RESOURCE ALLOCATION MODE

The profit motive of maximizing $P = R - C$ is sometimes a function of a set of control variables, $x_1, x_2, \ldots, x_n$, and can be represented as nonlinear and linear mathematical models. *A nonlinear model* for maximizing net profit P is

$$\begin{aligned} &\text{Objective function: Maximize } z = f(\boldsymbol{x}) \\ &\text{Constraint set:} \quad g(\boldsymbol{x})\{=, \geq, \leq\}\boldsymbol{b} \\ &\qquad\qquad -\boldsymbol{x} \geq \boldsymbol{0} \quad \text{(nonnegativity condition)} \end{aligned}$$

where z = measure of effectiveness ($z = P = R - C$)

$\boldsymbol{x}$ = a set of n control variables of $x_1, x_2, \ldots, x_n$

$f(\boldsymbol{x}) = f(x_1, x_2, \ldots, x_n)$ = function of control variables

$$\boldsymbol{g}(\boldsymbol{x}) = \begin{bmatrix} g_1(\boldsymbol{x}) \\ g_2(\boldsymbol{x}) \\ \vdots \\ g_m(\boldsymbol{x}) \end{bmatrix} = \text{vector of } m \text{ constraint equations of } \boldsymbol{x}$$

$$\boldsymbol{b} = \begin{bmatrix} b_1 \\ b_2 \\ \vdots \\ b_m \end{bmatrix} = \text{resource constraint vector}$$

A linear model for minimizing total costs $C = \boldsymbol{c}'\boldsymbol{x}$ is

$$\begin{aligned} &\text{Objective function: minimize } z = \boldsymbol{c}'\boldsymbol{x} \\ &\text{Constraint set:} \quad \boldsymbol{A}\boldsymbol{x}\{=, \geq, \leq\}\boldsymbol{b} \\ &\qquad\qquad \boldsymbol{x} \geq \boldsymbol{0} \end{aligned}$$

where $\boldsymbol{x}$ and $\boldsymbol{b}$ are the same as for nonlinear models

z = measure of effectiveness

c = unit cost vector consisting of n unit costs, $c_1, c_2, \ldots, c_n$

$\boldsymbol{A}$ = $m \times n$ constraint parameter matrix

Systems Analysis for Civil Engineers

Lori Mueller
143 Courtney St
454-1523

Systems Analysis for Civil Engineers

Paul J. Ossenbruggen
University of New Hampshire
Durham, New Hampshire

JOHN WILEY & SONS
New York Chichester Brisbane Toronto Singapore

Library of Congress Cataloging in Publication Data:

Ossenbruggen, Paul J. (Paul John)
Systems analysis for civil engineers.

Includes bibliographies and indexes.
1. System analysis. 2. Civil engineering.
I. Title.
T57.6.084 1983 620.7′2 83-14595
ISBN 0-471-09889-2

Printed in the United States of America

10 9 8 7 6 5 4 3 2 1

Preface

The purpose of this textbook is to combine the principles of economics and engineering into an unified approach for solving civil engineering design, planning, and management problems. In the first portion of the book, Chapters 1 through 3, the fundamental principles and methodology of systems analysis are presented. The concepts of optimization and selection of competing design alternatives are covered. After completing this part of the book, the student or practicing engineer should realize the important role that financial, physical, and institutional factors play in the overall process. More importantly, they should know how to incorporate these factors into the analysis. In the remaining portion of the book, a more in-depth study of economic and mathematical methods is undertaken.

Every attempt has been made to stimulate interest in the subject by illustrating the use of the systems analysis approach with practical engineering problems. Illustrative examples and problem assignments from various disciplines of civil engineering have been chosen. They include structural, geotechnical, and environmental engineering, water resource planning, construction, and transportation. When practical engineering examples are inappropriate simple mathematical examples are employed.

To formulate a systems analysis problem properly, the principles of economics must be understood. For this reason, engineering economics and the basic principles of microeconomics are covered. The effects of production cost and pricing are important in most practical problems.

Systems analysis has become an important decision-making tool largely owing to the advances that have been made in the computer hardware and software industries. With low-cost microcomputers, it is now possible for small engineering firms to perform analyses that were not practical a few years ago. When preparing this textbook, the importance of computer application was recognized. As a result, matrix methods are emphasized.

It is realized that some people tend to overemphasize the importance of mathematical methods and lose sight of the primary purpose of systems analysis. Whenever possible, graphical methods of solution are used. Graphical solutions generally provide more insight into the overall problem than do mathematical methods. As a consequence, if a choice can be made between mathematical and graphical methods, the graphical methods should be chosen. In order to illustrate the features of a mathematical approach and its shortcomings, illustrative problems using both mathematical and graphical methods are presented and compared.

This book is intended for individuals, who have sufficient engineering analysis and design background. It has been my experience that a typical civil engineering major entering his or her second semester of the junior year has adequate knowledge to utilize the concepts presented in the book. At the University of New Hampshire, the first half of the textbook is used in a required senior-level course, and the second half is used in an elective senior–graduate-level course.

I am appreciative of the support that I have received from my family and to all the people who have helped me in the preparation of this book. I am practically indebted to the reviewers, Dr. William P. Darby, Department of Technology and Human Affairs, Washington University in St. Louis, Dr. Michael Demetsky, Department of Civil Engineering, University of Virginia, and Dr. L. David Meeker, Department of Mathematics, University of New Hampshire. Special thanks goes to Mrs. Nan Collins, who patiently typed many revisions of this book. Finally, I would like to thank the students who used this book in its draft form, Nancy Di Mauro, Timothy Grant, and James Downs, who solved some of the problems given as text examples and homework problems, and my son, Paul, who developed special computer graphics and other algorithms to make the programs provided with this text more valuable to the students.

Author's Note on Computer Programs

With hand calculations, some of the methods described in this textbook are, quite frankly, time consuming, repetitious, and monotonous to apply. It is advocated that hand calculations be performed to learn the method and then computer programs be used in application. The following programs are most helpful:

Program	Applicable in Chapters
Nonlinear function $y = f(x)$ plot	2, 5, 8, 9
Matrix methods	Appendix, 6, 7, 9, 10
Histogram plot	4
Linear programming	6, 7
Newton's method	9
Newton/gradient projection method	9
Method of least squares	10
Scatter gram plot	10

With the exception of the Newton/gradient projection method, these programs are available for most large main-frame, mini-, and microcomputers.

Since graphical and matrix methods are stressed in this book, the first two programs have the broadest application. The matrix methods program should include the operations of transpose, multiplication, inversion, and the calculation of a determinant. These operations are of interest in themselves but most often are used as functions of larger computer algorithms. For instance, Apple II BASIC computer programs (that were written by the author to complement this textbook) call these matrix operations.

The author's linear programming algorithm is a special interactive program. In order for the user to obtain an answer, it is necessary for him to know the rules of the simplex method, a method of solving linear optimization problems as described in Chapter 6. The simplex method is an iterative method calling for matrix inversion and multiplication operations in each step.

Listings of all the programs listed above are available by contacting the author or the publisher, John Wiley. It is not imperative that the reader of this book use the author's programs. Furthermore, owing to computer language incompatibility problems, it may be necessary for the user to rewrite portions of the program to operate on his or her equipment.

Author's Note on Computer Programs

Contents

Systems Analysis for Civil Engineers

CHAPTER 1

The Process

After the completion of this chapter, the student should:

1. Understand the basic systems-analysis approach to formulating and solving engineering problems with mathematical models.
2. Appreciate how labor power, physical, and financial limitations as well as institutional requirements such as indicated in building and design codes affect engineering design, planning, and management processes.
3. Understand the advantages and limitations of systems analysis for solving practical engineering problems.

Systems analysis is a coordinated set of procedures that can be used to address issues of project planning, engineering design, and management. Systems analysis is a decision-making tool. An engineer can use it for determining how resources can be used most efficiently and most effectively to achieve a specified goal or objective. For successful decision making, *both technological* and *economic* considerations must be employed in the analysis. The premise is followed throughout this textbook.

Since systems analysis can be applied to a broad range of decision-making and engineering problems, we shall illustrate its application to problems in structural, geotechnical, environmental, transportation, water resources, and construction engineering. In each illustration we attempt to show how the principles of engineering can be combined with the principles of economics to achieve an optimum solution.

In the fields of economics, mathematics, and business, systems analysis is commonly referred to as *operations research*. In this textbook, we are concerned with the application of these principles to the solution of design, planning, and management problems in civil engineering.

1.1 RESOURCE ALLOCATION AND MATHEMATICAL MODELS

Systems analysis is an approach for allocating *resources* in an effective manner. Resources can be broadly classified as: *labor power*, *money*, and *materials*. Since resources have market value, that is, can be bought and sold, and since money is generally in short supply, the allocation of resources is extremely important. This is especially true when civil engineers are involved in large-scale public-works projects that cost millions of dollars.

The Goal and Objective

In order to allocate resources efficiently and effectively, we must have a clearly established *goal*, or *objective*. Generally, our goal will be to allocate resources in a manner that will maximize profit for a firm or maximize a societal benefit for the public. The resources of labor power, money, and materials are assumed to be used for producing goods and services. Since social benefits are difficult to measure and will only tend to confuse our introductory remarks, we utilize the profit motive and other simple measures to illustrate fundamental concepts of systems analysis. For example, the profit P is defined as the difference between the revenue or monetary benefit R received for a good or service and the cost C to provide the good or service. The following mathematical expression summarizes this definition.

$$P = R - C$$

If the revenue is assumed to be fixed, the greatest profit is obtained by minimizing the cost of production. Typically, the role of the engineer is to achieve this goal.

Constraints

Finding ways to minimize cost for the purpose of maximizing profit is easier said than done. *Financial, physical,* and *institutional constraints* must be considered. Financial constraints are brought about, generally, by the limited supply of money and the costs of borrowing it. This money may be used to obtain resources, such as by buying material or hiring workers. Physical constraints refer generally to the limitations of the properties of materials. A material has certain properties that can be measured, such as strength, elasticity, and other engineering characteristics. Institutional constraints are generally rules, laws, or guidelines specified by society, government, and the engineering profession. The rules as specified in design and building codes are examples of institutional constraints that must be considered by the engineer. Clearly, if an engineering design satisfies all constraints at the same time that it satisfies the given objective to maximize profit or minimize cost, the design can be considered an *optimum solution.* The purpose in this textbook is to discuss the ways of achieving this end.

The Optimum Solution

In systems analysis, a *mathematical model* is an important element of the decision-making process. The mathematical model is an exact and explicit statement of the objective, or goal, to be achieved. In addition, it consists of a set of financial, physical, and institutional constraint conditions that must be satisfied. The solution to a problem, the *optimum solution,* is a statement of how resources should be used in the most efficient and effective manner.

In this textbook we investigate problems from many areas of civil engineering. The goal, or objective, will vary from problem to problem. For example, our objective may be to minimize the cost of production, minimize the weight of a structure or choose the best alternative design that satisfies a public need. All these problems may be structured as systems-analysis problems using mathematical models. The important point to realize is that a systems-analysis model, regardless of the particular discipline within civil engineering concerned or the goal of the project, is always formulated using the same approach. The resulting mathematical model can be presented in a standard mathematical form with an objective function and a set of constraint equations.

In this chapter we are primarily concerned with the statement of the project goal, the statement of constraint conditions, and the formulation of mathematical models. The solution of systems-analysis or optimization problems is the subject of the following chapters of the book. We shall see that there are different algorithms for solving different types of mathematical models. We shall use graphical methods and methods of calculus to illustrate basic principles for determining the optimum solution.

The graphical method is a very powerful means of solving a certain class of optimization problems. Generally, it gives better insight into understanding the problem and evaluating alternative solutions than do mathematical algorithms. Whenever possible, graphical methods will be utilized to complement the discussion of mathematical approaches.

Mathematical Models

A typical systems analysis model will consist of a single objective function and a set of constraint equations. Models consisting of multiple objective functions are discussed in Section 1.4. Our discussion, for the most part, will be restricted to models with a single objective function.

The objective function is assumed to be a function of a set of *design*, *decision*, or *control variables*. In this textbook, the term "control variable" is generally used. The objective function may be expressed as a mathematical relationship.

$$z = f(x_1, x_2, \ldots, x_n)$$

where $x_1, x_2, \ldots, x_n$ are designated as a set of n control variables. The control variables, which may typically represent the assignment of the number of workers, an amount of money, and a volume of material, are all nonnegative values; they are introduced into the mathematical model as

$$\begin{aligned} x_1 &\geq 0 \\ x_2 &\geq 0 \\ &\vdots \\ x_n &\geq 0 \end{aligned}$$

The financial, physical, and institutional limitations are represented by a set of m constraint equations.

$$\begin{aligned} g_1(x_1, x_2, \ldots, x_n)\{=, \leq, \geq\}b_1 \\ g_2(x_1, x_2, \ldots, x_n)\{=, \leq, \geq\}b_2 \\ \vdots \qquad\qquad \vdots \qquad\qquad \vdots \\ g_m(x_1, x_2, \ldots, x_n)\{=, \leq, \geq\}b_m \end{aligned}$$

The set $\{=, \leq, \geq\}$ represents the possibility of having an equal to, less than or equal to, or greater than or equal to constraint condition. For example, in the construction of a steel building, the right-hand side of the equation set may consist of: $b_1 =$ \$10,000, the amount of money that may be expended on fabrication; $b_2 = 500$ ft, the number of linear feet of steel available for this project; and $b_3 = 20{,}000$ psi, an allowable stress limitation permitted in structural support members. By utilizing vector notation, we may simplify the model into a compact form:

$$z = f(\mathbf{x})$$

subject to

$$g_i(\mathbf{x})\{=, \leq, \geq\}b_i$$

where

$$i = 1, 2, \ldots, m \qquad \text{and} \qquad \mathbf{x} \geq \mathbf{0}$$

In vector notation the set of control variables $x_1, x_2, \ldots, x_n$ is represented by the *control vector* $\mathbf{x}$ or

$$\mathbf{x} = \begin{bmatrix} x_1 \\ x_2 \\ \vdots \\ x_n \end{bmatrix}$$

The constraint set may be represented as a vector also. Let

$$\mathbf{g}(\mathbf{x}) = \begin{bmatrix} g_1(x_1, x_2, \ldots, x_n) \\ g_2(x_1, x_2, \ldots, x_n) \\ \vdots \\ g_m(x_1, x_2, \ldots, x_n) \end{bmatrix} = \begin{bmatrix} g_1(\mathbf{x}) \\ g_2(\mathbf{x}) \\ \vdots \\ g_m(\mathbf{x}) \end{bmatrix}$$

and

$$\mathbf{b} = \begin{bmatrix} b_1 \\ b_2 \\ \vdots \\ b_m \end{bmatrix}$$

The model becomes

$$z = f(\mathbf{x})$$
$$\mathbf{g}(\mathbf{x})\{=, \leq, \geq\}\mathbf{b}$$
$$\mathbf{x} \geq 0$$

The function $f(\mathbf{x})$ and the set of functions represented by $\mathbf{g}(\mathbf{x})$ may be either linear or nonlinear set of functions of $\mathbf{x}$.

Control Variables and Vectors

A control variable is a term used to designate any parameter that may vary in the design, planning, or management process. A control variable may be either a discrete or continuous variable. In this textbook, the letter x_i, where the subscript $i = 1, 2, \ldots, n$, is generally used to identify the particular control variable.

An example of a discrete control variable is the number of workers assigned to a particular task. Let x_1 be equal to the number of assigned workers, where $x_1 = 1, 2, 3, \ldots$. Since the number of workers is always a positive number or equal to zero, it is represented as

$$x_1 \geq 0$$

We shall see that in most formulations of physical systems the control variables are nonnegative values. This simple constraint will have an important impact in our search for an optimum solution.

An example of the use of continuous control variables is illustrated with the use of Figure 1.1. In this case, the beam length is fixed and the beam width and height are considered the design variables. Since the length is a constant, it is not designated as a control variable. We define x_1 and x_2 as control variables, where x_1 is the beam width b and x_2 is the beam height h. In this case x_1 and x_2 are nonnegative values.

$$x_1 \geq 0$$

$$x_2 \geq 0$$

Utilizing vector notation, this set of variables can be written as

$$\boldsymbol{x} \geq \boldsymbol{0}$$

where

$$\boldsymbol{x} = \begin{bmatrix} x_1 \\ x_2 \end{bmatrix} = \begin{bmatrix} b \\ h \end{bmatrix} \geq \begin{bmatrix} 0 \\ 0 \end{bmatrix} = \boldsymbol{0}$$

The Objective Function

The objective function $f(\boldsymbol{x})$ is a single-valued function of the set of control variables or the control vector $\boldsymbol{x}$. The objective function is a mathematical statement of the goal and a measure of how effectively the goal is met. The selection of the measure

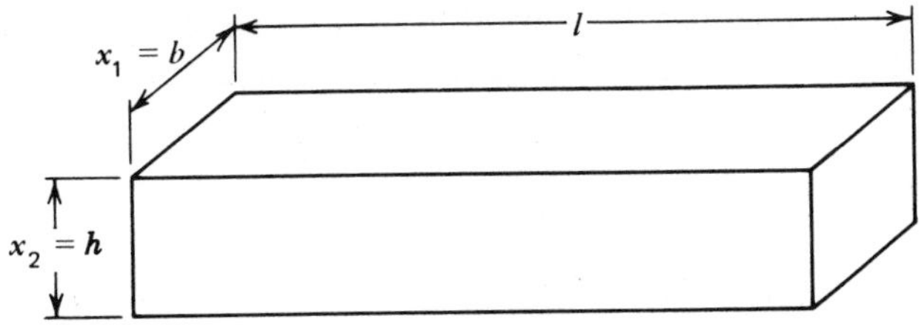

Figure 1.1 Rectangular beam.

of effectiveness will have an important effect on the outcome or solution. In some problems the selection of the measure of effectiveness and $f(\boldsymbol{x})$ is obvious. In other situations the choice may be extremely difficult. More will be said about measures of effectiveness in Section 1.2.

The objective function may either be maximized or minimized. For example, in a problem to maximize profits P, the objective function, is written as

$$\text{Maximize } z = \text{Maximize } P = f(\boldsymbol{x})$$

The symbol z represents the scalar quantity of the function $f(\boldsymbol{x})$. In this case, z is a scalar measure of profit P expressed in dollars and is assumed to be a function of the control variables $x_1, x_2, \ldots$.

The objective function may also be a minimization of z. For instance, the goal of a structural designer of aircraft is to minimize total weight W. In this case, the measure of effectiveness is a nonmonetary measure in units of pounds. The objective function for this case is

$$\text{Minimize } z = \text{Minimize } W = f(\boldsymbol{x})$$

Constraint Equations

A constraint equation is a mathematical equation expressing a financial, physical, or institutional limitation placed upon the problem. Generally, it is derived from fundamental principles of engineering or economics. A constraint equation may be stated as an equation with a *strict equality* condition

$$g(\boldsymbol{x}) = b$$

with a *less than or equal* condition

$$g(\boldsymbol{x}) \leq b$$

or a *greater than or equal* condition

$$g(\boldsymbol{x}) \geq b$$

For instance, in the planning of a two-room building as shown in Figure 1.2, the contract states that the total floor area of the building must have a minimum area of 5000 ft^2 and that each room of the building must have a maximum specified area of 5000 and 3000 ft^2, respectively. With these specifications, it is possible that the optimum solution will be a single room of 5000 ft^2. Let us state these specifications as a set of constraint equations. Let the control variables x_1 and x_2 represent the floor area of rooms 1 and 2, respectively. The constraint set is equal to

$$\begin{aligned} x_1 + x_2 &\geq 5000 \quad \text{(Total floor area)} \\ x_1 &\leq 5000 \quad \text{(Room 1)} \\ x_2 &\leq 3000 \quad \text{(Room 2)} \\ x_1 &\geq 0 \\ x_2 &\geq 0 \end{aligned}$$

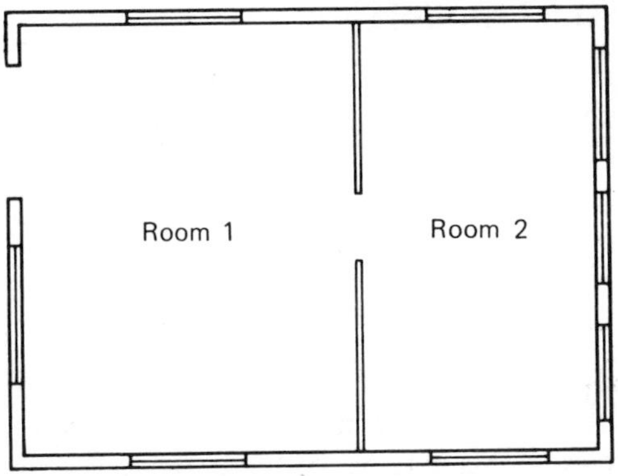

Figure 1.2 Floor plan.

The total floor area is equal to $x_1 + x_2$. According to specifications, it must be equal to or greater than 5000 ft^2. The other constraints, called *side constraints*, limit the room sizes, 1 and 2, to be less than or equal to 5000 and 3000 ft^2, respectively. Since the room sizes cannot be negative in value, we state that x_1 and x_2 are restricted to be positive numbers.

Feasible and Optimum Solutions

Any combination of control variables that satisfies the set of constraint conditions is called a *feasible solution.* A solution that does not satisfy all constraint equation conditions is called an *infeasible* solution. An *optimum solution* is a feasible solution that satisfies the goal of the objective function as well.

In the floor plan example the goal is to maximize total revenue. Assume that \$50/ft^2 and \$60/ft^2 are the unit revenues received for rooms 1 and 2 respectively. The maximum total revenue received, R, is a function of the floor area x_1 and x_2 and can be represented by the following objective function and constraint set.

$$\text{Maximize } R = \$50x_1 + \$60x_2$$

subject to

$$x_1 + x_2 \geq 5000$$
$$x_1 \leq 5000$$
$$x_2 \leq 3000$$
$$x_1 \geq 0$$
$$x_2 \geq 0$$

This mathematical model consists of a linear objective function and a set of linear constraint equations. A problem with this mathematical form is called a *linear mathematical model.* A model with either a nonlinear objective function or one or more nonlinear constraint equations is called a *nonlinear mathematical model.*

Linear Mathematical Models

Vector and matrix notation offer a convenient way to represent classes of mathematical models. As previously shown, the mathematical model is represented as

$$z = f(\boldsymbol{x})$$
$$\boldsymbol{g}(\boldsymbol{x})\{=, \geq, \leq\}\boldsymbol{b}$$
$$\boldsymbol{x} \geq \boldsymbol{0}$$

In this text linear mathematical models are represented in two ways. A model consisting of a set of constraint equations having equality, greater than or equal to, and less than or equal to constraints may be written in the following manner.

$$z = f(\boldsymbol{x}): \quad z = c_1x_1 + c_2x_2 + \cdots + c_nx_n$$

$$\boldsymbol{g}(\boldsymbol{x})\{=, \leq, \geq\}\boldsymbol{b}: \quad a_{11}x_1 + a_{12}x_2 + \cdots + a_{1n}x_n\{=, \leq, \geq\}b_1$$
$$a_{21}x_2 + a_{22}x_2 + \cdots + a_{2n}x_n\{=, \leq, \geq\}b_2$$
$$\vdots$$
$$a_{m1}x_1 + a_{m2}x_2 + \cdots + a_{mn}x_n\{=, \leq, \geq\}b_m$$

$$\boldsymbol{x} \geq 0: \begin{bmatrix} x_1 \\ x_2 \\ \vdots \\ x_n \end{bmatrix} \geq \begin{bmatrix} 0 \\ 0 \\ \vdots \\ 0 \end{bmatrix}$$

where $c_j = c_1, \ldots, c_n$, and $b_i = b_1, b_2, \ldots, b_m$ and $a_{ij} = a_{11}, a_{12}, \ldots, a_{mn}$ are constant parameters. Using matrix notation, linear mathematical models will have the following compact form.

$$z = \boldsymbol{c}'\boldsymbol{x}$$
$$\boldsymbol{a}_i'\boldsymbol{x}\{=, \leq, \geq\}b_i \qquad \text{where } i = 1, 2, \ldots, m$$
$$\boldsymbol{x} \geq \boldsymbol{0}$$

and $\boldsymbol{a}_i$, **c**, and b_i are vectors of the elements of the objective and constraint equations. The elements of c_j and b_i are generally called the unit cost and resource parameters, respectively. The elements of $\boldsymbol{a}_i$ are derived from technological and economic considerations; therefore, they are called technological parameters. The objective function may be represented as

$$z = [c_1 \; c_2 \cdots c_n] \cdot \begin{bmatrix} x_1 \\ x_2 \\ \vdots \\ x_n \end{bmatrix}$$

Likewise, the ith constraint equation may be represented as

$$[a_{i1}\ a_{i2} \cdots a_{in}] \cdot \begin{bmatrix} x_1 \\ x_2 \\ \vdots \\ x_n \end{bmatrix} \{=, \le, \ge\} \begin{bmatrix} b_1 \\ b_2 \\ \vdots \\ b_n \end{bmatrix}$$

where $i = 1, 2, \ldots, m$.

In Chapter 6 the linear model will be written in the so-called *standard form*, the second way of representing linear models:

$$\text{Minimize } z = \boldsymbol{c}'\boldsymbol{x}$$

$$\boldsymbol{A}\boldsymbol{x} = \boldsymbol{b}$$

$$\boldsymbol{x} \ge \boldsymbol{0}$$

All elements of $\boldsymbol{b}$ are positive values, $b_i \ge 0$, where $i = 1, 2, \ldots, m$. All contraint equations are strict equality constraints. The constraint set in expanded matrix form is

$$\boldsymbol{A}\boldsymbol{x} = \boldsymbol{b} \quad \text{or} \quad \begin{bmatrix} a_{11} & a_{12} \cdots a_{1n} \\ a_{21} & a_{22} \cdots a_{2n} \\ \vdots & \\ a_{m1} & a_{m2} \cdots a_{mn} \end{bmatrix} \cdot \begin{bmatrix} x_1 \\ x_2 \\ \vdots \\ x_n \end{bmatrix} = \begin{bmatrix} b_1 \\ b_2 \\ \vdots \\ b_n \end{bmatrix}$$

where the number of control variables n is equal to or greater than the number of equations m, $n \ge m$. We distinguish between vectors and matrices by reserving lower-case letters, $\boldsymbol{x}$, $\boldsymbol{c}$, $\boldsymbol{b}$, and $\boldsymbol{a}_i$, for vectors and upper-case letters, $\boldsymbol{A}$, for matrices. All vectors in this book are column vectors. A row vector is denoted as a transpose of a column vector. For example,

$$\boldsymbol{x} = \begin{bmatrix} x_1 \\ x_2 \\ \vdots \\ x_n \end{bmatrix}$$

The row vector of $\boldsymbol{x}$ is

$$\boldsymbol{x}' = [x_1\ x_2 \cdots x_n]$$

EXAMPLE 1.1 A Statically Determinate Minimum-Weight Truss

Consider the truss shown in Figure 1.3*a*. Formulate a mathematical model to design a simple truss of minimum weight. The critical buckling and maximum allowable tensile stresses of compression and tension members are 10 ksi and 20 ksi, respectively. The truss is to be constructed of steel. All compression and tension members are assumed to have the same cross-sectional area.

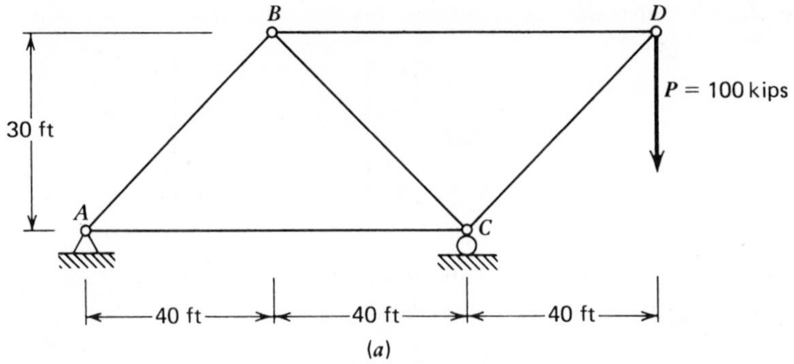

Figure 1.3*a*

Solution

Control Variables The structural members will be sized according to the type of member force, either compression or tension. Thus, the control variables are defined as

$$x_1 = A_1 = \text{Cross-sectional area of a compression member (in.}^2)$$

$$x_2 = A_2 = \text{Cross-sectional area of a tension member (in.}^2)$$

The control vector $\boldsymbol{x}$ is $\boldsymbol{x}' = [A_1\ A_2]$.

Reactions

The support reactions may be determined from Newton's law of static equilibrium. The free-body diagram for the reactions is shown in Figure 1.3*b*.

$$\sum F_x = 0: \quad H_a = 0$$

$$\sum F_y = 0: \quad -V_a + V_c - 100 = 0$$

$$\sum M_A = 0: \quad 80V_c - 120 \cdot 100 = 0$$

Solving this set of reactions results in $H_a = 0$, $V_a = 50$ kips, and $V_c = 150$ kips. (1 kip = 1000 pounds.)

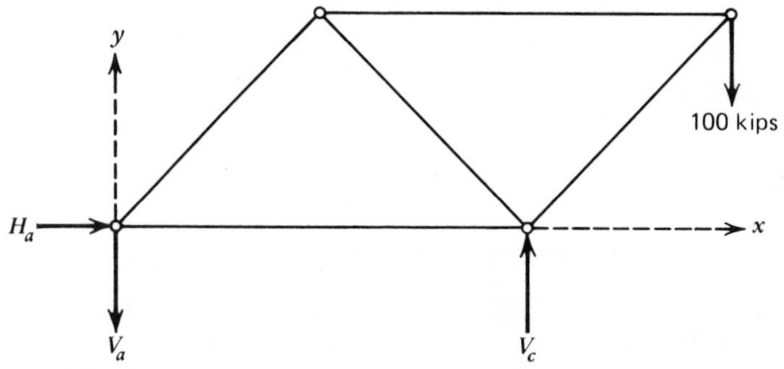

Figure 1.3*b*

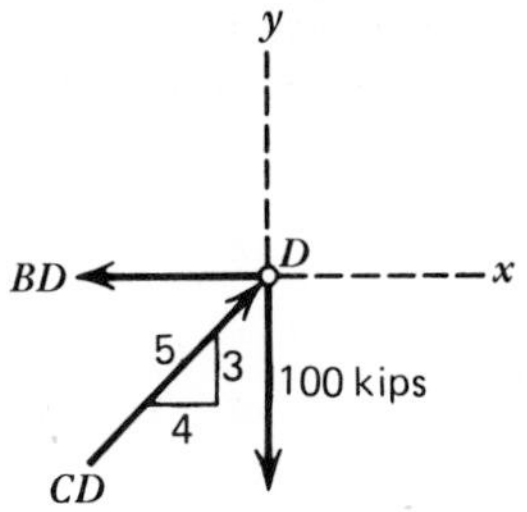

Figure 1.3c

Member Forces and Stresses We next use the method of joints to determine forces in each member. The free-body diagram for joint D is shown in Figure 1.3*c*.

$$\sum F_x = 0: \quad -BD + \tfrac{4}{5}CD = 0$$

$$\sum F_y = 0: \quad \tfrac{3}{5}CD - 100 = 0$$

The member forces are $CD = 167$ kips (compression) and $BD = 125$ kips (tension). The stress in the members will be equal to the member force divided by the cross-sectional area of the member. In this case, the tension member *BD* will have a stress σ_{BD} equal to

$$\sigma_{BD} = \frac{125}{A_2}$$

The compression member *CD* will have a stress equal to

$$\sigma_{CD} = \frac{167}{A_1}$$

The method of joints was used to determine the forces in members *BC*, *AB*, and *AC*. The member forces and reactions are summarized in Figure 1.3*d*. The stress in remaining members *AB*, *AC*, and *BC* will be equal to

$$\sigma_{AB} = \frac{83.3}{A_2}; \qquad \sigma_{AC} = \frac{66.7}{A_1}; \qquad \sigma_{BC} = \frac{83.3}{A_1}$$

respectively.

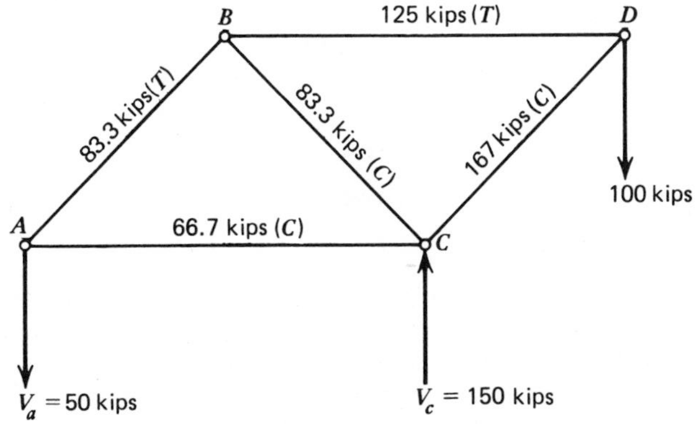

Figure 1.3*d*

Constraint Equations The stresses in each member may be equal to, but must never exceed, the critical buckling stress or allowable tensile stress. Thus, we may express these restrictions by the set of equations:

$$\text{Member } AB\text{:} \quad \frac{83.3}{A_2} \le 20 \qquad \text{or} \quad A_2 \ge 4.17$$

$$\text{Member } AC\text{:} \quad \frac{66.7}{A_1} \le 10 \qquad \text{or} \quad A_1 \ge 6.67$$

$$\text{Member } BC\text{:} \quad \frac{83.3}{A_1} \le 10 \qquad \text{or} \quad A_1 \ge 8.33$$

$$\text{Member } BD\text{:} \quad \frac{125}{A_2} \le 20 \qquad \text{or} \quad A_2 \ge 6.25$$

$$\text{Member } CD\text{:} \quad \frac{167}{A_1} \le 10 \qquad \text{or} \quad A_1 \ge 16.7$$

Objective Function The weight of each member is equal to the density of steel times the volume of each member. The density of steel is approximately 490 lb/ft^3 or 3.40 lb/ft-in.^{2}. The equation for weight of the truss is the sum of the weight of each individual member; thus,

$$z = 3.40[V_{AB} + V_{AC} + V_{BC} + V_{BD} + V_{CD}]$$

where V = volume of each member or the member length multiplied by the member cross-sectional area. The objective function is

$$z = 3.40[50A_2 + 80A_1 + 50A_1 + 80A_2 + 50A_1]$$

or

$$z = 612A_1 + 442A_2$$

Mathematical Model The problem formulation is complete. Summarizing the equations results in the following mathematical model:

$$\text{Minimize } z = 612A_1 + 442A_2$$

subject to the constraints

$$A_2 \ge 4.17$$

$$A_1 \ge 6.67$$

$$A_1 \ge 8.33$$

$$A_2 \ge 6.25$$

$$A_1 \ge 16.7$$

Remarks

The formulation of this problem utilizes basic principles of engineering mechanics and structural design. In most instances, the formulation of the mathematical model will be derived from basic principles of engineering or economics. All physical laws, such as Newton's law, must be satisfied, and all rules of good engineering practice must be incorporated in the model.

Systems analysis gives us a different perspective in solving engineering problems. In this problem, for instance, we have established the limitations on our design with structural engineering principles and combined these limitations in a set of constraint equations with a goal of minimizing the weight of the truss. The optimum solution satisfies the goal and the constraints at the same time. Systems analysis adds a new dimension to the design process that should lead us to a better design.

PROBLEMS

Problem 1

A contractor may purchase material from two different sand and gravel pits. The unit cost of material including delivery from pits 1 and 2 is \$5 and \$7 per cubic yard, respectively. The contractor requires 100 yd^3 of mix. The mix must contain a minimum of 30 percent sand. Pit 1 contains 25 percent sand, and pit 2 contains 50 percent sand.

The objective is to minimize the cost of material.

(a) Define the control variables.
(b) Formulate a mathematical model.

Problem 2

An aggregate mix of sand and gravel must contain no less than 20 percent nor more than 30 percent gravel. The in situ soil contains 40 percent gravel and 60 percent sand. Pure sand may be purchased and shipped to site at \$5.00/yd^3. A total mix of 1000 yd^3 is needed. There is no charge for using in situ material.

The goal is to minimize cost subject to the mix constraints.

(a) Define the control variables.
(b) Formulate the mathematical model.

Problem 3

There are two suppliers of pipe.

SOURCE	UNIT COST (\$/LINEAR FOOT)	SUPPLY (LINEAR FOOT)
1	\$100	100 ft maximum
2	\$125	Unlimited

Nine hundred feet of pipe is required.

The goal is to minimize the total cost of pipe.

(a) Define the control variables.
(b) Formulate a mathematical model.

Problem 4

A company requires at least 4.0 Mgal/day more water than it is currently using. A water-supply facility can supply up to 10 Mgal/day of extra supply. A local stream can supply an additional 2 Mgal/day. The company requires that the water pollution concentration be less than 100 mg/l. BOD, the biological oxygen demand. The water from the water-supply facility and from the stream has a BOD concentration of 50 mg/l and 200 mg/l, respectively. The cost of water from the water supply is \$100/Mgal and from the local stream is \$50/Mgal.

The goal is to minimize the cost of supplying extra water that meets water quality standards.

(a) Define all control variables.
(b) Formulate a mathematical model.

Problem 5

The unit selling price p of an item is \$150/unit, thus the total revenue is $R = 150q$. The production cost C is a function of output level q, $C = 100q^{0.8}$. The maximum output of the firm is 50 units/yr.

The objective is to maximize yearly profit subject to the production constraint.

(a) Define the control variable.
(b) Formulate a mathematical model.

Problem 6

Determine the maximum volume of a sphere subject to an internal pressure of 4000 psi (lb/in.^{2}). The volume of a sphere is $V = (\pi/6)d^3$, where d equals the mean diameter of the tank. The allowable stress of the material is equal to 20,000 psi. The hoop stress is a function of the tank diameter d and tank thickness t.

$$\sigma = \frac{pd}{4t}$$

The objective is to maximize tank volume subject to the limitation on stress.

(a) Define the control variable.
(b) Formulate a mathematical model.

1.2 THE SYSTEMS-ANALYSIS APPROACH

The systems-analysis approach consists of the following steps.

1. Establish an objective and an appropriate measure of effectiveness.
2. Formulate a mathematical model.
3. Determine the optimum solution.

In this chapter, we implicitly utilized steps 1 and 2 in formulating the mathematical model for a minimum-weight truss. In some real-world engineering problems, however, the establishment of the objective and the formulation of the mathematical models will not be so easy. We shall see in our development of systems analysis that most of our discussion in this textbook is devoted to steps 2 and 3. It might appear that step 1 is not as important as the others. But in this section we see that the establishment of a single goal or objective and the establishment of an appropriate measure of effectiveness will have a most important bearing upon the solution to a problem.

Multiobjective Problems

The types of systems-analysis problems we consider here are limited to a single objective. One difficulty in satisfying this restriction is establishing a mathematical model with unlike measures of effectiveness. Consider the purchase of an automobile. Most people decide which automobile to buy through comparison shopping. We weigh the attributes or features of one automobile against the others. All automobiles provide transportation, but it is the selling price of the vehicle, its performance, and styling features that are ultimately used to determine the final selection of the vehicle. We can measure these attributes by vehicle price, speed, fuel economy, reliability, comfort, and prestige of ownership. If we rank each automobile by each one of these attributes, it will be very unlikely that one vehicle would be our first choice in every category. In other words, the lowest-priced vehicle will rank number one in the price category, but it is unlikely that it would rank number one in speed, comfort, and prestige of ownership as well. In order to make a decision, we have to weigh these attributes. If price is important and prestige of ownership is unimportant, a greater weight will be placed on price than on owner prestige. By this *subjective reasoning process*, we can make a choice among the vehicles. It is possible that the vehicle chosen does not receive a top rating in any one category.

If we use a mathematical model, we shall have to quantify those attributes and place them in an objective function. Let the control variables $x_1, x_2, \ldots$ represent automobile types 1, 2, . . . , and so on.

Here, the control variables are discrete control variables taking on the values of 0 or 1, the automobile type is not selected or is selected for purchase, respectively. Thus, $x_1 \leq 0$ or 1, $x_2 = 0$ or 1, ... Unlike the subjective reasoning process utilized in the comparative shopping discussion, here we must explicitly assign numerical *weights* or a *measure of effectiveness* to each attribute. For instance, the purchase price is measured in dollars, the speed in miles per hour, and the fuel economy in miles per gallon. These are objective measures of effectiveness because they have economic value and performance rating that are distinguishable. The attributes of reliability, comfort, and prestige of ownership are personal preference items. They are considered subjective measures because they have psychological values that are difficult to quantify.

The objective measures of effectiveness are more desirable to use for engineering problems than are subjective measures. Even when all measures of effectiveness are objective measures, they may be *unlike measures of effectiveness* that are difficult to combine into a single objective function. For example, if selling price in dollars per car and fuel economy in miles per gallon are the only two objective measures of effectiveness, they cannot be added together nor meaningfully introduced into the same single objective function, because they do not have the same units. Therefore, they are considered unlike measures of effectiveness. This problem formulation requires two objective functions: one to minimize the selling price and the other to maximize fuel economy.

Obviously, multiobjective problems are important. We suggest that the optimum solution for each single-value objective be determined independently. Each optimum solution can then be compared in the context of the original problem. Subjective reasoning, trade-off analysis, and other methods can be used to make the final decision. The textbook by Giocoechea, et al (see bibliography) is recommended for further reading.

Establishing an Appropriate Objective Function

Once a single measure of effectiveness has been chosen, extreme care must be used in establishing an appropriate objective function. For profit-maximization and cost-minimization problems, money measured in dollars, for example, will be the obvious choice of the unit of measure. However, consider the total cost of an engineered structure. There are design, material, shipping, construction, and labor costs to consider. In lieu of considering all costs in a minimum-cost model, let us assume that minimum-cost models are established and solutions for each individual cost category are found. The optimum solution to a minimum-construction-cost problem may not be the same optimum solution to a minimum-material-cost problem, even though the constraints imposed upon the two problems are the same. The solution to the minimum-material-cost model may result in a very light and delicate structure. The cost to construct this structure may be prohibitively expensive. The solution to the minimum-construction-cost model may result in easy construction with materials being wasted. To eliminate these problems, a model considering all costs, designs, materials, shipping, construction, and labor will probably be most effective. Since the same measure of effectiveness, dollars, is used to measure each attribute, the objective function incorporating all attributes can be established by adding them together.

Cash-Flow Problems

Costs and revenues may be affected by time. The cost to construct a system is an initial cost and may be considered as a one-time capital cost. On the other hand, the cost to maintain and operate a system must be paid over the life of the project. When payments are made at different time periods, we cannot simply add them together as we did in establishing a minimum-total-cost model of an engineered structure. For instance, the value of a payment made today versus 10 years from now is different. Clearly, the payment that we receive today has more value than one we expect to receive in the future. Problems of this type are called *cash-flow problems* and will be evaluated with *time value of money relationships* to be discussed in detail in the chapter on engineering economics.

For illustration, consider two different types of construction plans. Alternative **A** calls for the construction of the entire system at the present time. Alternative **B**, a staged-construction plan, calls for construction of part of the system now and the remaining part in year 10. The cash flow of benefits and costs are shown in Figure 1.4.

The design life of the project is 20 years. For plan **A**, the capital cost of construction is \$15M, $C_0 = \$15M$, and the monetary annual benefit and annual operating costs are shown to be uniform over the entire life of the project. The benefits are $B_1 = B_2 = \cdots = B_{20} = B = \$5M/yr$, and the annual operating and maintenance costs are $A_1 = A_2 = \cdots = A_{20} = A = \$3M/yr$. The subscript refers to the year the payment is made. The annual net benefit $B - A$ is \$2M/yr. For plan **B**, construction takes place in years 0 and 10, $C_0 = \$9M$ and $C_{10} = \$8M$. The annual benefits and costs are not uniform over the 20-year life of the project as shown. For the first 10 years the annual net benefit $B - A$ is \$1M/yr, and for the last 10 years $B' - A'$ is \$3M/yr.

If we study the two cash-flow diagrams, we can see different advantages for the two plans. Plan **B** has the advantage of lower initial construction cost; however, plan **A** offers a greater net annual benefit for the first 10 years. In later years, plan **B** offers a greater net annual benefit. These advantages are derived by making the *same-year comparisons*. Moreover, it is meaningless to say that the total cost of construction of plan **B** is \$9M + 8M = \$17M. This addition should not be performed, because the cash flow, or payments, takes place in different time periods, years 0 and 10. Thus, we need a method of combining present and future cash flows.

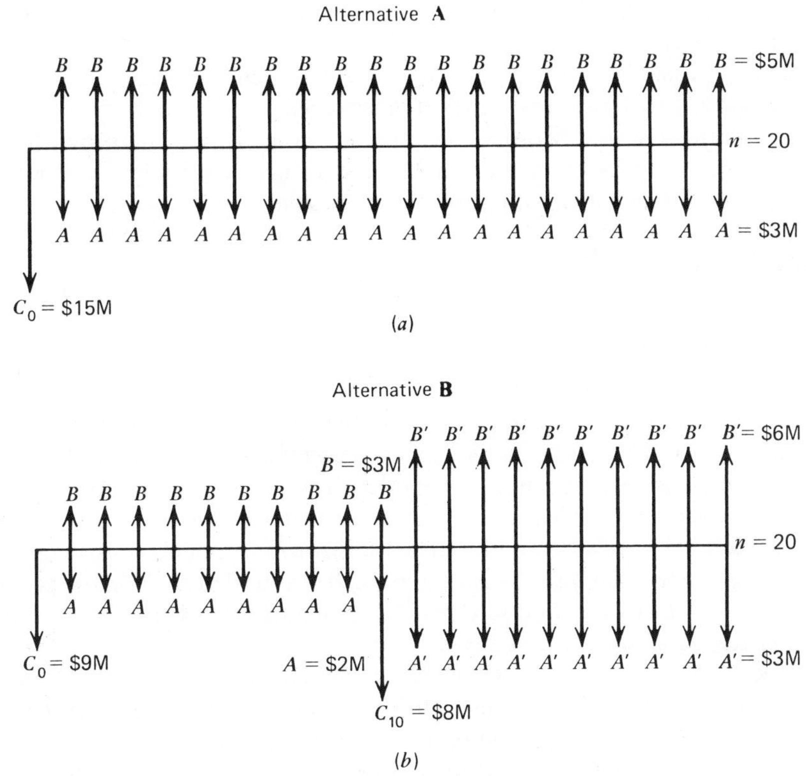

Figure 1.4 Cash-flow diagrams. (*a*) Alternative **A**, construction in year 0. (*b*) Alternative **B**, construction in year 0 and year 10.

Time of value of money relationships, incorporating factors of *time of payment* and *value of money*, which is measured in terms of an interest rate, will permit the transferral of a future payment to an equivalent present-worth payment. Now, simple arithmetic calculations can be performed. Moreover, all present and future benefits and costs can be combined into new value called the *net present-worth* value.

$$NPW = B_0 - A_0 - C_0$$

where A_0, B_0, C_0 represent the present worth of all annual costs, all annual benefits, and all capital costs, respectively. The net present worth of the competing alternatives $NPW^{\mathbf{A}}$ and $NPW^{\mathbf{B}}$ can be compared and the one offering the greater net present worth will be selected. If $NPW^{\mathbf{A}} > NPW^{\mathbf{B}}$, then construction plan **A** should be selected. If not, plan **B** is the better one. This is the essence of the *net present-worth selection method*. to be introduced in Chapter 3.

Nonmonetary measures of effectiveness are also important. In some instances, a physical measure, such as weight, is more important than a monetary measure. In the aircraft industry, reducing the overall weight of the frame and other components is of prime importance. Safety is of prime concern in all civil engineering designs.

Esthetics, comfort, and the protection of the environment are examples of other important objectives.

Monetary measures and engineering measures, such as weight and pollution emissions, are called objective measures of effectiveness. These quantities are measured in well-established scales. Measures of esthetics, comfort, and pleasure are called subjective measure because they are influenced by personal preference. Generally, subjective measures are avoided in engineering design.

Mathematical Model Types

We broadly classify system-analysis problems as

1. Resource allocation problems.
2. Alternative selection problems.

The same three-step procedure is used to solve these problems. Mathematical models consisting of a set of control variables, an objective function, and a set of constraint equations are called resource allocation models. The solution to this problem results in finding an *optimum combination of resources* that satisfies the set of constraint conditions and achieves a stated objective. Mathematical models consisting of a set of *mutually exclusive design alternatives* are called alternative selection problems. The net present-worth selection method is a technique for selecting the *best design alternative* from competing alternatives.

Now let us consider how resource allocation and alternative selection procedures can complement one another to achieve the best optimum design solution. There is a distinction between best design and optimum design. Consider a design of a bridge. Different technology and design configurations may be used to satisfy the basic requirement to span a given distance safely. If a steel truss and a concrete beam are considered the only alternatives, the one with the least combined construction and maintenance costs is considered the better choice. However, it may not be the best optimum design! Only if each design alternative is a minimum-cost design will the use of the alternative selection method lead to the best optimum design. It is possible to determine the least-cost designs by formulating separate resource allocation models for the steel truss and the concrete beam designs. These models will consist of an objective function to minimize cost subject to a set of constraint equations to ensure bridge safety and performance. Thus, the methods of solving resource allocation problems and the alternative selection method are complementary procedures that can be used to achieve the best optimum design.

A major difficulty of utilizing mathematical models is that not all of them can be easily solved. As a result, the use of combined methods of resource allocation and alternative selection may not be fully realized in the mathematical sense. Engineering is as much an art as an applied science. The best optimum design may be obtained by experimentation and trial and error. Thus, sometimes it is only practical to use the conceptual aspects of resource allocation and alternative selection to formulate problems and to seek solutions by mathematical means, by experimentation, or by trial and error. This book deals only with the mathematical methods of solution. Chapters 2 and 5 through 10 deal with methods of solving resource allocation problems, and Chapters 3 and 4 deal with methods of selecting the best design alternative. Chapter 5 also shows how cash-flow considerations are incorporated in resource allocation problems.

EXAMPLE 1.2 Bridge Location Study

Establish an objective function and measure of effectiveness to satisfy the goals of the City of Kingsbury. Owing to industrial growth, the City of Kingsbury is experiencing new demands for housing. City officials want to encourage the development of new housing in the West End. Since the economic future looks bright, the city government wants to improve the accessibility of the West End and find the best location for a new bridge. Two bridge sites have been chosen and are shown on the map in Figure 1.5.

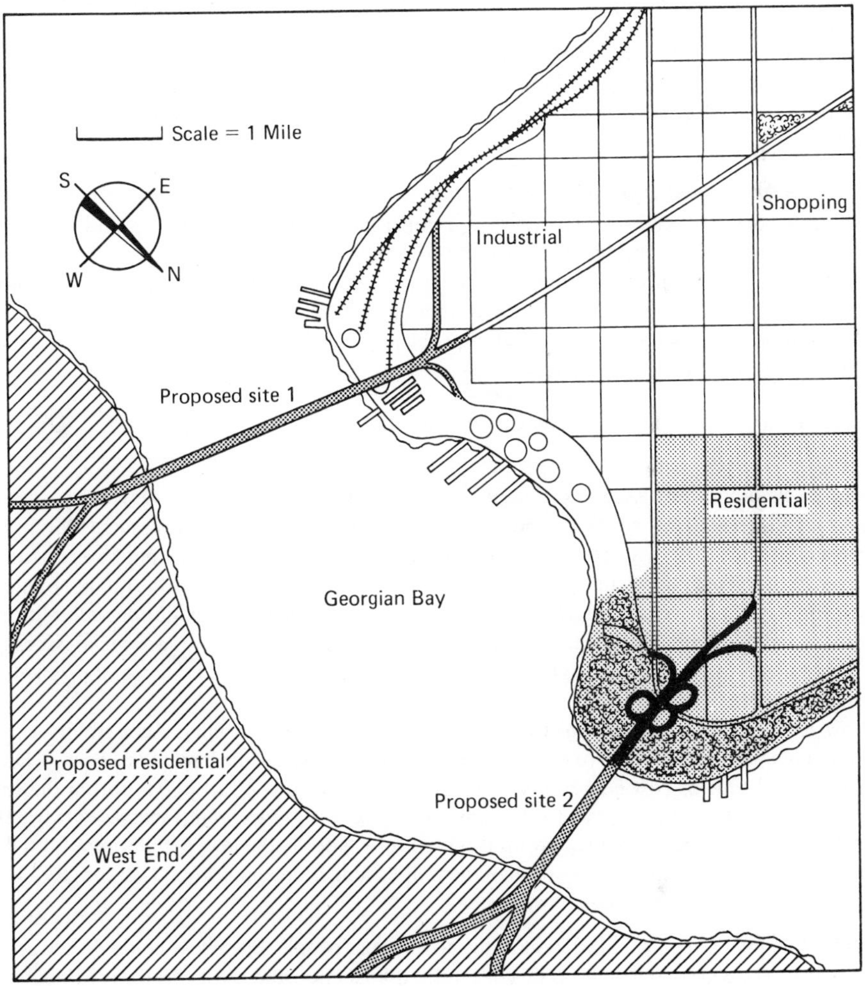

Figure 1.5

Discuss the use of systems analysis in relation to the following goals:

(a) Select a bridge site that will be the least costly to construct.
(b) Select a bridge site that will provide maximum societal benefit to the city.

Solution

(a) Minimize Construction Cost Minimum-cost problems are representative of the type of problem that a structural engineer faces when he designs a bridge. He must find the most economical means to construct the bridge. In this case, the measure of effectiveness will be specified in units of money or dollars. Costs associated with the design, the fabrication, and the construction of the bridge structure and foundation will be considered. Environmental conditions such as soil strength and fluid flow about the bridge piers will be investigated as part of the design process. The total cost of design, fabrication, construction, and material is considered an appropriate measure of effectiveness. The total cost for each site can be compared, and the least-cost alternative can be selected.

From a systems-analysis point of view, the selection of the least-cost design is straightforward. Acquiring, estimating, and establishing engineering data and costs will require a detailed study demanding much effort and time. The cost of this work is generally included in the design cost of the bridge.

(b) Maximize Societal Benefits Facility siting is representative of the type of problem that faces a transportation engineer. The issues are much more difficult to establish and quantify as compared to the least-cost bridge construction problem. A single-valued objective function cannot be established for this problem; therefore, it is not as amenable to the mathematical model presented in this text book. More important, there are other issues that make the mathematical modeling approach ill-suited for this kind of problem.

Consider the following two impacts.

1. Accessibility to West End from the City of Kingsbury.
2. Quality of life in the city.

Impact 1 An important issue in evaluating the impacts is *who* is affected. Obviously, the new population that moves into the West End will receive positive benefits. We can assume that the new bridge will provide them with good access to the city. Some city dwellers may not be as fortunate, however. If they live in the vicinity of the bridge ramp, we can anticipate that the quality of life for these people will be adversely affected. Traffic congestion, noise and air pollution, demolition of neighborhood buildings, and the taking of land may result. These problems are social issues that are very difficult to solve in the engineering sense. Political solutions to this type of problem are generally sought.

Now let us discuss different measures of effectiveness of accessibility. Accessibility can be measured in units of travel time or cost. However, how do we assess the impact on society? Do we measure travel time or cost between West End and the city only, or all points within the study region? How do we weigh reduced travel time and cost to one group against increased travel time and cost to another group? Which group is more important? Is the new population of the West End more important than the population of the well-established neighborhoods of the city? There are no clear-cut answers to these questions.

Impact 2 Quality of life means different things to different people. Even if we could decide upon a set of issues to be considered, how do we collect this information and use it effectively? Possibly we could interview people, asking them which bridge site they prefer, how the new bridge will affect them personally, and how they think it will affect their neighborhood. Obviously, incorporating this information into a model will be subjective at best.

Clearly, we have only scratched the surface of this problem. It should be evident that mathematical modeling is not well suited to the solution of this type of problem. Incidently, an environmental impact statement and public hearings are required for most major projects such as the one being considered in this hypothetical example. Local, state, and federal laws and regulations specify this need.

Systems analysis is an important decision-making tool but not a panacea. When suitable, it provides insight and understanding into optimization problems. Its advantages and shortcomings must be understood for it to be used effectively.

PROBLEMS

Problem 1

The statistics shown in Table 1.1 have been compiled for cities operating major transit systems.

(a) In your opinion which one measure of effectiveness most appropriately reflects the transit systems' overall efficiency? Why? Define overall efficiency.

(b) Are there other measures of effectiveness that reflect a transit system's overall efficiency? List them.

(c) In your opinion which one measure of effectiveness most appropriately reflects the transit systems' productivity? Why? Define productivity.

(d) Are there other measures of effectiveness that reflect the transit system's productivity? List them.

(e) Prepare a table, and rank the transit systems by city as being effective and productive. Use the measure of effectiveness from parts a, b, c, and d for the ranking.

(f) Which transit system ranks the best for overall efficiency and productivity? The worst? Why?

(g) Try to establish a single objective function to reflect overall efficiency utilizing all the measures of effectiveness listed under overall efficiency. List the problems associated with establishing this measure.

(h) Repeat step g for productivity.

Problem 2

Consider a cube, a cylinder, and a sphere as shown in Figure 1.6. They are to be used for storage. For each of the following, state the measure of effectiveness and determine:

(a) Which one has the maximum capacity.

(b) Which one utilizes the most material to construct.

(c) Which one offers the maximum capacity while utilizing the least building material.

(d) Which one has the maximum useable floor area.

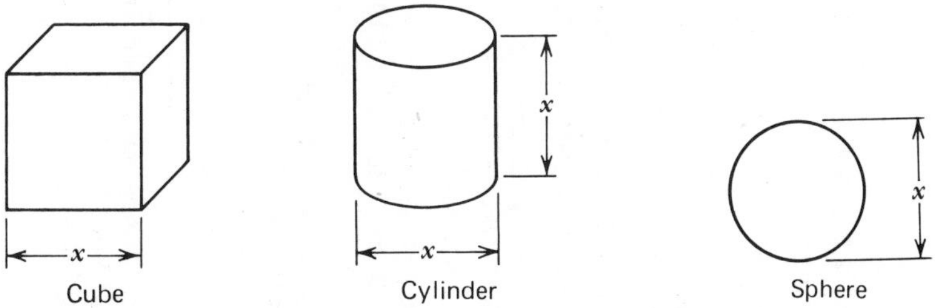

Figure 1.6

Table 1.1 Public Transit Systems[a]

	BOSTON	TORONTO	CHICAGO	NEW YORK	PHILADELPHIA	CLEVELAND	LOS ANGELES	WASHINGTON	SAN FRANCISCO
NUMBER OF EMPLOYEES	5,999	8,405	13,245	36,655	5,904	2,459	6,661	6,145	2,994
OPERATING BUDGET	$235.6M	$213.8M	$394.4M	$972.7M	$163.1M	$80.4M	$206.7M	$168.2M	$85.9M
RIDERSHIP (ANNUAL)	151.4M	337.6M	374.6M	1334.4M	212.9M	115.5M	230.5M	149.9M	120.2M
COST PER MILE OF VEHICLE OPERATION	$6.22	$2.18	$2.96	$3.22	$3.13	$3.05	$1.98	$2.82	$3.39
COST PER RIDER	$1.56	$0.63	$1.05	$0.72	$0.77	$0.69	$0.90	$1.12	$0.71
COST PER VEHICLE OWNED BY THE SYSTEM	$143,982	$92,960	$112,692	$103,984	$73,424	$73,531	$96,403	$76,945	$85,729
MILES OF OPERATION PER EMPLOYEE	6,333	11,672	10,057	8,242	8,825	10,736	15,646	9,690	8,466
WAGE COST PER MILE OF OPERATION	$4.16	$1.54	$2.35	$3.01	$2.20	$1.94	$1.55	$2.13	$2.70

Source: American Public Transit Association, Toronto Transit System, Massachusetts Bay Transportation Authority, Compiled by *Boston Globe*, Nov. 16, 1979.

[a] M = Millions

Problem 3

Identify five factors that may affect land developers in building the following types of structures. In other words briefly explain why:

New high-rise buildings are usually constructed in cities.

New fabrication plants tend to be located in suburbs.

Problem 4

Refer to Example 1.2.

(a) For each land classification—residential, shopping, and industrial—determine three different impacts that may affect the quality of life of the region owing to the construction of the new bridge.

(b) Discuss the problems associated with formulating a single measure of effectiveness to objective function that weighs both positive and negative impacts on the region.

Summary

The systems-analysis approach consists of the following steps.

1. State a goal, establish an appropriate measure of effectiveness, and develop an objective function.
2. Determine the financial, physical, and institutional limitations, and establish a set of constraint equations.
3. Determine a solution, the so-called optimum solution, that achieves the stated goal and satisfies all constraint conditions.

The mathematical model can be written as

Minimize or maximize	$z = f(\boldsymbol{x})$	(objective)
subject to	$\boldsymbol{g}(\boldsymbol{x})\{=, \leq, \geq\}\boldsymbol{b}$	(constraint set)
and	$\boldsymbol{x} \geq \boldsymbol{0}$	

where the constraint set consists of m constraint equations and $\boldsymbol{x}$ consists of a set of n control variables, $x_1, x_2, \ldots, x_n$. A control variable is a parameter that the engineer is free to vary during the design, planning, or management process.

The primary attribute of systems analysis is that it provides a systematic step-by-step procedure for formulating mathematical models and obtaining the optimum solution to these models. In this textbook, we discuss graphical and mathematical methods for solving various linear and nonlinear mathematical models. The systems-analysis approach may always be used to formulate problems. However, it is not a panacea; there is no guarantee that an optimum solution will be found. In some cases, the mathematics becomes so complex that a solution cannot be easily determined. For problems consisting of unlike measures of effectiveness or multiple objectives, the formulation of the problem as a mathematical model is difficult. In any event, thinking of an engineering problem in terms of satisfying a goal or objective, subject to a set of financial, physical, and institutional limitations, does provide a convenient framework for formulating engineering design, planning, and management problems.

Bibliography

Ambrose Giocoechea, Don R. Hansen, and Lucien Duckstein, *Multiobjective Decision Analysis with Engineering and Business Applications*, John Wiley, New York. 1982.

CHAPTER 2

An Overview of Optimization Methods

After the completion of this chapter, the student should be able to:

1. Solve optimization problems limited to one or two control variables by graphical means.
2. Recognize that linear mathematical models have the special property that the optimum solution will always occur at an extreme point.
3. Utilize the methods of calculus to find the critical-point solution of nonlinear optimization problems consisting of one control variable and classify the critical point as a minimum, maximum, or point of inflection.

The essential concepts for finding the optimum solution of resource allocation models can be illustrated with graphical means and models limited to two control variables, called *bivariate* models. That is, the control vector $\boldsymbol{x}$ consists of x_1 and x_2 only. When models consist of more than two control variables, graphical methods cannot be used and mathematical methods will be utilized. First, our discussion will be limited to models with linear objective and linear constraint equations. This is an extremely important class of problems, both from a mathematical and an engineering point of view.

Next, the graphical method will be used to solve bivariate nonlinear problems. Finally, calculus will be used to solve *univariate* mathematical models, models with one control variable. The section on calculus is intended to be a review.

The graphical method is an extremely important method because we can visualize the overall problem more clearly. We not only find the solution to a problem in a straightforward manner, we usually obtain insights that are not apparent by studying the mathematical model alone or by solving with mathematical methods. It is truly unfortunate that the graphical method of solution is not applicable to models with more than two control variables. The conclusions that we draw from solving univariate and bivariate models, however, are incorporated in solving multivariate linear models and the nonlinear programming found in later chapters of this book.

2.1 GRAPHICAL SOLUTION TO LINEAR MODELS

A bivariate linear mathematical model may be written as

$$z = c_1x_1 + c_2x_2$$
$$a_{11}x_1 + a_{12}x_2\{=, \leq, \geq\}b_1$$
$$a_{21}x_1 + a_{22}x_2\{=, \leq, \geq\}b_2$$
$$\vdots$$
$$a_{m1}x_1 + a_{m2}x_2\{=, \leq, \geq\}b_m$$

The number of constraint equations m will depend upon the problem statement.

The graphical procedure consists of the following steps.

1. Establish the feasible region from the set of constraint equations.
2. Assume a solution of z^0 and establish the slope of the line $c_1x_1 + c_2x_2 = z^0$.
3. Determine the optimum solution z^* by establishing a line that is parallel to z^0 and lies on the boundary of the feasible region.

Consider the mathematical model of Section 1.1. for maximizing total revenue for the building shown in Figure 1.2. The first step is to establish the feasible region. Let us establish the feasible region by investigating each constraint equation separately, and then we shall combine these results to construct the feasible region for the entire constraint set of equations.

The constraints $x_1 \geq 0$ and $x_2 \geq 0$ restrict the feasible region to be all points in the positive quadrant or quadrant I. In Figure 2.1*a*, we see that $x_1 > 0$ restricts the solution to be in quadrants I and IV. Similarly, in Figure 2.1*b*, we see that $x_2 \geq 0$ restricts the solution to be in quadrants I and II. The intersection of both constraints, $x_1 \geq 0$ and $x_2 \geq 0$, will always limit the solution to lie within the positive quadrant, quadrant I, as shown in Figure 2.1*c*.

The constraints $x_1 + x_2 \geq 5000$, $x_1 \leq 5000$, and $x_2 \leq 3000$ are shown in Figures 2.2*a*, 2.2*b*, and 2.2*c*, respectively. The arrows show the half-plane where the feasible region must lie. The intersection of these planes results in the feasible region as shown

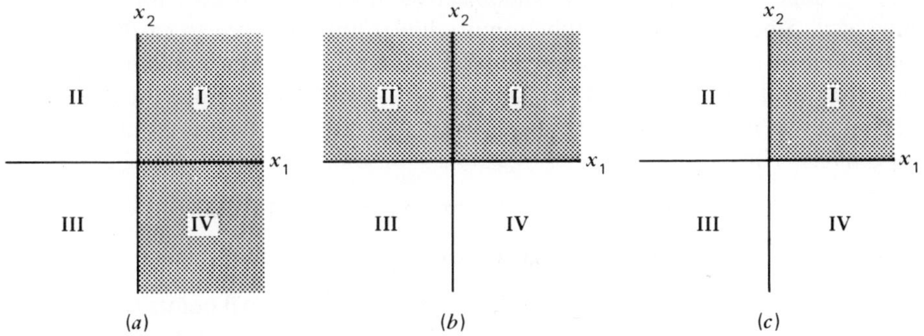

Figure 2.1 Nonnegative constraints $x_1 \geq 0$ and $x_2 \geq 0$.

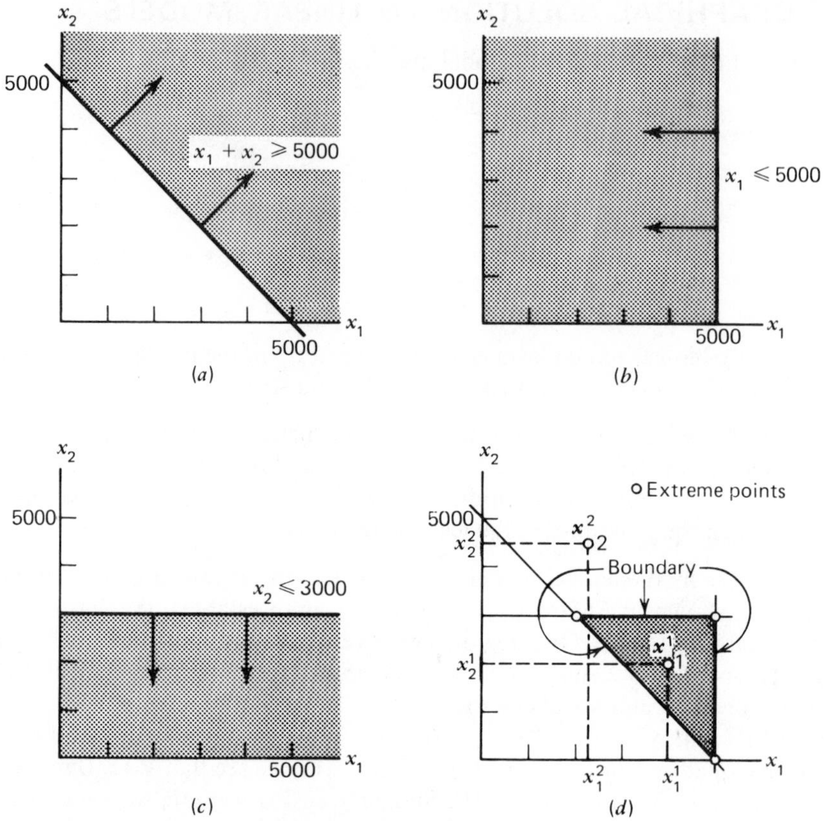

Figure 2.2 Establishing a feasible region.

by the shaded portion of the graph in Figure 2.2*d*. All points that lie within or on the boundary of the feasible region are called *feasible solutions*. All points that lie outside the feasible region are called *infeasible solutions*. Points $\boldsymbol{x}^1 = (x_1^1, x_2^1)$ and $\boldsymbol{x}^2 = (x_1^2, x_2^2)$ are examples of feasible and infeasible solutions, respectively.

The second step requires that an objective function be drawn for an assumed value of z^0. Since the objective function is a linear function, we can easily establish a *contour line* or locus of points satisfying $50x_1 + 60x_2 = z^0$. Let us arbitrarily choose z^0 to be equal to \$150,000. Thus, $50x_1 + 60x_2 = \$150{,}000$. For convenience, we determine the points $(x_1, 0)$ and $(0, x_2)$. These points lie on the x_1 and x_2 axes, respectively. Thus,

$$50x_1 + 60 \cdot 0 = 150{,}000 \quad \text{or} \quad x_1 = 3000 \text{ or } (3000, 0)$$

and

$$50 \cdot 0 + 60x_2 = 150{,}000 \quad \text{or} \quad x_2 = 2500 \text{ or } (0, 2500)$$

The z^0 line must pass through points (3000, 0) and (0, 2500). All points that lie on the line $z^0 = \$150{,}000$ and do not intersect as shown in Figure 2.3 are called infeasible solutions. None of these points can be the optimum solution.

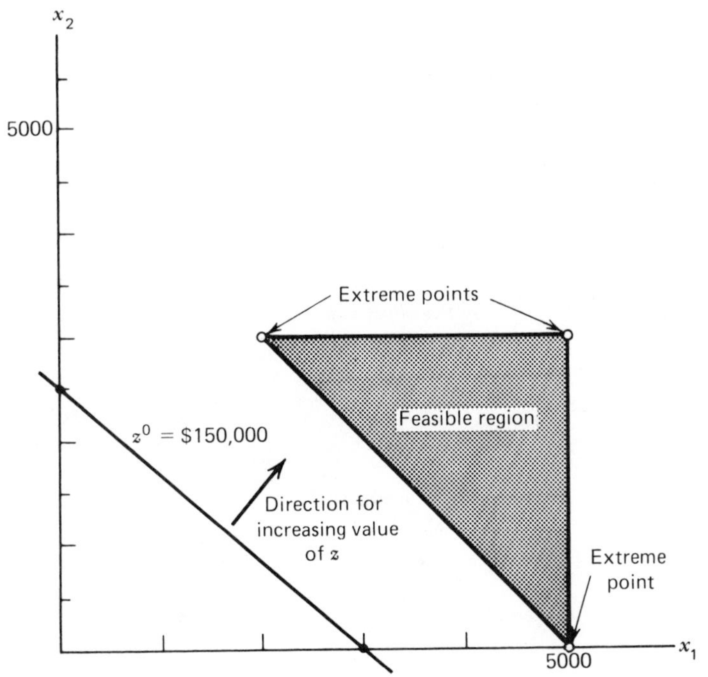

Figure 2.3 Searching for an optimum solution.

Utilizing the third step, the optimum solution may be found by drawing a contour line parallel to z^0 that lies at an extreme point of the feasible region. The direction for increasing the values of z is shown in Fig. 2.3. Any new contour line of z must intersect the feasible solution to be a candidate for an optimum solution. For linear models, an *extreme point* is defined to be the intersection of two or more constraint equations. An extreme point will lie on the boundary of the feasible region; therefore, it is a feasible solution and a candidate for the optimum solution. The contour line marked z^* is parallel to z^0 and it passes through the extreme point x^*, the location of the optimum solution for this maximization problem as shown in Figure 2.4. The point x^* is unique because z is a maximum and satisfies the conditions established in the constraint set of the problem. Thus, the optimum solution to this problem is

$$x_1^* = 5000 \text{ ft}^2 \text{ and } x_2^* = 3000 \text{ ft}^2$$

with optimum or maximum total revenue equal to $z^* = \$50 \cdot 5000 + \$60 \cdot 3000 =$ $430,000.

This is a straightforward approach to solving linear mathematical models. In step 3, we are able to establish the optimum point z^* because the slope of the objective function is always parallel to the contour line z^0 regardless of the assumed value of z^0. This is a property of linear functions. This approach can be used for any linear mathematical model restricted to two control variables with a minimum or maximum objective function.

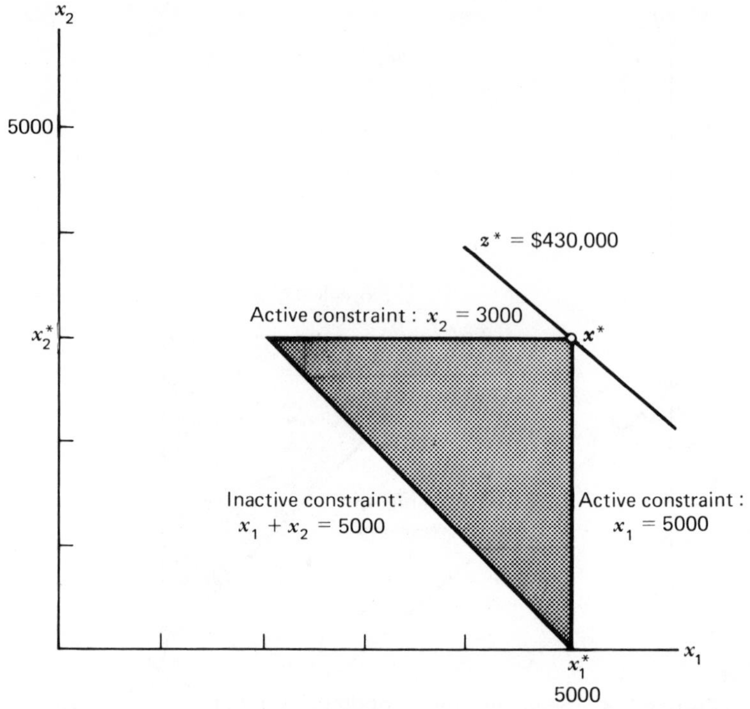

Figure 2.4 The optimum solution.

Active and Inactive Constraints

When identifying whether or not a solution is located on the boundary or at an extreme point of the feasible region, the terms active and inactive constraints are used. By definition, an *active constraint* will occur on the *boundary line* or at *an extreme point* of the feasible region. The constraint equation where $g(\boldsymbol{x})$ is either a linear or non-linear function of $\boldsymbol{x}$,

$$g(\boldsymbol{x}) \leq b$$

is said to be active constraint when the values of $\boldsymbol{x}$ satisfy the strict equality as:

$$g(\boldsymbol{x}) = b$$

On the other hand, when the solution $\boldsymbol{x}$ is such that the inequality is satisfied,

$$g(\boldsymbol{x}) < b$$

the constraint equation is said to be an *inactive* constraint. In Figure 2.4 the active and inactive constraints are indicated.

EXAMPLE 2.1 Minimum-Weight Truss

In Example 1.1, the following mathematical model was derived.

$$\text{Minimize } z = 612A_1 + 442A_2$$

$$A_2 \geq 4.17$$

$$A_1 \geq 6.67$$

$$A_1 \geq 8.33$$

$$A_2 \geq 6.25$$

$$A_1 \geq 16.7$$

where A_1 and A_2 represent the cross-sectional areas of the compression and tension members, respectively. Find the optimum member sizes using a graphical method of solution.

Solution

The first step is to establish the feasible region. Each equation has been plotted in Figure 2.5. Arrows have been placed upon each constraint equation to show the location of the feasible region for each one. The intersection of these constraints specify the feasible region. The feasible region is shown as the shaded region. It is bounded by the equations $A_1 = 16.7$ and $A_2 = 6.25$ as shown in Figure 2.6.

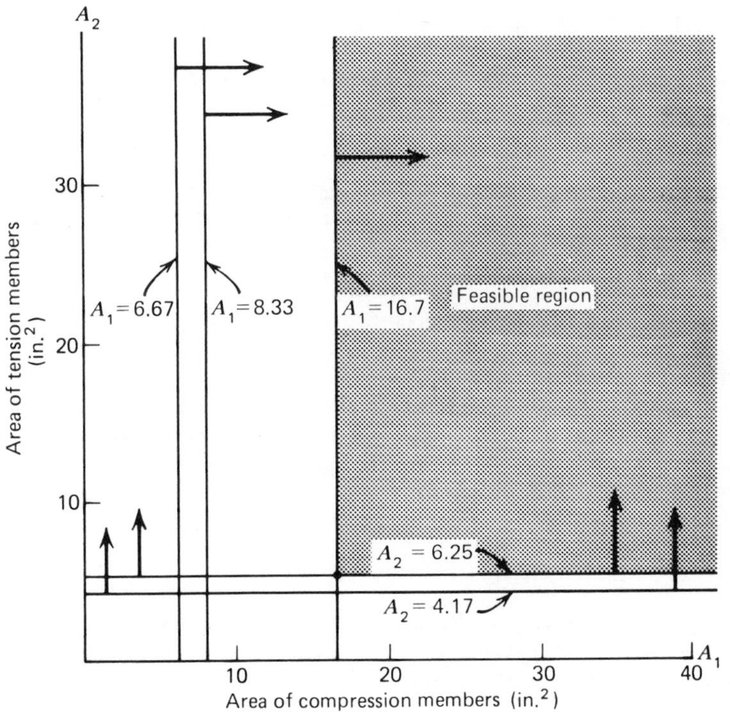

Figure 2.5

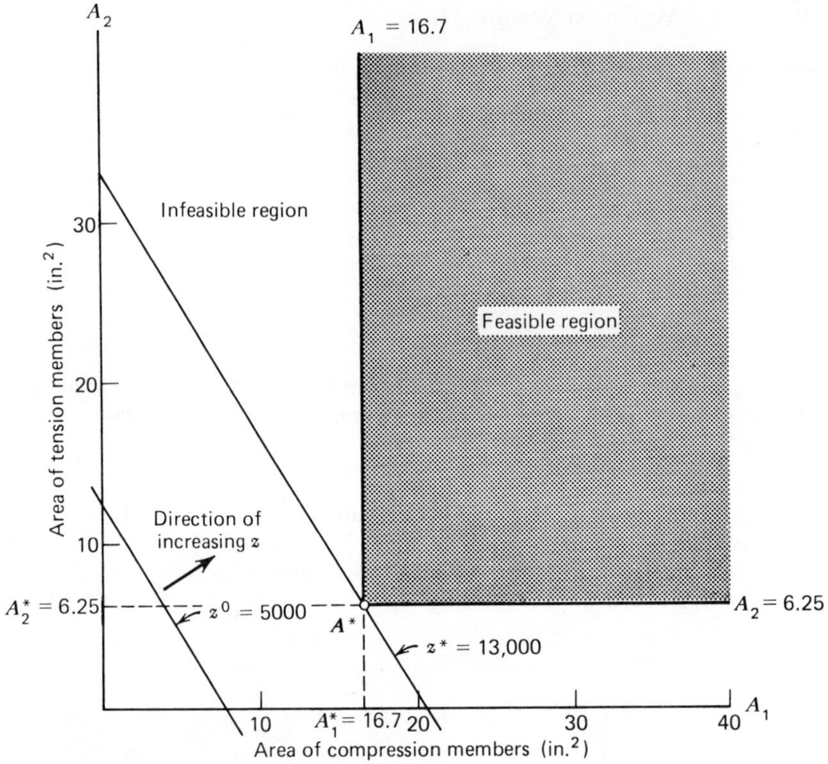

Figure 2.6

The next step is to determine the slope of the objective equation for an arbitrary value of z^0. Let $z^0 = 5000$. We shall find the intercepts of $612A_1 + 442A_2 = 5000$ on the A_1 and A_2 axes.

$$612A_1 + 442 \cdot 0 = 5000; \qquad A_1 = 8.16 \text{ or } (8.16, 0)$$

$$612 \cdot 0 + 442A_2 = 5000; \qquad A_2 = 11.31 \text{ or } (0, 11.31)$$

The line for $z^0 = 5000$ is shown in Figure 2.6.

Since all the points of z^0 lie in the infeasible region, no solution exists along the z^0 line. By increasing z, we can find the extreme point where z is a minimum. The arrow on $z^0 = 5000$ show the direction in which the optimum solution lies. The optimum solution $z^* = 13{,}000$ is shown in the figure. Note that lines z^* and z^0 are parallel.

The optimum solution is

$$A_1^* = 16.7 \text{ in.}^2, \quad A_2^* = 6.25 \text{ in.}^2$$

Thus, all compression members, AC, BC, and CD, have cross sectional areas of 16.7 in.^{2} and all tension members, AB and BD, have cross sectional areas of 6.25 in.^{2}. The total weight of the minimum truss is $z^* = 162 \cdot 16.7 + 442 \cdot 6.25$ or

$$z^* = 13{,}000 \text{ lb}$$

EXAMPLE 2.2 Minimum-Cost Aggregate Mix Model

A contractor is considering two gravel pits from which he may purchase material to supply a project. The unit cost to load and deliver the material to the project site is \$5.00/yd^3 from pit 1 and \$7.00/yd^3 from pit 2. He must deliver a minimum of 10,000 yd^3 to the site.

The mix that he delivers must consist of at least 50 percent sand, no more than 60 percent gravel, nor more than 8 percent silt. The material at pit 1 consists of 30 percent sand and 70 percent gravel. The material at pit 2 consists of 60 percent sand, 30 percent gravel, and 10 percent silt.

(a) Formulate a minimum-cost model.
(b) Determine the optimum solution by the graphical method.
(c) Determine the active and inactive constraint equations for the optimum solution.
(d) Determine the proportions of sand, gravel, and silt in the optimum solution.

Solution

(a) Formulation Since the gravel from pit 1 does not contain the minimum amount of sand to meet project requirements, the contractor may not utilize the cheaper material exclusively. He must mix the material from pits 1 and 2 to produce the required proportions.

We define the control variables to be

$$x_1 = \text{amount of material taken from pit 1 (in cubic yards)}$$
$$x_2 = \text{amount of material taken from pit 2 (in cubic yards)}$$

The cost function is

$$\text{minimize } c = \$5.00x_1 + \$7.00x_2$$

Let $x_1 + x_2$ equal the total amount of standard gravel mix delivered to the project site. The contractor must deliver at least 10,000 yd^3, thus the delivery constraint is

$$x_1 + x_2 \geq 10{,}000$$

The mixture must contain at least 50 percent sand. The contractor may obtain the desired amount of sand by combining the materials from each pit.

$$0.3x_1 + 0.6x_2 \geq 0.5(x_1 + x_2)$$

The products $0.3x_1$ and $0.6x_2$ are the amounts of sand taken from pits 1 and 2, respectively. The term $0.5(x_1 + x_2)$ is the amount of sand in the mix. Similarily, the constraint on the amount of gravel to be delivered is

$$0.7x_1 + 0.3x_2 \leq 0.6(x_1 + x_2)$$

Finally, the constraint equation for silt is

$$0.1x_2 \leq 0.08(x_1 + x_2)$$

The minimum cost model may be written as

$$\text{Minimize } c = 5x_1 + 7x_2$$

$$x_1 + x_2 \geq 10{,}000 \qquad \text{(delivery)}$$

$$0.3x_1 + 0.6x_2 \geq 0.5(x_1 + x_2) \qquad \text{(sand)}$$

$$0.7x_1 + 0.3x_2 \leq 0.6(x_1 + x_2) \qquad \text{(gravel)}$$

$$0.1x_2 \leq 0.08(x_1 + x_2) \qquad \text{(silt)}$$

$$x_1 \geq 0$$

$$x_2 \geq 0$$

or in standard form:

$$\text{Minimize } c = 5x_1 + 7x_2$$

$$x_1 + x_2 \geq 10{,}000 \qquad \text{(delivery)}$$

$$-2x_1 + x_2 \geq 0 \qquad \text{(sand)}$$

$$-x_1 + 3x_2 \geq 0 \qquad \text{(gravel)}$$

$$4x_1 - x_2 \geq 0 \qquad \text{(silt)}$$

$$x_1 \geq 0$$

$$x_2 \geq 0$$

The second cost model will be utilized for plotting. The first cost model will be used to answer part d and for checking the results.

(b) Graphical Solution

Step 1. *Establish the feasible region.* Since $x_1 \geq 0$ and $x_2 \geq 0$, the feasible region is restricted to be in a positive quadrant. Since only two points are needed to determine a line, each constraint equation is assumed to be a strict equality, and then the boundary of the constraint equation is found. For instance, for the equation:

$$x_1 + x_2 = 10{,}000$$

when $x_1 = 0$, x_2 must be equal to 10,000 or $x_2 = 10{,}000$ and when $x_2 = 0$, x_1 must be equal to 10,000 or $x_1 = 10{,}000$. Thus, the points (0, 10,000) and (10,000, 0) are sufficient to determine the boundary of the constraint, $x_1 + x_2 \geq 10{,}000$. Next, the boundary of the feasible region for the inequality constraints of sand, gravel, and silt is determined. In Figure 2.7*a* the arrows show the direction of the feasible region.

Since the gravel constraint $-x_1 + 3x_2 \geq 0$ lies outside the feasible region, it will not be considered in the search for the optimum solution.

Steps 2 *and* 3. Estimate an optimum solution, and test the condition of optimality. The minimum cost was assumed to be equal to \$80,000 or $c^0 = \$80{,}000$. Figure 2.7*b* shows this estimate to be too high. This estimate does not satisfy the condition of optimality.

The optimum-cost line will be parallel and less than the initial estimate of $c^0 = \$80{,}000$. Furthermore, for linear mathematical models, we should search for the optimum solution at an extreme point. For this problem, the optimum point is equal to

$$x_1^* = 3300 \text{ ft}^3, \quad x_2^* = 6700 \text{ ft}^3$$

with minimum cost $c^* = \$63{,}400$.

Note that the optimum-cost line is parallel to the initially estimated cost line, and it satisfies the conditions of optimality. In addition, the optimum solution occurs at an extreme point.

(c) Active and Inactive Constraints Since the optimum solution passes through the intersection of the lines marked delivery and sand, the equations $x_2 - 2x_1 = 0$ and $x_1 + x_2 = 10{,}000$ are both active constraints. The remaining equations are inactive constraints. These are the lines labeled silt and gravel. Since the constraint equation for gravel lies outside the feasible region, it will always be an inactive constraint.

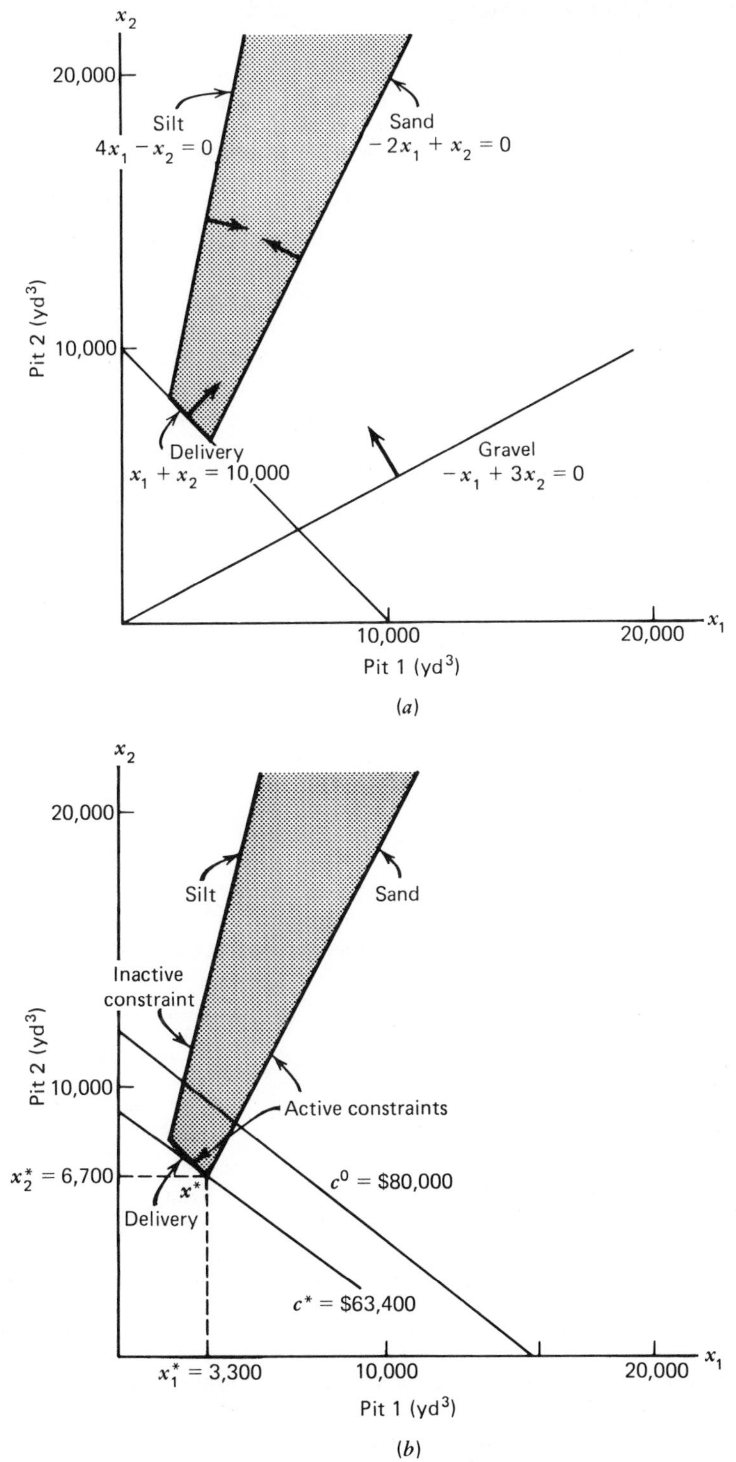

Figure 2.7 (*a*) Feasible region. (*b*) Optimum solution.

(d) The Optimum Mix The amount of sand, gravel, and silt that is the optimum mix is most conveniently determined from the first cost model. The amount of sand is

$$0.3x_1^* + 0.6x_2^* = 0.3 \cdot 3300 + 0.6 \cdot 6700 = 5010 \text{ yd}^3$$

The minimum amount required is $0.5(x_1^* + x_2^*) = 5000$, where $x_1^* + x_2^*$ is the total amount delivered.

The amount of gravel is

$$0.7x_1^* + 0.3x_2^* = 0.7 \cdot 3300 + 0.3 \cdot 6700 = 4320 \text{ yd}^3$$

This is less than $0.6(x_1^* + x_2^*) = 6000$, the maximum amount of gravel permitted in the mix.

The amount of silt is

$$0.1x_2^* = 0.1(6700) = 670$$

This is less than $0.08(x_1^* + x_2^*) = 800$, the maximum amount of silt permitted in the mix.

The constraint conditions are satisfied. The minimum cost to deliver the mix is $63,400.

PROBLEMS

Problem 1

The constraint set is

$$x_1 + x_2 \leq 3 \tag{1}$$

$$1 \leq x_2 \leq 2 \tag{2}$$

$$x_1 \geq 0 \tag{3}$$

(a) Clearly show the feasible region on a graph for the constraint set.
(b) Label all extreme points.
(c) If the less than or equal to constraint of Eq. (1) is changed to a strict equality, show the feasible region and label the extreme points.

Problem 2

Consider the following constraint sets:

$$x_1 - 2x_2 \leq 2$$

$$2x_1 + x_2 \leq 9$$

$$-3x_1 + 2x_2 \leq 3$$

$$x_1 \text{ is unrestricted in sign}$$

$$x_2 \geq 0$$

(a) Clearly show the feasible region on a graph for the following constraint set.
(b) Label all extreme points.
(c) If $x_1 \geq 0$ is imposed, show the feasible region and label the extreme points.

Problem 3

By graphical means, determine the location of the optimum solution.

$$\text{Maximize } z = 3x_1 + 2x_2$$

$$2x_1 + 4x_2 \leq 21$$

$$5x_1 + 3x_2 \leq 18$$

$$x_1 \geq 0$$

$$x_2 \geq 0$$

(a) Clearly show the feasible region.
(b) Determine the magnitude and location of the optimum point.
(c) Label the active and inactive constraints.

Problem 4

By the graphical approach, determine the optimum solution to

$$\text{Maximize } z = 5x_1 - 2x_2$$

$$x_1 - 2x_2 \leq 0$$

$$2x_1 + x_2 \leq 9$$

$$-3x_1 + 2x_2 = 3$$

$$x_1 \text{ is unrestricted in sign}$$

$$x_2 \geq 0$$

(a) Clearly show the feasible region.
(b) Determine the magnitude and location of the optimum point.
(c) Label the active and inactive constraints.

Problem 5

$$\text{Minimize } z = 3x_1 + x_2$$

$$2x_1 + 2x_2 \leq 9$$

$$2x_1 - 4x_2 = 6$$

$$x_1 \geq 0,$$

$$x_2 \geq 0$$

(a) Clearly show the feasible region.
(b) By graphical means determine the optimum solution.
(c) If the nonnegative restriction on x_2 (i.e., $x_2 \geq 0$) were removed, determine the effect on the optimal value of z.

Problem 6

$$\text{Maximize } z = 2x_2$$

$$x_1 \leq x_2 \quad (1)$$

$$-x_1 + 2x_2 \leq 2 \quad (2)$$

$$2x_1 + 2x_2 = 3 \quad (3)$$

x_1 is unrestricted in sign

$$x_2 \geq 0$$

(a) Clearly show the feasibility region.
(b) Determine the magnitude of the optimum point.
(c) Label the active and inactive constraints.
(d) How is the optimum point affected if constraint (2) is removed. Why?

Problems 7, 8, 9, and 10 were formulated in Section 1.1 as problems 1, 2, 3, and 4, respectively. The question here deals with the solution of the problems.

Problem 7

A contractor has two sand and gravel pits where he may purchase material. The unit cost including delivery from pits 1 and 2 is \$5 and \$7 per cubic yard, respectively. The contractor requires 100 yd^3 of mix. The mix must contain a minimum of 30 percent sand. Pit 1 contains 25 percent sand, and pit 2 contains 50 percent sand.

The object is to minimize the cost of material.

(a) Draw the feasible region.
(b) Determine the optimum solution by the graphical approach.
(c) Label the active and inactive constraints.

Problem 8

An aggregate mix of sand and gravel must contain no less than 20 percent nor more than 30 percent gravel. The in situ soil contains 40 percent gravel and 60 percent sand. Pure sand may be purchased and shipped to site at \$5.00/yd^3. A total mix of 1000 yd^3 is needed. There is no charge to use in situ material.

The goal is to minimize cost subject to the mix constraints.

(a) Draw the feasible region on the graph.
(b) Determine the optimum solution.
(c) Label the active and inactive constraints.

Problem 9

There are two suppliers of pipe:

SOURCE	UNIT COST (\$/LINEAR FOOT)	SUPPLY (LINEAR FOOT)
1	\$100	100 ft maximum
2	\$125	Unlimited

Nine hundred feet of pipe is required. The goal is to minimize the total cost of pipe.

(a) Find the optimum solution.

(b) Formulate a mathematical model with the supply of pipe from source No. 2 limited to 700 linear feet.

(c) Does a solution to part b exist? Use the graph to prove your answer.

Problem 10

A company requires at least 4.0 Mgal/day more water than it is currently using. A water-supply facility can supply up to 10 Mgal/day of extra supply. A local stream can supply an additional 2 Mgal/day. The concentration of pollution must be less than 100 mg/l BOD, the biological oxygen demand. The water from the water-supply facility and from the stream has a BOD concentration of 50 mg/l and 200 mg/l, respectively. The cost of water from the water supply is $100/Mgal and from the local stream is $50/Mgal. The goal is to minimize cost of supplying extra water that meets water quality standards.

By graphical means, show the feasible region, and determine the location of the optimum solution.

Problem 11

A treatment plant has a capacity of 8 Mgal/day and an operating efficiency of 80 percent. The operating efficiency is defined as the amount of BOD_5, the 5-day biological oxygen demand, removed by the treatment plant facility. For example, if 200 mg/l of BOD_5 enters the plant, the amount of BOD_5 leaving it is

$$200(1 - 0.8) = 40 \text{ mg/l of } BOD_5$$

Owing to regional population growth, the rate of flow of wastewater has increased to 20 Mgal/day. The BOD_5 of the wastewater is 200 mg/l. Since the plant can only treat Mgal/day, 12 Mgal/day is being diverted from the plant and is entering the river without treatment (Figure 2.8*a*).

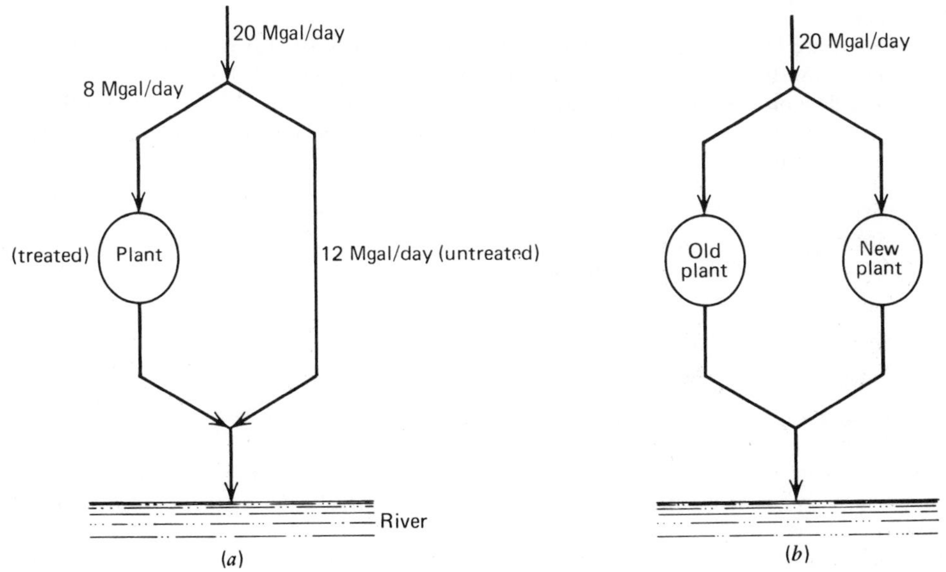

Figure 2.8

(a) Determine the BOD_5 of the combined flow of treated and untreated water entering the river (BOD_5 in mg/l).

(b) The cost of building a new treatment plant (Figure 2.8*b*) is

$$C = \$2.88 \times 10^6 Q \qquad (Q = \text{flow})$$

Define all control variables and formulate a mathematical model to treat 20 Mgal/day of wastewater which has 200 mg/l of BOD_5. The combined flow of treated water entering the river from the old and new plants must not exceed 25 mg/l of BOD_5. The efficiency of the existing plant is 80 percent, and of the new plant is 90 percent.

(c) Determine the minimum-cost solution by graphical means.

Problem 12

A 30-mile stretch of roadway (Figure 2.9*a*) is considered to have a poor level of service. Twenty-five million dollars ($25M) has been allocated for the project. The cost to improve the roadway is $1M/mile; therefore, the entire roadway cannot be improved.

The existing and proposed upgraded roadways are assumed to have a speed–density relationship as shown in Figure 2.9*b*.

(a) Derive speed–density functions for the existing and upgraded roads. The speed u is assumed to be a linear function of density k.

$$k = \text{density (vehicles/mile)}$$

$$u = \text{speed (miles/hr)}$$

(b) Derive a flow-density function for the existing and upgraded roads, $q = ku$, and draw the function for the existing and upgraded roadways, where flow is defined as q = flow (vehicles/hr).

(c) What are the maximum capacities of the existing and upgraded roads? Call them q_{max}.

(d) Determine the density at capacity k_{max} for the existing and upgraded roads.

(e) The flow on the existing roadway is 800 vehicles/hr between point A and intersection C, and 1200 vehicles/hr between C and B. Show that the vehicular speeds for uncongested flow, $q \le q_{max}$, are as follows:

Existing road
Segment A–C $\quad u_{AC} = 31.5$ miles/hr
Segment B–C $\quad u_{BC} = 20$ miles/hr

Upgraded road
Segment A–C $\quad u_{AC} = 52.4$ miles/hr
Segment B–C $\quad u_{BC} = 47.4$ miles/hr

(f) Determine the total travel time speed between A and B for existing road or $u_{AC} = 31.5$ miles/hr and $u_{BC} = 20$ miles/hr.

(g) Formulate a mathematical model to minimize total travel time between A and B. Assume a budgetary constraint of $25M. Clearly define control variables for mileage of roadway to be constructed.

(h) Solve for the optimum solution in part g by graphical means.

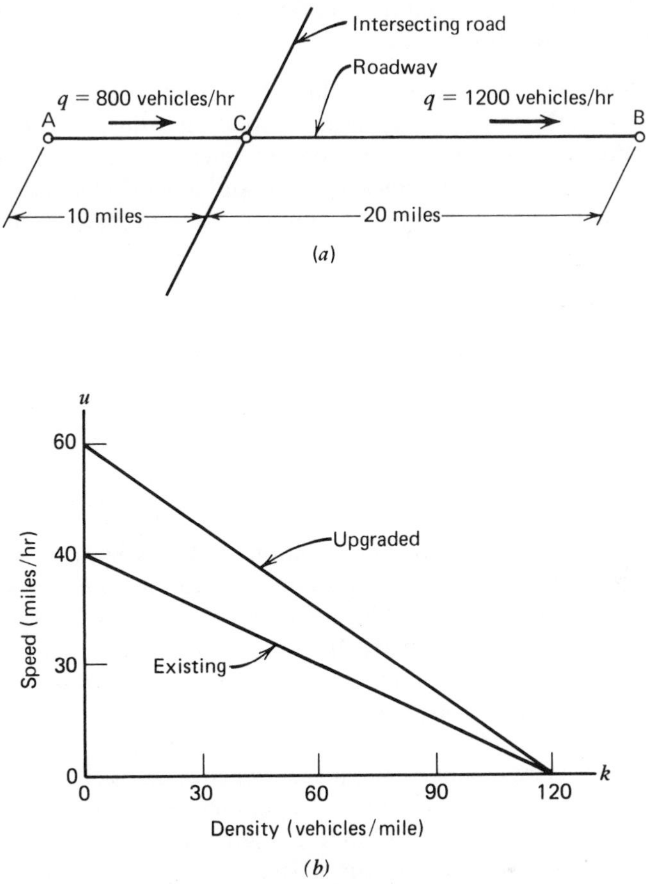

Figure 2.9

2.2 GRAPHICAL SOLUTION TO NONLINEAR MODELS

We shall extend our method for finding the optimum solution to nonlinear design problems limited to two control variables. For optimization problems of the general form, the mathematical model form is

$$z = f(x_1, x_2)$$

$$g_i(x_1, x_2)\{=, \leq, \geq\}bi \qquad i = 1, 2, \ldots, m$$

The graphical procedure consists of the following steps.

1. Establish the feasible region from the set of constraint equations.
2. Estimate an optimum solution, $z = z^0$, and draw the function, $z^0 = f(x_1, x_2)$.

3. Test to determine if the conditions of optimality are satisfied. If not, repeat steps 2 and 3 for a new estimate of z.

These steps are the same ones used to find the solution of linear models. The major difference between the search for an optimum solution of linear and nonlinear models is that the search for an optimum solution of a linear model is *restricted* to *extreme* points, whereas the search for an optimum solution of a nonlinear model must consider the *whole feasible region.* The optimum solution may occur at an *interior, boundary* or *extreme point.*

Conditions of Optimality

Compare the following functions

$$z_1 = c_1x_1 + c_2x_2$$

$$z_2 = x_1x_2$$

$$z_3 = (x_1 - 3)^2 + (x_2 - 4)^2$$

The functions z_1 and z_2 are monotonic functions and z_3 is a nonmonotonic function. The nonmonotonic function has one or more locations where local minima or maxima exist. The function z_3 has a local minimum at $x_1 = 3$ and $x_2 = 4$. Functions z_1 and z_2 do not possess this property. If x_1 and x_2 are restricted to a certain feasible region, it is possible that the nonmonotonic function z_3 will possess the property of a monotonic function. For instance, z_3 is a monotonic function for the region, $0 \leq x_1 \leq 1$ and $0 < x_2 \leq 2$.

The conditions of optimality are satisfied when a solution (x_1^*, x_2^*) satisfies all constraint equations and, at the same time, minimizes or maximizes the objective function of z. For a monotonically objective function of z, the condition of optimality is satisfied when the contour line of z is a boundary or extreme point of the feasible region. For nonmonotonically functions of z, the condition of optimality is satisfied when the contour line of z is a boundary point, extreme point, or an interior point of the feasible region.

Monotonic Objective Functions

A mathematical model is defined as being a nonlinear model when either the objective function is a nonlinear function or when one or more of the constraint equations are nonlinear functions. For example, the following nonlinear cost model consists of a linear objective function with a nonlinear constraint equation.

$$\text{Minimize } z = c_1x_1 + c_2x_2$$

$$g(x_1, x_2) = b$$

$$x_1 \geq 0$$

$$x_1 \geq 0$$

The linear objective function is a monotonically increasing function of $\boldsymbol{x}$.

The first step to solve this problem by graphical means is to draw the contour of the constraint equation on the graph. Since the constraint equation is a strict equality, any combination of x_1 and x_2 that lies on the constraint line is a feasible solution and a candidate for the optimum solution. The next step is to estimate the total cost z^0 and draw the cost function.

After steps 1 and 2 of the graphical procedure are performed, three possible outcomes may occur. They are illustrated in Figure 2.10.

Let us investigate each possible outcome with regard to step 3, that is, the testing of the condition of optimality. The solid lines marked z^0 are the initial estimates. The dashed lines marked z^* are the optimal solutions.

In Figure 2.10*a* the initial estimate z^0 is the optimum solution. This point is unique. No other solution will meet the constraint condition at smaller cost. At point 1, the *tangents* of the cost and constraint lines are equal. This is an important finding. For nonlinear models, it is possible that an optimum solution will occur at a *tangency point*.

Figure 2.10*b* shows the initial estimate z^0 to be below the constraint line $g(x_1, x_2) = b$. No feasible solution exists for this estimate. Since the feasible solution must lie on the constraint line, any combination of x_1 and x_2 that we select on the cost line will not fall in the feasibility region, therefore, it cannot satisfy the constraint requirement. Another interpretation of this result is that insufficient funds z^0 have been provided to achieve the minimum required output b. More money is needed. The new estimate of the cost line must be made as indicated by the arrows in Figure 2.10*b*. In lieu of repeating the estimating and testing for the conditions of optimality, we may utilize the properties of the linear cost function and conditions of optimality to solve this problem. The initial estimate of the cost line z^0 has established the slope of the cost function. All other cost lines must be parallel to this initial estimate. Since the conditions of optimality state that the objective function must be *tangent* to the constraint line of the feasible region, the optimum solution for this example must occur at a point where the cost line is tangent to the boundary $g(x_1, x_2) = b$, shown as point 1 in Figure 2.10*b*. The optimum is located at (x_1^*, x_2^*).

In Figure 2.10*c*, the initial guess is too high. Feasible solutions occur at point 2 and 3. The amount output at points 2 and 3 is the same $q(x_1^2, x_2^2) = g(x_1^3, x_2^3) = b$. We

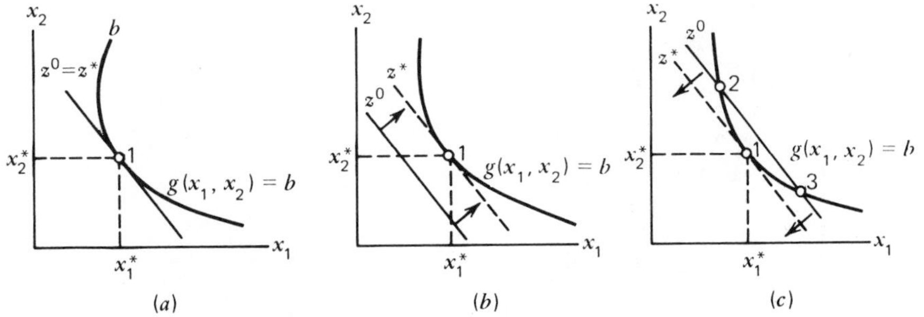

Figure 2.10 A linear objective and nonlinear constraint equation.

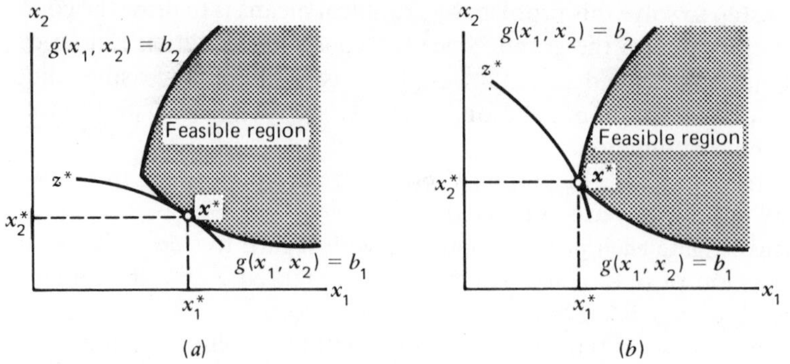

Figure 2.11 Conditions of optimality for a typical minimum-cost model.

are wasting money by producing at either one of these combinations of inputs of x_1 and x_2. If we change the values of x_1 and x_2 in such a manner as to remain on the constraint line, we can reduce the cost and remain in the feasible region. Since the slope of all cost lines must be parallel to the cost line z^0, we determine the optimum solution to be at point 1. Clearly, the optimum solution is located at (x_1^*, x_2^*). Furthermore, note the output $\mathbf{x}^*$ is the same at $\mathbf{x}^2$ or $\mathbf{x}^3$, $g(x_1^*, x_2^*) = g(x_1^2, x_2^2) = g(x_1^3, x_2^3) = b$.

Consider the optimum solutions to another nonlinear mathematical model consisting of two control variables and two constraint equations.

$$\text{Minimize } z = f(x_1, x_2)$$
$$g_1(x_1, x_2) \geq b_1$$
$$g_2(x_1, x_2) \geq b_2$$

If the mathematical model has a nonlinear, monotonic, increasing objective function and two nonlinear constraint equations, the optimum solution may occur at a tangent point or at an extreme point as shown in Figures 2.11*a* and 2.11*b*, respectively.

If the objective function is nonlinear and the two constraints are linear, we may obtain the solution at either a point tangent to the boundary or an extreme point as shown in Figure 2.12*a* and 2.12*b*.

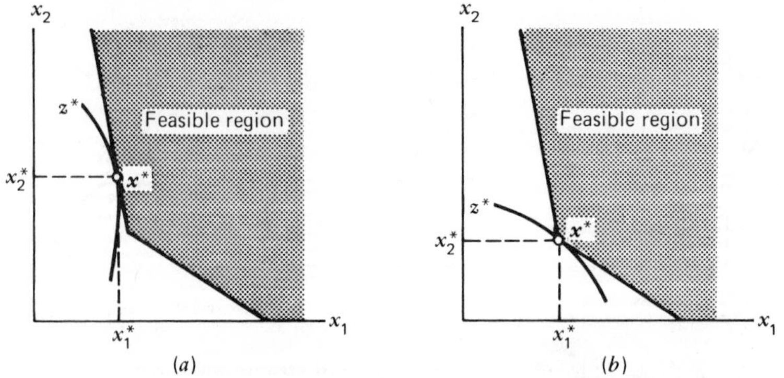

Figure 2.12 A nonlinear objective and linear constraint equation.

Nonmonotonic Objective Functions

Figures 2.10 through 2.12 show the optimum solutions to nonlinear mathematical models with various combinations of linear and monotonic nonlinear objective and constraint equations. We can see from these figures that the optimum solution will be located at a point that is either a *tangent* point or an *extreme point*. If the objective function is a nonmonotonic function as illustrated in Figure 2.13*a* and 2.14*a* the optimum solution of

$$\text{Maximize } z = f(x_1, x_2)$$
$$g_1(x_1, x_2) \le b_1$$
$$g_2(x_1, x_2) \le b_2$$
$$x_1 \ge 0,$$
$$x_2 \ge 0$$

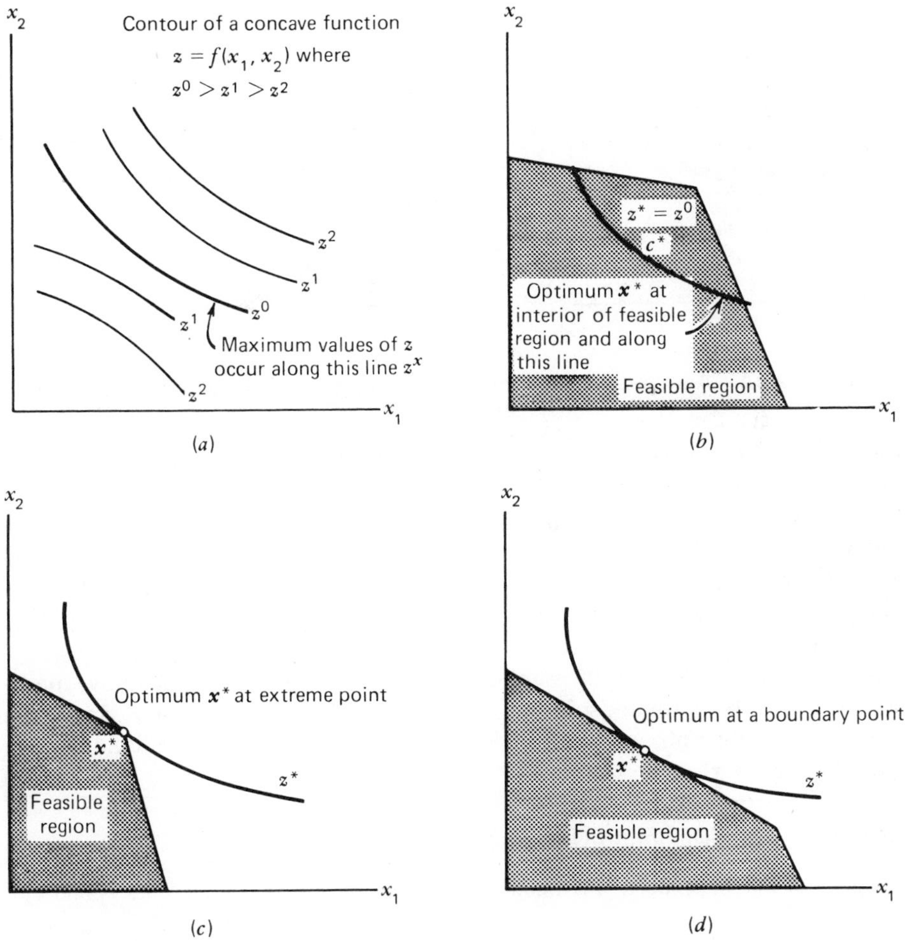

Figure 2.13 Nonmonotomic objective functions.

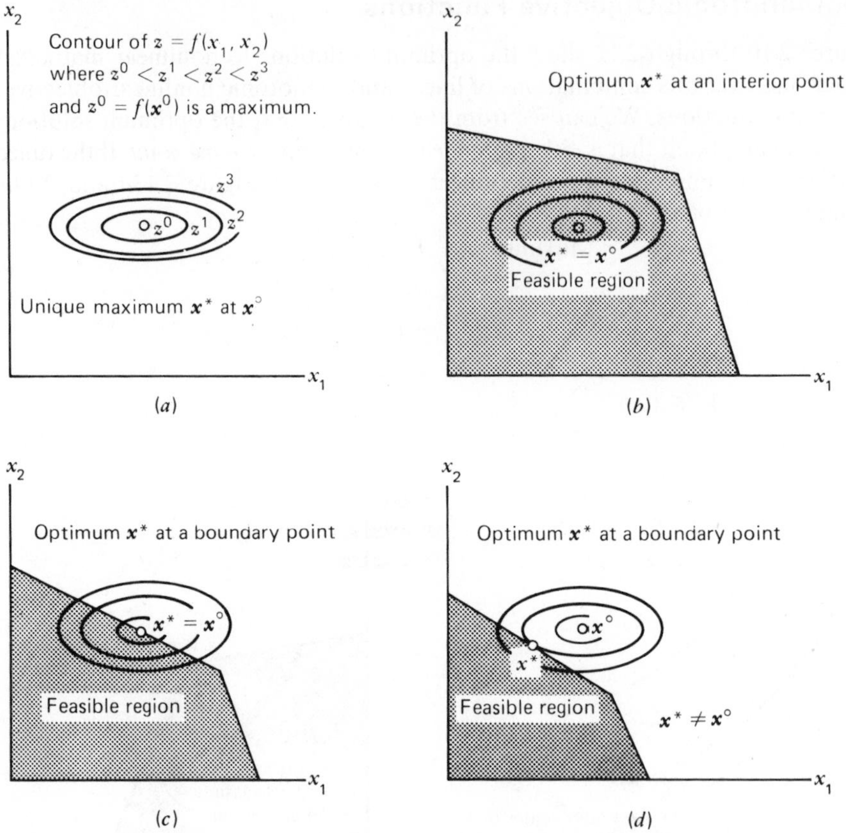

Figure 2.14 An objective function with a unique minimum at x^0.

may occur at an *interior*, *boundary*, or *extreme point*. Various linear constraint equations are shown in Figures 2.13 and 2.14.

In Figure 2.14*a*, a contour mapping of the function $z = f(x_1, x_2)$ is shown. It is a concave function with an unique maximum point at x°. Functions that have a single minimum or maximum point have special mathematical properties that will be used to help locate the solution of unconstrained and constrained optimization problems. Those features will be discussed in greater detail in the following section and in Chapter 8. For now, we use the graphical method of solution. For the above mathematical model, the optimum solution x^* will not necessarily occur at x°. In Figures 2.14*b* and 2.14*c*, the optimum solution is $x^* = x^\circ$, but in Figure 2.14*d* it is a boundary point where $x^* \neq x^\circ$.

EXAMPLE 2.3 Maximum Output Model

The output of a given production is a nonlinear function of two control variables x_1 and x_2.

$$q = x_1 x_2^{1.5}$$

The cost of production is

$$C = x_1^2 + x_2^{1.5}$$

(a) Formulate a maximum output model given that the production cost is equal to $100.
(b) Determine the optimum solution utilizing a graphical method of solution.

Solution

(*a*) ***Formulation*** The mathematical model is

$$\text{Maximize } q = x_1 x_2^{1.5}$$

$$x_1^2 + x_2^{1.5} = 100$$

$$x_1 \geq 0, \quad x_2 \geq 0$$

(*b*) ***Graphical Solution*** The maximum output will occur when the strict equality $x_1 + x_2^{1.5} = 100$ is imposed. Since the objective function is a monotonically increasing function, the solution will occur at a tangency point. The feasible solution is drawn and is shown as $C^* = 100$. Several different estimates of output were made. The level of output q^* is 375.5 units. This is shown as the tangency point on the graph. The optimum level of production is

$$x_1^* = 5 \text{ units}$$

$$x_2^* = 17.8 \text{ units}$$

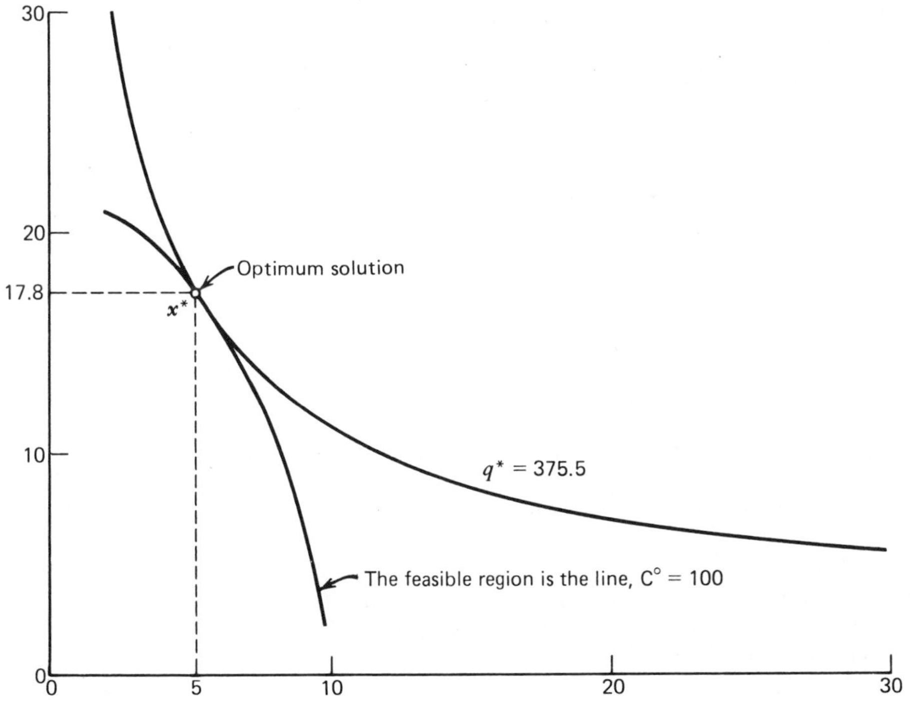

Figure 2.15

PROBLEMS

Problem 1

$$\text{Minimize } z = 4x_1 + x_2$$

$$x_2 \leq 1$$

$$x_1 x_2 \geq 2$$

$$x_1 + x_2 \leq 4$$

$$x_1 \geq 0$$

$$x_2 \geq 0$$

(a) On a graph show the feasible region.
(b) Show the location of optimum solution (x_1^*, x_2^*) and determine z^*.
(c) Classify each constraint as an active or inactive constraint.

Problem 2

(a) By graphical means, find the optimum solution to the minimum-cost model.

$$\text{Minimize } z = 20x_1 + 10x_2$$

$$x_1 + x_2 = 100$$

$$x_1 \geq 0$$

$$x_2 \geq 0$$

Label the optimum point as being an extreme, boundary, or interior point.

(b) Repeat part a with the objecture function equal to

$$z = 20x_1^{0.8} + 10x_2$$

(c) Repeat part a with the objective function equal to

$$z = 20x_1^{1.2} + 10x_2$$

Problem 3

The output q is a function of the inputs x_1 and x_2. The cost production is

$$q = x_1 x_2^{1.5}$$

$$C = x_1^{0.5} + x_2^{0.8}$$

(a) Formulate a minimum-cost model given that the level of production q is 375 units and the resources x_1 and x_2 are restricted to a maximum of 10 and 25 units, respectively.
(b) Show the feasible region.
(c) By graphical means determine the minimum-cost solution.

Problem 4

Design a chemical reactor of minimum cost (Figure 2.16). The flow of liquid is steady, $Q =$ 2 Mgal/day. The tank volume will be a function of the flow rate Q and retention time t. The retention time is equal to

$$t = \frac{V}{Q}$$

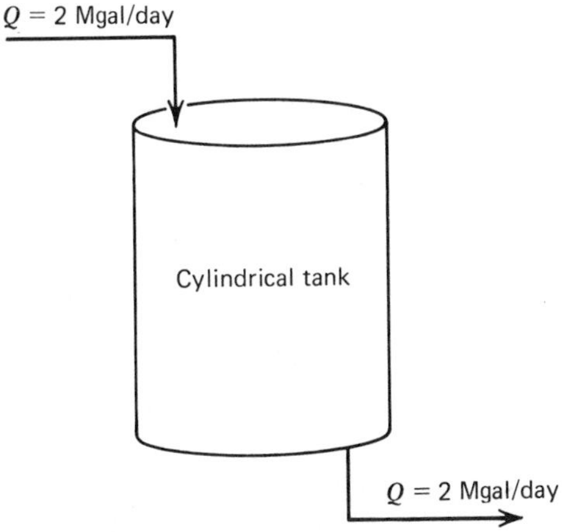

Figure 2.16

where Q is the flow rate and V is the volume. The retention time t is 90 mins. The cost to construct the walls and flooring is \$2.60/ft^2 of wall surface area and \$1.20/ft^2 of floor surface area, respectively. (1 gallon = 231 cubic inches.)

(a) Define the control variables.

(b) Formulate a minimum-cost model.

(c) Determine the optimum solution by graphical means.

Problem 5

A structural engineering consulting firm requires a large number of wide flange members for a particular structure. Each wide flange member is to be designed as a simply supported beam subject to a maximum load of 100 kips (1 kip = 1000 lb). In lieu of purchasing standard shapes, they will fabricate the wide flange sections by cutting $\frac{1}{2}$ in. steel plate and welding it together. The cross section of the member will be bd, where b and d are design variables. The sheet steel will be cut in a single pass with two torches.

(a) Formulate a mathematical model to minimize cost, or equivalently, the amount of steel to construct a wide flange member from flat plate. All calculations should be done in kips and inches. Consider the following AISC Building Code provisions.

A. Bending (AISC Section 1.5.1.5.1)

$$F_b = 0.66\ F_y$$

$$F_b = \text{allowable bending stress}$$

B. Shear (AISC Section 1.5.1.2)

$$F_s = 0.4\ F_y$$

The cross section may be taken as the product of the overall depth d and the web thickness t.

$$F_s = \text{allowable shear stress}$$

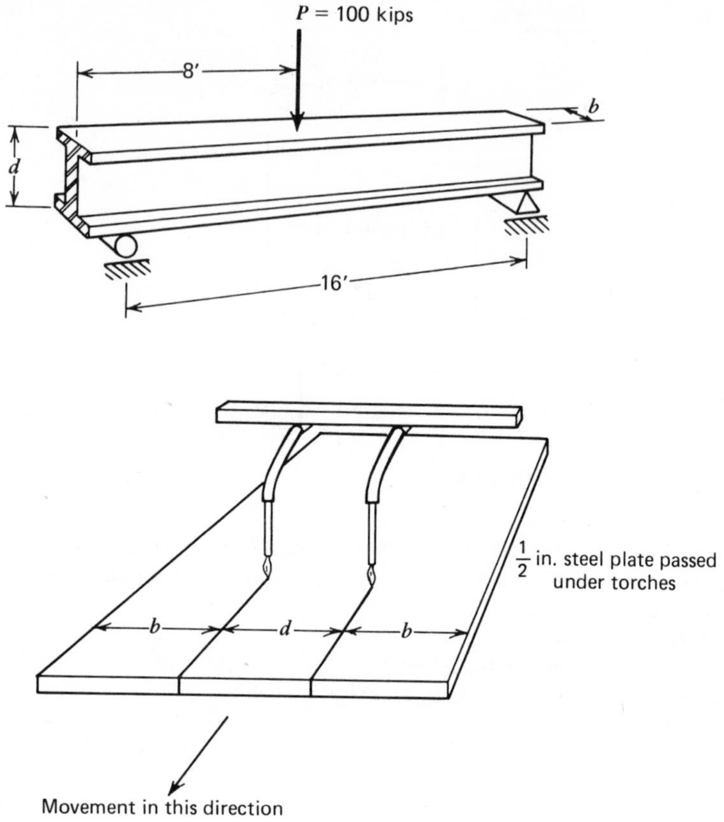

Figure 2.17

C. Lateral Stability (AISC Section 1.5.1.4.1e)

$$\frac{l}{b} \leq \frac{76b}{\sqrt{F_y}}$$

The compression flange shall be laterally supported at intervals not to exceed $76b/\sqrt{F_y}$ Assume no lateral support.

D. Material strength

$$F_y = 36 \text{ ksi}$$

E. For bending simplify by assuming the moment of inertia about the web and center line of the flanges is zero. That is, assume the moments of inertia about the centerline of the member is simply

$$I_f \simeq \frac{bd^2}{2}.$$

(b) Determine the feasible region.
(c) By graphical means, determine the optimum solution.

2.3 THE USE OF CALCULUS

The shape of the objective function and each equation of the constraint set has important bearing upon where to search for an optimum solution. If the function can be classified as a convex or concave function, the search can be simplified. In this section, univariate models or models with a single control variable are investigated. This section is intended to be a review of calculus methods and to show how basic principles of mathematics can be applied to optimization problems consisting of models with and without constraints.

The Unconstrained Model

A univariate mathematical model is expressed as a single variable function or

$$z = f(x)$$

In this discussion, the control variable x is assumed to be an unrestricted variable that allows for the location of the optimum value of $f(x)$ to be anywhere between plus and minus infinity. Let x° be a candidate for maximum, minimum, or point of inflection of $f(x)$; it is also called a *critical point*. If x° is the location of a maximum value of $f(x)$, the so-called *local* maximum, then $f(x^\circ)$ must be greater than or equal to values of the function $f(x)$ in the neighborhood of x°, or

$$f(x^\circ) \geq f(x) = f(x^\circ + \Delta x)$$

The neighborhood of x° will be defined as the set of points of the form

$$x = x^\circ + \Delta x$$

for small (positive or negative) Δx. Likewise, the minimum value of $f(x)$, a *local minimum*, is defined to be

$$f(x^\circ) \leq f(x) = f(x^\circ + \Delta x).$$

These relationships and the Taylor series expansion will be used to establish the necessary and sufficient conditions for the local maximum and minimum conditions of $f(x)$.

The *Taylor series expansion* for $f(x)$ evaluated at x, a point in the neighborhood of x°, is

$$f(x) = f(x^\circ) + \left.\frac{df}{dx}\right|_{x=x^\circ}(x - x^\circ) + \left.\frac{d^2f}{dx^2}\right|_{x=x^\circ}\frac{(x - x^\circ)^2}{2!} + \cdots$$

It is assumed that the function $f(x)$ is a continuous function and that first and second derivatives exist. The higher-order terms of the Taylor series are neglected. Since $x = x^\circ + \Delta x$, or $\Delta x = x - x^\circ$, the equation may be written as:

$$f(x) = f(x^\circ + \Delta x) = f(x^\circ) + f'(x^\circ)\,\Delta x + f''(x^\circ)\frac{(\Delta x)^2}{2!} + \cdots$$

where the derivatives

$$f'(x^\circ) = \left.\frac{df}{dx}\right|_{x=x^\circ} \quad \text{and} \quad f''(x^\circ) = \left.\frac{d^2f}{dx}\right|_{x=x^\circ}$$

are defined to be the first and second derivatives evaluated at the critical point x°, respectively. If x° is a minimum point, all points of $f(x)$ in the neighborhood of x° must be greater than $f(x^\circ)$, or

$$f(x^\circ + \Delta x) \geq f(x^\circ)$$

For this condition to be satisfied, the first derivative evaluated at $x^\circ, f'(x^\circ)$ must be equal to zero, or $f'(x^\circ) = 0$. This is the *necessary condition* for a minimum point. Since Δx is not restricted in sign, the term $f''(x^\circ)(\Delta x)^2$ will be equal to or greater than zero for a minimum to exist. Since $(\Delta x)^2$ is always positive, the only possible way for $f(x^\circ)$ to be a local minimum is that the first nonzero derivative of $f(x)$ evaluated at x° must be an even-order derivative and positive. This is the *sufficient condition* for a local minimum. A similar argument may be used to establish the necessary and sufficient condition for an univariate maximum.

The *necessary condition* for a local minimum, maximum, or point of inflection of $z = f(x)$ is that the first derivative of $f(x)$ evaluated at x° be equal to zero.

$$\left.\frac{dz}{dx}\right|_{x=x^\circ} = \left.\frac{df}{dx}\right|_{x=x^\circ} = f'(x^\circ) = 0$$

The second- and higher-order derivative of $f(x)$ is used to determine if the critical point x° is a minimum, maximum, or point of inflection. The *sufficient conditions* are:

Minimum

$$\left.\frac{d^nz}{dx^n}\right|_{x=x^\circ} = \frac{d^nf(x^\circ)}{dx^n} = f^n(x^\circ) > 0 \text{ and } n \text{ is an even integer}$$

Maximum

$$\left.\frac{d^nz}{dx^n}\right|_{x=x^\circ} = \frac{d^nf(x^\circ)}{dx^n} = f^n(x^\circ) < 0 \text{ and } n \text{ is an even integer}$$

Point of Inflection

$$\left.\frac{d^nz}{dx^n}\right|_{x=x^\circ} = \frac{d^nf(x^\circ)}{dx^2} = f^n(x^\circ) \neq 0 \quad \text{and } n \text{ is an odd integer}$$

For example, consider the function $z = x^3$. The necessary condition for a critical point is

$$\frac{dz}{dx} = 3x^2 = 0 \quad \text{or} \quad x = 0$$

Thus, the critical point is $x^\circ = 0$. The higher-order derivatives evaluated at $x^\circ = 2$ are

$$\frac{d^2z}{dx^2} = f''(x_0) = 6x^\circ = 0$$

$$\frac{d^3z}{dx^3} = f'''(x_0) = 6 > 0$$

Since $f'''(x^\circ)$ is nonzero, the function $z = x^3$ has a point of inflection.

Global Optima In order for the local optimum to be a *global minimum*, $x^\circ = x^*$, $f(x) \geq f(x^\circ) = f(x^*)$ for all values of x between plus and minus infinity. Likewise, the *global maximum* is $f(x) \leq f(x^\circ) = f(x^*)$ for all values between plus and minus infinity.

For a constrained mathematical model the location of the global optimum x^*, which is either a maximum or minimum, must satisfy the side constraint condition, $a \leq x \leq b$, and the objective function $z = f(x)$. The model is

$$z = f(x)$$

$$a \leq x \leq b$$

The optimum solution x^* may occur at an interior or at an extreme point. Consider the function shown in Figure 2.18. All points lying between a and b, $a < x < b$, are called *interior* points. The points $x = a$ and $x = b$ are called *extreme* points. The search for the location of the global optimum x^* will require a search of the feasible region or all points between a and b, $a \leq x \leq b$. The search may be limited to a finite number of points by utilizing the necessary and sufficient conditions for a local optimum. The search for the global minimum or maximum of the function $z = f(x)$

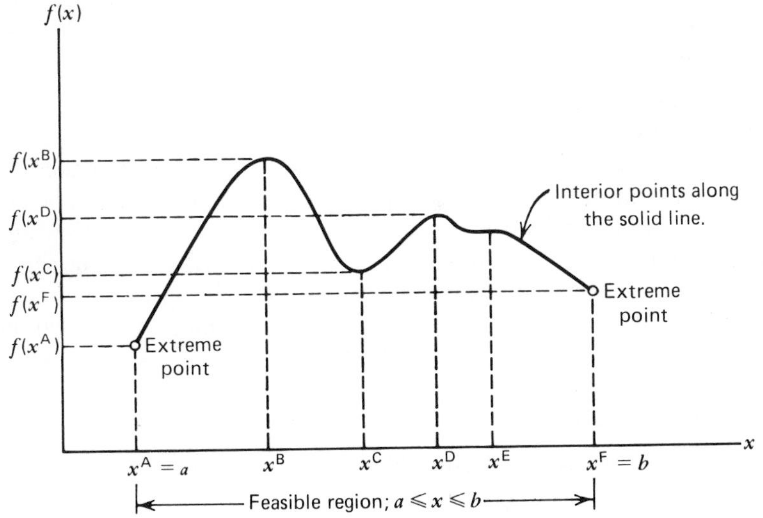

Figure 2.18 Local optima and extreme points.

shown in Figure 2.18 will be limited to the following six critical and extreme point locations.

x^A = extreme point

x^B = local maximum because $f''(x^B) < 0$

x^C = local minimum because $f''(x^C) > 0$

x^D = local maximum because $f''(x^D) < 0$

x^E = point of inflection because $f''(x^E) = 0$ and $f'''(x^E) \neq 0$

x^F = extreme point

If the minimum is sought,

$$\text{Minimize } z = f(x)$$
$$a \leq x \leq b$$

the location of the optimum solution may occur at the interior point x^C, or at the extreme points x^A or x^F. Compare the local minimum and extreme points $f(x^A)$, $f(x^C)$, and $f(x^F)$. From Figure 2.18, the global minimum occurs at the extreme point $x^* = x^A$, thus $z^* = f(x^A)$.

If the maximum is sought,

$$\text{Maximize } z = f(x),$$
$$a \leq x \leq b$$

the location of the optimum solution may occur at x^A, x^B, x^D, and x^E. Comparison of the local maxima and extreme points, $f(x^A)$, $f(x^B)$, $f(x^D)$ and $f(x^E)$, shows that $f(x^D)$ is the global maximum. Thus $x^* = x^B$ and $z^* = f(x^B)$. The location of the global maximum occurs at an interior point.

In the determination of the maximum and minimum value of $z = f(x)$, the point of inflection is not considered. Clearly, only local optima that satisfy the constraint $a \leq x \leq b$ and the two extreme points are considered.

Linear Mathematical Models If the objective function is a *linear function*, the search for optimum solution will be *restricted to extreme points* only. This is an important observation because it eliminates the search of the interior points. It makes the search for an optimum solution a more efficient procedure. This property will be utilized in searching for optima of multivariate linear models with the simplex method algorithm. More important, for linear functions the necessary condition cannot be satisfied and the second derivative of $f(x)$ is always equal to zero. The net result is that no information is obtained. This important conclusion will be extended to multivariate linear models also.

Unimodal Functions

Let us consider the constrained mathematical model once again.

$$\text{Maximize } z = f(x)$$
$$a \leq x \leq b$$

If the shape of the objective function is a *unimodal* function, the search is simplified even further. At most, a comparison of a single interior point and two extreme points must be investigated. The functions shown in Figure 2.19 are unimodal functions. In Figure 2.19*a*, since $f(a) > f(b) > f(x^\circ)$ the minimum value of $z = f(x)$ occurs at the interior point $x = c$. The maximum value of $z = f(x)$ occurs at the extreme point $x = a$. In Figure 2.19*b*, the critical value x° lies outside the feasible range of $a \leq x \leq b$, thus, the minimum and maximum search is located at either of the two extreme points. Since $f(b) > f(a)$, the values of the global maximum occur at $x = b$ and the global minimum occur at $x = a$.

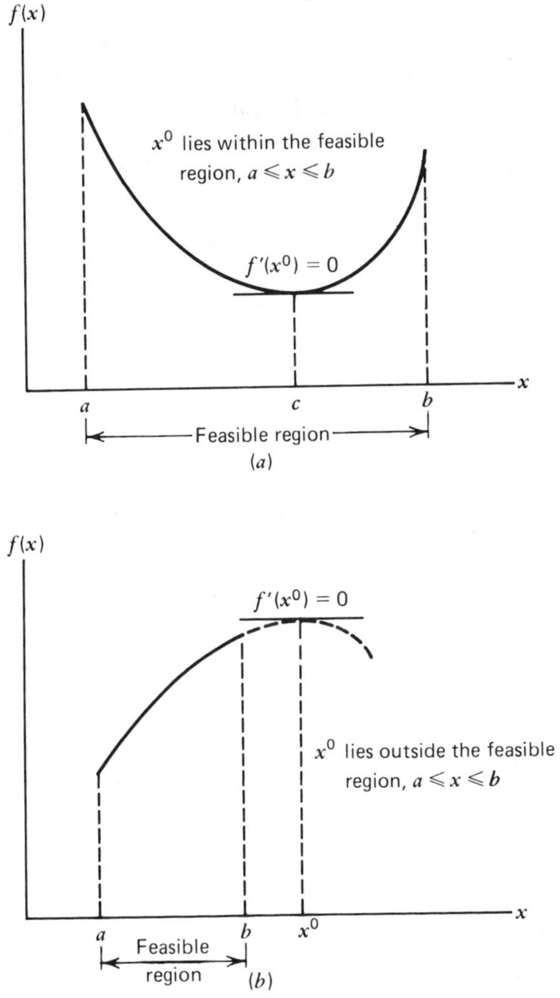

Figure 2.19 (*a*) Convex and (*b*) concave functions.

Test for Convex and Concave Functions A function is defined to be a convex function over a given range of x, $a \le x \le b$, if for any value of θ, between zero and one, $0 \le \theta \le 1$, the following inequality holds.

$$f(\theta a + (1 - \theta)b) \le \theta f(a) + (1 - \theta) f(b)$$

The test for a convex function may be established with this definition and the aid of the convex function shown in Figure 2.20. The value of x may be defined in terms of a, b, and θ.

$$x = \theta a + (1 - \theta)b, \qquad 0 \le \theta \le 1$$

The value of $f(x)$ or $f(\theta a + (1 - \theta)b)$ is depicted as point B. The relationship $\theta f(a) + (1 - \theta) f(b)$ is represented by a line drawn between points $f(a)$ and $f(b)$. The magnitude of $\theta f(a) + (1 - \theta) f(b)$ is shown as point A and $f(\theta a + (1 - \theta)b)$ is shown as point B. Since the magnitude at point A is greater than the magnitude at point B, the condition for a convex region is satisfied for the given value of x. Varying θ from 0 to 1 permits points between a and b to be evaluated. Thus, the function of $f(x)$ satisfies the definition for all values of x between a and b; therefore, the function of $f(x)$ is a convex function.

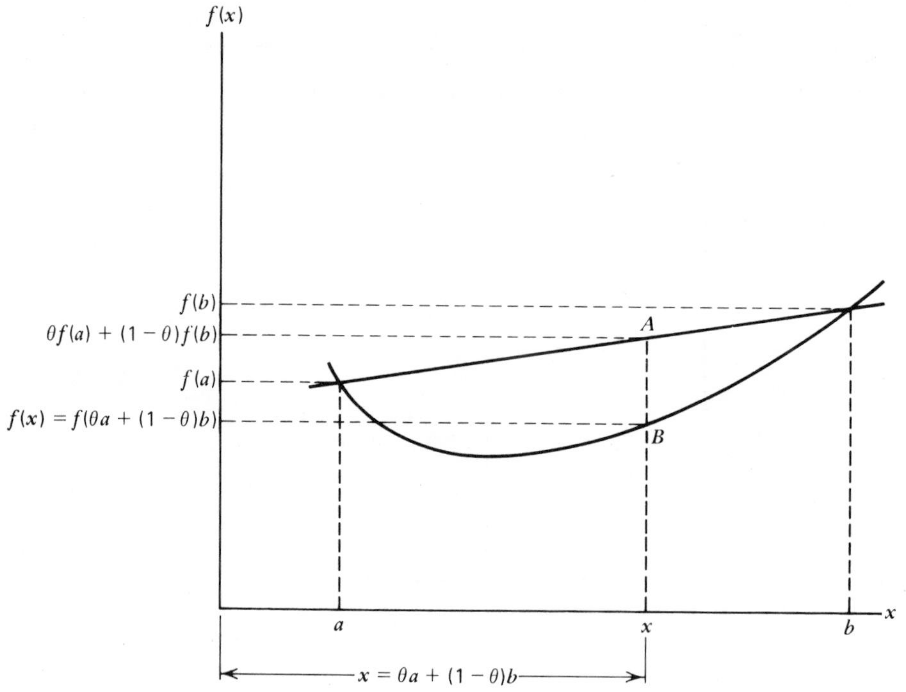

Figure 2.20 Test for a convex function.

A function is defined to be a concave function over a given range of x, $a \leq x \leq b$, if for any value of θ, $0 \leq \theta \leq 1$, the following inequality holds.

$$f(\theta a + (1 - \theta)b) \geq \theta f(a) + (1 - \theta) f(b)$$

Observe that if $f(x)$ is a convex function then $-f(x)$ is a concave function, and vice versa. For linear functions another important property emerges. A linear function satisfies the condition for the definition of a convex and a concave function, thus a linear function can be considered either a convex or a concave function.

If a function is known to be a convex, concave, or linear function, the search for an optimum solution is simplified. For linear functions only extreme points must be evaluated. For a convex function or concave function, local optima and the extreme points are evaluated.

EXAMPLE 2.4 The Sample Mean

To ensure safety of a concrete structure, three samples of compressive strength at 28 days are taken. The average or mean value of this sample must exceed the design strength of the concrete. Since the mean is a weighted sum of numbers, this procedure allows for one or two samples to be less than the design strength and still pass the strength test.

The following samples of 28 day strength were obtained: 3882, 3990, and 4512 psi.

(a) The sum of squared deviations can be calculated about any value. Show that when it is calculated about the mean, the sum of the square deviations is a minimum.

(b) Determine if this sample meets design specifications for 4000-psi concrete.

Solution

(a) Formulation Let $\bar{x}$ = unknown average or mean value of the three samples. Let x_1 = 3882 psi, x_2 = 3990 psi, and x_3 = 4512 psi, the sample 28 day strengths.

The deviation about the mean e is defined to be the difference between the observed and the mean value. Thus,

$$e_1 = x_1 - \bar{x} = 3882 - \bar{x}$$

$$e_2 = x_2 - \bar{x} = 3990 - \bar{x}$$

$$e_3 = x_3 - \bar{x} = 4512 - \bar{x}$$

The sum of the square deviations about the mean s is

$$s = e_1^2 + e_2^2 + e_3^2$$

$$s = (3882 - \bar{x})^2 + (3990 - \bar{x})^2 + (4512 - \bar{x})^2$$

The objective is to minimize s where $\bar{x}$ is the control variable.

$$\text{Minimize } s = (3882 - \bar{x})^2 + (3990 - \bar{x})^2 + (4512 - \bar{x})^2$$

The necessary and sufficient conditions for a minimum are

$$\frac{ds}{d\bar{x}} = 2(3882 - \bar{x})(-1) + 2(3990 - \bar{x})(-1) + (4512 - \bar{x})(-1) = 0$$

$$\frac{d^2 s}{d\bar{x}^2} = 2 + 2 + 2 > 0$$

Since $d^2s/d\bar{x}^2 > 0$, the conditions for a minimum are satisfied. Simplifying $ds/d\bar{x} = 0$ gives:

$$\bar{x} = \frac{3882 + 3990 + 4512}{3}$$

$$\bar{x} = 4128 \text{ psi}$$

The definition of the statistical mean and variance of sample size n is

$$\bar{x} = \frac{1}{n}\sum_i x_i$$

$$s_x^2 = \frac{1}{n-1}\sum_i (x_i - \bar{x})^2 \quad \text{or} \quad s_x^2 = \frac{1}{n-1}\sum_i e_i^2$$

Thus, the sum of square deviation s or, equivalently, the variance is a minimum about the mean $\bar{x}$.

(b) Adequacy Test If the average strength of the sample exceeds 4000 psi, we may assume that the entire batch exceeds the minimum strength requirement. Since $\bar{x} > 4000$ psi, the design requirement is met. Note that two out of three samples are less than 4000 psi.

EXAMPLE 2.5 Minimum-Weight Pressure Piping

The cost of piping is directly proportional to the amount of material used (Figure 2.21). Determine the minimum cost, or equivalently, the minimum-weight pipe to substain a pressure of 2500 psi. To accommodate the flow, the cross-sectional area of the pipe must be 10 in.^{2}. The allowable stress of the steel piping material is 30,000 psi. Formulate a mathematical model and solve for the minimum weight per unit length of pipe.

Solution

The total weight of piping will be equal to

$$W = \rho A l$$

where ρ is the unit weight of steel, 0.283 lb/in.^{3}; $A = 2\pi R t$ is the area of steel piping material; R is the pipe radius; t is the pipe wall thickness; and l is the length of pipe. Since the pipe is of uniform thickness along its entire length, the weight per unit length w will provide an equivalent objective. Thus, the objective is to minimize $w = W/l$. Since the cross section is 10 in.^{2}, the radius of the pipe may be determined.

$$\pi R^2 = 10$$

$$R = \left(\frac{10}{\pi}\right)^{1/2} = 1.78 \text{ in.}$$

Thus, the weight per unit length is

$$w = \rho A = \rho(2\pi R t)$$

$$w = 0.283 \cdot 2\pi \cdot 1.78t$$

$$w = 3.17t$$

Utilizing thin-wall pipe theory, the stress σ in the pipe is

$$\sigma = \frac{pR}{t}$$

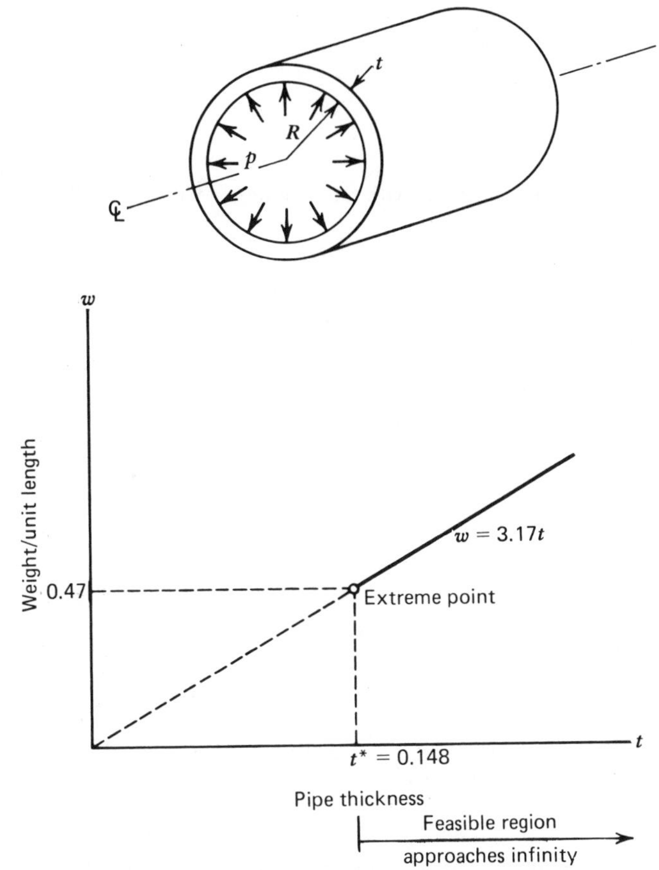

Figure 2.21

where $p = 2500$ psi = internal pressure

$$\sigma = \frac{2500(1.78)}{t} = \frac{4450}{t}$$

The mathematical model is

$$\text{Minimize } w = 3.17t$$

$$\sigma = \frac{4450}{t} \leq 30{,}000$$

$$t \geq 0$$

This is a nonlinear model because $4450/t$ is a nonlinear function. The constraint equation may be rearranged as

$$\frac{4450}{30{,}000} \leq t$$

$$0.148 \leq t$$

The model becomes

$$\text{Minimize } w = 3.17t$$

$$0.148 \leq t$$

$$t \geq 0$$

This is a linear model. Since w increases with t, the minimum value of t will occur at an extreme point, or

$$t^* = 0.148 \text{ in.}$$

The minimum weight per unit length is

$$w^* = 3.17t^*$$

$$w^* = 0.470 \text{ lb/in.}$$

PROBLEMS

Problem 1

(a) By graphical means, show that the function $|x|$ is a convex function over the range $-2 \leq x < 2$.

(b) Utilize the definition of a convex function,

$$f(\theta a + (1 - \theta)b) \leq \theta f(a) + (1 - \theta)f(b)$$

to prove $|x|$ is a convex function for the range of x between -2 and 2.

(c) For $x \geq 0$, is the function $|x|$ a convex function? Is it a concave function? Explain your answer.

Problem 2

Find all critical points and identify the critical points as local minima, local maxima, or points of inflection.

$$f(x) = x^3 + 6x^2 + 3x - 10$$

Problem 3

A cost function is equal to

$$C = 10q^{0.8}$$

$$q \geq 0$$

where q is the quantity of material produced or output.

(a) Use the definition of a convex function and a graphical plot to prove the function is a concave function in the feasible range.

(b) Use the sufficient condition for a global maximum to prove the function is a concave function for $q \geq 0$.

Problem 4

A moving load is to be placed upon a statically indeterminate beam as shown in Figure 2.22. The moment diagram is as shown.

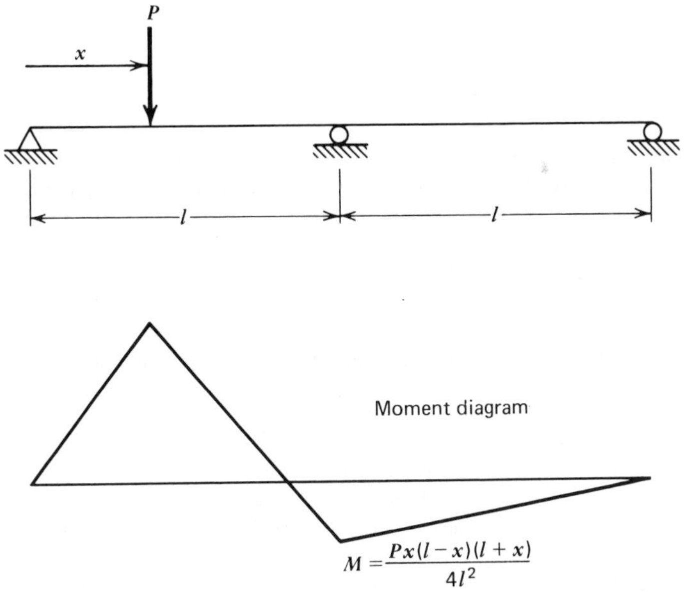

Figure 2.22

(a) Formulate the problem as a constrained optimization problem to determine the maximum moment at the middle support. The moment at the middle support is a function of load location x.

$$M = \frac{Px(l - x)(l + x)}{4l^2}$$

(b) Determine the location x where the moment at the support M is a maximum.
(c) Is the critical point determined in part b the location of a global maximum? Why?

Problem 5

The unit selling price p of an item is \$150/unit, thus the total revenue is $R = 150q$. The production cost C is a function of output level q, $C = aq^b$. The maximum ouput of the firm is 50 units per year.

(a) Formulate a mathematical model to maximize profit subject to the production constraint and production cost of $C = 100q^{1.2}$. Find the optimum level of production. Prove your answer is a maximum.
(b) Repeat part a for a production cost of $C = 100q^{0.8}$.

Problem 6

Determine the maximum volume of a sphere subject to an internal pressure of 4000 psi (pounds per square inch). The volume of a sphere is $V = (\pi/6)d^3$, where d is the mean diameter of the tank. The allowable stress of the material is equal to 20,000 psi. The hoop stress is a function of the pressure p, tank diameter d, and tank thickness t.

$$\sigma = pd/4t$$

The cost to purchase and fabricate the steel is \$200/lb. The surface area of a sphere is

$$A = \frac{3\pi}{2} d^2$$

Fabrication cost must not exceed \$6000.

(a) Formulate a mathematical model to maximize tank volume subject to the limitation on stress,

(b) Assume that both contraint equations are active constraints. Determine the tank thickness and tank diameter to maximize the tank volume. Note that there are two active constraints and two control variables; therefore, an extreme point will be evaluated.

(c) Use the graphical method of solution to prove the answer found in b is a global maximum.

Problem 7

The inflow rate to an industrial water treatment plant varies as a function of time t where t is in days (Figure 2.23).

$$Q_{\text{in}} = 2 + \sin(2\pi t) \text{ Mgal/day}$$

(a) Determine the total volume of wastewater treated each day.

(b) The wastewater is treated at a constant rate Q_{out} equal to 2 Mgal/day. Determine the size of the holding tank. The holding tank volume must be equal to the total cumulative volume of fluid being removed for treatment. Formulate a model with t as a control variable. Determine the tank size and prove that the tank is of maximum volume for continuous flow.

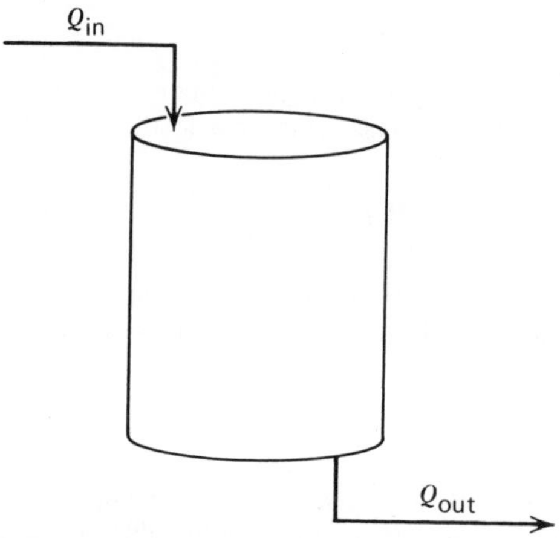

Figure 2.23

Summary

In this chapter, searching for an optimum solution for two-dimensional linear and nonlinear mathematical models is achieved by utilizing a graphical approach. A two-dimensional model consists of only two control variables of x_1 and x_2. A graphical method of solution provides a straightforward method for finding the optimum solution. The graphical method contains all the essential properties of searching for an optimum solution of higher dimension utilized in formal mathematical algorithms. When two-dimensional models are formulated, graphical methods should be used because the optimum solution can be obtained in a straightforward manner, and at the same time, the graphical representation gives better insight and better overall understanding of the problem than mathematical methods.

CHAPTER 3

Engineering Economics

After the completion of this chapter, the student should:

1. Be able to utilize the net present worth method to select the best design alternative among several candidate designs.
2. Understand the basic principles of capital and consumer theories to determine benefits, costs, interest rate, and design life to evaluate candidate alternatives.
3. Appreciate the special problems and issues of evaluating public projects.

The competitive market offers an investor many different opportunities to earn money. After weighing possible earnings offered by various companies and associated risks, the investor will then decide to make an investment in one of these companies. A transaction will only occur if the investor feels that he will earn a sufficient income in the future. If conditions are not to his liking, he will not invest at all. Very important facts emerge from this example. First, the selection process is competitive, and not all opportunities are worthy of investment. Second, there are risks that must be considered. Third, the anticipated earnings or benefits are not received at the time of the exchange of money but are received at a later time. In other words, the investment is an investment for future gain. The same issues and basic principles apply to selecting the best engineering design among various design alternatives.

Unlike the market, anticipated benefits from engineering facilities are not always marketable items. The benefits received from the use of a bridge, a water irrigation facility, and most other civil engineering systems are not sold in the market place. Methods to analyze these benefits in monetary terms must be established. The concepts of consumer theory will be used to help answer the question of how to evaluate benefits from engineering projects.

Civil engineering facilities will deliver services for many years. These facilities must be managed and maintained over their service lives. The costs of operating and maintaining the facility play an important role in the selection process. An equitable means of evaluating and comparing these current and future benefits and costs will be established utilizing the concepts of capital theory and time value of money.

Engineering economics provides a mechanism for incorporating these important issues into a decision-making selection method. The net present worth method will be introduced and used to evaluate and select the best design alternative. Particular attention will be given to the evaluation of public projects and to the effects of inflation upon the decision-making process.

3.1 THE SELECTION PROCESS

The problem facing a design engineer can be described by considering the following situation. A client hires an engineer to design a structural frame. Design specifications, building code requirements, and safety standards are provisions that must be met. The engineer is obligated to assure his client that his final design meets these standards for reasons of safety and performance. The successful engineer does not stop here. He assures the client that he is a cost-conscious individual by providing an economical design. Typically, a structure that meets all the design requirements and is economical to construct and maintain over its intended design life is the most desirable design. Generally, the design engineer considers different alternative designs. For example, concrete and steel structural frames with different geometric configurations may be considered. The most economical design will be the one that the engineer will recommend for construction.

This does not mean that the facility will be built. Only if the client feels that the anticipated costs to construct and to maintain the structure are less than the benefits will the client choose to build. Thus, it is important for the engineer to consider the benefits as well as the construction and maintenance costs when planning, designing, and recommending a final design to the client.

A similar situation occurs when a client hires an engineer to plan and design an industrial waste-treatment plant. Here, the benefit is an integral part of the design. The operating performance will vary with the choice of the treatment process. Thus, the engineer's final recommendation must consider the quality of the water and the cost to construct, operate, and maintain the facility. Again, both benefits and costs must be evaluated.

Two important points must be addressed. First, the engineer must have a mechanism for weighing benefits and costs on an equitable basis. Costs are inherently monetary measures, whereas the benefits cited in the two examples are not. In order to utilize engineering economic methods, *all* benefits and costs will be evaluated on a *monetary basis*. Thus, the benefits received from the structure frame and the waste-treatment facility will have to be transformed into dollar measures. The concepts of consumer economics are utilized to address this problem. This topic is discussed in Section 3.3. Until then, problems dealing with marketable monetary benefits will be utilized.

The second point is that the engineer must have a method of combining present and future costs and benefits. The value of future money is dependent upon the time at which it is received and its future purchasing power. It is more desirable to have a dollar today than a year or more from now. The dollar in hand today can be used for immediate consumption. Moreover, the future dollar may not maintain its purchasing power due to economic factors and other risks and uncertainties. Comparing dollars in hand today with future dollars usually requires that the worth of future dollars be *discounted*. Thus, the time value of money is an important issue that must be considered before the engineering economic selection methods can be appreciated. We use the following nomenclature.

P = present value of a single sum of money.

F_n = future sum of P at time period n.

i = interest rate.

Typically, i is given as annual interest rate in percent per year and n is given in years.

Interest

For simplicity, suppose that stock market investment opportunities are available and that payment of benefits will be made at the end of one year. Let P equal the amount of money or capital invested in the initial year and F_1 equal the anticipated earnings plus capital received at the end of the year. Here, $n = 1$ and F_1 is also called the benefit. The market rate of return is defined to be an annual interest rate i:

$$i = \frac{F_1 - P}{P}$$

The annual interest rate is a measure of the yield of capital or net productivity. An investor seeks to maximize i. If an investor is offered two different investment opportunities, one offering a 9 percent annual return and the other one 10 percent, the investment with the 10 percent rate of return will be chosen. This is the essence of the selection method, called the market rate of return method.

The market return rate i has many different names. Here, it is most commonly referred to as the *interest rate* and *opportunity cost interest rate*. It is also called the *discount rate* and *social opportunity-cost interest rate*.

Alternative Selection and Risk

The interest rate i may be used as a measure of the worth of an investment. Suppose two bonds are offered for sale at the same price and bond **A** has a higher rate of return than bond **B**, $i^{\mathbf{A}} > i^{\mathbf{B}}$. If these two investments are considered to be *risk free*, then bond **A** is a better choice.

Bonds that are offered by the U.S. Federal Government are considered to be risk-free because it is virtually 100 percent certain that the Government will make payment. Businesses, states, and local governments offer bonds and businesses offer stocks and bonds that may have greater elements of risk associated with them; however, the net profit from these investments may be far greater than those received from risk-free U.S. Bonds. The important point is that the investment process is subjective decision making where both anticipated net profit and risk must be weighed. The same considerations are incorporated in the engineering economics selection process.

Suppose that an investor is offered two investment opportunities for the same amount of money. One is an 8 percent risk-free government bond, and the other is the opportunity to purchase stock in a new high-technology firm that might have a rate of return as high as 40 percent per year. There is also the possibility that the firm may go bankrupt, making the stock worthless. In this case, the rate of return may be equal to zero. A gambler will speculate and purchase the stock. A person averse to risk will take the safer choice and purchase the bond. The decision involves many different social and economic factors. In any case, investments that involve greater risks usually offer higher rates of return than investments with lesser risk.

Opportunity-Cost Interest Rate

The interest rate is considered the rent, or price, paid for borrowed money. Since the lender surrenders the ability to purchase goods and services during the period of the loan, the lender has lost an opportunity to consume with his money. The interest payment the lender receives is considered payment for this lost opportunity. The interest rate i is an *opportunity-cost interest rate.*

The interest demanded by the lender will depend upon the risks involved. The risk averter will lend his money to a borrower whom he feels will honor the loan. A risk taker or gambler will speculate, and he will receive a greater return on his money than a risk averter. The risk averter will place his money in a savings account; whereas the gambler will speculate his money on a horse race. The earnings that the risk averter receives from the bank will be less than the payoff the gambler hopes to receive. Thus, lenders willing to invest in projects of greater risk usually demand a higher annual interest rate.

Banks adjust their interest rates to their business customers. A corporation will receive a bank loan with an interest rate based upon (1) the amount of inventory held by the firm, (2) the rate of business spending upon plant and equipment, and (3) the performance of corporate cash flow. In other words, the bank will adjust its loan interest based upon the risk involved.

The Net Present Worth Method

Throughout this introductory discussion, it has been implied that the market rate of return will always lead to the same selection of design alternatives. This is true only for the simplest evaluations as discussed thus far. For evaluation of projects with design lives greater than one year and with high annual costs relative to capital cost, the market rate of return tends to artificially inflate the interest rate and makes these projects appear more desirable. For this reason, and because of the mathematical difficulties associated with it, the market rate of return will not be used or discussed further. The net present worth, net annual worth, and benefit–cost methods are the most popular methods used in engineering practice. Since the benefit–cost ratio method suffers from the same difficulties cited for the market rate of return (plus other ones), it will not be discussed. We focus our discussion upon the net present worth method and to lesser extent the net annual worth method.

The net present worth method utilizes the same conceptual principles as the market rate of return selection method; however, instead of calculating the annual interest rate i, it is assumed to be known. For a simple time stream of one year, the net present worth is equal to

$$NPW = F_0 - P$$

In present worth calculations, *all benefits and costs are evaluated in the same period*, at time period 0, or at the present time. Thus, F_0 represents the present worth of the benefit received in year 1, F_1. With time-value-of-money equations for a 1-year time period, it will be shown that F_0 is equal to

$$F_0 = F_1(1 + i)^{-1}$$

and $NPW = F_1(1 + i)^{-1} - P$. Since $i > 0$, we see that F_0 is a discounted value of F_1. The investment will be considered *feasible* or *viable* only when the benefit received is greater than the cost. $F_0 > P$, or equivalently, $NPW \geq 0$. If $NPW < 0$, the alternative is infeasible and should not be selected. In the selection among competing alternatives, the alternative with the maximum net present worth will be the best choice. These principles will be used for time streams of greater than 1 year. The net annual worth method uses the same principle as the net present worth, except the benefits and costs are expressed as annual benefits and costs instead of as present values.

Time Value of Money

In this discussion, cash-flow diagrams for single-payment and uniform-payment series are used to determine the future amount of money F_n that one will receive from an interest-bearing account. Furthermore, it will be shown how these future sums F_n may be transformed into an equivalent present worth sum P.

Future Worth of a Single Payment The cash-flow diagram of a single-payment interest-bearing account is shown in Figure 3.1. Generally, in cash-flow diagrams, an arrow shown as a broken line will be utilized to indicate the unknown. In this case, the future payment F_n is the unknown. The payment will be made in the time period or year n. The future worth of a single payment is equal to the sum of money placed in an interest-bearing account at time 0 plus the accrued interest. We will define

P = a single sum of money placed in an interest-bearing account in period 0.

F_n = future sum, principal plus interest, withdrawn from the account in period n.

i = interest rate per period.

n = number of periods.

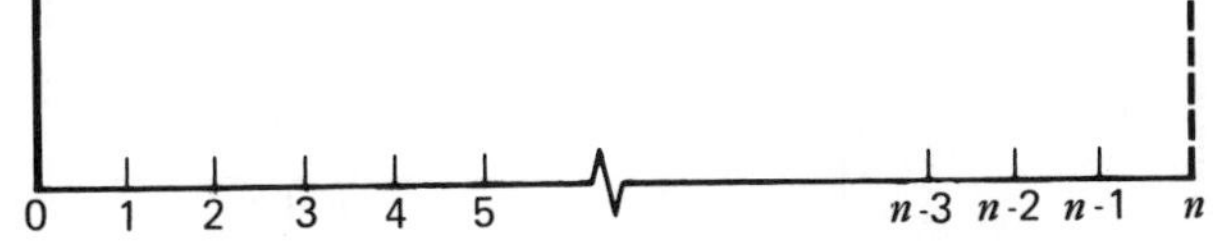

Figure 3.1 Future worth F_n of a single payment P.

Assume that the future worth F_n is unknown and all other values are known. If we place the original sum P into an interest-bearing account, the total worth after one time period will be equal to the original sum plus interest, or

$$F_1 = P + iP = P(1 + i)$$

If this sum of money F_1 is left in the account to accrue interest for another time period, the sum of money after the second period F_2 will be equal to

$$F_2 = F_1(1 + i) = P(1 + i)^2$$

If we continue to leave the money in the account for n periods F_n, the future worth of P will be equal to

$$F_n = P(1 + i)^n$$

The multiple compounding relationship $(1 + i)^n$ is called the single-payment compound amount factor.

Present Worth of a Single Payment The present value of a future sum of money F_n is called the present worth of a single payment P. The quantities P and F_n are considered to be *equivalent sums of money*. We may easily determine P with a known value of F_n by rewriting the future worth of a single payment equation as

$$P = F_n \frac{1}{(1 + i)^n} = F_n(1 + i)^{-n}$$

The multiple compounding relationship $(1 + i)^{-n}$ is called the single-payment present worth factor. The cash-flow diagram is shown in Figure 3.2.

Future Worth of a Uniform Series The cash-flow diagram for a uniform series interest account is shown in Figure 3.3. The future worth of a uniform payment series at time n is the total sum of n individual payments plus interest. We will define

A = a fixed sum of money placed in an interest-bearing account at the end of each payment period. (In the present context A may be considered to be a dummy parameter indicating either a operating and maintenance cost A or benefit B as previously defined.)

F_n = future sum of all A payments plus interest, withdrawn in period n

i = interest rate per period.

n = number of periods.

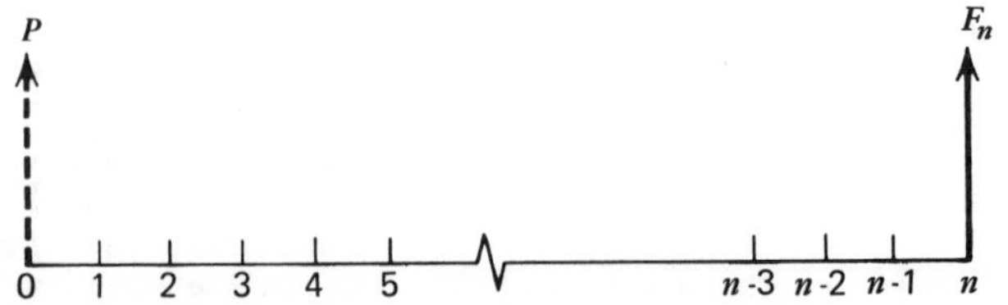

Figure 3.2 Present worth P of a future payment F_n.

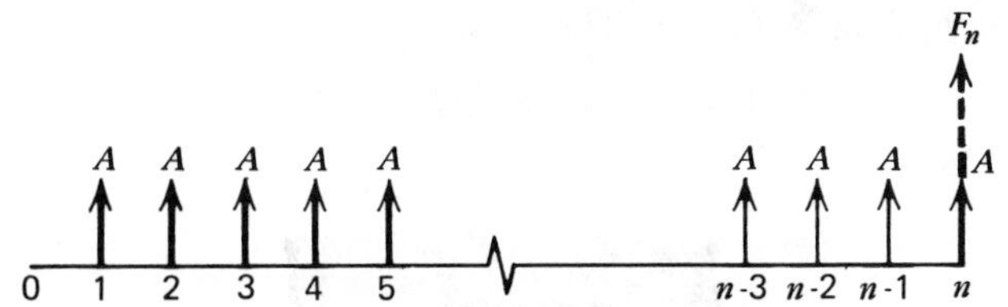

Figure 3.3 Future worth F_n of a series of uniform payments of A.

The return from a uniform series interest-bearing account can be derived by summing the returns from n single-payment interest-bearing accounts. The cash-flow diagram for an equivalent sum of single-payment diagrams is depicted in Figure 3.4.

The total sum from n future payments of A is

$$F_n = A(1 + i)^{n-1} + A(1 + i)^{n-2} + \cdots + A(1 + i) + A$$

or

$$F_n = A[1 + (1 + i) + \cdots + (1 + i)^{n-2} + (1 + i)^{n-1}]$$

This term may be simplified by multiplying both sides of the expression for F_n by $(1 + i)$.

$$F_n(1 + i) = A[(1 + i) + (1 + i)^2 + \cdots + (1 + i)^{n-1} + (1 + i)^n]$$

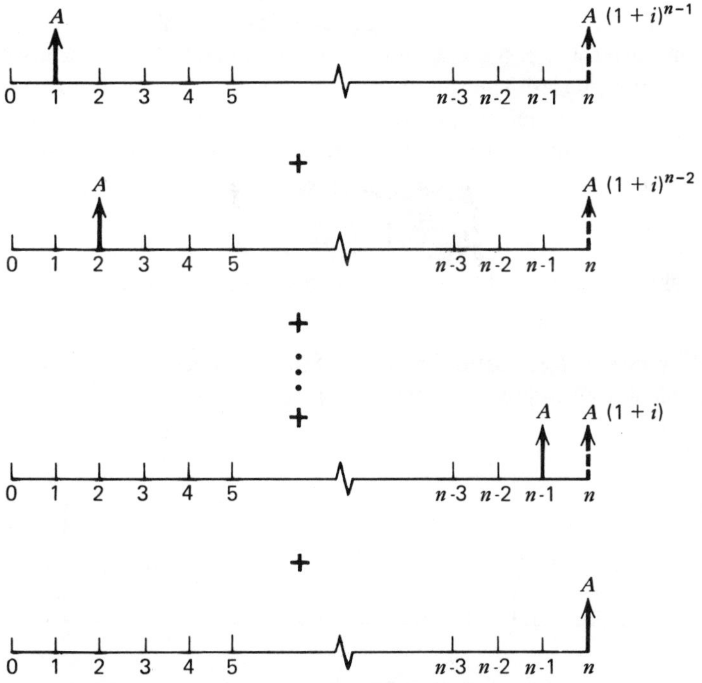

Figure 3.4 The sum of n future payments of A.

and substracting F_n from $F_n(1 + i)$. The result is

$$F_n(1 + i) - F_n = A[(1 + i)^n - 1]$$

or

$$F_n = A\left[\frac{(1 + i)^n - 1}{i}\right]$$

The multiple compounding relationship $[(1 + i)^n - 1]/i$ is called the uniform series compound amount factor.

Sinking Fund A sinking fund is an account where a future sum of F_n is known and the amount deposited A is unknown. This sum is obtained by placing fixed sums of money A into an interest-bearing account at regular intervals and withdrawing the total amount, principal plus interest, in time period n. We may utilize the future worth of a uniform series equation to determine the unknown A.

$$A = F_n\left[\frac{i}{(1 + i)^n - 1}\right]$$

The relationship $i/[(1 + i)^n + 1]$ is known as the sinking-fund deposit factor. The cash-flow diagram for a sinking fund is similar to the uniform series, interest-bearing account. The cash flow diagram is shown in Figure 3.5.

Present Worth of a Uniform Series The cash-flow diagram for the present worth of a uniform series is shown in Figure 3.6. The present worth P of a future sum of money A placed into an interest-bearing account at regular time intervals is called the present worth of a uniform series. The quantities P and F_n are considered to be equivalent sums of money. We can derive the present worth P by utilizing equations for future worth of uniform payment series and the present worth of a single payment.

$$P = A\left[\frac{(1 + i)^n - 1}{i(1 + i)^n}\right]$$

The relationship $[(1 + i)^n - 1]/i(1 + i)^n$ is known as the uniform series present worth factor.

Capital Recovery Fund A capital recovery fund is an account where a fixed sum of money is placed in an interest-bearing account and withdrawn in equal amounts for

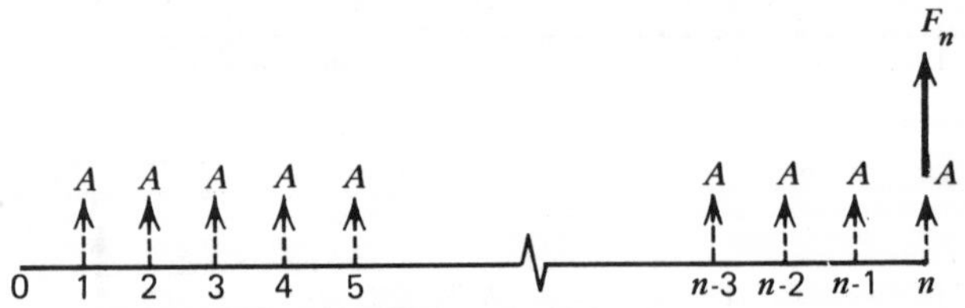

Figure 3.5 Future worth F_n of n uniform payments of A.

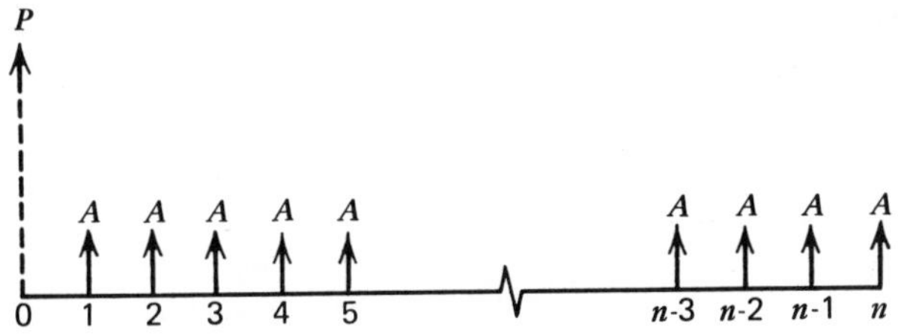

Figure 3.6 Present worth P of n future payments of A.

n payment periods. We may determine A from the present worth of a uniform series equation.

$$A = P\left[\frac{i(1+i)^n}{(1+i)^n - 1}\right]$$

The relationship $i(1 + i)^n/[(1 + i)^n - 1]$ is known as the capital recovery factor. The cash-flow diagram for capital recovery is similar to the cash-flow diagram for uniform series present worth. The cash-flow diagram is shown in Figure 3.7.

The Effect of Time and Interest Rate

In an earlier discussion, a feasible alternative was one where the present worth of the benefit exceeded the present worth of the cost, or $NPW = F_0 - C > 0$. For the sake of illustration suppose $F_n = \$100$, $P = \$50$, and $i = 5$ percent per annum. These are held constant, and the time n in which the \$100 benefit is to be received is changed. The present value of F_n and the NPW for $n = 5$, 10, 15, 20, and 50 yr are

n	$F_0 = F_n(1+i)^{-n}$	$NPW = F_0 - P$	THE ALTERNATIVE IS
5	\$78	\$28	Feasible
10	61	11	Feasible
15	48	−2	Infeasible
20	38	−12	Infeasible
50	9	−41	Infeasible

Clearly, the longer we must wait for the \$100, the less desirable the investment becomes.

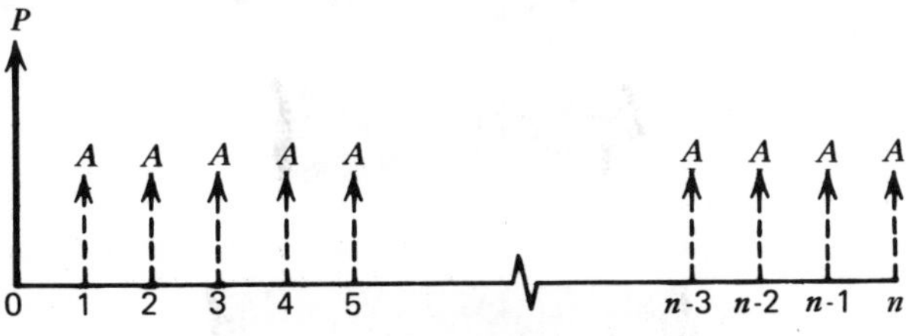

Figure 3.7 The annual worth A of a present worth P.

Now, let us hold $F_n = \$100$, $P = \$50$, and $n = 10$ yr constant and change the interest rate i to 3, 5, 10, and 15 percent.

i	$F_0 = F_n(1+i)^{-n}$	$NPW = F_0 - P$	THE ALTERNATIVE IS
3	74	\$24	Feasible
5	61	11	Feasible
10	39	−11	Infeasible
15	25	−25	Infeasible

With the NPW method, the future sum is a *fixed* cash payment that is discounted to a present value F_0. Thus, the larger the value of the interest rate or discount rate i, the smaller the value of F_0 and the less desirable the investment becomes.

If similar analyses are made for uniform series payments or a combination of single payment and uniform series payments, the same conclusions will be drawn. The longer the design life n or the larger the interest rate becomes, the more undesirable is the investment.

Multiple and Continuous Compounding

Not all interest rates are compounded annually; they may be compounded biannually, quarterly, monthly, daily, or even continuously. The formulas for single and uniform series as previously presented may be used for any discrete number of interest-bearing periods. Let $\tilde{i}$ equal the annual interest rate that is compounded m times per year. The interest rate i per period is

$$i = \frac{\tilde{i}}{m}$$

If the design life or investment period is equal to p years, the number of compounding periods is equal to

$$n = p \cdot m$$

The future sum of a single payment is $F_n = P(1 + i)^n$ and for a uniform series is

$$F_n = A\left[\frac{(1+i)^n - 1}{i}\right].$$

Making appropriate substitutions, the future worth of a single and uniform series in period n will be respectively equal to

$$F_n = P\left(1 + \frac{\tilde{i}}{m}\right)^{pm}$$

and

$$F_n = A\left[\frac{(1 + \tilde{i}/m)^{pm} - 1}{\tilde{i}/m}\right]$$

The other single and uniform series formulas may be used by appropriately substituting $i = \tilde{i}/m$ and $n = pm$ into these formulas.

An effective annual interest rate i_{eff} may be calculated by utilizing the future worth of a uniform series compounded m times during a year. For $p = 1$ yr, $n = pm = m$, the number of compounding periods per year, or F_{n}, $= P(1 + \tilde{i}/m)^m$. The future worth of a single payment F_1 is equal to $F_1 = P(\tilde{i} + i_{\text{eff}})$. The following equality must be satisfied.

$$F_1 = P(1 + i_{\text{eff}}) = P\left(1 + \frac{\tilde{i}}{m}\right)^m$$

or the effective yearly interest rate in percent per year is

$$i_{\text{eff}} = \left(1 + \frac{\tilde{i}}{m}\right)^m - 1$$

For continuous compounding, the number of interest periods per year is assumed to approach infinity.

$$\lim_{m \to \infty} F_{pm} = P \lim_{m \to \infty} \left(1 + \frac{\tilde{i}}{m}\right)^{pm}$$

or the future sum in year p, F_p, is

$$F_p = Pe^{\tilde{i}p}$$

The effective annual rate of interest is determined with the following equality.

$$P(1 + i_{\text{eff}}) = Pe^{\tilde{i}}$$

where p equals 1 yr, $p = 1$, or

$$i_{\text{eff}} = e^{\tilde{i}} - 1$$

EXP 1 ∧ ĩ — Continuous Compounding

The effective annual interest rates for different numbers of interest periods show the effect of multiple and continuous compounding for $\tilde{i} = 10$ percent/yr.

n	1	2	4	6	12	24	365
i_{eff} (%/yr)	10.00	10.25	10.38	10.43	10.47	10.49	10.52

For continuous compounding, $i_{\text{eff}} = 10.52$ percent.

EXAMPLE 3.1 Investment Selection: The Effect of Payment Timing

Consider two different \$2000.00 investment opportunities. The schedule of benefits is

INVESTMENT	YEAR 1	YEAR 2
A	\$150.00	\$300.00
B	\$250.00	\$195.00

Assume an opportunity-cost interest rate of 10 percent per annum. The principal of \$2000.00 is returned at the end of the investment period. Use the net present worth method to select the better alternative.

Solution

The cash-flow diagrams for the investments are shown in Figure 3.8. The cash-flow diagram of costs shows the benefits and cost as F_1, F_2 and C, respectively. Since the initial \$2000 investment is returned in year 2, the total benefit F_2 is the sum of \$2000 plus the scheduled benefit obtained from the table above. The present worth of alternative **A** is the present worth of benefits minus cost. The single-payment present worth factor $(1 + i)^{-n}$ is used in this analysis.

$$NPW^A = F_1(1 + i)^{-1} + F_2(1 + i)^{-2} - C$$

$$NPW^A = 150(1 + 0.1)^{-1} + 2300(1 + 0.1)^{-2} - 2000$$

$$NPW^A = 136 + 1901 - 2000 = \$37$$

Since $NPW^A > 0$, alternative **A** is a viable alternate and should be considered for possible selection.

For alternative **B**, the present worth is

$$NPW^B = F_1(1 + i)^{-1} + F_2(1 + i)^{-2} - C$$

$$NPW^B = 250(1 + 0.1)^{-1} + 2195(1 + 0.1)^{-2} - 2000$$

$$NPW^B = 227 + 1814 - 2000 = \$41$$

Alternative **B** is also a viable alternate. Since $NPW^B > NPW^A$, alternative **B** is the better alternative and the better choice.

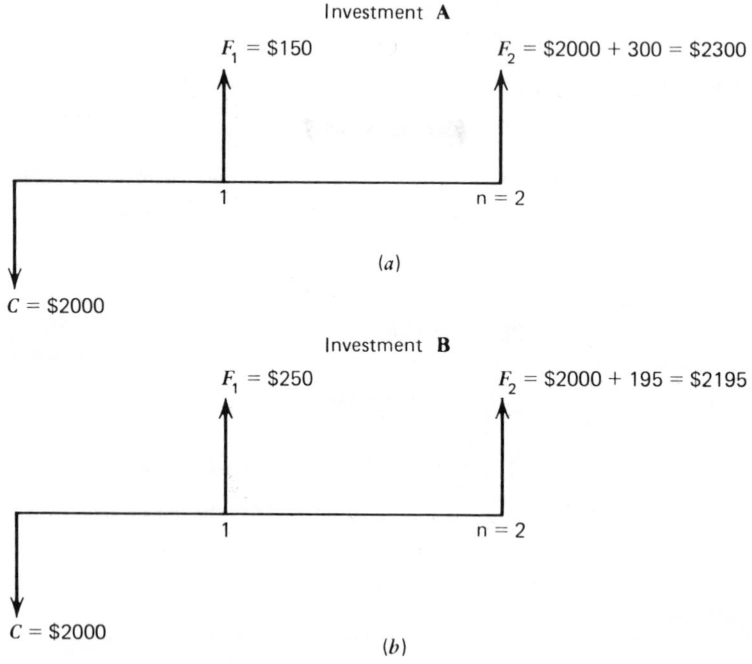

Figure 3.8

Remarks

It is interesting to observe that the total sum of nondiscounted money received from alternative **A** is slightly greater than alternative **B**: \$450 > \$445. However, the timing of payments has an important impact on the final selection. Alternative **B** gives more spendable income after the first year than alternative **A**, both in terms of scheduled benefits and present worth. Consequently, the larger return from alternative **B** in year 1, a \$100 difference in scheduled benefits, favors it as the better choice.

Benefits are considered positive cash flow and will be shown as up arrows in cash-flow diagrams, whereas costs are considered negative cash flow and will be shown as down arrows.

EXAMPLE 3.2 Loan Payment

(a) What is the annual payment on a \$10,000 loan at an interest rate of 10 percent per annum for 3 years?

(b) After each year determine the remaining principal to be paid each year on the loan.

Solution

(a) The annual payment is determined as a capital recovery fund where $C = \$10{,}000$.

$$A = C\left[\frac{i(1+i)^n}{(1+i)^{n-1}}\right]$$

$$A = \$10{,}000\left[\frac{(0.1)(1+0.1)^3}{(1+0.1)^3 - 1}\right] = \$4021$$

At end of each year, a payment of \$4021 must be made.

(b) In order to calculate the remaining principal after each year C_j, we calculate the amount of money paid in interest charges I_j for the year under consideration, $j = 1, 2$, or 3.

$$I_j = i_j C_{j-1}$$

Let $C = C_0 = \$10{,}000$.

The reduction in principal after each year R_j is the annual payment made per year, $A - I$. The new principal C_j is the old principal minus the reduction in principal.

$$C_j = C_{j-1} - R_j$$

After the first year:

$$I_1 = iC_0$$

$$I_1 = \$10{,}000(0.1) = \$1000$$

$$R_1 = A - I_1$$

$$R_1 = \$4021 - \$1000 = \$3021$$

$$C_1 = C_0 - R_1$$

$$C_1 = \$10{,}000 - \$3021 = \$6979$$

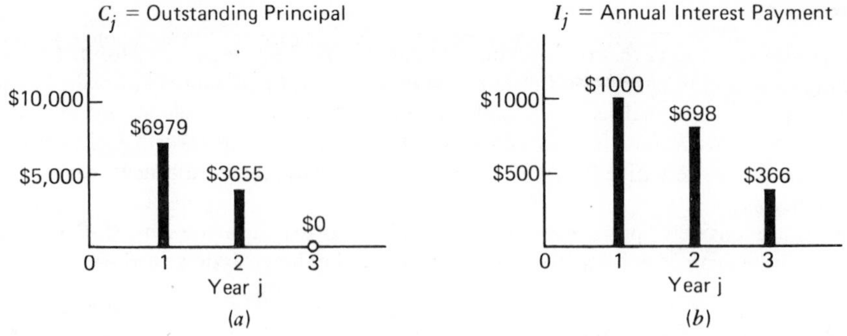

Figure 3.9

After the second year;

$$I_2 = iC_1$$

$$I_2 = \$6979(0.1) = \$698$$

$$R_2 = A - I_2$$

$$R_2 = \$4021 - \$698 = \$3323$$

$$C_2 = C_1 - R_2$$

$$C_2 = \$6979 - \$3323 = \$3656$$

After the third year:

$$I_3 = iC_2$$

$$I_3 = \$3656(0.1) = \$366$$

$$R_3 = A - I_3$$

$$R_3 = \$4021 - \$366 = \$3655$$

$$C_3 = C_2 - R_3$$

$$C_3 = \$3656 - \$3655 \approx 0.00$$

The diagrams in Figure 3.9 show the outstanding principal at the end of each year and the amount of interest paid each year for the fixed yearly payment of $4021.15.

PROBLEMS

Problem 1

A uniform series of payments is $100 per year for 10 years. The annual interest rate is 8 percent.

(a) Draw a cash-flow diagram.
(b) Determine the present worth of the uniform series P.
(c) Determine the future worth of the uniform series in year 10, F_{10}.
(d) Determine the future worth of the uniform series in year 5, F_5. Use all 10 $100 payments shown in the cash flow diagram of part a.
(e) Show that the present worth of F_{10} is equal to P, the present worth of the uniform series.
(f) Repeat e for F_5.

Problem 2

In 5 years, a \$4000 bond will mature, and the bond holder will receive \$5000, interest payments plus principal. No other payments are received.

(a) Draw a cash-flow diagram.

(b) Determine the annual interest rate.

Problem 3

Which of the following banks offers the greatest return on savings?

Bank A Simple annual interest rate of 5 percent.

Bank B An annual interest of 4.75 percent compounded quarterly.

Bank C An annual interest of 4.6 percent compounded continuously.

Problem 4

Calculate the present worth of 24 monthly payments of \$1000 per month. Assume an annual interest rate at 10 percent.

Problem 5

A \$1000 investment is expected to yield \$600 in years 1 and 2. There is no salvage value.

(a) Draw a cash-flow diagram.

(b) If $i = 5$ percent, is the investment viable? In other words, is the present worth *NPW* greater than zero?

(c) If $i = 10$ percent, is the investment viable?

Problem 6

A \$1000 investment yields \$350 and \$200 in two consecutive years. The salvage value at the end of year 2 is \$500. Is the investment a viable one at an annual interest rate of 8 percent?

(a) Draw a cash-flow diagram.

(b) Utilize the net present worth method.

Problem 7

Select the alternative that yields the greatest net benefit. Each alternative has a purchase price of \$1000. The \$1000 is returned after one year.

Alternative A A lump sum revenue of \$200 is received at the end of one year.

Alternative B Quarterly payments of \$40 are received.

(a) Draw a cash-flow diagram.

(b) Determine the better alternative for $i = 5$ percent annual interest rate.

(c) Repeat step b for $i = 15$ percent annual interest rate.

3.2 CHOOSING A DESIGN ALTERNATIVE

In order to distinguish among different types of cash flow for engineering projects, the following nomenclature will be employed.

B = user benefit.

C = capital investment.

A = operating and maintenance costs.

The capital cost is generally a single payment made in the initial year of the project and is generally designated as C or C_0. There are exceptions. For example, in stage construction, capital investments are made at times other than in the initial year. They will be denoted with subscript notation C_k, where k is the time period over which the capital is paid. The benefits B and operating and maintenance costs A are generally uniform series payments over the entire life of the project. If the benefits and annual costs are not uniform series payments, subscripted notation will be used.

In Section 3.1 the expression $NPW = F_0 - P$ is used to find the net present worth value, where F_0 represents the present worth of future payments of F and P is the present value payment. Utilizing the new notation, the present worth of benefits, annual operating and maintenance costs, and capital investments as B_0, A_0, and C_0, respectively, can be classified as F_0 and P. Thus, the conclusions drawn in Section 3.1 are applicable here. The decision-making process for selecting the best alternative among the competing alternatives **A**, **B**, **C**, ... is as follows:

1. Calculate $NPW = B_0 - A_0 - C_0$ for each alternative **A**, **B**, **C**, ... or $NPW^{\mathbf{A}}$, $NPW^{\mathbf{B}}$, $NPW^{\mathbf{C}}$, ...
2. Consider only feasible alternatives satisfying the inequality condition, $NPW \geq 0$.
3. Select the feasible alternative with the maximum value of NPW.

In this section the computational and theoretical details of the net present worth are investigated. Particular attention is paid to user benefits, the do-nothing alternative, establishing equivalence when comparing projects with different project design lives, and the effects of inflation. Once these principles are understood, then the net present worth method may be used with confidence.

Direct User Benefits

In evaluating different alternatives the following premises are employed.

1. Capital investment leads to a *direct increase in production efficiency*.
2. This increased production efficiency leads to a lower production cost that is passed on to the users in the form of *lower prices* or *savings*.

For example, when a manufacturer purchases a new machine to reduce production costs, this is considered a capital investment with direct user benefit. In a competitive market it is reasonable to assume that the company will pass its production cost savings on to its consumers in the form of lower prices. Building a new highway may be considered to be a capital investment project with direct user benefit. Here the user benefits are not in the form of cash savings. Instead, the benefits are time savings, improved safety benefits, and better accessibility. By assigning unit savings measures to the benefits of time, safety, and accessibility the total benefit may be determined in monetary terms.

Contrast these two examples with a government subsidy to reduce consumer prices. This action may be meritorious, but it is not considered a direct benefit in terms of our definition. Production efficiencies did *not* cause the reduction in price. There is a transfer of money from one source to another one. In other words, some people gain while others lose purchasing power. With direct benefits we assume the society as a whole is better served.

The Do-Nothing Alternative

Over the past generation, the public has been increasingly concerned with the protection of the natural environment. Bitter social arguments have occurred over the construction of nuclear and coal-fired power plants, urban highways, dams, and other large-scale civil engineering projects. As a result, a study of the *do-nothing* alternative should be considered a part of the decision-making process. The do-nothing alternative represents the status quo because in it no capital is expended and, in turn, no direct user benefits are received. The do-nothing alternative is not restricted to the protection of the environment only; it has applicability to planning, management, and design problems.

In engineering economic evaluations, the do-nothing alternative is not considered a separate alternative. Since $C = 0$, we assume that the production efficiency is not changed and no direct benefits can be received from it. This assumption does not exclude the possibility of receiving benefits from the do-nothing alternative. For example, a neighborhood may operate an old school building. Education benefits are received by the students who attend it even though no new capital investments are made. In addition, school operation and maintenance costs exist. The alternative to construct a new school building should be undertaken only if the new structure provides *additional* educational benefits. Thus, B equals the total educational benefits received from the new school minus the total educational benefits received for the old school. The operation and maintenance cost A equals the total operation and maintenance cost of the new school minus the total operation and maintenance cost of the old school. The capital cost of the new school is C. With time value of money relationships the net present worth of the new school alternative NPW may be determined. The new school is a viable alternative if $NPW \geq 0$.

Projects with Different Project Lives

We cannot over emphasize that all design-alternative comparisons must be made on an equal basis. Our principal assumption is that all engineering design alternatives will be equally functional over the *entire* design life of the project. If the set of design alternatives includes those with different design lives, the selection process must be modified.

For example, suppose we have two alternatives that use trucks. The first alternative uses diesel-powered trucks, which have been shown to remain operational for three times longer than gasoline-powered trucks. The second alternative uses the gasoline-powered trucks. For the sake of simplicity, let us suppose that a diesel-powered truck will remain operational for 300,000 miles. If the truck is to be driven an estimated

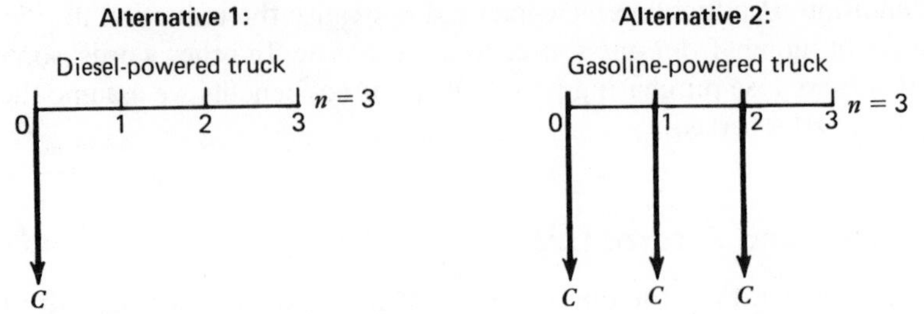

Figure 3.10 Comparison of projects using the principle of replacement.

100,000 miles per year it will provide service for 3 years. If a gasoline-powered truck remains operational for 100,000 miles, it will have to be replaced every year. In order to make equal comparisons between the two alternatives we assume that the design lives of the two projects are 3 years each. The cost C is the purchase price of a new diesel-powered truck and gasoline-powered truck for each respective alternative. We may represent each alternative with cash-flow diagrams as shown in Figure 3.10. The capital costs for the new gasoline-powered truck are assumed to be the same in years 0, 1, and 2; thus $C_0 = C_1 = C_2 = C$, as shown in the cash-flow diagram in Figure 3.10.

Another approach to this problem is to use the same design life for each alternative, but to estimate the salvage value of the alternative with the longer life. In the case of the diesel- and gasoline-powered trucks example, the cash-flow diagrams are based upon 1 year, as shown in Figure 3.11. The salvage value of the diesel truck after 1 year, B_1, may be estimated. This might be accomplished by comparing the present price of a used item of the same design. For example, if we were selling a truck we might obtain the price for a truck of similar design in the used-car listings of the newspaper or from a used-car salesman. Generally, estimating the value of used civil engineering projects is a difficult task. The first technique is recommended.

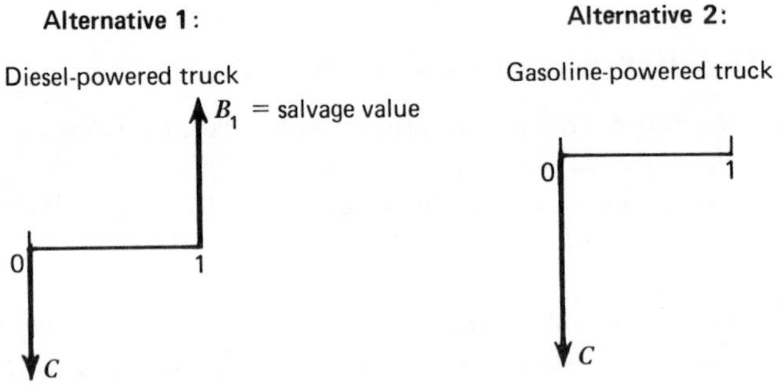

Figure 3.11 Comparison of projects using the principle of salvage value.

Real and Inflationary Prices

The price that a consumer pays for a good or service is assumed to be at least equal to the level of satisfaction that the consumer received from it. If the selling price is too high, no transaction will take place. The dollar exchange price is assumed to reflect the worth of the good or service or its *utility*. The *real price* is assumed to be the *exchange price* at the time the purchase is made and a measure of utility or its value in use.

Real prices are affected by supply and consumer demand. A consumer who purchases a gallon of gasoline receives a fixed amount of energy that can perform a given amount of work. Since the amount of energy is fixed, its utility is the same no matter when it is used. In other words, the utility is constant over time. If consumer demand and the supply of gasoline remain constant over this time period, the real price of gasoline remains constant over time. Hence, the real price is also referred to as *constant dollars*. A shortage will cause the real price to rise, even though its utility remains the same. An over abundance causes real prices to fall. If supply and demand are fixed and the price increases or decreases, the selling prices are called *inflationary* and *deflationary prices*, respectively. The following discussion covers inflation. The same principles apply to deflation.

By convention, inflation is treated as exponential growth and expressed in terms of percentage per year. For example, if inflation is 10 percent per year, an item that costs \$10 today will cost \$11 next year. Mathematically, we shall discover that there is no difference in how one does calculations whether the exponential growth measures inflation or opportunity cost of money. The concepts, however, are very different!

Now consider a simple example that shows how inflation affects calculations involving time value of money. Assume an individual borrows \$100 to buy a suit. The terms of the loan call for the borrower to pay back the principal and interest, calculated at 7 percent per year, at the end of 5 years. The total amount to be paid at the end of 5 years is

$$F_5 = \$100(1 + 0.07)^5$$
$$= \$140.26$$

One might assume that the lender will make a profit of \$40.26 from the loan. That is, when the loan is repaid with interest, the lender could buy his own suit for \$100.00 and have \$40.26 left as a reward for waiting 5 years to buy it. But when the lender goes to buy his own suit, he finds that the identical suit, which cost \$100.00 5 years ago, now costs \$150.00. By making the loan, the lender has actually lost \$9.74. This occurred because inflation lowered the buying power, or the value, of the dollars that were repaid by the borrower.

In problems dealing with inflation we must carefully distinguish between exchange prices and the actual time at which the monetary transfer is made. A suit bought in year 0 has an exchange price of \$100, which will be represented in *in-hand year-0 dollars* as $P = \$100$. The terminology "*in-hand year-0 dollars*" is used to emphasize the fact that there is an actual money transfer being made at time 0. The identical suit bought with in-hand year-5 dollars has an exchange price of \$150 and is represented as $\hat{P}$ = \$150. The circumflex ˆ will be used to indicate that the monetary exchange is made in a nonzero year, $n = 1, 2, \ldots$. The average rate of inflation, f, can be determined with the following expression

$$\hat{P} = P(1 + f)^n$$

where f represents the average annual rate of inflation and n is the year in which the transaction takes place. The rate of inflation for the suit problem is

$$\$150 = \$100(1 + f)^5$$

or

$$f = 0.0845 = 8.45 \text{ percent}$$

The rate of inflation is 8.45 percent per year.

The consumer price index and producer price index are measures of inflation. The consumer price index is determined by evaluating the changes in prices of a specific group of goods and services. Similarly, the producer price index reflects the change in wholesale prices of crude, intermediate, and finished goods. These government-supplied statistics show the trend of inflation over time. If the annual rates of inflation are known over time, $f_1, f_2, \ldots, f_n$, the *in-hand year*-0 price P may be expressed as an in-hand year-n price $\hat{P}$:

$$\hat{P} = P(1 + f_1)(1 + f_2) \cdots (1 + f_n)$$

In engineering problems, the inflation rates will have to be predicted for future years. Since there are many uncertainties in this prediction, a uniform inflation rate is assumed, $f_1 = f_2 = \cdots = f_n = f$. Thus, $\hat{P} = P(1 + f)^n$ is used in most calculations.

Let us investigate the suit problem further. If the lender knew the inflation rate over the next five years is going to be 8.45 percent per year and he wants a future return of \$50, he will demand a payment of \$200 (\$50 + \$150). The \$50 is considered a real return on the loan. This money will be received in year 5 and can be spent at that time. The \$150, on the other hand, is necessary to keep pace with inflation. The required interest rate i on the \$100 loan may be calculated as

$$\hat{F}_5 = P(1 + i)^5$$

or

$$\$200 = \$100(1 + i)^5$$

$$i = 0.1487 = 14.87 \text{ percent}$$

The interest rate on the loan equals 14.87 percent per year. If the lender could have known the future behavior of suit prices when he made the loan, he could merely have specified 14.87 percent per year as an interest rate. The interest rate i is an opportunity-cost interest rate that accounts for the real return of \$50 plus the declining value of the dollar over the period of the loan.

Using $\hat{P}$ = \$150 gives a different point of view of the problem. The payment in year 5 must satisfy the relationship

$$\hat{F}_5 = \hat{P}(1 + r)^5$$

where r is a newly defined *real-return interest rate* and $\hat{P}$ is in-hand year-5 dollars, or $\hat{P} = \$150.00$:

$$\$200.00 = \$150.00(1 + r)^5$$

or

$$r = 0.0592 = 5.92 \text{ percent}$$

The real rate of return is 5.92 percent per year. The interest rate r is an opportunity-cost interest rate like i, but it is based upon the cheaper in-hand year-5 dollars. Let us investigate the relationship among i, f, and r.

For a project design life of n yr, we can express the future sum F_n in terms of P and i as

$$\hat{F}_n = P(1 + i)^n$$

and in terms of $\hat{P}$ and r as

$$\hat{F}_n = \hat{P}(1 + r)^n$$

By substitution of $P(1 + f)^n$ for $\hat{P}$ and equating the above expressions, we find that

$$(1 + i) = (1 + r)(1 + f)$$

or

$$i = (1 + r)(1 + f) - 1$$

The opportunity-cost interest rate i is a function of the inflationary and real interest rates, f and r, respectively. The real-return interest rate may be calculated as

$$r = \frac{i - f}{1 + f}$$

For the lender to receive a net gain, r must be greater than zero, $r > 0$. If $i = f$, then the rate of return r equals zero and the lender keeps pace with inflation only.

Effects of Fixed Cash and Inflationary Price Payments Let us consider the evaluation of two different types of investments. The first one deals with an initial capital cost of \$1000 and a fiscal return in *fixed cash payments* of \$180 per year for 10 years. The second one has an initial \$1000 investment, but the return on investment is the form of *revenue* from sales of 100 items selling at \$1.80 per item. Since revenue is equal to the selling price times quantity sold, $R = pq$, $R = B = pq = \$1.80$ per year where B is in-hand year-0 dollar. Furthermore, we assume that the selling price p keeps pace with inflation. The cash flows for those two cases are the same, but the kind of payments, cash or revenue, will affect our analyses. Let us calculate the net present worth values for these two cash flows. For these purposes, we assume a real rate of return r of 10 percent per year and an inflation rate f of 5 percent per year. The opportunity-cost interest rate is

$$i = (1 + r)(1 + f) - 1$$

$$i = (1 + 0.10)(1 + 0.05) - 1$$

$$i = 0.155 = 15.5 \text{ percent}$$

For the *fixed cash payment problem*, the future payments are 10 annual yearly cash payment of $B = \$180$. These annual payments do not keep pace with inflation. The net present worth of this investment is

$$NPW = -1000 + 180\left[\frac{(1 + 0.155)^{10} - 1}{0.155(1 + 0.155)^{10}}\right]$$

$$NPW = \$-113.58$$

In terms of net present value, there is a net loss.

For the *revenue-with-inflationary-prices problem*, we assume that the selling price follows the inflation rate.

$$\hat{p} = p(1 + f)^k$$

In in-hand year-k dollars, the annual benefit or annual revenue is

$$\hat{B}_k = pq(1 + f)^k$$

or

$$\hat{B}_k = B(1 + f)^k$$

The present worth of a single payment $\hat{B}$ received in year k is

$$PW = \frac{\hat{B}_k}{(1 + i)^k}$$

Since $(1 + i) = (1 + r)(1 + f)$, the relationship becomes

$$PW = \frac{B(1 + f)^k}{(1 + r)^k(1 + f)^k}$$

or

$$PW = B\left[\frac{1}{(1 + r)^k}\right]$$

The effect of inflation drops out. The problem reduces to the evaluation of PW in terms of in-hand year-0 dollars B and the real interest rate r. The net present worth of this investment is

$$NPW = -\$1000 + 180\left[\frac{(1 + 0.10)^{10} - 1}{0.1(1 + 0.10)^{10}}\right]$$

$$NPW = \$106.02$$

There is a net gain for this investment.

Comparing the evaluation of these two cases shows the importance of the kinds of payments. Most engineering projects involve future benefits and costs that keep pace with inflation because the price of goods and services bought and sold may be estimated as $\hat{p}_k = p(1 + f)^k$. As a result, we can often estimate capital costs, benefits, and annual costs in terms of year-0 dollars and utilize the real rate of return r in our calculations. When this assumption cannot be justified, the analysis must be conducted as a fixed-cash payment problem.

Mathematically, the real interest rate r is easy to determine if i and f are known. The difficulty arises in determining i and f. As previously discussed, the opportunity-cost interest rate i is dependent upon the rate of inflation f. Inflation may change the buying behavior of consumers, businesses, and government, and may influence the government economic policy. These factors affect the investment process, which, in turn, affects the interest rates. It is observed that during periods of high inflation, interest rates tend to be high. For example, when the loan agreement is made for the suit, the future rate of inflation f is not known. It will have to be forecast to perform the above analysis. The economic conditions at the time the loan agreement is made will have an important bearing upon the interest rate that the lender offers and the borrower accepts. In any event, borrowers *favor low interest rates and long investment periods* because they hope to pay back the loan with inflated dollars and cheaper money. Lenders favor the opposite, *higher interest rates and shorter loan periods.*

In engineering economics, judgment must be used in assigning not only the interest and inflation rates, but also the cost and benefits as well. Up-to-date sources of cost data for various types of engineering facilities are listed in Table 3.1. Consumer price and producer price indices are tabulated in the U.S. Department of Commerce document, *Statistical Abstracts of the United States* and in the *Engineering News Record.*

It has been common practice in engineering to ignore the rate of inflation and assume it is equal to zero, $f = 0$. This approach avoids the problem of forecasting the inflation rate. Since $r = (i - f)/(1 + f)$, $r = i$. This assumption tends to make alternatives less attractive and the do-nothing alternative a more likely outcome. This practice tends to favor the do-nothing alternative. As a result, it is considered to be a conservative approach by some. Although the assumption leads to a simpler analysis, it is not recommended, especially in periods of high inflation. When the rate of inflation is small, it has little impact upon the final selection. In this case, inflation can be safely ignored.

EXAMPLE 3.3 A Least-Cost Selection

A company must replace malfunctioning equipment in order to remain in operation. It has two options that have the following characteristics.

Machine A purchase price: $100,000
estimated annual maintenance cost: $10,000
design life: 6 yr

Machine B purchase price: $75,000
estimated annual maintenance cost: $13,000*
major overhaul of equipment in year 3: $20,000*
design life: 6 yr

Utilize the principles of the net present worth method to select the better choice between machines **A** and **B**. Assume that the cost of money is 12 percent per annum.

* The total maintenance cost in year 3 is the sum of the annual maintenance and overhaul costs, or $33,000.

Table 3.1 Sources of Cost Estimation Data

BUILDING COSTS

1	Dodge	*Manual for Building Construction Pricing and Scheduling*
2	McKee, Berger & Mansueto	*Building Cost File*
3	Means	*Building Construction Cost Data*

These documents, published annually, contain unit costs for general building construction items.

HEAVY CONSTRUCTION COSTS

1	Dodge	*Guide to Public Works and Heavy Construction Costs*
2	Coert & Engelsman	*Heavy Construction Cost File*

These documents, published annually, contain unit prices of labor, equipment, and materials for public works and heavy construction.

BUILDING SYSTEMS COSTS

1	Dodge	*Construction Systems Costs*
2	McKee, Berger & Mansueto	*Design Cost File*
3	Means	*Building Systems Costs*

These documents, published annually, contain prices for installed in-place and composite cost for complete building systems.

PRICE INDICES

1 *Statistical Abstracts of the United States*, Department of Commerce, Bureau of Census
2 *Engineering News Record*, McGraw-Hill, a monthly publication
3 *Chemical Engineering*, McGraw-Hill, a monthly publication
4 *Water Pollution Control Federation Journal*

SPECIAL REPORTS

Agencies of the U.S. Government compile cost information for particular engineering facilities. This technical information may be obtained by contacting the appropriate government agency directly or by investigating the reference section of the library. The *Government Reports Annual Index* prepared by the U.S. Department of Commerce, National Technical Information Service (NTIS), is a good source of information about engineering reports. The report is listed under the Subject Index, as "Construction Costs," of the *Government Reports Annual Index 1978*. This document contains abstracts, author, title, and publisher information. For example, a U.S. Environmental Protection Agency (EPA) report on construction cost of wastewater treatment and sewers is listed as:

Dames and Moore, Denver Colo.
Construction Costs for Municipal Wastewater Treatment Plants: 1973–1977. Jan. 1978, 214 p EPA/430/19-77/013, EPA/MCD-37

Solution

For net present worth selection, the alternative with the largest *NPW* value is considered the best choice. For this problem, the benefit *B* is equal to the revenue received from sales of the product produced with the equipment. The sales revenue, or benefit, *B* will be the same no matter which loan plan is selected. The key issue is to replace the worn equipment and get it back on line. As a result, the benefit *B* does not have to be utilized in this analysis. The machine of lesser capital and maintenance costs *C* will be considered the better choice. A problem of this type is called a *least-cost selection problem.* For least-cost selection it is assumed that benefits *B* will always exceed the costs *C* or $B > C$.

The time stream of payments is shown in the cash-flow diagrams in Figure 3.12. For machine **B**, the \$20,000 major overhaul payment required in year 3 is considered a single-payment capital cost and is designated as C_3.

The present worths of these alternatives are determined with the time value of money equations for the present worth of single payments and uniform series. For machine **A**, the present worth of costs is

$$PW^{A} = 100{,}000 + 10{,}000\left[\frac{(1 + 0.12)^6 - 1}{0.12(1 + 0.12)^6}\right]$$

$$PW^{A} = 100{,}000 + 41{,}100 = \$141{,}100$$

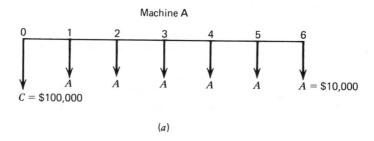

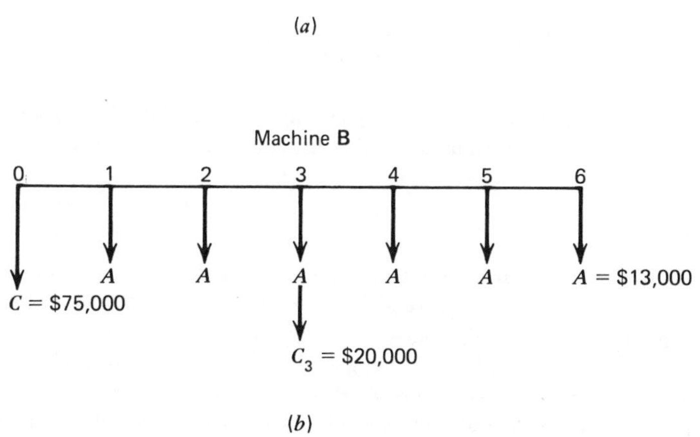

Figure 3.12

For machine **B**, the present worth of costs is:

$$PW^{\mathbf{B}} = 75{,}000 + 13{,}000\left[\frac{(1 + 0.12)^6 - 1}{0.12(1 + 0.12)^6}\right] + \frac{20{,}000}{(1 + 0.12)^3}$$

$$PW^{\mathbf{B}} = 75{,}000 + 53{,}400 + 14{,}200 = \$142{,}600$$

Since $PW^{\mathbf{A}} < PW^{\mathbf{B}}$, the better choice is to purchase machine **A**.

EXAMPLE 3.4 Incorporating a Do-Nothing Alternative into the Selection Process

A contractor has decided to buy a new four-wheel-drive truck costing \$6500. The truck will serve him for a period of 6 years before he must replace it. The contractor does not work during the winter months if it snows. If he plows snow he can gross an estimated \$1500 each winter. The contractor wants to know if he should buy a \$1000 plow. If he plows snow, the truck will have to be replaced after 3 years. The truck salvage value is assumed to be zero. He can borrow money at an interest rate of 10 percent per annum. Solve by using the net present worth method.

Solution

The two alternatives are summarized as follows.

Buy the truck at \$6500 with no plow.

Buy the truck at \$6500 with \$1000 plow and earn \$1500/year.

For this example, we compare the expected present values of the net profit from plowing or not plowing snow. The first alternative is considered the *do-nothing* alternative. Our only concern here is to determine if the contractor should buy a snow plow, not if he should buy a truck. He must purchase the truck to continue to operate his contracting business. We can assume if he does not buy the truck he may go out of business. Thus, we are investigating the benefit of plowing snow only. We compare the *extra* costs and benefits associated with the snow plow alternatives.

The cash-flow diagram is shown in Figure 3.13. This cash-flow diagram shows that the truck and plow must be replaced after 3 years. The net present worth is

$$NPW = 1500\left[\frac{(1 + 0.1)^6 - 1}{0.1(1 + 0.1)^6}\right] - 1000 - 7500\left[\frac{1}{(1 + 0.1)^3}\right] = -\$102$$

Since $NPW < 0$, the alternative is not viable. The contractor should buy a truck without a plow because the negative present worth indicates that the contractor can expect to lose money if he does otherwise.

Remarks

It is interesting to note that if the interest rates were 5 percent, then purchasing the snow plow would be the better selection. At 5 percent, $NPW = \$135$. The interest rate has an important bearing upon the final choice. The benefits or revenue received from the snow plowing business are expected or anticipated earnings. They are not guaranteed. If there is a drought the earnings from snow plowing will not be realized. Likewise, heavy snows will produce revenues greater than expected. Lenders will take risk into consideration when offering a loan and assigning an interest rate. For projects of high risk, the interest rates will be greater than ones with little risk. A high interest rate puts more weight on present and near future returns; whereas, with a low interest rate, future returns are given a higher weight. The contractor may be offered a loan at an annual interest

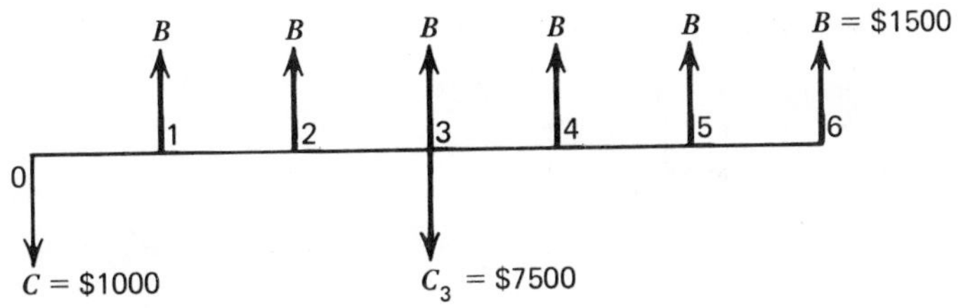

Figure 3.13 Plan B.

rate of 5 percent if the contractor lives in a snow belt. On the other hand, if the contractor lives in regions where snow droughts occur, an annual interest rate of 10 percent may be offered. Decision analysis may be used for projects involving uncertainty. This topic will be discussed in the next chapter.

EXAMPLE 3.5 Projects with Different Design Lives

Consider two projects with the estimated revenues shown below. There is no guarantee that the revenue will be received, and certain risks are associated with each investment.

PROJECT	CAPITAL COST	TOTAL ANTICIPATED REVENUE	ESTIMATED YEARLY REVENUE
A	\$1000	\$2000 in 10 yr	\$200
B	\$1000	\$2500 in 20 yr	\$125

Project **B** earns more money over a longer period; however, alternative **B** has a smaller yearly revenue. Utilize the net present worth method with an annual interest rate of 10 percent to determine the better alternative.

Solution

The yearly revenues of \$200 and \$125 will be used because they reflect the time when the revenue is received. The revenues are assumed to be equally distributed over the life of project.

The cash-flow diagrams for capital cost and yearly revenue are shown in Figures 3.14*a* and *b*.

If we compare these projects as they stand, we are not making a fair comparison. For project **A**, revenue is not earned after year 10. In order to continue to earn money, reinvestment is necessary. It is assumed that the future investment will be made in the same project. Thus, the cash-flow diagram for a design life of 20 years is as shown in Figure 3.14*c* for alternative **A**. The alternative with the maximum net profit will be the best investment. The net present worths of alternatives **A** and **B** are

$$NPW^{A} = 200\left[\frac{(1+0.1)^{20}-1}{0.1(1+0.1)^{20}}\right] - 1000 - 1000\left[\frac{1}{(1+0.1)^{10}}\right] = \$317.17$$

and

$$NPW^{B} = 125\left[\frac{(1+0.1)^{20}-1}{0.1(1+0.1)^{20}}\right] - 1000 = \$64.20$$

respectively. Since $NPW^{A} > NPW^{B}$, the better investment is project **A**.

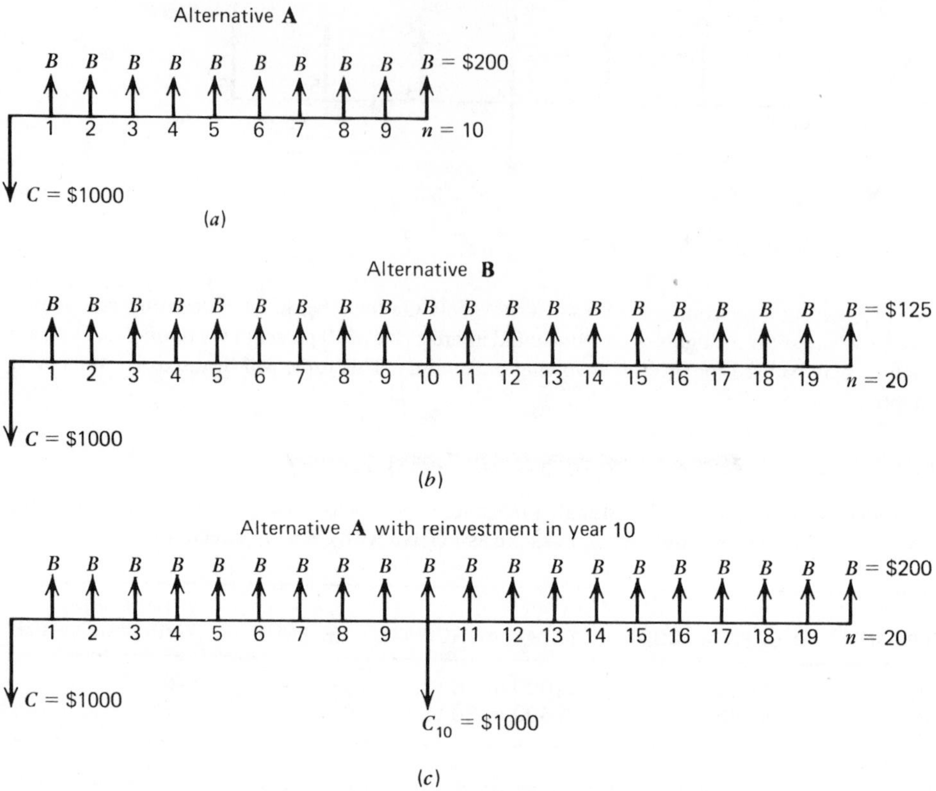

Figure 3.14

EXAMPLE 3.6 The Effect of Inflation Upon a Business Plan

A firm has signed a three-year labor contract that promises to pay an average annual increase of 10 percent to its workers. This contract is considered to be inflationary because it is assumed that the worker productivity will remain the same and no production cost savings will be realized. The total annual labor cost in present dollars is $150,000.

(a) Determine the amount of money in present dollars that must be placed in a fund to pay future labor costs. Assume the fund yields no annual interest.

(b) Determine the amount of money in present dollars that must be placed in a fund yielding 7 percent per annum to pay future labor costs.

Solution

(a) Let $f = 0.10 =$ the annual rate of inflation and
$A = \$150{,}000 =$ the in-hand year 0 dollar labor cost.
$\hat{A}_k =$ the in-hand year k dollar labor cost, where $k = 1, 2, 3$.

The amount of money that must be paid to labor in any given k year is

$$\hat{A}_k = A(1 + f)^k$$

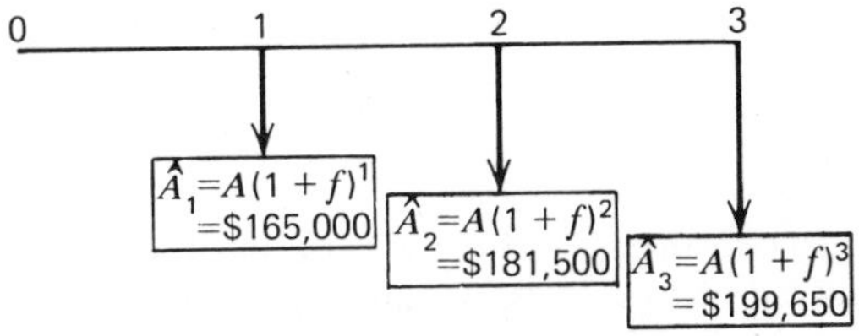

Figure 3.15

Substituting the appropriate value of A, f, and k into this relationship, the values of $\hat{A}_k$ are calculated. The annual payments are shown in the cash flow diagram in Figure 3.15. The total amount of money in the fund in year-0 dollars P is

$$P = \hat{A}_1 + \hat{A}_2 + \hat{A}_3 = 165{,}000 + 181{,}500 + 199{,}000 = \$546{,}150$$

(b) The amount of money that must be paid to labor each year was found in part a in terms of in-hand k dollars.

$$\hat{A}_k = A(1 + f)^k$$

Define P_k as the portion of the fund money in in-hand year-0 dollars to pay labor in year k. Since we pay labor in year-k in-hand dollars, the following relationship must hold.

$$\hat{A}_k = P_k(1 + i)^k$$

where $i = 0.07$ = annual interest rate from the fund. Rearranging the equation and substituting $\hat{A}_k = A(1 + f)^k$ into the equation, we obtain

$$P_k = \frac{\hat{A}_k}{(1 + i)^k} = \frac{A(1 + f)^k}{(1 + i)^k}$$

Since $(1 + i) = (1 + r)(1 + f)$, P_k may be written in terms of the real rate of return r.

$$P_k = \frac{A}{(1 + r)^k}$$

where

$$r = \frac{i - f}{1 + f} = \frac{0.07 - 0.10}{1 + 0.10} = -0.0273$$

The total amount of money in the fund is

$$P = P_1 + P_2 + P_3 = A\left[\frac{1}{(1 + r)^1} + \frac{1}{(1 + r)^2} + \frac{1}{(1 + r)^3}\right]$$

$$P = 150{,}000\left[\frac{1}{1 - 0.0273} + \frac{1}{(1 - 0.0273)^2} + \frac{1}{(1 - 0.0273)^3}\right]$$

$$P = \$475{,}710$$

Remarks

The determination of P, the amount of money placed in a fund yielding 7 percent per year, may be found with use of the relationship for determining the present worth of a uniform series.

$$P = A\left[\frac{(1 + r)^n - 1}{r(1 + r)^n}\right]$$

where $A = \$150{,}000$, $n = 3$, and $r = -0.0273$.

The real interest rate is determined with the following relationship

$$r = \frac{i - f}{1 + f}$$

It is not the simply difference between i and f, $r \neq i - f$. Furthermore, if the interest rate i and inflation rate f are equal, the present worth of a uniform series equation cannot be used. Since the present value of inflated dollars for any year k can be determined with

$$\hat{A}_k = \frac{A(1 + f)^k}{(1 + i)^k}$$

and $i = f$, then $\hat{A}_k = A$. Thus, the amount of money that must be placed in an interest-bearing account, where $i = f$, is $P = nA$. For this problem, if $i = f$, then $P = 3 \times \$150{,}000 = \$450{,}000$.

EXAMPLE 3.7 The Effect of Inflation Upon Investment

A 5-year bond having a face value of $5000 yields 12 percent per year. Assume that the investor is in the 25 percent marginal tax bracket. The annual rate of inflation is assumed to be 9 percent per year.

(a) Determine the present worth of this investment, and estimate the annual rate of return on investment after taxes.

(b) Determine the bond yield if the investor wants a real rate of return on investment after taxes of 3 percent per year.

Solution

(a) The effective bond yield rate i after taxes is

$$i = \hat{i}(1 - t)$$

$$i = 0.12(1 - 0.25) = 0.09$$

where $\hat{i}$ is the before-tax annual yield rate of the bond and t is the tax bracket rate of 25 percent. Thus, the effective yield is 9 percent per annum. The annual payments to the investor after taxes are the same each year.

$$B = \$5000 \cdot 0.07 = \$450$$

The investor's initial investment of $5000 is returned to him or her in year 5. The cash flow diagram is shown in Figure 3.16.

All future annual benefits B and the fixed payments B_5 are fixed-cash payments and do not keep pace with inflation. Thus, the opportunity-cost interest rate i is equal to 9 percent, the

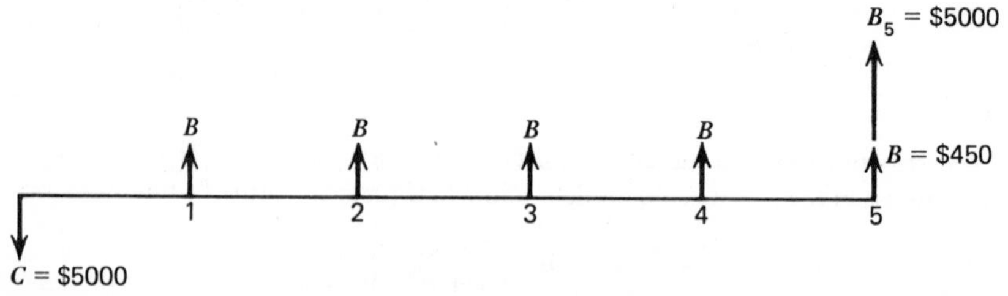

Figure 3.16

after-tax annual yield from the bond. The present worth of the investment is equal to the present worth of all future benefits minus the investment cost C or

$$P = B\left[\frac{(1+i)^n - 1}{i(1+i)^n}\right] + \frac{B_5}{(1+i)^n} - C$$

Since the term of the bond is 5 years, $n = 5$ and

$$P = 450\left[\frac{(1+0.09)^5 - 1}{0.09(1+0.09)^5}\right] + \frac{5000}{(1+0.09)^5} - \$5000$$

or $P = \$0$.

The investor receives a zero return on this investment. Since $P = 0$, we can deduce that $i = f$ and $r = (i - f)/(1 + f) = 0$, or $r = 0$ percent per year.

(b) In order for the investor to receive a real rate of return of 3 percent per year, the effective annual yield i must satisfy the following relationship:

$$i = (1 + r)(1 + f) - 1$$

where $r = 0.03$ and $f = 0.09$.

$$i = (1 + 0.03)(1 + 0.09) - 1 = 0.1227$$

The before-tax annual yield rate for the bond i must satisfy the relationship

$$i = \hat{i}(1 - t)$$

or

$$\hat{i} = \frac{i}{1 - t} = \frac{0.1227}{1 - 0.25} = 0.1636$$

The before-tax annual yield rate on the bond must be 16.36 percent for the investor to receive a real rate of return to 3 percent per annum.

Remarks

Bond purchasers are at a disadvantage during periods when inflation increases over time because they receive payment in future dollars which have lost purchasing power. The organization selling the bond benefit because they pay dividends based on the face value of the bond. These payments are fixed cash payments. Under these conditions, the organization selling the bond favor long-term borrowing, whereas the bond purchasers, favor the opposite. During periods of high inflation, organizations must offer higher rates of interest to attract purchasers. Thus, inflation affects the cost of money or discount rates and will, in turn, affect the interest rates used in the engineering economic selection process.

EXAMPLE 3.8 The Effect of Inflation Upon the Selection Process

A firm is considering the purchase of machines. The alternatives being considered are shown in the following table.

MACHINE	CAPITAL COST	ANNUAL O & M COST[a]	DESIGN LIFE
A	\$20,000	\$3,000	8 yr
B	\$15,000	\$3,500	4 yr

[a] O & M = operation and maintenance

Owing to advances in technology, it is assumed that machine **B** will be replaced with an improved model, model **C**. It has the following characteristics.

MACHINE	CAPITAL COST	ANNUAL O & M COST	DESIGN LIFE
C	$10,000	$2,500	4 yr

The annual sales revenue is $8000, and the opportunity cost of money is 12 percent per annum.

Determine the better alternative utilizing the net present worth method assuming that *all prices* and *expenses will increase at the same inflationary rate* of 7 percent per year.

Solution

Since all prices and expenses are assumed to keep pace with inflation, $p_k = p(1 + f)^k$, the analysis is conducted using in-hand year-0 dollar estimates for future benefits B, annual operation and maintenance costs A, and capital costs C. The real return interest is used as the discount rate.

$$r = \frac{i - f}{1 + f} = \frac{0.12 - 0.07}{1 + 0.07} = 0.0467$$

The cash flow diagrams for alternatives **A** and **B** are shown in Figure 3.17. The annual net benefit is equal to the annual revenue minus annual cost, $B - A$. For machine A, the annual net benefit is

$$B - A = \$8000 - 3000 = \$5000$$

For machine **B**, the annual net benefit for the first 4 years is

$$B - A = \$8000 - 3500 = \$4500$$

and the last 4 years is

$$B - A' = \$8000 - 2500 = \$5500$$

The net present worths of alternatives **A** are thus:

$$NPW^{\mathbf{A}} = -\$20{,}000 + 5000\left[\frac{(1 + 0.0467)^8 - 1}{0.0467(1 + 0.0467)^8}\right]$$

$$NPW^{\mathbf{A}} = -20{,}000 + 32{,}752$$

$$NPW^{\mathbf{A}} = 12{,}752$$

For alternative **B**

$$NPW^{\mathbf{B}} = -\$15{,}000 + 4500\left[\frac{(1 + 0.0467)^4 - 1}{0.0467(1 + 0.0467)^4}\right]$$

$$+ \left[-10{,}000 + 5500\left[\frac{(1 + 0.0467)^4 - 1}{0.0467(1 + 0.0467)^4}\right]\right](1 + 0.0467)^{-4}$$

$$NPW^{\mathbf{B}} = -\$15{,}000 + 16{,}080 + [-10{,}000 + 19{,}653](1 + 0.0467)^{-4}$$

$$NPW^{\mathbf{B}} = \$1080 + 8042 = \$9122$$

Since $NPW^{\mathbf{A}} > NPW^{\mathbf{B}}$, machine **A** is a better choice.

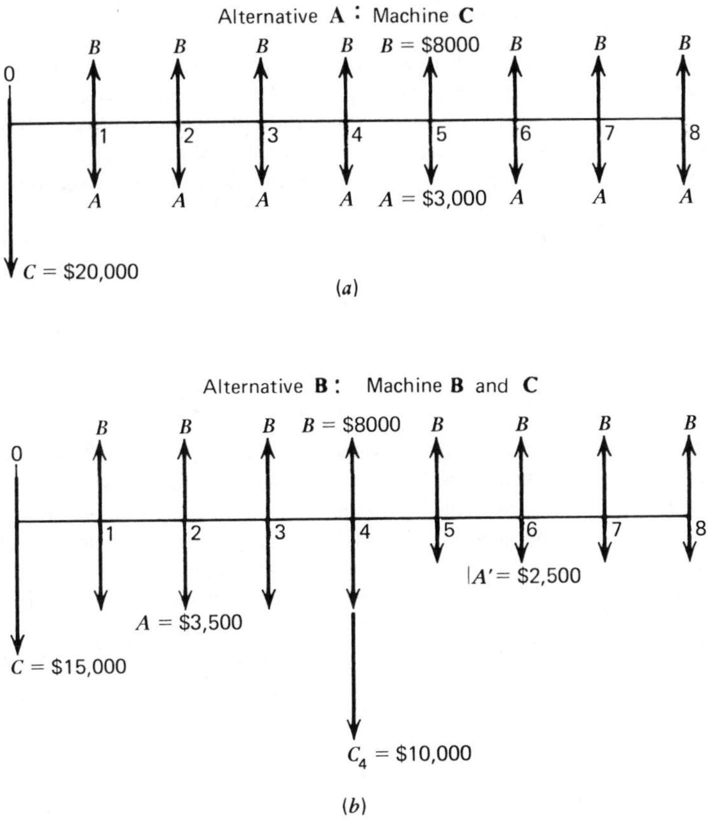

Figure 3.17

$-20000 + 5000\left[\frac{(1.12)^8 - 1}{.12(1.12)^8}\right]$

Remarks

If the inflation rate is ignored, conservative values of the present worth values of net benefit are obtained. The above analysis has been repeated for $i = 0.12$ and $f = 0.00$. The results are tabulated in the following table.

ALTERNATIVE	r = 4.67 PERCENT PER ANNUM	i = 12 PERCENT PER ANNUM
A	$12,375	$4,840
B	$9,122	$2,930

For $i = 12$ percent per annum, the better choice is machine **A**. This agrees with the result for $r = 4.67$ percent per annum. The major difference is the present worth of net benefits for $i =$ 12 percent is less than the present worth of net benefits for $r = 4.67$ percent.

EXAMPLE 3.9 Stage Construction Planning

A town operates a modern secondary wastewater treatment plant that has sufficient capacity to treat waste generated by a population of 25,000 people. The 22-in. main interceptor of the town's sanitary sewer system is 30 years old and is scheduled for replacement in 10 years. It has sufficient capacity to transport flow for an estimated 20,000 people. The Master Plan calls for the development of a new suburban residential area that requires the immediate construction of a sanitary sewer. The sewer system under consideration is shown on the map in Figure 3.18*a*.

Develop a sewer system construction plan to meet the immediate development need and the anticipated need of future population growth. (See Figure 3.18*b*.) Municipal bonds yielding a real annual interest rate of 8 percent will be sold. Utilize the principle of net present worth to select the best sewer system construction plan. The service life of all sewer pipe is assumed to be equal to 40 years.

Solution

All sewer system alternatives will be designed to meet the anticipated growth in the dense residential and suburban regions as specified in the Town Master Plan. Since all design strategies must meet minimum wastewater flow requirements for a given population demand, *all* alternatives offer the *same* benefit B to the town's population. As a result, it is not necessary to determine explicitly the benefit for each alternative. The *least-cost alternative* is considered the best selection.

The construction cost and pipe size will be estimated with the information from the EPA *Construction Costs for Municipal Wastewater Conveyance Systems*: 1973–1977 document. The design discharge flow will be estimated by assuming an average per capita flow of 100 gallons per person-day.

The following estimates are utilized in developing the alternative sanitary sewer system plans.

POPULATION	ESTIMATED DISCHARGE (Mgal/day)	PIPE DIAMETER (in.)	CONSTRUCTION COST ($/LINEAR FOOT)		
			AVERAGE	SUBURBAN[a]	DENSE RESIDENTIAL[a]
2,500	0.25	10	47.00	43.00	51.00
5,000	0.50	15	73.00	66.00	79.00
10,000	1.0	18	94.00	86.00	102.00
15,000	1.5	21	118.00	107.00	127.00
25,000	2.5	24	124.00	113.00	134.00

[a] The construction costs are estimated by multiplying the average construction cost by the appropriate multiplier: open country—0.67; suburban residential—0.91; dense residential (within city limits)—1.08; and commercial industrial—1.20.

A plan design life of 50 years, $n = 50$, is assumed. A present worth factor for $n = 50$ yr with an annual interest rate of 8 percent is

$$\frac{1}{(1+i)^n} = \frac{1}{(1+0.8)^{50}} = 0.0213$$

The discount value of payments for periods greater than 50 yr will be smaller than 0.0213; thus, they are assumed to be sufficiently small and equal to zero.

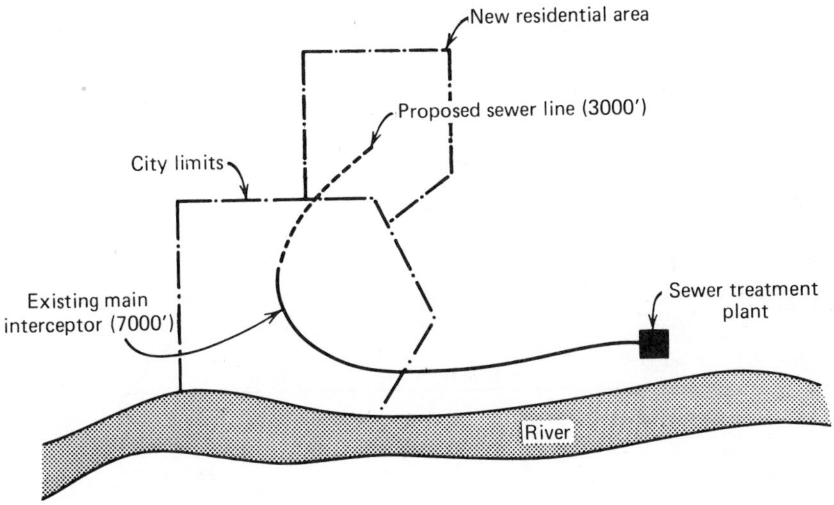

Figure 3.18*a*

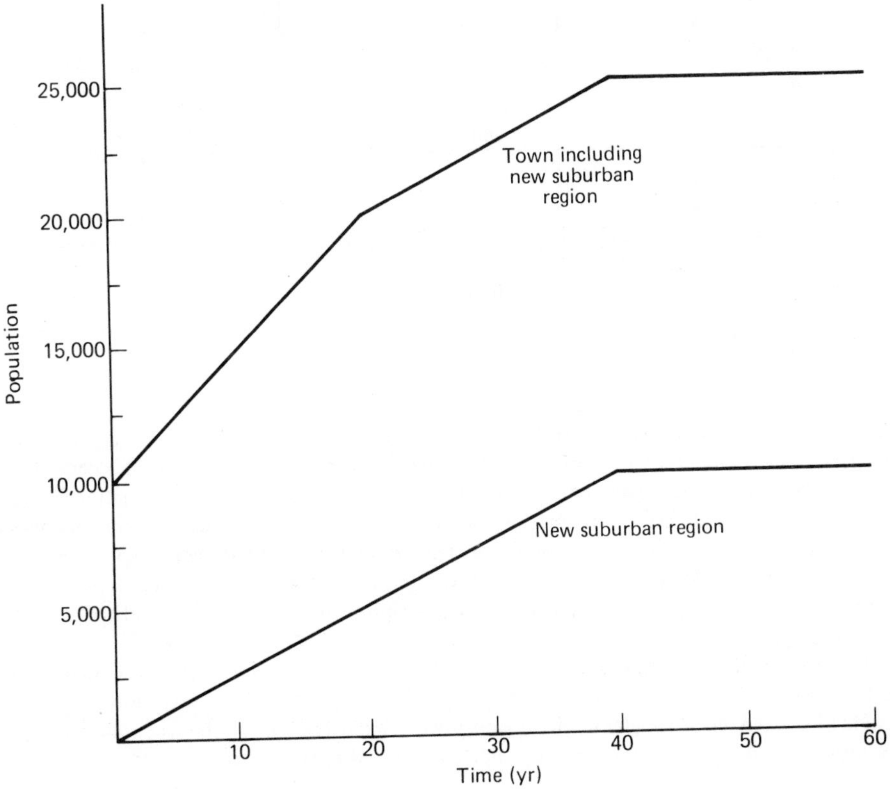

Figure 3.18*b*

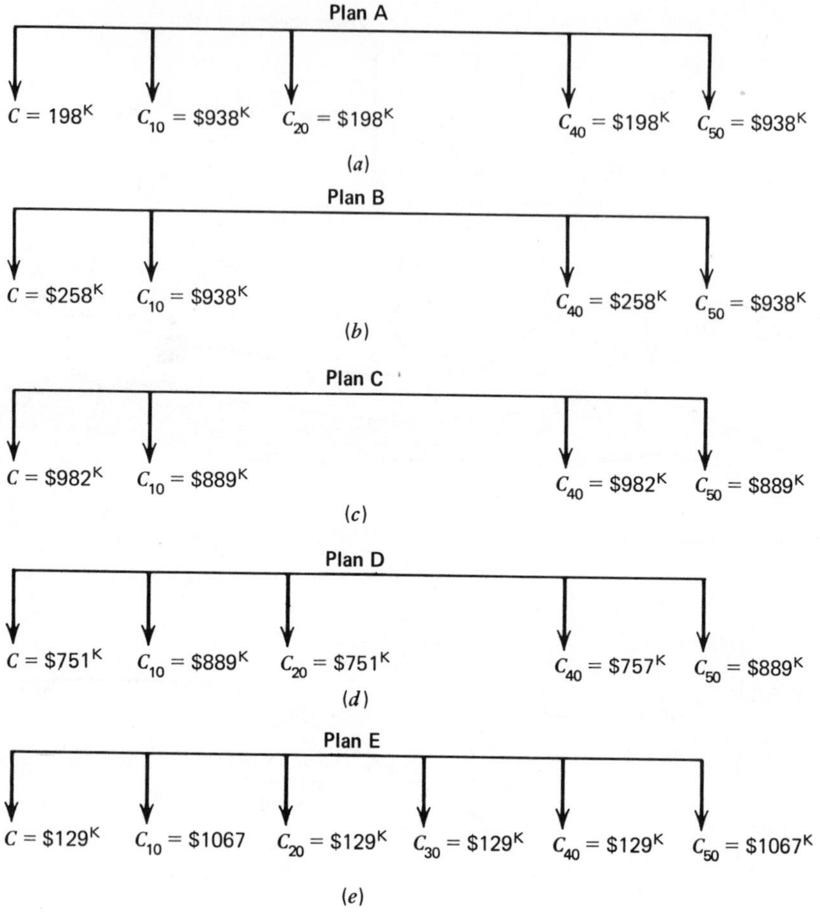

Figure 3.19

In each of the following construction schedules the capacity of the sewer line is specified in terms of the population served instead of the flow discharge or pipe diameter. This permits us to compare system capacity to service demand, which is given in Figure 3.18*b* as population forecasts.

There are several different combinations of building new sewer lines and replacing old ones or building new sewer lines next to existing ones. The key to a successful design is supplying a sewer capacity at least equal to the service demand. Five alternatives are considered. For each alternative, the construction schedule is accompanied with a description of the sewer system and how it changes over time. (See Figure 3.19.)

The capital construction cost is a function of the unit construction cost and sewer length. A sewer constructed in year k is denoted as $C_k = cl$, where c is the unit construction cost per unit length for a given pipe-diameter service region, town or suburban, and l is the sewer pipe length. For example, a 3000-ft 15-in. suburban sewer line constructed in year 20 is

$$C_{20} = cl = \$66 \cdot 3000 = \$198{,}000 = \$198\text{K} \qquad (\text{K} = 1000)$$

Plan A *Construction Schedule*

Year 0 Build a 15-in. suburban sewer line.
$C_0 = 66 \cdot 3000 = \$198{,}000 = \$198K$

Year 10 Replace the town interceptor line with a new 24-in. pipeline.
$C_{10} = 134 \cdot 7000 = \$938{,}000 = \$93K$

Year 20 Build a new 15-in. suburban sewer line in parallel to the existing 15 in. sewer line.
$C_{20} = 66 \cdot 3000 = 198{,}000 = 198K$

Year 40 Replace 15-in. suburban line constructed in year 0.
$C_{40} = C_0 = \$198K$

Year 50 Replace 24-in. town sewer line constructed in year 10.
$C_{50} = C_{10} = \$938K$

CONSTRUCTION YEAR *k*	PERIOD (yr)	REGION	SEWER CAPACITY (POPULATION)	SERVICE DEMAND FOR YEAR *k* + 10	PIPE SIZE (in.)	PIPE AGE (yr)
0	0–10	Suburban	5,000	2,500	15	New
		Town	20,000	15,000	22	30
10	10–20	Suburban	5,000	5,000	15	10
		Town	25,000	20,000	24	New
20	20–30	Suburban	10,000	7,500	15	New
					15	20
		Town	25,000	22,500	24	10
30	30–40	Suburban	10,000	10,000	15	10
					15	30
		Town	25,000	25,000	24	20
40	40–50	Suburban	10,000	10,000	15	New
					15	20
		Town	25,000	25,000	24	30
50	50–60	Suburban	10,000	10,000	15	10
					15	30
		Town	25,000	25,000	24	New

The present worth is determined from

$$PW^A = \$198K + \frac{938K}{(1.08)^{10}} + \frac{198K}{(1.08)^{20}} + \frac{\$198K}{(1.08)^{40}} + \frac{938K}{(1.08)^{50}}$$

$$PW^A = \$198K + 434K + 42.4K + 9.1K + 20.0K$$

$$PW^A = \$704K = \$704{,}000$$

Plan B *Construction Schedule*

Year 0 Build a 18-in. suburban sewer line.
$C_0 = 86 \cdot 3000 = 258{,}000 = \$258K$

Year 10 Replace the town interceptor line with a new 24-in. pipeline.
$C_{10} = 134 \cdot 7000 = 938{,}000 = \$938K$

Year 40 Replace 18-in. suburban line constructed in year 0.
$C_{40} = C_0 = \$258K = \$258K$

Year 50 Replace the 24-in. town sewer line constructed in year 10.
$C_{50} = C_{10} = \$938K$

CONSTRUCTION YEAR k	PERIOD (yr)	REGION	SEWER CAPACITY (POPULATION)	SERVICE DEMAND FOR YEAR $k+10$	PIPE SIZE (in.)	PIPE AGE (yr)
0	0–10	Suburban	10,000	2,500	18	New
		Town	20,000	15,000	22	30
10	10–20	Suburban	10,000	5,000	18	10
		Town	25,000	20,000	24	New
20	20–30	Suburban	10,000	7,500	18	20
		Town	25,000	22,500	24	10
30	30–40	Suburban	10,000	10,000	18	30
		Town	25,000	25,000	24	20
40	40–50	Suburban	10,000	10,000	18	New
		Town	25,000	25,000	24	30
50	50–60	Suburban	10,000	10,000	18	10
		Town	25,000	25,000	24	New

The present worth is

$$PW^B = 258K + \frac{938K}{(1.08)^{10}} + \frac{258K}{(1.08)^{40}} + \frac{938K}{(1.08)^{50}}$$

$$PW^B = 258K + 434.5K + 11.9K + 20.0K$$

$$PW^B = \$724K = \$724{,}000$$

Plan C *Construction Schedule*

Year 0 Build an 18-in. suburban sewer line.
Build a 18-in. sewer line parallel to existing interceptor sewer.
$C_0 = 86 \cdot 3000 + 102 \cdot 7000 = 982{,}000 = \$982K$

Year 10 Replace old 22-ft town sewer line with a 21-in. pipe.
$C_{10} = 107 \cdot 7000 = 889{,}000 = \$889K$

Year 40 Replace old sewer lines constructed in year 0.
$C_{40} = C_0 = 982K$

Year 50 Replace old sewer lines constructed in year 10.
$C_{50} = C_{10} = \$889K$

CONSTRUCTION YEAR k	PERIOD (yr)	REGION	SEWER CAPACITY (POPULATION)	SERVICE DEMAND FOR YEAR $k+10$	PIPE SIZE (in.)	PIPE AGE (yr)
0	0–10	Suburban	10,000	2,500	18	New
		Town	25,000	15,000	18	New[a]
					22	30
10	10–20	Suburban	10,000	5,000	18	10
		Town	25,000	20,000	18	10[a]
					21	New[a]
20	20–30	Suburban	10,000	7,500	18	20
		Town	25,000	22,500	18	20
					21	10
30	30–40	Suburban	10,000	10,000	18	30
		Town	25,000	25,000	18	30
					21	20
40	40–50	Suburban	10,000	10,000	18	New
		Town	25,000	25,000	18	New
					21	30
50	50–60	Suburban	10,000	10,000	18	10
		Town	25,000	25,000	18	10
					21	New

[a] Pipes in parallel.

The present worth is

$$PW^c = 982\text{K} + \frac{889\text{K}}{(1.08)^{10}} + \frac{982\text{K}}{(1.08)^{40}} + \frac{889\text{K}}{(1.08)^{50}}$$

$$PW^c = 982\text{K} + 412\text{K} + 44.7\text{K} + 79.0$$

$$PW^c = \$1458\text{K} = 1{,}458{,}000$$

Plan D *Construction Schedule*

Year 0 Build a 15-in. suburban sewer line.
Build a 15-in. sewer line parallel to existing interceptor line.
$C_0 = 66 \cdot 3000 + 79 \cdot 7000 = \$751{,}000 = \$751\text{K}$

Year 10 Replace the inceptor line with a new 21-in. pipeline.
$C_{10} = 127 \cdot 7000 = 889{,}000 = \$889\text{K}$

Year 20 Build a 15-in. suburban and 15-in. town sewer lines.
$C_{20} = 66 \cdot 3000 + 79 \times 7000 = 751{,}000 = \$751\text{K}$

Year 40 Replace sewer lines built in year 0.
$C_{40} = C_0 = \$751\text{K}$

Year 50 Replace sewer lines built in year 10.
$C_{50} = C_{10} = \$889\text{K}$

CONSTRUCTION YEAR k	PERIOD (yr)	REGION	SEWER CAPACITY (POPULATION)	SERVICE DEMAND FOR YEAR $k+10$	PIPE SIZE (in.)	PIPE AGE (yr)
0	0–10	Suburban	5,000	2,500	15	New
		Town	25,000	15,000	15	New
					22	30
10	10–20	Suburban	5,000	5,000	15	10
		Town	20,000	20,000	15	10
					21	New
20	20–30	Suburban	10,000	7,500	15	New
					15	20
		Town	25,000	22,500	15	New
					15	20
					21	10
30	30–40	Suburban	10,000	10,000	15	10
					15	30
		Town	25,000	25,000	15	10
					15	30
					21	20
40	40–50	Suburban	10,000	10,000	15	New
					15	20
		Town	25,000	25,000	15	20
					15	New
					21	30
50	50–60	Suburban	10,000	10,000	15	10
					15	30
		Town	25,000	25,000	15	30
					15	10
					21	New

The present worth is

$$PW^{D} = \$751\text{K} + \frac{889\text{K}}{(1.08)^{10}} + \frac{751\text{K}}{(1.08)^{20}} + \frac{751\text{K}}{(1.08)^{40}} + \frac{889\text{K}}{(1.08)^{50}}$$

$$PW^{D} = \$751\text{K} + \$411.8\text{K} + \$161\text{K} + 34.6\text{K} + 19.0\text{K}$$

$$PW^{D} = 1378\text{K} = 1{,}378{,}000$$

Plan E *Construction Schedule*

Year 0 Build a 10-in. suburban sewer line.
$C_0 = \$43 \cdot 3000 = 129{,}000 = \$129\text{K}$

Year 10 Build a 10-in. sewer line parallel to existing sewer line built in year 0. Replace the interceptor sewer line with a 24-in. pipeline.
$C_{10} = \$43 \cdot 3000 + 134 \times 7000 = 1{,}067{,}000 = \$1067\text{K}$

Year 20 Build a 10-in. pipeline parallel to existing line.
$C_{20} = \$43 \cdot 3000 = 129{,}000 = \$129\text{K}$

Year 30 Build a 10-in. sewer line in the suburban region.
$C_{30} = \$43 \cdot 3000 = 129{,}000 = \$129\text{K}$

Year 40 Replace the 10-in. suburban sewer lines constructed in year 0.
$C_{40} = C_0 = \$129\text{K}$

Year 50 Replace the 10-in. suburban and 24-in. town lines constructed in year 10.
$C_{50} = C_{10} = \$1067\text{K}$

CONSTRUCTION YEAR k	PERIOD (yr)	REGION	SEWER CAPACITY (POPULATION)	SERVICE DEMAND FOR YEAR $k+10$	PIPE SIZE (in.)	PIPE AGE (yr)
0	0–10	Suburban	2,500	2,500	10	New
		Town	20,000	15,000	22	30
10	10–20	Suburban	5,000	5,000	10	10
					10	New
		Town	25,000	20,000	24	New
20	20–30	Suburban	7,500	7,500	10	New
					10	10
					10	20
		Town	25,000	22,500	24	10
30	30–40	Suburban	10,000	10,000	10	New
					10	10
					10	20
					10	30
		Town	25,000	25,000	24	20
40	40–50	Suburban	10,000	10,000	10	New
					10	10
					10	20
					10	30
		Town	25,000	25,000	24	30
50	50–60	Suburban	10,000	10,000	10	New
					10	10
					10	20
					10	30
		Town	25,000	25,000	24	New

The present worth is

$$PW^E = 129\text{K} + \frac{1067}{(1.08)^{10}} + \frac{129}{(1.08)^{20}} + \frac{129}{(1.08)^{30}} + \frac{129}{(1.08)^{40}} + \frac{1067}{(1.08)^{50}}$$

$$PW^E = 129\text{K} = 49.40\text{K} + 27.7\text{K} + 13.0\text{K} + 5.9\text{K} + 22.8\text{K}$$

$$PW^E = 692\text{K} = \$692{,}000$$

The present worth of each plan is

$$PW^{\mathbf{A}} = \$704{,}000$$

$$PW^{\mathbf{B}} = \$724{,}000$$

$$PW^{\mathbf{C}} = \$1458{,}000$$

$$PW^{\mathbf{D}} = \$1377{,}000$$

$$PW^{\mathbf{E}} = \$692{,}000$$

The best (least-cost) alternative is therefore plan **E**.

Remarks

Closer evaluation of the five alternatives shows that the three top choices—plans **A**, **B**, and **E**—require a minimum amount of construction activity in the town region over the 50-year design life. Plans **C** and **D** call for the parallel construction of new pipe next to the old pipe. This construction takes place in the town, where it is the most costly to perform construction work.

This policy of minimizing the amount of construction activity over the design life of the project is not a general rule of thumb that applies to all stage construction problems. Of the three lowest-cost alternatives, plan **E** calls for the maximum amount of construction in the suburban region. The planning strategy here is to design for the smallest pipe size to meet the near-term 10-year population demand and to add to the flow capacity by adding more pipe at 10-year increments. Comparing the columns labeled Sewer Capacity and Service Demand for the suburban region shows that plan **E** does not provide overcapacity. Plans **A** and **B**, however, do. Even though plan **E** is the minimum-cost alternative; there may be additional costs of construction and intangible costs, due to community dissatisfaction with the disruption caused by the frequent construction schedule. Plan **B** utilizes a strategy of designing the pipe to provide for capacity for the 40th-year population demand. This plan requires the replacement of pipe at 40-yr intervals. Plan **A** is considered an immediate plan requiring construction in the suburban region at 20-yr intervals.

PROBLEMS

Problem 1

A bank offers an annual interest rate on savings of 6 percent compounded quarterly. The money is withdrawn after 1 yr.

(a) Determine the present worth of the $1000 savings after 1 yr. Assume an annual opportunity-cost interest rate of money for the saver is zero, $i = 0$ percent.
(b) Repeat step a, for $i = 5$ percent.
(c) Repeat step a, for $i = 10$ percent.

Problem 2

It is assumed that the annual rate of inflation f is 8 percent. For a 3-year period, $100,000 is to be withdrawn at the end of each year.

(a) Draw a cash-flow diagram of inflated dollars for $n = 3$.
(b) Determine the amount of money that must be placed in an interest-bearing account with an annual interest rate of 6 percent to keep pace with inflation. Use $f = 8$ percent for annual inflationary rate and $n = 3$.
(c) Repeat b for an annual rate of interest of 9 percent.

Problem 3

The annual inflation rate f is assumed to be 9 percent per year. The annual net revenue and all future costs from an investment are assumed to keep pace with inflation. The annual interest rate is 16 percent.

ALTERNATIVE	DESIGN LIFE	CAPITAL COST	NET BENEFIT
A	10	$100,000	$15,000
B	5	$60,000	$17,000

(a) Draw cash-flow diagrams for each alternative.
(b) Determine an equivalent interest rate r.
(c) Utilize r and the net present worth method to determine the better alternative.
(d) If $f = 0$, determine the better alternative.
(e) If $i = f = 16$ percent determine the better alternative.
(f) Prepare a table to show the effect of inflation f upon the net present worth values found in parts c, d, and e.

Problem 4

The sale of construction equipment is estimated to be 50 units per year, and the unit selling price is $10,000 per unit. The annual operating and maintenance costs to produce them are $400,000 per annum. Assume an annual interest rate of 10 percent for a 10-yr design life of the production equipment.
(a) Draw a cash-flow diagram of annual net revenue.
(b) Determine the present worth of annual net revenue for a 10 percent annual interest rate.
(c) If the annual rate of inflation is 7 percent, determine the equivalent annual interest rate. Assume that the unit selling price of the construction equipment keeps pace with inflation.
(d) Utilize the result in part c to determine the net present worth value.

Problem 5

The annual sales are $50,000 with an annual production cost of $35,000. Two alternatives for reducing annual production costs are available.

ALTERNATIVE	CAPITAL COST	ESTIMATED ANNUAL PRODUCTION COST	DESIGN LIFE	SALVAGE
A	$50,000	$25,000	5	$15,000
B	$40,000	$30,000	3	$15,000

Assume an annual interest rate of 10 percent. Furthermore, assume annual sales remain the same over the design life of the project.
(a) Draw a cash-flow diagram.
(b) Utilize the net present worth method to determine the better alternative.

(c) The annual inflation rate is assumed to be 8 percent. Determine the equivalent annual interest rate. Assume all sales keep pace with inflation.
(d) Utilize the results of part c to answer b.

Problem 6

Consider the following alternatives.

ALTERNATIVE	CAPITAL COST	ANNUAL REVENUE	ANNUAL O & M COSTS[a]	DESIGN LIFE
A	$150,000	$50,000	$30,000	10
B	$130,000	$50,000	$30,000	5
C	$175,000	$50,000	$25,000	10
D	$160,000	$50,000	$28,000	10
	$50,000 in year 5		$20,000 after year 5	

[a] O & M = operation and maintenance

Alternative **D** will be installed in two stages. In the first stage, the firm will operate equipment requiring $28,000 a year to operate and maintain. In year 5, the equipment will be upgraded and the annual operating and maintenance cost will be reduced to $20,000. Assume an annual interest rate of 7 percent. Assume no salvage value.
(a) Draw cash-flow diagrams for each alternative.
(b) Use the net present worth method to determine the best alternative.

Problem 7

Determine which alternative should be chosen. The annual interest rate is 15 percent. Assume no salvage value.

ALTERNATIVE	DESIGN LIFE	CAPITAL COST	ANNUAL REVENUE	ANNUAL COST
A	10	$100,000	$110,000	$90,000
B	5	$50,000	$10,000 for years 1, 2, 3	0
		$20,000 in year 3	$15,000 for years 4 and 5	0

Alternative **B** requires an initial investment of $50,000, and an additional investment at the beginning of year 3. For the first three years, $10,000 annual revenue is received at the end of each year. For years 4 and 5, $15,000 is received.
(a) Draw the cash-flow diagrams for alternatives **A** and **B**.
(b) Use the net present worth method to determine the better alternative.

Problem 8

For each of the following, assume a design life of one year.

ALTERNATIVES	PRINCIPAL OR INITIAL INVESTMENT	ANNUAL O & M[a] COSTS	BENEFIT OR INTEREST PAYMENT	SALVAGE VALUE	TAX RATE
1. Savings account	$1,000		$5\frac{3}{4}$% on principal	$1,000	20% on interest
2. Bond	$1,000		14% on principal	$1,000	20% on interest
3. Stock	$1,000		10% (est.) on principal	$1,300 (est.)	20% on interest 40% on net gain[b]
4. Savings certificate	$1,000		11%	$1,000	No tax
5. Machine	$1,000	$7,000 (est.)	$10,000 (est.)	$500 (est.)	8% on net profit 40% on net gain[b]

[a] O & M = operation and maintenance
[b] Net gain (loss) = Salvage value − Investment. Assume no tax is paid on losses.

(a) Determine the net benefit or loss after 1 yr for each alternative.
(b) Excluding risk considerations, which alternative is the best choice?
(c) Based on your knowledge of investments, list these alternatives in order of risk, the lowest risk at the top of the list. Explain the reason behind your ranking.

Problem 9

Determine which alternative investment plan is more profitable. The principal will be returned at the end of the investment period.

ALTERNATIVE	PRINCIPAL	ESTIMATED YEARLY REVENUE AS PERCENT OF PRINCIPLE	INCOME TAXES
A	$100,000	12% for 3 yr	3% on revenue
B	$95,000	8% for 4 yr	No taxes

Use the net present worth method of comparison at an annual interest rate of 10 percent to determine the investment plan and an inflation rate of 7 percent per year. Assume all future cash payments are fixed cash payments.

Problem 10

The current operating and maintenance cost is $50,000 per year. If new equipment costing $100,000 is installed, the annual operating and maintenance costs would be reduced to $30,000. The equipment has a design life of 6 yr with no salvage value. Assume an annual interest rate of 5 percent.
(a) Draw a cash-flow diagram.
(b) Use the net present worth method to determine if the investment is feasible.

Problem 11

An investor is considering either installing a heat pump or retrofitting a hydroelectric plant and selling the power to industry.

ALTERNATIVE	CAPITAL COST	DESIGN LIFE (yr)	ANNUAL OPERATING AND MAINTENANCE	ANNUAL OUTPUT
A. Heat pump	\$193,000	15	\$10,000 plus fuel oil cost[a]	5×10^9 Btu
B. Hydroelectric	\$226,000	30	5,000	525,000 kW-hr

[a] The heat pump extracts heat by burning solid waste material and fuel oil. It has a 3.3 coefficient of performance rating, the ratio of energy output to energy consumed in fuel oil. Assume there is no cost for the solid waste. The unit cost for power to operate the heat pump is 6.4 cents/kW-hr (1 kW-hr = 3413 Btu).

The power will be sold at 7.7 cents/kW-hr. The investor may obtain a loan at an annual interest rate of 10 percent.

(a) Draw a cash-flow diagram for each alternative.
(b) Use the net present worth method to determine the better alternative.
(c) If the annual inflation rate of $f = 5$ percent is assumed to affect all prices, determine the better alternative by the net present worth method.

Problem 12

A city operates a sports stadium that has a seating capacity of 25,000 people for baseball and football games. The average ticket price is \$4.00 per person. The annual attendance for all events is 1.5 million people per year. The annual operating cost is \$4M. The city is considering three plans.

A. Continue to operate the existing facility.

B. Destroy the old one, and construct a new facility seating 50,000 people.

C. Add 10,000 new seats to the existing stadium.

PLAN	ANNUAL MAINTENANCE COST	CAPITAL COST	EXPECTED ANNUAL ATTENDANCE
B	\$6M	\$15M	2.0M
C	\$5M	\$6M	1.7M

The annual opportunity-cost interest rate is 10 percent. Ignore inflationary effects. The design life is 20 yr. Consider construction and all additional revenues and costs in your selection.

(a) Draw cash-flow diagrams for the options considered.
(b) Utilize the net present worth method to determine the best plan.

Problem 13

A town must construct a 1-mile sewer line. In order to ensure the wastes do not settle in the pipe, the design velocity of 5 ft/sec is selected. Town engineers have evaluated the population and industrial growth of the area and have forecast the sewer flow to increase as shown in Figure 3.20. Determine

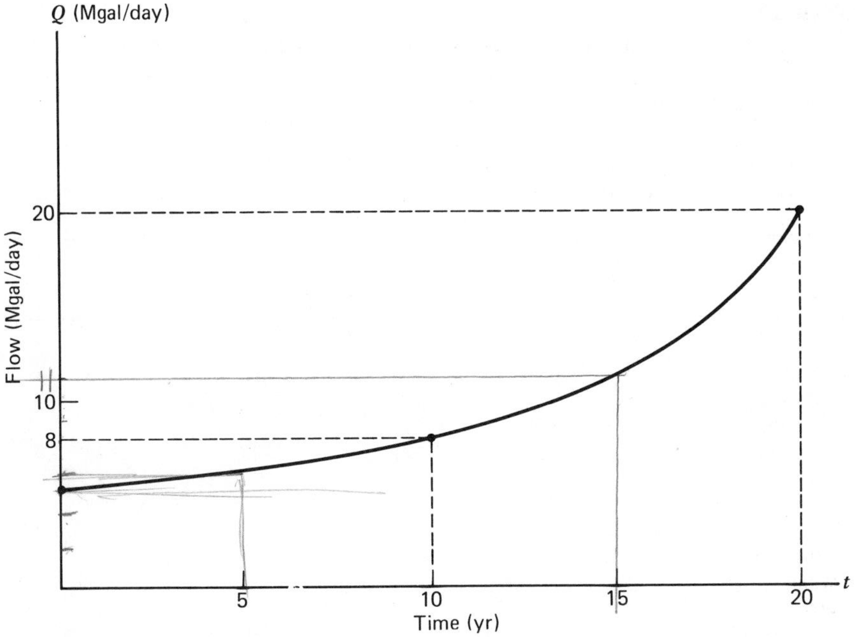

Figure 3.20

the least-cost alternative for plans to:

A. Construct a pipeline to adequately handle the forecasted 20-yr design flow.

B. Construct a pipeline in two stages. The first-stage plan calls for constructing a pipeline to handle the forecast 10-yr design flow and then calls for constructing a second pipe to handle additional flow for the forecast 20-yr design flow.

The construction cost of installing the pipe is a function of pipe diameter and length, $c = \$7$ per inch of pipe diameter per linear foot of pipe. Determine the least-cost alternative.

(a) Draw cash-flow diagrams.

(b) Use an annual interest rate of 7 percent and ignore the effects of inflation. Use the net present worth method.

Problem 14

A municipal storm sewer culvert several blocks in length has become inadequate owing to a commercial development. Two alternative plans have been designed to protect the area against flooding.

Alternative A replace the existing 36-in. corrugated pipe with a new 54-in. pipe.

Alternative B add a parallel 20-in. pipe next to the existing 36-in. corrugated pipe.

The construction cost for alternative **A** is $35,000 with a projected life of 50 yr. The annual maintenance cost is $1000. The construction cost for alternative **B** is $25,000. The existing 36-in. pipe will have to be replaced in 20 yr at a cost of $40,000. It will have a salvage value of $10,000 in 50 years. The maintenance cost will be $800/yr for the first 20 yr and $1000/yr for the next 30 yr.

(a) For each alternative show the cash-flow diagrams.

(b) Determine the best alternative by use of the net present worth method. The annual interest rate is 8 percent.

Problem 15

Use the net present worth alternative selection method to determine the best house plan. (See table.)

A. House with solar panels and aluminum siding.
B. House with solar panels and wood siding.
C. Conventional house with aluminum siding.
D. Conventional house with wood siding.

	PLAN[a]			
	A	B	C	D
Initial costs:				
Furnace	—	—	$2K	$2K
Solar				
Installation	$15K	$15K	—	—
House insulation	$2K	$2K	$3K	$3K
Aluminum siding	$5K	—	$5K	—
Wood siding	—	$3K	—	$3K
Maintenance Costs:				
Solar panel overhaul	$2K after 20 yr	$2K after 20 yr	— —	— —
Replace siding	$6K every 20 yr	—	$6K every 20 yr	—
Replace furnace	—	—	$2K after 20 yr	$2K after 20 yr
Annual heating oil	—	—	$1.5K	$1.5K
Painting	—	$2K every 5 yr	—	$2K every 5 yr

[a] (K = $1000)

Assume the design life of each house equals 40 yr and the annual interest rate is 6 percent. The following information should be considered. Assume no salvage value after 40 yr. Use cash-flow diagrams for each alternative. Assume all homes are equally comfortable.

Problem 16

Use the present worth least-cost alternative selection method to determine the best foundation design. A 10-m wide and 30-m long building will be constructed on a clay material. The design pressure of the building is assumed to be 24 kN/m^2. The clay has a specific weight of 100 pcf. Three alternatives are to be evaluated.

A. Pile foundation—piles are driven to bedrock.
B. Floating foundation.
C. Structure placed upon fill.

Alternative A

Use the following material and costs.

$$\text{Piles: } 8 \times 8\text{BP } 36 \qquad \sigma_y = 36{,}000 \text{ psi} \qquad \text{Factor of safety} = 2$$

$$\text{Construction cost} = \$10.20/\text{vertical linear foot}$$

(a) Determine the number of piles required and the construction cost.

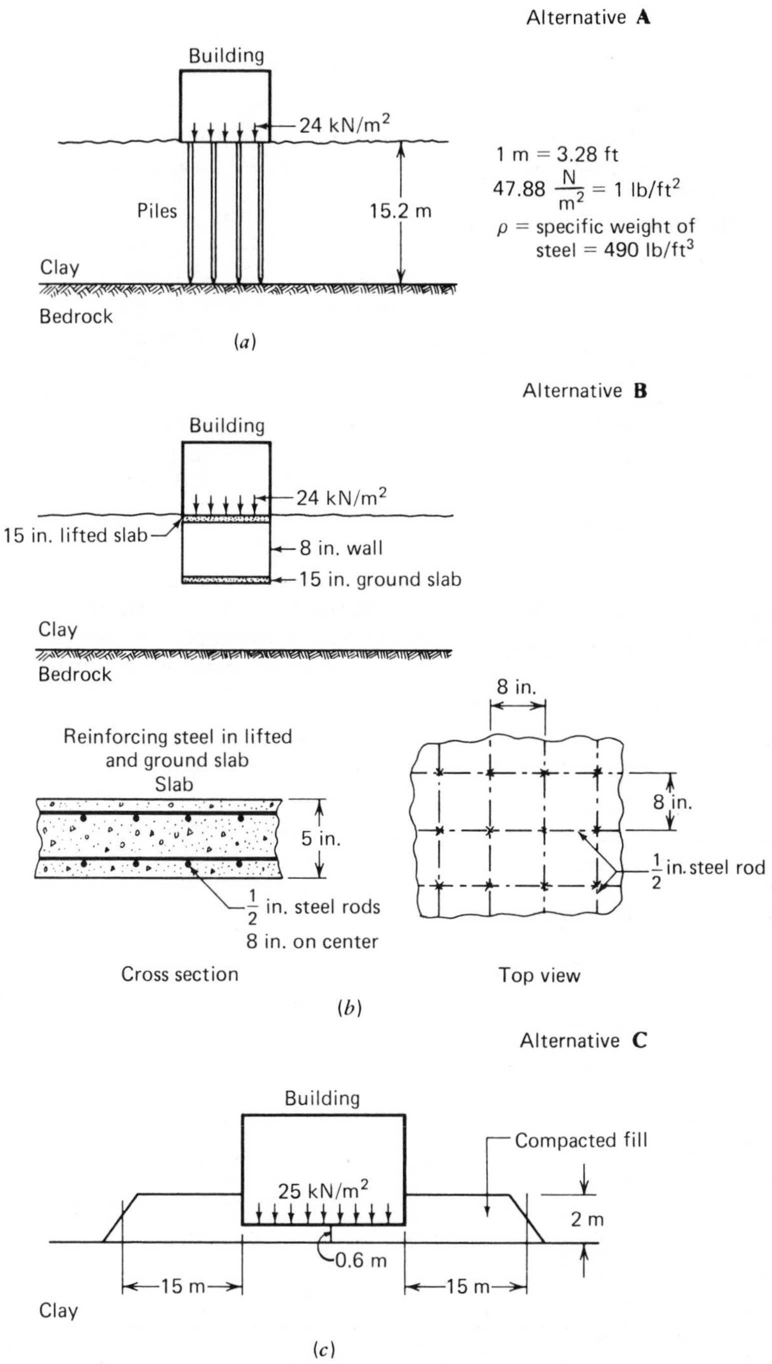

Figure 3.21 (*a*) Alternative 1, pile foundation. (*b*) Alternative 2, floating foundation. (*c*) Alternative 3, compacted fill foundation.

Alternative B

Define:

$$h = \text{depth of foundation (unknown)}$$

(b) Find the cost of the water-tight floating foundation. Assume the total weight equals the buoyancy force. Ignore the weight of the 8-in. walls in estimating the weight of the floating foundation. The specific weight of concrete is 145 lb/ft^3. Reinforcing steel is used in the lifted and ground slabs.

Assume the following costs:

Excavation:	$\frac{1}{2}$-yd^3 capacity shovel	(30-yd^3/hr): \$2.09/yd^3
8-in. walls	to 8-ft high	\$2.10/SFCA
	8 to 16-ft high	\$2.65/SFCA
	16-ft or higher	\$3.20/SFCA
Formwork:	15-in. ground slabs	\$1.50/ft^2
Reinforcing steel in slabs:		\$750/ton
Cast in Place Concrete: 8-in. walls		\$320/yd^3
15-in. lifted slabs (including formwork)		\$675/ft^2
15-in. ground slabs		\$2.55/ft^2

SFCA = square foot contact area

Alternative C

The total cost of compacted fill is \$4.40/yd^3. The building settles at a rate shown by the settlement curve in Figure 3.22. Forty-five meters of 8-in. water-supply pipe and 8-in. sewer pipe must be repaired after the building settles 0.15 m. Consider repair costs over the entire 30-yr design life of the building. Consider the following costs:

Excavation	(8-in. pipe):	\$0.75/linear ft
Pipe	(8-in. pipe):	\$3.70/linear ft

(c) Draw a cash flow diagram for alternative **C**.

(d) Use the net present worth method to determine the best alternative. Assume an annual interest rate of 10 percent per year.

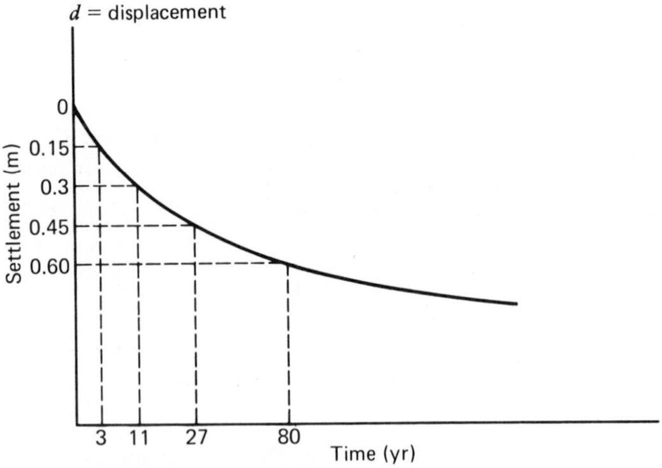

Figure 3.22 Water supply and sewer line repair made in years 3, 11, and 27.

Problem 17

The purchase price of a dump truck is

$$c = \$2000q^{1.1}$$

where q is the size of the truck load in cubic yards. The annual operating cost is estimated to be equal to 10 percent of the truck purchase price. The selling price of the material is $\$6/\text{yd}^3$. In one year 200 deliveries will be made. The maximum size of a truck is no greater than 30 yd^3. Assume a truck design life of 5 years and an annual interest rate of 15 percent.

(a) Formulate a mathematical model to maximize the present worth of net profit. Define the control variable. Draw and label a cash-flow diagram.
(b) Give the necessary and sufficient conditions for the optimum truck size.
(c) Solve for the optimum truck size.
(d) The annual rate of inflation is 8 percent. Do you think the future costs and benefits are better classified as cash payments keeping pace with inflation or fixed cash payments. Why?
(e) Use your answer from part d to reformulate and solve the optimization model given in parts a through c.

3.3 EVALUATING PUBLIC PROJECTS

Thus far in our discussion, mathematical procedures for selecting the best alternative among several candidates have been the prime topic. The benefits, costs, interest rates, and design life were assumed to be known values. In this section we direct our discussion to answering questions about which benefits and costs should be included in the analysis and what interest rate and design life should be assigned. Basic principles of consumer theory are introduced and used to help answer these questions.

The Supply Function

Investments by private business and government are made to increase the efficiency of delivering goods or services. The new efficiencies are assumed to drive down the cost of production, which, in time, will be passed on to consumers in the form of lower prices for goods and services. In private business the benefits received from investment are in the form of increased profits. In government the benefits received are generally in the form of improved service to society. For example, an investment in the highway system brings about higher levels of service, and an investment in a water project may bring increased flood protection or increased recreational benefits. Generally, the benefits received from government projects are not in the form of direct monetary payments as in the case of private enterprise. Thus, the term *social benefit* is used to describe these nonmonetary returns of investment.

A supply curve shows the relationship between price p and volume of output q. The supply–price relationship is also called a production function. Imagine that we are in the business of fabricating automobiles. The selling price of the automobile is designated by the variable p and the number of automobiles is designated by the variable q. The supply function S^0 in Figure 3.23 shows an increase in automobile price as volume of production increases.

If new fabricating equipment replaces the old equipment, its improved efficiency will cause a decrease in fabrication cost, which is passed on to consumers in the form of

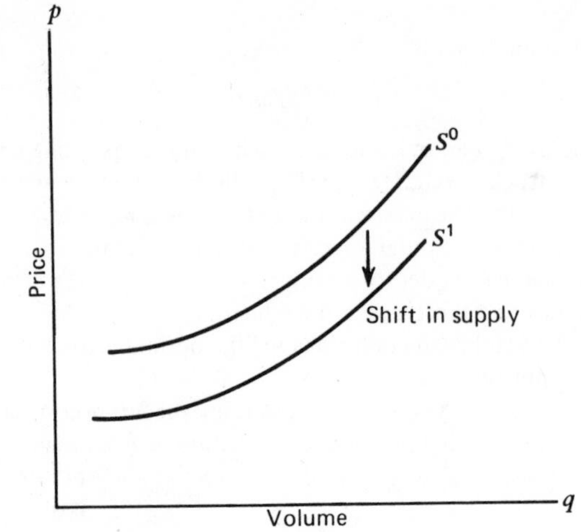

Figure 3.23 The effect of unproved production efficiency.

lower prices. In Figure 3.23 the price–volume relationship, the old and new equipment scenarios, are represented by the curves S^0 and S^1, respectively. Economists call the reduction in price caused by an investment in production a downward shift in supply.

Consumer Demand

The number of automobiles purchased will depend upon consumer demand. The demand is assumed to be a decreasing function of price, as shown in Figure 3.24. The demand is the willingness of consumers to purchase goods. For our example, the

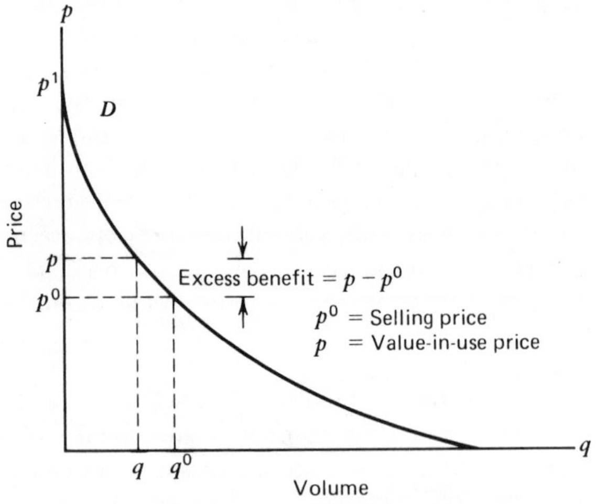

Figure 3.24 Consumer demand.

demand curve D shows the relationship between price and sales volume of automobiles.

The value that a consumer places upon a good is the maximum sacrifice the consumer is willing to make in order to acquire the use of the good or service. At price p consumers are willing to sacrifice or exchange p dollars to receive q automobiles. The *utility* or *value in use* to the consumers is assumed to be equal to the price p they are willing to pay. This concept was discussed previously in Section 3.2 on "Real and Inflationary Prices." The actual price paid p^0 will be less than or equal to the value in use, $p^0 \leq p$. Thus, the qth vehicle sold has the value of p dollars as shown in Figure 3.24. The difference between the value-in-use price and the selling price $(p - p^0)$ is a measure of consumer satisfaction and is called the *excess benefit*.

The level of consumption of goods will depend upon consumer demand and the level of production. *Market equilibrium* is the point where consumer demand equals supply. It is shown as point a in Figure 3.25 at a level of consumption q^0 and the exchange price of p^0. At price p^0, q^0 consumers are willing to exchange their money for obtaining the use of an automobile. At volumes greater than q^0 the unit value in use of the additional automobiles is less than the selling price, thus sales greater than q^0 are not made.

The consumer demand curve shows that not all purchasers of automobiles receive the same satisfaction of ownership. Excess benefit $(p - p^0)$, a measure of owner satisfaction, varies with q. At q^0, a consumer received an automobile that he considered to have a value in use equal to the price paid or $p = p^0$. There is no excess benefit at q^0. At $q = 1$, this consumer is considered to receive maximum satisfaction or an excess benefit of $(p^1 - p^0)$. The total satisfaction obtained by all consumers is called the *consumer surplus*. It is shown as the shaded area p^1ap^0 in Figure 3.25.

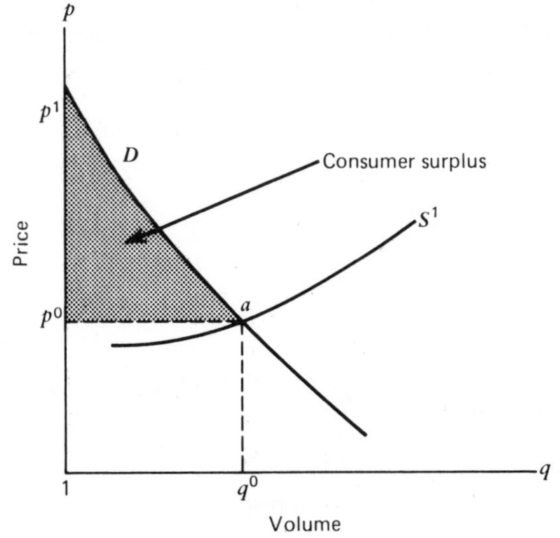

Figure 3.25 Market equilibrium and consumer surplus.

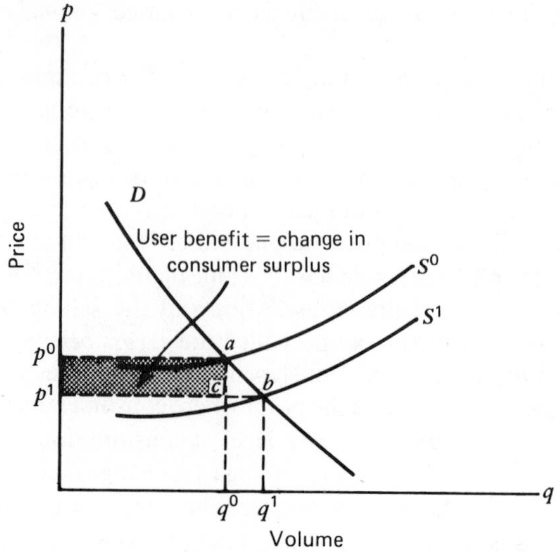

Figure 3.26 Change in consumer surplus.

Since investment is assumed to cause an increase in efficiency and a downward shift in the supply curve, a new market equilibrium level is achieved. A new market equilibrium is assumed to occur at q^1 as shown in Figure 3.26. This investment brought about an added benefit to consumers equal to the change in consumer surplus, which is shown as the shaded region p^0abp^1.

The change in consumer surplus reflects the increased consumer satisfaction, value in use, or utility brought about by the price reduction from p^0 to p^1. The *user benefit B* is assumed to be equal to the *change in consumer surplus.* The user benefit B is approximated to be equal to the area of a trapezoid p^0abp^1 or

$$B = q^0(p^0 - p^1) + \tfrac{1}{2}(q^1 - q^0)(p^0 - p^1)$$

The quantity $q^0(p^0 - p^1)$ is the added benefit from lower prices for the original sales and $\frac{1}{2}(q^1 - q^0)(p^0 - p^1)$ is the added benefit from increased sales. The equation is simplified to

$$B = \frac{(p^0 - p^1)(q^0 + q^1)}{2}$$

Demand Elasticity

The demand elasticity is defined to be the ratio of the percentage change in quantity demanded to the percent change in price, or

$$w = \frac{\%\Delta \text{ in demand}}{\%\Delta \text{ in price}}$$

$$w = \frac{\Delta q/q}{\Delta p/p} = \frac{p}{q}\left(\frac{\Delta q}{\Delta p}\right) = \frac{p}{q}\left(\frac{dq}{dp}\right)$$

In engineering and economic studies, it is common to use the exponential demand function

$$p = \alpha q^{\beta}$$

where the coefficient α is a unit price and β is a negative coefficient, $\beta < 0$. The elasticity of this function may be determined by calculating dp/dq and substituting it into the definition of elasticity.

$$\frac{dp}{dq} = \alpha\beta q^{\beta - 1}$$

Thus, from the definition of elasticity, we obtain

$$w = \frac{p}{q} \frac{1}{dp/dq} = \frac{\alpha q^{\beta}}{q(\alpha\beta q^{\beta - 1})} = \beta^{-1}$$

The demand elasticity is a constant value for an exponential demand function. Since $\beta < 0$, the elasticity ω is always negative. This is true for all demand functions.

If a slight change in price causes a large change in demand, it is called an elastic demand when $w \leq -1$. Inelastic demand occurs when the demand remains relatively unchanged no matter what change in price is made, $-1 \leq w \leq 0$. Generally, items required for survival such as fuel oil and water have an inelastic demand.

If the demand is inelastic, the change in demand Δq will be very small as shown in Figure 3.27. When $\Delta q \approx 0$ the consumer surplus is estimated to be $B = q^0(p^0 - p^1)$. For elastic demand, the same simplification cannot be made.

Monetary Return and Social Benefit

For simplicity, assume that a private firm consists of a group of stockholders or investors whose sole purpose is receiving a maximum return on investment. The government, on the other hand, consists of taxpayers who pay taxes to receive such diverse services or benefits as education, public health protection, military defense, environmental protection, transportation, and communication. The taxpayer expects the government to provide these services most efficiently. Obviously, the roles of the private firm and government are not the same. A private firm is motivated to maximize net profits in order that it may pay its stockholders the highest dividend. The government, on the other hand, is motivated to maximize social benefits or consumer surplus.

If a stockholder does not feel that his investment is being optimized, he may sell his shares in the firm and invest the money in another firm. If a taxpayer does not feel that his taxes are being used most appropriately, he may voice his objection in the voting booth. This is the democratic process. Through the market system of buying and selling stocks and the voting system, the stockholders and taxpayers have mechanisms for ensuring that their desires to receive maximum net profits and maximum social benefits are met. This simplistic view of the motivation of private firms and the government illustrates a fundamental difference between them.

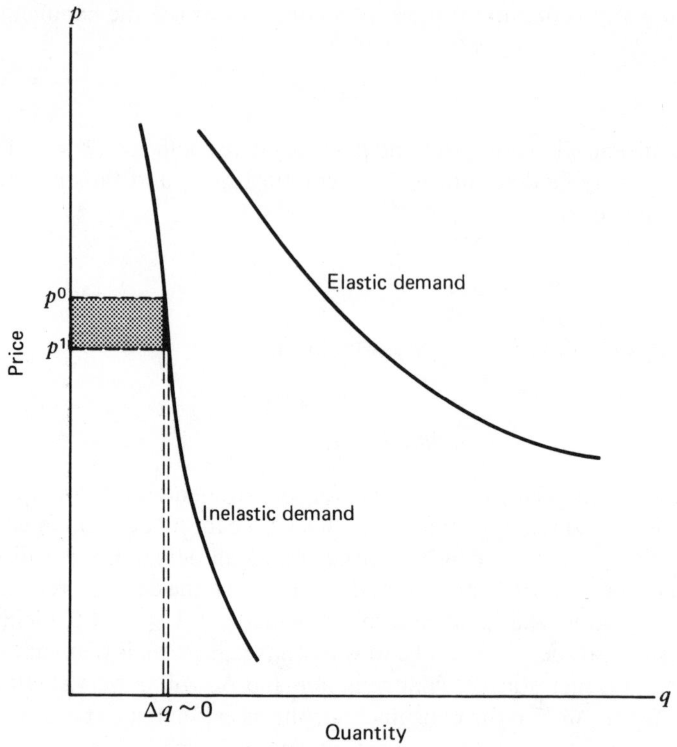

Figure 3.27 Elastic and inelastic demand.

Table 3.2 lists the similarities and differences between private and public sectors in an ideal alternative selection process.

Social Benefits

Social benefits may be defined as tangible and intangible benefits. A *tangible benefit* is a benefit that can be measured in dollars. For example, highways tolls are considered tangible benefits because the revenues received are in the form of cash. In the case of the toll-free highway the benefit received by the motorists is in the form of

Table 3.2 Public versus Private Alternative Selection

ISSUES	PRIVATE FIRM	GOVERNMENT
Motivation	Maximize net profits	Maximize social benefits
Output	Sale of goods and services	Public services
Benefit	Revenue	Tangible and intangible
Costs	Capital	Capital
	Operation and maintenance	Operation and maintenance

a time savings. Since time savings can be calculated in terms of vehicle operating costs, that is, cost of fuel, oil, and so on, time savings are considered tangible benefits. An *intangible social benefit* is defined as a benefit that cannot be assigned a dollar value. One of the outcomes of building a highway may be to reduce noise pollution. It is considered an intangible social benefit because its value is impossible to assess meaningfully in dollars. Social benefits may be considered as negative benefits as well as positive ones. If a highway causes noise pollution or has a deteriorating effect upon the neighborhood it bisects, it may be considered a negative intangible benefit.

Private versus Public Projects

The private firm has an obligation to its stockholders to maximize net profit $P = R - C$. This objective may be stated as

$$\text{Maximize } P = pq - C$$

where R = revenue = pq and C = cost of production, the sum of the capital, operating and maintenance costs evaluated as annual payments. This objective does not consider consumer surplus. The firm is interested in consumer demand only to determine the increase in profit it can make by its investment.

The producer receives an extra revenue from a capital investment by increasing production from q^0 to q^1. The net increase in revenue is equal to $(p^1q^1 - p^0q^0)$. Since the net profit is the difference between revenue and cost, the change in net profit is

$$(p^1q^1 - C^1) - (p^0q^0 - C^0) = (p^1q^1 - p^0q^0) - (C^1 - C^0)$$

where C^0 and C^1 represent the production costs before and after the investment. The firm will be interested in the capital investment only if net revenue $(p^1q^1 - p^0q^0)$ exceeds net production costs $(C^1 - C^0)$.

Government has a mission to maximize social benefit to the public. Since society pays for the cost to provide these services, they should receive the benefit from them. Public projects should be considered as viable alternatives for construction only if the consumer surplus exceeds costs. That is, $B > C$. The costs are considered to be the total sum of capital, operating and maintenance costs. With the net present worth method, the alternative with the largest present value difference of consumer surplus and cost, or $B_0 - C_0$, is the best selection.

Evaluating Transportation and Water Resource Projects

Engineering economics methods may be utilized to examine various public projects including water supply, transportation, health, military, land usage, and education systems. Since engineering economic analysis has been most commonly applied to water and highway projects, Tables 3.3 and 3.4 have been prepared. These tables list the primary social benefits and costs that are considered for these types of projects.

Consumer surplus theory is used to determine *all direct benefits* as a result of public investment. Conceptually, direct consumer surplus is determined by

$$B = \frac{(p^0 - p^1)(q^0 + q^1)}{2}$$

Table 3.3 Water Projects

SOCIAL BENEFITS	EVALUATION OF SOCIAL BENEFIT
1. Direct: Irrigation	Forecast additional output and prices of agricultural goods
Secondary: Increased sales by grain merchants, transport concerns, millers, bakers, etc.	The secondary benefit is assumed to be in direct proportion to the increase in agricultural goods
Special considerations	Farm price subsidies
2. Direct: Flood control	(a) Forecast the savings from losses to damage of property, furnishings, crops, etc. to different owner types: individuals, businesses, farmers, etc. (b) Avoidance of deaths due to drowning (c) Avoidance of temporary costs from flood victim evacuation, sanitation breakdowns, etc.
3. Direct: Hydroelectric power	Evaluate the savings realized by not having to buy from an alternative source
Externalities:	Effects upon operations of existing or potential dams downstream
Intangible benefits and disbenefits	(a) Recreation benefits (b) Environmental effects on wildlife, etc.

Table 3.4 Highway Transportation Projects

SOCIAL BENEFITS	EVALUATION OF SOCIAL BENEFITS
1. Direct: Net annual savings	Forecast vehicle operating cost savings for (a) Existing traffic users (b) Generated traffic users (c) Diverted traffic users (d) Normal-growth traffic users
2. Direct: Reduction in accidents	Forecast the reduction in the number of accidents and the cost of accidents
3. Direct: Effect upon public transportation	Forecast benefits or costs associated by competing public transport modes
Externalities	Added income to roadside businesses
Costs	Capital, annual operating and maintenance costs
Intangible items.	(a) Noise and air pollution (b) Effects upon neighboring communities

For transportation analysis, p^0 and p^1 may be estimated with the use of running-cost-of-motor-vehicle data and with q^0 and q^1 as estimates of current and future traffic, respectively. Likewise, for water projects similar prices and outputs can be estimated. This looks simple enough, but there are complexities that make this process more difficult than it first appears.

For example, the direct benefit from an irrigation project is an intangible benefit. As a result, agricultural output and the cost of agricultural products are used as a surrogate measure of the social benefit. Owing to farm subsidies the amount of money a farmer receives for his crops may be vastly different from the market prices paid by the consumers, the recipients of the benefits. Judgment must be used in assigning prices to these items.

Externalities and Secondary Benefits

When applying engineering economics methods, *double counting* must be avoided. Public investment projects should take account of external effects, *externalities*, when they have a *direct* impact upon productivity and should not consider externalities that affect prices only. If an irrigation dam project causes a reduction in available down-stream water, the amount of water for hydropower and irrigation in this area may be adversely affected. These effects should be considered in the analysis, because the direct impact of the investment causes increased benefits to the upstream area and negative benefits to downstream area. Both positive and negative impacts should be included in the analysis in this case.

On the other hand, if higher benefits or profits are caused by an increase in profits in one location and a decrease in another, they should not be included as a user benefit if they are not a direct impact of productivity changes. Consider the effects of a new highway upon two roadside businesses; one is located at a busy location along the new road and the other one is located on the old roadway, which is no longer busy with customers. The roadside business on the new road now receives higher benefits. The other one reduces prices and receives lower profits. The higher profits that the business along the new road receives is not due to increased productivity, but traffic diversion. The net effect upon society is assumed to be zero. Some gain while others lose. Only accounting for increased profits at the new site in a net present worth analysis is considered *double counting* because the direct benefits plus externalities from price changes from a transfer of income artificially inflate the user benefit B.

Secondary benefits are brought about by improved efficiency in a primary industry that is passed on to a support industry. A primary industry produces a raw material; a support industry utilizes the raw material to produce a consumer product. For example, the primary industry sells grain, and the secondary industry sells bread. A support business may receive increased profits owing to an increase in productivity and a lowering of prices of the primary industry. In turn, the support business passes its benefits to its consumers in the form of lower prices. This spillover effect is called a secondary benefit. For irrigation projects this spillover effect will cause an increase in sales or benefits by corn merchants, transporters, food producers, and the like. Some claim the benefits received by all these beneficiaries should be used as the user benefit. Others feel that the secondary benefits are brought about by the more-efficient production in the primary industry and should be included as user benefits by them exclusively and not by support industries. No clear-cut resolution of this controversial issue is offered in the literature.

Demand Shifts

Consumer preferences for goods and services are brought about by many factors. They may be a direct result of the construction of a new facility or by other factors. Increased traffic may be caused by the improved accessibility from constructing a new roadway, or increased demand may be brought about by an improved automobile design that has nothing to do with the new roadway investment. The second type of growth is called naturally generated traffic demand. The shifts in consumer demand may be analyzed with the use of supply and demand curves as shown in Figure 3.28.

Consider the downward shifts in the supply function from S^0 to S^1 caused by the improved efficiency offered by the construction of a new roadway and the shift in demand from D^0 to D^1 caused by increased demand due to naturally generated traffic

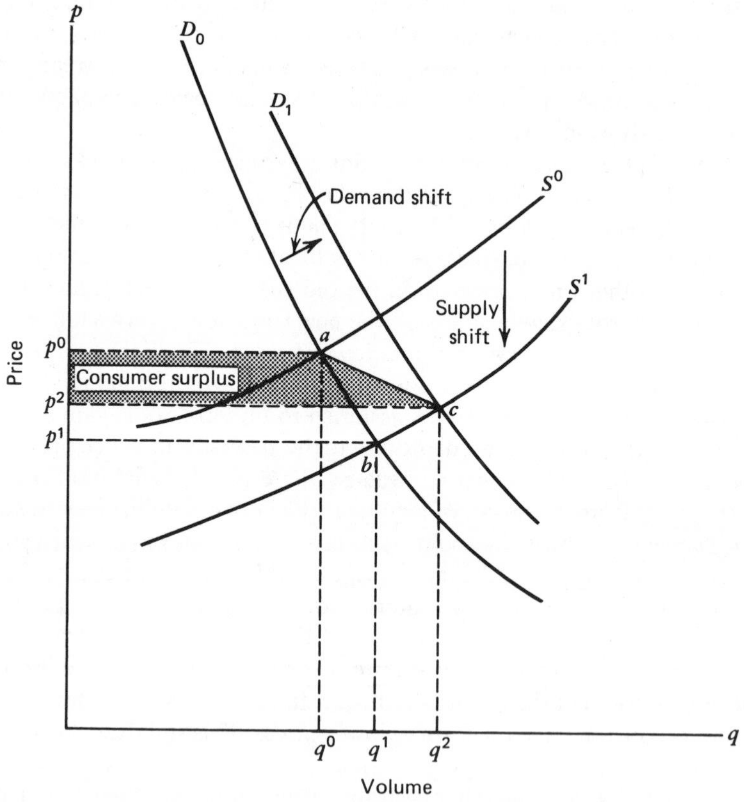

Figure 3.28 Effects of demand shift and increased productivity.

demand. Consider the supply and demand shifts as two separate processes. First, the new roadway is opened, which causes a shift in market equilibrium from point *a* to point *b*. The consumer surplus for this intermediate equilibrium point is the area p^0abp^1. Next, the increased demand results in a shift of market equilibrium from point *b* to point *c*. The consumer surplus for this equilibrium position is the shaded area p^0acp^2. The consumer surplus is estimated to be

$$B = \frac{(p^0 - p^2)(q^0 + q^2)}{2}$$

This user benefit should be used in economic analysis if increased demand is assumed to occur.

Consider the effects of the roadway construction and the increased demand upon the original q^0 users. The new roadway causes a downward shift in cost to the individual users from price p^0 to price p^1. Since $p^1 < p^0$, the original q^0 road users experience an increased level of service. The increased demand caused by improved accessibility and natural growth, however, adversely affects the level of service received by these users. When the demand shifts from q^1 to q^2, the motorists pay p^2, which is greater than the minimum price p^1. If the demand continues to grow, the price of travel could increase to a price level greater than p^0. If this occurs, the project may be considered a design failure.

Design Life

Estimation of the design life of a facility is a subjective process. Design life may be based upon the anticipated time of usable physical or economic service. A project may be deemed unusable because of the effects of physical deterioration, technological changes, demand shifts, economic considerations, and so on. The significance of the design life upon the present worth method will depend upon the assignment of the interest rate. Higher interest rates tend to diminish the effects of far future benefits and costs.

Social Opportunity-Cost Rates

There is common agreement among economists and engineers that the interest rate assigned to public projects, called the *social opportunity-cost interest rate*, should reflect the risk involved. Higher interest rates should be used on projects involving greater risks. The higher rates put more weight upon present and near-term benefits and costs. Long-term benefits and costs are given less weight. Higher rates favor the present generation of users, whereas lower rates favor future generations. Water projects are considered less risky than highway projects. Flood protection dams, hydropower and irrigation projects, have a high success rate of delivering services over the entire life of the project. These facilities remain serviceable for as long as 50 years. They serve both the present and future generations. Owing to unexpected traffic growth, some highways have become congested soon after opening. In this case the present generation gleans the full benefit for a short period of time. High interest rates should be assigned to high-risk projects of this type.

The actual assignment of a social opportunity-cost rate is a complex issue involving economic and political factors. No formula or clear-cut method for assigning it exists. Engineers and planners generally agree that the minimum annual interest rate for both private and public projects should be at least equal to the interest rate being paid on risk-free government bonds. The payment of government bonds is considered to be 100 percent certain; therefore, they are considered to be risk-free. The annual interest rate being paid on them is considered the lower limit. For government projects the social opportunity cost should be at least equal to the return on government bonds and adjusted upward for uncertainties and risk.

This is similar to the approach used by banks on business loans. Since taxes must be paid upon business profits, which can be as much as 50 percent of the profits, the annual interest rates used for the evaluation of private investment may be more than double the rate used for government projects. Most engineers and planners feel that the social opportunity-cost rate should be assigned a value between the lower limit of risk-free government bonds and the upper limit of business investment.

Individuals who buy government bonds are removing their money from the private sector. If this money is being used to pay for a public project, then it is only reasonable for the social opportunity-cost rate to be at least equal to the minimum risk-free annual interest rate. Furthermore, it is reasonable to assume that the annual interest rate on government bonds should be greater than the estimated annual inflation rate. If not, individuals will find other opportunities for investment.

These principles are not utilized in practice. Some critics claim that the social opportunity-cost rates are assigned values that are too low and the methods utilized for measuring benefits do not reflect the true worth of the project. They claim that legislation favors the construction of public projects. The annual interest rate assigned by Federal law to certain water resource projects has risen from $3\frac{1}{8}$ percent in 1967 to a proposed 7 percent in 1980. The inflation rate as measured by the consumer price and producer price indices has risen dramatically. The consumer price index has risen from 2.9 percent in 1967 to more than 13 percent in 1979. The producer price index has risen from 2 percent in 1967 to more than 13 percent in 1979. It peaked at a value of more than 18 percent in 1974.

Distribution of User Benefits

The questions of who receives benefits from a public project and who pays for a public project are important. They are not explicitly considered in the mathematical formulation of the engineering economic methods. Since demand curves do not distinguish between users or beneficiaries, the rich and poor are assumed to receive equal benefits. Furthermore, public projects are financed by selling government bonds and by general tax revenues. These are totally different methods of financing public projects. Unfortunately, since the question of who pays and who receives benefits is so complex, no completely satisfactory answer exists.

Owing to the difficulty of measuring tangible benefits and not explicitly introducing intangible benefits into the engineering economic selection methods, selecting the best design alternative among candidates offering different societal benefits is not recommended, and the methods should be limited to projects offering the same societal benefits. For example, selecting the best highway design among different highway designs is considered appropriate because the primary recipients are highway users. However, selecting the best design among highway and flood protection dams projects is not recommended. The benefits received from these projects are not like quantities. The highway offers benefits in the form of time savings, and the dam offers benefits in the form of risk aversion. Even though these benefits may be measured as tangible benefits, they measure different types of benefits. The road user receives a benefit every time that he or she uses the road. It is possible that a flood occurrence will not take place during the design life of the project, and no tangible benefit is actually received. The increased security associated with risk aversion, however, is an intangible benefit that cannot be ignored.

Here, even though the public may not always agree with the final decision, the political process proves to be more effective in selecting projects offering different societal benefits than are mathematical methods. Clearly, engineering economics and the methods of resource allocation as discussed in other parts of this book have their advantages and disadvantages and must be used where appropriate.

EXAMPLE 3.10 Water Consumption

Residential and commercial users are charged different rates as shown in the following table.

USER	WATER RATE	SALES	DEMAND ELASTICITY COEFFICIENT
Residential	\$0.50/unit	10M units	−0.225
Commercial	\$0.40/unit	12.5M units	−0.10

This dual rate system has been challenged as being unfair to the residential user. A fixed rate of \$0.46/unit has been proposed for all users.

(a) Estimate the total demand for water. Does the new policy encourage water conservation?
(b) Estimate the total change in revenue due to the price change.
(c) Estimate the consumer surplus for residential and commercial users.

Solution

(a) The change in water demand may be estimated with the demand elasticity relationship.

$$w = \frac{\Delta q/q}{\Delta p/p}$$

or

$$\Delta q = w \frac{\Delta p}{p} q$$

For residential users:

$$\Delta q = -0.225 \cdot \frac{(0.46 - 0.50)}{0.50} \cdot 10\text{M} = 0.18\text{M}$$

For commercial users:

$$\Delta q = -0.10 \cdot \frac{(0.46 - 0.40)}{0.40} \cdot 12.5\text{M} = -0.188\text{M}$$

Since $\Delta q = q^1 - q^0$ or $q^1 = q^0 + \Delta q$, the new residential demand q^{r1} for water is

$$q^{r1} = 10\text{M} + 0.18\text{M} = 10.18\text{M units}$$

Similarly, for commercial users the water use q^{c1} is

$$q^{c1} = 12.5\text{M} - 0.188\text{M} = 12.31\text{M units}$$

The total demand is $q^1 = q^{r1} + q^{c1}$

$$q^1 = 10.18\text{M} + 12.31 = 22.49\text{M}$$

The new price structure does not encourage water conservation. The demand will remain approximately the same.

(b) The new total revenue R^1 is

$$R^1 = p^1 q^1 = \frac{\$0.46}{\text{unit}} \cdot 22.49\text{M} = \$10.34\text{M}$$

The old revenue is

$$R^0 = p^{r0} q^{r0} + p^{c0} q^{c0} = 0.50 \cdot 10 + 0.40 \cdot 12.5 = \$10\text{M}$$

The net change in revenue ΔR is estimated to be

$$\Delta R = r^1 - r^0 = \$10.34 - 10\text{M} = \$0.34\text{M}$$

$$\Delta R = \$350{,}000$$

The price structure will bring an estimated increase of \$350,000 in revenue.

(c) The consumer surplus diagram is shown in Figure 3.29. The consumer surplus is equal to

$$B = (p^0 - p^1)\bar{q}$$

where

$$\bar{q} = \frac{q^0 + q^1}{2}$$

For residential users the consumer surplus B^r is estimated to be

$$B^r = (p^{r0} - p^{r1})\bar{q} = (0.50 - 0.46)\frac{(10.18\text{M} + 10.00)}{2}$$

$$B^r = \$0.40\text{M}$$

For commercial users the consumer surplus B^c is estimated to be

$$B^c = (P^{c1} - p^{c0})\bar{q}$$

$$B^c = (0.4 - 0.46) \times \frac{(12.31 + 12.5)}{2}$$

$$B^c = -\$0.74\text{M}$$

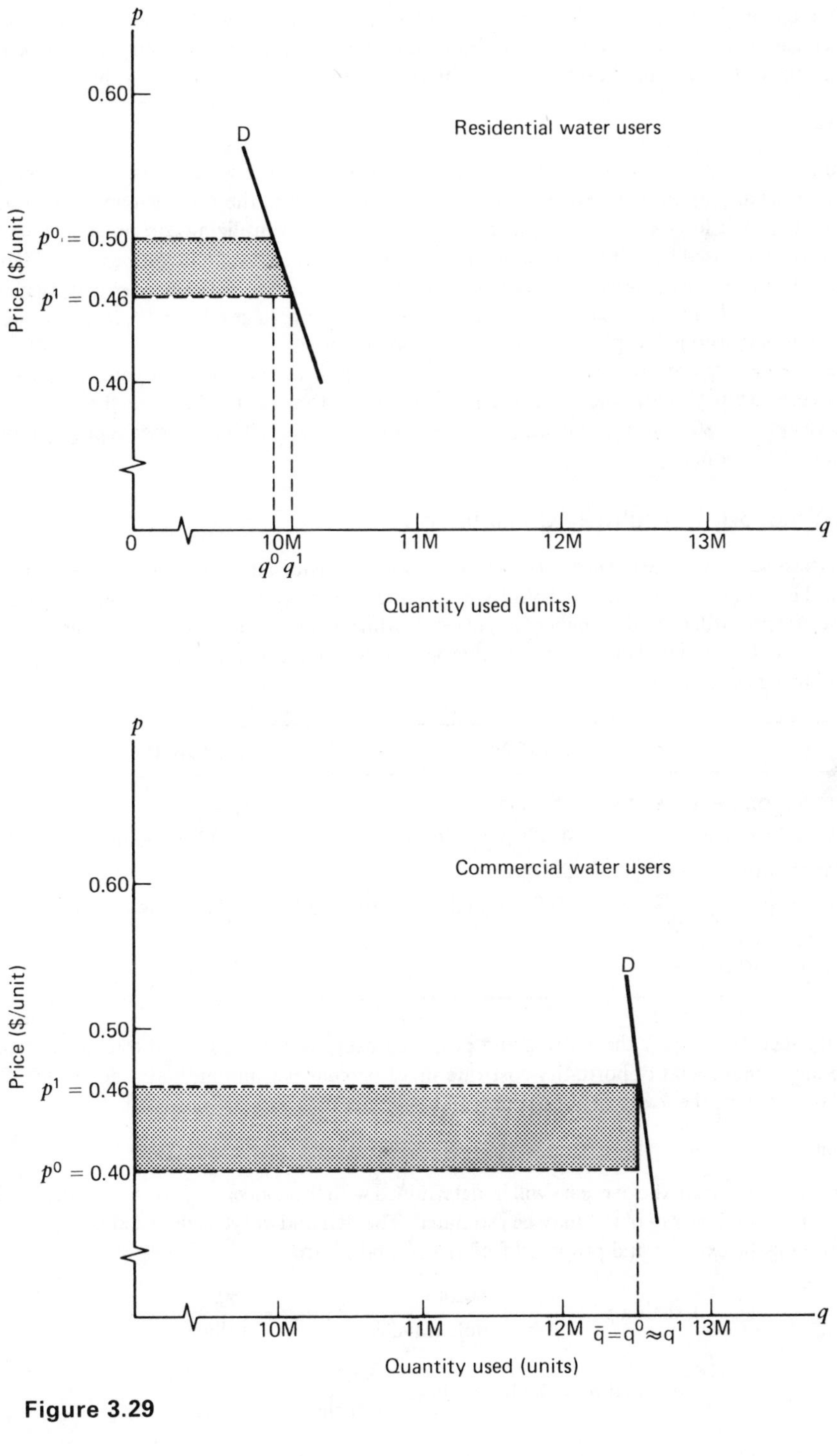
p
0.60
D
Residential water users
p⁰ = 0.50
p¹ = 0.46
0.40
Price ($/unit)
0
10M
11M
12M
13M
q
q⁰ q¹
Quantity used (units)
p
Commercial water users
0.60
D
0.50
p¹ = 0.46
p⁰ = 0.40
Price ($/unit)
10M
11M
12M
q̄ = q⁰ ≈ q¹
13M
q
Quantity used (units)

Figure 3.29

The change in price causes an increased burden in price to commercial users and only a slight decrease in commercial water use, as reflected in the negative consumer surplus. The benefits accrue to the residential users as shown by the positive consumer surplus value.

Remarks

Overall, since the proposed fixed rate does not appear to promote water conservation and puts more of a penalty upon commercial users than residential users, the proposal appears to be ineffective in establishing a fair price system. The basic reason for utilizing consumer surplus is to estimate the increased benefit for selection of design alternatives. Capital investment should, by its nature, cause a *decrease in price due to an improvement in productivity and reduced cost of production.* Here, the *price change was due to a policy change, not increased productivity.* In any case, the calculation of consumer surplus is conducted as shown above.

The revenues that are received are generally used to pay for the operation, maintenance, and capital recovery payments; therefore, the price to the customers should cover these costs. The breakeven price is \$0.444/unit. It is the price that should be charged if the water supply utility is a nonprofit operation.

EXAMPLE 3.11 Highway Construction

An overpass will eliminate the heavy traffic congestion and reduce the accident rate at a busy intersection. The cost of construction is \$6M for a new interchange and land acquisition. It is assumed that the overpass will provide a higher level of service during the 4-hour peak period over the 2-mile design length. It is assumed that no appreciable benefits will be received during the off-peak periods. The following estimates have been made.

	EXISTING	PROPOSED
Vehicular operating speed	16 miles/hr	40 miles/hr
Peak period flow	4,000 vehicles/hr	5,000 vehicles/hr
Accident rate	500/yr	25/yr
Cost	\$0.24/vehicle mile traveled	\$0.18/vehicle mile traveled
Cost per accident (fatal and property)	\$2,000	\$2,000

The annual costs to repair the existing and proposed overpasses are assumed to equal \$250,000. Utilize an annual social opportunity-cost rate of 12 percent per annum and a design life of 15 years, to determine the feasibility of the project by the net present worth method.

Solution

The direct benefits from the overpass will be determined with the concepts of consumer surplus. The price of travel and user demand may be estimated. The demand curve is depicted in Figure 3.30. The price for the existing and proposed facilities p^0 and p^1 are

$$\text{Existing } p^0 = \frac{\$0.24}{\text{vehicle mile traveled}} \cdot 2 \text{ miles} = \frac{\$0.48}{\text{vehicle}}$$

$$\text{Proposed } p^1 = \$0.18 \cdot 2 \text{ miles} = \frac{\$0.36}{\text{vehicle}}$$

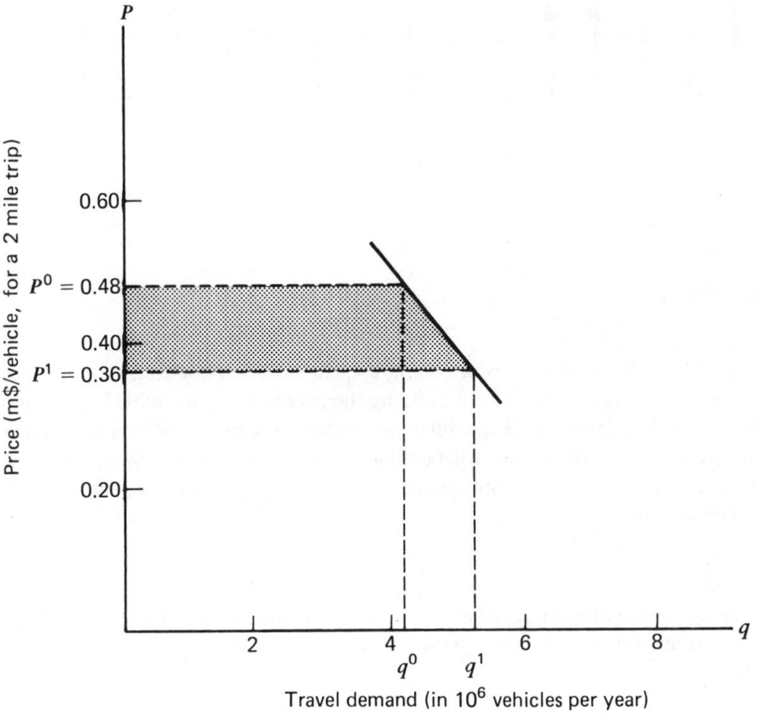

Figure 3.30

The annual numbers of users, q^0 and q^1, are estimated to be

$$\text{Existing } q^0 = 4000 \frac{\text{vehicles}}{\text{hr}} \cdot \frac{4 \text{ hr}}{\text{day}} \cdot \frac{5 \text{ days}}{\text{week}} \cdot \frac{52 \text{ weeks}}{\text{yr}} = 4.16 \times 10^6 \frac{\text{vehicles}}{\text{yr}}$$

$$\text{Proposed } q^1 = 5000 \frac{\text{vehicles}}{\text{hr}} \cdot \frac{4 \text{ hr}}{\text{day}} \cdot \frac{5 \text{ days}}{\text{week}} \cdot \frac{52 \text{ weeks}}{\text{yr}} = 5.2 \times 10^6 \frac{\text{vehicles}}{\text{yr}}$$

The consumer surplus as annual savings B_s is

$$B_s = \tfrac{1}{2}(p^0 - p^1)(q^0 + q^1) = \tfrac{1}{2}(0.48 - 0.36)(4.16 + 5.20) \times 10^6$$

$$B_s = \$561{,}600/\text{yr} = \$0.562\text{M/yr}$$

The annual savings for accident avoidance B_a is

$$B_a = \$2000(500 - 25) = \$950{,}000/\text{yr} = \$0.950\text{M/yr}$$

The total annual saving or benefit B as shown on the cash-flow diagram is estimated to be

$$B = B_s + B_a = \$1.51\text{M/yr}$$

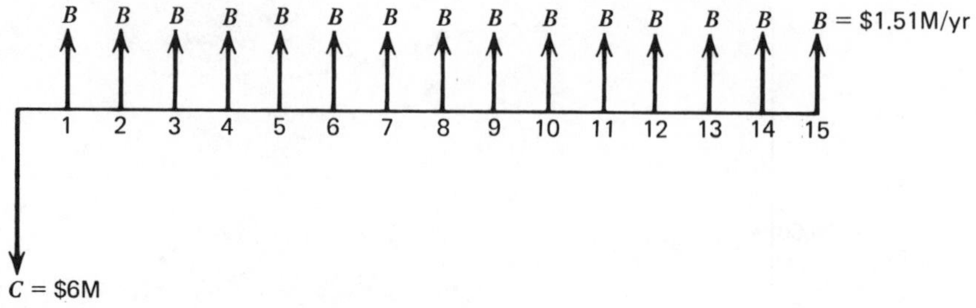

Figure 3.31

In this problem, two alternatives are being considered, the build and no-build options. The added benefits or savings received from building the overpass are accounted for in the change of consumer surplus B. The savings is the difference between the proposed overpass and the existing road. Likewise, the cost of the project will be the added cost of the project. Since there are no capital costs associated with the no-build option, the added capital cost will be equal to the cost to construct the new overpass.

$$C = \$6\text{M}$$

Since the existing road and proposed overpass have the same maintenance costs of $250,000 per year, the added maintenance cost is equal to zero.

$$A = 0$$

The cash-flow diagram shows the distribution of benefits and cost over time.

The net present worth of the proposed overpass is

$$NPW = \$1.51\text{M}\left[\frac{(1 + 0.12)^{15} - 1}{0.12(1 + 0.12)^{15}}\right] - \$6\text{M}$$

$$NPW = \$10.3\text{M} - \$6\text{M} = \$4.3\text{M}$$

Since $NPW > 0$, the proposed overpass is a viable alternative and is recommended for construction.

Remarks

All the selection methods in this chapter require that the measures of effectiveness be monetary measures. In order to analyze this problem, social benefits received by the users of the overpass were considered during the peak period only. The benefits are assumed to be time savings and a reduction in the accident rate. Consider the difficulty in determining these estimates.

Placing a monetary measure upon time involves very careful thought. Who is receiving the time saving? Do business, shopping, and recreational trips receive the same weight? What is the traffic mix? How many business, shopping, and recreational trips occur each day? Issues such as these make the assignment of a monetary value to a trip very difficult. Obviously, the monetary value assigned to the time saving will have an important bearing upon the selection of the overpass. For simplicity, we have assumed that all trips are of equal importance and that the average driver will save approximately 12 cents for each trip through the 2-mile area.

The costs of fatal and property accidents were combined into a single estimate. Arriving at these values is difficult because accident costs are generally determined from insurance data. This value may not accurately reflect the cost if a life is lost. Furthermore, not all accidents, especially lesser property-damage accidents, are reported.

Other quantifiable economic impacts that are included in a highway improvement study are relocation and parking costs. There are other impacts that are both quantifiable noneconomic impacts and nonquantifiable ones. The quantifiable noneconomic measures include air pollution, persons displaced by construction, businesses displaced, and accessibility improvement. Nonquantifiable impacts include land development patterns and esthetic value. Clearly, these impacts are important issues. These nonmonetary measures are not introduced into the economic analysis method; they are considered through the political process of public hearings. Here, issues are discussed and resolved in the best way possible.

PROBLEMS

Problem 1

(a) Label each of the following as a tangible or intangible benefit or cost. If your answer is tangible, give the method that is normally used to evaluate it.
  (i) Highway tolls.
  (ii) Highway level of service.
  (iii) Water irrigation project.
  (iv) Water recreation project.
  (v) Flood protection project.

(b) When are "externalities" considered in the evaluation of public projects?

(c) Why, in general, is a water project assigned a lower social opportunity-cost rate than a highway project?

Problem 2

The current sales of an item at a price of \$2.00 per unit are 100 units. A price change to \$2.10 per unit is being contemplated. Estimate the total sales if the demand elasticity is -0.20.

Problem 3

The demand elasticities and demand for transit service are:

	ELASTICITY, w	DEMAND
Work trips	$-1/3$	3000 riders/day
Shopping trips	-1.11	1500 riders/day

The fare is \$0.50 per rider.

(a) Determine the unknown parameters α and β for the demand function $p = \alpha q^{\beta}$ for work and shopping trips.

(b) Draw demand curves for work and shopping trips on the same figure.

(c) Determine the expected demand for work and shopping trips if the fare is reduced to \$0.45/rider.

(d) Determine the consumer surplus for work and shopping trips.

Problem 4

It has been observed that a 1 percent increase in transit fare will result in a 3 percent decrease in ridership.

(a) Determine the elasticity of demand.

(b) A transit company carries 50,000 passengers/day at a fare of \$0.70/ride. The fare is to be increased to \$0.95/ride. Estimate the ridership at the new fare.

(c) Assume that demand has the following exponential relationship, where number of riders q is a function price or fare p.

$$p = \alpha q^{\beta}$$

Determine the unknown constant α and demand elasticity β.

(d) Utilize the demand function $p = \alpha q^{\beta}$ to determine the expected ridership for a fare of \$0.95/ride. Compare this answer to one from part b.

(f) Draw the demand curve, $p = f(q)$. Show that the demand elasticity is a negative value.

Problem 5

A demand function for bus service between two cities is

$$p = \$1000 - 5q$$

where q = number of riders/day and p = price/ride. The fare or cost to provide the service is \$25/ride.

(a) Draw a supply–demand curve.

(b) Determine the daily and annual ridership.

(c) Determine the annual consumer surplus.

(d) For a \$750,000 capital investment in new buses, the fare may be reduced to \$22 per ride. Determine the daily and annual ridership for this price.

(e) For part d, determine the change in consumer surplus.

(f) If the design life of the new buses is 5 years and loans can be obtained at an annual interest rate of 10 percent, is the \$750,000 investment feasible? Assume that the bus company is operated as a public facility. Use the net present worth method.

Problem 6

The estimated flood damage to an area is \$150,000 per year. Two alternative flood management schemes have been proposed.

ALTERNATIVE	CAPITAL COST	ESTIMATED ANNUAL FLOOD DAMAGE	ANNUAL O & M[a] COSTS
A	\$120,000	\$40,000	\$20,000
B	\$160,000	\$15,000	\$10,000

[a] O & M = operation and maintenance

Assume no salvage value, and an annual interest rate of 10 percent and a 60-year design life. Utilize the net present worth method to determine the better alternative.

Problem 7

A sea wall is to be built to protect the coast from damage caused by high waves. A $0.3 million sea wall of 3 ft will be built. It is estimated that an annual savings of $100,000 in flood damage avoidance will be realized. If a $0.35 million sea wall of 4-ft is constructed the estimated annual savings is estimated to be $105,000. If a $0.37 million dollar, 5 ft sea wall is constructed, it is estimated that the annual savings will be $112,000. Assume a 10 percent per annum interest rate and a design life of ten years.

(a) Use the net present worth method to show that the 3-ft sea wall is a feasible alternative.

(b) Use the net present worth method to determine if a sea wall of 3.5 ft or 4 ft should be constructed instead of the 3 ft sea wall. Assume that the 3 ft sea wall is the so-called do-nothing alternative. Draw the cash flow diagrams for those alternatives.

Problem 8

An industrial building is located in an area subject to flooding. In order to protect the building, construction of a canal, project **A**; a single dam, project **B**; or a series of two dams, project **C**, is contemplated. (See Figure 3.32.)

PROJECT	CONSTRUCTION COST	ANNUAL MAINTENANCE COST	ESTIMATES ANNUAL FLOOD COST
A Constructional	$120,000	$20,000	$40,000
B Build dam 1	$160,000	$10,000	$15,000
C Build dams 1 and 2	$200,000	$11,000	$10,000

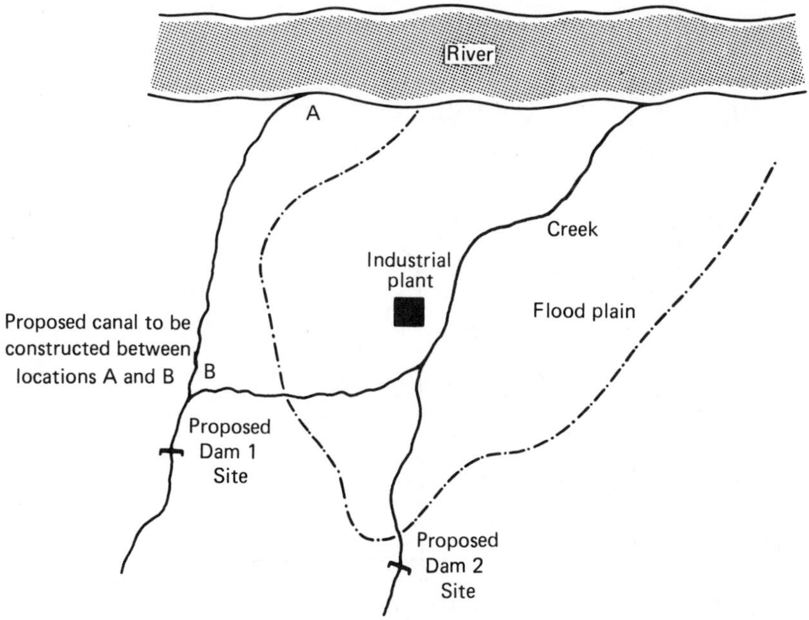

Figure 3.32

If project **A** is constructed, the excess runoff from the creek will be diverted to up-river via the canal. If project **B** or **C** is constructed, the excess runoff will be stored behind the dam or dams and gradually released. Additional flood protection will be provided if project **C** is constructed. Project **C** consists of the dam 1 plus the additional dam 2. Dam 2 will not be constructed unless dam 1 is built. Assume a discount rate of 10 percent for a design life of 60 years. Use the net present worth method to determine which option should be constructed. If no flood protection system is built the annual flood cost is estimated to be $150,000.

Problem 9

At one time, a major source of electrical power was small hydroelectric plants. These sites have been abandoned. Owing to shortages in supply of coal, oil, and other nonreplenishable energy sources, retrofitting of hydroelectric dam sites is now being considered as a source of power.

(a) What is the opportunity cost to retrofit these sites?

(b) The supply–demand curve for electricity before retrofitting is shown in Figure 3.33. Retrofitting and producing electricity from these dam sites will cause a shift in the total electrical energy supply as shown in the diagram.

  (i) Label the new price of electricity p^1.

  (ii) Label the new electricity demand q^1

  (iii) Indicate the consumer surplus.

(c) Is the new supply of electricity, a direct benefit, secondary benefit, or externality? Explain your answer. (Use definitions or examples if you like.)

(d) One of the reasons for retrofitting hydroelectric sites is to conserve nonreplenishable fuel sources. Use your answer shown in the supply–demand curve from part b to indicate if conservation is achieved. Explain your answer.

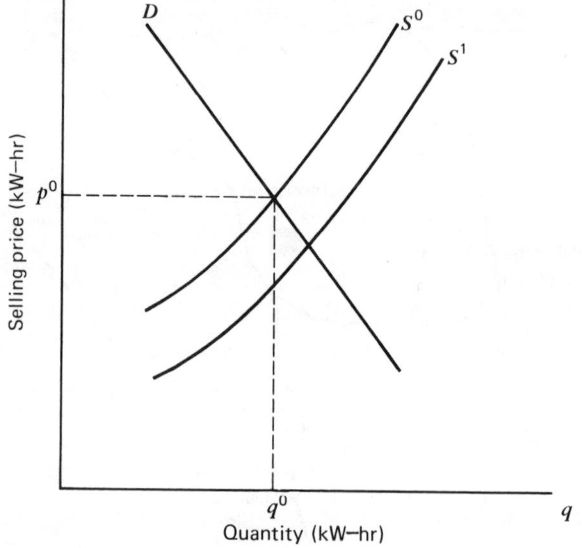

Figure 3.33

(e) If electricity supplied from hydropower is used as a substitute for electricity supplied for nonreplenishable fuel sources, the supply of electricity from part b will remain at q^0.
    (i) Determine the price of electricity.
    (ii) What is the consumer surplus? Explain.
    (iii) How would you measure the benefit of conserving nonreplenishable fuel sources in an engineering economic analysis?

(f) Most water resource projects are considered to be relatively safe investments. In your opinion, do you feel that retrofitting of small hydroelectric plants will be of great risk to the investor? Explain your answer.

(g) How are risks incorporated into engineering economic evaluations?

Summary

Utilizing engineering economics requires that

1. All benefits and costs be monetary measures.
2. All present and future benefits and costs be compared on an equal basis.

Construction, maintenance, and operating costs are monetary measures. For private projects, revenues are clearly pecuniary measures. For public projects, benefits take many forms. Time savings from improved highway service and the avoidance of risk from flood protection dams are examples of tangible benefit that may be estimated in monetary terms. Convenience, comfort, air, water, and noise pollution are intangible factors and cannot be reasonably evaluated with the methods discussed in this chapter. Subjective evaluation is required. Even though there are conceptual and practical problems, engineering economics has proved to be a worthwhile decision-making tool.

The concepts of time value of money are used to combine present and future costs into measures that can be compared equitably. The value of a future payment will depend upon the cost of money, the annual interest rate, and the time of payment. All current and future payments may be transformed to present worth payments. Equitable comparisons of present worth payments may be made. It has been shown that inflation will affect the interest rate, and during periods of high inflation interest rates tend to be high. The net present worth uses the principle of selecting the best project that provides the greatest net gain to the user or beneficiary.

The assignment of the interest rate has an important bearing on the final selection. For public projects many complex economic and political factors and matters of risk and uncertainty make the choice difficult. High interest rates tend to make investment and engineering construction less attractive and the do-nothing alternative more attractive. An effective interest rate, which accounts for the opportunity cost of capital and inflation, is used in the alternative selection process.

The principles of consumer surplus are used to provide a basis for measuring societal benefit. Only direct benefits should be considered in the evaluation. Externalities and secondary benefits that have a direct influence upon productivity, not price, should be counted as benefits.

In Chapters 1 and 2, systems analysis models are formulated as resource allocation problems by establishing an objective and a set of constraints. The solution to resource allocation problems results in an optimum combination of resources that satisfies the set of constraint conditions and achieves a stated objective. In this chapter, engineering economic methods are used to evaluate the performance or benefits and costs associated with a set of design alternatives. Each design is considered to be a *mutually exclusive feasible alternative*.

Now, let us consider how systems analysis and engineering economics can complement one another to achieve the best selection among optimally designed solutions. Consider the design of a bridge. Different technology and design configurations may be used to satisfy the basic requirement to span the given distance safely. If a steel truss and a concrete beam are considered the only alternatives, the one with the least combined construction and maintenance costs is considered the best choice. However, it may not be an *optimum design*! Only if each design alternative is a minimum-cost design will the use of engineering economics aid in selecting the *best optimum design*. It is possible to determine the least-cost designs by formulating separate resource allocation models for the steel truss and the concrete beam designs. These models will consist of an objective function to minimize cost subject to the set of constraint equations to ensure bridge safety and performance. Thus, the systems analysis approach and engineering economics methods are complementary procedures that can be used to achieve the optimal design.

In Chapter 1, the advantages and shortcomings of systems analysis were discussed. A major difficulty of utilizing mathematical models is that not all of them can be easily solved. As a result, the combined use of system analysis and engineering economics methods may not be fully realized in the mathematical sense. Engineering is as much an art as an applied science. Optimum designs are obtained by experimentation and trial and error. Thus, sometimes it is only practical to use the concepts of systems analysis and engineering economics to formulate problems and to seek solutions by mathematical means, by experimentation, or by trial and error.

Bibliography

Otto Eckstein, *Water-Resource Development: The Development of Project Evaluation*, Harvard University Press, Cambridge, Mass. 1958.

Frederick F. Frye, "Alternative Multimodal Passenger Transportation Systems: Comparative Economic Analysis," *Highway Research Board*, National Cooperative Highway Research Program, Report 146, 1973.

Eugene Grant, W. Grant Ireson, Richard Leavenworth, *Principles of Engineering Economy*. John Wiley, 7th ed., New York. 1982.

Ian G. Heggie, *Transport Engineering Economics*, McGraw-Hill, London, England. 1972.

Jack Hirshleifer, James C. DeHaven, and Jerome W. Milliman, *Water Supply: Economics, Technology*, and *Policy*, The University of Chicago Press, Chicago, Ill. 1960.

Roland McKean, *Efficiency in Government Through Systems Analysis with Emphasis on Water Resources*, John Wiley, New York. 1958.

A. R. Prest and R. Turvey, "Cost-Benefit Analysis: A Survey," *The Economic Journal*, No. 300, Vol. LXXV, December 1965, pp. 683–735.

Lloyd Sawchuk, "Declining Block Rates Can Encourage Water Conservation," *American Water Works Association Journal*, January 1981, pp. 13–15.

George A. Taylor, *Managerial and Engineering Economy*, D. Van Nostrand, 3rd ed., New York. 1980.

Robley Winfrey, *Economic Analysis for Highways*, International Textbook Co., Scranton, Penn. 1969.

John White, Marvin H. Agee, and Kenneth Case, *Principles of Engineering Economic Analysis*, John Wiley, New York. 1977.

CHAPTER 4

Decision Analysis

After the completion of this chapter, the student should:

1. Be able to utilize the fundamental principles of probability and statistics to analyze various civil engineering systems where the performance is dependent upon uncertain states of nature or random phenomena.
2. Understand the principle of expected monetary value and couple it with the principles of engineering economics to form a procedure to select engineering design alternatives under uncertainty.
3. Understand the principles of utility theory and its importance in the selection process.

Floods, earthquakes, strong winds, and other natural phenomena are called *random events*. Since these events occur with no specific pattern, they might be best described as being haphazard events. The engineer cannot control the outcome of a natural event or *state of nature*, or even forecast it perfectly. Thus, engineering structures are designed to withstand the impact of a specific event or are designed to protect the public against a specific level of devastation. The uncertainty associated with forecasting some future state of nature affects the engineering design process. In structural design, for instance, it is common practice to provide a margin or factor of safety to account for uncertainties in loading magnitude and frequency and other uncertainties in structural and material reliability. Although this practice prevents loss of property and life, it adds to the capital cost of the system. The question of how much extra protection should be provided and at what cost is difficult to answer. Decision analysis offers a rational approach to aid in this decision-making process. It uses the principles of engineering economics, described in Chapter 3, and probability theory, a mathematical tool used in the study of random phenomena. Decision analysis couples these disciplines into a framework that closely resembles the selection process of net present worth, the process of choosing the alternative offering the maximum net present worth. The important concepts and methodology employed in decision theory can be best described by example.

Suppose a building is subject to earthquake loading. Three alternatives using different philosophies are proposed:

A. Build an inexpensive structure with no resistance to earthquake loading.
B. Build a moderately priced structure with some earthquake resistance.
C. Build an expensive earthquake-resistant structure.

For alternative A, no capital money is provided for earthquake protection. If a major earthquake occurs, it is assumed that the structure will suffer serious structural damage resulting in building collapse and possible loss of life. The property damages may be estimated as the cost to replace the building and its contents. Placing a monetary value on human life is an extremely difficult moral and social question to be answered. We assume that these cost assignments can be made. These costs are future costs. At the opposite extreme, for alternative **C** it is assumed that substantial capital, a present cost, is provided to avoid all possible future losses to property and life. The moderately priced structure, alternative **B**, provides sufficient capital to avoid future loss of life, but it is anticipated that future dollars will be spent upon minor repair of the structure.

A *decision tree* as shown in Figure 4.1 offers a convenient way to visualize the evaluation and selection process. The square block □ represents a decision node, and the circles ○ represent the random event nodes. The branches between the decision node and chance or random nodes represent the alternatives for selections **A**, **B**, and **C**. The branches between the random event node and consequence blocks represent the states of nature that the design engineer cannot control. They are random events

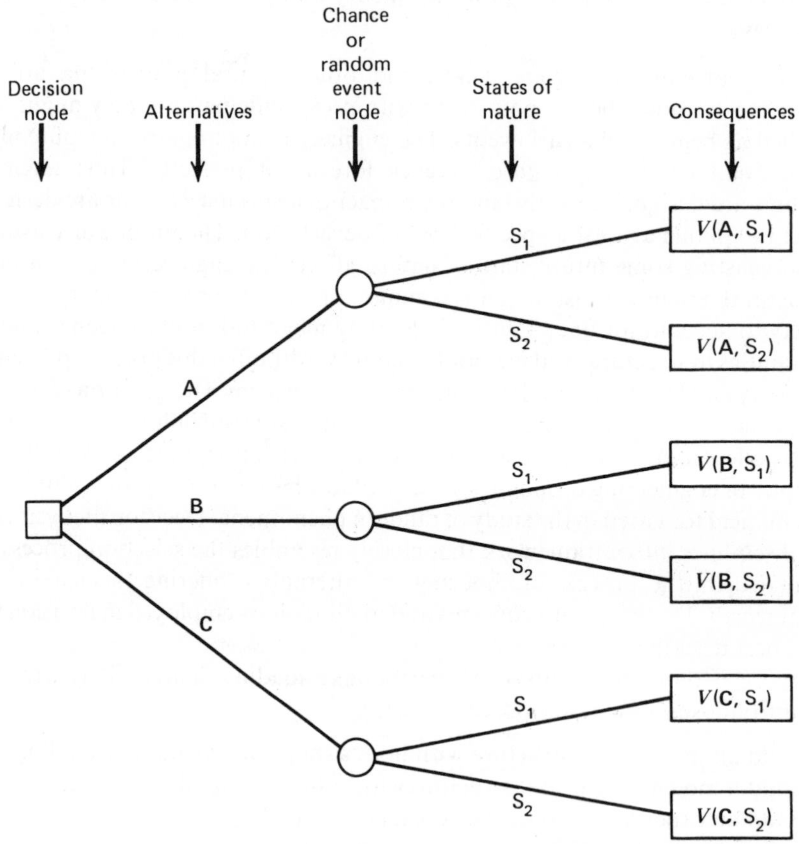

Figure 4.1 A decision tree for alternative selection with uncertainty.

that will be evaluated with statistics and probability theory. For this example, the states of nature are the random events:

S_1 No major earthquake occurs.

S_2 A major earthquake occurs.

The monetary value of the consequence is placed in the blocks as shown in the figure under the label "Consequences." The consequence V will depend upon the design alternative **A**, **B**, or **C** and the state of nature S_1 or S_2. For this example, the consequence is the financial loss in property and human life plus the extra capital used for earthquake resistance. The consequences are as follows:

$V(\mathbf{A}, S_1)$ = \$0 because no extra capital is provided for earthquake resistance.

$V(\mathbf{A}, S_2)$ = loss of property and life given a major earthquake.

$V(\mathbf{B}, S_1)$ = capital cost for earthquake resistance.

$V(\mathbf{B}, S_2)$ = capital cost for earthquake resistance and future repair costs owing to property damage from a major earthquake.

$V(\mathbf{C}, S_1)$ = capital cost for earthquake resistance.

$V(\mathbf{C}, S_2)$ = capital cost for earthquake resistance and \$0 future loss if a major earthquake occurs.

In this development we have assumed that under normal loading each structural alternative offers the same user benefits; thus we need only to investigate the *extra* capital needed for earthquake protection and the potential losses if a major earthquake occurs. The present worth values of capital cost and future monetary value of loss to property and human life are used as the measure of the consequence. For example,

$V(\mathbf{A}, S_2)$ = present worth of capital cost for alternative A_1 + future damage iosses given state S_2, a major earthquake occurs.

This is a least-cost selection problem. For state of nature S_1 all structures, alternatives **A**, **B**, and **C**, are assumed to be structurally sound, and no loss of property or life will occur, thus no loss is assessed to each of these alternatives. However, the extra capital cost associated with providing an earthquake-resistant structure is indicated in the consequence blocks for state S_1. For S_2, the present worth of capital plus future damage costs caused by the earthquake is indicated in the consequence blocks.

For alternative **C**, it is assumed that the structure will withstand a major earthquake without damage, thus there are assumed to be no losses associated with a major earthquake. The decision tree shown in Figure 4.2 therefore shows that the only loss is the capital cost of construction, which is independent of the state of nature.

In order to make a decision, the likelihood of earthquake damage must be considered. There are many possibilities. Let us consider two possibilities. If the frequency of occurrence is great, alternative **C** may be the best choice because it will probably offer the best protection in the long run. On the other hand, if the likelihood of a major earthquake is remote, alternative **A** may be the best choice. Earthquakes in the United States most frequently occur along the major fault lines in California. In the eastern

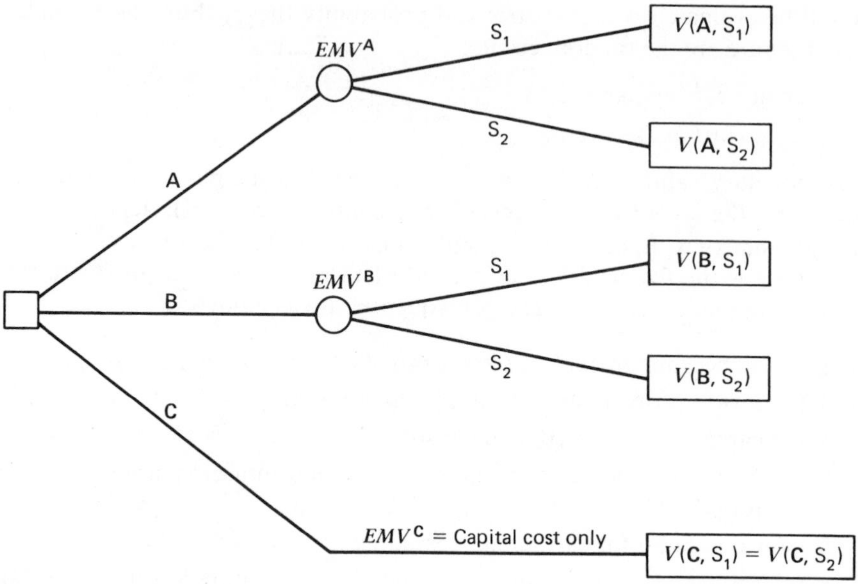

Figure 4.2 Alternative C is unaffected by a major earthquake; thus, $V(C_1 S_1) = V(C_1 S_2)$ = capital costs only.

and central parts of the country, where major faults do not exist, earthquakes occur less frequently. However, the most severe earthquakes, measured in terms of magnitude and affected area, occurred in New Madrid, Missouri in 1811 and 1812. Earthquakes of this magnitude have not occurred in this area since that time. It is uncertain whether an earthquake like these will ever occur again in this area. Furthermore, in areas where no history of major earthquake activity exists, there is a chance that a major earthquake may occur at some future time. Even if a good long-term historical record exists, there is a possibility that it may not be adequate for predicting the future. Our selection approach, however, will attempt to consider these uncertainties as well as economic trade-offs among alternatives.

Decision theory provides a framework to aid in making a difficult choice. Factors of probability of occurrence of an earthquake and economic loss, both present and future losses, are *weighted*. The concept of *expected monetary value*, EMV, and time value of money relationships are used to combine these factors. Probability theory offers a way to perform this calculation. For now, assume that the expected monetary values for each alternative are known. The values of EMV^A, EMV^B, and EMV^C are recorded on the decision tree as shown in Figure 4.2. The alternative with the minimum magnitude of EMV will be considered the best choice. Once the EMV values are known, the selection process is the same as used in Chapter 3 for net present worth selection under certainty.

Since this decision-making process requires that future events be forecast, historical records are commonly employed to estimate the probability of a future state of nature. When records do not exist, subjective reasoning and experimentation may be employed. In order to utilize decision theory, a knowledge of probability and statistics is required; therefore, a major portion of this chapter will be devoted to this topic. These concepts will be used to determine the value of experimentation and testing for the short-term decision-making problems most often found in construction planning.

4.1 FUNDAMENTALS OF PROBABILITY THEORY AND STATISTICS

Suppose 20 samples are taken from the same batch of concrete, a carefully designed experiment is developed where each sample is loaded to failure under the same test conditions, and the failure load is measured. Although the tests are done under the same conditions, the failure loads for each sample are different. How do we specify the failure load of the concrete? Probability theory and statistics may be used to answer this and similar questions that are encountered in civil engineering.

Random Events

In probability theory and statistics, *statistical experiments* are used to generate information about a specific outcome. The failure load test is an example of a statistical experiment. The measured failure loads are called *random events*, *outcomes*, or *states of nature*. The terms have the same meaning and will be used interchangeably. In order to demonstrate simple statistical experiments, rolling a die and tossing coins will be used to describe basic principles.

Roll a die and observe the outcome. A die has six sides with the numbers 1 through 6 printed on them. For illustration, we shall assume that the outcome of this experiment is the number 3. The random outcome, designated with capital letter A, will be described by using set notation.

$$\mathrm{A} = \{x : x = 3\}$$

This statement is read, "A is the random outcome of an experiment x where the observed value of x is equal to 3." Lower-case letters are used to designate the outcomes of statistical experiments, and capital letters are used to designate random events.

The set of *all* possible outcomes is called the *sample space*. For the roll of a die, the sample space Z is represented as the set

$$\mathrm{Z} = \{x : x = 1, 2, 3, 4, 5, 6\}$$

This is read, "Z is the set of all x such that the values of x are equal to 1, 2, 3, 4, 5, or 6."

Sample space Z and event A are illustrated by the Venn diagram in Figure 4.3. We can see that the random event A is a subset of Z. The probability of the specific outcome A from a set of all possible outcomes of Z is defined as the *chance likelihood*; it is measured as the ratio of the number of events in A to the total number of possible outcomes in Z. If we assume that the die is fair (that is, not "loaded"), there is an

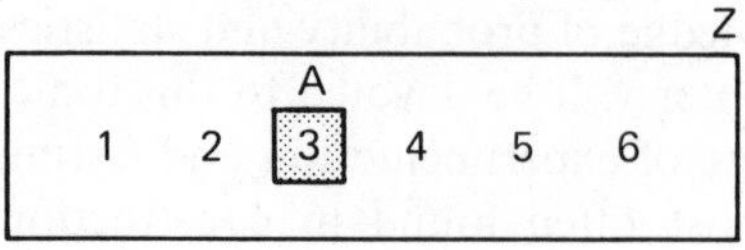

Figure 4.3 The outcome of the experiment is the number 3.

equally likely chance that 1, 2, 3, 4, 5, or 6 will appear on a single roll; the probability of the number 3 showing is represented as the probability of the event A.

$$P[\mathrm{A}] = \tfrac{1}{6}$$

In this example, since all numbers are equally likely to occur in a single roll, we simply count the number of sample points in A and divide by the number of sample points in Z.

If we are interested in determining the probability that a fair die will come up with an odd number on a single roll, we proceed in a similar manner. Define the outcome of the experiment as

$$\mathrm{B} = \{\text{roll of die is an odd number}\}$$

or

$$\mathrm{B} = \{x : x = 1, 3, 5\}$$

The sample space is Z as previously defined. The Venn diagram for this experiment is shown in Figure 4.4. Since the die is assumed to be fair, the probability of B is simply determined by counting the number of sample points in B and Z and determining the ratio of these numbers. Thus

$$P[\mathrm{B}] = \tfrac{3}{6} = \tfrac{1}{2}$$

The probability of obtaining an odd number may be obtained by using a different approach. Here, we use the concept of *mutually exclusive events.* First, we define the following states of nature for a single roll of the die and their associated probabilities as:

$$\mathrm{E}_1 = \{x : x = 1\} \qquad P[\mathrm{E}_1] = \tfrac{1}{6}$$
$$\mathrm{E}_2 = \{x : x = 2\} \qquad P[\mathrm{E}_2] = \tfrac{1}{6}$$
$$\mathrm{E}_3 = \{x : x = 3\} \qquad P[\mathrm{E}_3] = \tfrac{1}{6}$$
$$\mathrm{E}_4 = \{x : x = 4\} \qquad P[\mathrm{E}_4] = \tfrac{1}{6}$$
$$\mathrm{E}_5 = \{x : x = 5\} \qquad P[\mathrm{E}_5] = \tfrac{1}{6}$$
$$\mathrm{E}_6 = \{x : x = 6\} \qquad P[\mathrm{E}_6] = \tfrac{1}{6}$$

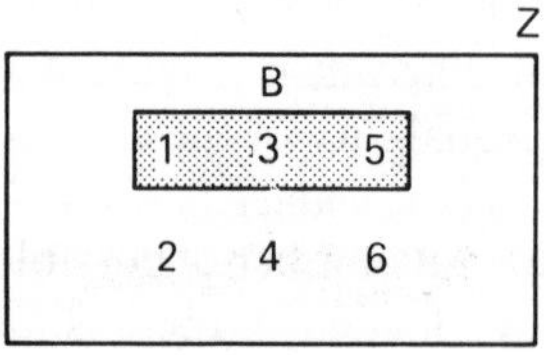

Figure 4.4 The outcome of the experiment is an even number.

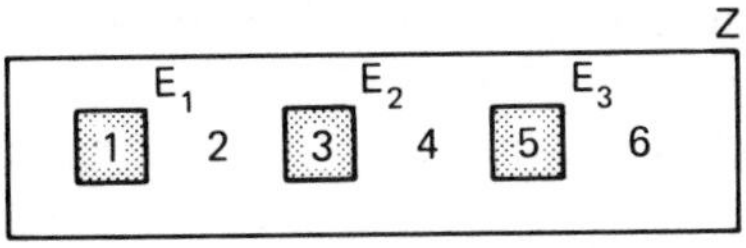

Figure 4.5 E_1, E_3, and E_5 are mutually exclusive events.

The probability of an odd number has been previously defined as the event B. Here, event B is the outcome that event E_1, E_3, or E_5 is observed. It is also called the *union* where

$$B = E_1 \text{ or } E_3 \text{ or } E_5$$

Using the symbol $\cup$ to represent the word "or," the event becomes

$$B = E_1 \cup E_3 \cup E_5$$

Since on a single roll of a die, only one number 1, 2, 3, 4, 5, or 6 can occur, these events are called mutually exclusive events. The Venn diagram for mutually exclusive events shows the subsets as non-overlapping events, as shown in Figure 4.5.

The *intersection* of two events A and B is the event C containing all elements that are common to both A and B.

$$C = A \text{ and } B$$

Using the symbol $\cap$ to represent the word "and," the intersection is written as

$$C = A \cap B$$

Figure 4.6 shows that intersection C may contain sample points as in Figure 4.6*a* or may not contain sample points as in Figure 4.6(*b*). When A and B are mutually exclusive, $C = A \cap B = \varnothing$, the *null set*, therefore,

$$P[C] = P[A \cap B] = P[\varnothing] = 0.0$$

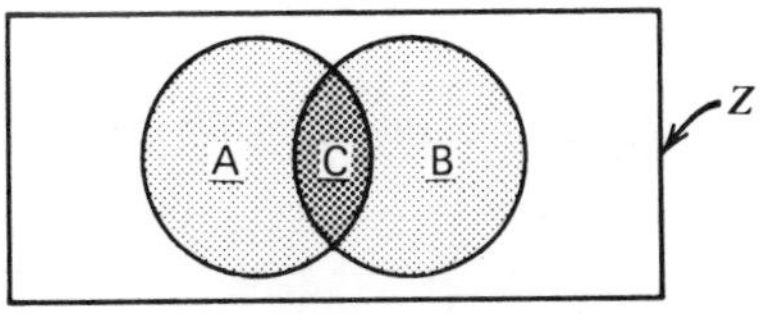

$C = A \cap B$ is the intersection of A and B

(*a*)

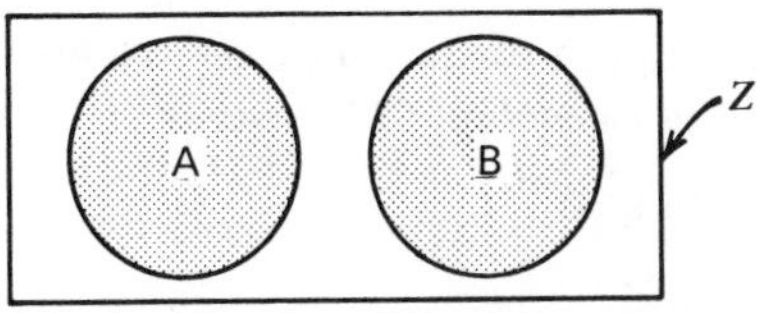

$C = A \cap B = \phi$, the null set

(*b*)

Figure 4.6 These are possible outcomes for $C = A \cap B$. (*a*) Event C contains sample points when A and B are any two random events. (*b*) Event C is the null set when A and B are mutually exclusive. If A and B are mutually exclusive events, then the intersection of A and B is the null set, $C = A \cap B = \varnothing$.

Thus, from Figures 4.3 and 4.4 we can see that the intersection C for the die problem is

$$C = A \cap B = \{x : x = 3\}.$$

The probability of C is $P[C] = \frac{1}{6}$.

In Figure 4.5 the intersection of E_1 and E_3 for the rolling of the die is the null set.

$$E_1 \cap E_3 = \varnothing$$

The events of E_1 and E_3 occurring on the same roll is impossible. The probability of this event occurring is zero or

$$P[E_1 \cap E_3] = 0.$$

Thus, this simple example shows intersection of mutually exclusive events is always the null set $\varnothing$.

The probability of two mutually exclusive events is simply the sum of the probabilities of these two events.

$$P[A \cup B] = P[A] + P[B]$$

The definition may be extended to the problem of determining the probability of obtaining an odd number on a single roll of a fair die. Thus,

$$P[B] = P[E_1 \cup E_3 \cup E_5]$$

$$P[B] = P[E_1] + P[E_3] + P[E_5]$$

$$P[B] = \tfrac{1}{6} + \tfrac{1}{6} + \tfrac{1}{6}$$

$$P[B] = \tfrac{1}{2}$$

In the determination of $P[B]$, we utilized the axioms of probability without calling attention to them.

Axioms of Probability

The axioms of probability are:

AXIOM 1 The probability of an event A is a number greater than or equal to zero and less than or equal to one.

$$0 \leq P[A] \leq 1$$

AXIOM 2 The probability of the sample space Z is

$$P[Z] = 1$$

AXIOM 3 The probability of an event that is the union of two mutually exclusive events, A and B, $A \cup B$, is the sum of the probabilities of each of these events.

$$P[A \cup B] = P[A] + P[B]$$

AXIOM 4 The probability of the null set $\varnothing$ is equal to zero.

$$P[\varnothing] = 0$$

Consider an experiment where two coins are flipped. The sample space representing the toss of one coin consists of the outcome of a head h and a tail t. The sample space for two coins consists of the following set of points.

$$Z = \{(h, h), (h, t), (t, t)\}$$

The sample outcomes from this experiment are:

$$E_1 = \{\text{two heads}\} = \{(h, h)\}$$

$$E_2 = \{\text{a head and a tail}\} = \{(h, t)\}$$

$$E_3 = \{\text{two tails}\} = \{(t, t)\}$$

E_1, E_2, and E_3 are mutually exclusive events, thus

$$P[E_1] + P[E_2] + P[E_3] = P[Z] = 1$$

In this experiment, the outcomes E_1, E_2, and E_3 are not equally likely; there is twice the chance of having the outcome E_2. Assign weights of one to E_1 and E_3 and a weight of two to E_2. The probability of an event is equal to the ratio of the weight assigned to the event to the sum of the weights. If we assign weights of one to E_1 and E_3 and a weight of 2 to E_2, we obtain the following probability assignments.

$$P[E_1] = \tfrac{1}{4}$$

$$P[E_2] = \tfrac{2}{4} = \tfrac{1}{2}$$

$$P[E_3] = \tfrac{1}{4}$$

Let us prove this result by evaluating the coin-toss problem a little differently. In the above evaluation we did not distinguish between the coins. Now, let us assume that the two coins are a dime and a quarter. The sample space for this experiment is

$$Z' = \{(h_d, h_q), (h_d, t_q), (t_d, h_q), (t_d, t_q)\}$$

where the subscripts d and q represent the dime and quarter, respectively. The sample outcomes are:

$$E_1 = \{\text{two heads}\} = \{(h_d, h_q)\}$$

$$E_2 = \{\text{a head and a tail}\} = \{(h_d, t_q), (t_q, h_d)\}$$

$$E_3 = \{\text{two tails}\} = \{(t_d, t_q)\}$$

These are the same definitions used for the undesignated coin toss experiment. All outcomes of the sample space Z' are equally likely; thus equal weights are assigned to each chance outcome. Therefore, since E_1 consists of only one sample point (h_d, h_q), then

$$P[E_1] = \tfrac{1}{4}$$

E_2 consists of two sample points, (h_d, t_q) and (t_d, h_q), then

$$P[E_2] = \tfrac{2}{4} \text{ or } \tfrac{1}{2}$$

and E_3 consists one sample point, (t_d, h_q), then

$$P[E_3] = \tfrac{1}{4}$$

This concludes our proof. We conclude by stating the following theorem:

THEOREM 1 If an experiment consists of n equally likely events and the outcome A consists of m of these events, then the probability of A is

$$P[A] = \frac{m}{n}$$

This theorem satisfies all axioms of probability. Observe that the probability of an event A is a relative frequency measure.

Empirical Frequency Distributions

In most engineering applications, the sample space generally does not consist of equally likely outcomes. This theorem has limited application. For example, a motor has a design life of one year of continuous operation. The sample consists of two outcomes: either the motor is operating after one year, the success event S, or it is not, the failure event F. The sample space is

$$Z = \{S, F\}$$

Clearly, these events are not equally likely outcomes. Typically, historical records or results from laboratory or field tests are used to estimate the probability of the random outcomes.

In the case of the motor, let us assume that a field experiment is conducted. Suppose that 100 motors are tested. After one year, 98 are still operating and only 2 are not. Using relative frequency estimates, the probabilities of success and failure are estimated by

$$P[S] = \tfrac{98}{100} = 0.98$$

$$P[F] = \tfrac{2}{100} = 0.02$$

To obtain the *true* probability of events S and F *all* motors must be tested. There may be many thousands produced and operated each year. Since this is impossible to achieve a test of all motors, the *population*, a *sample* consisting of 100 motors is used. Thus, $P[S]$ and $P[F]$ are *estimates* of the true probabilities of success and failure. If the population of all motors were available, the true probability of success and failure would be calculated by counting the successes and failures and determining the relative frequency. We will use sampling and the relative frequency estimation techniques to calculate the probability distributions of random variables.

A Discrete Random Variable

In engineering, we are generally concerned with quantitative measures. The sample space for count data such as the number of vehicular arrivals per time, the number of traffic fatalities, and machine failures per year may be represented by the sample space

$$Z = \{x : x = 0, 1, 2, \ldots\}$$

where x is a set of positive integers. For certain applications, the sample space may take on negative integers.

The probability that the random variable X for the outcome of the statistical experiment is equal to x is written as

$$P[X = x] = p_X(x)$$

where $p_X(x)$ is called the *probability function* of the *discrete random variable* X. The capital letter X, in this case, is used to represent the random variable and the lower-case letter x represents the numerical value that X takes on after experimentation.

Since random variables are random events, the probability function $p_X(x)$ must satisfy the axioms of probability. By deduction we state the following properties

1. $0 \leq p_X(x) \leq 1$
2. Since $P[Z] = 1$, then $\sum_i p_X(x_i) = 1$
3. $P[(X = x_i) \cup (X = x_j)] = p_x(x_i) + p_x(x_j)$ where x_i and x_j represent two different outcomes in Z.
4. $P[(X = x_i) \cap (X = x_j)] = P[\varnothing] = 0.$

The coin flip of two coins example may be analyzed in terms of a discrete random X where

$$X = \{\text{sum of the number of heads in a flip of two coins}\}$$
$$X = \{x : x = 0, 1, 2, \ldots\}$$

The random events E_1, E_2, and E_3 may be defined in terms of the random variable X.

$$E_1 = \{\text{two heads}\} = \{(h, h)\} \text{ or } [X = 2]$$
$$E_2 = \{\text{one head}\} = \{(h, t)\} \text{ or } [X = 1]$$
$$E_3 = \{\text{no heads}\} = \{(t, t)\} \text{ or } [X = 0]$$

From our previous analysis we obtained

$$P[E_1] = P[X = 2] = \tfrac{1}{4}$$
$$P[E_2] = P[X = 1] = \tfrac{1}{2}$$
$$P[E_3] = P[X = 0] = \tfrac{1}{4}$$

Thus, the probability functions are

$$P[X = 0] = p_X(0) = \tfrac{1}{4}$$
$$P[X = 1] = p_X(1) = \tfrac{1}{2}$$
$$P[X = 2] = p_X(2) = \tfrac{1}{4}$$

Typically, the probability function is determined with the use of historical, laboratory, or field test data. Suppose we are interested in determining the number of vehicles that enter a parking lot each morning at 8:00 A.M. Let

$$X = \{\text{the number of vehicles observed in a period of 1 min}\}$$

$$X = \{x: x = 0, 1, 2, \ldots\}$$

For a period of 30 days a traffic study is performed in exactly the same manner. The number of vehicles between the times of 8:00 A.M. and 8:01 A.M. are counted and recorded. The results of the experiment indicate that in 3 days no vehicles are counted, in 21 days one vehicle was counted, and in 6 days two vehicles are counted. A *histogram* is a graph that relates the number of observations of a particular outcome to that outcome. The histogram for the observations of the parking lot experiment is given in Figure 4.7. The function $p_X(x)$ is estimated by determining the relative frequencies:

$$p_X(0) = \tfrac{3}{30} = 0.10$$

$$p_X(1) = \tfrac{21}{30} = 0.70$$

$$p_X(2) = \tfrac{6}{30} = 0.20$$

$$p_X(3) = p(4) = \cdots = 0$$

Note that the properties derived from the axioms of probability are satisfied with the use of these sampling and estimation techniques.

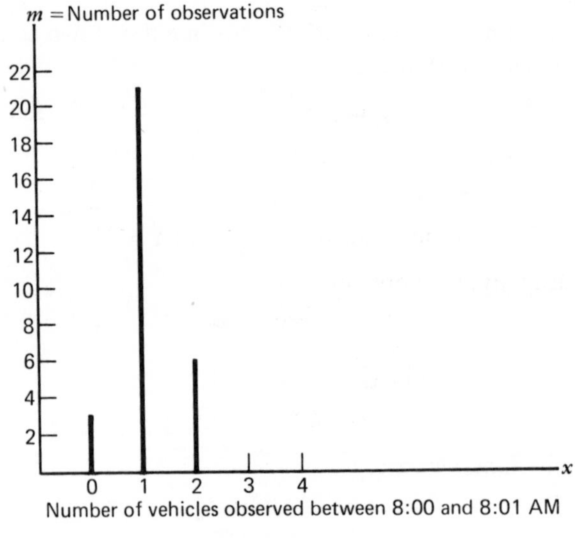

Figure 4.7 The number of vehicles observed between 8:00 and 8:01 A.M.

The Continuous Random Variable

For civil engineering problems dealing with failure load, strength of material, flow rates, and time, for example, a *continuous random variable* is adopted. A continuous random variable may be equal to any number on the real number line. For most engineering applications, the random variable X is equal to positive real numbers.

$$X = \{x: x \geq 0\}$$

The random variable for certain applications may be defined in the negative region. The outcome for a continuous random variable X is described with the *probability density function* $f_X(x)$. From the axioms of probability we may make the following observations

1. $f_X(x) \geq 0$ where $P[a \leq X \leq b] = \int_a^b f_X(x)\,dx$

2. Since $P[Z] = 1$, thus $\int_{-\infty}^{\infty} f_X(x)\,dx = 1$ where $x = \{x: -\infty \leq x \leq \infty\}$

In dealing with a continuous random variable, envision the probability $P[a \leq X \leq b]$ as the area under the graph of the function $f_X(x)$ as shown in Figure 4.8. It is equal to the integral

$$P[a \leq X \leq b] = \int_a^b f_X(x)\,dx$$

From this interpretation, we can conclude that the continuous random variable is equal to a particular value of X, say, $X = a$, is equal to zero

$$P[X = a] = \int_a^a f_X(x)\,dx = 0$$

There are several probability density functions used in engineering to describe random processes. Here, we illustrate its use with two examples.

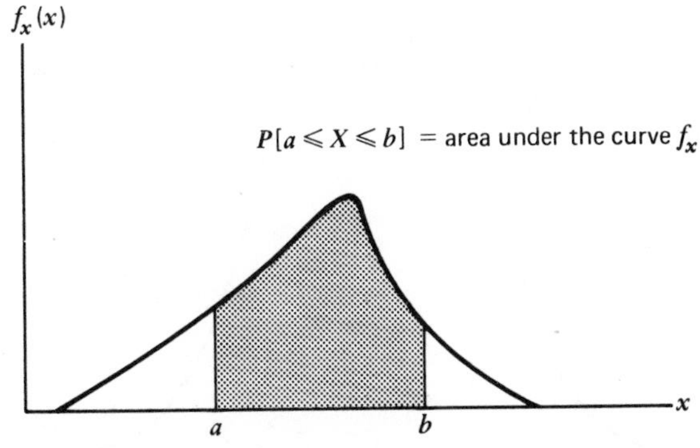

Figure 4.8 An interpretation of the probability that X lies between a and b.

The arrival rate of vehicles may be represented by a shifted exponential distribution, where the random variable X is defined as the interarrival time of vehicles in seconds.

$$f_X(x) = \lambda e^{-\lambda(x-\tau)} \qquad \text{where } x \geq \tau$$

The arrival rate is 0.3 vehicles per second and τ, the minimum time between arrivals, is 1 sec. The probability that the interarrival time is more than 4 sec is

$$P[X > 4] = \int_4^{\infty} 0.3e^{-0.3(x-1)}\, dx = 0.41$$

There is a 41 percent chance that the interarrival time is greater than 4 sec.

The normal distribution, $N(\mu, \sigma^2)$, is used to represent various random phenomena.

$$f_X(x) = \frac{1}{\sigma\sqrt{2\pi}} e^{-(x-\mu)/2\sigma^2}$$

where $-\infty < x < \infty$, μ is the mean, and σ^2 is the variance of X. The variance is a measure of dispersion. The standard deviation is the square root of the variance, $\sigma = \sqrt{\sigma^2}$. The normal distribution $f_X(x)$ is shown in Figure 4.9.

The effect of variance σ^2 upon the normal probability density functions is shown in Figure 4.10. If there is little dispersion in the data, the probability distribution will be peaked as shown by curve A. If there is much variation in the data, the distribution will be depicted as shown by curve B.

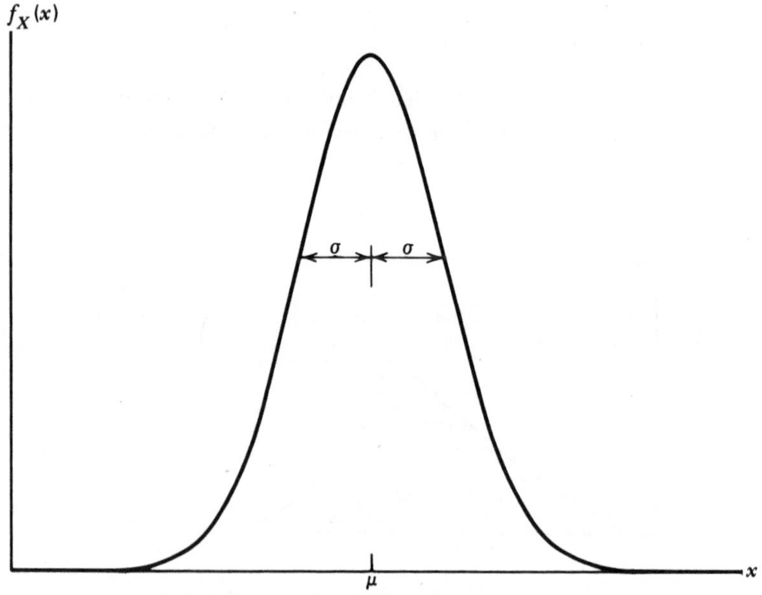

Figure 4.9 The normal distribution, $N(\mu, \sigma^2)$.

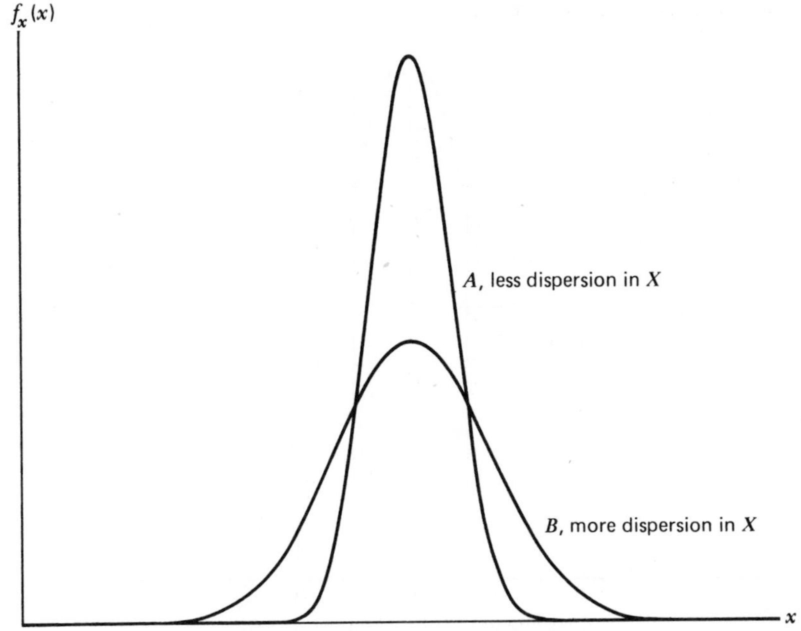

Figure 4.10 The effect of variance upon the normal probability distribution.

Suppose that the mean annual rainfall $\bar{x}$ is 50 inches with standard deviation σ equal to 10 in. The probability of a drought is defined to be the state of nature when the total annual rainfall is less than 20 in. The probability of drought is

$$P[X \leq 20] = \int_0^{20} \frac{1}{10\sqrt{2\pi}} e^{-(x-50)^2/(2.10^2)}\, dx$$

Since the evaluation of this integral is not easily determined, tables of unit normal probability have been prepared. See Appendix B. Presently, we shall discuss the steps used in this calculation of the integral. For now, we can assume the area under the curve is equal to 0.0013 or

$$P[X \leq 20] = 0.0013$$

There is a likelihood of less than 1 percent, or 0.13 percent, that a drought will occur.

In these two examples, the probability density function is assumed to be known. In most engineering applications, we rely on historical, field, and laboratory data to estimate the probability density function $f_X(x)$. The approach to estimate probability density function $f_X(x)$ is similar to the one used for $p_X(x)$. The relative frequency of the data set is used as estimates of $f_X(x)$. With a continuous variable, class intervals are established by dividing the range of X into 5 to 20 intervals. The choice of the number of intervals is dependent upon the number of observations. A histogram is a graph

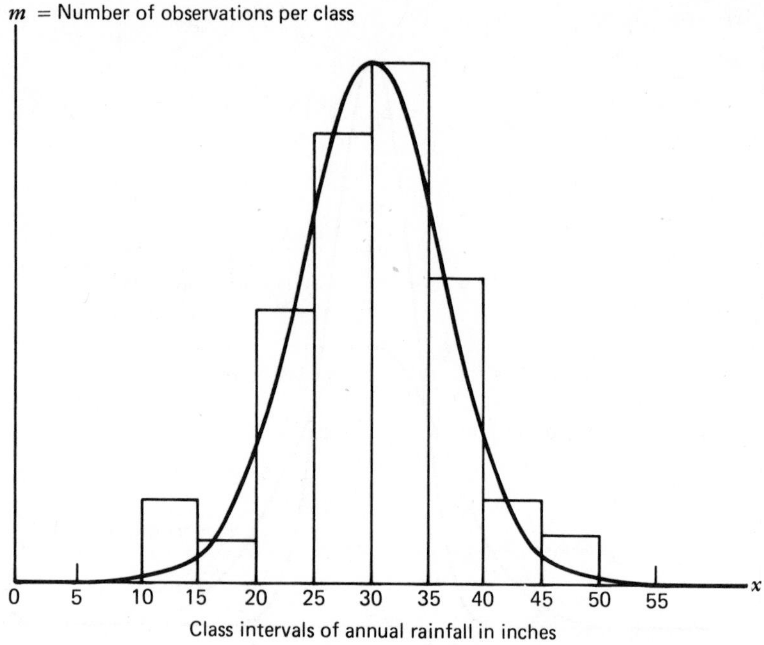

Figure 4.11 Estimating the normal probability density function.

that relates the number of observations of a particular outcome to the class interval. The histogram is useful in determining the shape of $f_X(x)$. In Figure 4.11, the histogram of the annual rainfall data shows that it has a shape that closely resembles the normal probability distribution. Thus, the use of normal probability distribution, $N(\mu, \sigma^2)$, appears to be justified. The theoretical mean μ and variance σ^2 of the normal probability density function may be estimated with *statistical mean* $\bar{x}$ and *variance* s_x^2.

$$\bar{x} = \frac{1}{n} \sum_i x_i$$

$$s_x^2 = \frac{1}{n-1} \sum_i (x_i - \bar{x})^2 = \frac{1}{n-1} \left[\sum_i x_i^2 - n\bar{x}^2 \right]$$

where n equals the number of observations in the record and $i = 1, 2, \ldots, n$.

The histogram of the interarrival data used in determining should have a shape that closely resembles an exponential function. The use of statistical methods is used to estimate the parameters of the models of $f_X(x)$. The discussion of these methods is beyond the scope of this textbook. We shall use histograms and the relative frequency estimation technique to estimate the distributions of $p_X(x)$ and $f_X(x)$. The use of class intervals, histograms, and relative frequencies for estimating a probability density function $f_X(x)$ is illustrated in Example 4.1, Estimates of Flood Maintenance and Damage Costs.

Cumulative Probability Functions

The cumulative probability function is used to calculate the probability of various events described with discrete and continuous random variables. The cumulative probability distribution $F_X(x)$ is defined as

$$F_X(x) = P[X \leq x]$$

or the probability that the random event X will be equal to or less than the value of x. Using a graphical interpretation, $F_X(x)$ is the area under the $p_X(x)$ curve or $f_X(x)$ curve evaluated from minus infinity to the upper limit of x as shown in Figure 4.12.

The formulas for calculating $F_X(x)$ for discrete and continuous random variables are:

Discrete $\quad F_X(x) = \sum_{-\infty}^{x} f_X(t)$

Continuous $\quad F_X(x) = \int_{-\infty}^{x} f_X(t)\, dt$

where t is a dummy variable.

The cumulative function for the parking lot experiment is calculated as follows:

$$F_X(0) = f_X(0) = 0.10$$

$$F_X(1) = f_X(0) + f_X(1) = 0.10 + 0.70 = 0.80$$

$$F_X(2) = f_X(0) + f_X(1) + f_X(2) = 0.10 + 0.70 + 0.20 = 1.00$$

$$F_X(x) = 1 \text{ for other values of } x > 2$$

The cumulative probability function for the interarrival time experiment, the shifted exponential distribution, is

$$F_X(x) = \int_{1}^{\infty} 0.3e^{-0.3(t-1)}\, dt = 1 - e^{-0.3(x-1)} \qquad \text{for } x \geq 1$$

$$F_X(x) = 0 \qquad \text{for } x < 1$$

The cumulative normal probability distribution is

$$F_X(x) = P[X \leq x] = \int_{-\infty}^{x} \frac{1}{\sigma\sqrt{2\pi}} e^{-(t-\mu)/2\sigma^2}\, dt$$

Since the normal probability distribution is described by the parameters μ and σ, the normal distribution is written as $N(\mu, \sigma^2)$. For this example, we use the notation $N(50, 100)$. The table given in Appendix B is the cumulative probability function for the unit normal probability function, $N(0, 1)$, or

$$F_U(u) = \int_{-\infty}^{u} f_U(z) dz$$

where $\quad f_U(z) = \frac{1}{\sqrt{2\pi}} e^{-z^2/2} \qquad \text{for } -\infty < z < \infty$

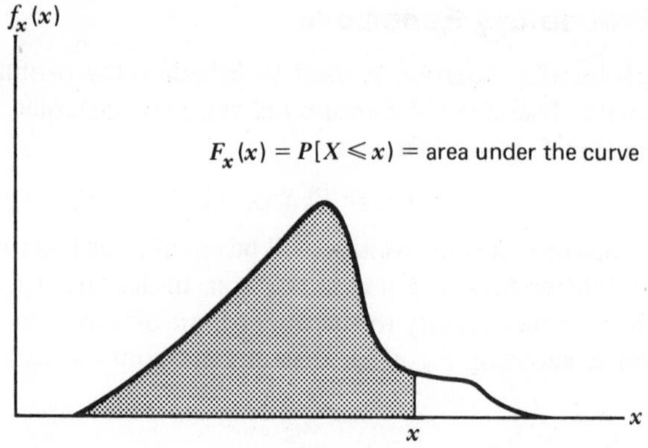

Figure 4.12 The cumulative probability of the continuous random variable X.

In order to use the cumulative unit normal distribution of Appendix B, the following transformation is made.

$$z = \frac{x - \mu}{\sigma}$$

Furthermore, the following relationship holds for $N(\mu, \sigma^2)$ and $N(0, 1)$.

$$F_X(x) = F_U(u)$$

Now that we know the properties of the cumulative normal and unit normal distributions, the probability of a drought occurring in the rainfall example may be calculated as

$$P[X \leq 20] = F_X(20) \qquad \text{for } N(50, 100)$$

The value of u is

$$u = \frac{x - \mu}{\sigma} = \frac{20 - 50}{10} = -3$$

Thus, $P[X \leq 20] = F_X(20) = F_U(-3) = 0.0013$. See Appendix B.

Expectation

It is convenient to summarize a data set with the mean and standard deviation. In our discussion of the normal distribution we gave formulas for the statistical mean and variance. We denoted these values as $\bar{x}$ and s^2, respectively. In addition, for bell-shaped distributions, we stated that $\bar{x}$ and s^2 are approximations of the theoretical mean μ and variance σ^2. The mean and variance parameters are not restricted to the normal probability distribution only; the mean and variance may be determined for any probability distribution of $p_X(x)$ and $f_X(x)$. The mean, m_X is a measure of the central tendency of X and the variance, σ^2, is a measure of the dispersion of X.

Knowing $p_X(x)$ or $f_X(x)$, the value of m_X and σ^2 may be calculated by determining the expected value of X. The mean is equal to

$$m_X = E[X]$$

where $E[X]$ is called the first moment of X.

The variance is equal to

$$\sigma^2 = E[(X - m_X)^2] = E[X^2] - m_X^2$$

where $E[X^2]$ is called the second moment of X.

The first and second moments are

$$E[X] = \sum_x x p_X(x)$$

and

$$E[X^2] = \sum_x x^2 p_X(x)$$

if X is a discrete random variable and

$$E[X] = \int_{-\infty}^{\infty} x^2 f_X(x)\, dx$$

$$E[X^2] = \int_{-\infty}^{\infty} x^2 f_X(x)\, dx$$

if X is a continuous random variable. Knowing $E[x^2]$ and m_x, the variance σ^2 may be determined as $\sigma^2 = E[x^2] - m_x^2$.

If the random variable Y is a function of the random variable X, $Y = g(X)$, and the probability of X is known, the expected value of Y for a discrete random variable is

$$E[Y] = E[g(X)]$$

If X is a discrete random variable,

$$E[Y] = \sum_x g(x) p_X(x)$$

If X is a continuous random variable

$$E[Y] = \int_{-\infty}^{\infty} g(x) f_X(x)\, dx$$

The Expected Monetary Value

In decision analysis, the state of nature will be described as a set of random events, $S_1, S_2, \ldots$, or by a random variable of X. In either case, we are interested in calculating the long-term benefit or cost, that is, the expected monetary value, EMV. The concept of expectation will be used to find the EMV. We shall describe the analysis by example.

Suppose we found that the probability of a motor operating continuously without stopping for a period of one year is 0.98. We define this random event as a success S, and $P[S] = 0.98$ and the event that the pump will stop before the end of the 1 year period as a failure F and $P[F] = 0.02$. An engineer using a pump of this kind finds that if the pump fails, it will cost his client \$15,000 owing to production delays, repair and lost profits, or $V(F) = \$15{,}000$. If the pump does not fail, there are no monetary losses, or $V(S) = \$0$. The expected loss or long-term loss is calculated as the expected monetary value EMV.

$$\text{EMV} = V(\text{S})P[\text{S}] + V(\text{F})P[\text{F}]$$

$$\text{EMV} = \$0 \cdot 0.98 + \$15{,}000 \cdot 0.02 = \$300$$

The expected long-term loss is \$300. It should be clear that if the pump is used and it fails, the *actual cost to the client is* \$15,000, *not* \$300, the long term average. In the next section, we shall use the EMV in the alternative selection process.

The cost of operating a waste-water treatment system is a function of its flow. $C = 2x^{0.8}$, where C is the annual maintenance cost in millions of dollars and x is the flow in million gallons per day. The flow is characterized by a uniform probability density distribution.

$$f_X(x) = \tfrac{1}{5} \qquad 5 \le x \le 10 \text{ Mgal/day}$$

The average daily flow, variance of flow, and expected annual operating cost are sought. Expected values will give us the desired information.

$$m_X = \text{average daily flow}$$

$$m_X = E[X] = \int_{-\infty}^{\infty} x f_X(x)\,dx = \int_{5}^{10} x(\tfrac{1}{5})\,dx = 7.5 \text{ Mgal/day}$$

$$\sigma^2 = \text{variance of daily flow}$$

$$\sigma^2 = E[X^2] - m_X^2$$

where

$$E[X^2] = \int_{-\infty}^{\infty} x^2 f_X(x)\,dx = \int_{5}^{10} x^2(\tfrac{1}{5})\,dx = 58.33 \text{ (Mgal/day)}^2$$

Thus,

$$\sigma^2 = 58.33 - 7.5^2 = 2.08 \text{ (Mgal/day)}^2$$

or

$$\sigma = \text{standard deviation} = \sqrt{\sigma^2} = 1.44 \text{ Mgal/day}$$

Since the flow is assumed to be a random variable of X, the given cost relationship $C = 2X^{0.8}$ may be written as a function of X, or

$$C = 2X^{0.8} = g(X)$$

Let

$$\text{EMV} = E[C] = \text{expected annual operating cost}$$

$$\text{EMV} = E[g(X)] = \int_{-\infty}^{\infty} g(x) f_X(x)\, dx$$

$$\text{EMV} = \int_{5}^{10} 2x^{0.8}(\tfrac{1}{5})\, dx = \$9.99\text{M/yr}$$

For alternative selections, present and future benefits and costs are combined. The time value of money as presented in Chapter 3 is used for this purpose. If the design life of the waste-water treatment plant system is 15 years and the annual interest rate is 12 percent, the present worth of annual operating costs is

$$P = \text{EMV}\left[\frac{(1 + i)^n - 1}{i(1 + i)^n}\right] = 9.99\left[\frac{(1 + 0.12)^{15} - 1}{0.12(1 + 0.12)^{15}}\right]$$

$$P = \$68.0\text{M}$$

Conditional and Independent Random Events

Thus far, we have limited our discussion to problems dealing with mutually exclusive random events. We have shown when two random events A and B are mutually exclusive, their intersection is the null set, $A \cap B = \varnothing$, and the probability of random events A or B occurring is simply the sum of their respective probabilities.

$$P[A \cup B] = P[A] + P[B].$$

In this section we shall investigate engineering problems that are not mutually exclusive random events.

A civil engineer is concerned with the possibility of structural failure of a machine part due to overloads, loads placed upon the structural component that exceed the design load. In order to investigate this problem, an experiment is conducted in which 10 structural components are deliberately overloaded to a fixed percentage above the design load, and the number of structural failures are counted. In order to aid in the analysis we define the events

$$A = \{\text{the structure component is overloaded}\}$$

$$B = \{\text{the structure fails}\}$$

In this experiment event A is not strictly a random variable; the loading conditions are known with certainty. Event B is a random variable. The number of structures that fail is not known and can only be determined by experimentation. The goal of this experiment is to determine the *conditional probability* $P[B|A]$. The notation $P[B|A]$ is read "the probability of B occurs given that A occurs." For this problem $P[B|A]$ is the probability that the structural component fails given that the component is overloaded.

The probability $P[B|A]$ will be estimated by calculating the relative frequency. The sample space for this experiment is

$$Z = \{F, S\}$$

where F refers to a structural component failure and S refers to a structurally secure component. During testing, 5 out of the 10 components fail, thus with the use of relative frequency estimation,

$$P[B|A] = \tfrac{5}{10} = 0.50$$

Fifty percent of the structural components that are overloaded are expected to fail.

Based upon this structural test information alone, it appears that the company selling the machine using this component will experience a large number of complaints from its customers. The company conducts a separate survey of its customers to determine how many of them use their equipment in overload situations. They find only 1 percent of their customers overload the machine, or $P[A] = 0.01$. Based upon this information and the structural component failure tests, $P[B|A] = 0.50$, the probability that a machine will fail under actual working conditions can be estimated. In other words, the probability $P[B]$ can be determined. Before we find this number we shall introduce other properties of conditional probability.

Definition 1 The conditional probability of B given A is defined as the ratio of the probability of the intersection of A and B to the probability of A and

$$P[B|A] = \frac{P[A \cap B]}{P[A]}$$

Definition 2 If random event B does *not* depend upon A, the events A and B are *independent random events* and

$$P[B|A] = P[B]$$

From definition 1, the intersection of $A \cap B$ may be formed by rewriting the conditional probability expression as

$$P[A \cap B] = P[B|A]\,P[A]$$

The probability of the intersection of two independent events is the simple product

$$P[A \cap B] = P[B]\,P[A]$$

Shortly, we shall illustrate the use of this important property by example.

THEOREM 1 If two events A and B occupy the entire sample space $Z = \{A, B\}$, then these two events are considered to be *complementary events*.

THEOREM 2 Where the complement of event A, written as A^c, the event B, and

$$P[A^c] = P[B] = 1 - P[A]$$

Definition 3 The *marginal probability* of B is defined as

$$P[\mathrm{B}] = \sum_i P[\mathrm{B} \cap \mathrm{A}_i] \qquad \text{for all } i$$

where the probabilities of $P[\mathrm{B} \cap \mathrm{A}_i]$ are the intersection of event B with all A_i events.

With the theorems and definitions of marginal probability, complementary events, and conditional probability, we can now determine $P[\mathrm{B}]$, the probability that the company's machine will fail under actual working conditions. From the company survey data we found $P[\mathrm{A}] = 0.01$. The complement of A is the event A^c, the customer does not run the machine in the overloaded condition. Thus,

$$P[\mathrm{A}^c] = 1 - P[\mathrm{A}] = 1 - 0.01 = 0.99$$

Ninety nine percent of the company's customers are estimated to use the equipment as designed.

Since there are only two events, A and A^c, the *marginal probability* is equal to

$$P[\mathrm{B}] = P[\mathrm{B} \cap \mathrm{A}] + P[\mathrm{B} \cap \mathrm{A}^c]$$

where

$$P[\mathrm{B} \cap \mathrm{A}] = P[\mathrm{B}|\mathrm{A}]P[\mathrm{A}]$$

and

$$P[\mathrm{B} \cap \mathrm{A}^c] = P[\mathrm{B}|\mathrm{A}^c]P[\mathrm{A}^c]$$

or

$$P[\mathrm{B}] = P[\mathrm{B}|\mathrm{A}]P[\mathrm{A}] + P[\mathrm{B}|\mathrm{A}^c]P[\mathrm{A}^c]$$

Except for $P[\mathrm{B}|\mathrm{A}^c]$ all other information is known. Since the company does not expect the machine part to fail under proper loading conditions, the assignment $P[\mathrm{B}|\mathrm{A}^c] = 0$ is made. The probability of B is determined as

$$P[\mathrm{B}] = 0.5 \cdot 0.01 + 0 \cdot 0.99 = 0.005$$

Thus, less than 1 percent of the machines in use are expected to fail owing to the failure of the structural component. If the assignment $P[\mathrm{B}|\mathrm{A}^c] = 0$ is not expected to be correct, a laboratory test of the machine under normal operating conditions can be undertaken to obtain a more accurate estimation.

THEOREM 3 If A and B are any two random events, then the probability of union of A and B

$$P[\mathrm{A} \cup \mathrm{B}] = P[\mathrm{A}] + P[\mathrm{B}] - P[\mathrm{A} \cap \mathrm{B}]$$

The Venn diagram of Figure 4.13 shows the intersection of A and B, $\mathrm{A} \cap \mathrm{B}$, as the overlap of A and B. Since the probability is considered a relative frequency and can be considered to be a simple counting process, care must be taken to avoid double counting of the number of sample points in $\mathrm{A} \cap \mathrm{B}$. Thus $P[\mathrm{A} \cup \mathrm{B}]$ is the sum of the relative frequency of all points in A and in B minus the relative frequency of all points in $\mathrm{A} \cap \mathrm{B}$.

If a system employs two motors, A and B, and these motors have probabilities of failure, for a 1-year period of 0.02 and 0.05, respectively; the probability that the system will fail in 1 year may be determined as

$$P[\mathrm{A} \cup \mathrm{B}] = P[\mathrm{A}] + P[\mathrm{B}] - P[\mathrm{A} \cap \mathrm{B}]$$

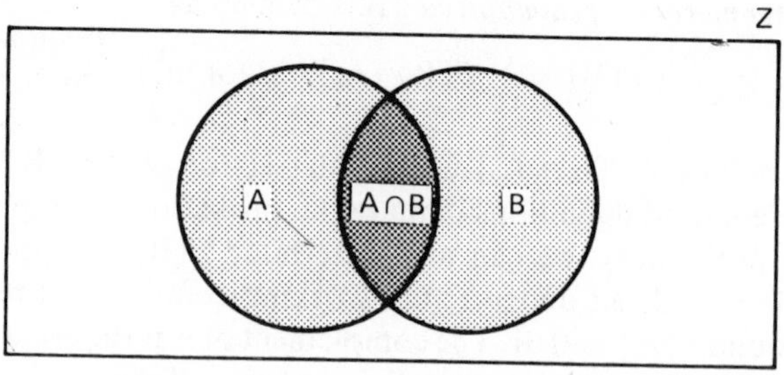

Figure 4.13 Venn diagram for the union of A and B, A ∪ B.

Since it is reasonable to assume that the failure of one motor to operate is independent of the other one, the probability of both motors failing is

$$P[A \cap B] = P[A]P[B]$$

Thus,

$$P[A \cup B] = 0.02 + 0.05 - 0.02 \cdot 0.05 = 0.069$$

There is a 6.9 percent likelihood of system failure.

EXAMPLE 4.1 Estimates of Flood Maintenance and Damage Costs

The maintenance and damage of a storm sewer system depends upon the "flood stage height," or the number of feet that a river rises above its banks. From 1975 through 1979, serious flooding with stage heights of 1.50 feet or more has caused millions of dollars in damage. In order to evaluate the effectiveness of different flood control designs, a 100-year record of stage height data has been collected . The historical record of maximum stage heights is as follows.

	STAGE HEIGHT IN FEET									
	YEAR									
DECADE	0	1	2	3	4	5	6	7	8	9
1880	2.25	1.76	0.26	2.30	2.55	0.82	0.88	0.90	0.92	0.55
1890	1.05	0.71	1.80	2.00	1.10	1.20	1.24	0.77	0.70	1.76
1900	1.20	1.05	0.40	1.83	0.72	0.76	0.82	0.71	0.78	2.40
1910	2.65	0.45	1.82	2.45	0.55	0.20	0.56	0.82	0.80	2.45
1920	0.27	0.79	0.81	1.90	1.10	0.55	0.80	0.89	0.91	1.20
1930	0.49	0.82	0.76	1.95	0.52	0.79	1.21	1.22	1.05	1.04
1940	0.42	0.83	0.77	1.89	0.60	0.82	1.01	1.05	1.15	1.20
1950	0.74	0.41	0.62	0.30	0.60	0.83	1.12	1.11	1.20	1.76
1960	0.24	0.49	0.63	1.26	0.69	0.24	0.81	0.79	1.45	0.77
1970	1.30	1.35	1.40	0.70	0.49	2.10	0.81	2.05	2.02	1.20

(a) Estimate and plot the frequency and cumulative probability distributions for the flood data.

(b) A proposed flood-control facility has the capacity to handle flood waters with stage heights of 1.50 ft or less. Determine the probability that in any given year the facility will fail to protect the area.

(c) Utilize the frequency distribution to estimate the mean annual flood stage height.

(d) Estimate the expected annual maintenance and damage costs due to flooding. The flood maintenance and damage costs are assumed to vary exponentially with stage height. $C = 10x^3$, where C is the cost in millions of dollars and x is the stage height.

Solution

(a) Define a continuous random variable as

$$X = \{\text{maximum annual stage height in feet}\} = \{x : x \geq 0\}$$

The historical data are arranged in 12 class intervals in the following table and are then plotted as shown in the histogram in Figure 4.14. The class intervals have ranges between 0 and 0.25 ft, 0.25 and 0.50 ft, and so on. The probability density function $f_X(x)$ is estimated as a relative frequency to be equal to the number of observations per class interval divided by the total number of occurrences, $n = 100$.

$$f_X(x_i) = \frac{m_i}{n}$$

The cumulative frequency is estimated as

$$F_X(x_i) = \sum_{t \leq x_i} f_X(t)$$

Table of Estimates

i	x_i Class Interval	Tally	m_i No. of observations per class interval	$f_X(x_i)$	$F_X(x_i)$
1	1-0.25		= 3	0.03	0.03
2	0.26-0.50		= 10	0.10	0.13
3	0.51-0.75		= 16	0.16	0.29
4	0.76-1.00		= 25	0.25	0.54
5	1.01-1.25		= 21	0.21	0.75
6	1.26-1.50		= 5	0.05	0.80
7	1.51-1.75		0	0.00	0.80
8	1.76-2.00		= 10	0.10	0.90
9	2.01-2.25		= 3	0.03	0.93
10	2.26-2.50		= 5	0.05	0.98
11	2.51-2.75		= 2	0.02	1.00
12	2.76-3.00		0	0.00	1.00
			$n = 100$		

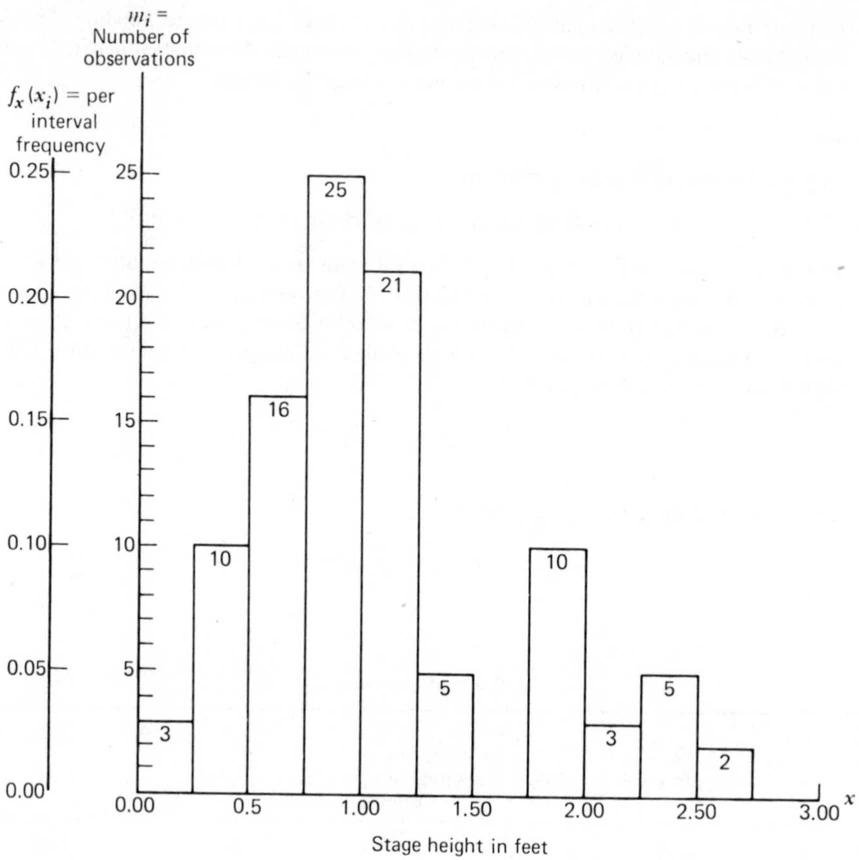

Figure 4.14

For example,

$$F_X(x_3) = f_X(x_1) + f_X(x_2) + f_X(x_3)$$

$$F_X(x_3) = 0.03 + 0.10 + 0.16 = 0.29$$

The graph $F_X(x_i)$ is shown in Figure 4.15.

(b) The failure event F is defined as

$$\text{F} = \{\text{stage height exceeds 1.5 feet}\}$$

or, in terms of the random variable X,

$$\text{F} = \{x : x > 1.5 \text{ ft}\}$$

$$P[\text{F}] = P[X > 1.5 \text{ ft}] = \sum_{x_i \geq 1.75} f_X(x_i) = 0.20$$

This answer may be found from the cumulative frequency diagram.

$$P[\text{F}] = P[X > 1.5] = 1 - \text{F}_X(1.5) = 1 - 0.80 = 0.20$$

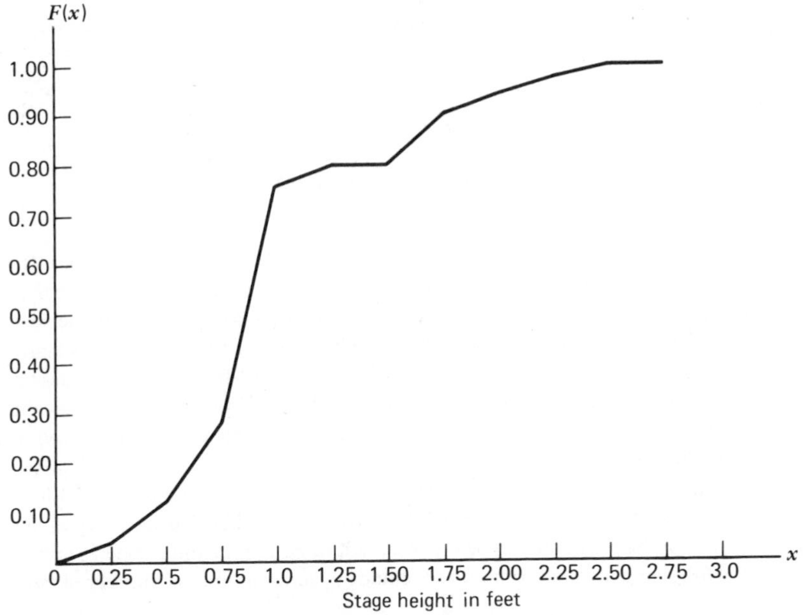

Figure 4.15

(c) The mean annual flood stage height $E[X]$ is estimated with the following relationship.

$$E[X] = \sum_x x f_X(x) = x_1 f_X(x_1) + x_2 f_X(x_2) + \cdots + x_{12} f_X(x_{12})$$

$$E[X] = (0.125)(0.03) + (0.375)(0.10) + \cdots + (2.875)(0.00)$$

$$E[X] = 1.09\text{ft}$$

The expected annual stage height is 1.09 ft.

(d) The expected annual maintenance and damage cost, or $E[\text{A}]$, is

$$E[\text{A}] = \sum_x 10x^3 f_X(x) = 10x_1^3 f_X(x_1) + 10x_2^3 f_X(x_2) + \cdots + 10x_{12}^3 f(x_{12})$$

$$E[\text{A}] = 10[(0.125)^3(0.03) + (0.375)^3(0.10) + \cdots + (2.875)^3(0.00)]$$

$$E[\text{A}] = \$26.2\text{M}$$

Remarks

From 1945 through 1975, the maximum annual stage height was less than the expected maximum height of 1.09 ft, as calculated in part c. In addition, since the stage height was less than 1.50 ft except in 1959 when the maximum stage height reaches 1.76 ft, a sense of security prevailed in the region and land development took place. During 1975, 1977, and 1978, when serious flooding took place, the annual maintenance and damage costs far exceeded the expected annual cost of \$26.2M. It is a weighted average, reflecting the costs associated with *all* maximum stage height conditions, both small and large. The single events when serious flooding occurred far exceed the expected annual cost.

YEAR	1975	1977	1978
Flood cost	\$92.6M	\$86.2M	\$82.4M

Our analysis for selection will use expected values for forecasting future outcomes. It must be remembered that the expected value represents a long-term average. These recent incidences have caused the demand for a better flood-control system. This problem is discussed in the next section.

In this example, we are relying on old data, a 100-year historical record. If the new land development caused the recent flooding, then this historical data cannot be used. We are assuming that the land development has not affected the river flow in the flood plain area and that the serious flooding, which occurred during the 1970s, is a random phenomenon like the flooding that is characterized by the 100-year record.

In this example, the number of intervals was selected to be 12. The interval width of 0.25 in. seemed to be a convenient range. The shape of the distribution will be dependent upon the interval size and the nature of the phenomenon being observed. In Benjamin and Cornell's textbook, they suggest that the following formula be used as a guide for determining the number of intervals k.

$$k = 1 + 3.3 \log_{10} n$$

where n = number of observations. They point out if the number of observations is small, then the choice of the precise point at which the interval divisions are to occur may affect the shape of the frequency distribution. Using this formula with 100 data points, eight intervals are needed for this problem.

EXAMPLE 4.2 Water-Supply Pump System Reliability

Consider three pump system configurations that are all capable of delivering 20 Mgal/day of water at an effective pressure head of 40 ft. (See Figure 4.16.)

System A Two 20-Mgal/day, 20-ft head pumps in series.

System B One 20-Mgal/day, 40-ft head pump with a 20-Mgal/day, 40-ft head backup pump.

System C Two 10-Mgal/day, 40-ft head pumps with a 10-Mgal/day 40-ft head backup pump.

The probability that a single pump will fail to operate over a period of 1 year is 5 percent. Pump failures are assumed to be independent events. Determine the most reliable system. The criterion for selection is a pump system with minimum probability of failing to deliver 20 Mgal/day.

Solution

Individual pump failure is defined to be event when zero flow is delivered. Thus, the following random events are defined:

$$F_i = \{\text{pump } i \text{ fails to operate for } i = 1, 2, 3\}$$

Thus $P[F_1] = P[F_2] = P[F_3] = 0.05$, where we define the discrete random variable Q as the system flow rate or

$$Q = \{\text{delivered flow rate in Mgal/day}\} = \{q : q = 0, 10, 20\}$$

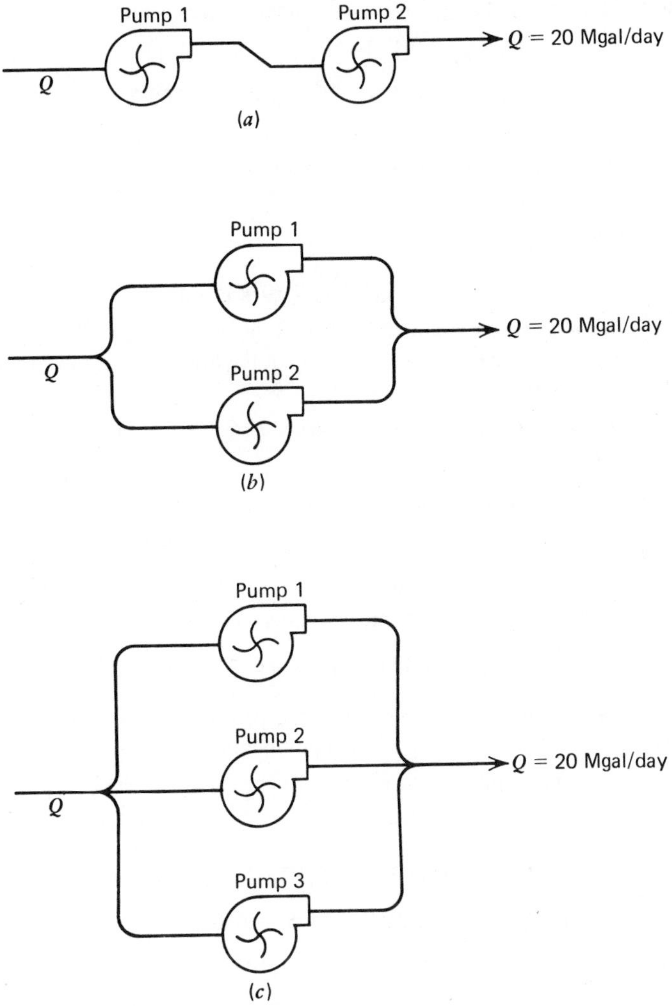

Figure 4.16 (*a*) System **A**, two-pump series system. (*b*) System **B**, primary pump with backup. (*c*) System **C**, two primary pumps with backup.

System failure for configurations **A**, **B**, and **C** is defined as the random events

$$F_{\mathbf{A}}, F_{\mathbf{B}}, F_{\mathbf{C}} = \{\text{systems } \mathbf{A}, \mathbf{B}, \text{ and } \mathbf{C} \text{ fail to deliver 20 Mgal/day of water}\}$$

For system **A**, failure occurs when either pump 1 or 2 fails to operate. The probability of event $F_{\mathbf{A}}$ is the union of F_1 and F_2.

$$P[F_{\mathbf{A}}] = P[Q = 0]$$

$$P[F_{\mathbf{A}}] = P[F_1 \cup F_2]$$

$$P[F_{\mathbf{A}}] = P[F_1] + P[F_2] - P[F_1 \cap F_2]$$

Since the failure of pumps 1 and 2 is assumed to be independent events,

$$P[F_A] = P[F_1] + P[F_2] - P[F_1]P[F_2]$$

$$P[F_A] = 0.05 + 0.05 - (0.05)^2 = 0.0975$$

For system **B**, the failure event occurs when both pumps fail to operate at the same time. The probability of the event F_B is the intersection of F_1 and F_2.

$$P[F_B] = P[Q = 0] = P[F_1 \cap F_2]$$

Since the events are independent,

$$P[F_B] = P[F_1]P[F_2] = (0.05)^2 = 0.0025$$

For system **C**, the failure event occurs when two of the three pumps fail to operate or when all three pumps fail to operate. Here, failure occurs when the flow is equal to zero or 10 Mgal/day. The probability of failure is

$$P[F_C] = P[(Q = 0) \cup (Q = 10)] = P[Q = 0] + P[Q = 10]$$

because $Q = 0$ and $Q = 10$ are mutually exclusive events. Let F_1^c, F_2^c, and F_3^c be complementary events that represent the events that pumps 1, 2, and 3 are operational, respectively.

$$P[F_1^c] = 1 - P[F_1], \quad P[F_2^c] = 1 - P[F_2] \quad \text{and} \quad P[F_3^c] = 1 - P[F_3]$$

All events are assumed to be independent, thus $P[F_1] = P[F_2] = P[F_3] = P[F]$.

The probability that no flow is delivered by system **C** is probability that all pumps fail to operate.

$$P[Q = 0] = P[F_1]P[F_2]P[F_3] = P[F]^3.$$

The probability that the flow is 10 Mgal/day is the probability that one out of three pumps operate. All possible combinations of individual pump failure must be considered, thus

$$\begin{aligned} P[Q = 10] &= P[F_1]P[F_2^c]P[F_3^c] + P[F_1^c]P[F_2^c]P[F_3] + P[F_1^c]P[F_2]P[F_3^c] \\ &= 3P[F]^2(1 - P[F]) \end{aligned}$$

$$P[F_C] = 3P[F]^2(1 - P[F]) + P[F]^3$$

$$P[F_C] = 3(0.05)^2(1 - 0.05) + (0.05)^3$$

$$P[F_C] = 0.0071 + (0.125 \times 10^{-5}) \approx 0.0071$$

The results are summarized as follows:

	DEFINITION OF FAILURE	PROBABILITY OF FAILURE
System A	Q = 0 Mgal/day	0.0975
System B	Q = 0 Mgal/day	0.0025
System C	Q = 0 or 10 Mgal/day	0.0071

The most reliable system is system **B**; therefore, it is recommended.

Remarks

The most unreliable system is plan A, the two-pump series system. The parallel systems are much more reliable.

Let us investigate the criterion for selection in greater detail. System **B** has a small probability of failure, $Q \leq 20$ Mgal/day; however, if both pumps fail, no water is delivered. System **C**, on the other hand, may deliver water at a rate of 10 Mgal/day when two of the three pumps fail to operate. The probabilities of delivering no water and 10 Mgal/day are

$$P[Q = 0] = P[F_1 \cap F_2 \cap F_3] = P[F]^3 = 0.125 \times 10^{-3}$$

$$P[Q = 10] = P[F_c] - P[Q = 0] = 0.0071 - 0.125 \times 10^{-3} = 0.007$$

It is very unlikely that no water will be delivered for system **C**. If economic losses that are associated with pump system failure are incorporated into the decision making, it may prove that the system **C** is more desirable than system **B**. These economic considerations are discussed in the next section.

PROBLEMS

Problem 1

A construction company has collected 100 samples. Each sample consists of an investigation of 30 consecutive days. The following discrete random variable is defined.

X = {the number of days in the period of 30 consecutive days that construction is stopped due to bad weather, late deliveries, and other delays} = $\{x: x = 0, 1, 2, \ldots, 30\}$.

A histogram of X is shown in Figure 4.17.

(a) What is the probability function of X?

(b) What is the expected number of days construction is stopped in 30 consecutive days?

(c) What is the probability that there are no days that construction is stopped in 60 consecutive days? (*Hint*: Define the random events.

$$S_1 = \{\text{no delays occur in the first 30 days}\}$$

$$S_2 = \{\text{no delays occur in the second 30 days}\}$$

Assume S_1 and S_2 are independent events.)

(d) For any reason if construction is delayed, the construction company must pay a penalty. The penalty function is

$$C = 0 \qquad \text{for } x = 0, 1$$

$$C = 10{,}000x \qquad \text{for } x = 2, 3, \ldots$$

Determine the expected penalty cost for a 30-consecutive-day period.

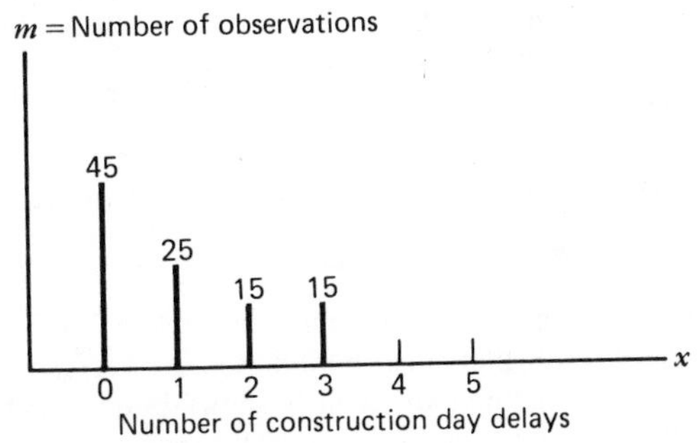

Figure 4.17

Problem 2

A construction job consists of the following tasks that must be done in the following order:

Task 1 Prepare site for placement of concrete forms.

Task 2 Construction of concrete forms.

Task 3 Pour concrete.

The time to complete these tasks is described by the following discrete random variables.

$T_1 = \{\text{time in days to complete task 1}\} = \{t: t = 0, 1, 2 \ldots\}$

$T_2 = \{\text{time in days to complete task 2}\} = \{t: t = 0, 1, 2, \ldots\}$

$T_3 = \{\text{time in days to complete task 3}\} = \{t: t = 0, 1, 2, \ldots\}$

The contractor will remain on schedule if $T_1 + T_2 + T_3 \leq 4$ days. The probabilities of completing each task are:

$$P[T_1 \leq 1] = 0.60$$

$$P[T_2 \leq 2] = 0.90$$

$$P[T_3 \leq 1] = 0.98$$

If a task exceeds any one or more of these limits, the construction job will be delayed by one or more full days. Tasks 1, 2, and 3 are assumed to be independent events.

(a) Determine the probability that the contractor will remain on schedule.

(b) If the contractor does not meet the schedule of 4 days, he must pay a \$1000 penalty. No penalty is levied if the schedule is met. Determine the expected penalty charge.

(c) If task 1 is completed in one or less days, what is the expected penalty charge?

Problem 3

The following historical record for the number of snowfalls per year has been tabulated. A snowfall is defined as the event that sufficient accumulation, usually a depth of $\frac{1}{2}$ in. or more, has fallen to warrant the call for snow plows to remove the snow from roadways.

	YEAR									
DECADE	0	1	2	3	4	5	6	7	8	9
1940	6	8	2	18	2	14	3	4	8	20
1950	6	9	4	5	3	6	8	9	3	4
1960	8	1	10	15	2	8	6	7	10	5
1970	3	5	12	13	1	9	7	3	12	8

(a) Plot a histogram and probability distribution for the random event:

$$X = \{\text{Number of snowfalls per year}\} = \{x: x \geq 0\}$$

where the interval k is determined with the relationship $k = 1 + 3.3 \log_{10} n$.

(b) Determine a cumulative probability distribution.

(c) Determine the probability that no more than 10 snowfalls occur per year.

(d) Determine the probability that 15 or more snowfalls occur per year.

(e) Determine the expected mean m_X and standard deviation σ.

Problem 4

The compressive strengths of 20 concrete cylinders are:

2200 psi	1950
1860	1860
2440	1780
2100	2350
1960	2260
2020	2100
2230	1985
1980	2300
2100	2100
2260	2210

(a) Determine the statistical mean $\bar{x}$ and variance s_x^2 of the data.
(b) Plot a histogram and frequency distribution using the class intervals: 1700–1799, 1800–1899, 1900–1999, ...
(c) Determine the expected mean m_X and variance σ^2 using the probability definitions. Compare the means, m_X and $\bar{x}$ and the variances, σ^2 and s_x^2.
(d) Plot a cumulative frequency diagram.

Problem 5

The probability density function $f_T(t)$ for rainfall is

$$f_T(t) = 1/6e^{-t/6} \qquad \text{for } t \geq 0$$

where $T = \{\text{time in days between rainfall}\} = \{t: t \geq 0\}$
(a) Plot a cumulative probability density function.
(b) Determine the probability that rainfall will occur in one day or less.
(c) Determine the expected time between rainfall events.
(d) Determine the probability of drought, no rain will occur in 20 or more days.
(e) Crop damage on a farm area is assumed to be

$$D = 0 \qquad \text{for } t \leq 20$$

$$D = \$11{,}000(t - 20) \qquad \text{for } t > 20$$

Determine the expected damage.

Problem 6

See Figure 4.18. The pumps are numbered items 1 and 2 and the stirring motor is numbered item 3.
(a) Determine the probability that the system will fail during a 1-yr period, where

$$\text{F}_i = \{\text{Failure of item in a 1-yr period, } i = 1, 2, 3\}$$

$$P[\text{F}_1] = P[\text{F}_2] = 0.10$$

$$P[\text{F}_3] = 0.05$$

Assume all events are independent

(b) Determine the expected monetary loss if the cost of system failure is \$10,000.
(c) Repeat part a for $P[\text{F}_2|\text{F}_1] = 0.13$, $P[\text{F}_1] = 0.10$, and $P[\text{F}_3] = 0.05$

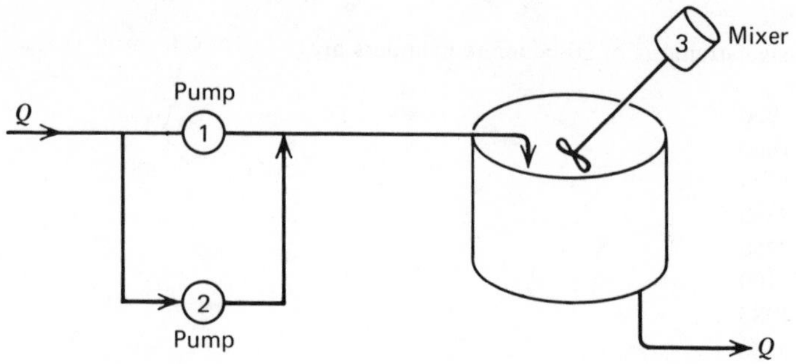

Figure 4.18

Problem 7

A probability density distribution of live load W is assumed to have a triangular distributed between 0 and 100 lb per linear foot. (See Figure 4.19.) The density function is

$$f_W(w) = a\left(1 - \frac{w}{100}\right) \qquad \text{for } 0 \leq w \leq 100$$

(a) Determine the unknown coefficient a.
(b) Determine the cumulative frequency function of W.
(c) What is the probability that the live load is 50 lb per linear foot or greater?
(d) Determine the expected mean load m_W.
(e) The maximum moment of a beam is

$$M = \frac{wl^2}{12}$$

where l is the beam length. What is the maximum expected moment m_M?

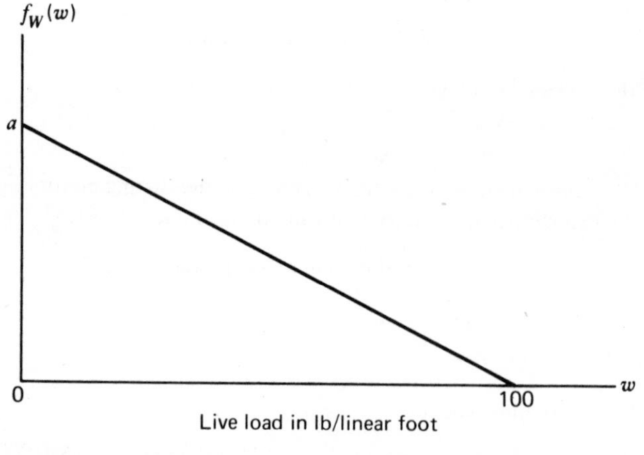

Figure 4.19

Problem 8

The total cost y is a sum of a fixed cost α and variable cost βx. Thus, the total cost function is

$$Y = \alpha + \beta x$$

Assume the probability function $f_X(x)$ is known.

(a) Show by using the definitions of the mean that the expected cost $E[Y] = E[\alpha + \beta X]$ is

$$E[Y] = \alpha + \beta E[X] = \alpha + \beta m_X$$

(b) Show the variance of y is

$$\sigma_Y^2 = \beta^2 \sigma_X^2$$

(c) The cost is $y = 10 + 8x$. The mean and standard deviations of X are 16 and 8, respectively. Determine the mean and variance of Y.

Problem 9

The histogram for hourly wind speeds on a mountain top location is shown in Figure 4.20.

(a) Define a random variable for wind speed and estimate the probability function for wind speed.
(b) Estimate the cumulative frequency function for wind speed.
(c) Estimate the mean wind speed.
(d) A wind turbine will generate electric power between 10 and 35 miles/hr. Below 10 miles/hr there is insufficient wind velocity to overcome the internal friction of the turbine, and above 35 miles/hr the turbine will be damaged by the strong winds. Determine the probability that the wind turbine will deliver power.

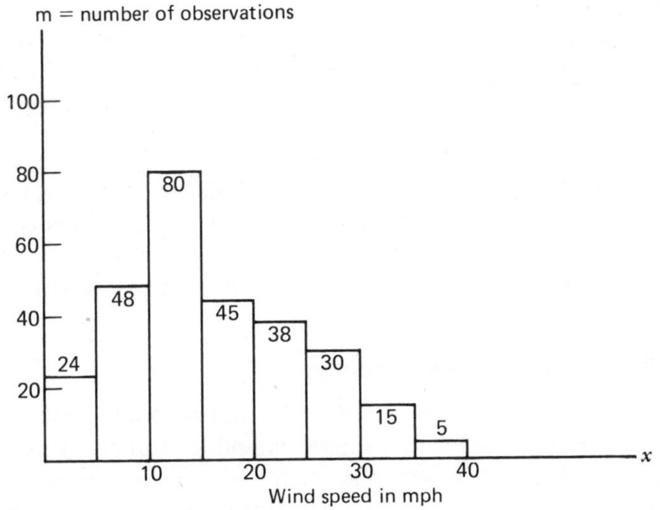

Figure 4.20

(e) Determine the mean power output in kilowatts that the turbine may generate, where power is a function of the cube of the wind speed x.

$$P = kx^3$$

The constant $k = 5 \times 10^{-3}$ is a function of air density, sweep area of turbine blades, and efficiency.

Problem 10

The probability of failure of pumps 1, 2, and 3 shown in Figure 4.21 is

$$P[F_1] = P[F_2] = [F_3] = 0.01$$

where F = {pump failure}. The random events of pump failure are independent.

(a) Define the random event for system failure S in terms of F_1, F_2, and F_3.
(b) Draw a Venn diagram describing system failure in terms of F_1, F_2, and F_3.
(c) Determine the probability of system failure.

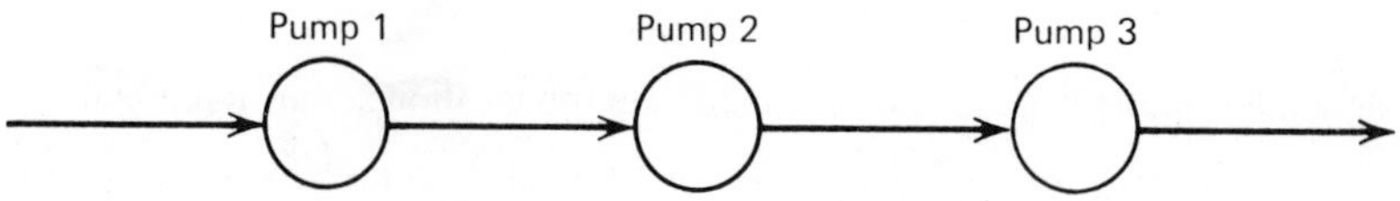

Figure 4.21

4.2 DECISION MAKING UNDER UNCERTAINTY

The alternative selection procedure is straightforward. The steps are:

1. Determine the consequence for each alternative **A**, **B**, **C**, ... and each state of nature, S_1, S_2, ...
2. Determine the probability for the states of nature, $P[S_j]$ for $j = 1, 2, 3, \ldots$
3. Calculate the expected monetary value for each alternative, EMV^A, EMV^B, EMV^C, ... and select the alternative with the maximum utility; or, equivalently, select the alternative with the minimum opportunity cost.

Step 1 generally requires the use of time value of money relationships and other considerations dealing with present and future costs and benefits as discussed in Chapter 3. Although the procedure is simple to apply, there are important conceptual considerations with regard to the selection process that must be discussed.

Expected Monetary Value Theory

The concept of expected monetary value EMV and its implication for alternative selection can best be described with a simple gambling example. A prize of \$10 will be given if we correctly guess the outcome of a flip of a coin as a head or tail. We may elect to pay \$4 to play the game or elect not to play. The decision tree is shown in Figure 4.22.

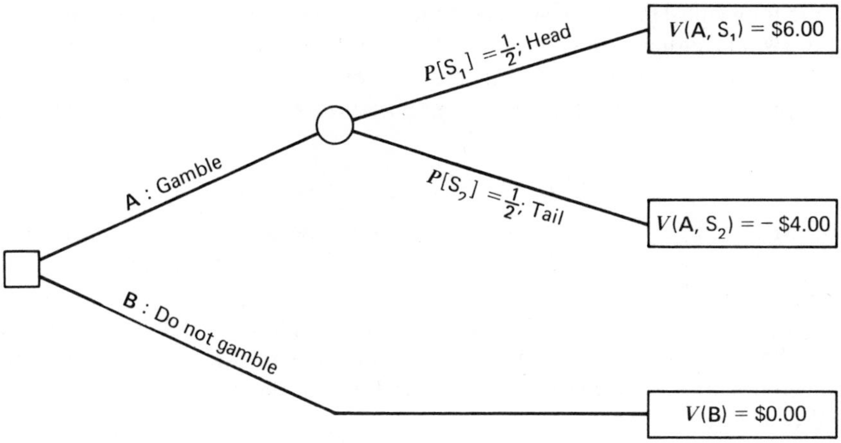

Figure 4.22 The coin-flip gamble decision tree.

The consequences of our action are as follows.

Alternative A Since we pay \$4 to play, the total gain is the prize minus the cost to play. If we choose a head and the outcome of the coin flip is a head, we obtain a net benefit of

$$V(\mathbf{A}, \mathrm{S}_1) = \$10 - \$4 = \$6$$

If the outcome is a tail, the net negative benefit or loss is

$$V(\mathbf{A}, \mathrm{S}_2) = \$0 - \$4 = -\$4$$

Alternative B The value of the do-nothing alternative is

$$V(\mathbf{B}) = \$0$$

The expected monetary value will be used to aid us in determining if it is worthwhile to gamble or not. For alternative **A**,

$$\mathrm{EMV}^{\mathbf{A}} = P[\mathrm{S}_1]V(\mathbf{A}, \mathrm{S}_1) + P[\mathrm{S}_2]V(\mathbf{A}, \mathrm{S}_2)$$

$$\mathrm{EMV}^{\mathbf{A}} = \tfrac{1}{2}(\$6) + \tfrac{1}{2}(-\$4)$$

$$\mathrm{EMV}^{\mathbf{A}} = \$1$$

Since there is no chance involved in the do-nothing alternative, the expected monetary value for alternative **B** is

$$\mathrm{EMV}^{\mathbf{B}} = \$0$$

According to this theory, we choose the alternative with the maximum expected monetary value. Since $\mathrm{EMV}^{\mathbf{A}} > \mathrm{EMV}^{\mathbf{B}}$, alternative A is considered the better choice.

According to probability theory, if the coin is flipped a large number of times, the expected outcome will be 50 percent heads and 50 percent tails. Since a net benefit of \$1 is associated with the gamble, we are at an advantage because the odds are with us in the long run. However, we must remember that only *one play is made* and the outcome will be either a \$6 gain or a \$4 loss, not the expected gain of \$1. With EMV theory we select the gamble and hope that the long-term odds are with us.

If the price of play were changed to \$6, the expected monetary value of alternative **A** would be

$$EMV^{A} = \tfrac{1}{2}(\$4) + \tfrac{1}{2}(-\$6) = -\$1$$

Since $EMV^{A} < EMV^{B} = 0$, we shall not gamble, because the long-term odds are against us. In this case, according to the EMV theory, alternative **B**, the do-nothing alternative, is the better choice.

The probabilities will also affect the selection. Suppose that the price to play is changed back to \$4, but we win if the roll of a die is either 1 or 2, S_1, and we lose if the roll comes up 3, 4, 5, or 6, S_2. The probabilities for the states of nature of winning S_1 and of losing S_2 are

$$P[S_1] = \tfrac{2}{6} = \tfrac{1}{3}$$

$$P[S_2] = \tfrac{4}{6} = \tfrac{2}{3}$$

The EMV for alternative **A** is

$$EMV^{A} = \tfrac{1}{3}(\$6) + \tfrac{2}{3}(-\$4) = -\$0.67$$

Since $EMV^{B} = 0$ and $EMV^{B} > EMV^{A}$, the better choice is not to gamble.

Risk Aversion

These examples demonstrate how important both the economic preference and the likelihood of occurrence are in the decision-making process. The probabilities of the states of nature are weighing factors. Let us investigate one more example, where the cost to play the game is \$5. The EMV^{A} for the gamble is

$$EMV^{A} = \tfrac{1}{2}(\$5) + \tfrac{1}{2}(-\$5)$$

$$EMV^{A} = \$0$$

For alternative **B**, $EMV^{B} = \$0$. There is no clear-cut choice, $EMV^{A} = EMV^{B}$. This is called the *indifference* condition.

According to the EMV theory either choice will lead to equally acceptable outcomes. The amount of the stake will influence our choice. When the stake becomes too high we shall avoid the gamble because a loss may destroy us financially. The EMV theory for high stakes is not a suitable selection criterion. For the relatively small stakes presented in the gambles described thus far, the consequences of winning or losing are assumed not to affect our overall financial situation seriously; thus the EMV selection criterion is considered acceptable.

If the stakes are increased to \$50,000 to play, with \$100,000 for calling the flip of the coin correctly and \$0 for calling it incorrectly, the EMV for the gamble is

$$EMV^{A} = \tfrac{1}{2}(\$100{,}000 - \$50{,}000) + \tfrac{1}{2}(\$0 - \$50{,}000)$$

$$EMV^{A} = \$0$$

For alternative **B**, $EMV^B = 0$. According to the EMV theory, we have an indifference condition, $EMV^A = EMV^B$, and either choice, the gamble or the do-nothing alternative, is equally acceptable. For most people, choosing a gamble like this one is too risky, and it will not be chosen. The financial hardship of losing the gamble will have a devastating impact upon our financial well being. For this gamble we are called *risk averters*. The EMV theory does not adequately describe the risk averse situation. The concepts of utility theory will be used for this purpose.

Utility Theory

In Chapter 3, we define *utility* as the *value in use* of a good or service. It is also the measured satisfaction that a consumer derives from various quantities of commodities. In our discussion of consumer demand, we define excess benefit as the difference between the value-in-use price, usually obtained from the demand function, minus the selling price. It is also a measure of utility. Since the selling price is determined from the conditions of market equilibrium, the excess benefit is a function of both demand and supply. These concepts help explain why water, which has a high utility and a plentiful supply, has a low selling price and gold, which has a low utility and a limited supply, has a high selling price.

Utility used in decision theory adds new insight and meaning to this definition. We shall investigate the behavior of a single individual in contrast to the broader interpretation of excess benefit and the effects of supply and demand. In decision theory, we concentrate upon the satisfaction derived from possessing money and the dissatisfaction of not having money. The new function, called a *utility function* $U(q)$, will be used to measure the preference for different quantities of money q. The units of $U(q)$, a psychological measure of user satisfaction, are called a "utiles." They do not have physical or economic meaning in the strict sense to which engineers are accustomed. They are best described by example.

For our illustration, we use a range of −\$50,000 to \$100,000 in the high-stake coin-flip problem. The utility curves $U(q)$ as shown in Figure 4.23 are for a gambler and a risk averter. The risk averter and the gambler place the same utility upon the extremes: $U(-\$50{,}000) = 0$ and $U(\$100{,}000) = 100$. However, the risk averter places different utility upon all intermediate gains and losses as compared to the gambler, as shown in the figure. Both the gambler and the risk averter have the same preference; they prefer \$100,000 to \$50,000.

$$\$100{,}000 > \$50{,}000$$

In terms of utilities, they still have the same preference.

$$U(\$100{,}000) > U(\$50{,}000)$$

However, the gambler derives more satisfaction from the \$100,000 than the risk averter. This preference is measured in terms of utiles.

Gambler Preference $\quad U(\$100{,}000) > U(\$50{,}000)$

$$100 > 66.7$$

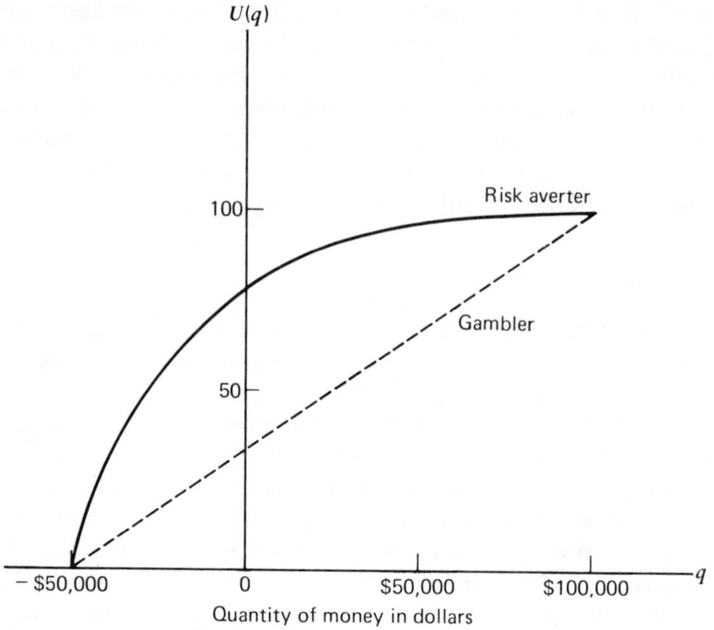

Figure 4.23 Utility functions for a gambler and risk averter.

Risk Averter Preference $U(\$100{,}000) > U(\$50{,}000)$

$$100 > 98$$

The relative magnitude of utility, not the absolute magnitude, is important in weighing the preference.

The properties of utilities are summarized by the following axioms.

Axioms of Utility

1. If q_1 is preferred to q_2 or $q_1 > q_2$, then $U(q_1)$ is preferred to $U(q_2)$ or $U(q_1) > U(q_2)$.
2. The transitive property. If $q_1 > q_2$ and $q_2 > q_3$, then $q_1 > q_3$, then $U(q_1) > U(q_2)$ and $U(q_2) > U(q_3)$, then $U(q_1) > U(q_3)$.
3. If choices are indifferent, $q_1 \sim q_2$, then they can be substituted for each other or $U(q_1) \sim U(q_2)$.
4. Utilities can be combined by using basic probability theory and the definition of expectation: $U(\mathbf{A}) = p(S_1)U(q_1) + p(S_2)U(q_2)$, where S_1 and S_2 represent two mutually exclusive random states for alternative **A**.
5. If two combined utilities lead to the same utility, $U(\mathbf{A}) \sim U(\mathbf{B})$, choose the alternative with the greater probability of success. For instance, if alternative **A** involves risk, $U(\mathbf{A}) = p(S_1)U(q_1) + p(S_2)U(q_2)$, alternative **B** involves no risk, and $U(\mathbf{A}) \sim U(\mathbf{B})$, choose alternative **B**.

Now let us return to the high-stake coin-flipping example and assume we are risk averters. Since we want to maximize our satisfaction, we maximize our utility. The decision tree for this problem is shown in Figure 4.24, where the consequences are measured in terms of utilities of the net gain and losses taken from Figure 4.23. For alternatives **A** and **B**, the utilities are

$$U(\mathbf{A}, S_1) = U(\$100{,}000 - \$50{,}000) = U(\$50{,}000) = 98$$

$$U(\mathbf{A}, S_2) = U(\$0 - \$50{,}000) = U(-\$50{,}000) = 0$$

$$U(\mathbf{B}) = U(\$0) = 75$$

From axiom 4, the utility for alternative **A** is

$$\begin{aligned} U(\mathbf{A}) &= P(S_1)U(\mathbf{A}, S_1) + P(S_2)U(\mathbf{A}, S_2) \\ &= \tfrac{1}{2}(98) + \tfrac{1}{2}(0) = 49 \end{aligned}$$

Since $U(\mathbf{B}) > U(\mathbf{A})$, the better choice is not to gamble.

Assume we are gamblers. The utilities are

$$U(\mathbf{A}, S_1) = U(\$50{,}000) = 66.7$$

$$U(\mathbf{A}, S_2) = U(-\$50{,}000) = 0$$

$$U(\mathbf{B}) = U(0) = 33.3$$

$$U(\mathbf{A}) = \tfrac{1}{2}(66.7) + \tfrac{1}{2}(0) = 33.3$$

Since $U(\mathbf{A}) = U(\mathbf{B})$ the indifference condition occurs. From axiom 5, we shall choose alternative **B**, the do-not-gamble alternative. According to utility theory, we play safe and avoid risk in this case.

For this high-stakes problem, the indifference condition was obtained using EMV theory. There is no difference between EMV and utility theories for the gambler.

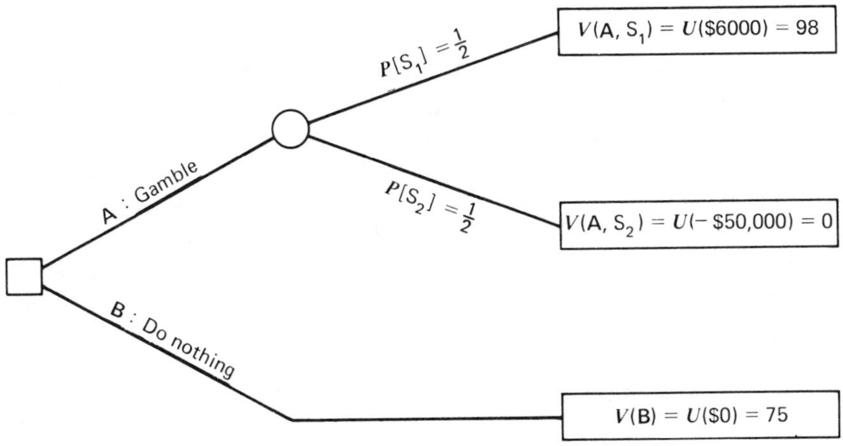

Figure 4.24 A high-stake coin-flipping problem.

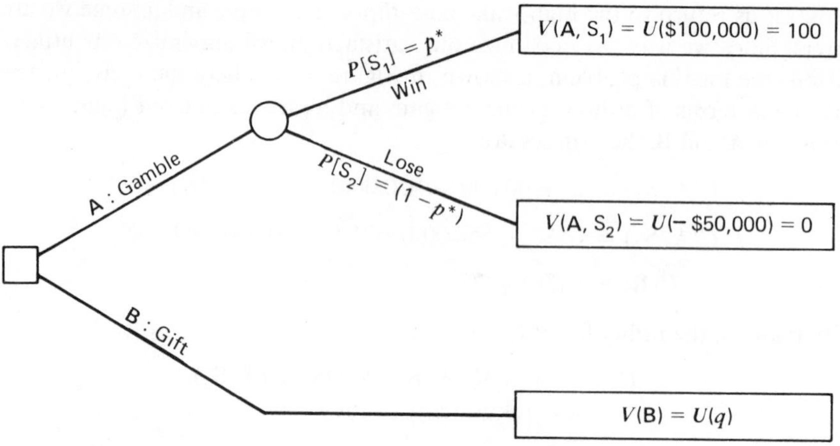

Figure 4.25 Decision tree for establishing a utility curve.

The same alternative will be chosen no matter what kind of problem is being analyzed. Since there is no new information gained by transforming monetary measures to utility measures for the gambler, utility theory is not used, and the EMV approach is sufficient. Incidentally, utility theory requires extra calculations that are avoided when EMV is used.

Throughout this discussion we assumed that the utility function was known. It is derived by offering an individual two alternatives: one to gamble and the other one is the do nothing choice. Here we express the alternative **B** as a gift. The magnitude of the gift q will be assumed to a variable. The decision tree is shown in Figure 4.25, where the utilities of receiving a \$100,000 winning and paying a \$50,000 loss are:

$$U(\$100{,}000) = 100$$
$$U(-\$50{,}000) = 0$$

The purpose of offering these two alternatives is to determine the indifference probability p^* and to find a point on the person's utility curve for a certain sum of money q. Assume an individual is offered a \$20,000 gift. At this point, $U(\$20{,}000)$ is unknown also. Through a bargaining-type discussion, the person is offered to accept the gamble with different winning probabilities. If the person does not choose the gamble and chooses the gift, the probability of winning is changed until the person becomes indifferent to the gamble and gift choices. For example, suppose a gift of \$20,000 is offered and the indifference probability p^* for this gift offer is 0.90, $p^* = 0.90$. The utility of \$20,000 is

$$U(\$20{,}000) = 0.9U(\$100{,}000) + 0.1U(-\$50{,}000)$$

or

$$U(\$20{,}000) = 0.9(100) + (0.1)(0) = 90$$

This value $U(\$20,000)$ is represented as one point on the utility curve shown in Figure 4.23. This process is repeated for different gift amounts until the continuous function as shown in Figure 4.23 is approximated.

Obviously, this curve-fitting process is extremely difficult to perform. In practical engineering applications and for many other practical reasons, the utility function is not derived. As a result, the expected monetary-value approach is used in engineering. As previously mentioned, a gambler utilizes the EMV approach to decision making. Obviously, engineers and their clients are most likely risk averters and opposed to gambling. Since EMV theory does not incorporate risk aversion, the EMV selection process must be used with care. Thus, we will use the concepts of utility as listed in the axioms of utility for the special case of gambling.

Opportunity Costs

The theoretical concepts of EMV and utility are important to the understanding of the selection process when both uncertainty and risk aversion are involved. In order to emphasize its importance, these concepts will be used in our discussion of opportunity cost.

Consider the situation with two alternatives and two states of nature as shown in the payoff matrix.

States of nature		Alternatives A_1	A_2
	S_1	u_{11}	u_{12}
	S_2	u_{21}	u_{22}

The consequences shown in the payoff matrix are monetary measures. The preferences are

$$u_{11} > u_{12} \qquad \text{and} \qquad u_{22} > u_{21}$$

If we knew the state of nature, we would choose

$$u_{11} \quad \text{if } S = S_1 \qquad \text{and} \qquad u_{22} \quad \text{if } S = S_2$$

Since the state of nature is unknown, there is a distinct possibility that we shall not receive as much satisfaction as hoped. If the outcome does not correspond to our choice, a penalty or loss of opportunity results:

CHOICE	OUTCOME	OPPORTUNITY COST, l
A_1	S_2	$l_{21} = u_{22} - u_{21}$
A_2	S_1	$l_{12} = u_{12} - u_{11}$

Obviously if the outcome corresponds to the choice, there is no lost opportunity, $l_{11} = l_{22} = 0$. The payoff matrix for opportunity cost for each state of nature is

		Alternatives $\mathbf{A}_1$	Alternatives $\mathbf{A}_2$
States of nature	S_1	$l_{11} = 0$	$l_{12} = u_{12} - u_{11}$
	S_2	$l_{21} = u_{22} - u_{21}$	$l_{22} = 0$

When using the decision-analysis approaches described in this section, the same choice will result whether we maximize EMV or minimize the expected opportunity cost. For some engineering applications that involve net profits or savings, a maximum EMV formulation is most appropriate. In some cases, it is easier to minimize the cost, thus the minimum expected opportunity-cost approach is better.

EXAMPLE 4.3 Selection of Water-Supply Pump System

This example is a continuation of Example 4.2, the Water-Supply Pump System Reliability example. In this example, only the parallel pumping systems will be considered. In addition, the capital costs of the pump systems and economic losses of not delivering 20 Mgal/day will be incorporated into the analysis. The alternative plans are

Plan A Two 20-Mgal/day pumps in parallel operation.

q, Flow condition (Mgal/day)	$P[Q = q]$
0	0.0025
10	0.0
20	0.9975

Plan B Three 10-Mgal/day pumps in parallel operation.

q, Flow condition (Mgal/day)	$P[Q = q]$
0	0.000125
10	0.007
20	~0.993

The probabilities cited here are derived from plans **B** and **C** in Example 4.2.

The capital cost for construction of a pumping station, in millions of dollars, is a function of flow q in million gallons per day (Mgal/day).

$$C = 0.035q^{1.25}$$

Failure to deliver 20 Mgal/day of water is assumed to cause inconvenience and economic losses to domestic, business, and agricultural users. In addition, public safety may be seriously compromised because medical, health-care delivery, and fire-protection services all rely on the continuous supply of drinking water. The annual losses are estimated to be

q, FLOW (Mgal/day)	ANNUAL LOSS
0	\$250M
10	\$25M
20	0

Determine the better pump system. Use a social opportunity-cost interest rate of 5 percent and a system design life of 15 years.

Solution

1. Define the states of nature S_1, S_2, and S_3 in terms of the discrete random variable Q.

$$Q = \{q:q = 0, 10, 20 \text{ Mgal/day}\}$$

or

$$S_1 = \{\text{pump system delivers no flow}\} = \{q:q = 0\}$$

$$S_2 = \{\text{pump system delivers 10 Mgal/day}\} = \{q:q = 10\}$$

$$S_3 = \{\text{pump system delivers 20 Mgal/day}\} = \{q:q = 20\}$$

The probability assignments are

Plan A

$$P[S_1] = P[Q = 0] = 0.0025$$

$$P[S_2] = P[Q = 10] = 0.0$$

$$P[S_3] = P[Q = 20] = 0.9975$$

Plan B

$$P[S_1] = P[Q = 0] = 0.0001$$

$$P[S_2] = P[Q = 10] = 0.0070$$

$$P[S_3] = P[Q = 20] = 0.9929$$

2. The capital costs for the systems are

Plan A $C = 2(0.035)(20)^{1.25} = \$2.96M$

Plan B $C = 3(0.035)(10)^{1.25} = \$1.86M$

A minimum opportunity-cost selection approach will be utilized. The annual benefit received from both systems is the same. The benefit is assumed to be greater than the costs. The present worth of annual losses PW for each state of nature or flow is

$$S_1: \text{PW} = 250\left[\frac{(1 + 0.05)^{15} - 1}{0.05(1 + 0.05)^{15}}\right] = \$2{,}595M$$

$$S_2: \text{PW} = 25\left[\frac{(1 + 0.05)^{15} - 1}{0.05(1 + 0.05)^{15}}\right] = \$260M$$

$$S_3: \text{PW} = 0$$

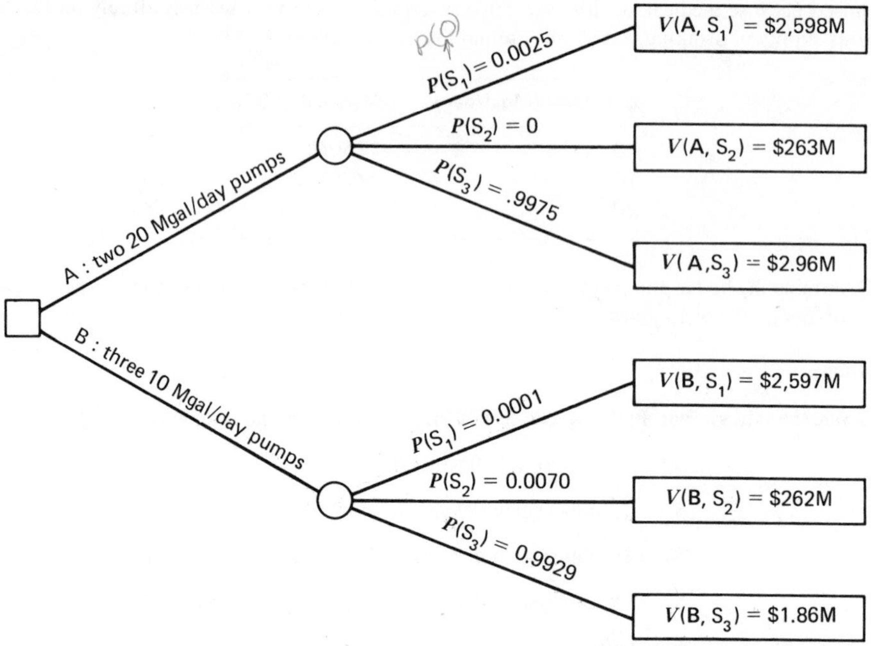

Figure 4.26

The decision tree for the two proposed systems is shown in Figure 4.26. The present worth of capital plus economic losses is shown in the boxes. They are calculated as follows:

	STATE OF NATURE q	PRESENT WORTH OF CAPITAL AND LOSSES
Plan A	S_1	$V(\mathbf{A}, S_1) = \$2.96M + \$2595M = \$2598M$
	S_2	$V(\mathbf{A}, S_2) = \$2.96M + \$\ 260M = \$\ 263M$
	S_3	$V(\mathbf{A}, S_3) = \$2.96M + \quad 0 = \$2.96M$
Plan B	S_1	$V(\mathbf{B}, S_1) = \$1.86M + \$2595M = \$2597M$
	S_2	$V(\mathbf{B}, S_2) = \$1.86M + \$\ 260M = \$\ 262M$
	S_3	$V(\mathbf{B}, S_3) = \$1.86M + \quad = \$1.86M$

The EMV for each alternative are

Plan A

$$\text{EMV}^{\mathbf{A}} = P[S_1]V(\mathbf{A}, S_1) + P[S_2]V(\mathbf{A}, S_2) + P[S_3]V(\mathbf{A}, S_3)$$

$$\text{EMV}^{\mathbf{A}} = (0.0025)(2{,}598\text{M}) + (0.0)(263\text{M}) + (0.9975)\cdot(2.96\text{M})$$

$$\text{EMV}^{\mathbf{A}} = \$6.50\text{M} + 0 + \$2.95\text{M} = \$9.45\text{M}$$

Plan B

$$\text{EMV}^{\mathbf{B}} = P(S_1)V(\mathbf{B}, S_1) + P(S_2)V(\mathbf{B}, S_2) + P(S_3)V(\mathbf{B}, S_3)$$

$$\text{EMV}^{\mathbf{B}} = (0.0001)(2{,}597\text{M}) + (0.0070)(262\text{M}) + 0.9929(1.86\text{M})$$

$$\text{EMV}^{\mathbf{B}} = \$0.325\text{M} + \$1.83\text{M} + \$1.85\text{M} = \$4.01\text{M}$$

Since $EMV^A > EMV^B$, the better selection is plan **B**, the three 10-Mgal/day parallel pumps system.

EXAMPLE 4.4 A Flood Protection System

In Example 4.1, Estimates of Flood Maintenance and Damage Costs, the annual flooding damage is estimated to be \$26.2M. Two different flood-control systems are proposed consisting of different design configurations of a reservoir, levees, and channel works. Plan **A**, a \$150M facility, can manage flows of water with stage heights of 1.5 ft or less. Plan **B**, a \$200M facility, is designed to handle stage heights of 2 ft or less. The annual damage cost A due to flooding is assumed to be a function of stage height.

Plan A

$$A = 0 \qquad \text{for } x \leq 1.5 \text{ ft}$$
$$A = 6(x - 1.5) \qquad \text{for } x > 1.5 \text{ ft}$$

Plan B

$$A = 0 \qquad \text{for } x \leq 2 \text{ ft}$$
$$A = 5(x - 2) \qquad \text{for } x > 2 \text{ ft}$$

where x is the stage height. The annual costs are in millions of dollars. Use the frequency distribution estimates $f_X(x)$ in Example 4.1 to determine the probability of the various states of nature. The design lives of the two systems are assumed to be 40 years. The social opportunity cost interest rate is 10 percent per annum. Utilize decision analysis under uncertainty to determine the choices among plan **A**, plan **B**, and the do-nothing alternative.

Solution

The same approach for plans **A** and **B** will be used. First, the states of nature will be defined. Let

$$S_1 = \{\text{Stage height causing no flood damage}\}$$
$$S_j = \{\text{Stage height causing flood damage where } j = 2, 3, \ldots\}$$

For plan **A**,

$$P[S_1] = P[X \leq 1.5\text{ft}] = F_X(x_6) = 0.80$$

This value is obtained from the table in Example 4.1. For stage heights above 1.5 ft, $X > 1.5$ ft, the extent of damage is a function of x, thus the following states of nature are estimated with the relative frequencies of the Table of Estimates from Example 4.1.

$$P[S_2] = f_X(x_7) = 0.00$$
$$P[S_3] = f_X(x_8) = 0.10$$
$$P[S_4] = f_X(x_9) = 0.03$$
$$P[S_5] = f_X(x_{10}) = 0.05$$
$$P[S_6] = f_X(x_{11}) = 0.02$$
$$P[S_7] = f_X(x_{12}) = 0.00$$

These probabilities are tabulated in the table marked plan **A**. For plan **B**

$$P[S_1] = P[X \leq 2 \text{ ft}] = F_X(x_8) = 0.90$$

The probabilities for $X \geq 2$ ft, states of nature, S_2, S_3, S_4, S_5, are obtained from the table in Example 4.1 and recorded in the table marked plan **B**.

Next, the cost associated with each of these states of nature, $C(S_j)$, will be determined with the annual cost function given in the problem statement. Then the annual savings $B(S_j)$ is determined, and, finally, the net present worth, $NPW(S_j)$, of benefits minus costs is determined. For illustration, the state of nature S_3 for plan **A** will be described in detail. The other values are recorded in the table marked plan **A** and **B**.

For S_3, the stage interval is between 1.76 ft and 2.00 or its midpoint is $x_8 = 1.875$, thus

$$C(S_3) = 6(x_8 - 1.5)$$

$$C(S_3) = 6(1.875 - 1.5) = 2.25\text{M}$$

The estimated damage with no flood protection was estimated to be \$26.2M. The savings are

$$B(S_3) = \$26.2\text{M} - C(S_3)$$

$$B(S_3) = \$26.2\text{M} - \$2.25\text{M} = \$24.0\text{M}$$

The savings are assumed to accrue over the entire project life as depicted in the cash-flow diagram in Figure 4.27.

The net present worth is equal to the present worth of the benefit minus the capital cost of construction, or

$$NPW(S_3) = B(S_3)\left[\frac{(1+i)^n - 1}{i(1+i)^n}\right] - C$$

where $C = \$150\text{M}$, $i = 10$ percent and $n = 40$ yr, or

$$\left[\frac{(1+0.1)^{40} - 1}{0.1(1+0.1)^{40}}\right] = 9.78$$

$$NPW(S_3) = 9.78 \cdot 24.0\text{M} - \$150\text{M} = \$84.2\text{M}$$

the value $V(\mathbf{A}, S_3) = NPW(S_3) = \$84.2\text{M}$.

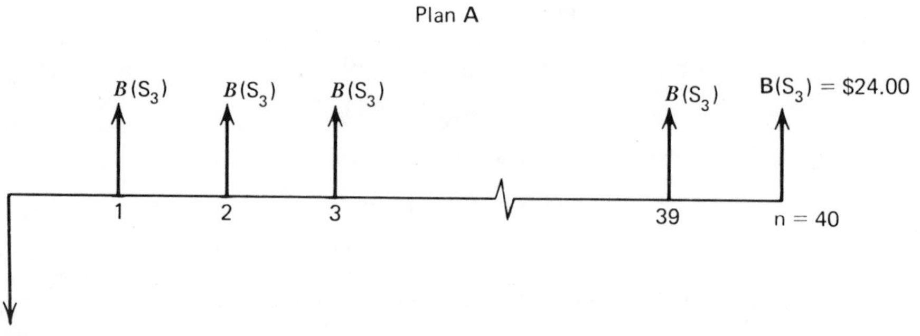

Figure 4.27

Plan A

STATE OF NATURE	MIDPOINT OF CLASS INTERVAL	PROBABILITY $P[S_j]$	DAMAGE COST $C(S_j) = A$	SAVINGS $B(S_j) = \$26.2M - C(S_j)$	$V(\mathbf{A}, S_j) = NPW(S_j) = 9.78B(S_j) - \$150M$
S_1	—	0.80	0	$26.2M	$106.2M
S_2	1.625	0.00	$0.75M	25.4M	98.4M
S_3	1.875	0.10	2.25M	24.0M	84.2M
S_4	2.125	0.03	3.75M	23.5M	69.5M
S_5	2.375	0.05	5.25M	21.0M	54.9M
S_6	2.625	0.02	6.75M	19.5M	40.2M
S_7	2.875	0.00	8.25M	18.0M	25.5M

Plan B

STATE OF NATURE	MIDPOINT OF CLASS INTERVAL	PROBABILITY $P[S_j]$	DAMAGE COST $C(S_j) = A$	SAVINGS $B(S_j) = \$26.2M - C(S_j)$	$V(\mathbf{B}, S_j) = NPW(S_j) = 9.78B(S_j) - \$200M$
S_1	—	0.90	0	$26.2M	$56.2M
S_2	2.125	0.03	$0.63M	25.6M	50.1M
S_3	2.375	0.05	1.88M	24.3M	37.9M
S_4	2.625	0.02	3.13M	23.1M	25.7M
S_5	2.875	0.00	4.38M	21.8M	13.5M

The decision tree is shown in Figure 4.28. The expected monetary values are

$$EMV^{A} = P[S_1]V(\mathbf{A}, S_1) + P[S_2]V(\mathbf{A}, S_2) + \cdots + P[S_7]V(\mathbf{A}, S_7)$$

$$EMV^{A} = 0.80 \cdot 106.2M + 0.0 \cdot 98.4M + \cdots + 0 \cdot 25.6M$$

$$EMV^{A} = \$99.0M$$

$$EMV^{B} = P[S_1]V(\mathbf{B}, S_1) + P[S_2]V(\mathbf{B}, S_2) + \cdots + P[S_5]V(\mathbf{B}, S_5)$$

$$EMV^{B} = 0.90 \cdot \$56.2M + 0.03 \cdot \$50.1M + \cdots + 0.00 \cdot \$13.5M$$

$$EMV^{B} = \$54.5M$$

Since $EMV^{A} > EMV^{B}$, choose plan **A**, the \$150M facility.

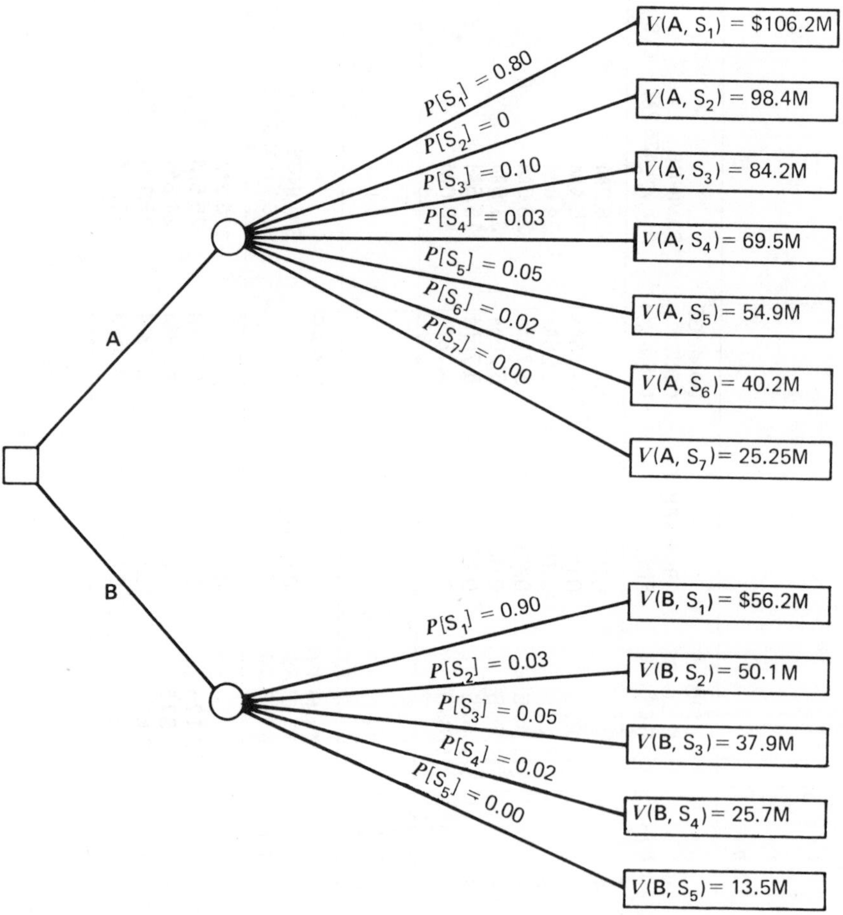

Figure 4.28

Remarks

The do-nothing alternative has been included in the analysis of plans **A** and **B**. If $EMV^A < 0$ and $EMV^B < 0$, they are both infeasible alternatives and no flood-protection system should be built. Here, both alternatives are feasible, but plan **A** offers a greater advantage over **B**; thus it should be constructed.

Weighing all factors of costs, benefits, and risks, plan **A** is more desirable than either plan **B** or the do-nothing alternative. Comparing plans **A** and **B** for given stage height, say $x = 2.125$, the estimated damage is \$3.75M for plan **A** and \$0.63M for plan **B**. Thus, if a flood of this magnitude occurs, plan **B** occurs greater protection. However, during the decision-making process we weigh all factors of benefits, costs, and risk. In this case, we find the extra capital cost required to build plan **B** outweighs the extra benefit of flood protection offered by this plan, thus plan **A** is chosen.

PROBLEMS

Problem 1

A building will be subject to wind loads. The probability that the wind will exceed 100 miles/hr is 0.01 percent in any year. It is estimated that the building damage will be \$50,000 if it is subject to a loading of this magnitude. No damage occurs for wind less than 100 miles/hr. In order to protect the building against 100-mile/hr or more winds a capital investment of \$25,000 is needed. If the wind exceeds 100 miles/hr, an estimated building damage of \$5,000 is expected. Utilize decision theory to determine the best alternative;

A. No capital investment.

B. \$25,000 capital investment.

The design life is 20 years and the opportunity-cost interest rate is 10 percent. Draw a decision tree.

Problem 2

Dice are rolled. The sum of the numerical values on the dice will determine if the person wins or loses the gamble. The winning outcomes of the roll of the dice, winnings, and losses are

CHOICE	WINNING OUTCOME	WINNINGS	LOSSES
A	Even number	\$500	−\$500
B	7	\$2500	−\$500
C	11	\$8500	−\$500
D	7 or 11	\$1875	−\$500

A decision tree for this problem is shown in Figure 4.29.
(a) Show the sample space for the roll of the die.
(b) Define the random events

$$S_1 = \{\text{Person wins}\}$$

$$S_2 = \{\text{Person loses}\}$$

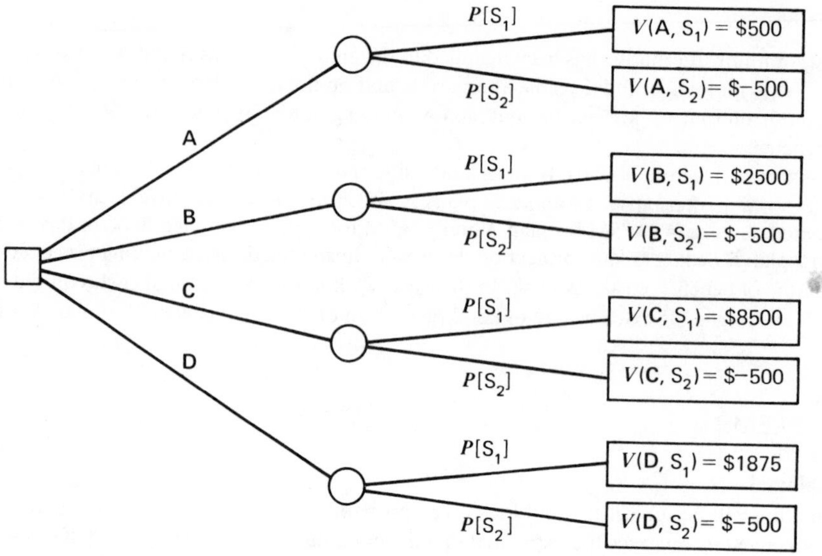

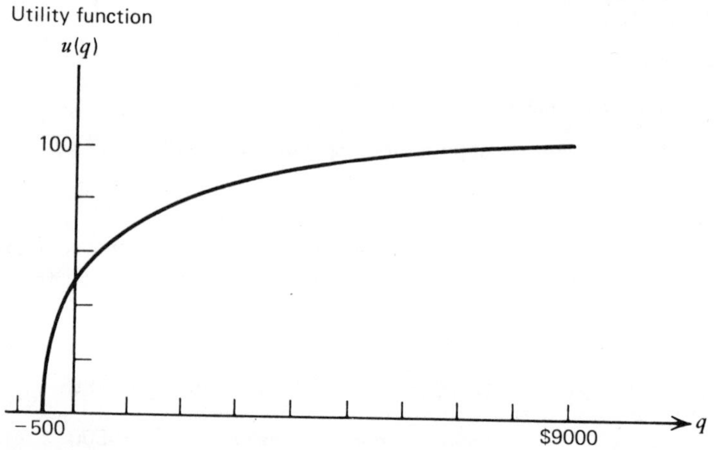

Figure 4.29

Determine the $P[S_1]$ and $P[S_2]$ for each gamble **A**, **B**, **C**, and **D**.

(c) Use EMV to determine the best gamble.

(d) Use the utility function and decision analysis to determine the best gamble.

(e) From part d, rank the gambles in order from the highest risk aversion to the least.

Problem 3

Two pumping systems supplying 10 Mgal/day as shown in Figure 4.30 are proposed. If the system fails, it is anticipated that the area receiving this water will suffer a \$50M annual loss. $P[F] = P[\text{pump failure}] = 0.03$. All failure events are assumed to be independent events.

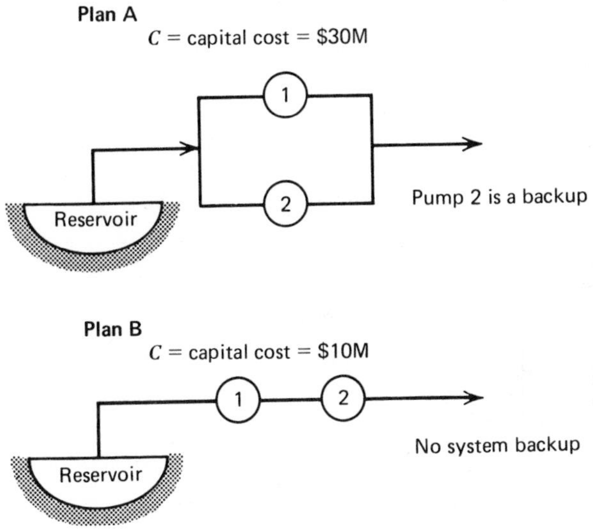

Figure 4.30

(a) Define the states of nature S_1 and S_2 for plans **A** and **B**.
(b) Determine the probabilities of S_1 and S_2 for plans **A** and **B**.
(c) Determine the present worth of losses including capital costs $V(\mathbf{A}, S_1)$, $V(\mathbf{A}, S_2)$, $V(\mathbf{B}, S_1)$, and $V(\mathbf{B}, S_2)$. Assume a design life of 15 years and an annual interest rate of 5 percent.
(d) Use decision analysis to determine the least-cost alternative.

Problem 4

Two mountain-top locations, sites **A** and **B**, have been selected for the possible installation of wind turbines to generate electricity to be sold at $0.05 kW-hr. A loan for wind turbines may be secured at an annual interest rate of 15 percent. The design life is 8 years. Wind speed data have been collected and analyzed and are presented in the form of frequency distributions in Figure 4.31. The two turbines will operate between 10 and 40 miles/hr. Below 10 miles/hr, there is insufficient wind to generate power, and above 40 miles/hr the machine is shut down to avoid wind damage.

(a) Estimate the percentage of time the turbines will be operating at each site.
(b) Estimate the number of hours per year the turbines will be operating at each site.
(c) Determine the percentage of time the turbines will be shut down because of high winds, that is, wind speeds greater than 40 miles/hr.
(d) Owing to the high percentage of time the turbine at site **B** is exposed to high winds, a more rugged machine will be installed there. The turbine at site **B** costs $40,000 and has a rated output of 40 kW. The turbine at site **A** has a cost of $50,000 and a rated output of 50 kW. The operation and maintenance costs are estimated at 2 percent of the capital cost at site **A** and 5 percent at site **B**. Estimate the annual revenue for operating the turbines at each site.
(e) Use decision analysis to determine the best alternative. Do not forget the do-nothing alternative.

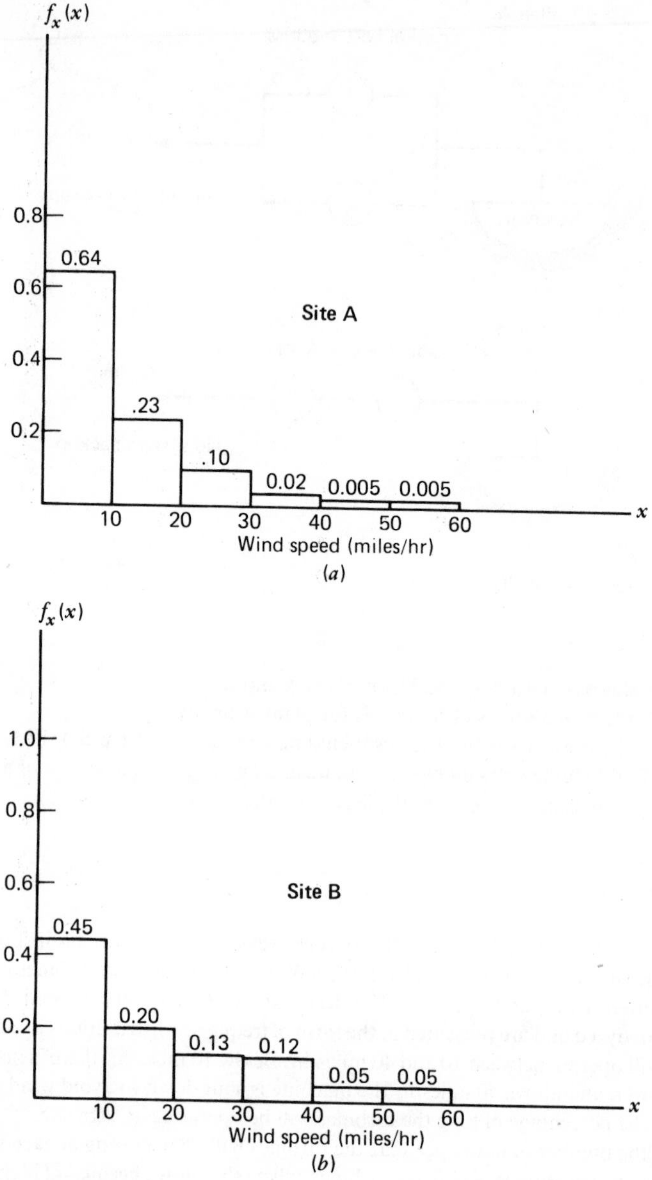
$f_x(x)$
0.8
0.64
0.6
Site A
0.4
.23
0.2
.10
0.02
0.005
0.005
10
20
30
40
50
60
x
Wind speed (miles/hr)
(a)
$f_x(x)$
1.0
0.8
0.6
Site B
0.45
0.4
0.20
0.2
0.13
0.12
0.05
0.05
10
20
30
40
50
60
x
Wind speed (miles/hr)
(b)

Figure 4.31

Problem 5

Two dams are being considered for construction.

A. $4M construction cost.

B. $6M construction cost.

For a serious flood it is assumed that there is a 10 percent chance the dam **A** will fail causing $10M in damages and a 5 percent chance that dam **B** will fail causing $15M in damages. If a major flood occurs, it is assumed that there is a 15 percent chance dam **A** will fail, causing $15M in damages, and there is an 8 percent chance that dam **B** will fail, causing $17M in damages. The probabilities of serious and major floods are estimated to be 5 percent and 1 percent, respectively. The design life of the project is 40 years, and the social opportunity-cost interest rate is 8 percent.

(a) Draw a decision tree and label it appropriately.

(b) Determine the better alternative to minimize losses.

Hint: Let $P[F_S] = P[\text{serious flood occurs in a 1-year period}] = 0.05$

$$P[F_M] = P[\text{major flood occurs in a 1-year period}] = 0.01$$

$$P[\text{Dam } \mathbf{A} \text{ fails} \mid F_S] = P[F_{\mathbf{A}} \mid F_S] = 0.1$$

$$P[\text{Dam } \mathbf{B} \text{ fails} \mid F_S] = P[F_{\mathbf{B}} \mid F_S] = 0.05$$

$$P[\text{Dam } \mathbf{A} \text{ fails} \mid F_M] = P[F_{\mathbf{A}} \mid F_M] = 0.15$$

$$P[\text{Dam } \mathbf{B} \text{ fails} \mid F_M] = P[F_{\mathbf{B}} \mid F_M] = 0.08$$

Problem 6

A town is currently being served by a water-pumping system supplying water at a rate of 10 Mgal/day. Population growth projections show that the town will require 20 Mgal/day. (See Figure 4.32.) The consequences of not delivering 20 Mgal/day are estimated to be an annual loss of:

$$\$250\text{M} \quad \text{if } Q = 0 \text{ Mgal/day}$$

$$\$25\text{M} \quad \text{if } Q = 10 \text{ Mgal/day}$$

$$\$0\text{M} \quad \text{if } Q = 20 \text{ Mgal/day}$$

Pump house 3 will be constructed to bring the total supply to a minimum of 20 Mgal/day. Two alternatives are considered.

A. Build a water-pumping system supplying 10 Mgal/day at $4M.

B. Build a water-pumping system supplying 20 Mgal/day at $7M.

For plan **A**, the new and old systems can deliver a total of 20 Mgal/day. For plan **B**, the new and old systems can deliver a total of 30 Mgal/day. Plan **B** offers an oversupply and security against failure of the old system. The likelihoods of failure are

$$P[F_1] = P[F_2] = 0.02$$

$$P[F_3] = 0.01$$

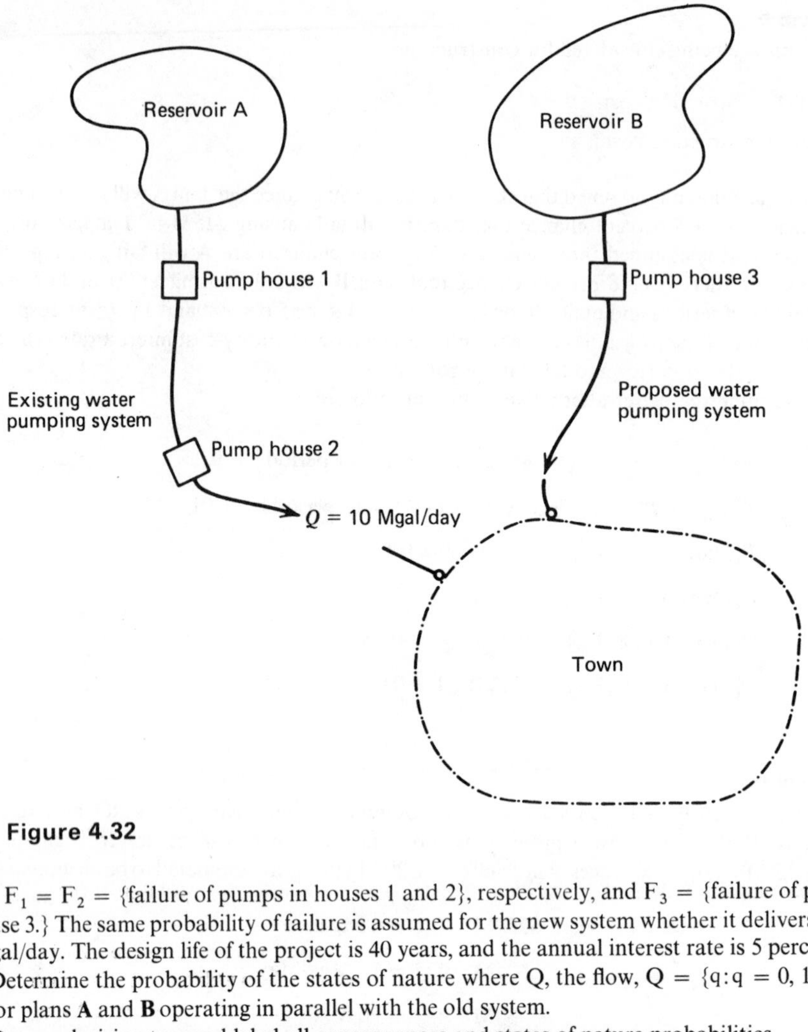

Figure 4.32

where $F_1 = F_2$ = {failure of pumps in houses 1 and 2}, respectively, and F_3 = {failure of pumps in house 3.} The same probability of failure is assumed for the new system whether it delivers 10 or 20 Mgal/day. The design life of the project is 40 years, and the annual interest rate is 5 percent.

(a) Determine the probability of the states of nature where Q, the flow, $Q = \{q : q = 0, 10, 20\}$ for plans **A** and **B** operating in parallel with the old system.

(b) Draw a decision tree and label all consequences and states of nature probabilities.

(c) Utilize decision theory to determine the better alternative system.

Problem 7

A contractor has the opportunity to bid on supplying sand and gravel for a construction job. The volume of available material in the borrow pit is not known. Assume that the borrow pit is a rectangular solid as shown in Figure 4.33. The contractor, however, estimates that there is a uniform probability that the depth of the borrow is between 10 and 20 ft.

$$p_X(x) = \tfrac{1}{10} \quad \text{for } 10 \leq x \leq 20$$

The contractor plans to make one of the following bids.

A. Bid on supplying 400 yd^3 of material.

B. Bid on supplying 600 yd^3 of material.

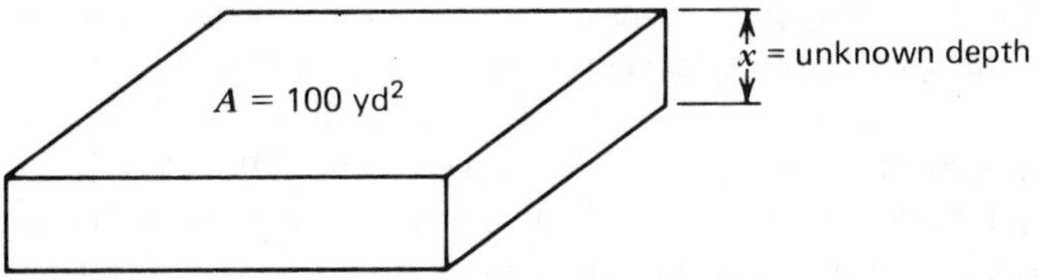

Figure 4.33

The profit received from alternative **A** depends on borrow pit depth. The contractor receives a \$1000 profit if the depth is 4 yd or 12 ft or more.

$$P = \$1000 \qquad x \geq 12 \text{ ft}$$

He receives a profit equal to revenue received from the sale of sand, \$284x, minus a fixed cost of \$2400.

$$P = 284x - 2400 \qquad x \leq 12 \text{ ft}$$

The profit received from alternative **B** is

$$P = \$1500 \qquad x \geq 18 \text{ ft}$$

$$P = 284x - 3600 \qquad x \leq 18 \text{ ft}$$

(a) Plot the profit versus depth for each alternative. Note that bid **A** is risk averse and bid **B** is not.
(b) Use decision analysis to determine the better alternative bid.

4.3 CONSIDERATIONS IN PROJECT PLANNING

In the preceding section, our discussion centered on the selection of alternative designs that provide a long-term benefit over some given design life. In this section and the following one we focus our attention upon the short-term decision-making problems most often found in construction and project planning.

Perfect and Imperfect Information

In construction and project planning, the use of historical records, laboratory and field experiments, and good engineering judgment can lead to the avoidance of penalty costs. For example, if insufficient geological data are obtained prior to construction, the design may prove to be unacceptable, resulting in additional project costs due to redesign, reconstruction, and delay. Our goal is to minimize penalties or opportunity costs such as these.

Field and laboratory experimentation may be used to estimate the *true state of nature*. If the true state of nature is known, an appropriate design may be constructed with no penalty charges to the project. For example, by test drilling the exact depth to firm strata may be determined, the proper bearing pile length may be ordered, and the piles may be driven at the construction site without penalty and delay costs. This kind of test is considered to give *perfect information* because there is no doubt or risk associated with selecting the proper pile length.

Less expensive tests may give *imperfect information*, but they may prove to be more cost effective than the more expensive perfect tests. Seismic testing is less expensive and less time consuming to conduct than test drilling. However, there are risks involved. The seismic test may give a false depth reading. If so, the cost of seismic tests and the redesign, reconstruction, and delay costs are the penalty. If the test results are correct, only the cost of the test must be paid.

On some occasions, no testing may prove to be more cost effective than perfect or imperfect testing. For the pile length determination, an investigation of similar foundations in the vicinity of the building site may give sufficient information. A construction engineer may also rely upon his experience and "gut feeling" in deciding upon the depth to bed rock. The engineer may assign odds or a probability value that he is correct within a certain tolerance. These probability assignments are called *subjective probability* measures. Obviously, the value of this information may be questionable. By use of imperfect test results, the confidence in these assignments may be improved. Our goal is to determine the following conditional and marginal probabilities.

Define the random events

$$S_i = \{\text{True state of nature } i\}$$

$$Z_j = \{\text{State of nature } j \text{ as indicated by the experiment}\}$$

Conditional probability

$P[S_i|Z_j]$ = Probability that the true state of nature is S_i given the outcome of an experiment Z_j

Marginal probability

$$P[Z_j] = \text{probability that the experimental outcome is } Z_j$$

These probabilities will be determined with the *Bayes theorem.* In order to choose among the perfect, imperfect, and no-test alternatives, the cost of testing, the value of the test information, and the penalty costs must be weighed also. The Bayes theorem will be discussed in this section and then introduced into the decision-making process in Section 4.4.

Bayes Theorem

The terms *prior* and *posterior probabilities* are used to designate the probability of the state of nature before and after a test or experiment. In this discussion, the words, "test" and "experiment" are used interchangeably. The following definitions will be used to define the random outcomes of the true state of nature and of experiments or tests.

Where the true state of nature and experiment outcomes are represented by the random variables:

Prior probability

$P[S_i]$ = The marginal probability that the true state of nature is the event S_i

Posterior probability

$P[S_i|Z_j]$ = The conditional probability the true state of nature is the event S_i given the outcome of an experiment or test is Z_j.

The prior probability is generally a subjective probability assignment. As previously stated, it is our objective to improve our confidence in these assignments by testing.

The Bayes theorem is derived from the definition of the probability of an intersection and the use of the definition of conditional probability.

$$P[S_i \cap Z_j] = P[Z_j \cap S_i]$$

$$P[S_i|Z_j]P[Z_j] = P[Z_j|S_i]P[S_i]$$

Rearranging the equation, the posterior probability is obtained.

$$P[S_i|Z_j] = \frac{P[Z_j|S_i]P[S_i]}{P[Z_j]}$$

From this equation, we can see that the posterior probability is a function of the *prior probability* $P[S_i]$, the probability that the true state of nature is S_i before observing the outcome of an experiment Z.

The conditional probability $P[Z_j|S_i]$ is called the *sample likelihood*. It is the probability that the test event Z_j occurs given the true conditional state of nature S_i.

The sample likelihood is generally determined by calibrating an instrument. Since the true state of nature is known during calibration, the sample likelihood is estimated as a relative frequency as discussed in Section 4.1.

The denominator of Bayes equation $P[Z_j]$ is called a marginal probability. It is determined with the use of the definition of marginal probability and the prior probabilities and sample likelihoods.

$$P[Z_j] = \sum_i P[Z_j \cap S_i] = \sum_i P[Z_j|S_i]P[S_i]$$

For decision making the conditional probability distribution $P[S_i|Z_j]$ and the marginal probability distribution $P[Z_j]$ are required. In order to determine these distributions, a prior or subjective probability $P[S_i]$ must be assigned and a sample likelihood distribution $P[Z_j|S_i]$ must be known. Once this information is known, the value of the test or experiment may be determined in terms of probability measures. In Section 4.4 the value of this information will be evaluated in terms of expected monetary values EMV. The use of the Bayes theorem is best illustrated by example.

EXAMPLE 4.5 Geological Field Exploration

To reduce the cost of construction, a civil engineer has investigated "as-built" drawings for pile foundations in a vicinity surrounding the construction site. Sixty percent of the existing sites have a depth to bedrock of 40 ft or less. The construction calls for eight piles as depicted in Figure 4.34. Based upon this information and his experience, the engineer assigns the following subjective probabilities that the depth to bedrock at locations A through H are 40 ft or less.

$$P[\text{depth} \leq 40] = 0.60$$

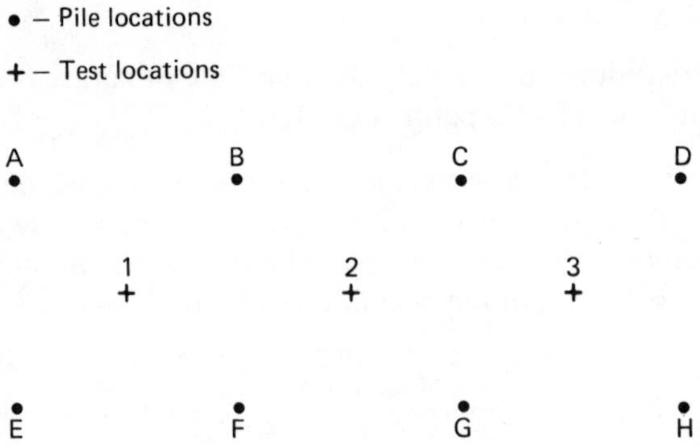

(*a*) Plan View of Site

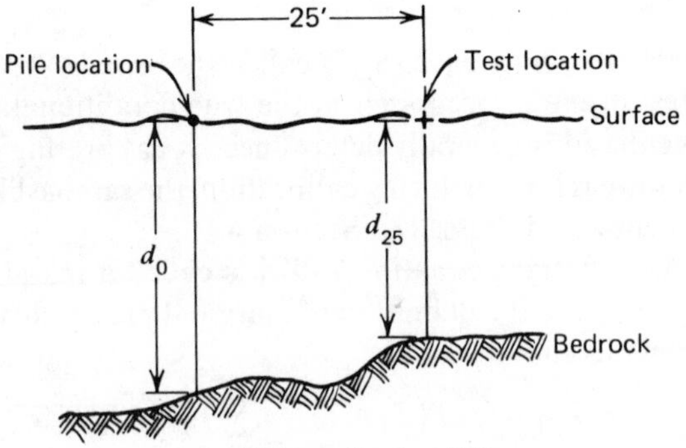

(*b*) Cross Sectional View

Figure 4.34

Based upon this information alone, there is a substantial risk that pile lengths of 40 or more feet are required. The engineer investigates the existing foundations and finds that there is a strong likelihood that the depth of bedrock at a 25-ft radius of the pile is the same. The following sample likelihoods were determined.

	$P[Z_i \mid S_i]$	
	S_1	S_2
Z_1	0.85	0.10
Z_2	0.15	0.90

where the states of nature and test outcomes are:

$$S_1 = \{\text{Depth of bedrock at a pile location is 40 ft or less}\} = \{D_0 \le 40 \text{ ft}\}$$
$$S_2 = \{\text{Depth of bedrock at a pile location is more than 40 ft}\} = \{D_0 > 40 \text{ ft}\}$$
$$Z_1 = \{\text{Depth of bedrock at a test location is 40 ft or less}\} = \{D_{25} \le 40 \text{ ft}\}$$
$$Z_2 = \{\text{Depth of bedrock at a test location is more than 40 ft}\} = \{D_{25} > 40 \text{ ft}\}$$

let D_0 be the random variable to describe the depth of the bedrock at the pile location

$$D_0 = \{d_0 : d_0 \ge 0\}$$

and D_{25} be a random variable to describe the depth of the bedrock at test location

$$D_{25} = \{d_{25} : d_{25} \ge 0\}$$

Three soil borings are made at the test locations marked 1, 2, and 3. These locations are 25 ft from the nearest pile locations. The following observations were made.

LOCATION	DEPTH, d_{25} (ft)	OUTCOME, Z_i
1	35	Z_1
2	32	Z_1
3	30	Z_1

(a) Given these test results, determine the probability that the depth to bedrock at point A is 40 ft or less.

(b) Given these test results determine the probability that the depth to bedrock of point B is 40 ft or less. Utilize the test at adjacent test locations in this determination.

Solution

(a) The Bayes theorem will be used to calculate whether the depth of bedrock at location A is 40 ft or less. Since the depth to bedrock at test location 1 is less than 40 ft, the conditional probability $P[S_1|Z_1]$ will be determined. From the Bayes theorem,

$$P[S_1|Z_1] = \frac{P[Z_1|S_1]P[S_1]}{P[Z_1]}$$

or

$$P[S_1|Z_1] = \frac{P[Z_1|S_1]P[S_1]}{P[Z_1|S_1]P[S_1] + P[Z_1|S_2]P[S_2]}$$

Since $P[S_2] = 1 - P[S_1] = 1 - 0.60 = 0.40$, the condition probability is

$$P[S_1|Z_1] = \frac{(0.85)(0.60)}{(0.85)(0.60) + (0.10)(0.40)} = 0.927$$

(b) The results from two test locations may be used to find the probability that the bedrock depth at location B is 40 ft or less because test locations 1 and 2 are 25 ft from pile location B. Utilizing the Bayes theorem and the fact that the test results at locations 1 and 2 have depths less than 40 ft, the following relationship may be utilized.

$$P[S_1|Z_1 \cap Z_1] = \frac{P[Z_1 \cap Z_1|S_1]P[S_1]}{P[Z_1 \cap Z_1]}$$

However, the experimental results are assumed to be independent; thus,

$$P[Z_1 \cap Z_1 | S_1] = P[Z_1 | S_1]P[Z_1 | S_1] = P[Z_1 | S_1]^2$$

$$P[Z_1 \cap Z_1 | S_1] = (0.85)^2 = 0.7225$$

The marginal probability is

$$P[Z_1 \cap Z_1] = P[Z_1 \cap Z_1 | S_1]P[S_1] + P[Z_1 \cap Z_1 | S_2]P[S_2]$$

Again, since the two test borings at locations 1 and 2 are assumed to be independent observations, the marginal probability of intersection ($Z_1 \cap Z_1$) is

$$P[Z_1 \cap Z_1] = P[Z_1 | S_1]P[Z_1 | S_1]P[S_1] + P[Z_1 | S_2]P[Z_1 | S_2]P[S_2]$$

$$P[Z_1 \cap Z_1] = P[Z_1 | S_1]^2 P[S_1] + P[Z_1 | S_2]^2 P[S_2]$$

$$P[Z_1 \cap Z_1] = (0.85)^2(60) + (0.10)^2(0.40)$$

$$P[Z_1 \cap Z_1] = 0.4335 + 0.0040 = 0.4375$$

Thus,
$$P[S_1 | Z_1 \cap Z_1] = \frac{(0.7225)(0.60)}{0.4375} = 0.991$$

This test program shows that there is a better than 99 percent chance that the bedrock depth will be less than 40 ft.

EXAMPLE 4.6 A Traffic-Counting Experiment

A total of 170 observations were made by an individual who is assumed not to have miscounted. The observer is assumed to give perfect test information. At the same time, a mechanical traffic counter, which has a pneumatic tube placed across the road, records the traffic also. (See Figure 4.35.) The following random variables have been assigned.

$$S_i = \{\text{The person counts } x \text{ vehicles in a 30-sec time period}\} = \{x: x = 0, 1, 2, 3\}$$

$$Z_j = \{\text{The machine records } y \text{ vehicles in a 30-sec time period}\} = \{y: y = 0, 1, 2, 3\}$$

Corresponding 30-sec time period record counts of the human observer and machine are compared. The observer counted one vehicle S_1, in 30 time periods, $n_1 = 30$. The machine, on the other hand, recorded only one vehicle 26 times, Z_1. It undercounted 4 times, Z_0. These results are

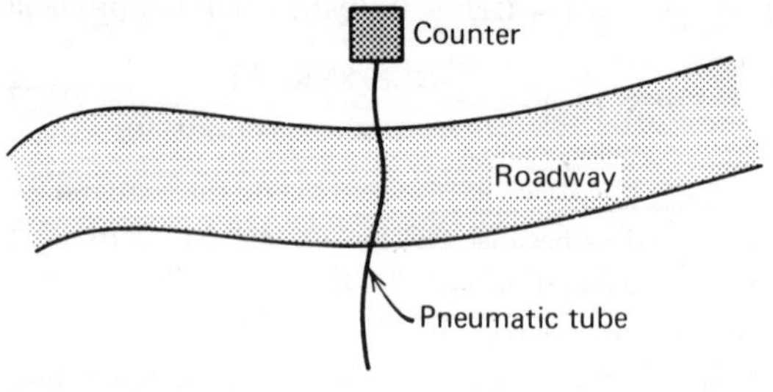

Figure 4.35

shown in column 1 of the following table. The data show that the machine is not fully reliable for S_2 and S_3. The counter will give the correct count or undercount, but it will never overcount.

		Person count		
		S_1	S_2	S_3
Traffic counter	Z_0	4	3	0
	Z_1	26	7	5
	Z_2	0	50	15
	Z_3	0	0	60
	Totals	30	60	80

(a) Estimate the sample likelihood distribution $P[Z_j|S_i]$. Assume $P[Z_0|S_0] = 1.0$ and $P[Z_1|S_0] = P[Z_2|S_0] = P[Z_3|S_0] = 0$.
(b) Estimate the probability distribution of the true count given that 1 vehicle is observed. Assume that there is an equally likely chance that the true count is 0, 1, 2, or 3 vehicles.
(c) In order to increase the confidence in the forecast, two traffic counters are used. One traffic counter records zero vehicles, and the second one records one vehicle. Determine the probability that one vehicle is observed given these observations.

Solution

(a) The sample likelihood distribution is estimated as a relative frequency measure. For example, the conditional probability of recording 1 vehicle, given that one person counts 1 vehicle, is determined as

$$P[Z_1|S_1] = \frac{26}{30} = 0.867$$

The sample likelihood distribution for all outcomes is found in a similar manner. They are

	$P[Z_j|S_i]$			
	S_0	S_1	S_2	S_3
Z_0	1	0.133	0.050	0.000
Z_1	0	0.867	0.117	0.063
Z_2	0	0.00	0.833	0.187
Z_3	0	0.00	0.000	0.750

(b) The probability of true count given a traffic counter recording is determined with the Bayes theorem.

$$P[S_i|Z_j] = \frac{P[Z_j|S_i]P[S_i]}{P[Z_j]} \text{ for } i = 0, 1, 2, 3 \text{ and } j = 0, 1, 2, 3$$

where the marginal probability is

$$P[Z_j] = \sum_{i=0}^{3} P[Z_j|S_i]P[S_i]$$

For a machine count of 1 vehicle, $Z_1 = 1$, the marginal probability is

$$P[Z_1] = P[Z_1|S_0]P[S_0] + P[Z_1|S_1]P[S_1] + P[Z_1|S_2]P[S_2] + P[Z_1|S_3]P[S_3]$$

where the prior probabilities are the subjective probabilities

$$P[S_0] = P[S_1] = P[S_2] = P[S_3] = \tfrac{1}{4} = 0.25$$

Thus, $$P[Z_1] = (0.000)(0.25) + (0.867)(0.25) + (0.117)(0.25) + (0.063)(0.25)$$

$$P[Z_1] = 0.262$$

The probability distribution given that one vehicle is recorded is

$$P[S_0|Z_1] = \frac{P[Z_1|S_0]P[S_0]}{P[Z_1]} = \frac{(0.000)(0.25)}{0.262} = 0.000$$

$$P[S_1|Z_1] = \frac{P[Z_1|S_1]P[S_1]}{P[Z_1]} = \frac{(0.867)(0.25)}{0.262} = 0.828$$

$$P[S_2|Z_1] = \frac{P[Z_1|S_2]P[S_2]}{P[Z_1]} = \frac{(0.117)(0.25)}{0.262} = 0.112$$

$$P[S_3|Z_1] = \frac{P[Z_1|S_3]P[S_3]}{P[Z_1]} = \frac{(0.063)(0.25)}{0.262} = 0.060$$

Even though the traffic counter is an imperfect testing machine, we are certain that the true traffic count is not 0 vehicles. There is a chance that the machine undercounted and the actual count is 2 or 3 vehicles. Obviously, our confidence has increased dramatically from a probability of 25 percent to 82.7 percent. There is a value in utilizing the traffic counter even though it is not a perfect counter.

(c) Assuming that both traffic counters behave identically, the conditional probability that the true count is 1 vehicle given that the traffic counters record 0 and 1 vehicle is

$$P[S_1|Z_0 \cap Z_1] = \frac{P[Z_0 \cap Z_1|S_1]P[S_1]}{P[Z_0 \cap Z_1]}$$

The traffic count observations are independent events, thus,

$$\begin{aligned} P[Z_0 \cap Z_1|S_1] &= P[Z_0|S_1]P[Z_1|S_1] \\ &= (0.133)(0.867) = 0.115 \end{aligned}$$

The marginal probability is

$$\begin{aligned} P[Z_0 \cap Z_1] = {} & P[Z_0|S_0]P[Z_1|S_0]P[S_0] + P[Z_0|S_1]P[Z_1|S_1]P[S_1] \\ & + P[Z_0|S_2]P[Z_1|S_2]P[S_2] + P[Z_0|S_3]P[Z_1|S_3]P[S_3] \end{aligned}$$

$$\begin{aligned} P[Z_0 \cap Z_1] = {} & (1)(0)(0.25) + (0.133)(0.867)(0.25) \\ & + (0.050)(0.117)(0.25) + (0.000)(0.063)(0.25) \end{aligned}$$

$$P[Z_0 \cap Z_1] = 0.030$$

The conditional probability is

$$P[S_1 | Z_0 \cap Z_1] = \frac{(0.115)(0.25)}{0.030} = 0.958$$

Remarks

Even though the two traffic counters give inconsistent readings, there is value in using two recorders instead of one. From parts a and b, the probabilities are

$$P[S_1 | Z_1] = 0.867$$

$$P[S_1 | Z_0 \cap Z_1] = 0.958$$

One traffic recorder undercounted; therefore, more weight is placed upon the higher count. If both recorders give the same reading of 1 vehicle, the probability increases slightly.

$$P[S_1 | Z_1 \cap Z_1] = 0.979$$

The $P[S_1 | Z_1 \cap Z_1]$ is not equal to 1, because there is a chance that both recorders undercounted on the same event. The traffic recorder will never overcount, but there is always the chance it will undercount. These results show the advantage of repeat experiments.

PROBLEMS

Problem 1

The seismic test is an imperfect test with a sample likelihood of:

	$P[Z_j \| S_i]$	
	S_1	S_2
Z_1	0.60	0.30
Z_2	0.30	0.70
Z_3	0.10	0.00

where Z_j = {test outcome 1, 2, 3} and S_i = {true state of nature, 1, 2}

(a) Are the probabilities $P[Z_1 | S_1]$, $P[Z_2 | S_1]$, and $P[Z_3 | S_1]$ mutually exclusive or independent random events?

(b) Given S_1, what is the probability that the seismic test is either Z_1 or Z_2?

(c) Determine the marginal probability of Z_2 given that the subjective probabilities of S_1 and S_2 are equal to 0.25 and 0.75, respectively.

Problem 2

Show that for perfect testing the posterior probability function is:

$$P[S_i | Z_j] = \begin{cases} 1.0 & \text{for } i = j \\ 0.0 & \text{for } i \neq j \end{cases}$$

Problem 3

A chemical test used in the environmental laboratory has been calibrated. The sample likelihood is:

	$P[Z_i \mid S_j]$		
	S_1	S_2	S_3
Z_1	0.90	0.15	0.10
Z_2	0.10	0.70	0.30
Z_3	0.00	0.15	0.60

S_1, S_2, S_3 = {low, medium, and high concentration levels of samples, respectively.}

Z_1, Z_2, Z_3 = {low, medium, and high concentration test readings, respectively.}

(a) There is an equally likely chance that the true state of nature is S_1, S_2, or S_3. Assign subjective probabilities to these states.
(b) A sample is analyzed by the chemical test, and the test indicates a high concentration level. Determine the posterior probability that the true state of nature is a high concentration.
(c) A second sample is analyzed, and it indicates a high concentration level. Determine the posterior probability assuming the two tests are independent events.

Problem 4

Use the problem description and sample likelihoods of Example 4.5 to answer this question. The soil borings at the locations marked + are as follows:

LOCATION	DEPTH, d_{25}
1	45 ft
2	42 ft
3	35 ft

(a) Calculate the probability that the depth at pile location A to bedrock is 40 ft or less given the above boring test results.
(b) Determine the probability that the depth to bedrock at pile location B is 40 ft or less.
(c) Determine the probability that the depth to bedrock at pile location C is 40 ft or less.
(d) Prepare a table comparing the probabilities that the depth to bedrock is 40 ft or less for all pile locations A through H.

Problem 5

Use the problem description and sample likelihoods given in Example 4.6 to answer this question. Assume $P[S_0] = P[S_1] = P[S_2] = P[S_3] = 0.25$.

(a) Given that no vehicles are recorded by the traffic counter, determine the probabilities that the true number of vehicles is 0, 1, 2, and 3.
(b) Assume two traffic counters are used. One recorder records 0, and the other one records 2. Determine the probabilities that 0, 1, 2, and 3 are the true state of nature.

4.4 THE VALUE OF EXPERIMENTATION AND TESTING

In the preceding sections, we found that testing can build our confidence in a design or construction plan. In this section we determine the value of this information and whether an experiment or test should be incorporated into an engineering project. Both test cost and penalty costs are important. Our problem is to determine if a test should be conducted or not and, if it should, which test is the most cost-effective one. Our selection method is an extension of the EMV and utility theory methods discussed in Section 4.2. Here our objective is to select the alternative with the minimum opportunity cost.

Decision Tree

Consider three types of test alternatives.

E_1 No test or experiment.

E_2 Perform an imperfect test or experiment.

E_3 Perform a perfect test or experiment.

In addition, we must select from competing design alternatives, **A** and **B**. The decision tree for three test alternatives E_1, E_2, and E_3, two design alternatives **A** and **B**, and two states of nature S_1 and S_2 are shown in Figure 4.36. The outcomes of the experiments are shown as Z_1 and Z_2.

In order to demonstrate how a sequential decision tree is used in alternative selection we describe each of its components in the following subsections. Our procedure consists of the following steps.

1. Determine the opportunity costs V.
2. Calculate the subjective probabilities, and posterior probabilities when appropriate, for the states of nature S_1 and S_2.
3. Based upon the probabilities of step 2, determine the EMV for each decision node marked d_1, d_2, and d_3.
4. At the first decision nodes d_1, d_2, and d_3, select the better alternative **A** or **B**. These are the minimum opportunity-cost selections.
5. Use the probabilities for test states Z_1 and Z_2 to determine the EMV for the better alternatives found at nodes d_1, d_2, d_3, and for E_3.
6. By using the minimum opportunity cost criteria, determine the best alternative E_1, E_2, or E_3.

Opportunity Costs

The opportunity costs shown in the sequential decision tree and required by step 1 consist of the cost of testing plus the penalty costs that are associated with time delay, redesign, and reconstruction. For the no-experiment alternative E_1, there are no costs associated with tests, only penalty costs. For the imperfect testing alternative E_2, the opportunity costs consist of both testing and penalty costs. For alternative E_3, the opportunity cost is the test cost only. Generally, it is easier to determine a payoff matrix and then determine the opportunity-cost matrix.

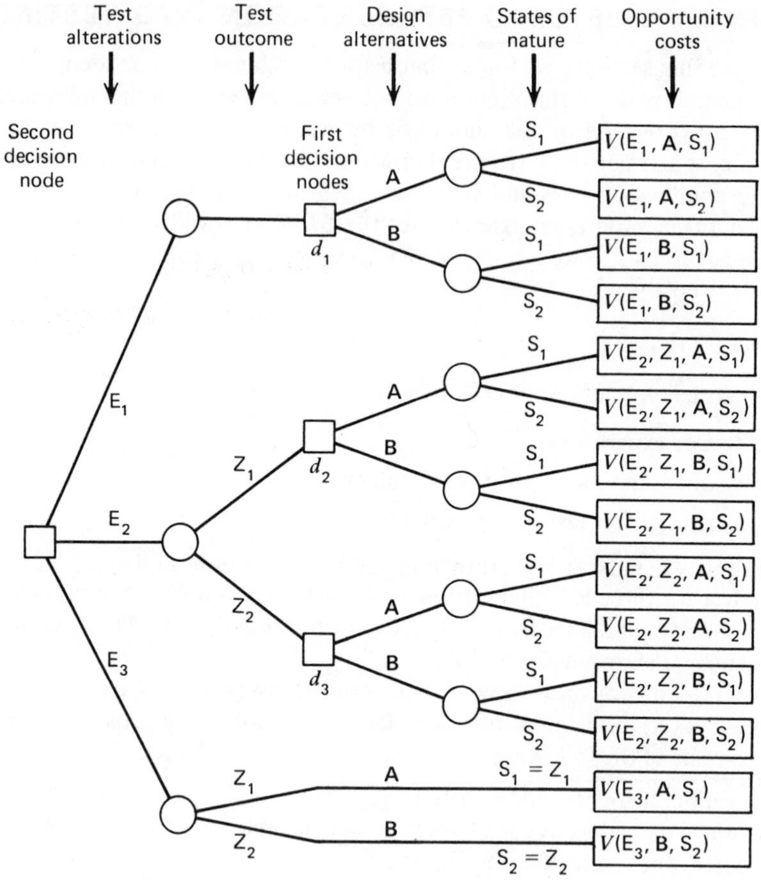

Figure 4.36 Sequential decision tree incorporating test and design alternative selection.

Expected Monetary Value

In step 2, the subjective probabilities and posterior probabilities are calculated as described in Section 4.3. For alternative E_1 and decision node d_1, the EMVs for **A** and **B** are

$$\text{EMV}^{\mathbf{A}} = V(\text{E}_1, \mathbf{A}, \text{S}_1)P[\text{S}_1] + V(\text{E}_1, \mathbf{A}, \text{S}_2)P[\text{S}_2]$$

$$\text{EMV}^{\mathbf{B}} = V(\text{E}_1, \mathbf{B}, \text{S}_1)P[\text{S}_1] + V(\text{E}_1, \mathbf{B}, \text{S}_2)P[\text{S}_2]$$

The alternative having the minimum EMV value is the better choice.

$$\text{EMV}^{\text{E}_1} = \min(\text{EMV}^{\mathbf{A}}, \text{EMV}^{\mathbf{B}})$$

For alternative E_2 and decision node d_2, the EMVs are based upon the subjective probabilities of $P[S_1|Z_1]$ and $P[S_2|Z_1]$. Thus,

$$EMV^{\mathbf{A}|Z_1} = V(E_2, Z_1, \mathbf{A}, S_1]P[S_1|Z_1] + V(E_2, Z_1, \mathbf{A}, S_2)P[S_2|Z_1]$$

$$EMV^{\mathbf{B}|Z_1} = V(E_2, Z_1, \mathbf{B}, S_1)P[S_1|Z_1] + V(E_2, Z_1, \mathbf{B}, S_2)P[S_2|Z_1]$$

The alternative having the minimum EMV is the better choice.

$$EMV^{E_2|Z_1} = \min(EMV^{\mathbf{A}|Z_1}, EMV^{\mathbf{B}|Z_1})$$

The same procedure is followed at node d_3.

$$EMV^{\mathbf{A}|Z_2} = V(E_2, Z_2, \mathbf{A}, S_1)P[S_1|Z_2] + V(E_2, Z_2, \mathbf{A}, S_2)P[S_2|Z_2]$$

$$EMV^{\mathbf{B}|Z_2} = V(E_2, Z_2, \mathbf{B}, S_1)P[S_1|Z_2] + V(E_2, Z_2, \mathbf{B}, S_2)P[S_2|Z_2]$$

The better choice is

$$EMV^{E_2|Z_2} = \min(EMV^{\mathbf{A}|Z_2}, EMV^{\mathbf{B}|Z_2})$$

For the perfect test, alternative E_3, there are theoretically no first decision nodes; therefore, no evaluation is made for alternative E_3. However, the opportunity costs $V(E_3, \mathbf{A}, S_1)$ and $V(E_3, \mathbf{B}, S_2)$ are equal because they equal the cost of testing only and are independent of the states of nature S_1 and S_2.

$$EMV^{E_3} = V(E_3, \mathbf{A}, S_1) = V(E_3, \mathbf{B}, S_2)$$

The results of performing steps 1 through 4 are summarized in Figure 4.37. The selection steps 5 and 6 for the decision tree are the same ones followed in Section 4.2. For alternative E_2, the EMV is

$$EMV^{E_2} = EMV^{E_2|Z_1}P[Z_1] + EMV^{E_2|Z_2}P[Z_2]$$

where $P[Z_1]$ and $P[Z_2]$ are the marginal probabilities.

The best selection is the one associated with minimum expected opportunity cost EMV*. EMV* = $\min(EMV^{E_1}, EMV^{E_2}, EMV^{E_3})$. If EMV = EMV^{E_1}, then the value of experimental tests does not provide sufficient information to make them cost effective or competitive. If EMV* = EMV^{E_3}, we avoid all risk by employing perfect testing. Here the risks exceed the cost of perfect testing. If EMV* = EMV^{E_2}, imperfect tests are employed.

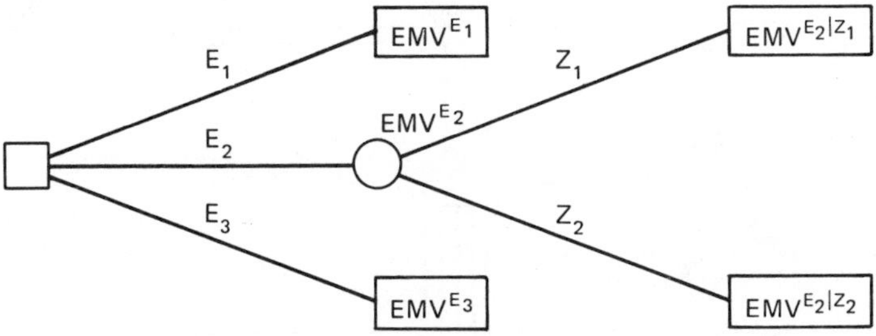

Figure 4.37 Intermediate decision tree.

Again, we must remind ourselves that our selection process utilizes expected values. Only perfect testing avoids risk. If either no testing or imperfect testing is determined as the best choice using EMV theory, there is possibility a major penalty or opportunity cost will be paid. The EMV value is a weighted value and is always less than the opportunity costs found in the payoff matrix.

EXAMPLE 4.7 A Bridge Pier Foundation

The central pier foundation of a bridge is to be supported on a foundation composed of 10 bearing piles. (See Figure 4.38.) Two pile types are available.

Precast concrete pile at $8/ft including driving.

Steel H pile at $10/ft including driving and splicing.

From geological field data, it is estimated that there is equal likelihood that the depth to firm strata is between 35 and 40 ft or between 40 and 50 ft. A 40-ft pile will be sufficient for a depth range of 35 to 40 ft and a 50-ft pile will be sufficient for a range of 40 to 50 ft. For simplicity, no other possibilities are assumed to exist.

Precast concrete piles must be ordered in advance. If the piles are too short, the delay cost is estimated to be $7000. In order to reduce the difficulty of shipping long steel piles, 25-ft piles will be ordered and spliced on site. If the 50-ft steel and concrete piles are too long, they may be cut at an estimated cost of $1000 and $2000, respectively.

Field testing may be employed. Consider the following three test alternatives:

E_1 No testing.

E_2 Seismic testing at $500.

E_3 Test borings at $4000.

The seismic tests are imperfect tests with the following sample likelihood.

	SAMPLE LIKELIHOOD $P[Z_i \mid S_j]$	
	S_1	S_2
Z_1	0.85	0.20
Z_2	0.15	0.80

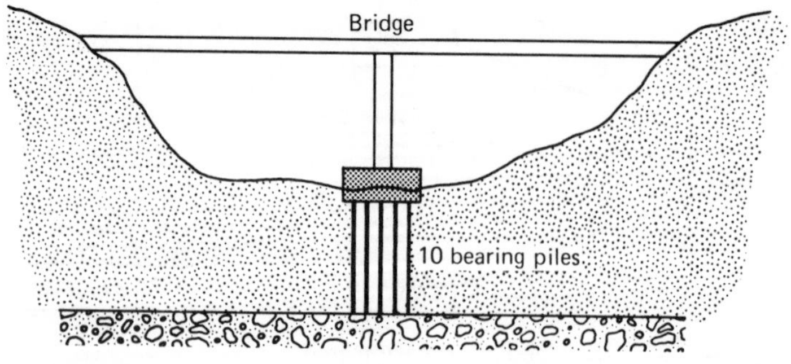

Figure 4.38

where the states of nature and test outcomes are

$$S_1 = \{\text{Actual depth to firm strata is 35 to 40 ft}\}$$

$$S_2 = \{\text{Actual depth to firm strata is 40 to 50 ft}\}$$

$$Z_1 = \{\text{Indicated depth to firm strata is 35 to 40 ft}\}$$

$$Z_2 = \{\text{Indicated depth to firm strata is 40 to 50 ft}\}$$

Consider three alternatives.

A. Order ten 40-ft precast concrete members at $3200. ($8/ft · 10 piles · 40 ft)
B. Order ten 50-ft precast concrete members at $4000. ($8/ft · 10 piles · 50 ft)
C. Order ten 50-ft steel H piles at $5000. ($10/ft· 10 piles · 50 ft)

(a) Draw a sequential decision tree including the test alternatives of no testing, imperfect seismic testing, and perfect test boring and the three pile length alternatives **A, B,** and **C**.
(b) Establish a payoff matrix of total construction, delay, and testing costs.
(c) Establish a payoff matrix of opportunity costs.
(d) Use the Bayes theorem to estimate posterior probabilities and the other likelihood needed in the analysis.
(e) Use decision theory to determine the best alternative.

Solution

(a) A decision tree has been constructed as shown in Figure 4.39. The monetary value of the consequence V has three or four arguments to emphasize that it is a function of the cost of experiment E_1, E_2, or E_3, the outcome of the experiment Z_1 or Z_2, the alternative **A**, **B**, or **C**, and the true state of nature S_1 or S_2. Opportunity costs are given in the decision tree. The costs are found in Table 4.3 and are calculated in the following parts (b) and (c).

Table 4.1 Itemized Costs

CONSEQUENCE	COMMENTS PILES ARE	COSTS DELAY	CONSTRUCTION	TEST	TOTAL
$V(E_1, \mathbf{A}, S_1)$	Correct length	—	$3,200	—	$3,200
$V(E_1, \mathbf{A}, S_2)$	Too short	$7,000	$3,200	—	$10,200
$V(E_1, \mathbf{B}, S_1)$	Too long	$2,000	$4,000	—	$6,000
$V(E_1, \mathbf{B}, S_2)$	Correct length	—	$4,000	—	$4,000
$V(E_1, \mathbf{C}, S_1)$	Too long	$1,000	$5,000	—	$6,000
$V(E_1, \mathbf{C}, S_2)$	Correct length	—	$5,000	—	$5,000
$V(E_2, Z_1, \mathbf{A}, S_1)$	Correct length	—	$3,200	$500	$3,700
$V(E_2, Z_1, \mathbf{A}, S_2)$	Too short	$7,000	$3,200	$500	$10,700
$V(E_2, Z_1, \mathbf{B}, S_1)$	Too long	$2,000	$4,000	$500	$6,500
$V(E_2, Z_1, \mathbf{B}, S_2)$	Correct length	—	$4,000	$500	$4,500
$V(E_2, Z_1, \mathbf{C}, S_1)$	Too long	$1,000	$5,000	$500	$6,500
$V(E_2, Z_1, \mathbf{C}, S_2)$	Correct length	—	$5,000	$500	$5,500
$V(E_2, Z_2, \mathbf{A}, S_1)$	Correct length	—	$3,200	$500	$3,700
$V(E_2, Z_2, \mathbf{A}, S_2)$	Too short	$7,000	$3,200	$500	$10,700
$V(E_2, Z_2, \mathbf{B}, S_1)$	Too long	$2,000	$4,000	$500	$6,500
$V(E_2, Z_2, \mathbf{B}, S_2)$	Correct length	—	$4,000	$500	$4,500
$V(E_2, Z_2, \mathbf{C}, S_1)$	Too long	$1,000	$5,000	$500	$6,500
$V(E_2, Z_2, \mathbf{C}, S_2)$	Correct length	—	$5,000	$500	$5,500
$V(E_3, Z_1, \mathbf{A}, S_1)$	Correct length	—	$3,200	$4,000	$7,200
$V(E_3, Z_2, \mathbf{B}, S_2)$	Correct length	—	$4,000	$4,000	$8,000

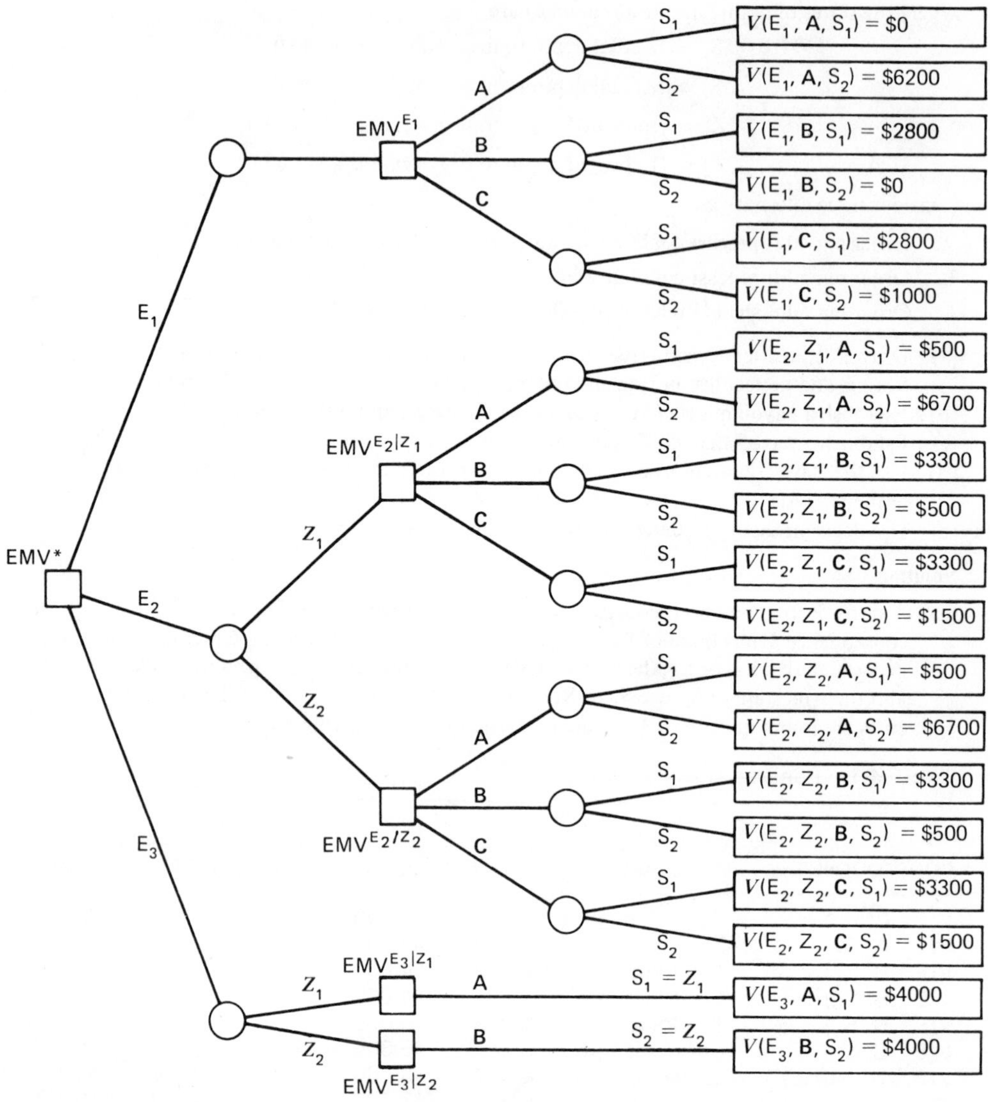

Figure 4.39

(b) The total cost figures found in Table 4.1 have been rearranged and placed in the appropriate column S_1 or S_2 of the payoff matrix of total costs marked as Table 4.2.

(c) The payoff matrix of opportunity cost of Table 4.3 is determined by subtracting \$3200 from every entry in column 1 of Table 4.2 and by subtracting \$4000 from every entry of column 2 of Table 4.2.

(d) The preposterior or subjective probabilities are equally likely; therefore, they are assumed to be

$$P[S_1] = 0.5, \qquad P[S_2] = 0.5$$

Table 4.2 Total Costs

	STATES OF NATURE	
CONSEQUENCE	S_1	S_2
$V(E_1, \mathbf{A}, S_i)$	3,200	10,200
$V(E_1, \mathbf{B}, S_i)$	6,000	4,000
$V(E_1, \mathbf{C}, S_i)$	6,000	5,000
$V(E_2, Z_1, \mathbf{A}, S_i)$	3,700	10,700
$V(E_2, Z_1, \mathbf{B}, S_i)$	6,500	4,500
$V(E_2, Z_1, \mathbf{C}, S_i)$	6,500	5,500
$V(E_2, Z_2, \mathbf{A}, S_i)$	3,700	10,700
$V(E_2, Z_2, \mathbf{B}, S_i)$	6,500	4,500
$V(E_2, Z_2, \mathbf{C}, S_i)$	6,500	5,500
$V(E_3, Z_1, \mathbf{A})$	7,200	—
$V(E_3, Z_2, \mathbf{B})$	—	8,000

Table 4.3 Opportunity Cost

	STATES OF NATURE	
CONSEQUENCE	S_1	S_2
$V(E_1, \mathbf{A}, S_i)$	$0	$6,200
$V(E_1, \mathbf{B}, S_i)$	2,800	0
$V(E_1, \mathbf{C}, S_i)$	2,800	1,000
$V(E_2, Z_1, \mathbf{A}, S_i)$	500	6,700
$V(E_2, Z_1, \mathbf{B}, S_i)$	3,300	500
$V(E_2, Z_1, \mathbf{C}, S_i)$	3,300	1,500
$V(E_2, Z_2, \mathbf{A}, S_i)$	500	6,700
$V(E_2, Z_2, \mathbf{B}, S_i)$	3,300	500
$V(E_2, Z_2, \mathbf{C}, S_i)$	3,300	1,500
$V(E_3, Z_1, \mathbf{A})$	4,000	—
$V(E_3, Z_2, \mathbf{B})$	—	4,000

The probability outcomes of the imperfect experiment Z_1 and Z_2 are estimated with marginal probabilities.

$$P[Z_1] = P[Z_1|S_1]P[S_1] + P[Z_1|S_2]P[S_2]$$

$$P[Z_1] = (0.85)(0.5) + (0.20)(0.5) = 0.525$$

$$P[Z_2] = 1 - P[Z_1] = 1 - 0.525 = 0.475$$

From the Bayes theorem, the posterior or conditional probabilities can be determined.

$$P[S_1|Z_1] = \frac{P[Z_1|S_1]P[S_1]}{P[Z_1]} = \frac{(0.85)(0.5)}{0.525} = 0.810$$

$$P[S_2|Z_1] = 1 - P[S_1|Z_1] = 1 - 0.810 = 0.190$$

$$P[S_1|Z_2] = \frac{P[Z_2|S_1]P[S_1]}{P[Z_2]} = \frac{(0.15)(0.5)}{0.475} = 0.158$$

$$P[S_2|Z_2] = 1 - P[S_1|Z_2] = 1 - 0.158 = 0.842$$

(e) The decision tree is read from left to right, the same way as you read the page of this book. The analysis, however, is conducted from right to left. To simplify the analysis, parts of the complete decision tree shown in Figure 4.38 have been redrawn in Figures 4.40, 4.41 and 4.42. In each case, the opportunity costs are shown as consequences, and the probabilities of the state of nature or test outcome are indicated. The EMV principle is utilized, and these values are indicated at the random event node.

The EMV analysis for no experimentation E_1, imperfect testing E_2 given Z_1, and imperfect testing E_2 given Z_2 are shown in the appropriate decision trees. Figures 4.40, 4.41, and 4.42, respectively. For example, the EMV for the no-experiment alternative, Figure 4.40, is:

$$EMV^{\mathbf{A}|E_1} = (0.5)(0) + (0.5)(6200) = \$3100$$

$$EMV^{\mathbf{B}|E_1} = (0.5)(2800) + (0.5)(0.0) = \$1400$$

$$EMV^{\mathbf{C}|E_1} = (0.5)(2800) + (0.5)(\$1000) = \$1900$$

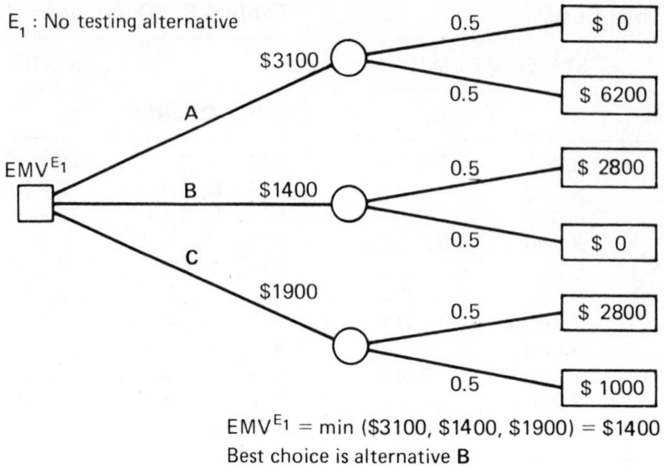

Figure 4.40

The best choice is the least expected cost, or alternative **B**, as indicated in Figure 4.40.

The results of these subanalyses are recorded on the decision tree in Figure 4.43, and the EMV analysis is performed. Note that $EMV^{E_3|z_1} = EMV^{E_3|z_2} = \$4000$, thus $EMV^{E_3} = \$4000$. From this decision tree we see the best choice is experiment E_2. The analysis reveals that seismic testing should be undertaken. If the test gives a depth recording between 35 and 40 ft, Z_1, we should order a precast pile 40 ft in length, plan **A**. See Figure 4.41. If the test result is between 40 and 50 ft, Z_2, we should order a precast pile of 50 ft, plan **B**. See Figure 4.42.

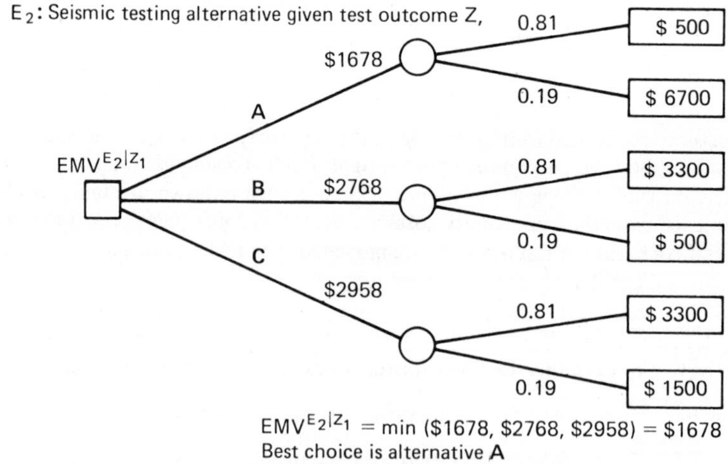

Figure 4.41

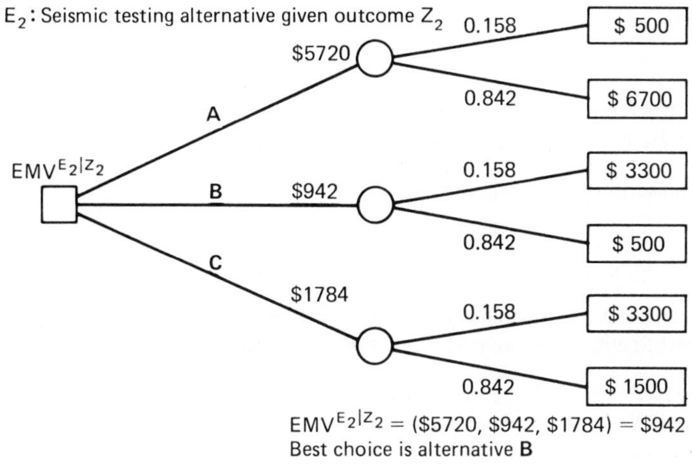

Figure 4.42

Remarks

From the investigation of these results we can deduce that the steel is not considered a viable alternative because the unit cost of steel is greater than the unit cost of the concrete, \$10/ft > \$8/ft.

The value of information obtained from \$500 seismic test is the most cost effective. There is too much risk involved with the no-test alternative, and the \$4000 test boring alternative is too expensive.

If the seismic test should prove to be incorrect, the penalty or opportunity cost as shown in Table 4.3 is \$6700 when the precast pile is too short and \$3300 when the pile is too long. These penalty costs include the cost of testing, delay and reconstruction costs for piles that prove to be too short, and costs from material wasted for piles that prove to be too long.

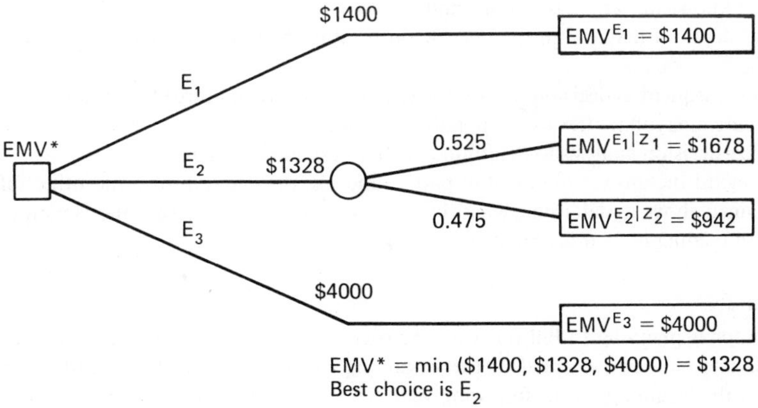

Figure 4.43

PROBLEMS

Problem 1

A contractor has a borrow pit from which he may remove sand and gravel and supply to a construction site. He is contemplating offering the following bids:

Bid A Bid on supplying 400 yd^3 of material.

Bid B Bid on supplying 600 yd^3 of material.

Bid C Do not bid.

The contractor's profit will depend upon the depth of material in the borrow pit x. He does not know the exact depth, but estimates his profit P as follows:

Bid A $P = \$900$ if $9 \text{ ft} \le x \le 15 \text{ ft}$
$P = \$1000$ if $15 \text{ ft} < x \le 21 \text{ ft}$

Bid B $P = -\$200$ if $9 \text{ ft} \le x \le 15 \text{ ft}$
$P = \$2000$ if $15 \text{ ft} < x \le 21 \text{ ft}$

Before submitting a bid the contractor is contemplating having a \$500 seismic test performed. The seismic test is an imperfect test with a sample likelihood of

	$P[Z_i \mid S_j]$	
	$S_1 = \{x: 9 \le x \le 15 \text{ ft}\}$	$S_2 = \{x: 15 < x \le 21 \text{ ft}\}$
$Z_1 = \{x: x = 9 \text{ ft}\}$	0.10	0.00
$Z_2 = \{x: x = 12 \text{ ft}\}$	0.65	0.10
$Z_3 = \{x: x = 15 \text{ ft}\}$	0.20	0.20
$Z_4 = \{x: x = 18 \text{ ft}\}$	0.05	0.60
$Z_5 = \{x: x = 21 \text{ ft}\}$	0.00	0.10

The contractor is also considering submitting a bid without relying on seismic testing. He has some prior knowledge about the depth, and he estimates that there is an equally likely chance that the actual depth of material in the borrow pit is either between 9 and 15 ft or between 15 and 21 ft.

(a) Draw a sequential decision tree for the two test options and three bid options.
(b) Prepare a payoff matrix for all possible outcomes.
(c) Utilize decision analysis to determine the best test and bid plan.
(d) Disregard the answer obtained in part c, and use the decision tree and other information obtained above to determine what the best bid should be if a seismic test is performed and a depth reading of 18 ft is obtained.

Problem 2

A contractor is planning to bid on two construction jobs, each yielding an estimated profit of \$250,000. The bids are accepted at the beginning of the year, and the award announcement is made at the beginning of the following year. The \$250,000 profit is received at that time. The contractor is competing on the bids with three other firms that have the same experience as her firm. In other words, each firm is equally likely to be awarded the \$250,000 contract.

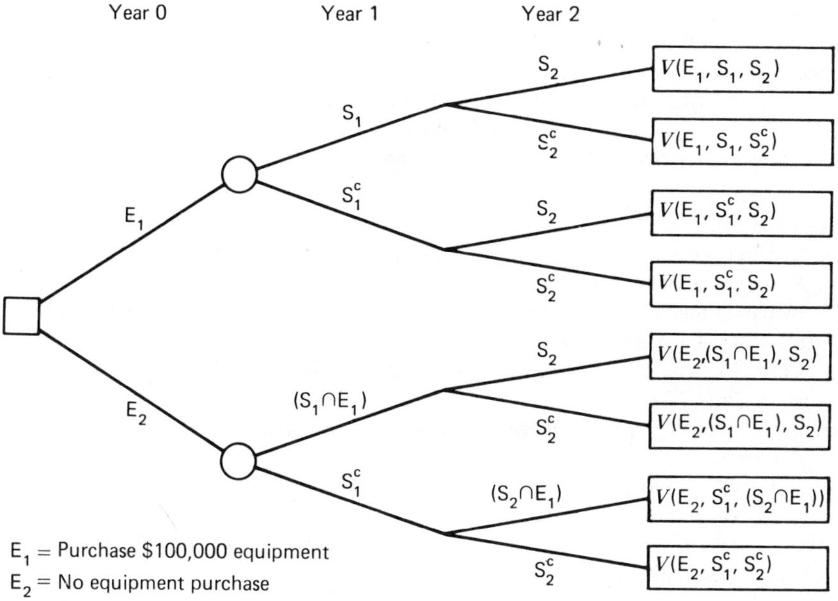

E_1 = Purchase \$100,000 equipment
E_2 = No equipment purchase
S_1 = \$250,000 contract is awarded in year 1
S_2 = \$250,000 contract is awarded in year 2
$(S_1 \cap E_1)$ = \$250,000 contract is awarded and \$100,000 equipment purchase is made in year 1
$(S_2 \cap E_1)$ = \$250,000 contract is awarded and \$100,000 equipment purchase is made in year 2

Figure 4.44

The contractor anticipates that if she owns the \$100,000 equipment her chances of receiving the first bid will double. In the second year the contractor feels her firm and the three competing firms will have an equal chance of an award. The contractor wants to decide now whether or not to purchase the equipment. Her plan of action is depicted in the decision tree in Figure 4.44.

The contractor realizes there is a danger of purchasing the equipment and not being awarded the contract. She assumes that her opportunity-cost interest rate is 12 percent per annum.

Decision analysis may be used to aid the contractor in deciding whether to purchase the equipment or not.

(a) Determine the consequence in present worth dollars of each possible outcome. Draw cash-flow diagrams for each outcome.
(b) Determine whether the contractor should purchase the equipment or not.

Summary

In this chapter, the elements of chance are incorporated into the selection process. By use of statistics, probability, the expected monetary value EMV of alternative designs may be compared on an equitable basis. Factors of economic worth, utility, or opportunity cost, and frequency of occurrence are incorporated into the process. The selection criterion is to maximize the EMV of utility or minimize the EMV of opportunity cost.

The concepts of utility, risk aversion, and gambling were discussed. Since utility functions are virtually impossible to determine for individuals or firms, they are not explicitly entered into the selection process; however, the concepts offer meaningful insight to the practical decision-making process. Again, we emphasize that the methods presented in this textbook are decision-making aids and are not a substitute for good judgment.

Subjective reasoning, field and laboratory tests, and historical records are used in the decision-making problems with uncertainty. The Bayes theorem provides a mechanism for incorporating this kind of information into the analysis. It is especially valuable for evaluating construction plans when a choice among various design alternatives and various field test alternatives must be made.

Bibliography

Alfredo Ang and Wilson Tang, *Probability Concepts in Engineering Planning and Design, Volume I Basic Principles*, John Wiley, New York. 1975.

Jack R. Benjamin and C. Allen Cornell, *Probability, Statistics and Decision for Civil Engineers*, McGraw-Hill, New York. 1970.

Gary C. Hart, *Uncertainty Analysis, Loads, and Safety in Structural Engineering*, Prentice Hall, Englewood Cliffs, N.J. 1982.

Frederick, Hiller and Gerald J. Lieberman, *Introduction to Operations Research*, 3rd ed., Holden Day, San Francisco, Calif. 1980.

Dale Meredith, Kam Wong, Ronald Woodhead, and Robert Wortman, *Design and Planning of Engineering Systems*. Prentice Hall, Englewood Cliffs, N.J. 1973.

John White, Marvin Agee, and Kenneth Case, *Principles of Engineering Economic Analysis*, John Wiley, New York. 1977.

CHAPTER 5

Economic Considerations for Resource Allocation

After the completion of this chapter, the student should:

1. Be familiar with the basic concepts and definitions of economics for evaluating systems engineering problems in terms of cost and performance.
2. Understand the underlying principles of market equilibrium and its effect upon selling price, output, and profit maximization.
3. Understand the effects of competition and the lack of competition upon private and public pricing and output.
4. Be able to optimize the allocation of input resources to maximize profits while satisfying conditions of the limitations of technology.

An engineering system consists of a set of components that are *all* necessary for the system to operate and function properly. For example, a truss consists of compression members, tension members, connections, and a foundation, all of which are required to support a given load. The absence or underdesign of any one component will cause failure of the system. The task of the design engineer is to select the proper combination of components, that is, member sizes, in order that the entire structure perform its job safely. There are an infinite number of designs that will satisfy this function. In the context of this example, the design that performs at minimum cost will be considered the best one. Our objective in systems analysis is to find a solution to a design, planning, or management problem that allocates resources in the most efficient and effective manner. In this chapter we consider problems with objectives stated in monetary measures of effectiveness. Maximize-net-profit, minimize-cost, and maximize-output problems will be solved subject to technological performance constraints and other restrictions encountered in design, planning, and management.

In systems analysis we evaluate the *overall performance* and *total cost* of the entire system by the *individual performance* and *individual cost* of its components. Since the entire system is optimized, each individual component must perform its task at optimum efficiency also. As a result, an optimum design will consist of a set of components all sized to work as a compatible unit. The overdesign of individual components will tend to be eliminated.

The maxim, "a chain is only as strong as its weakest link," is helpful to illustrate the meaning of an optimally designed system versus an overdesigned system. Suppose a chain consists of a set of links that are all of equal strength P. One link in the chain has the strength of P', which is less than the other links, $P' < P$. The ultimate load of the chain is P'. All links of the chain of strength P are considered overdesigned. In contrast, an optimally designed chain will consist of links of equal strength. Theoretically, all links of the chain will fail at the same load, since they are all working to their maximum capacity. As a result, the optimally designed chain performs the same task as the overdesigned chain while requiring less material.

Since both resource performance and resource cost are utilized in the system-analysis decision-making process, the principles of the *theory of the firm* can help us conceptualize, formulate, and solve engineering design problems. The theory of the firm is from the branch of economics called microeconomics.

A *firm* is a general term used to describe an enterprise that produces a commodity or service for profit. If the total revenue, the product of the unit selling price times the number of items sold, is greater than the cost to produce them, the firm makes a profit; otherwise, it will experience a loss. Consequently, the efficiency in which a commodity is produced is a very important consideration in the success of a firm.

Furthermore, the selling price has a direct bearing upon the level of production of the firm and its ability to realize a profit. The total demand for an item or service will depend upon many factors including consumer demand, the production cost, and the number of competing firms. Supply and demand functions, which were so important in engineering economic decision making, will be used to address these questions of market equilibrium. These factors are used primarily for finding the selling price of a good or service. Once known, the firm can allocate its resources to maximize its profit while ensuring that overall performance considerations are satisfied.

5.1 COST AND PRODUCTION

A private firm will continue to manufacture and market a good or service as long as it is profitable. The level of production will depend upon consumer demand and the cost of production. The net profit P is equal to the revenue R minus cost C, or

$$P = R - C = pq - c(q)$$

where p is the sale price of the good or service, q is the quantity sold, and $C = c(q)$ represents all fixed and variable production costs including the cost of capital, operating and maintenance. It is common to use the capital letter C to represent capital cost and production cost. Here, $C = c(q) =$ production cost. In the chapter on engineering economics we use C to indicate a capital cost; here we use C_0 to avoid confusion with C, production cost.

The necessary condition to maximize the net profit P of an individual firm is

$$\frac{dP}{dq} = \frac{dR}{dq} - \frac{dC}{dq} = 0$$

or

$$\frac{dP}{dq} = \frac{d(pq)}{dq} - \frac{dc(q)}{dq} = 0$$

thus

$$\frac{dR}{dq} = \frac{dC}{dq}$$

$$\frac{d(pq)}{dq} = \frac{dC}{dq}$$

Let MR equal the marginal revenue, MR $= dR/dq$, and MC equal the marginal cost, MC $= dc(q)/dq$. Thus, the marginal revenue must equal the marginal cost, or

$$\text{MR} = \text{MC}$$

The sufficient condition for a maximum is

$$\frac{d^2P}{dp^2} = \frac{d^2(pq)}{dq^2} - \frac{d^2C}{dq^2} < 0$$

Market Equilibrium and Perfect Competition

The total production, or *aggregate production* q, is equal to the sum of the output of each individual firm q_i, where the subscript i represents the firm number $i = 1, 2, \ldots, n$.

$$\hat{q} = q_1 + q_2 + \cdots + q_n = \sum_i^n q_i$$

To distinguish between aggregate production and production by individual firms, the circumflex ^ will be used to designate aggregate production. When necessary, the subscript i, where $i = 1, 2, \ldots, n$, will be used to distinguish among firms. Thus, $q_1, q_2, \ldots, q_n$ represents the output from n different firms and $\hat{q}$ represents the aggregate production. When discussing a single firm no distinction is needed and the subscript i will not be used.

In a perfectly competitive market it is assumed that there are many firms that all produce the same identical good or provide the same identical service. Furthermore, all consumers are assumed to have *perfect information* about all firms competing in this market. As a result, the firms, by necessity, will sell their good or service at the same price p^0. Mathematically, the market equilibrium levels of output $\hat{q}^0$ and price p^0 may be determined by equating the demand and supply functions as depicted in Figure 5.1.

$$S = D$$

The demand D is equal to the price as function of aggregate supply $\hat{q}$.

$$D = p(\hat{q})$$

Any firm that raises its price above p^0 will lose sales because the buyers are cognizant of its higher sale price and the lower prices offered by the other firms. The firm will be forced to reduce its price to p^0 to be competitive.

If a firm is able to reduce its production cost, it will be able to pass this saving along in terms of lower prices and at the same time keep a favorable profit margin. The

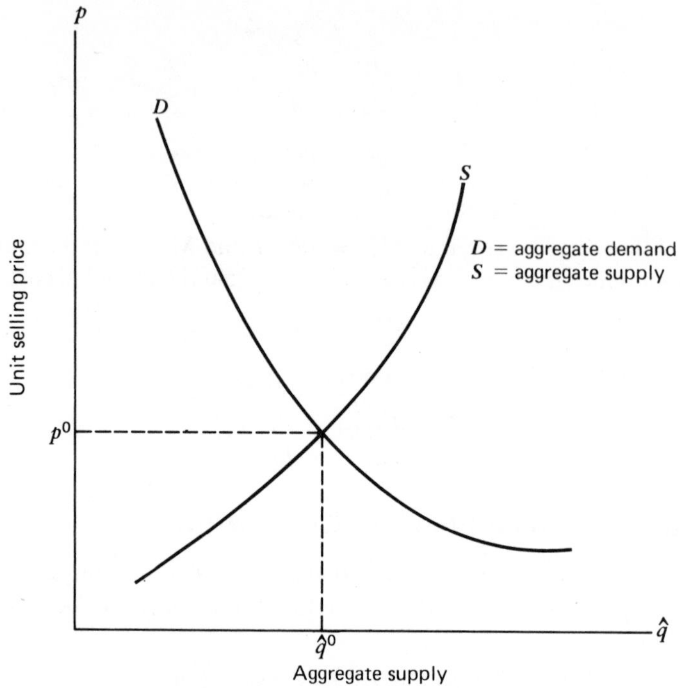

Figure 5.1 Market equilibrium, $S = D$.

individual economics realized by the one firm will cause an overall downward shift in the supply curve. As a result, the selling price charged by each individual firm will be forced to a lower level. Prices will stabilize to a new market equilibrium price p^1 that is lower than the original price p^0, $p^1 < p^0$. We discussed this principle in Chapter 3. See Figure 3.26. For the remaining discussion, the market equilibrium price will be denoted as p^0 instead of p^1. Thus, in a perfectly competitive market, each firm will sell its output at price p^0 or $p_1 = p_2 = \cdots = p_n = p^0$.

The Cost Function

The cost function is defined as a mathematical relationship of cost to magnitude of output q:

$$C = c(q) = \alpha + \beta(q)$$

where the parameter α, equals the fixed cost of production and the function $\beta(q)$ equals the variable cost of production. The fixed cost may be considered overhead; it is the cost a firm must pay regardless of the level of output q. The rent on buildings, loan payments, and utility charges are examples of fixed costs. The variable cost $\beta(q)$ is the direct cost associated with producing an output q. The variable cost may consist of charges for material, labor, equipment, and the like.

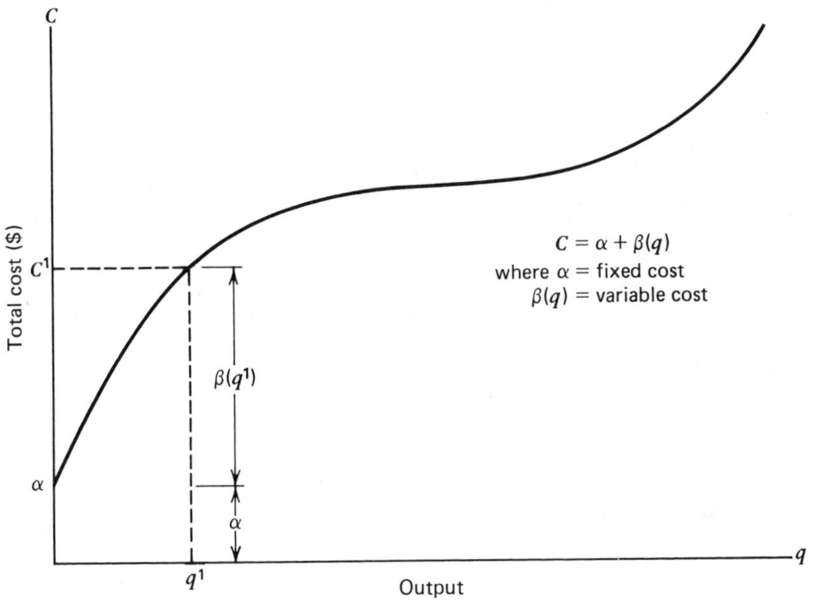

Figure 5.2 Theoretical cost curve.

Cost functions take many forms. Figure 5.2 shows a cost function where the cost increases with increasing output, levels off, and then increases at higher levels of production. At an output of q^1, the cost is $C^1 = \alpha + \beta(q^1)$.

Economy of Scale

The average total cost $\bar{c}$ of each item produced is equal to the total cost divided by the output.

$$\bar{c} = \frac{C}{q} = \frac{c(q)}{q} = \frac{\alpha + \beta(q)}{q} = \frac{\alpha}{q} + \frac{\beta(q)}{q} = AC$$

The average fixed cost is α/q, and the average variable cost is $\beta(q)/q$. In Figure 5.3 the average cost $\bar{c}$ is shown as the slope of the line OA, which is drawn from the origin O to a point A on the cost line.

The relationships of the total and average cost functions are shown in Figure 5.3. As the output q increases, the average cost of production decreases, and then increases at higher levels of production q. The production level, where the average cost is a minimum $\bar{c}'$, is shown at output level q'. *Economy of scale* is defined as a decrease in average cost as output increases. In Figure 5.3, an economy of scale exists for production levels between 0 and q'. Beyond q' no economy of scale exists.

In engineering design, economy of scale can be used to help decide whether overcapacity of a system should be provided. If economy of scale exists, it may be wise for the engineer to anticipate future growth and provide extra capacity in the system. If

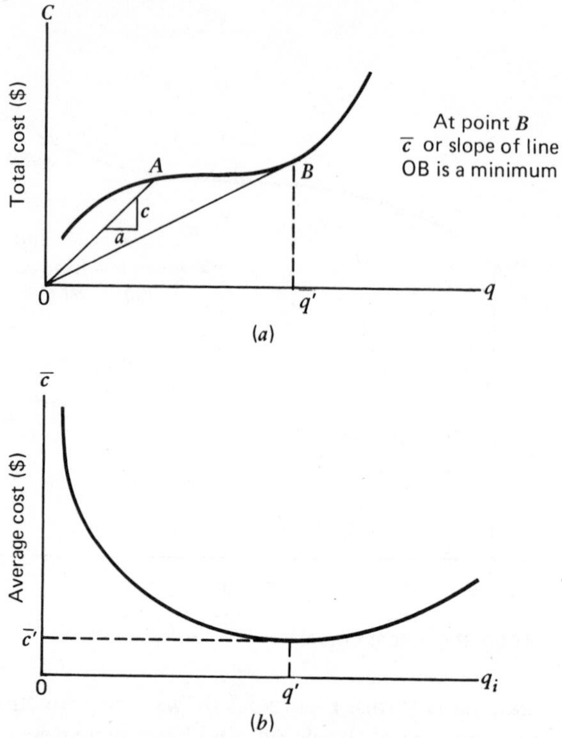

Figure 5.3 Average cost function.

the future growth is realized, the average cost of production will be lessened, and greater profits will be obtained.

Necessary and Sufficient Conditions for Maximum Profit

In order for an individual firm to maximize its profit it should adjust its production to a level q to satisfy the condition of profit maximization.

$$\text{MR} = \text{MC}: \qquad \frac{d(p^0 q)}{dq} = \frac{dC}{dq}$$

Since p^0 is a constant, the necessary condition for maximum profit is reduced to

$$p^0 = \frac{dC}{dq}$$

The sufficient condition for maximum profit is satisfied when

$$\frac{d^2(p^0 q)}{dq^2} - \frac{d^2 C}{dq^2} \leq 0$$

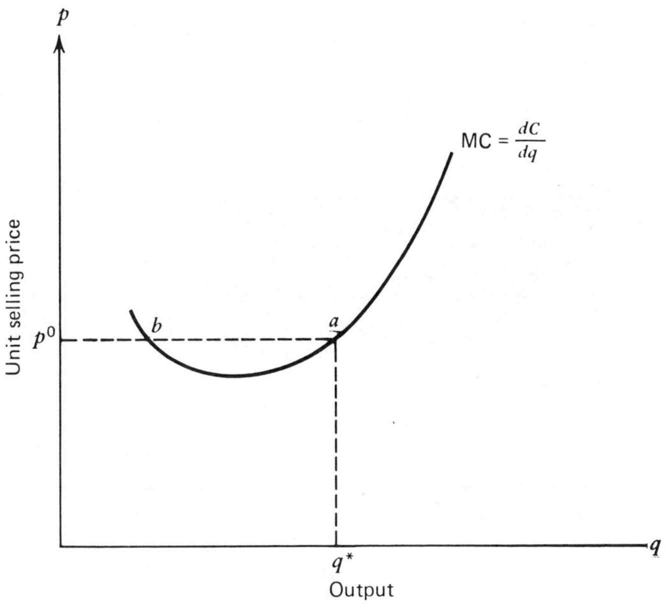

Figure 5.4 Perfect competition.

This simplifies to $-d^2C/dq^2 \leq 0$, or equivalently, the *slope of the marginal cost curve must be positive*. When the sufficient condition is satisfied, the location of the maximum profit, depicted as point a in Figure 5.4, the intersection of p^0 with the marginal cost curve MC. The maximum profit occurs at the production level q^*, point a. If the slope of MC is negative, the solution to the necessary conditions represent condition of minimum net loss for the firm, point b.

The maximum profit P^* is:

$$P^* = p^0q^* - c(q^*)$$

The equation may be written in terms of the average cost. The maximum profit is

$$P^* = p^0q^* - \bar{c}q^*$$

or

$$P^* = (p^0 - \bar{c})q^*$$

The profit will be positive when the price is greater than the average cost of production, $p^0 > \bar{c}$. An important result emerges. Profits are directly related to *production efficiency* as measured by the average production cost AC at the production level q. This result is consistent with the assumptions made in supply–demand relationships considered in engineering economics.

In Figure 5.5, the conditions for maximum profit P^* are shown for two different selling prices. In Figure 5.5*a*, the selling price exceeds the average cost $\bar{c}$ or $p^0 > \bar{c}$. In Figure 5.5*b*, the selling price is less than $\bar{c}$, or $p^0 < \bar{c}$. In both cases the sufficient

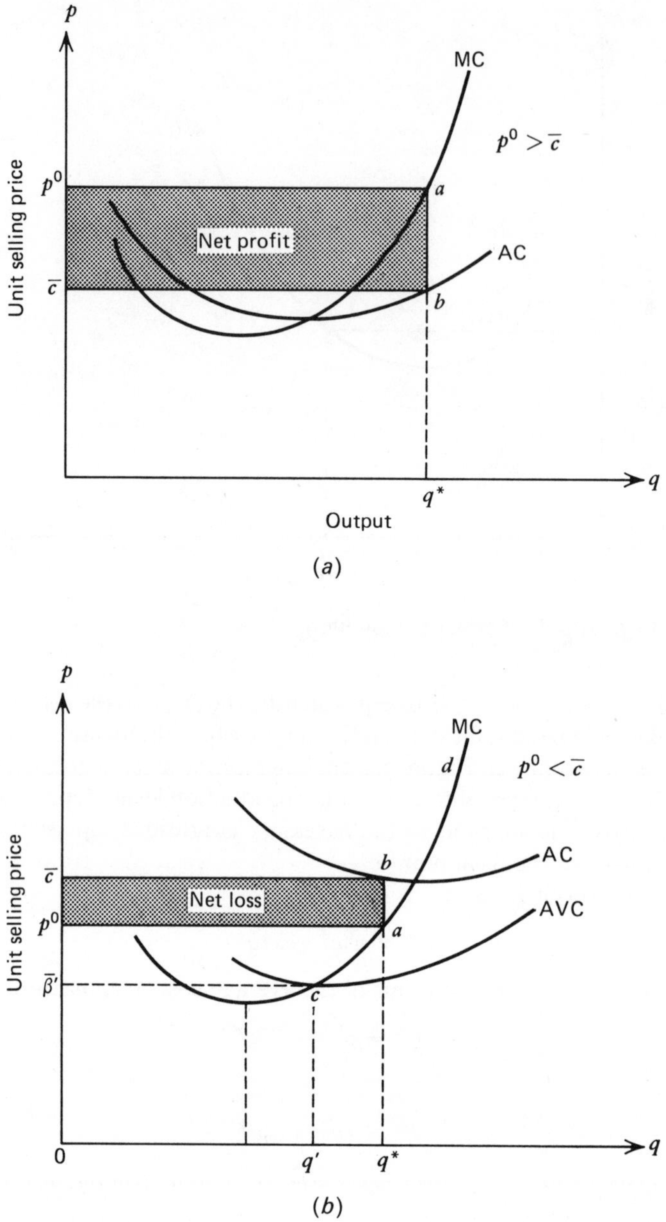

Figure 5.5 (*a*) Optimum level of output q^*. (*b*) Firm operating at a loss.

condition for profit maximization is satisfied; the MC curve is positive. For each case the maximum net profit or loss is equal to the shaded area $p^0ab\bar{c}$. Under the conditions of perfect competition, the firm cannot control the price p^0. Consequently, the only way it can improve its profits is by minimizing the cost of production $c(q)$.

Long- and Short-Term Operation

For the case when the selling price is less than the average cost, $p^0 < \bar{c}$, the firm is faced with the problem of deciding whether to continue to produce or shut down. In the *long term*, if the selling price is less than the average cost of production, the firm should discontinue operation. The loss is depicted in Figure 5.5*b* as the shaded area $p^0ab\bar{c}$.

In the *short term*, it may be advisable to continue production. The net profit $P^* = (p^0 - \bar{c})q^*$ for this cost function is

$$P^* = (p^0 - \bar{\beta})q^* - \alpha$$

where $\bar{\beta}$ equals the average variable cost. The average variable cost is shown in Figure 5.5 as AVC. Consider two conditions: shut down or $q = 0$, and produce at level $q = q^*$, where q^* represents the condition of maximum profit. For the first case, since the production is shut down, $q = 0$, and the net revenue equals zero.

$$P = (p_0 - \bar{\beta})\cdot 0 - \alpha = -\alpha$$

The net loss will be equal to the fixed cost α.

Consider the second condition, that the firm will produce at q^*. This production level satisfies the necessary and sufficient conditions for a maximum. In Figure 5.5*b*, p^0 intercepts MC at point a, and the slope of the MC curve is positive. When the price p^0 exceeds the minimum average variable cost $\bar{\beta}'$, where $\bar{\beta}'$ is shown in Figure 5.5*b* as point c, the loss will be a minimum.

$$P^* = (p^0 - \bar{\beta})q^* - \alpha$$

If the price p^0 falls below $\bar{\beta}'$, $p^0 < \bar{\beta}'$, the minimum loss occurs during the shut-down, $P^* = -\alpha$. The firm should continue production as long as MC is greater than AVC.

The results of this analysis are summarized as

$0 \le p^0 < \bar{\beta}'$	$q^* = 0$	Shut down operation
$\bar{\beta}' \le p^0 < \bar{c}$	$q^* > 0$	Minimize net loss in the short run
$\bar{c} \le p^0$	$q^* > 0$	Positive net profits

Cost Elasticity

The marginal cost is defined as the first derivative or slope of the cost function.

$$\text{MC} = \frac{dC}{dq} \quad \text{or} \quad \text{MC} = \frac{\Delta C}{\Delta q}$$

The cost elasticity ε is defined as the ratio of percentage change in cost C to the percentage change in supply q.

$$\varepsilon = \frac{\% \, \Delta \text{ in cost}}{\% \, \Delta \text{ in supply}}$$

$$\varepsilon = \frac{(\Delta C/C \times 100)}{(\Delta q/q \times 100)} = \frac{q}{C}\frac{\Delta C}{\Delta q}$$

In the limit $\Delta q \rightarrow 0$, the elasticity is equal to:

$$\varepsilon = \frac{q}{C}\frac{dC}{dq}$$

Rearranging, the cost elasticity may be written as the ratio of marginal cost to average cost.

$$\varepsilon = \frac{dC/dq}{C/q} = \frac{\text{MC}}{\text{AC}}$$

Exponential Cost Function

It has been found in various cost surveys of engineering projects that the exponential form of the cost function is applicable. For an individual firm, it is

$$C = c(q) = \delta q^{\gamma}$$

where δ and γ are constants. Generally, the cost function varies as an increasing function of q as shown in Figure 5.6 for different values of γ. When $\gamma = 1$ the variable cost function is a linear function.

The cost elasticity for the exponential cost function is

$$\varepsilon = \frac{q}{C}\frac{dC}{dq} = \left(\frac{q}{\delta q^{b}}\right)\frac{d}{dq}(\delta q^{\gamma}) = \left(\frac{q}{\delta q^{\gamma}}\right)^{\gamma} b\delta q^{\gamma-1} = \gamma$$

For the experimental cost function, the cost elasticity is constant, $\varepsilon = \gamma$, and independent of the level of production q.

The average cost function for the average cost function is

$$\bar{c} = \frac{C}{q} = \frac{\delta q^{\gamma}}{q} = \delta q^{\gamma-1}$$

The effect of cost elasticity upon average cost and economy of scale is shown in Figure 5.6. When $\gamma < 1$, an economy of scale exists. We shall discover that productions with economy of scale have very different optimum solutions from production functions that do not.

The Effects of Cost Elasticity

The exponential cost function $C = \delta q^{\varepsilon}$, is used to represent cost relationships of various engineering facilities. Consider the necessary and sufficient conditions for maximum net profit. The profit function is

$$\text{Maximize } P = p^{0}q - \delta q^{\varepsilon}$$

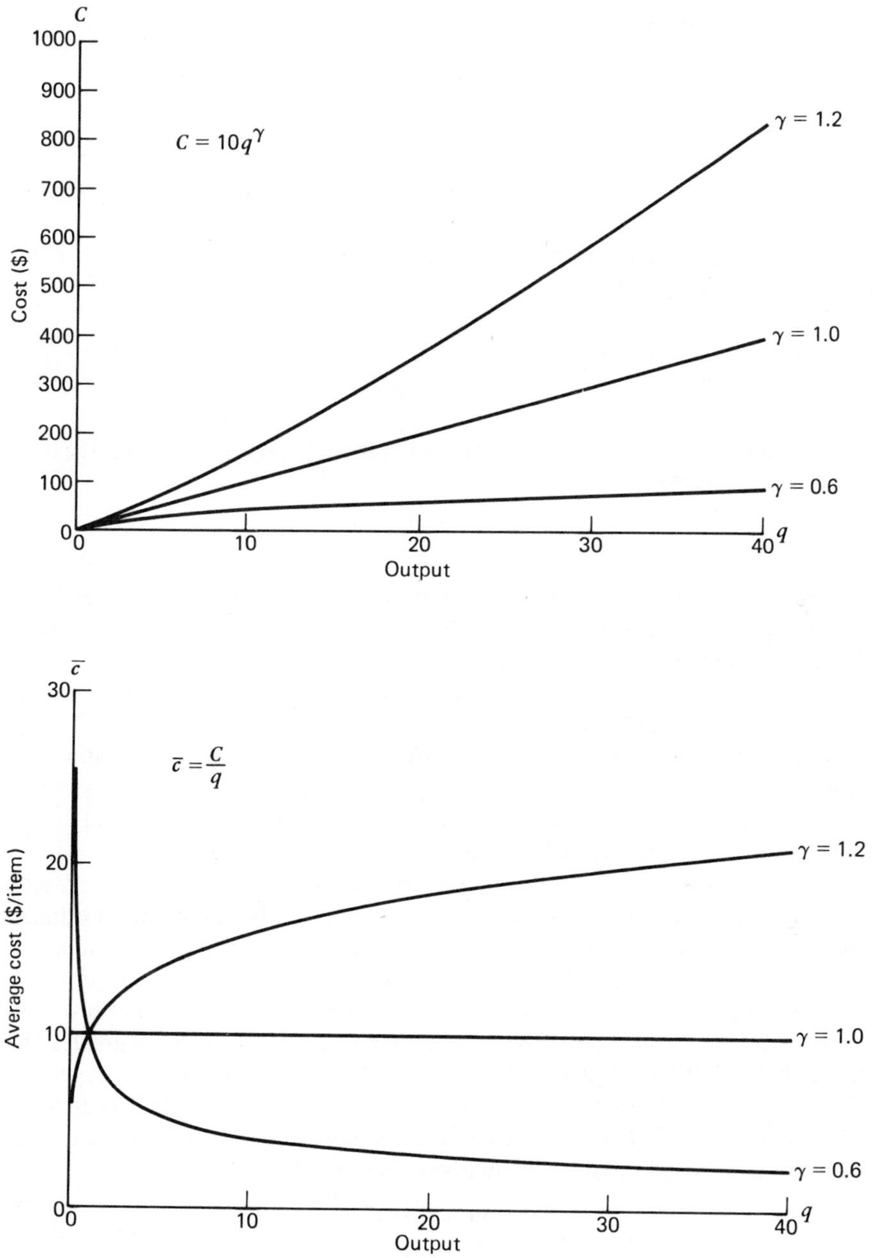

Figure 5.6 The effect of elasticity upon average cost functions.

The necessary condition is:

$$\frac{dP}{dq} = p^0 - \delta\varepsilon q^{\varepsilon-1} = 0$$

The sufficient condition for maximum profit is

$$\frac{d^2P}{dq^2} = -\delta\varepsilon(\varepsilon - 1)q^{\varepsilon-2} < 0$$

The average cost function for an exponential cost function is $\bar{c} = \bar{c}(q) = \delta q^\varepsilon/q = \delta q^{\varepsilon-1}$. The above equations will be rewritten in terms of the average cost.

$$\text{Maximize } P = (p^0 - \bar{c})q$$

$$\frac{dP}{dq} = p^0 - \varepsilon\bar{c}(q^*) = 0$$

The average cost function $\bar{c}(q^*)$ is used to emphasize that it is evaluated at the optimum production level q^*.

$$\frac{d^2P}{dq^2} = -\frac{\varepsilon(\varepsilon - 1)\bar{c}(q^*)}{q} < 0$$

When no economy of scale exists, $\varepsilon > 1$, the sufficient condition is satisfied because ε, $(\varepsilon - 1)$, $\bar{c}$, and q are positive values then d^2P/dq^2 must be negative. Thus, the net profit is a maximum positive value. Since $\varepsilon > 1$, the average cost function $\bar{c} = \delta q^{\varepsilon-1}$ is an increasing function and MC > 0. This situation is shown in Figure 5.7*a*. The maximum profit P^* is the shaded area $p^0ab\bar{c}$.

Contrast this situation when an economy of scale exists, $\varepsilon < 1$. The sufficient condition is not satisfied, $d^2P/dq^2 > 0$, and the solution q^0 to the necessary conditions represents a minimum value of P. Since ε, $\bar{c}$, and q are positive values and $(\varepsilon - 1)$ is a negative value, d^2P/dq^2 must be positive. The average cost function $\bar{c} = \delta q^{\varepsilon-1}$ is a decreasing function as shown in Figure 5.7*b*. The shaded area $p^0ab\bar{c}$ represents the minimum loss for output q^0. A profit may be achieved by increasing production.

A profit will be obtained when $p^0 > \bar{c}$. Since $\bar{c}$ is a decreasing function, increasing the level of production q will reduce the loss. At point c, the level of production is $q = q^1$, selling price p^0 equals the average cost, $p^0 = \bar{c}(q^1)$, and the net profit is zero, $P = (p - \bar{c}(q^1))q^1 = 0$. Increasing q to higher levels of production greater than q^1 will result in positive net profits. Since $\bar{c}$ is a decreasing function, the larger q is, the greater the profit will become. Theoretically, since q is not restricted, q can approach infinity, and the profit P approaches infinity. A more realistic situation occurs when a constraint is imposed upon q. The mathematical model becomes

$$\text{Maximize } P = p^0q - \delta q^\varepsilon$$

$$q \le \tilde{q}$$

The maximum will occur when the constraint is active, $q^* = \tilde{q}$. For the calculation of the maximum profit, the profit function utilizing $c = \delta q^\varepsilon$ will be used. Thus,

$$P^* = p^0\tilde{q} - \delta\tilde{q}^\varepsilon$$

Remember, this situation occurs only when an economy of scale exists, $\varepsilon < 1$.

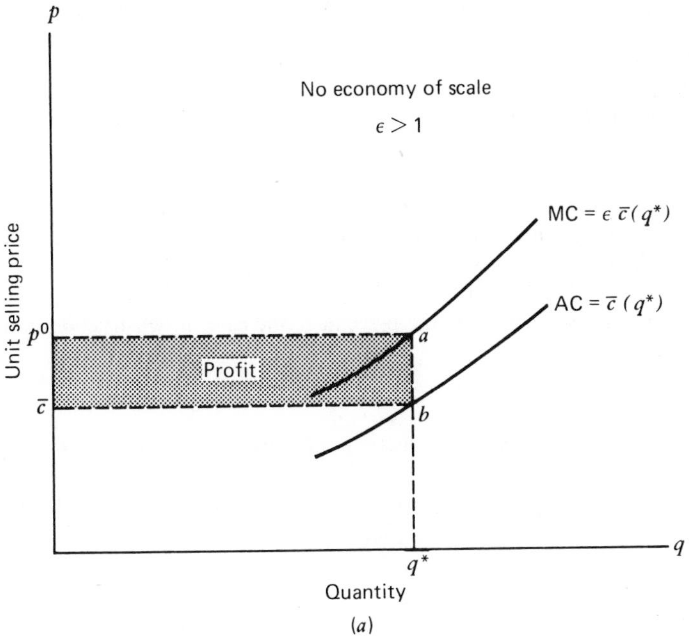

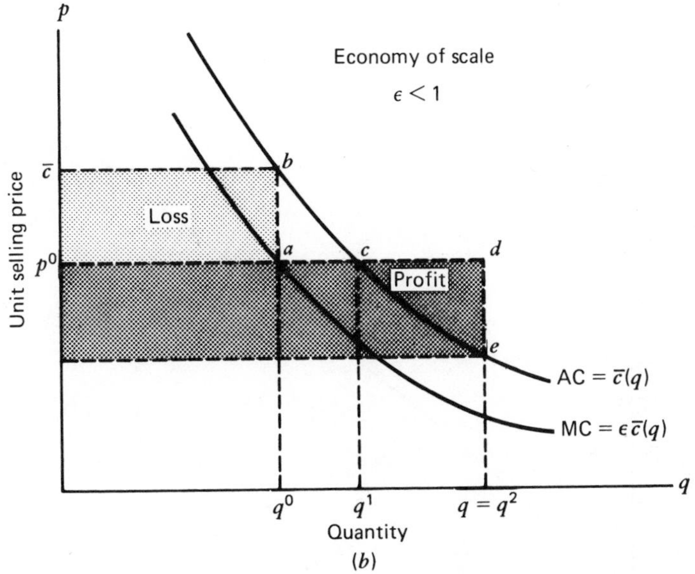

Figure 5.7 The effect of cost elasticity upon new profit.

EXAMPLE 5.1 Construction Costs For Waste-Water Treatment Plants

The U.S. Environmental Protection Agency has conducted a national survey of the construction cost of secondary water-treatment plants for the years 1973 through 1977. They have estimated a cost of construction to be equal to

$$C = c(q) = \$2 \times 10^6 \, q^{0.88}$$

where C equals the cost in dollars and q equals the plant output capacity in million gallons per day (Mgal/day). The largest plant constructed had an output capacity of 70 Mgal/day.

(a) Draw a graph of the cost function.

(b) Does an economy of scale exist?

(c) By use of cost elasticity estimate the percentage increase in construction cost from 50 to 55 Mgal/day.

Solution

(a) The graph of the construction cost function in Figure 5.8 shows a slight decrease in cost as q increases.

(b) An economy of scale exists when there is a decrease in average cost as output q increases. The average construction cost $\bar{c} = c(q)/q$, or

$$\bar{c} = \frac{2 \times 10^6 q^{0.88}}{q} = 2 \times 10^6 \, q^{-0.12}$$

Figure 5.9 is the average construction cost curve. Since the average cost $\bar{c}$ decreases with increasing output, economy of scale exists for the entire of range of output from 0 to 70 Mgal per day.

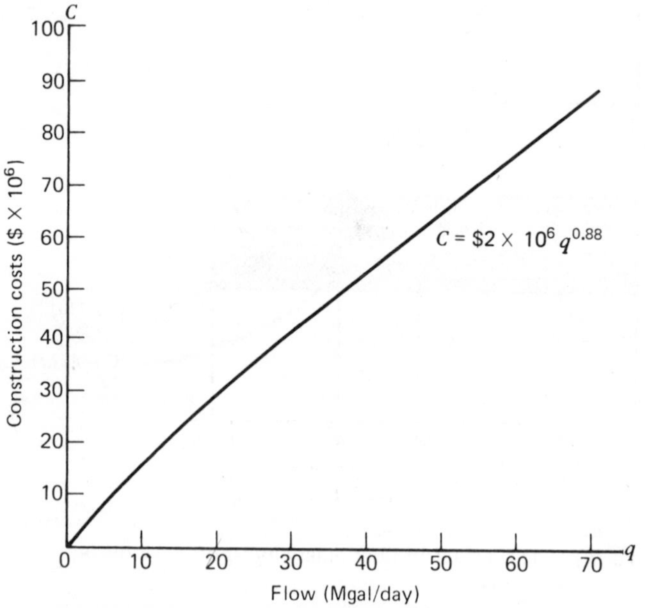

Figure 5.8 The construction cost of a secondary waste-water treatment plant.

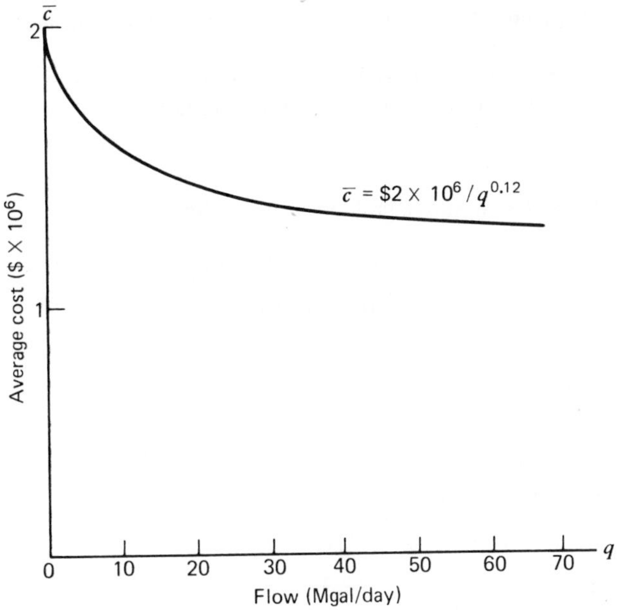

Figure 5.9 Average cost function for a given flow design.

(c) Since the cost function is an exponential function, the cost elasticity ε is constant and equals 0.88. The percentage increase in production from 50 to 55 Mgal per day is

$$\frac{\Delta q}{q} = \frac{55 - 50}{50} \cdot 100 = 10 \text{ percent}$$

The percentage increase in cost is estimated with the definition of cost elasticity. Thus

$$\frac{\Delta C}{C} = \varepsilon\left(\frac{\Delta q}{q}\right)$$

$$\frac{\Delta C}{C} = 0.88(10) = 8.8 \text{ percent}$$

A 10 percent increase in production results in a 8.8 percent increase in production cost. Clearly, an economy of scale exists.

EXAMPLE 5.2 Production Efficiency and Economy of Scale

The selling price of an item in a perfectly competitive market is \$10/unit. A company is considering purchasing three different types of equipment to produce the item.

MACHINE	CAPITAL COST	PRODUCTION COST
Type A	$C_0 = \$10{,}000$	$C = 2q^{1.2}$
Type B	$C_0 = \$25{,}000$	$C = 2q^{0.8}$
Type C	$C_0 = \$15{,}000$	$C = 2q$

The maximum yearly demand is 2000 items.

(a) Determine the optimum level of production, the net maximum profit, and the average cost of production.
(b) Determine the net annual worth NAW to determine which machine is the best investment. Assume that a 5-year loan at an interest rate of 10 percent per annum is available.

Solution

The mathematical model to maximize profit for machine types is the same.

$$\text{Maximize } P = 10q - C$$

$$0 \le q \le 2000$$

The necessary condition is $dP/dq = 10 - dC/dq = 0$, or

$$\frac{dC}{dq} = 10$$

The sufficient condition for a maximum is

$$\frac{dC^2}{dq^2} > 0.$$

Machine Type A The necessary condition for an optimum, assuming an inactive constraint, is

$$\frac{dC}{dq} = 10$$

$$\frac{d(2q^{1.2})}{dq} = 2.4q^{0.2} = 10$$

$$q = q^{\circ} = \left(\frac{10}{2.4}\right)^5 = 1256 \text{ units}$$

The sufficient condition is

$$\frac{d^2(2q^{1.2})}{dq^2} = 0.48q^{-0.8} > 0$$

Since $0.48q^{-0.8} > 0$ for all positive values of q, the sufficient condition is satisfied and the optimum production level is $q_A^* = q^0 = 1256$. The maximum profit for machine type **A** is

$$P^* = 10q_A^* - 2q_A^{*1.2}$$

$$P^* = 10 \cdot 1256 - 2(1256)^{1.2} = \$2094$$

The average cost of production at optimum q_A^* is

$$\bar{c}^* = \frac{C}{q} = 2q_A^{*0.2}$$

$$\bar{c} = 2(1256)^{0.2} = \$8.33/\text{unit}$$

Machine Type B The necessary condition assuming the constraint is inactive is

$$\frac{dC}{dq} = 10$$

$$\frac{d(2q^{0.8})}{dq} = 1.6q^{-0.2} = 10$$

$$q = q^{\circ} = \left(\frac{1.6}{10}\right)^5 = 0.0001$$

The sufficient condition is

$$\frac{d^2(2q^{0.8})}{dq^2} = -0.32q^{-1.2} < 0$$

Since the $-0.32^{-1.2} < 0$ for all values of positive q, the necessary condition for a maximum is not satisfied. The value $q^{\circ} = 0.0001$ represents the location for a minimum profit.

The maximum profit P^* will occur when the constraint $q \le 2000$ is active. $q_B^* = 2000$.

$$P^* = 10q_B^* - 2q_B^{*0.8}$$

$$P^* = 10(2000) - 2(2000)^{0.8}$$

$$P^* = \$19{,}125$$

The average production cost at optimum q_B^* is

$$\bar{c} = \frac{C}{q} = 2q_B^{*-0.2}$$

$$\bar{c} = 2(2000)^{-0.2} = \$0.44/\text{unit}$$

Machine Type C The mathematical model is linear.

$$\text{Maximize } P = 10q - 2q = 8q$$

$$q \le 2000$$

$$q \ge 0$$

The production cost function is a linear function; therefore, the methods of calculus do not apply. The search for an optimum will occur at an extreme point. In other words, when $q = 2000$. $q_C^* = 2000$. The maximum profit P^* is

$$P^* = 8q_C^*$$

$$P^* = 8(2000) = \$16{,}000$$

The average cost per unit is

$$\bar{c} = \frac{C}{q} = \$2.00/\text{unit}$$

(b) Assuming a design life of 5 years and an annual interest rate of 10 percent, the annual benefits and costs of capital and operating costs for each alternative are determined with the capital recovery factor.

Machine Type A,

$$B = P^* = \$2094/\text{yr}$$

$$C_0 = 10{,}000 \cdot \left[\frac{0.1(1+0.1)^5}{(1+0.1)^5 - 1}\right] = \$2638/\text{yr}$$

The annual profit is

$$\text{NAW}^{\mathbf{A}} = (B - C_0) = \$2097 - \$2638 = -\$544/\text{yr (loss)}$$

This is not a viable alternative, because the capital investment C_0 exceeds the net profit B.

Machine B,

$$B = P^* = \$19{,}125/\text{yr}$$

$$C_0 = \$25{,}000\left[\frac{0.1(1+0.1)^5}{(1+0.1)^5 - 1}\right] = \$6595/\text{yr}$$

$$\text{NAW}^{\mathbf{B}} = (B - C_0) = \$19{,}125 - \$6595 = \$12{,}530/\text{yr (profitable)}$$

Machine C,

$$B = P^* = \$16{,}000/\text{yr}$$

$$C = \$1000\left[\frac{0.1(1+0.1)^5}{(1+0.1)^5 - 1}\right] = \$3957/\text{yr}$$

$$\text{NAW}^{\mathbf{C}} = (B - C^0) = \$16{,}000 - \$3957 = \$12{,}043/\text{yr (profitable)}$$

Since $\text{NAW}^{\mathbf{B}} > \text{NAW}^{\mathbf{C}}$, the best optimum choice is machine type **B**.

Remarks

An economy of scale exists for machine type **B** because it has a cost elasticity of 0.8. For machine types **A** and **C** no economy of scale exists, because they have cost elasticities of 1.2 and 1.0, respectively.

The production efficiency is the deciding factor in choosing machine type **B**.

MACHINE TYPE	OUPUT q^*	AVERAGE COST $\bar{c}(q^*)$	CAPITAL COST
A	1,256	\$8.33/unit	\$10,000
B	2,000	\$0.44/unit	\$25,000
C	2,000	\$2.00/unit	\$15,000

Clearly, machine type **B** is the most efficient and most costly machine. From the net annual worth calculations it becomes evident that its high initial cost is not an overriding factor in its selection. If its price were greater, machine type **C** might become the best choice. This example clearly illustrates the importance of production efficiency, production cost, and capital cost in the selection process.

PROBLEMS

Problem 1

The cost function is $C = 10q^{1.2}$.

(a) Determine the average cost.
(b) Determine the marginal cost function.
(c) Prove the cost elasticity is 1.2.
(d) Does an economy of scale exist? Why?

Problem 2

A cost function is given as $C = 6 + 8q$, where q is the output.

(a) Determine the marginal cost function.
(b) Determine the average cost function.
(c) Determine the elasticity function.
(d) Does an economy of scale exist? Why?
(e) Do you recommend providing additional capacity based upon economies of scale? Why?
(f) Repeat parts a through e for the cost function $C = 8q$.

Problem 3

The cost functions for two machines producing the same product are

Machine A $C = 110q^{0.95}$
Machine B $C = 70q^{1.05}$

The production level is estimated to be 100 units/day.

(a) Compare machines **A** and **B** by completing the following table:

	MACHINE A	MACHINE B
C = Production cost at q = 100 units		
ε = Cost of elasticity		
Economy of scale, yes or no		

(b) Utilize the definition of cost elasticity to estimate the production cost for machines **A** and **B** for a 10 percent production increase.
(c) Repeat part b for a 30 percent production increase.
(d) Utilize the cost function to determine the production cost for machines **A** and **B** for a 10 percent production increase.
(e) Repeat part d for a 30 percent production increase.
(f) Prepare a table to compare the results of parts b through e. Do the elasticity estimates give reasonable cost values for 10 percent and 30 percent production increases.

Problem 4

A water-treatment system consists of water-distribution and treatment systems. The construction costs as a function of flow q (in million gallons per day) are

Distribution	$\$350{,}000q$
Treatment	$\$2{,}000{,}000q^{0.88}$
Design cost (a fixed cost)	$\$200{,}000$

(a) Does an economy of scale exist for the distribution system? Draw the average cost function as a function of q.
(b) Does an economy of scale exist for the treatment system? Draw the average cost function as a function of q.
(c) What is the total design and construction cost of the entire project as a function of q? Draw an average cost function as a function of q.
(d) Does an economy of scale exist for the entire project? Explain your answer.

Problem 5

The construction cost of a wastewater filtration and chemical addition processor are

Filtration	$C = \$1.85 \times 10^5 q^{0.84}$ (q in million gallons per day)
Chemical Addition	$C = \$2.36 \times 10^4 q^{1.68}$

(a) Do economics of scale exist for each of these processes? Why?
(b) Using the definition of elasticity, estimate the increase in cost for a 10 percent, 50 percent, 100 percent increase in flow for each process.
(c) Use the cost function to determine the percentage increase in cost for an initial flow of $q = 20$ Mgal/day, and percentage increase in flow of 10 percent, 50 percent, and 100 percent for each process.
(d) Repeat part c for an initial flow of $q = 2$ Mgal/day.
(e) Prepare a table and compare the results of parts b, c, and d. Is the cost elasticity approach for estimating percentage increase in cost valid for all increases in production?

Problem 6

At capacity a machine may produce no more than 100 items per year. The production cost function is $150q^{0.5}$, where q equals the number of items produced per year. There is a fixed cost of \$250 per year.
(a) Formulate a mathematical model to determine the optimum level of production to maximize net profit.
(b) Determine the optimum level of production.
(c) Calculate the average cost of production.
(d) Should production be continued or shut down in the long and short terms? Why?

Problem 7

Answer Problem 6 for a production cost function of $50q^{1.1}$.

Problem 8

The following information has been gathered.

MACHINE	C_0 CAPITAL COST	n DESIGN LIFE	$\beta(q)$ ANNUAL PRODUCTION COST	α ANNUAL FIXED COSTS
A	\$1,000	2	$150q^{0.5}$	\$750
B	\$1,500	3	$130q^{0.6}$	750

The unit selling price is \$25. The opportunity-cost interest rate is 10 percent per year. In a period of one year, a total of 100 units may be produced with machines **A** and **B**.

(a) Determine the optimum level of production to maximize the net profit for each alternative.

(b) Use the present worth method to determine the better alternative.

5.2 PRICING STRATEGIES

When only one firm produces a good or service, the assumptions of perfect competition may no longer be valid. Since the firm has exclusive control over the market, it may act as a monopoly. The firm cannot be undersold by competitors; therefore, it may maximize its profit by controlling *both* the level of production and the selling price of p^0. Many public projects are owned or regulated by government agencies. Since it is the only supplier and has no competition, it could act as a monopoly. The charge of government is to maximize the public good, not profit; therefore, the principles of welfare economics and other pricing strategies are used to describe government behavior. In this section we compare the theories of perfect competition, monopoly, and welfare economics, and investigate how these principles may be applied to public and private operations.

Aggregate Supply Functions

An individual firm in a perfectly competitive market will adjust its level of production q to maximize its profit. The selling price p^0 that the firm charges is determined in the market place, or

$$S = D$$

In order to estimate the selling price, *all* firms producing the good or service must be considered. In addition, it is assumed that each firm will be charging the *same price* p^0; thus

$$p_i = p^0 \qquad i = 1, 2, \ldots, n$$

or

$$p_1 = p_2 = \cdots = p_n = p^0$$

The subscript i is used to designate each of the firms.

Since each firm is assumed to maximum its individual profit, the relationship $p^0 = MC$ must hold for each firm. For an exponential cost function,

$$C_i = \delta_i q_i^{\varepsilon_i}$$

$$\text{MC}_i = \delta_i \varepsilon_i q_i^{(\varepsilon_i - 1)}$$

Since $p^0 = \text{MC}_i$, the level of production q_i may be determined as a function of selling price p^0.

$$p^0 = \delta_i \varepsilon_i q_i^{(\varepsilon_i - 1)}$$

or

$$q_i = \left(\frac{p^0}{\delta_i \varepsilon_i}\right)^{1/(\varepsilon_i - 1)}$$

The *aggregate supply* q is equal to sum of the individual outputs of each firm; thus

$$\hat{q} = \sum_{i=1}^{n} q_i$$

$$\hat{q} = \sum_{i=1}^{n} \left(\frac{p^0}{\delta_i \varepsilon_i}\right)^{1/(\varepsilon_i - 1)}$$

This relationship shows that the total supply is a function of selling price p^0, the efficiency of production of each individual firm as measured by the constant δ_i and cost elasticity ε_i, and the number of firms n. If we assume that each firm uses the same production method, then $\delta_1 = \delta_2 = \cdots = \delta_n = \delta$ and $\varepsilon_1 = \varepsilon_2 = \cdots = \varepsilon_n = \varepsilon$. The aggregate supply function is simplified to

$$\hat{q} = n\left(\frac{p^0}{\delta\varepsilon}\right)^{1/(\varepsilon - 1)}$$

and the selling price p^0 may be written as function of total aggregate supply.

$$p^0 = \delta\varepsilon\left(\frac{\hat{q}}{n}\right)^{\varepsilon - 1}$$

The effects of cost elasticity ε and the number of firms n upon p^0 are shown in Figure 5.10. For a production with no economy of scale, increasing the number of firms n will tend to shift the supply curve downward. For a production with economy of scale, the opposite will occur, or the unit selling price increases with increasing n.

Imperfect Competition

In an imperfect market, an individual supplier or small group of suppliers can influence the selling price of an item. Monopolies, duopolies, and oligopolies are terms used to describe markets consisting of one, two, and a small number of firms, respectively, having this influence on prices. Since the delivery of public services, such as water treatment and supply, wastewater collection and treatment, and electrical power

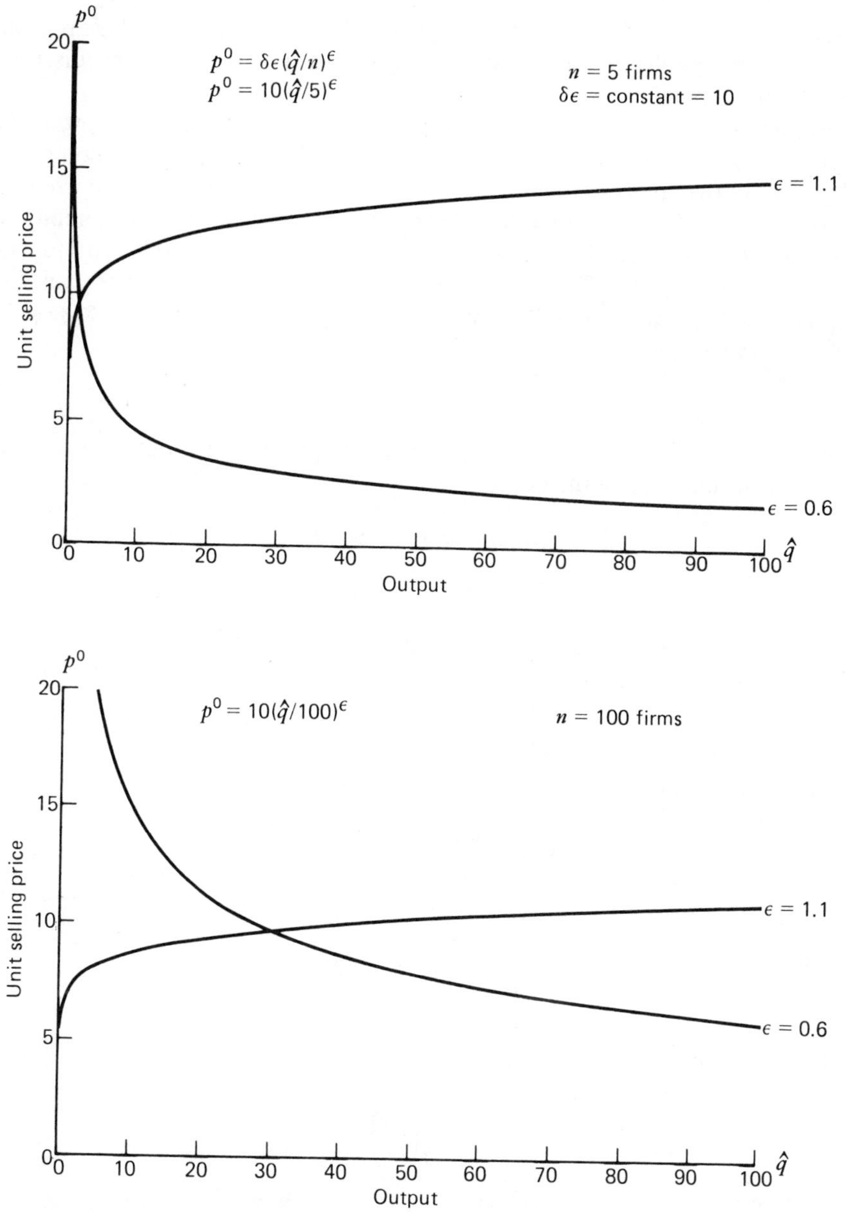

Figure 5.10 The effect of cost elasticity and number of firms upon selling price.

distribution are usually delivered by a single government agency or a firm regulated by government, public services may be considered a monopoly. The discussion focuses on monopolies.

The basic principles of calculus that have been used to maximize net profits in a perfectly competitive market apply to a monopoly in an imperfectly competitive market also. In that analysis the principle assumption is that the p^0 is independent of the production level q. For a monopoly it is assumed that a single firm can influence the selling price; therefore, the selling price p is assumed to be a variable and a function of q, the production level. The demand curve will be used to relate price p and production q. The superscript on selling price 0 is dropped to emphasize that the selling price is a variable; or

$$p = p(q)$$

The objective of the monopoly is to maximize profit where both p and C are assumed to be functions of the output q.

$$\text{Maximize } P = R - C$$

$$\text{Maximize } P = pq - C$$

The necessary condition for profit maximization is

$$\frac{dP}{dq} = \frac{dR}{dq} - \frac{dC}{dq} = 0$$

The marginal revenue is

$$\text{MR} = \frac{d(pq)}{dq} = p + q\frac{dp}{dq}$$

or

$$\text{MR} = p\left(1 + \frac{q}{p}\frac{dp}{dq}\right)$$

The demand elasticity w is equal to

$$w = \frac{p}{q}\left(\frac{dq}{dp}\right)$$

The expression for MR can be written as

$$\text{MR} = p\left(1 + \frac{1}{w}\right) = p\left(\frac{w+1}{w}\right)$$

The necessary condition for a monopolist's maximum profit is

$$\text{MR} = p\left(\frac{w+1}{w}\right) = \text{MC}$$

The sufficient condition for maximum profit of a monopoly is

$$\frac{d^2(pq)}{dq^2} - \frac{d^2C}{dq} < 0$$

or

$$\frac{d(\text{MR})}{dq} - \frac{d(\text{MC})}{dq} < 0$$

A maximum profit is realized if the slope of the marginal revenue curve MR is less than the slope of the marginal cost curve. Since the demand curve is negative, $w \leq 0$, the condition is satisfied. The optimum values of price p_m^* and q_m^* output are indicated in Figure 5.11.

The maximum net profit for a monopoly P^* is

$$P^* = p_m^* q_m^* - \bar{c}q_m^* = (p_m^* - \bar{c})q_m^*$$

where $\bar{c}$ is the average cost of production. The net profit is shown as the area $p_m^* ab\bar{c}$ in Figure 5.11. The same principles for determining whether to operate or shut down a firm, whether it is a perfectly competitive firm or a monopoly, will depend upon the average production costs. In the long term if $p_m^* > \bar{c}$, the monopoly will remain in operation; otherwise it will shut down.

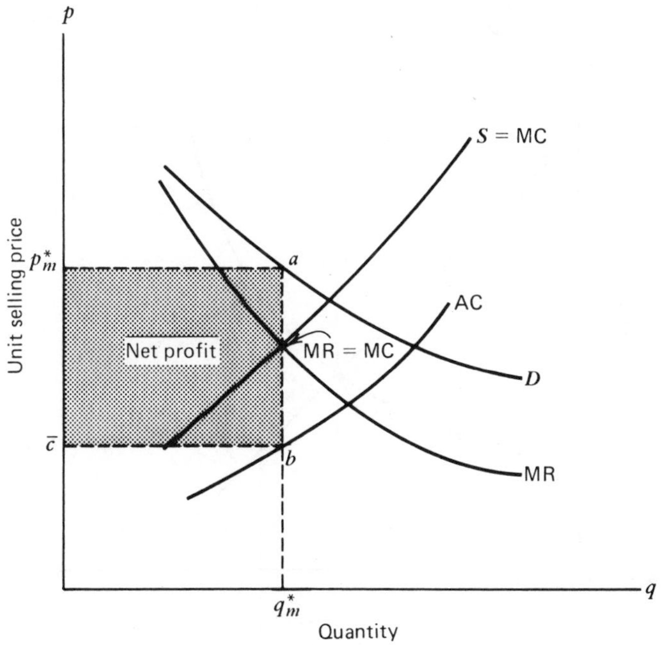

Figure 5.11 Maximum profit for a monopoly.

A monopoly is not forced to operate under conditions of imperfect competition. This is an option that is open to it. It may operate at other levels of production and selling prices. Remember that under those conditions it will not be maximizing its profit as described by the relationship MR = MC.

Compare a monopoly operating in a perfectly competitive market with the same monopoly operating in an imperfect market. In Figure 5.12, points *a* and *b* represent the necessary conditions for profit maximization in a perfect and an imperfect market, respectively. For perfect competition, the necessary condition for profit maximization is $p^0 = \text{MC}$ as indicated at point *b*. The selling price is determined from the conditions of market equilibrium, $S = D$, the intersection of the supply and demand curves. Since a monopolist controls price and production level, it will charge a higher price and provide a smaller quantity than a monopolist operating as a perfect competitor, $q^0 > q_m^*$ and $p^0 < p_m^*$.

Welfare Economics and Subsidy

The welfare of a society as a whole will depend upon the distribution of goods and services to each individual member of society. If a redistribution of goods or services is made to improve the welfare of one individual and the redistribution can be accomplished only be decreasing the welfare of both consumers and producers, then the condition of maximum welfare is achieved. When this condition is achieved, an optimum distribution of output to consumers and an optimum distribution of resources to producers are assumed. These conditions are called the condition of

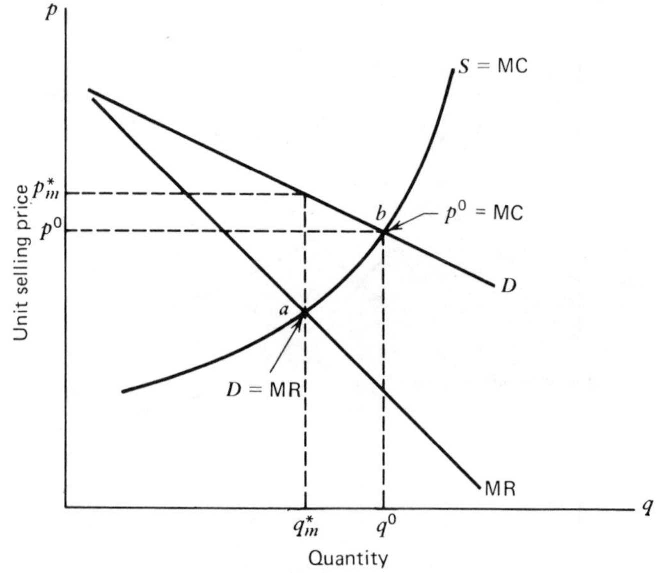

Figure 5.12 A monopoly operating under conditions of perfect and imperfect competition.

Pareto optimality. We assume the government will act as a monopoly and will distribute goods, services, and resources to maximize its distribution to *both* consumers and producers.

Society needs and desires are assumed to be quantifiable measures that can be described with supply and demand functions. This simplifying assumption makes it possible mathematically to describe an optimum welfare condition and Pareto optimality. Another critical assumption is that the conditions of perfect competition apply to a monopoly. Since the monopoly is assumed to have no influence on the selling price, the price p, or $p = p(q)$, the demand function $D = p(q)$, will be equal to the marginal cost of production. $D = \text{MC}$ as shown as the intersection of the supply and demand curves, point b of Figure 5.12.

$$p^0 - \frac{dC}{dq} = 0$$

When this condition is satisfied, it is assumed that the distribution of output to consumers and resources to producers is optimized.

The condition of Pareto optimality has important implications upon the maximum profit of the firm. The condition of Pareto optimality attempts to maximize the distribution of goods and services; therefore, the producer will *not* receive a maximum profit, and, in fact, it may have to operate at a loss. In the long term, with a firm operating at a loss the monopoly will cease operation. If the monopoly is a government agency, a *subsidy*, paid by taxes on income, real estate, and other items, may be made to the agency to keep it operating.

If the condition of Pareto optimality is relaxed, a private or government agency may establish a *nonprofit organization*. Since $P^* = (p^0 - \bar{c})q$, the level of production q and unit selling price p^0 will be established in such a manner that $P^* = 0$. No subsidy is required; however, the number of goods or services sold will not be as great, nor will the price be as low, compared with the condition of Pareto optimality.

The condition of Pareto optimality does not imply that all consumer needs will be satisfied. If the price of the good or service is reduced to a price lower than the intersection of the price and supply curves, more people will purchase it. Thus, the condition of Pareto optimality is a compromise, or neutral condition, for both consumers and suppliers.

EXAMPLE 5.3 Comparison of Different Pricing Strategies

A linear aggregate demand for a product is $p = 160 - q$. The cost of production is $C = 50q^{0.9}$. There are 10 firms that produce the product utilizing the same equipment. In other words, the same production function may be utilized for each firm. $C_i = 50q_i^{0.9}$, where $i = 1, 2, \ldots, 10$.

Determine the optimum level of production, the total net profit for the firm, and the selling price of the product for each of the following assumptions.

(a) Utilize the conditions of profit maximization for each of 10 competing firms.

(b) Utilize the conditions of imperfect competition for a monopoly, and assume a cost function of $C = 50q^{0.9}$.

(c) Utilize the condition of Pareto optimality for a monopoly for $C = 50q^{0.9}$.

Solution

(a) The selling price that each firm will charge is determined with the principle of market equilibrium or $S = D$, where $D = p = 160 - q$. A market supply curve is determined by summing the aggregate marginal costs for each firm. The marginal cost MC_i for each firm is

$$\text{MC}_i = \frac{dC_i}{dq_i} = \frac{d(50q_i^{0.9})}{dq_i} = 45q_i^{-0.1}$$

For conditions of perfect competition the selling price must equal the marginal cost, or

$$p^0 = \text{MC}_i$$

or

$$p^0 = 45q_i^{-0.1}$$

Solve for q_i in terms of the selling price p^0.

$$q_i = \left(\frac{p^0}{45}\right)^{-10} = \left(\frac{45}{p^0}\right)^{10}$$

Since each firm is assumed to be the same, the total supply $\hat{q}$ is

$$\hat{q} = \sum_{i=1}^{10} q_i = 10q_i$$

or

$$\hat{q} = 10\left(\frac{45}{p^0}\right)^{10}$$

The aggregate supply function is

$$p^0 = 45\left(\frac{\hat{q}}{10}\right)^{-0.1} = 45\left(\frac{10}{\hat{q}}\right)^{0.1} = \frac{56.7}{\hat{q}^{0.1}}$$

The total demand for the product is determined from the conditions of market equilibrium. Thus

$$S = D$$

$$\frac{56.7}{\hat{q}^{0.1}} = 160 - \hat{q}$$

The graphical solution gives the total market demand, $\hat{q} = 123$ units, and the market selling price as $p^0 = \$35.00$. (See Figure 5.13.)

The maximum net profit for each firm is determined by utilizing necessary and sufficient conditions for maximization. The net profit for each firm is

$$\text{Maximize } P_i = p^0 q_i - C_i = 35q_i - 50q_i^{0.9}$$

for $i = 1, 2, \ldots, 10$. The necessary condition is

$$\frac{dP}{dq_i} = 35 - 45q_i^{-0.1} = 0$$

The sufficient condition, or $q_i = (\frac{45}{35})^{10} = 12.3$ units for a maximum, is

$$\frac{d^2P}{dq_i} = +45q_i^{0.1} > 0$$

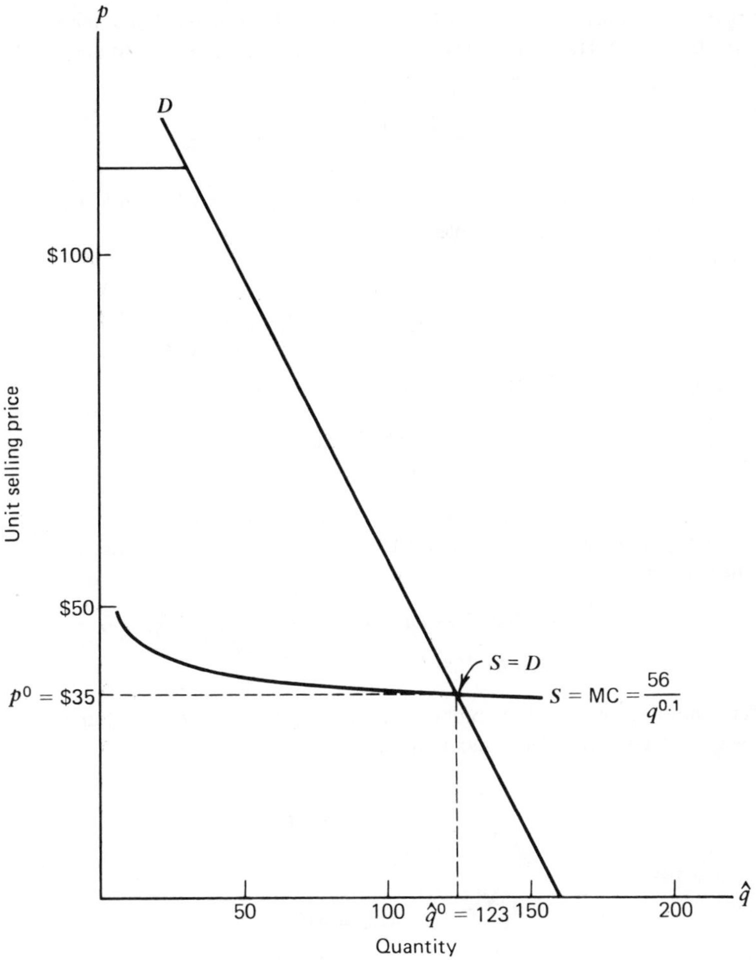

Figure 5.13 Market equilibrium for 10 competing firms.

Since $45q^{0.1} > 0$ for positive value of q, the condition for a maximum profit is not satisfied. This is the condition for a minimum.

The maximum net profit P^* is

$$P_i^* = \$35(12.3) - \$50(12.3)^{0.9}$$

$$P_i^* = \$430.50 - \$478.50$$

$$P_i^* = -\$48.00 \qquad i = 1, 2, \ldots, 10$$

Each firm will lose $48.00.

Since the cost elasticity of the production function is less than 1.0, $b = 0.9$, an economy of scale exists, which implies that the average cost production $\bar{c}_i$ decreases with increasing production. Furthermore, greater net profits will be obtained by increasing output q_i. However, in a perfectly competitive market, all 10 firms and all consumers know the selling price and the quantity of items to be sold in the entire market. It is reasonable to assume that all firms produce an equal amount, or

$$q_i = \frac{\hat{q}}{10} = \frac{123}{10} = 12.3 \text{ items/firm}$$

This value is equal to the production level q_i determined from $p^0 = MC_i$. There are *too many competing firms* to make it profitable for each of them.

The average cost function is

$$\bar{c}_i = \frac{50}{q_i^{0.1}} \qquad i = 1, 2, \ldots, 10$$

The average cost for each firm to produce 12.3 items is:

$$\bar{c}_i = \frac{50}{(12.3)^{0.1}} = \$38.90/\text{unit}$$

Note that the production cost exceeds the selling price, $\bar{c}_i > p^0$.

(b) For a monopoly under imperfect competition, the firm adjusts the price p and output q to maximize profit.

$$\text{Maximize } P = pq - C = (160 - q)q - 50q^{0.9}$$

or

$$\text{Maximize } P = 160q - q^2 - 50q^{0.9} \tag{1}$$

Under conditions of imperfect competition the price is assumed to be function of output q. The necessary condition for a maximum is

$$\frac{dP}{dq} = 160 - 2q - 45q^{-0.1} = 0$$

or equivalently,

$$MR = MC$$

$$160 - 2q = 45q^{-0.1}$$

A graphical solution (Figure 5.14) indicates that the optimum production level is $q_m^* = 65$ units. The selling price is obtained from the demand curve, or $p_m^* = \$95.00$.

The sufficient condition for a maximum is

$$\frac{d^2P}{dq^2} = -2 + 4.5q^{-1.1} < 0$$

$$\frac{d^2P}{dq^2} = -2 + 4.5(95)^{-1.1} = -1.97 < 0$$

Thus, the sufficient condition is satisfied.

At this level the average production cost is

$$\bar{c} = \frac{50}{q^{0.1}} = \frac{50}{(65)^{0.1}} = \$32.94/\text{unit}$$

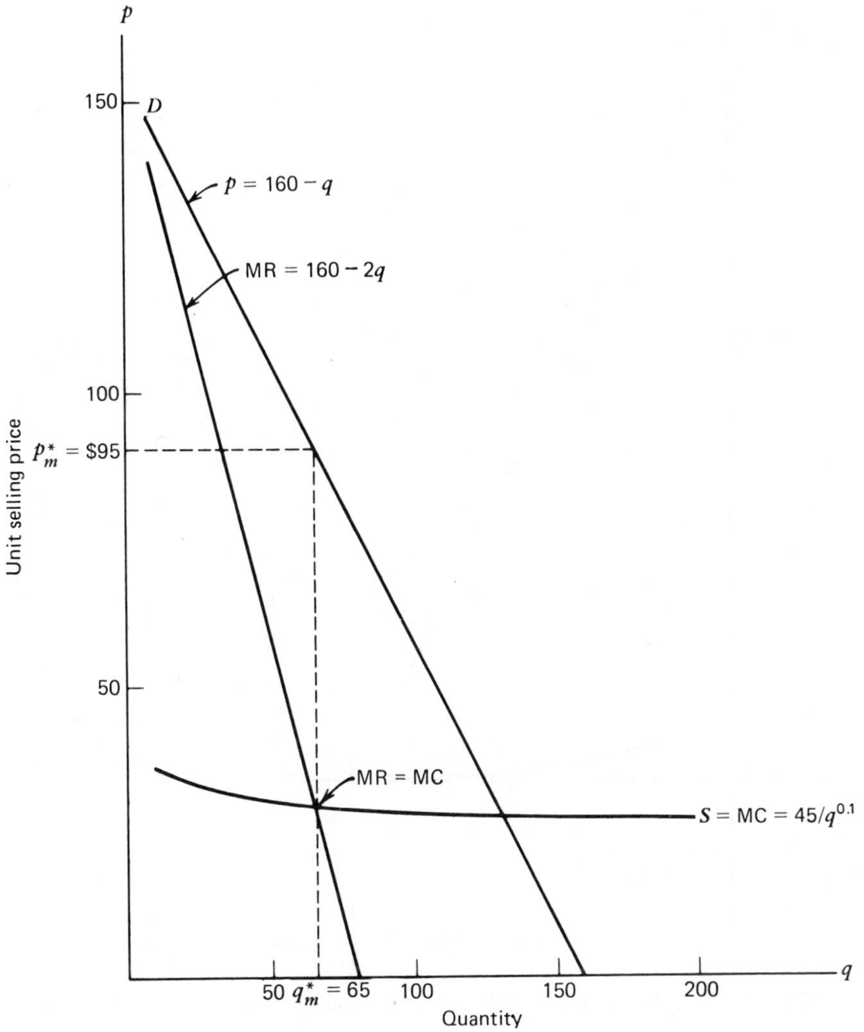

Figure 5.14 Conditions of imperfect competition for the monopoly.

The maximum net profit P^* may be obtained from Eq. (1); or

$$P^* = (p_m^* - \bar{c})q_m^*$$

$$P^* = (\$95.00 - \$32.94)65 = \$4034$$

Here, the selling price exceeds production costs $p_m^* > \bar{c}$.

(c) Under the conditions of welfare economics for a monopoly, perfect competition is assumed. (See Figure 5.15.)

$$D = \text{MC}$$

$$160 - q = 45q^{-0.1}$$

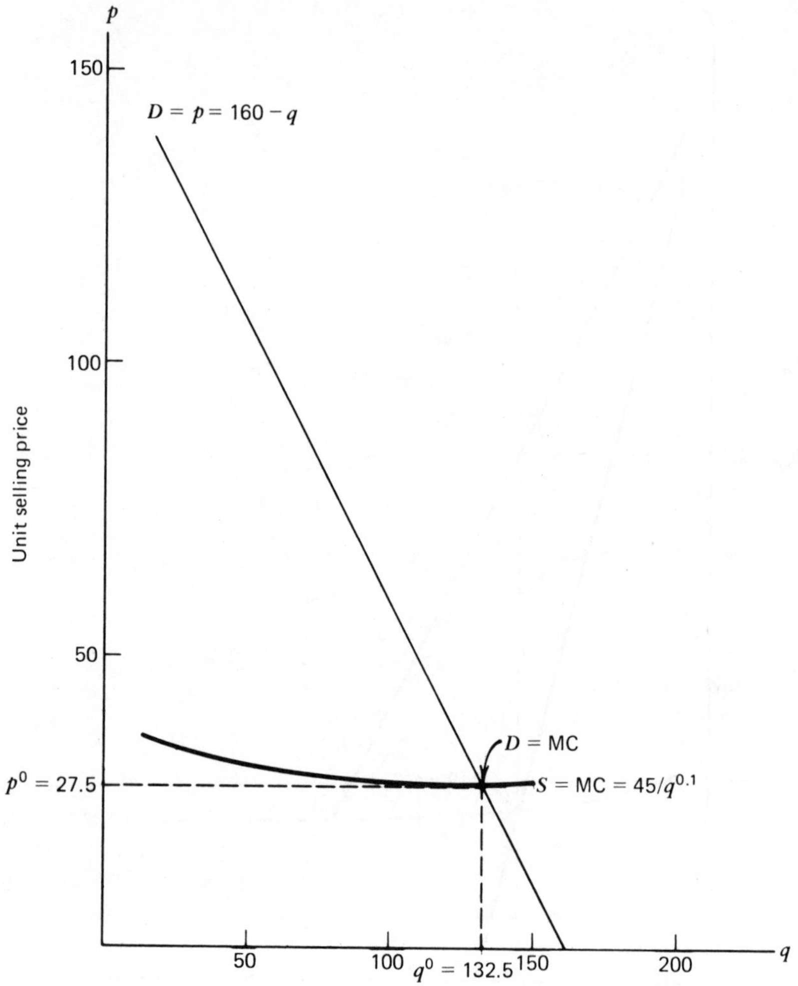

Figure 5.15 Conditions of perfect competition for the monopoly.

or

$$q^0 = 132.5 \text{ units and } p^0 = \$27.50$$

At this level of production the average cost of production is

$$\bar{c} = \frac{50}{q^{0.1}} = \frac{50}{(132.50)^{0.1}} = \$30.67/\text{unit}$$

The loss for the firm is

$$P^* = (p^0 - \bar{c})q^* = (\$27.50 - \$30.67)132.5$$
$$P^* = -\$420.02$$

The selling price is less than the average cost $p^0 < \bar{c}$.

Remarks

The following results were obtained:

CONDITION	NET PROFIT PER FIRM P^*	UNIT SELLING PRICE p^0	AVERAGE PRODUCTION COST UNIT $\bar{c}$	TOTAL OUTPUT q
10 firms perfect competition	−\$48.00	\$35.00	\$38.90	123
Monopoly: imperfect competition	\$4034	95.00	32.94	65
Monopoly: perfect competition	−\$420.03	27.50	30.67	132.5

From a consumer's point of view, the monopoly producing under the conditions of perfect competition offers the best price and the most plentiful supply of goods. The monopoly under imperfect competition offers the worst price and the least supply. Ten firms operating under perfect competition offer an intermediate price and supply.

From the producer's point of view, the monopoly producing under the condition of imperfect competition is the only one that is profitable. There are too many competing firms to make each of the 10 firms profitable. Since the selling price $p^0 = \$35.00$ and average production cost $\bar{c}_i$ is \$38.90, 10 firms cannot make a profit in the long term. In the long term, many of these firms will cease operation.

It will be possible for a smaller number of firms to produce the item at a profit. The minimum number of items that a firm must produce in order to make zero net profit occurs when the selling price equals the average cost of production. Since

$$P^* = (p^0 - \bar{c}_i)q = 0$$

or

$$\bar{c}_i = p^0$$

$$\frac{50}{q_i^{0.1}} = 35$$

$$q_i = (\tfrac{50}{35})^{10} = 35.4 \text{ items}$$

Since the total demand is $q = 123$, the number of firms n that will remain in production is estimated to be

$$n = \frac{123}{35.4} = 3.47 \text{ or } 3 \text{ firms}$$

The monopoly producing under the condition of Pareto optimality will not remain in operation in the long term. Since the condition of Pareto optimality is considered a goal that society is striving to achieve, a relaxation of the restriction upon the selling price is one way to keep the firm in operation. Another way is to subsidize the firm. If a cash payment of \$420.03 is made to the monopoly, it will break even, and conditions of Pareto optimality will be satisfied. The consumers will be able to purchase the item at a price of \$27.50. The subsidy of \$420.03 will be collected from society as a whole by charging buying taxes or through some other government fund-raising activity.

EXAMPLE 5.4 Analysis of an Irrigation System

The demand for water in a farming region is shown in Figure 5.16. Each farmer requires approximately 1 foot of water per acre per year. The demand for water is assumed to be equal to $p = 7.69 \times 10^{10} q^{-3.51}$.

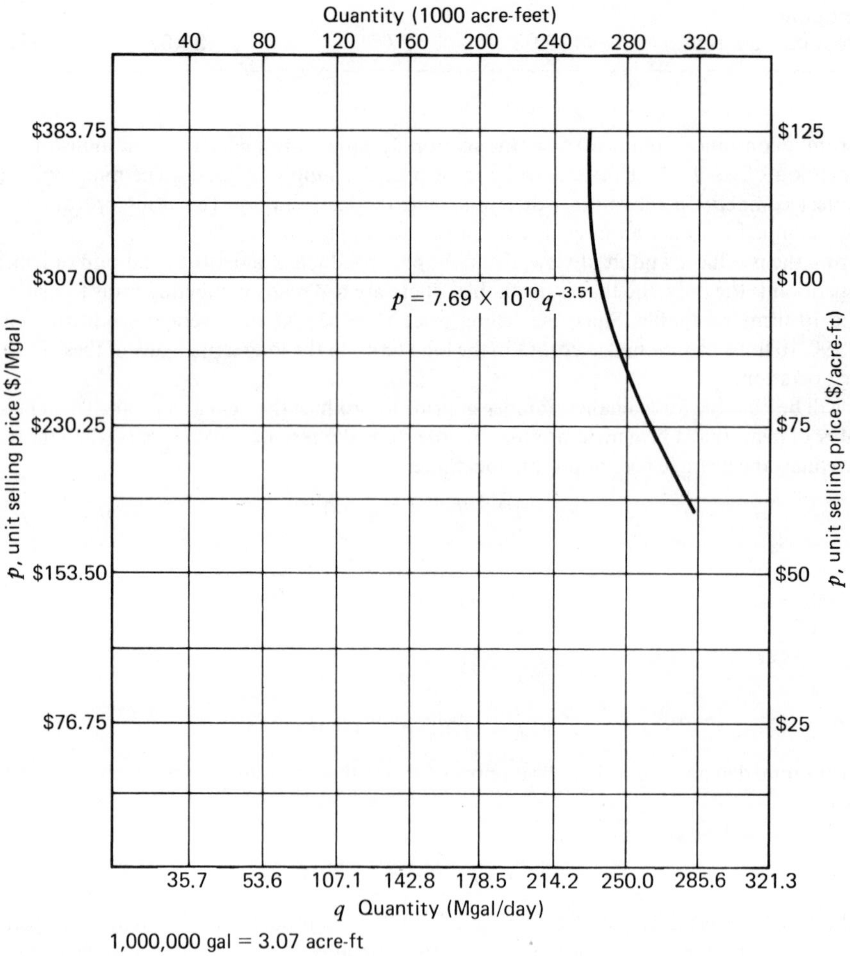

Figure 5.16

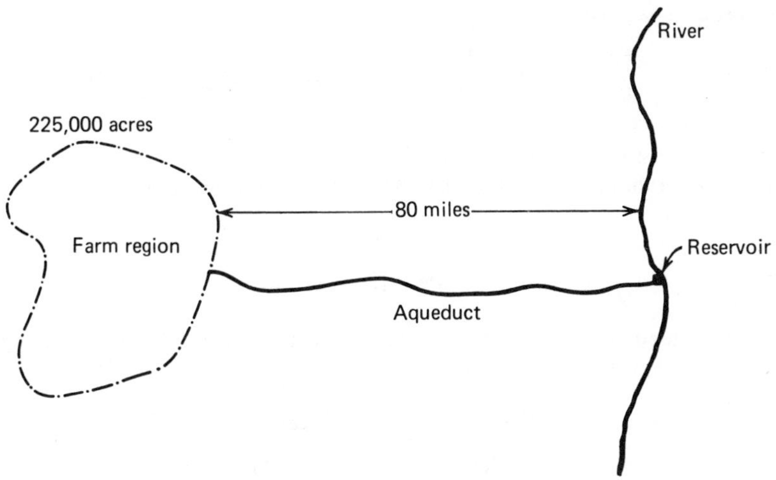

Figure 5.17

A reservoir located 80 miles away has sufficient capacity to supply the farm region. (See Figure 5.17.) The cost of constructing an aqueduct is

$$C = \$100q^{0.48}/\text{linear ft}$$

where q is measured in million gallons per day. Assume that a bond at 8 percent per annum will be sold to finance the project. Assume a design life of 40 yr.

(a) Determine the size of an aqueduct to deliver water, assuming that the water-supply agency is a nonprofit organization. Determine the unit selling price of water.
(b) Determine the size of an aqueduct, assuming the water-supply agency operates under the condition of Pareto optimality. Determine the unit selling price of water.
(c) Determine the size and unit selling price of water of an aqueduct assuming the condition of a monopoly in imperfect competition.

Solution

(a) The cost per mile of aqueduct is a function of flow

$$C = \$100q^{0.48} \cdot 5280 = \$0.528q^{0.48} \times 10^6/\text{mile}$$

and the cost for 80 miles is

$$C = 0.528q^{0.48} \cdot 80 = \$42.2q^{0.48} \times 10^6$$

The annual cost for the design life of 40 yr and an interest rate of 8 percent is determined by utilizing the capital recovery factor.

$$C = 42.2q^{0.48} \times 10^6 \left[\frac{0.08(1 + 0.08)^{40}}{(1 + 0.08)^{40} - 1}\right]$$

$$C = \$3.22q^{0.48} \times 10^6/\text{yr}$$

The cost on a daily basis is

$$C = \frac{3.22q^{0.48} \times 10^6}{365} = \$8800q^{0.48} \tag{1}$$

The average cost is $\bar{c} = C/q = 8800q^{-0.52}$.

For a nonprofit water-supply agency, the net profit P is equal to zero.

$$P^* = 0$$

$$P^* = (p - \bar{c})q = 0$$

or the average cost of supply must equal the unit selling price, $p = \bar{c}$. The unit selling price as a function of flow is equal to the demand function $p = 7.69 \times 10^{10}q^{-3.51}$. Thus

$$p = \bar{c}$$

$$7.69 \times 10^{10}q^{-3.51} = 8800q^{-0.52}$$

or

$$q = 219 \text{ Mgal}$$

The average cost and unit price is

$$\bar{c} = p = \frac{8800}{(219)^{0.52}} = \$535/\text{Mgal}$$

(b) For Pareto optimality, the demand and supply are equal or

$$D = \text{MC}$$

$$7.69q^{-3.51} \times 10^{10} = \frac{d}{dq}(0.008q^{0.48} \times 10^6)$$

$$7.69q^{-3.51} \times 10^{10} = 0.0042q^{-0.52} \times 10^6$$

$$q^{-2.99} = (5.46 \times 10^{-8})$$

$$q^0 = 268.5 \text{ Mgal}$$

The selling price of water per million gallons is

$$p^0 = 7.69q^{-3.51} \times 10^{10}$$

$$p^0 = \$229/\text{Mgal}$$

$$\bar{c} = \frac{0.0088q^{0.48} \times 10^6}{q} = 0.0088q^{-0.52} \times 10^6$$

$$\bar{c} = \$480/\text{Mgal}$$

The values of p^0, $\bar{c}$, and q^0 are shown in Figure 5.18.

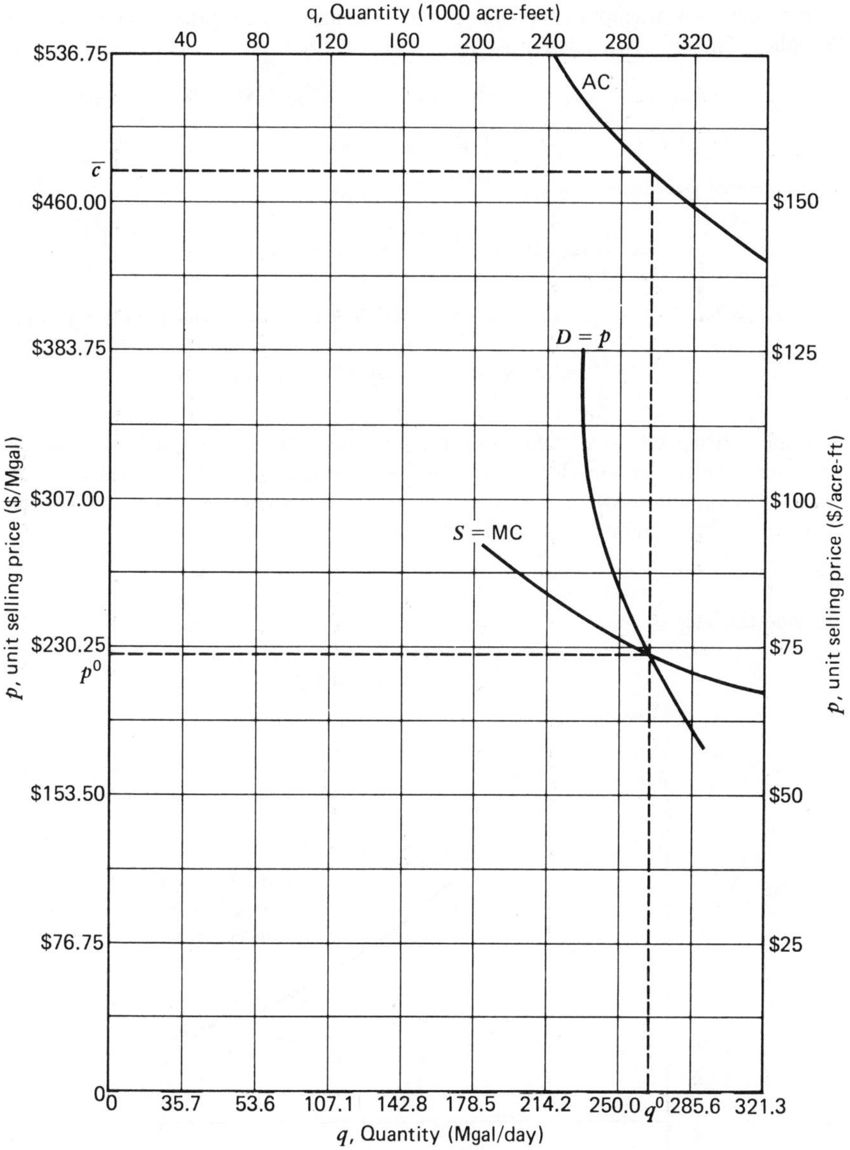

Figure 5.18 Conditions of Pareto optimality.

(c) A monopoly will attempt to maximize profits by controlling the price and quantity of water supplied. The objective function is

$$\text{Maximize } P = pq - C = (7.69q^{-3.51} \times 10^6)q - \$0.0088q^{0.48} \times 10^6$$

$$\text{Maximize } P = 7.69q^{-2.51} \times 10^{10} - 0.0088q^{0.48} \times 10^6$$

The necessary condition is

$$\frac{dP}{dq} = -19.3q^{-3.51} - 0.0042q^{-0.52} \times 10^6 = 0$$

The term $dP/dq = 0$ is equivalent to MR = MC. The sufficient condition for a maximum is

$$\frac{d^2P}{dq^2} = 67.75q^{-4.51} + 0.0022q^{-1.52} \times 10^6 > 0$$

For all q greater than 0, $d^2P/dq^2 > 0$; therefore, the sufficient condition is not satisfied. A monopoly would not consider this a viable investment at any level of q. Thus, no solution exists. The condition MR = MC cannot be satisfied, as shown in Figure 5.19, because the marginal revenue and marginal cost curves do not intersect.

Remarks

The demand elasticity is

$$w = \frac{1}{\beta} = \frac{1}{-3.51} = -0.285$$

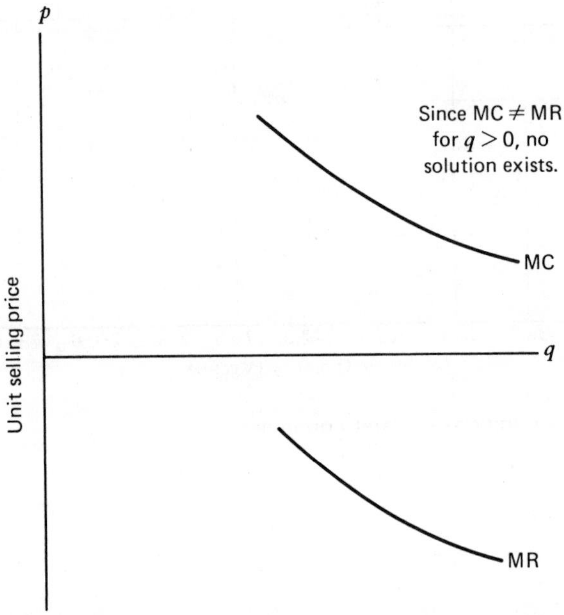

Figure 5.19

Since $-1 < w \leq 0$, it is an inelastic demand. The marginal revenue is

$$\mathrm{MR} = \frac{d(pq)}{dq} = p\left(1 + \frac{q}{p}\frac{dp}{dq}\right) = p\left(1 + \frac{1}{w}\right)$$

The marginal revenue is a positive value, $\mathrm{MR} > 0$ when $(1 + 1/w) > 0$, the rate of change of revenue will positively increase as the output q increases. When this condition is satisfied, the monopoly will maximize its profit. Contrast this result with the values in this illustrative example. The demand is inelastic, $w = -0.285$. The marginal revenue is

$$\mathrm{MR} = p\left(1 - \frac{1}{0.285}\right) = -2.51p$$

Since the marginal revenue is a negative value, the net change of revenue is negative for increasing price p. Clearly, a monopoly is interested in projects where its return of investment will result in increase revenue. Consequently, it will not be interested in the aqueduct project.

The demand for irrigation water in the farm region is large; approximately 250 Mgal is required each day. The willingness of the farmers to pay for water is reflected in the demand curve. The farmers, however, will not pay any price for water. The maximum amount they will pay is assumed to be \$383.75/Mgal. Above this price, farmers will drill for water or possibly use their land for ranching or some other activity requiring less water. Water selling at \$229/Mgal, the Pareto optimality price, is a very attractive price for farmers. However, the water supplier must pay \$480/Mgal to deliver it at that price. Since $p^0 < \bar{c}$, the supplier will require a subsidy to supply it at this price. The other alternative of changing a price equal to the average cost of delivery $p = \bar{c}$ is not attractive to the farmer. The selling price is \$563/Mgal. This is substantially greater than the maximum selling price of \$383.75/Mgal the farmers are willing to pay.

This problem illustrates why government provides certain services rather than private enterprise. Water is an essential commodity that the nation requires. The demand for it is strong, and the cost to provide it is great. In this example, it has been shown that the average cost exceeds the selling price $\bar{c} > p^0$. A private firm cannot provide water and be profitable. Government, since it has an obligation to its citizens, may have to construct the facility and the nation as a whole through its taxing system pay for it. That is, the farmers will be charged \$229/Mgal. Since this revenue will not pay for the cost of construction, a subsidy is required.

PROBLEMS

Problem 1

Consider a single firm that sells goods under two different market demand conditions as shown in Figure 5.20. Theoretically, the firm may operate as a

1. Monopoly at selling price p_m^* and at output level q_m^*.
2. Pareto optimality at selling price p^0 and at output level q^0.
3. Nonprofit organization at selling price p and at output level q.

(a) For the three theoretical operating conditions and the condition of low or high demand indicated on the graphs, p_m^*, p_m^*, and p^0, and q^0q, and q.

(b) Under what conditions does the firm operate at net profit?

(c) Under what conditions will the firm require a subsidy if the firm operates at a net loss?

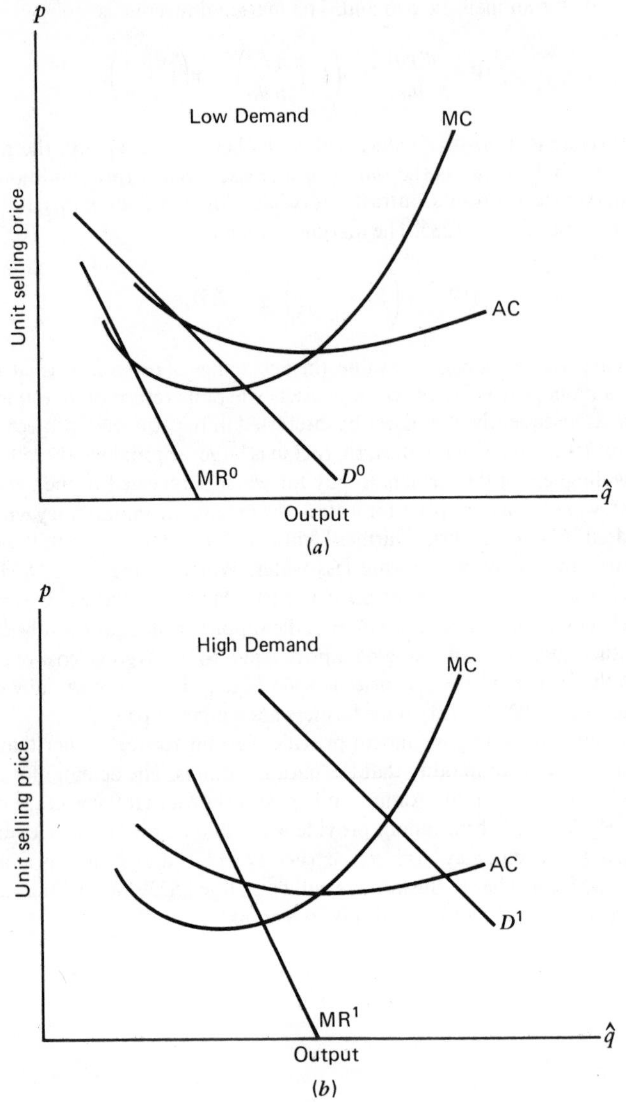

Figure 5.20

Problem 2

The demand function is $p = 800 - 4q$. The production cost function for firm A is $C = 3q^2$.

(a) Determine the selling price p^0 and quantity supplied q^0 under the condition of Pareto optimality. Draw supply and demand curves.

(b) A new firm, firm B, enters the market and competes with firm A. The production cost function for firm B is $C = 2q^2$. Determine the aggregate supply function, $p = f(q)$, for both firms A and B.

(c) Determine the new selling price p^1 and the new quantity supplied q^1 at market equilibrium for the two firms.
(d) Determine the change in consumer surplus B by introducing a new firm.
(e) Do you consider the new firm entering the market a direct social benefit, secondary benefit, or externality. Why?

Problem 3

The consumer demand for a product is given by the function $p = \$104 - 3q$. A firm's production cost is $C = 50 + 5q$. Three firms compete in this market. The total number of items produced or demanded is $\hat{q}$.

(a) Determine the aggregate supply for the three firms.
(b) Use the concept of market equilibrium to determine the selling price p and amount demanded $\hat{q}$. Draw the supply–demand curve.
(c) Determine the maximum profit for each firm in a perfectly competitive market.
(d) Determine the selling price, level of output, and profit for a monopoly under conditions of profit maximization.
(e) Determine the selling price, level of output, and profit for a monopoly under conditions of Pareto optimality.
(f) Compare the selling price, output level, average production cost, and profit for each of the three firms, a monopoly, and a monopoly under the condition of Pareto optimality.
(g) Determine the short-term price at which the firm will shut down.

Problem 4

The annual demand for prefabricated structural members is $p = 10^{10}q^{-2}$. There are 100 firms manufacturing them. They all use the same technology of prefabrication. The cost of manufacture for each firm is

$$C = 400q^{1.2}$$

(a) From principles of market equilibrium, determine the selling price p^0 and level of production q^0.
(b) Determine the average cost of production for one of these firms.
(c) What is the expected profit for one of the firms?
(d) One company has developed a new fabricating machine. The production cost function is

$$C = 450q^{1.1}$$

Determine the market equilibrium price and annual sales if the company introduces this new technology.

(e) Determine the level of production if the company operates the old and new machines.
(f) What is the total profit for the company?
(g) What is the average cost of production for the new and the old machine?
(h) Should the company abandon the old production method and utilize the new machine only?

Problem 5

The construction cost of a wastewater treatment plant is

$$C = 1.74 \times 10^6 q^{0.83}$$

where q is the flow rate in million gallons per day (Mgal/day). The plant must have the capacity to serve a town of 25,000 persons. Each person will produce approximately 100 gal/day of waste.

(a) Determine the annual user charge p^0. Assume the condition of Pareto optimality and that each person will pay an equal share of the annual cost. Assume that the plant is paid off in 10 years at an annual interest rate of 12 percent.
(b) Determine the total subsidy to construct the plant.
(c) Under Federal Law PL92-500, grant assistance for the construction of treatment plants is 75 percent. Does this pricing strategy, with user prices based upon Pareto optimality, appear to be in line with the Federal grant program?

Problem 6

The total cost of a sewer system is

$$C = 0.27q^{1.35}$$

where q is the number of persons served. The sewer system must serve 50,000 persons in an area of 30 square miles. (1 acre = 1/640 square miles.)
(a) Determine the annual user charge p^0 in dollars per acre. Assume the condition of Pareto optimality and that the sewer system is paid off in a period of 15 years at an annual interest charge of 13 percent.
(b) Will a subsidy be required?

5.3 THE ALLOCATION OF RESOURCES

We have seen how the selling price p^0, the number of firms, and economy of scale have influenced production level q. In this section we investigate how resources—workers, money, and material—are allocated to achieve the goal of maximizing profit while satisfying production and technology constraints. The condition of perfect competition will be assumed; thus the mathematical model will be written as

$$\text{Maximize } P = R - C = p^0 q - C$$

where q, the production level, and C, the cost function, are assumed to be functions of the resource vectors $\boldsymbol{x}$ or $q = g(\boldsymbol{x})$, and $C = c(\boldsymbol{x})$. In economics the relationship $q = g(\boldsymbol{x})$ is called the *production function.* In engineering the production function is called a *technological function.*

Production and Technological Functions

The production or technological function is a mathematical expression that relates the output q as a function of a set of input or control variables $\boldsymbol{x}$. It is a single-valued continuous function with first- and second-order derivatives. The output q and control variables $\boldsymbol{x}$ are restricted to positive values. These restrictions can be represented by the mathematical expressions:

$$q = g(\boldsymbol{x}) \geq 0$$

and

$$\boldsymbol{x} \geq 0$$

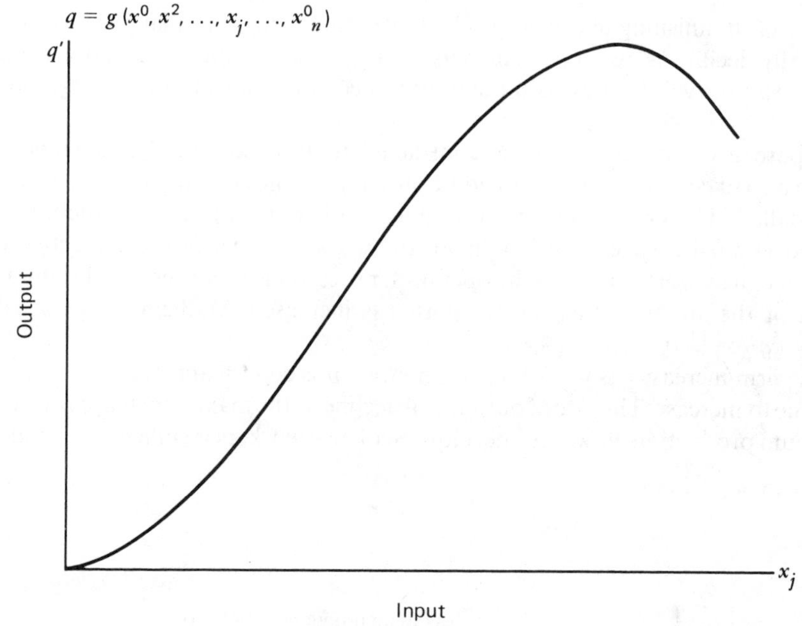

Figure 5.21 A theoretical production function.

If all but one control variable are held constant, a curve can be plotted; that is, $q = g(x_1^0, x_2^0, \ldots, x_j, \ldots x_n^0)$ where the superscript ^{0} represents a fixed value and the control variable x_j is assumed to be a variable. Figure 5.21 shows the theoretical shape of a production function.

Marginal Production

The marginal production MP_j is defined as the first partial derivative with respect to one of the input variables x_j.

$$MP_j = \frac{\partial g}{\partial x_j} \qquad j = 1, 2, \ldots, m$$

The marginal production is the slope of the production curve as shown in Figure 5.22. There are three possibilities: the slope may be positive, $MP_j > 0$; equal to zero, $MP_j = 0$; or negative, $MP_i < 0$. As the input x_j is increased, the slope of the production function changes from a positive slope to a negative slope. This implies that as the input increases the output will increase to some maximum value and then will decline. The maximum productivity occurs when the marginal productivity or slope of the productivity curve is equal to zero. The location of maximum productivity is designated by x_j'.

Law of Diminishing Marginal Productivity

The law of diminishing marginal productivity states that marginal productivity will eventually decline as an input x_j increases and the other inputs are held constant. A simple example will help explain the importance of the law of diminishing marginal returns.

Suppose a construction firm is contracted to transport sand and gravel from a pit to a road construction site. In addition, suppose the construction firm is initially understaffed. The firm decides to hire more workers to increase production. Since the understaffed crew cannot handle all the work required of them effectively, the addition of new workers makes the operation proceed more efficiently. The amount of output or the amount of earth transported is increased. Mathematically speaking, $MP_j = \partial g/\partial x_j > 0$.

If the firm increases its work force even more, it is conceivable that the output will continue to increase. The rate of output will decline as the maximum is approached. At maximum production, however, the addition of new workers is sufficient to hinder the

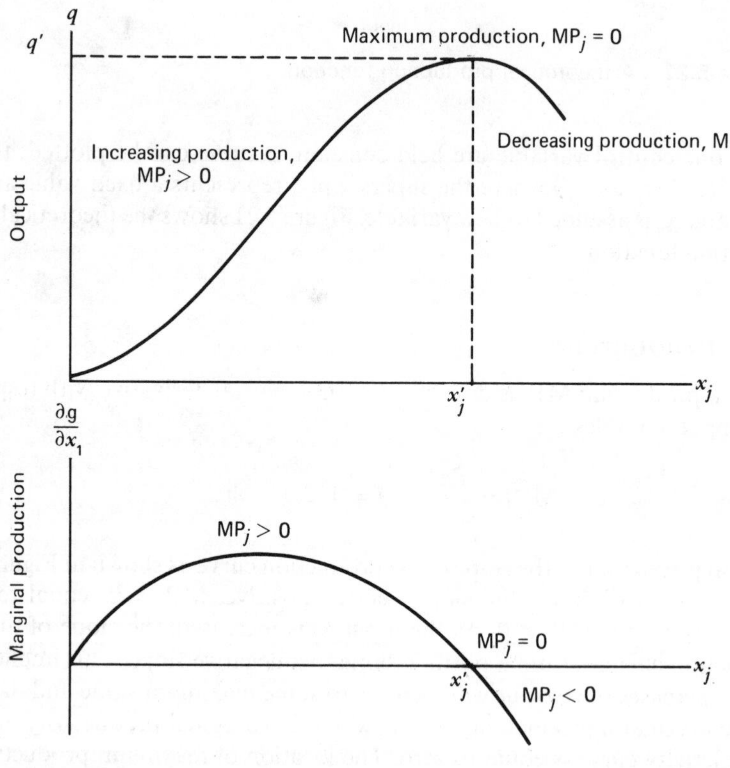

Figure 5.22 Marginal productivity.

other workers' performance. Consequently, the entire work force cannot efficiently handle their work as well anymore and production falls, $MP_j = \partial g/\partial x_j < 0$. In this range, the total output actually declines with the addition of a larger work force.

Isoquants

An *isoquant* shows the relationship among various combinations of inputs or control variables at a *fixed* production level q^0. The following relationship assumes that q^0 is fixed and the values of $\mathbf{x}$ are variables.

$$q^0 = g(\mathbf{x})$$

In two dimension, the function $q^0 = g(x_1, x_2)$ may be drawn as a contour line. Figure 5.23 shows the output as a fixed value of q^0 for various combinations of x_1 and x_2. At the output level q^0, all points on the isoquant line are equal. All combinations of x_1 and x_2 along the isoquant line produce the same output q^0. Thus, the outputs at any two points, shown as points 1 and 2 in Figure 5.23, must be equal; or

$$q^0 = g(x_1^1, x_2^1) = g(x_1^2, x_2^2)$$

Let us utilize the properties of an isoquant function as depicted in Figure 5.23 for another construction example. Suppose x_1 represents the number of bulldozers and x_2 represents the number of backhoes that are used to move earth. At point 1, the number of bulldozers is less than at point 2, $x_1^1 < x_1^2$, but the number of backhoes is

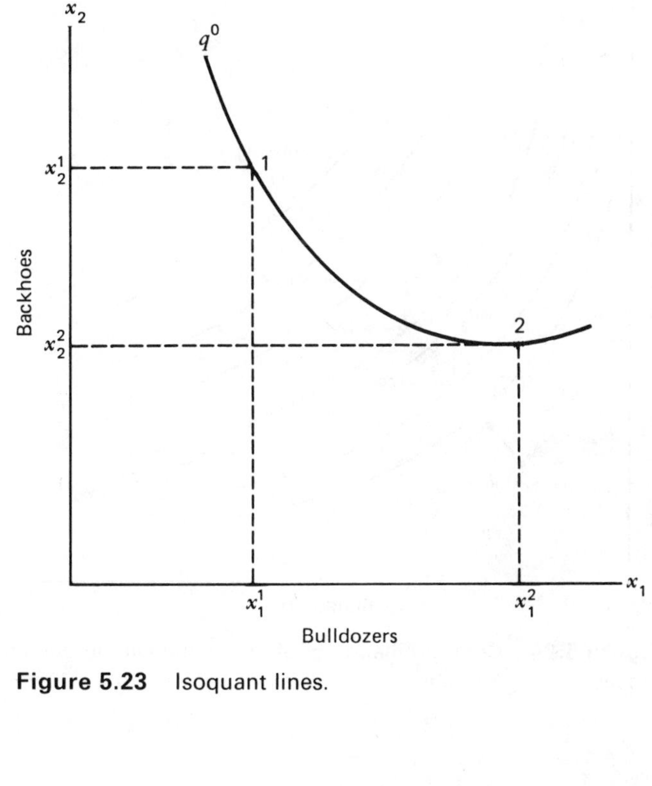

Figure 5.23 Isoquant lines.

more than at 2, $x_2^1 > x_2^2$. Either combination of (x_1^1, x_2^1) or (x_1^2, x_2^2) will result in the same amount of earth being moved. That is, $q^0 = g(x_1^1, x_2^1) = g(x_1^2, x_2^2)$.

In principle this concept can be applied to any number of input variables, $q^0 = g(x_1, x_2, \ldots, x_n)$, where n is any positive integer. Unfortunately, graphing of contour lines is restricted to two-dimensional functions. For higher-dimension production functions, mathematical procedures, not graphical procedures, will be used. This discussion will be reserved for later chapters.

The law of diminishing marginal returns can be illustrated with the use of isoquant lines. In Figure 5.24 we assume that the isoquant q' represents the maximum output. All other isoquants are less than the maximum output q'. This function is a non-monotonic function. If we position ourselves at the origin and move in the direction of the arrow, the slope of the contour is positive as we move toward the peak contour q' and then decreases on the other side of q'. Since marginal productivity is the partial derivative of productivity, as we proceed from the origin in the direction of the arrow the marginal productivities are positive, $MP_1 = \partial g/\partial x_1 > 0$ and $MP_2 = \partial g/\partial x_2 > 0$ below the maximum q'; at the maximum they are equal to zero, $MP_1 = \partial g/\partial x_1 = 0$ and $MP_2 = \partial g/\partial x_2 = 0$, and then they are less than zero on the other side of q', $MP_1 = \partial g/\partial x_1 < 0$ and $MP_2 = \partial g/\partial x_2 < 0$. In Chapter 2, we show that the solution of a constrained mathematical model involving nonmonotonic functions could have

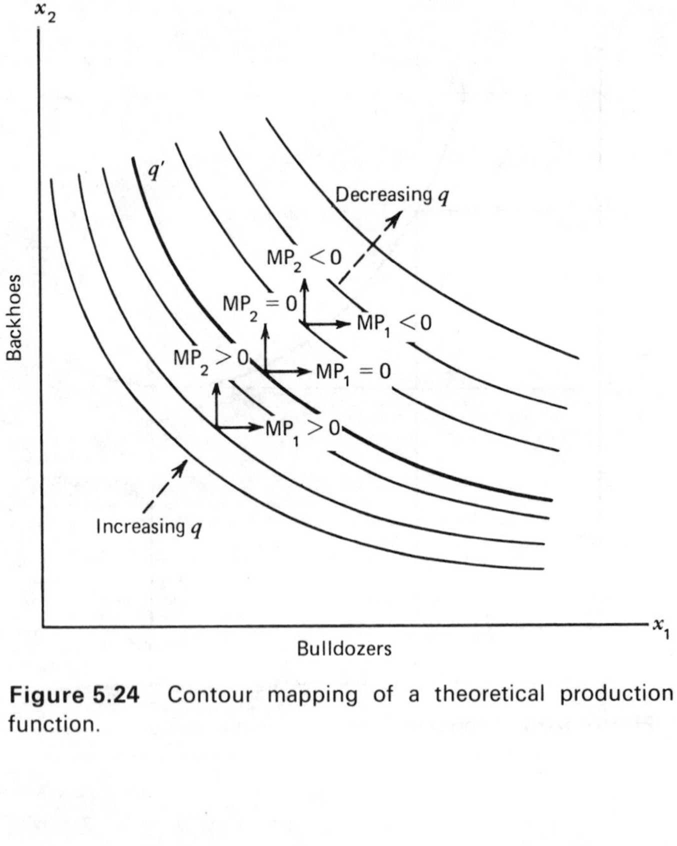

Figure 5.24 Contour mapping of a theoretical production function.

an optimum at an extreme point, boundary line, or an interior point. We shall show that the search for the optimum for profit maximization problems will be restricted to a search along the isoquant line.

Rate of Technical Substitution

Different combinations of resources x_1 and x_2 will provide a given output.

$$q^0 = g(x_1, x_2)$$

If the level of output q^0 is assumed to be held constant and one resource is changed, then there must be a corresponding change in the other resource to keep q^0 constant. This situation may be expressed with the total derivative.

$$dq^0 = \frac{dg(\mathbf{x})}{d\mathbf{x}} = \frac{\partial g(\mathbf{x})}{\partial x_1} dx_1 + \frac{dg(\mathbf{x})}{\partial x_2} dx_2 = 0$$

The *marginal productions* are defined to be the partial derivatives:

$$\text{MP}_1 = \frac{\partial g(\mathbf{x})}{\partial x_1} \quad \text{and} \quad \text{MP}_2 = \frac{\partial g(\mathbf{x})}{\partial x_2}$$

Thus, $\text{MP}_1 dx_1 + \text{MP}_2 dx_2 = 0$ along the isoquant q^0.

Rearranging this expression gives

$$\frac{-dx_1}{dx_2} = \frac{\text{MP}_2}{\text{MP}_1} = \text{RTS}$$

This ratio RTS is defined to be the *rate of technical substitution.* This expression will be used to search for maximum profit.

Optimum Allocation of Resources

Consider the unconstrained mathematical model:

$$\text{Maximize } P = p^0 g(\mathbf{x}) - c(\mathbf{x})$$

The necessary condition for profit maximization is that the partial derivations with respect to x_1 and x_2 must be equal to zero. Thus

$$\frac{\partial P}{\partial x_1} = 0 \quad \text{and} \quad \frac{\partial P}{\partial x_2} = 0$$

and

$$\frac{\partial P}{\partial x_1} = \frac{\partial}{\partial x_1} (p^0 g(\mathbf{x}) - c(\mathbf{x})) = 0$$

$$p^0 \frac{\partial g(\mathbf{x})}{\partial x_1} - \frac{\partial c(\mathbf{x})}{\partial x_1} = 0$$

Likewise

$$\frac{\partial P}{\partial x_2} = \frac{\partial}{\partial x_2}(p^0 g(\mathbf{x}) - c(\mathbf{x})) = 0$$

$$p^0 \frac{\partial g(\mathbf{x})}{\partial x_2} - \frac{\partial c(\mathbf{x})}{\partial x_2} = 0$$

The *marginal costs* are defined to be $\text{MC}_1 = \partial c(\mathbf{x})/\partial x_1$ and $\text{MC}_2 = \partial c(\mathbf{x})/\partial x_2$. The necessary condition in terms of marginal production and cost definitions is

$$p^0 \text{MP}_1 = \text{MC}_1 \quad \text{or} \quad p^0 = \frac{\text{MC}_1}{\text{MP}_1}$$

$$p^0 \text{MP}_2 = \text{MC}_2 \quad \text{or} \quad p^0 = \frac{\text{MC}_2}{\text{MP}_2}$$

Thus

$$\frac{\text{MC}_1}{\text{MP}_1} = \frac{\text{MC}_2}{\text{MP}_2}$$

or, rearranging, $\text{MP}_2/\text{MP}_1 = \text{MC}_2/\text{MC}_1$. Since the rate of technical substitution is defined to be $\text{RTS} = \text{MP}_2/\text{MP}_1$, the necessary condition for profit maximization is

$$\text{RTS} = \frac{-dx_1}{dx_2} = \frac{\text{MP}_2}{\text{MP}_1} = \frac{\text{MC}_2}{\text{MC}_1}$$

This condition is satisfied when the isoquant q is tangent to the cost function C as shown in Figure 5.25.

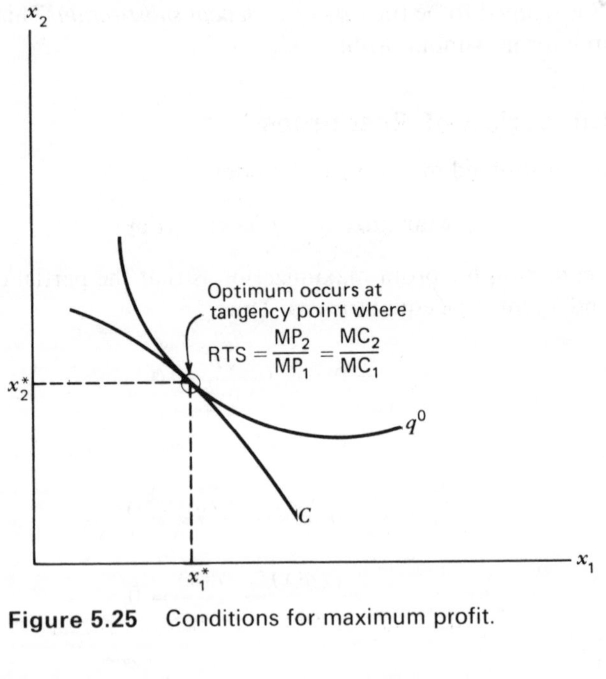

Figure 5.25 Conditions for maximum profit.

Minimum Cost Model

If the level of production is held constant at a level q^0, the mathematical model to maximize profit becomes the following constrained mathematical model:

$$\begin{aligned} \text{Maximize} \quad P &= p^0 q^0 - c(\boldsymbol{x}) \\ g(\boldsymbol{x}) &= q^0 \\ \boldsymbol{x}' &= [x_1 x_2] \end{aligned}$$

Substitute the constraint $q^0 = g(\boldsymbol{x})$ into the objective function. The model becomes

$$\text{Maximize } P = p^0 g(\boldsymbol{x}) - c(\boldsymbol{x})$$

This model is equivalent to one we just solved. The necessary condition for profit maximization is

$$\text{RTS} = \frac{-dx_1}{dx_2} = \frac{\text{MP}_2}{\text{MP}_1} = \frac{\text{MC}_2}{\text{MC}_1}$$

The optimum solution will occur at the tangency point of q^0 and C as shown in Fig. 5.25.

A different interpretation may be given to this problem. Let the revenue R be a constant, or $R = p^0 q^0$. Consequently, the only way to maximize profit is to minimize the production cost subject to the production constraint.

$$\begin{aligned} \text{Minimize} \quad C &= c(\boldsymbol{x}) \\ g(\boldsymbol{x}) &= q^0 \\ \boldsymbol{x}' &= [x_1, x_2] \end{aligned}$$

Since this is an equivalent model to the profit maximization model, the location of the minimum cost solution $\boldsymbol{x}^*$ will occur along the isoquant q^0 and will be tangent to cost function C.

Maximum Output Model

If the production cost is fixed C^0, the profit maximization model becomes the constrained mathematical model:

$$\begin{aligned} \text{Maximize} \quad P &= p^0 q - C^0 = p^0 q(\boldsymbol{x}) - C^0 \\ C^0 &= c(\boldsymbol{x}) \\ \boldsymbol{x}' &= [x_1 x_2] \end{aligned}$$

By substituting the cost constraint $C^0 = c(\boldsymbol{x})$ into the objective function, we obtain the same mathematical model described previously. Again, the solution to this problem is shown in Figure 5.25 where $C = C^0$.

Since the production cost is constant, an equivalent model is

$$\begin{aligned} \text{Maximize} \quad q &= g(\boldsymbol{x}) \\ C^0 &= h(\boldsymbol{x}) \\ \boldsymbol{x}' &= [x_1 x_2] \end{aligned}$$

Here, the output q is maximized subject to a fixed cost constraint. Since the cost is a constant, the total derivative of the cost $C = c(\mathbf{x})$ must be equal to zero.

$$dC = \frac{\partial h(\mathbf{x})}{\partial x_1} dx_1 + \frac{\partial h(\mathbf{x})}{\partial x_2} dx_2$$

or

$$dC = \text{MC}_1\, dx_1 + \text{MC}_2\, dx_2 = 0$$

Thus,

$$\frac{-dx_1}{dx_2} = \frac{\text{MC}_2}{\text{MC}_1}$$

Since RTS $= -dx_1/dx_2$ and RTS $= \text{MP}_2/\text{MP}_1$ the necessary condition for profit maximization becomes

$$\frac{\text{MP}_2}{\text{MP}_1} = \frac{\text{MC}_2}{\text{MC}_1}$$

Again, the optimum solution $\mathbf{x}^*$ will occur along the isocost line where $C = C^0$ and will be tangent to the isoquant line q.

It is implied here that the optimum solution to a profit maximization, least cost, and maximum production models will *always* occur at a point of tangency. But this is *not always* the case. Resource constraints and technological, economic, and other constraint conditions may be imposed also. These constraints may cause the optimum solution to occur either at a tangency point or at an extreme point. As a result, generalizations on where the optimum solution will be located cannot be made. A thorough investigation of the model must be made. In this section, since only two-dimensional models are being investigated, the use of graphical methods greatly enhances the investigation.

EXAMPLE 5.5 Strength of a Simply Supported Beam

Consider a simply supported beam like that in Figure 5.26. Assume that the cross section of the beam is rectangular and the control variables are the beam depth h and beam width b. The steel beam is stressed to its allowable yield strength of 30,000 psi.

(a) Use the basic principles of statics and strength of materials to derive a production function for a simply supported steel beam.

(b) Draw production functions as a function of beam depth and beam width, respectively.

(c) Graph the isoquant functions for $P = 10{,}000$ and 50,000 lb.

Solution

(a) The output q will be equal to the concentrated load placed on the beam at its midspan. The output will be expressed as a function of the control variables:

$$h = \text{beam depth (in.)}$$

$$b = \text{beam width (in.)}$$

All other properties are assumed to be constants. Thus, the allowable stress and beam length are

$$\sigma = 30{,}000 \text{ psi}$$

$$l = 10 \text{ ft} = 120 \text{ in.}$$

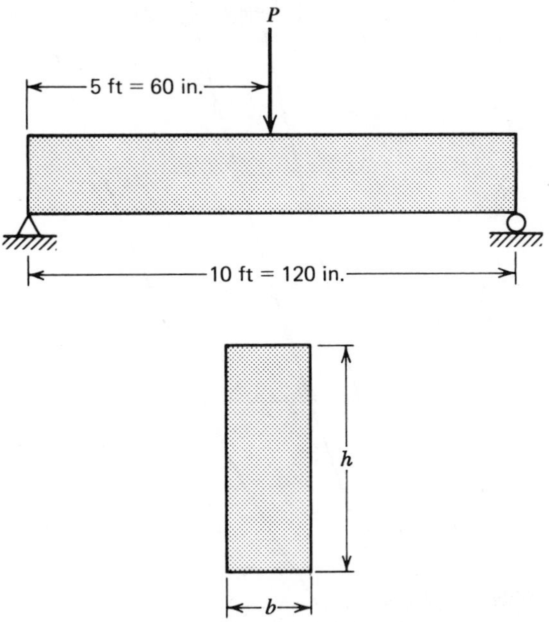

Figure 5.26 A simply supported beam.

From statics, the maximum moment at the midspan of the beam is equal to

$$M = \frac{Pl}{4} = \frac{P(120)}{4} = 30P$$

From strength of materials, the maximum bending stress is equal to

$$\sigma = \frac{Mc}{I}$$

where the area moment of inertia for a rectangular section is $I = bh^3/12$. The distance, from the neutral axis and extreme fiber c, is

$$c = \frac{h}{2}$$

Thus, the maximum stress is

$$\sigma = M\left(\frac{h}{2}\right)\frac{12}{bh^3} = \frac{6}{bh^2}$$

Substitute $M = 30P$ and the allowable stress $\sigma = 30{,}000$ psi into the equation and simplify.

$$\sigma = \frac{6}{bh^2}(30P) = 30{,}000$$

By solving for P, the production function $q = g(b, h)$ is

$$q = P = 167bh^2$$

(b) The maximum concentrated load that a beam of width b^0 can sustain is

$$q = P = 167b^0h^2$$

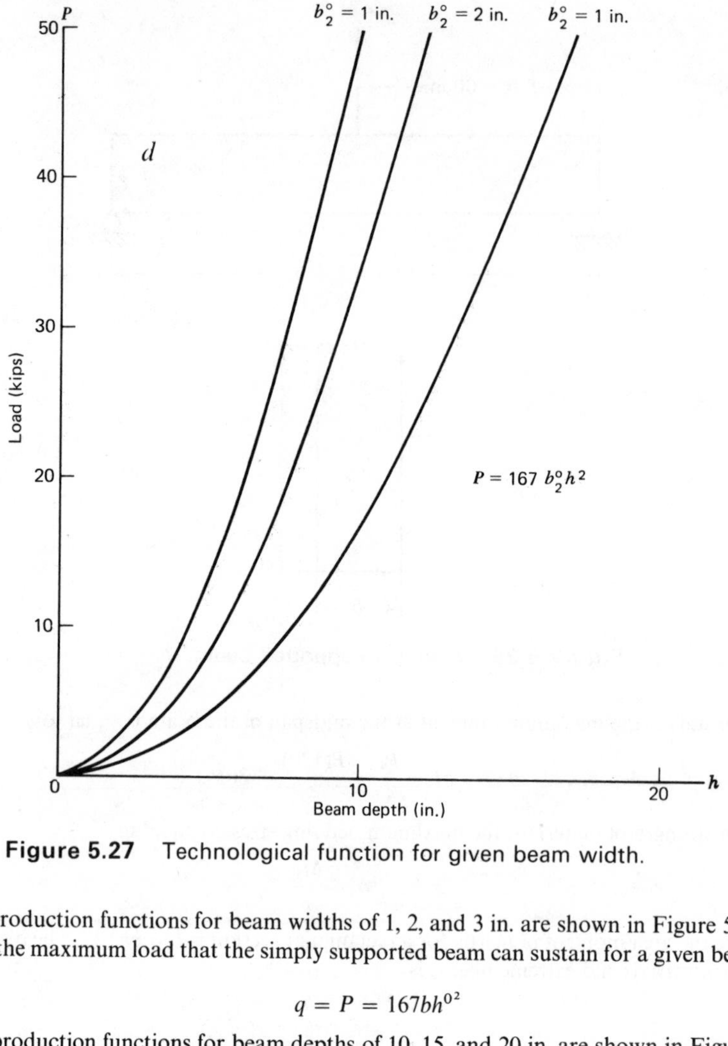

Figure 5.27 Technological function for given beam width.

The production functions for beam widths of 1, 2, and 3 in. are shown in Figure 5.27. Likewise, the maximum load that the simply supported beam can sustain for a given beam depth h^0 is

$$q = P = 167bh^{0^2}$$

The production functions for beam depths of 10, 15, and 20 in. are shown in Figure 5.28.

c) The isoquant for $q^0 = P = 10{,}000$ lb is

$$g(b, h) = q^0$$
$$167bh^2 = 10{,}000$$
$$bh^2 = 59.9$$

Likewise, the isoquant for $P = 50{,}000$ lb is

$$g(b, h) = q^0$$
$$167bh^2 = 50{,}000$$
$$bh^2 = 299.4$$

A graph of the isoquants is shown in Figure 5.29.

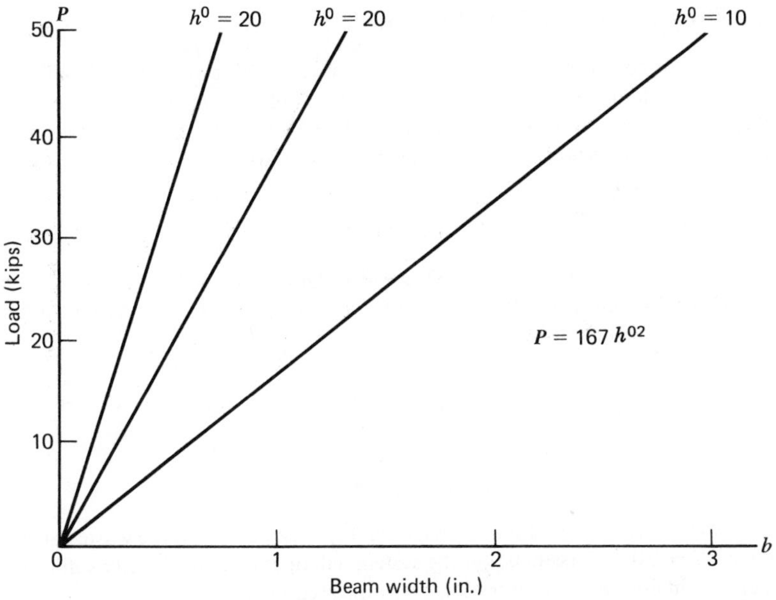

Figure 5.28 Technological function for given beam depth.

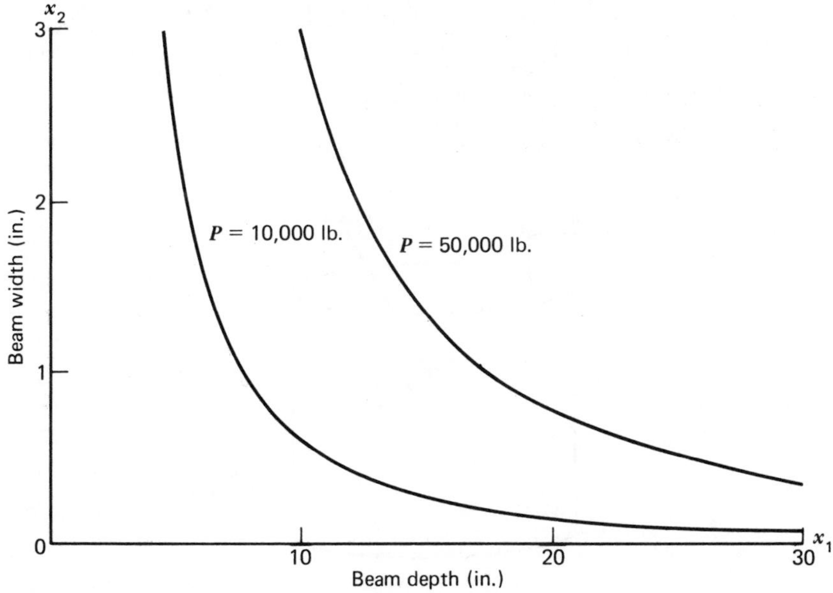

Figure 5.29

EXAMPLE 5.6 A Reactor Tank System

Significant savings in construction may exist if a chemical reactor system is placed in series as shown in the schematic diagram. A waste liquid passes through the system at a constant flow rate in million gallons per day. Chemicals are added to each tank to reduce the concentration in milligrams per liter of waste from S_0 to S. (See Figure 5.30 for the two systems under consideration.)

The cost of a single tank is

$$C = 0.1V^{0.6}$$

where C is measured in millions of dollars ($M), and V is measured in millions of gallons (Mgal). The flow rate Q is 1 Mgal/day and the specific reaction rate μ is 0.01/min. Assume a first-order chemical reaction, where the utilization rate is

$$\frac{dS}{dt} = -\mu S$$

or

$$S = S_0 e^{-\mu t}$$

The rate of removal is a function of μ and time t. Ninety-nine percent of the waste concentration will be removed, $S = 0.01S_0$. Assume that the system will operate under steady-state conditions. That is, the change in flow and concentration over time is equal to zero.

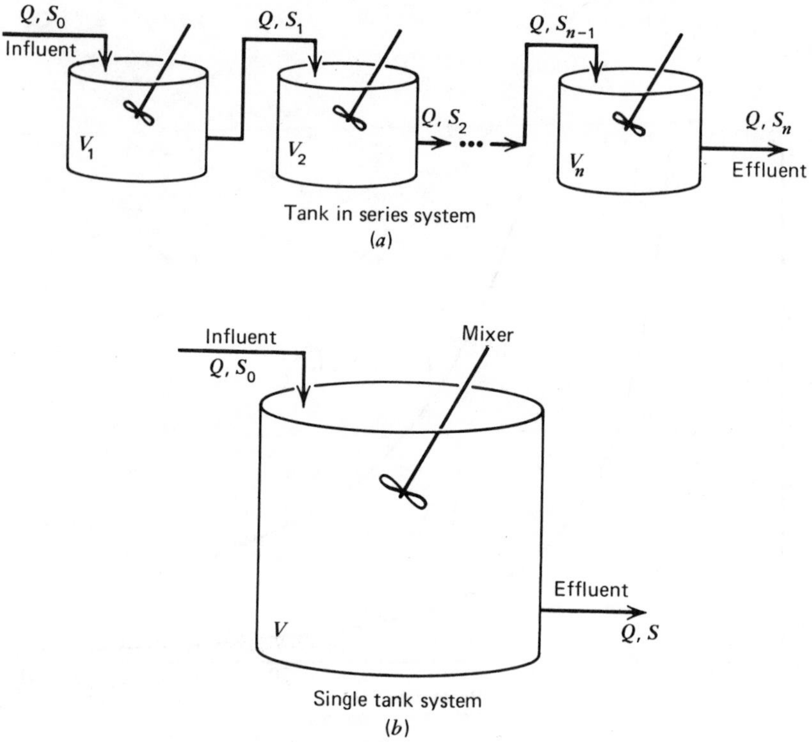

Figure 5.30

(a) Formulate a minimum cost model for a single tank and the optimum tank volume V.
(b) Formulate a minimum cost model for n tanks in series, and determine the optimum number of tanks n and the volume of all tanks in the system V.
(c) Compare the average cost for the single-tank and multiple-tank systems.

Solution

(a) A mass balance on the reactant or waste liquid may be written for a single tank as shown in Figure 5.30(b).

$$\text{Accumulation} = \text{Inflow} - \text{Outflow} - \text{Net utilization}$$

$$V\frac{dS}{dt} = QS_0 - QS - \mu SV$$

For steady-state conditions, $dS/dt = 0$.

$$QS_0 - QS - \mu V$$

or

$$V = \frac{Q}{\mu}\left(\frac{S_0 - S}{S}\right)$$

The minimum cost model is

$$\text{Minimize } C = 0.1V^{0.6}$$

$$V = \frac{Q}{\mu}\left(\frac{S_0 - S}{S}\right)$$

$$V \geq 0$$

Since Q, μ, S_0, and S are specified, the tank volume V may be determined.

$$V = \frac{1}{14.4}\left(\frac{S_0 - 0.01S_0}{0.01S_0}\right) = 6.87 \text{ Mgal}$$

where $$\mu = \frac{0.1}{\text{day}} \cdot 60\,\frac{\text{min}}{\text{hr}} \cdot 24\,\frac{\text{hr}}{\text{day}} = \frac{14.4}{\text{day}}$$

The cost is: $$C = 0.1 \cdot (6.87)^{0.6} = \$0.318\text{M} = \$318{,}000$$

No control over the design of the single-tank system is possible. The parameters Q, u, S_0, and S are specified; therefore, the tank volume V is specified and, in turn, the cost cannot be optimized.

(b) The control variables for tank reactors in series are

n = Number of tanks

V = Volume of the n tanks or the entire system, M/gal

Assume that each tank of the system has the same volume, or $V_i = V/n$. See Figure 5.31. The cost of the system in terms of n and V is

$$C = n\left(0.1\left(\frac{V}{n}\right)^{0.6}\right) = 0.1n^{0.4}V^{0.6}$$

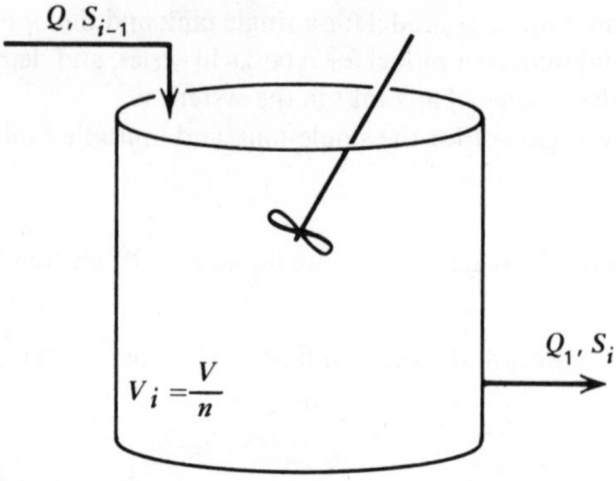

Figure 5.31

The constraint equation may be determined by writing a mass balance on one of the tanks in the system as shown in Figure 5.30(*a*). The mass balance for tank i is

$$\frac{V}{n}\left(\frac{dS_i}{dt}\right) = QS_{i-1} - QS_i - \mu\left(\frac{V}{n}\right)S_i \qquad i = 1, 2, \ldots, n$$

For steady-state conditions, $dS_i/dt = 0$.

$$Q(S_{i-1} - S_i) - \frac{\mu V}{n} S_i = 0$$

or

$$\frac{S_i}{S_{i-1}} = \frac{1}{1 + \mu V/nQ} \qquad i = 1, 2, \ldots, n$$

By repeated substitution, the relationship between inflow and outflow concentration may be obtained.

$$\frac{S_n}{S_0} = \frac{1}{(1 + uV/nQ)^n}$$

where $S_n = S = 0.01S_0$. The tank volume of the system is

$$\frac{\mu V}{n}\left[\left(\frac{S_0}{S_n}\right)^{1/n} - 1\right]^{-1} = Q$$

or for $Q = 1$ Mgal/day:

$$\frac{14.4V}{n}\left[\left(\frac{S_0}{0.01S_0}\right)^{1/n} - 1\right]^{-1} = \frac{14.4V}{n}[100^{1/n} - 1]^{-1} = 1$$

The minimum cost model is

$$\text{Minimize } C = 0.1n^{0.4}V^{0.6}$$

$$\frac{14.4V}{n}[100^{1/n} - 1]^{-1} = 1$$

$$V \geq 0 \qquad n = 1, 2, \ldots$$

The optimum solution may be obtained by substituting the active constraint equation into the objective function. The objective function may be written as a function of the single initial variable n. The necessary condition for minimum cost is $dC/dn = 0$. This approach is entirely adequate; however, the marginal analysis approach and the use of graphs give better insight into the nature of the optimization process.

Consider the constraint equation. Rewrite it that V is a function of n,

$$V = \frac{n}{14.4}[100^{1/n} - 1]$$

The relationship between V and n is shown as the isoquant or production function. This figure shows a sixfold reduction in tank volume for a system consisting of two or more tanks. (See Figure 5.32.)

The necessary condition for minimum cost is

$$\frac{\mathrm{MP}_2}{\mathrm{MP}_1} = \frac{\mathrm{MC}_2}{\mathrm{MC}_1}$$

where the production is equal to the flow. Thus $\mathrm{MP}_1 = \partial Q/dn$ and $\mathrm{MP}_2 = \partial Q/V$ and $\mathrm{MC}_1 = \partial C/\partial n$ and $\mathrm{MC}_2 = \partial C/\partial V$.

The optimum solution is shown in Figure 5.33 as the tangent of the isocost and isoquant functions. The optimum solution is

$$V^* = 0.600 \text{ Mgal} \qquad n^* = 4 \qquad C^* = \$128{,}000 = \$0.128\text{M}$$

The single-tank system cost \$318,000 and had a volume of 6.87 Mgal. Clearly, the tank system is the better choice.

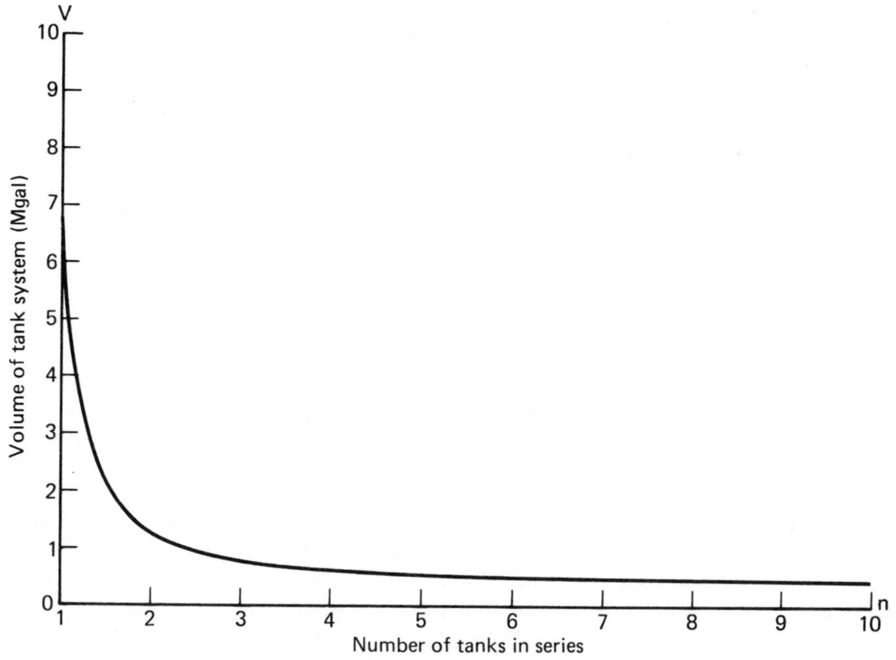

Figure 5.32 Isoquant for Q = 1 Mgal/day.

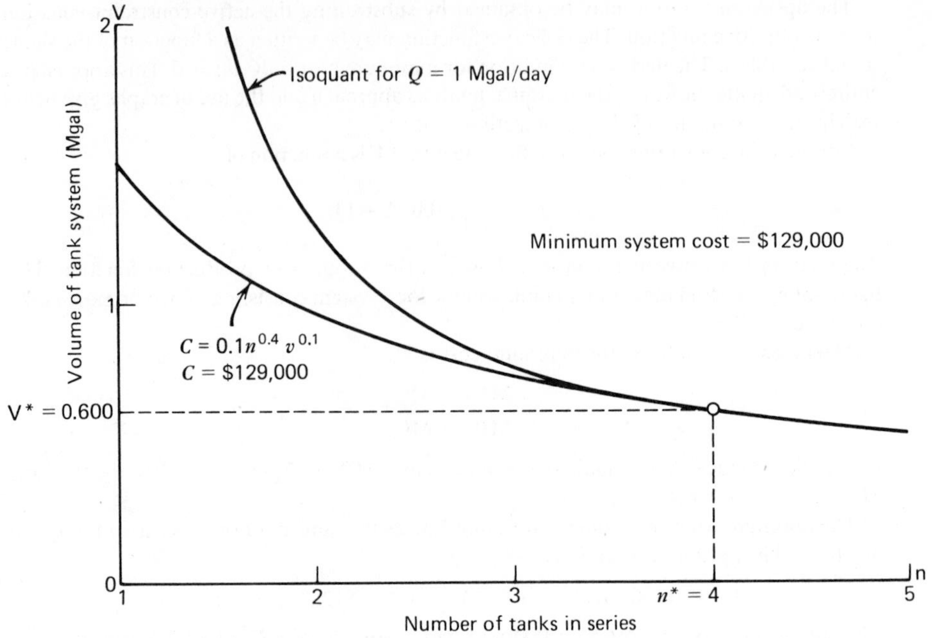

Figure 5.33

(c) The average costs for the single and five-tank systems are

$$\bar{c} = \frac{C}{Q}$$

or

$$\bar{c} = \frac{C}{Q} = \frac{\$0.318\text{M}}{1 \text{ Mgal}} = \frac{\$0.32}{\text{gal}}$$

for the single tank and

$$\bar{c} = \frac{C}{Q} = \frac{\$0.129\text{M}}{1 \text{ Mgal}} = \frac{\$0.13}{\text{gal}}$$

for the tank system. A saving of \$0.19/gal is realized with the tank system.

Remarks

In this problem, we have seen how *both technology* and *economics* may be used to reduce costs. The importance of the cost relationship cannot be overemphasized. In this example, an economy of scale was assumed. The cost elasticity of $C = 0.1V^{0.6}$ is 0.6, which is less than 1. If the cost elasticity is greater than 1, no economy of scale exists and an entirely different result occurs. For optimality, n approaches infinity. This is the condition for a plug flow reactor. This is a different type of process consisting of a single long tank with a high length-to-width ratio. In this instance, the continuous flow-stirred tank reactor in series would be abandoned for a plug flow reactor.

EXAMPLE 5.7 A Minimum-Construction-Cost Pump–Pipeline System

Design a pump–pipeline system to deliver 6 Mgal/day (9.3 ft^3/sec) of water from a river to a reservoir. The distance between the river and reservoir is 500 ft, and the elevation head is 20 ft. Minimize the cost of construction. The pump efficiency is assumed to be 85 percent. Only pumps that are rated 32 hp or less are available from suppliers. See Figure 5.34.

(a) State the basic assumptions.
(b) Establish technological functions for the individual components of the system.
(c) Establish cost functions for the individual components of the system.
(d) Formulate a mathematical model to minimize capital costs.
(e) By graphical means determine the minimum cost solution.

Solution

(a) The pump must have sufficient power to deliver a volume flow rate Q of 9.3 ft^3/sec, elevate the water 20 ft, and overcome pipe friction. In other words, the pump must deliver a pump head equal to the elevation head h plus the head loss h_l in the pipe. See Figure 5.35. Let

h = elevation head = 20 ft

h_l = Pipe head due to friction loss (ft)

$(h + h_l)$ = Pump head

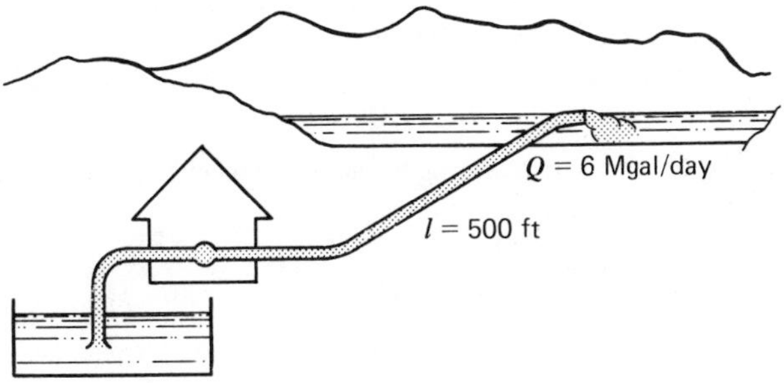

Figure 5.34 Pump-pipeline system.

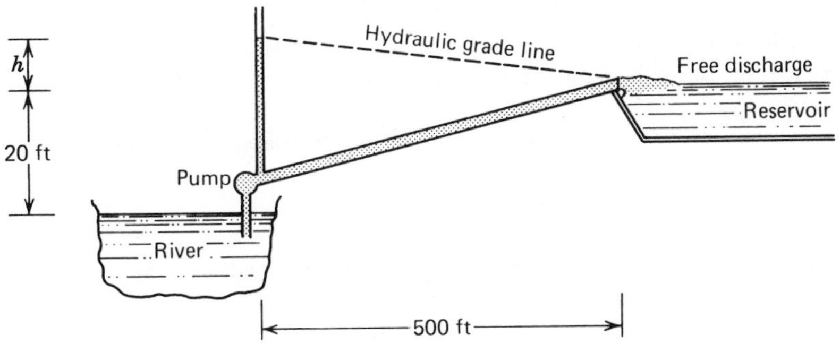

Figure 5.35 Hydraulic considerations.

Since the volume flow rate Q and head loss are functions of the pump power and pipe diameter, we must determine the sizes of pump and pipe so they work as a compatible system. Thus, the control variables are defined to be

$p =$ Brake horsepower of the pump (hp)

$d =$ Diameter of circular pipe (in.)

Our goal is to select the optimum pump size p and pipe diameter d such that the total cost of construction C is minimized and all technological conditions are met. The technological conditions are written in the form of constraint equations. At this time we do not know the details of the mathematical model, only its generic form, which can be stated as

$$\text{Minimize } C = h(p, d)$$

$$g_i(p, d)\{=, \leq, \leq\}q_i \qquad i = 1, 2, \ldots, m$$

After establishing the production and cost functions for the pipe and pump, we shall return to the formulation stage and determine explicitly a least-cost mathematical model.

(b) Basic principles from fluid mechanics will be used to determine the technological functions for the pump and pipe.

The pump efficiency e may be written as

$$e = \frac{Q\gamma(h + h_l)}{550p}$$

where e, is the pump efficiency, equals 85 percent, or 0.85; Q is the volume rate of flow in cubic feet per second; and γ is the specific weight of water, which is 62.4 lb/ft^3.

The pump technological function Q will be expressed as a function of the pipe head loss h_l and the pump brake horsepower p:

$$q_1 = g_1(h_l, p_1) = Q$$

The function may be obtained by rearranging the pump efficiency equation and substituting all known constants into it:

$$Q = \frac{550ep}{\gamma(h + h_l)} = \frac{550(0.85)p}{62.4(20 + h_l)} = \frac{7.49p}{20 + h_l}$$

A graph of the pump technological function for various values of pipe head loss h_l is shown in Figure 5.36.

The Hazen–Williams equation for pipe flow will be used to determine the pipe technological function. The Hazen–Williams empirical formula for mean velocity in feet per second is

$$V = 1.318C_1R^{0.63}S^{0.54}$$

where C_1 is the pipe roughness coefficient and equals 100 (the pipe is assumed to be relatively rough); R is the hydraulic radius, in feet; and S is the slope of the hydraulic grade line, which is h_l/l.

The hydraulic radius is assumed to be equal to the ratio of pipe area to pipe perimeter:

$$R = \frac{\pi d^2}{4\pi d} = \frac{d}{4}$$

where d, the pipe diameter, is measured in units of feet. The diameter d may be written in terms of the control variable d measured in units of inches, thus

$$R = \frac{d}{48}$$

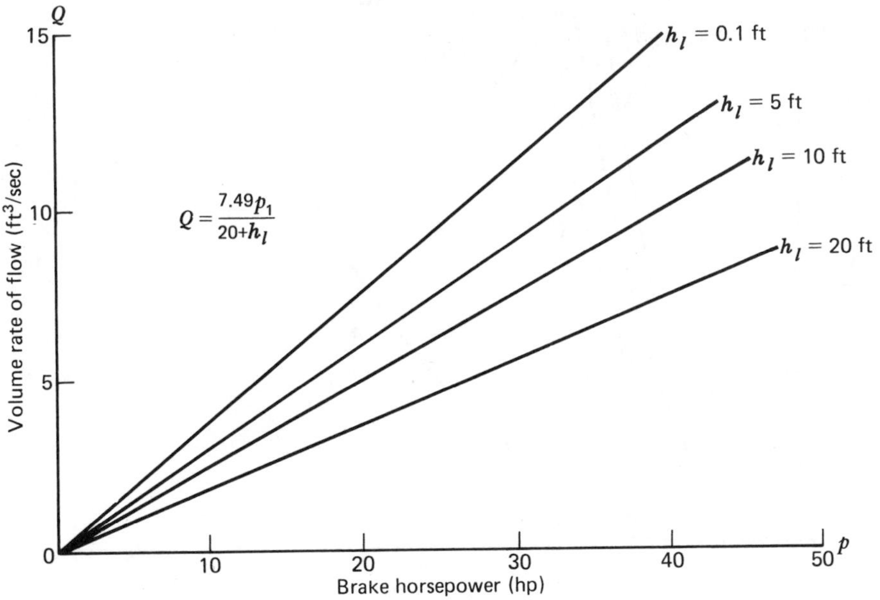

Figure 5.36

The slope of the hydraulic grade line is

$$S = \frac{h_l}{500}$$

The output q may be determined by using the flow equation $Q = VA$ where A is the pipe area in square feet

$$A = \frac{\pi}{4}\left(\frac{d}{12}\right)^2$$

The relationship is expressed as a function of the output q the pipe friction head loss h_L, and the pipe diameter d.

$$q_2 = g_2(h_L, d) = Q$$

Substituting the Hazen–Williams equation into the flow equation and simplifying will result in the pipe technological function:

$$Q = VA = 131.8\left(\frac{d}{48}\right)^{0.63}\left(\frac{h_l}{500}\right)^{0.54}\frac{\pi}{4}\left(\frac{d}{12}\right)^2$$

$$Q = 0.00219h_l^{0.54}d^{2.63}$$

A graph of the pipe technological function for various values of pipe head loss h_l is shown in Figure 5.37.

(c) The following cost data were obtained from the Environmental Protection Agency Reports prepared by Dames and Moore entitled: *Construction Costs for Municipal Wastewater Treatment Plants: 1973–1977* and *Construction Costs for Municipal Wastewater Conveyance Systems: 1973–1977.*

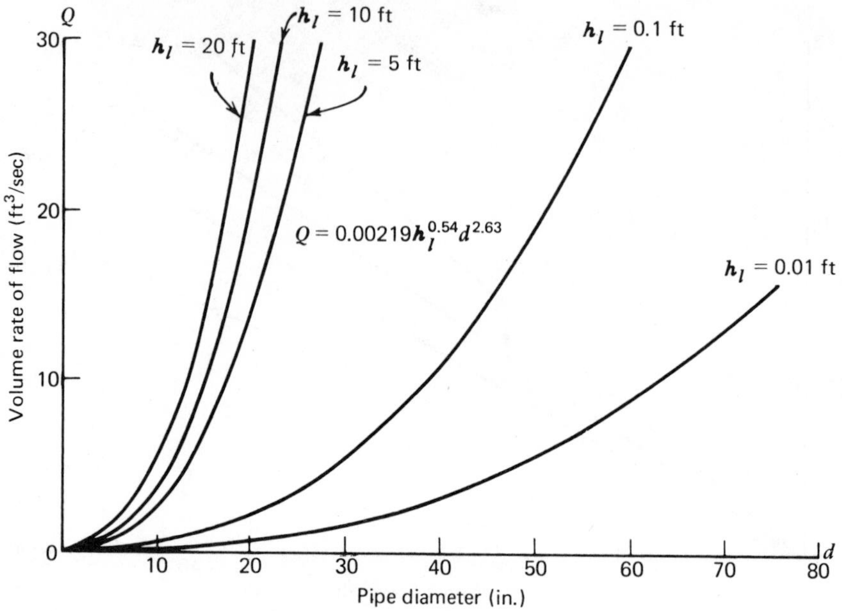

Figure 5.37 Production function for a given pipe diameter.

The pump station cost functions were developed from construction cost bid data that include equipment concrete, steel, and piping costs for a fully operational pump station. The pump station cost function is an exponential function:

$$C_1 = 33{,}200Q^{1.26}$$

A graph of the cost function is shown in Figure 5.38, where Q is the volume rate of flow in million gallons per day, not in terms of the control variable d. The construction cost survey data reported the pump station size in terms of volume rate of flow only. They did not include the size of the pump in terms of brake horsepower.

Since the cost elasticity $\varepsilon = 1.26$ is greater than one, economies of scale do not exist. As a result, it is not advisable to provide a pump station greater than 6 Mgal/day in anticipation of greater future demand for water.

The unit cost per foot of pipe is

$$c_2 = -31.1 + 7.06d \qquad \text{for } d \geq 4.41 \text{ ft}$$

Since the unit cost function is a function of pipe diameter d, the total cost of pipe is equal to the product of unit cost of pipe c_2 times the length of pipe $l = 500$ ft:

$$C_2 = p_2 l$$

or

$$C_2 = (-31.1 + 7.06d_2)500 = -15{,}500 + 3530d$$

A graph of the pipeline cost function for $l = 500$ ft. is shown in Figure 5.39.

(d) The total cost of construction of the system is the sum of the costs to construct the pump station and pipeline:

$$\text{Minimize } C = C_1 + C_2 = 33{,}200Q^{1.26} - 15{,}500 + 3530d$$

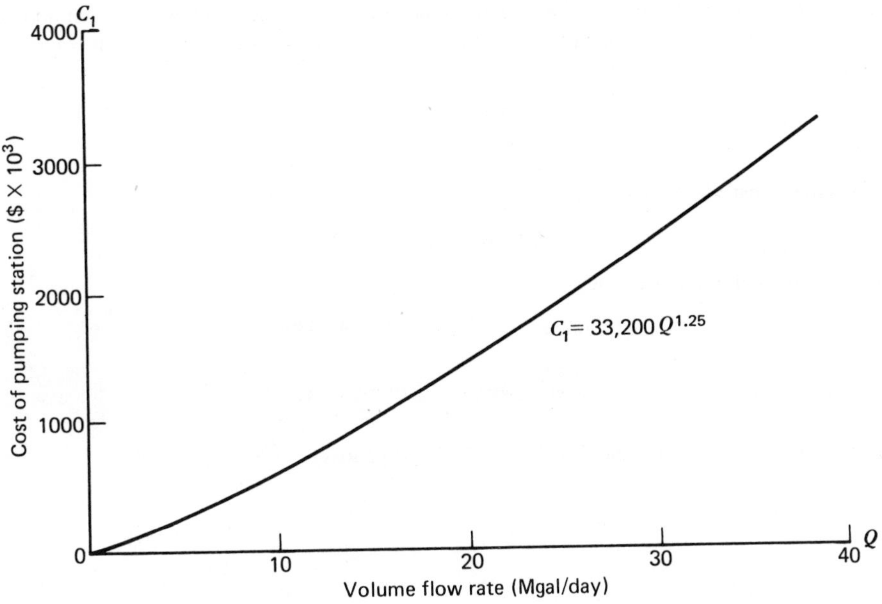

Figure 5.38 Construction cost function for pump station.

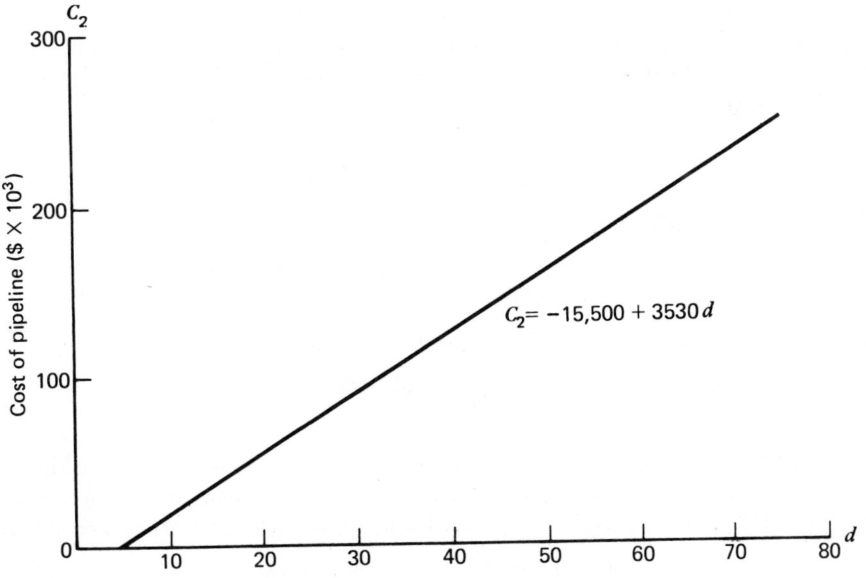

Figure 5.39 Cost function for 500-ft pipeline.

Since $Q = 6$ Mgal/day (9.3 ft^3/sec) must be delivered, the pump station and pipeline technological equations may be written as constraint equations. Thus, the pump station constraint equation is equal to

$$\frac{7.49p}{20 + h_l} = 9.3$$

The pipeline constraint equation is

$$0.00219h_l^{0.54}d^{2.63} = 9.3$$

The mathematical model becomes

$$\text{Minimize } C = 33{,}200q^{1.26} - 15{,}550 + 3530d$$

$$\frac{7.49p}{20 + h_l} = 9.3 \qquad \text{(Pump efficiency)}$$

$$0.00219h_l^{0.54}d^{2.63} = 9.3 \qquad \text{(Pipe loss)}$$

$$p \leq 32$$

$$d \leq 72$$

$$p \geq 0$$

$$d \geq 4.41$$

$$h_l \geq 0$$

The constraints $p \leq 32$ and $d \leq 72$ were added to the mathematical model because the suppliers can only supply pumps of 32 hp or smaller and the cost estimate equation for pipe construction was derived with the survey data limited to 72-in.-diameter pipe. Since p, d, and h_l must be positive values to have a valid solution, they are explicitly stated in the model.

Our original goal was to have the mathematical model as a function of p and d only. We may achieve the desired form by eliminating h_l in the constraint set. In the pipe loss equation, solve for h_l in terms of d:

$$h_l = \left(\frac{9.3}{0.00219d^{2.63}}\right)^{1/0.54} = \frac{5{,}209{,}000}{d^{4.87}}$$

and substitute it in the pipe efficiency constraint.

$$p = \frac{9.3(20 + h_l)}{7.49} = 1.24\left(20 + \frac{5{,}209{,}000}{d^{4.87}}\right)$$

The two constraint equations simplify to the single constraint:

$$p = 24.8 + \frac{6{,}470{,}000}{d^{4.87}}$$

Since the cost function for the pump station is given as a function of flow, we can substitute $Q = 6$ Mgal/day directly into the objective function. The cost of the pump station is dependent on volume flow rate Q and not brake horsepower p. The total cost is a function of pipe size only. Thus

$$C = 33{,}200(6)^{1.26} - 15{,}550 + 3530d$$

$$C = 302{,}000 + 3530d$$

Thus, our mathematical model becomes

$$\text{Minimize } C = 302{,}000 + 3530d$$

$$p = 24.8 + \frac{6{,}470{,}000}{d^{4.87}}$$

$$p \leq 32$$

$$d \leq 72$$

$$p \geq 0$$

$$d \geq 4.41$$

(e) The determination of the optimum solution is determined as follows:

STEP 1 Now that the mathematical model is established, we may utilize the graphical method for finding the optimum solution. A graph of the feasible solution will be determined from the constraint equations. The feasible region must lie on the isoquant, $Q = 9.3$ ft^3/sec or $p = 24.8 + 6{,}470{,}000d^{-4.87}$, and satisfy the condition that p and d is less than 32 and 72, respectively. See Figure 5.40.

STEPS 2 AND 3 Since the cost function is a function of d only, the optimum cost function must be parallel to the p axis and satisfy the condition of optimality. The optimum solution is located at an extreme point as shown in Figure 5.40. The optimum solution is

$$p^* = 32 \text{ hp}$$

$$d^* = 17\text{-in.-diameter pipe}$$

$$C^* = \$362{,}000$$

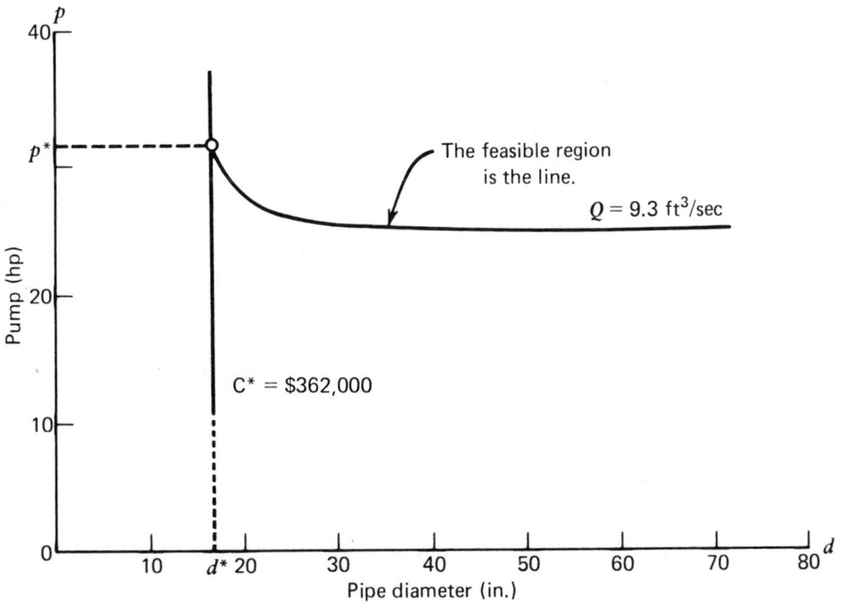

Figure 5.40

The pump station must utilize a pump of 32 brake horsepower at 85 percent efficiency and a pipe of 17-in. diameter to obtain the minimum cost. We can see from the isoquant that we may utilize a larger pipe and smaller pump to deliver the 9.3 ft^3/sec. However, any other combination will result in higher cost than C^*. The friction head at optimum is $h_f^* = 5.30$ ft.

EXAMPLE 5.8 Pump–Pipeline System: Minimize Capital Plus Operating Cost

Design a pump–pipeline system to deliver 6 Mgal/day (9.3 ft^3/sec) of water from a river to a reservoir. Refer to Example 5.7 for further details.

(a) Minimize the cost of construction and operating costs. Assume the cost of electricity is equal to \$0.05/kW-hr and the system has a design life of 20 years. Assume a discount rate of 10 percent.
(b) Use engineering economics to select either the minimum capital cost solution found in Example 5.7 or the minimum capital cost plus energy cost solution found in part a.
(c) Compare the construction and annual operating costs for both the two minimum-cost models.

Solution

(a) All the information in Example 5.7 is pertinent here. The objective function of the mathematical model for Example 5.7 must be modified to include the annual cost of operation. The cash-flow diagram for operating cost is shown in Figure 5.41(a). We assume the pump will operate continuously over the entire year. The annual cost A can be determined as a function of brake horsepower:

$$A = 0.05 \frac{\$}{\text{kW-hr}} \cdot 24 \frac{\text{hr}}{\text{day}} \cdot 365 \frac{\text{days}}{\text{yr}} \cdot 0.7457 \frac{\text{kW}}{\text{hp}} \, p \text{ (hp)}$$

$$A = \$326.60 \, p/\text{yr}$$

Let C_3 be the present worth of the annual operating cost A:

$$C_3 = \left[\frac{(1+i)^n - 1}{i(1+i)^n}\right] A$$

The discount rate is 10 percent, thus C_3 simplifies to

$$C_3 = \left[\frac{(1+0.1)^{20} - 1}{0.1(1+0.1)^{20}}\right] 326.6p = 2791p$$

The total cost C will be equal to the sum of the construction plus the operating cost:

$$C = C_1 + C_2 + C_3$$

where $C_1 + C_2$ is obtained from the mathematical model of Example 5.7. Thus, the objective function becomes

$$\text{Minimize } C = 302{,}000 + 353d + 2781p$$

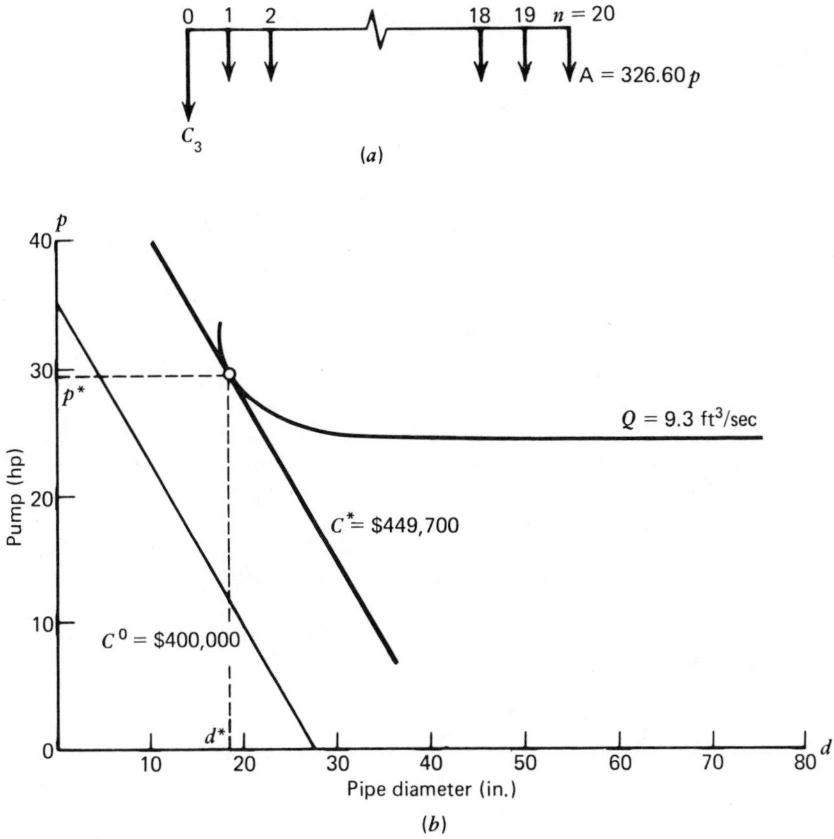

Figure 5.41

The mathematical model is

$$\text{Minimize } C = 302{,}000 + 3530d + 2781p$$

$$p = 24.8 + \frac{6{,}470{,}000}{d^{4.87}}$$

$$p \leq 32$$

$$d \leq 72$$

$$p \geq 0$$

$$d > 0$$

(a) The feasible region is the same one as utilized in Example 5.7. Since the cost function is linear, we may easily determine an isocost line by estimating an initial cost C^0. The isocost line for $C^0 = \$400{,}000$ is shown in Figure 5.41(*b*).

The optimality criterion requires that the solution occur at a point tangent to the feasible region or at an extreme point. An isocost line that is parallel to the initially estimated isocost line C^0 and is tangent to the feasible isoquant q is shown in Figure 5.42. Thus the optimum solution is

$$p^* = 29 \text{ hp}$$

$$d^* = 19 \text{ in.}$$

$$C^* = \$449{,}700$$

(b) Since the two designs, Example 5.7 and the design found in part a, are independent alternatives that satisfy the design requirement to deliver 6 Mgal/day of water to the reservoir from the river, the least-cost alternative selection method of Chapter 3 will be employed to choose the least-cost alternative. We consider both capital costs and operating costs. The present worth of costs will be utilized in the selection process.

The cash-flow diagram for the 32-hp pump and 17-in. pipe system, Example 5.7 or alternative **A**, is shown in Figure 5.42(*a*). The initial costs are equal to the construction cost; thus, the cost to build this system is \$362,000, as previously calculated as the optimum solution for Example 5.7. The operating cost was not considered in the analysis of Example 5.7. The annual operating cost is a function of the pump size, $A = \$326.60p$. Thus, the annual operating cost for a 32 hp pump is

$$A = \$326.60p = \$326.60(32) = \$10{,}450.00/\text{yr}$$

The present value of capital and operating cost PW^A for a discount rate of 10 percent and a design life of 20 yr is

$$PW^A = C_0 + \left[\frac{(1+i)^n - 1}{i(1+i)^n}\right]A = 362{,}000 + \left[\frac{(1+0.1)^{20} - 1}{0.1(1+0.1)^{20}}\right]10{,}450 = \$451{,}000$$

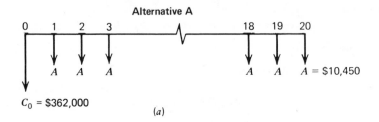

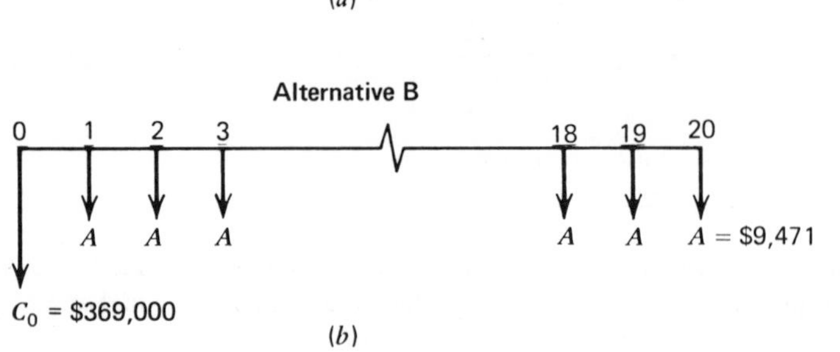

Figure 5.42

The cash-flow diagram for the 29-hp pump and 19-in. pipe is the same one shown in part a. Compare the cash-flow diagram in Figure 5.43*a* to the cash-flow diagram of alternative **A** in Figure 5.43*b*. The capital cost is determined from the cost equation of Example 5.7. Thus

$$C_0 = \$302{,}000 + \$3530d = \$302{,}000 + \$3530(19) = \$369{,}000$$

The annual operating cost is

$$A = \$326.60p = \$326.60(29) = \$9471/\text{yr}$$

The present value of capital and operating cost is equal to the optimum solution of part a, thus

$$\text{PW}^{\mathbf{B}} = C^* = \$449{,}700$$

Since $\text{PW}^{\mathbf{B}} < \text{PW}^{\mathbf{A}}$, the better solution is to construct the 29-hp pump and 19-in. pipe system.

(c) Comparing the present worth and the optimum pipe and pump sizes of alternatives **A** and **B** reveals that there is little difference between the two designs. The difference in present worth between the two designs is hardly distinguishable. However, let us examine the construction cost and yearly operating costs.

Since the pumping station construction cost is constant, the total extra construction cost can be determined with the unit cost of pipe only. Thus, the extra construction cost ΔC is equal to

$$\Delta C = (-\$15{,}550 + \$3530(19)) - (-\$15{,}550 + \$3530(17)) = \$7000$$

The annual operating cost savings ΔA is the difference between annual costs to operate the two systems. The difference in annual operating costs is estimated to be

$$\Delta A = \$326.60(32) - \$326.60(29) = \$980/\text{yr}$$

The additional \$7000 construction cost represents a small increase in capital expenditure. The percentage increase is only 1.93 percent. The annual savings of \$980, however, represents a substantial savings. It is a 10.4 percent annual operating cost savings. From this perspective, there is a significant gain in using the 29-hp pump and 19-in. pipe system over the 32-hp pump and 17-in. pipe system, especially if the cost of energy increases in future years.

Remarks

This is a classical example of choosing between present and future gains. There is a \$7000 opportunity cost associated with the 29-hp-pump–19-in.-pipe system. It will take several years to regain this lost purchasing power. Our decision-making process has considered economic factors only. Energy conservation and environmental protection may also favor one design over the other.

PROBLEMS

Problem 1

Consider the lever mechanism in Figure 5.43. Assume the lever is a rigid body.

(a) Derive a function $W = f(F, l)$ to relate the weight lifted W to the applied force F and the fulcrum length l. Assume the control variables are F is the applied force in pounds; and l is the distance between fulcrum and applied force in feet.

(b) Draw the isoquant for $W = 100$ lb.

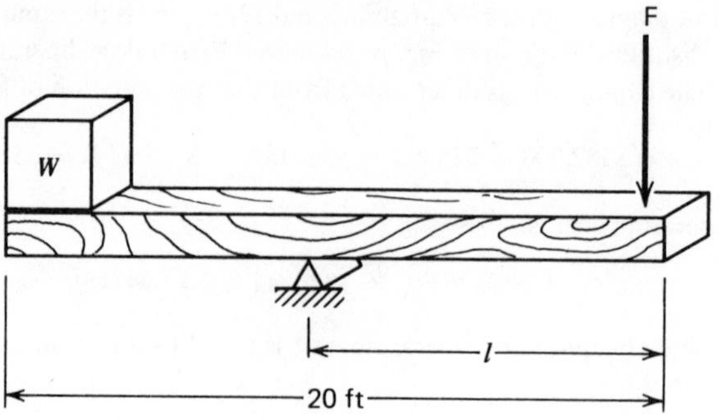

Figure 5.43

Problem 2

The simple supported beam in Figure 5.44 supports a load at its midspan.

(a) Determine a relationship $P = f(h, b)$, where P is the load that is supported at a maximum bending stress of 36,000 psi. The control variables are a function of the cross section of the beam, where h is the beam depth in inches, and b is the beam width in inches. The stress is equal to $\sigma = Mc/I$, where M is the maximum bending stress, I is the moment of inertia, and $c = h/2$.

(b) Draw an isoquant for $P = 100$ kips.

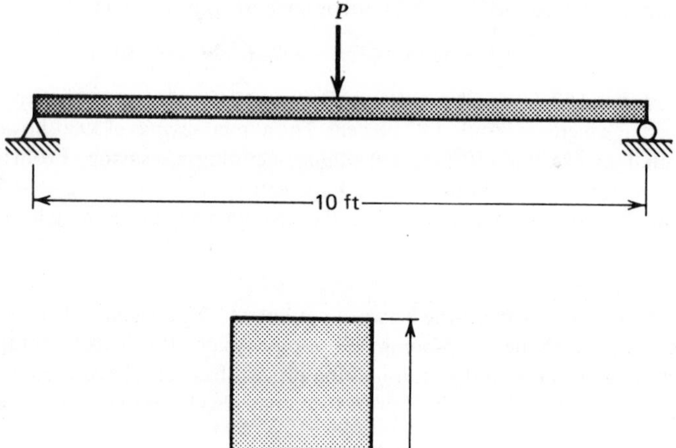

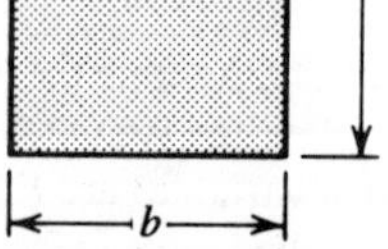

Figure 5.44

Problem 3

A wastewater aeration tank, constructed below ground level, is to be built. (See Figure 5.45.) Assume a square cross-sectional area A and the tank depth d to be the control variables.

Consider the following costs:

Excavation (using $\frac{1}{2}$ yd^3 capacity shovel)	\$2.09/yd^3
6 in. ground slabs (formwork and cost-in-place concrete)	\$4.00/ft^2
8 in walls (formwork)	\$2.60/ft^2
8 in. walls (cost in place)	\$3.20/ft^2
Reinforcing steel (total)	\$10,000

(a) Determine a total cost function $C = c(A, d)$.
(b) Draw isocost lines for $C^0 = \$50{,}000$ and \$100,000.
(c) Determine the maximum tank volumn for $C^0 = \$50{,}000$.

Problem 4

The cost to construct a continuous flow reactor tank is $C = 0.1V^{1.2}$, where V is measured in million gallons. Since the elasticity is 1.2, the exponent of the cost function, no economy of scale exists for a single tank. By placing tanks in series it is possible to obtain an economy of scale. Assume a first-order reaction where $k = 0.01$/min and

$$\frac{S_n}{S_0} = \frac{1}{(1 + kV/nQ)^n}$$

where V is the volume of the tank system, in million gallons; S_0 is the influent concentration of the reactant in milligrams per liter; S_n is the effluent concentration of the reactant in milligrams per liter; and n is the number of tanks. Q is the flow in million gallons per day. The required removal efficiency is 99 percent.

(a) Determine the volume and cost of the single tank.
(b) Determine the cost function for n tanks in series.
(c) Plot the average cost versus the tank volume, where the tank volume is the volume held by the entire tank system V. Show that no economy of scale exists.
(d) Draw an isoquant function for $Q = 1$ Mgal/day, where n and V are control variables.
(e) By graphical means determine the minimum cost of the tank system. (If n approaches infinity, a plug flow reactor should be constructed.)

Summary

In this chapter, basic concepts of microeconomic theory are applied to engineering design problems. The conditions of perfect and imperfect competition were shown to effect the selling price and quantity of goods or services produced. The efficiency of production, consumer demand, and the number of firms have a major impact upon production also.

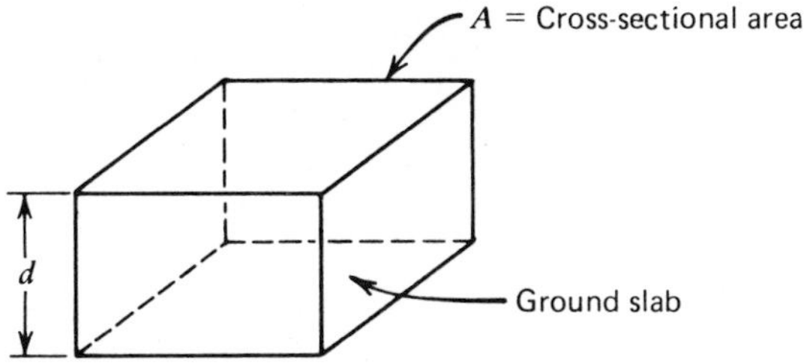

Figure 5.45

For perfect competition, aggregate supply functions were derived from cost functions of individual firms and used to find the market selling price and the total output in market equilibrium evaluations. Knowing the selling price, the output level of individual firms and optimum allocation of resources to maximize profit can be determined. For monopolies, they may maximize profit by controlling the selling price and quantity produced. Monopolies may, however, maximize-distribute the output of goods and services to maximize the welfare of society. Generally, this type of allocation will cause the monopoly to lose money in the long term. Thus, it may be necessary to subsidize to supplement the firm to keep it operational.

Our investigation of cost function shows that economy of scale and cost of elasticity can be helpful in determining the size of an engineering facility. If economics of scale exist, it may be wise to provide for additional capacity in anticipation of increased future demand. The average cost of production will be less for higher output levels if economies of scale exist.

Graphical procedures were used to find the optimum solution of various two-dimensional minimum cost and maximum-output models. For nonlinear models, we found the optimum solution to occur at the tangency point of the isocost and isoquant functions.

Bibliography

James M. Henderson and Richard E. Quandt. *Microeconomic Theory: A Mathematical Approach*, 2nd ed. McGraw-Hill, New York. 1958.

Richard de Neufville and Joseph H. Stafford. *Systems Analysis for Engineers and Managers.* McGraw-Hill, New York. 1971.

Paul A. Samuelson. *Economics: An Introductory Analysis*, 4th ed., McGraw-Hill, New York. 1958.

Edward K. Morlok. *Introduction to Transportation Engineering and Planning.* McGraw-Hill, New York. 1978.

Martin Wohl and Brian V. Martin. *Traffic Systems Analysis.* McGraw-Hill, New York. 1967.

CHAPTER 6

Optimization of Linear Mathematical Models

After completion of this chapter, the student should:

1. Be able to find the optimum solution of a linear programming problem with the simplex method.
2. Be able to find the optimum solution of a linear problem requiring artificial variables by the two-phase simplex method.
3. Be able to formulate linear mathematical models classified as resource allocation and net analysis problems.

In Chapter 2 linear mathematical models were formulated and the optimum solution for two-dimensional resource allocation problems were determined by graphical procedures. We concluded that the optimum solution of linear mathematical model will always occur at an extreme point. We utilize these concepts to establish mathematical methods to analyze the feasible region represented by a linear equation of the form:

$$\text{Minimize } z = \boldsymbol{c}'\boldsymbol{x}$$

$$\boldsymbol{a}_i'\boldsymbol{x}\ \{=, \leq, \geq\} b_i \qquad i = 1, 2, \ldots, m$$

$$\boldsymbol{x} \geq \boldsymbol{0}$$

Here we introduce surplus and slack variables to transform inequality constraint equations to strict equality constraint equations:

$$\text{Minimize } z = \boldsymbol{c}'\boldsymbol{x}$$

$$\boldsymbol{A}\boldsymbol{x} = \boldsymbol{b}$$

$$\boldsymbol{x} \geq \boldsymbol{0}$$

The new mathematical model is in the standard form of linear programming. With the use of basic and nonbasic variables, an extreme point of $\boldsymbol{A}\boldsymbol{x} = \boldsymbol{b}$ may be evaluated and tested as being the minimum z. An algorithm, called the *simplex method*, will be introduced to efficiently search and test extreme points as the location of the minimum value of z.

Matrix algebra is used in this chapter. An introduction of essential properties and methods of solving problems of the form $\boldsymbol{A}\boldsymbol{x} = \boldsymbol{b}$ is given in Appendix A.

6.1 THE STANDARD FORM

The most important conclusion that was drawn from the graphical determination of the optimum solution of a bivariate linear mathematical model is that the optimum solution will occur at an extreme point. Our search for the optimum will *always* be restricted to the evaluation of extreme points. In this chapter mathematical methods are used to solve linear models having n control variables of $\boldsymbol{x}$. A linear mathematical model is formulated and placed in the so-called standard form. New control variables called *slack* and *surplus variables* will be used to transform a set of linear equations with equality less-than-or-equal-to and greater-than-or-equal-to constraints to strict equality constraints only. Furthermore, procedures to search the extreme points are established. This discussion lays the groundwork for finding the optimum solution by the simplex method to be discussed in the following section.

Slack and Surplus Variables

The *standard form of a linear mathematical model* utilizing matrix notation is

$$z = \boldsymbol{c}'\boldsymbol{x}$$

$$\boldsymbol{A}\boldsymbol{x} = \boldsymbol{b}$$

$$\boldsymbol{x} \geq \boldsymbol{0}$$

where $\boldsymbol{x}$ is an n vector and $\boldsymbol{A}$ is an $m \times n$ matrix. The standard form requires that (1) all members of the $\boldsymbol{b}$ vector be *positive values* and (2) each constraint equation be a *strict equality*.

After the model is formulated utilizing inequality constraints, it can be placed in standard form. The first requirement that all elements of $\boldsymbol{b}$ be positive may be easily satisfied by multiplying a given equation with a negative value of $\boldsymbol{b}$ by -1. If the constraint equation has an inequality constraint it must be transformed from a less-than-or-equal-to constraint $\leq$ to a greater-than-or-equal-to constraint $\geq$ and vice versa. For example, a model contains the following set:

$$-x_1 + 2x_2 \geq -3$$

$$2x_1 - 6x_2 \leq -2$$

$$x_1 \geq 0, x_2 \geq 0$$

It may be transformed to positive values by multiplying each equation by -1 and changing the sense of the inequality constraints. The transformed equation set becomes

$$x_1 - 2x_2 \leq 3$$

$$-2x_1 + 6x_2 \geq 2$$

$$x_1 \geq 0, x_2 \geq 0$$

The second requirement for standard form is that the constraint set contain only equality constraints. Thus, the equation set does not satisfy this requirement and must be modified. New control variables called *slack* and *surplus variables* will be introduced into the model.

A slack variable is a *positive, independent, control variable* that is introduced into a constraint equation containing a less-than-or-equal-to constraint.

$$\boldsymbol{a}'\boldsymbol{x} \leq b$$

or

$$a_1x_1 + a_2x_2 + \cdots + a_nx_n \leq b$$

To transform this equation to one with a strict equality requires the introduction of a slack variable x_p. The transformed equation is

$$\boldsymbol{a}'\boldsymbol{x} + x_p = b$$

or

$$a_1x_1 + a_2x_2 + \cdots + a_nx_n + x_p = b$$

A surplus variable serves a function that is similar to a slack variable. It is a *positive, independent, control variable* that is introduced into the constraint equation containing a greater-than-or-equal-to constraint:

$$\boldsymbol{a}'\boldsymbol{x} \geq b$$

or

$$a_1x_1 + a_2x_2 + \cdots + a_nx_n \geq b$$

Let the surplus variable be designated as the positive control variable $x_r \geq 0$. The transformed equation is

$$\boldsymbol{a}'\boldsymbol{x} - x_r = b$$

or

$$a_1x_1 + a_2x_2 + \cdots + a_nx_n - x_r = b$$

Since surplus and slack variables are introduced into constraint equations with less-than-or-equal-to ($\leq$) and greater-than-or-equal-to ($\geq$) constraints, respectively, both a surplus and a slack variable *will never appear* in the same equation.

Furthermore, slack and surplus variables may be used to transform both linear and nonlinear equations. Thus, if $g(\boldsymbol{x})$ represents a nonlinear function, the following two inequality constraints may be transferred to equality constraints by utilizing a slack and surplus variable, respectively.

$$g(\boldsymbol{x}) \leq b \Rightarrow g(\boldsymbol{x}) + x_p = b$$

$$g(\boldsymbol{x}) \geq b \Rightarrow g(\boldsymbol{x}) - x_r = b$$

Consider the constraint set in the numerical example. In this chapter we are concerned with linear equations only. The first equation of the equation set requires a surplus variable, and the second equation requires a slack variable. Thus, the constraint set

$$x_1 - 2x_2 \leq 3 \Rightarrow x_1 - 2x_2 + x_3 = 3$$

$$-2x_1 + 6x_2 \geq 2 \Rightarrow -2x_1 + 6x_2 - x_4 = 2$$

$$x_1 \geq 0, x_2 \geq 0 \Rightarrow x_1 \geq 0, x_2 \geq 0, x_3 \geq 0, x_4 \geq 0$$

where x_3 is the slack variable and x_4 is a surplus variable. The requirement that all slack and surplus variable are positive values is explicitly expressed as $x_3 \geq 0$ and $x_4 \geq 0$.

Slack and surplus variables have physical meaning and may be defined just like the control variables used in the original formulation problem. Furthermore, since they are control variables, all mathematical rules that apply to the original set of control variables also apply to the slack and surplus variables. As a result, it is unnecessary to distinguish among the original, slack, and surplus control variables when performing mathematical manipulations on the mathematical model.

Evaluating Extreme Points

If a *unique* solution of $\boldsymbol{Ax} = \boldsymbol{b}$ exists, then the number of unknowns must equal the number of independent equations, an inverse of $\boldsymbol{A}$ must exist, and $\boldsymbol{x} = \boldsymbol{A}^{-1}\boldsymbol{b}$. Consequently, the $\boldsymbol{A}$ matrix must be a square $n \times n$ matrix.

For linear mathematical models in standard form, generally, the $\boldsymbol{A}$ matrix is not a square matrix; therefore, a *unique solution does not exist.*

$$z = \boldsymbol{c}'\boldsymbol{x}$$

$$\boldsymbol{Ax} = \boldsymbol{b}$$

$$\boldsymbol{x} \geq \boldsymbol{0}$$

For optimization problems, the number of control variables n will be greater the number of constraint equations m, or $n > m$. The discussion focuses upon the solution of the constraint set $\boldsymbol{Ax} = \boldsymbol{b}$, where $\boldsymbol{A}$ is an $m \times n$ matrix. The vector $\boldsymbol{x}$ contains n elements, $\boldsymbol{x}' = [x_1 \; x_2 \ldots x_n]$.

The concepts of basic and nonbasic variables, partitioned matrices, and matrix algebra will be used to determine the solution to the constraint set. Remember, the purpose of this work is to be able to search the extreme points of the constraint set for possible solutions to the optimization problem. The understanding of the concept of basic and nonbasic variables is an essential prerequisite for understanding the principles of the simplex method, which is discussed in the next section.

The Basis

A vector space is a set of vectors that has the property that if two vectors of the set are added or if a vector of the set is multiplied by a scalar, the sum or product is a member of the set. A *basis* is a set of independent vectors in a vector space that spans a vector space; that is, every vector in the vector space is a sum of scalar multiples of vectors from the basis. For example, $\boldsymbol{e}^{1'} = [1 \; 0 \; 0]$, $\boldsymbol{e}^{2'} = [0 \; 1 \; 0]$, and $\boldsymbol{e}^{3'} = [0 \; 0 \; 1]$ are a basis for the vector space of three-dimensional vectors. For the constraint equation $\boldsymbol{Ax} = \boldsymbol{b}$, where $\boldsymbol{A}$ is an $m \times n$ matrix and $n > m$, $\boldsymbol{A}$ may be represented as a set of column vectors, or

$$\boldsymbol{A} = [\boldsymbol{a}_1 \; \boldsymbol{a}_2 \cdots \boldsymbol{a}_n] = [\boldsymbol{a}_1 \; \boldsymbol{a}_2 \cdots \boldsymbol{a}_m \; \boldsymbol{a}_{m+1} \cdots \boldsymbol{a}_n]$$

where the set of all $\boldsymbol{a}_i$ vectors is assumed to be independent. According to the definition of a basis it is possible linearly to combine m column vectors of $\boldsymbol{A}$ in the following manner.

$$[\boldsymbol{a}_1\ \boldsymbol{a}_2 \cdots \boldsymbol{a}_m] \cdot \begin{bmatrix} x_1 \\ x_2 \\ \vdots \\ x_m \end{bmatrix} = \boldsymbol{b}$$

Here, the first m column vectors of $\boldsymbol{A}$ were selected in this illustration. Other combinations of $\boldsymbol{a}_i$ may be chosen. The matrix $[\boldsymbol{a}_1\ \boldsymbol{a}_2 \ldots \boldsymbol{a}_m]$ or some other combination is called the *basic matrix* $\boldsymbol{A}_B$ and $[x_1\ x_2 \ldots x_m]$ is called the basic vector $\boldsymbol{x}_B$. Thus

$$\boldsymbol{A}_B \boldsymbol{x}_B = \boldsymbol{b}$$

The elements of $\boldsymbol{x}_B$ are

$$\boldsymbol{x}_B = \boldsymbol{A}_B^{-1} \boldsymbol{b}$$

Since $\boldsymbol{A}_B$ is assumed to consist of a set of linearly independent $\boldsymbol{a}$ vectors, the inverse $\boldsymbol{A}_B^{-1}$ exists and a solution to $\boldsymbol{x}_B$ is obtained. Thus, the elements of $\boldsymbol{x}_B$ form a unique linear combination of the basic matrix vectors $\boldsymbol{a}_i$ that span the vector space.

Matrix partitioning allows the constraint set to be written as

$$\boldsymbol{A}\boldsymbol{x} = [\boldsymbol{A}_B\ \boldsymbol{A}_N] \cdot \begin{bmatrix} \boldsymbol{x}_B \\ \boldsymbol{x}_N \end{bmatrix} = \boldsymbol{b}$$

The vector $\boldsymbol{x}$ has been partitioned into two subvectors $\boldsymbol{x}_B$ and $\boldsymbol{x}_N$, the *basic* and *nonbasic vectors*, respectively. The number of control variables assigned to $\boldsymbol{x}_B$ is equal to the number of constraint equations m. The remaining control variables will be assigned to $\boldsymbol{x}_N$, thus it will contain $(n - m)$ variables. The matrix $\boldsymbol{A}_B$ is a square $m \times m$ matrix and $\boldsymbol{A}_N$ is an $m \times (n - m)$ matrix. A solution of $\boldsymbol{x}_B$ may be determined in terms of the nonbasic variables $\boldsymbol{x}_N$:

$$\boldsymbol{x}_B = \boldsymbol{A}_B^{-1}[\boldsymbol{b} - \boldsymbol{A}_N \boldsymbol{x}_N]$$

When evaluating an extreme point, all nonbasic vector elements are assigned values of zero, $\boldsymbol{x}_N = \boldsymbol{0}$.

Consider the following set of inequality constraints:

$$\begin{aligned} x_1 + 2x_2 &\leq 2 \\ x_2 &\leq \tfrac{1}{2} \\ x_1 + x_2 &\geq 1 \end{aligned}$$

Since there are only two control variables, this constraint set can be interpreted with the use of Figure 6.1.

The goal is to be able to determine each of the extreme points (1, 0), (2, 0), $(\frac{1}{2}, \frac{1}{2})$, and $(1, \frac{1}{2})$ with basic and nonbasic variables.

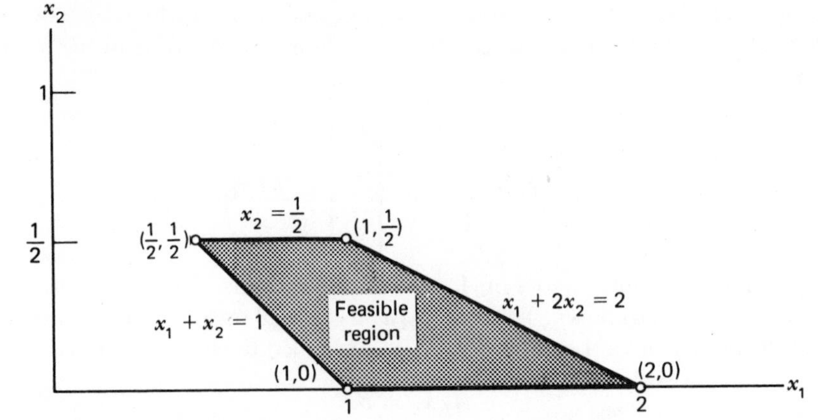

Figure 6.1 Evaluation of extreme points.

Consider the extreme point $(\frac{1}{2}, \frac{1}{2})$. This point occurs at the intersection of the equations $x_1 + x_2 = 1$ and $x_2 = \frac{1}{2}$. In other words, these equations are active constraints. Now consider the constraint set in standard form. It is

$$\begin{aligned} x_1 + 2x_2 + x_3 \qquad\qquad &= 2 \\ x_2 \qquad + x_4 \qquad &= \tfrac{1}{2} \\ x_1 + x_2 \qquad\qquad - x_5 &= 1 \end{aligned}$$

where x_3 and x_4 are slack variables and x_5 is a surplus variable.

In order for these equations to be active constraints, the following must hold.

$$\begin{aligned} x_2 + x_4 \qquad &= \tfrac{1}{2} \\ x_1 + x_2 \qquad - x_5 &= 1 \end{aligned}$$

Since $x_1 = \frac{1}{2}$ and $x_2 = \frac{1}{2}$,

$$\begin{aligned} \tfrac{1}{2} + x_4 &= \tfrac{1}{2} \\ \tfrac{1}{2} + \tfrac{1}{2} - x_5 &= 1 \end{aligned}$$

or $x_4 = 0$ and $x_5 = 0$. For $x_1 + 2x_2 + x_3 = 2$ to be an inactive constraint, the following equation must be satisfied.

$$\begin{aligned} x_1 + 2x_2 + x_3 &= 2 \\ \tfrac{1}{2} + 2(\tfrac{1}{2}) + x_3 &= 2 \end{aligned}$$

or $x_3 = \frac{1}{2}$. The solution in vector form is

$$\mathbf{x} = \begin{bmatrix} x_1 \\ x_2 \\ x_3 \\ x_4 \\ x_5 \end{bmatrix} = \begin{bmatrix} \frac{1}{2} \\ \frac{1}{2} \\ \frac{1}{2} \\ 0 \\ 0 \end{bmatrix}$$

Now, consider obtaining this solution using matrix partitioning, basic and non-basic variables, and matrix algebra. The constraint set is written in matrix form $\boldsymbol{Ax} = \boldsymbol{b}$ as

$$\begin{bmatrix} 1 & 2 & 1 & 0 & 0 \\ 0 & 1 & 0 & 1 & 0 \\ 1 & 1 & 0 & 0 & -1 \end{bmatrix} \cdot \begin{bmatrix} x_1 \\ x_2 \\ x_3 \\ x_4 \\ x_5 \end{bmatrix} = \begin{bmatrix} 2 \\ \frac{1}{2} \\ 1 \end{bmatrix}$$

Since there are three constraint equations and five control variables, the basic vector $\boldsymbol{x}_B$ will consist of three control variables, and the remaining control variables will be assigned to $\boldsymbol{x}_N$. Consider the following assignment:

$$\boldsymbol{x}_B = \begin{bmatrix} x_1 \\ x_2 \\ x_3 \end{bmatrix} \qquad \boldsymbol{x}_N = \begin{bmatrix} x_4 \\ x_5 \end{bmatrix}$$

The partitioned matrix is:

$$\boldsymbol{A}_B \boldsymbol{x}_B + \boldsymbol{A}_N \boldsymbol{x}_N = \boldsymbol{b}$$

$$\begin{bmatrix} 1 & 2 & 1 \\ 0 & 1 & 0 \\ 1 & 1 & 0 \end{bmatrix} \cdot \begin{bmatrix} x_1 \\ x_2 \\ x_3 \end{bmatrix} + \begin{bmatrix} 0 & 0 \\ 1 & 0 \\ 0 & -1 \end{bmatrix} \cdot \begin{bmatrix} x_4 \\ x_5 \end{bmatrix} = \begin{bmatrix} 2 \\ \frac{1}{2} \\ 1 \end{bmatrix}$$

Solving for $\boldsymbol{x}_B$ gives a solution of $\boldsymbol{x}_B$ as a function of $\boldsymbol{x}_N$:

$$\boldsymbol{x}_B = \boldsymbol{A}_B^{-1}[\boldsymbol{b} - \boldsymbol{A}_N \boldsymbol{x}_N]$$

$$\begin{bmatrix} x_1 \\ x_2 \\ x_3 \end{bmatrix} = \begin{bmatrix} 1 & 2 & 1 \\ 0 & 1 & 0 \\ 1 & 1 & 0 \end{bmatrix}^{-1} \cdot \left\{ \begin{bmatrix} 2 \\ \frac{1}{2} \\ 1 \end{bmatrix} - \begin{bmatrix} 0 & 0 \\ 1 & 0 \\ 0 & -1 \end{bmatrix} \cdot \begin{bmatrix} x_4 \\ x_5 \end{bmatrix} \right\}$$

$$\begin{bmatrix} x_1 \\ x_2 \\ x_3 \end{bmatrix} = \begin{bmatrix} 0 & -1 & 1 \\ 0 & 1 & 0 \\ 1 & -1 & -1 \end{bmatrix} \cdot \left\{ \begin{bmatrix} 2 \\ \frac{1}{2} \\ 1 \end{bmatrix} - \begin{bmatrix} 0 & 0 \\ 1 & 0 \\ 0 & -1 \end{bmatrix} \cdot \begin{bmatrix} x_4 \\ x_5 \end{bmatrix} \right\}$$

or

$$\begin{bmatrix} x_1 \\ x_2 \\ x_3 \end{bmatrix} = \begin{bmatrix} \frac{1}{2} \\ \frac{1}{2} \\ \frac{1}{2} \end{bmatrix} + \begin{bmatrix} 1 & 1 \\ -1 & 0 \\ 1 & -1 \end{bmatrix} \cdot \begin{bmatrix} x_4 \\ x_5 \end{bmatrix}$$

Expanding, it becomes:

$$x_1 = \tfrac{1}{2} + x_4 + x_5$$

$$x_2 = \tfrac{1}{2} - x_4$$

$$x_3 = \tfrac{1}{2} + x_4 - x_5$$

Since the nonbasic variables x_4 and x_5 equal zero, the solution is

$$x = \begin{bmatrix} x_B \\ x_N \end{bmatrix} = \begin{bmatrix} \frac{1}{2} \\ \frac{1}{2} \\ \frac{1}{2} \\ 0 \\ 0 \end{bmatrix}$$

This is the same result as obtained previously.

Note, if x_4 and x_5, the nonbasic vector x_N, are assigned values other than zero, the solution vector x does not contain the extreme point $(\frac{1}{2}, \frac{1}{2})$.

Obviously, the choice of variables that enter the x_B and x_N will determine the extreme point under investigation. The basic and nonbasic variables were carefully chosen to obtain the desired result. With the procedures discussed thus far, there is no guarantee that all basic variables are nonnegative, $x_B \geq 0$. Procedures to ensure that only nonzero variables enter the basis will be established in the following section.

EXAMPLE 6.1 A Construction Management Model

A construction company is contracted to excavate 6-ft and 18-ft-wide trenches. It may remove no more 10,000 yd^3/day of excavation material from a site because of a limited supply of dump trucks. To meet schedule requirements it must excavate at least 1600 yd^3/day from the 6-ft trench and at least 3000 yd^3/day from the 18-ft trench. The contractor has 12 backhoe operators that he may assign either to the $\frac{1}{2}$-yd^3 tractor backhoe or the $2\frac{1}{2}$-yd^3 hydraulic backhoe.

The performance and cost data for the backhoes are

BACKHOE	OPERATING COST	DAILY PERFORMANCE
$\frac{1}{2}$-yd^3 tractor	\$394/machine day	200 yd^3/day for 6-ft trench
$2\frac{1}{2}$-yd^3 hydraulic	\$1,110/machine day	1,000 yd^3/day for 18-ft trench

(a) Formulate the problem to minimize daily operating cost.
(b) Formulate the mathematical model in standard form.
(c) Define the slack and surplus control variables in terms of the original formulation.
(d) Assume that eight and four workers are assigned to the tractor and hydraulic backhoes, respectively. Interpret the meaning of the slack and surplus variables.

Solution

(a) Define the control variables as

x_1 = number of machine operators assigned to operate the $\frac{1}{2}$-yd^3 tractor backhoe.

x_2 = number of machine operators assigned to operate the $2\frac{1}{2}$-yd^3 hydraulic backhoe.

The total daily cost to operate is

$$C = \$394x_1 + \$1110x_2 \qquad \text{Operating cost}$$

The total amount that he may remove from the site each day because of the limited number of trucks will be equal to the cubic yards of material removed from each trench per day; or

$$200x_1 + 1000x_2 \leq 10{,}000 \qquad \text{Trucks}$$

In order to meet schedule requirements, an appropriate number of operators must be assigned to each trench:

$$200x_1 \geq 1600 \qquad \text{6-ft trench}$$

$$1000x_2 \geq 3000 \qquad \text{18-ft trench}$$

Since the contractor has only 12 workers, the contractor's assignment of operators must satisfy the following condition:

$$x_1 + x_2 \leq 12 \qquad \text{Workers}$$

Owing to the limited number of trucks, it may not be necessary to assign all 12 operators; hence the inequality constraint is specified. The mathematical model is

$$\begin{aligned}
\text{Minimize } C = \$394x_1 + \$1110x_2 & \qquad \text{Operating cost} \\
200x_1 + 1000x_2 \leq 10{,}000 & \qquad \text{Trucks} \\
200x_1 \qquad\qquad \geq 1600 & \qquad \text{6-ft trench} \\
1000x_2 \geq 3000 & \qquad \text{18-ft trench} \\
x_1 + \qquad x_2 \leq 12 & \qquad \text{Workers} \\
x_1 \geq 0, x_2 \geq 0 &
\end{aligned}$$

(b) To represent this constraint set in standard form, slack and surplus variables must be introduced. Since the first and fourth constraint equations have less-than-or-equal-to constraints, slack variables x_3 and x_6 will be introduced. The second and third constraint equations have greater-than-or-equal to constraints; thus, surplus variables x_4 and x_5 are introduced. The model in standard form is

$$\begin{aligned}
&\text{Minimize } C = 394x_1 + 1110x_2 \\
&200x_1 + 1000x_2 + x_3 = 10{,}000 \qquad \text{Trucks} \\
&200x_1 - x_4 = 1600 \qquad \text{6-ft trench} \\
&1000x_2 - x_5 = 3000 \qquad \text{18-ft trench} \\
&x_1 \quad x_2 + x_6 = 12 \qquad \text{Workers} \\
&x_1 \geq 0, x_2 \geq 0, x_3 \geq 0, x_4 \geq 0, x_5 \geq 0, x_6 \geq 0
\end{aligned}$$

In matrix form, the objection function and constraint set may be represented as

$$\text{Minimize } C = \boldsymbol{c}'\boldsymbol{x}$$

$$\boldsymbol{A}\boldsymbol{x} = \boldsymbol{b}$$

$$\boldsymbol{x} \geq \boldsymbol{0}$$

with

$$\boldsymbol{A} = \begin{bmatrix} 200 & 1000 & 1 & 0 & 0 & 0 \\ 200 & 0 & 0 & -1 & 0 & 0 \\ 0 & 1000 & 0 & 0 & -1 & 0 \\ 1 & 1 & 0 & 0 & 0 & 1 \end{bmatrix}$$

$$\boldsymbol{x}' = [x_1 \quad x_2 \quad x_3 \quad x_4 \quad x_5 \quad x_6]$$

$$\boldsymbol{b}' = [10{,}000 \quad 1600 \quad 3000 \quad 12]$$

(c) Let the slack and surplus variables be defined as:

x_3 = amount of excavation material below the maximum amount of 10,000 yd^3 that is removed from the site per day, in cubic yards per day.

x_4 = amount of excavation material above the minimum scheduled amount of 1600 yd^3 that is removed from the 6-ft trench per day, in cubic yards per day.

x_5 = amount of excavation material above the minimum scheduled amount of 3000 yd^3 that is removed from the 18-ft trench per day, in cubic yards per day.

x_6 = number of operators that are not assigned to work on this project.

(d) Let $x_1 = 8$ and $x_2 = 4$. For the truck constraint equation, solve for x_3 in terms of x_1 and x_2:

$$x_3 = 10{,}000 - 200x_1 - 1000x_2$$

$$x_3 = 10{,}000 - 200(8) - 1000(4) = 4400$$

Since $x_3 \neq 0$, the constraint equation is an inactive constraint. The total trucking capacity is not being used. An additional 4400 yd^3 of material may be removed from the site if more operators are assigned to backhoes.

From the 6 ft trench constraint equation, solve for x_4 in terms of x_1:

$$x_4 = 200x_1 - 1600$$

$$x_4 = 200(8) - 1600 = 0$$

Since $x_4 = 0$, the constraint equation is active. This may be interpreted as meaning that the minimum amount of material is being removed from the 6-ft trench to meet schedule requirements.

From the 18 ft trench constraint equation, solve for x_5 in terms of x_2:

$$x_5 = 1000x_2 - 3000$$

$$x_5 = 1000(4) - 3000 = 1000$$

Since $x_5 > 0$, the constraint is inactive. More than enough material is being excavated each day to meet the minimum schedule requirement. Excavating with four hydraulic backhoe operators produces an output 1000 yd^3/day greater than the minimum schedule requirement.

From the worker constraint equation, solve for x_6 in terms of x_1 and x_2:

$$x_6 = 12 - x_1 - x_2$$

$$x_6 = 12 - 8 - 4 = 0$$

Since $x_6 = 0$, all 12 operators are assigned to the project, and the constraint is active.

EXAMPLE 6.2 Formulating a Basis Model

(a) Use matrix partitioning and matrix algebra to solve for a basic vector consisting of x_1 and x_3 in terms of x_2 and x_4.

$$2x_1 + 3x_2 + 2x_3 + x_4 = 14$$

$$3x_2 + 2x_3 + x_4 = 6$$

(b) Use matrix partitioning and matrix inversion to solve for a basic vector consisting of x_2 and x_4 in terms of x_1 and x_3.

Solution

(a) Write the equation set in the matrix form $\boldsymbol{Ax} = \boldsymbol{b}$:

$$\begin{bmatrix} 2 & 3 & 2 & 1 \\ 0 & 3 & 2 & 1 \end{bmatrix} \cdot \begin{bmatrix} x_1 \\ x_2 \\ x_3 \\ x_4 \end{bmatrix} = \begin{bmatrix} 14 \\ 6 \end{bmatrix}$$

The solution $\boldsymbol{x}_B = \boldsymbol{A}_B^{-1}[\boldsymbol{b} - \boldsymbol{A}_N \boldsymbol{x}_N]$. The partition matrix is

$$[\boldsymbol{A}_B \ \boldsymbol{A}_N] \cdot \begin{bmatrix} \boldsymbol{x}_B \\ \boldsymbol{x}_N \end{bmatrix} = \boldsymbol{b}$$

where

$$\boldsymbol{x}_B = \begin{bmatrix} x_1 \\ x_3 \end{bmatrix} \quad \text{and} \quad \boldsymbol{x}_N = \begin{bmatrix} x_2 \\ x_4 \end{bmatrix}$$

Express the partition matrix as

$$\boldsymbol{A}_B \boldsymbol{x}_B + \boldsymbol{A}_N \boldsymbol{x}_N = \boldsymbol{b}$$

Rearrange the columns of $\boldsymbol{A}$ and the rows of $\boldsymbol{x}$ as

$$\left[\begin{array}{cc:cc} 2 & 2 & 3 & 1 \\ 0 & 2 & 3 & 1 \end{array}\right] \cdot \begin{bmatrix} x_1 \\ x_3 \\ \cdots \\ x_2 \\ x_4 \end{bmatrix} = \begin{bmatrix} 14 \\ 6 \end{bmatrix}$$

$$\begin{bmatrix} 2 & 2 \\ 0 & 2 \end{bmatrix} \cdot \begin{bmatrix} x_1 \\ x_3 \end{bmatrix} + \begin{bmatrix} 3 & 1 \\ 3 & 1 \end{bmatrix} \cdot \begin{bmatrix} x_2 \\ x_4 \end{bmatrix} = \begin{bmatrix} 14 \\ 6 \end{bmatrix}$$

Solve for $\boldsymbol{x}_B$.

$$\boldsymbol{x}_B = \boldsymbol{A}_B^{-1}\{\boldsymbol{b} - \boldsymbol{A}_N \boldsymbol{x}_N\}$$

$$\begin{bmatrix} x_1 \\ x_3 \end{bmatrix} = \begin{bmatrix} 2 & 2 \\ 0 & 2 \end{bmatrix}^{-1} \cdot \left\{ \begin{bmatrix} 14 \\ 6 \end{bmatrix} - \begin{bmatrix} 3 & 1 \\ 3 & 1 \end{bmatrix} \cdot \begin{bmatrix} x_2 \\ x_4 \end{bmatrix} \right\}$$

$$\begin{bmatrix} x_1 \\ x_3 \end{bmatrix} = \frac{1}{4} \begin{bmatrix} 2 & -2 \\ 0 & 2 \end{bmatrix} \cdot \left\{ \begin{bmatrix} 14 \\ 6 \end{bmatrix} - \begin{bmatrix} 3 & 1 \\ 3 & 1 \end{bmatrix} \cdot \begin{bmatrix} x_2 \\ x_4 \end{bmatrix} \right\}$$

$$\begin{bmatrix} x_1 \\ x_3 \end{bmatrix} = \begin{bmatrix} 4 \\ 3 \end{bmatrix} - \begin{bmatrix} 0 & 0 \\ \frac{3}{2} & \frac{1}{2} \end{bmatrix} \cdot \begin{bmatrix} x_2 \\ x_4 \end{bmatrix}$$

Expanding, the solution becomes

$$x_1 = 4$$

$$x_3 = 3 - \tfrac{3}{2}x_2 - \tfrac{1}{2}x_4$$

(b) The partitioned matrix where $\boldsymbol{x}_B' = [x_2 \ x_4]$ and $\boldsymbol{x}_N' = [x_1 \ x_3]$ is:

$$\begin{bmatrix} 3 & 1 \\ 3 & 1 \end{bmatrix} \cdot \begin{bmatrix} x_2 \\ x_4 \end{bmatrix} + \begin{bmatrix} 2 & 2 \\ 0 & 2 \end{bmatrix} \cdot \begin{bmatrix} x_1 \\ x_3 \end{bmatrix} = \begin{bmatrix} 14 \\ 6 \end{bmatrix}$$

Since the inverse of $\boldsymbol{A}_B$ does not exist, *no solution* of $\boldsymbol{x}_B$ exists when $x_1 = x_3 = 0$.

EXAMPLE 6.3 Minimum-Cost Aggregate Mix Model

A construction site requires a minimum of 10,000 yd^3 of sand and gravel mixture. The mixture must contain no less than 5000 yd^3 of sand and no more than 6000 yd^3 of gravel. Materials may be obtained from two sites:

SITE	DELIVERY COST (\$/yd^3)	PERCENT SAND	PERCENT GRAVEL
1	\$5	30	70
2	\$7	60	40

(a) Formulate and solve the problem by graphical means.
(b) Rewrite the mathematical model formulated in standard form.
(c) Define the slack and surplus variables.
(d) Identify the basic and nonbasic variables found in part a, and show that the same solution obtained graphically may be found utilizing $\mathbf{x}_B = \mathbf{A}_B^{-1}[\mathbf{b} - \mathbf{A}_N\mathbf{x}_N]$.
(e) Determine the minimum cost by use of matrix methods, where $C^* = \mathbf{cx}^*$.

Solutions

(a) The linear mathematical model is

$$\text{Minimum } C = 5x_1 + 7x_2$$

$$x_1 + x_2 \geq 10{,}000 \qquad \text{Delivery}$$

$$0.3x_1 + 0.6x_2 \geq 5000 \qquad \text{Sand}$$

$$0.7x_1 + 0.4x_2 \leq 6000 \qquad \text{Gravel}$$

$$x_1 \geq 0$$

$$x_2 \geq 0$$

where the control variables x_1 and x_2 are defined as

x_1 = amount of material taken from site 1 in cubic yards.

x_2 = amount of material taken from site 2 in cubic yards.

The feasible region is shown in Figure 6.2. The graphical solution gives

$$x_1^* = 3300 \text{ yd}^3$$

$$x_2^* = 6700 \text{ yd}^3$$

(b) Since the delivery and sand constraint equations have greater-than-or-equal-to constraints, surplus variables must be introduced. Since the gravel constraint equation has a less-than-or-equal-to constraint, a slack variable must be introduced. Thus, the linear mathematical model in standard form is

$$\text{Minimum } C = 5x_1 + 7x_2$$

$$x_1 + x_2 - x_3 = 10{,}000$$

$$0.3x_1 + 0.6x_2 - x_4 = 5000$$

$$0.7x_1 + 0.4x_2 + x_5 = 6000$$

$$x_1 > 0_1,\ x_2 \geq 0,\ x_3 \geq 0,\ x_4 \geq 0,\ x_5 \geq 0$$

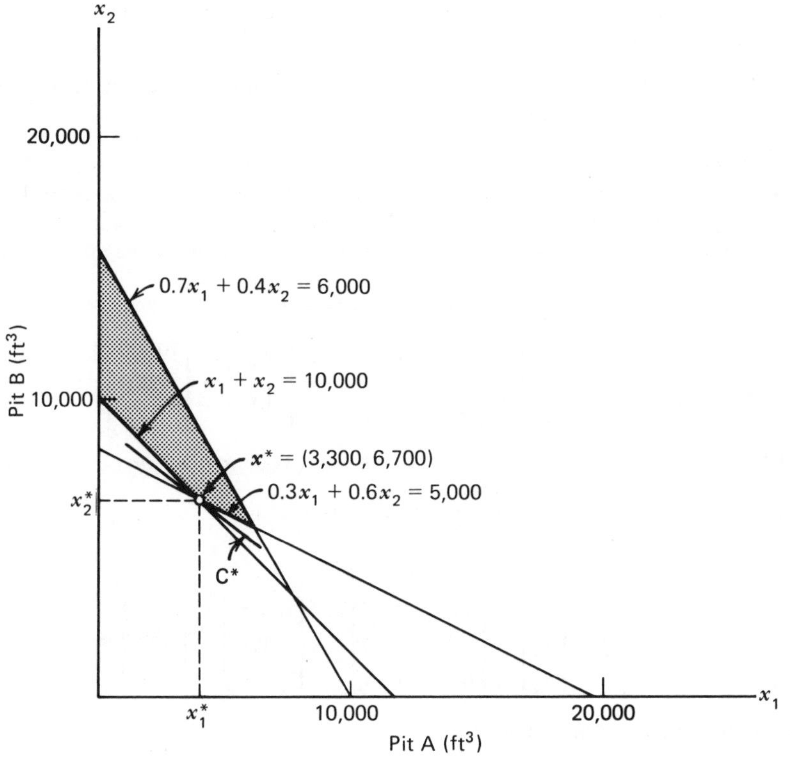

Figure 6.2

In matrix notation the model may be stated as

$$\text{Minimum } C = \boldsymbol{c}'\boldsymbol{x}$$
$$\boldsymbol{A}\boldsymbol{x} = \boldsymbol{b}$$
$$\boldsymbol{x} \geq \boldsymbol{0}$$

where

$$\boldsymbol{c}' = [5 \quad 7 \quad 0 \quad 0 \quad 0]$$

$$\boldsymbol{x}' = [x_1 \quad x_2 \quad x_3 \quad x_4 \quad x_5]$$

$$\boldsymbol{A} = \begin{bmatrix} 1.0 & 1.0 & -1 & 0 & 0 \\ 0.3 & 0.6 & 0 & -1 & 0 \\ 0.7 & 0.4 & 0 & 0 & 1 \end{bmatrix}$$

$$\boldsymbol{b}' = [10{,}000 \quad 5000 \quad 6000]$$

(c) The slack and surplus are defined as

x_3 = excess quantity of mix in cubic yards delivered from sites 1 and 2 to the construction site.

x_4 = amount of excess in cubic yards that is contained in the mix above the minimum 50 percent sand requirement.

x_5 = amount of gravel in cubic yards that is contained in the mix below the maximum 60 percent gravel requirement.

(d) From Figure 6.2, the optimum solution occurs at the intersection of $x_1 + x_2 = 10{,}000$ and $0.3x_1 + 0.6x_2 = 5000$. These equations thus being active constraints, the control variables x_3 and x_4, must therefore be members of the nonbasic vector $\boldsymbol{x}_N$. The remaining variables x_1, x_2, and x_5 are basic variables $\boldsymbol{x}_B$. The constraint set $\boldsymbol{Ax} = \boldsymbol{b}$ may be expressed as

$$\boldsymbol{A}_B \boldsymbol{x}_B + \boldsymbol{A}_N \boldsymbol{x}_N = \boldsymbol{b}$$

$$\begin{bmatrix} 1.0 & 1.0 & 0 \\ 0.3 & 0.6 & 0 \\ 0.7 & 0.4 & 1 \end{bmatrix} \cdot \begin{bmatrix} x_1 \\ x_2 \\ x_5 \end{bmatrix} + \begin{bmatrix} -1 & 0 \\ 0 & -1 \\ 0 & 0 \end{bmatrix} \cdot \begin{bmatrix} x_3 \\ x_4 \end{bmatrix} = \begin{bmatrix} 10{,}000 \\ 5000 \\ 6000 \end{bmatrix}$$

The solution is

$$\boldsymbol{x}_B = \boldsymbol{A}_B^{-1}[\boldsymbol{b} - \boldsymbol{A}_N \boldsymbol{x}_N]$$

Since $\boldsymbol{x}_N = \boldsymbol{0}$, the solution of $\boldsymbol{x}_B$ or optimum solution $\boldsymbol{x}^*$ is

$$\boldsymbol{x}_B = \boldsymbol{A}_B^{-1}\boldsymbol{b}$$

$$\boldsymbol{x}^* = \begin{bmatrix} x_1 \\ x_2 \\ x_5 \end{bmatrix} = \begin{bmatrix} 1.0 & 1.0 & 0 \\ 0.3 & 0.6 & 0 \\ 0.7 & 0.4 & 1 \end{bmatrix}^{-1} \cdot \begin{bmatrix} 10{,}000 \\ 5000 \\ 6000 \end{bmatrix} = \begin{bmatrix} 3000 \\ 6700 \\ 1000 \end{bmatrix}$$

(e) The solution vector $\boldsymbol{x}^*$ from part d is

$$\boldsymbol{x}^* = \begin{bmatrix} x_1^* \\ x_2^* \\ x_3^* \\ x_4^* \\ x_5^* \end{bmatrix} = \begin{bmatrix} 3300 \\ 6700 \\ 0 \\ 0 \\ 1000 \end{bmatrix}$$

The definition of the control variables may be used to interpret the solution $\boldsymbol{x}^*$. Since $x_3^* = 0$, a minimum amount of 10,000 yd^3 of material is delivered to the site. The amount of material taken from pits 1 and 2 is 3300 yd^3 and 6700 yd^3, respectively. Since $x_4^* = 0$, the minimum amount of sand (50 percent of the mix is sand) is delivered. Since $x_5^* = 1000$ yd^3, the mix contains less than the maximum amount of gravel permitted.

The minimum total cost may be obtained by using the cost function:

$$C^* = \boldsymbol{c}'\boldsymbol{x}^*$$

$$C^* = [5 \quad 7 \quad 0 \quad 0 \quad 0] \cdot \begin{bmatrix} 3300 \\ 6700 \\ 0 \\ 0 \\ 1000 \end{bmatrix}$$

$$C^* = \$63{,}400$$

PROBLEMS

Problem 1

Consider the following constraint set:

$$x_1 + x_2 + 2x_3 \geq 1$$
$$2x_1 \leq 4$$
$$x_3 \geq 2$$

(a) Remove the inequality constraints by adding slack and surplus variables x_4, x_5, and x_6. Identify x_4, x_5, and x_6 as being slack or surplus variables.
(b) Use matrix partitioning and matrix inversion methods to solve for a basis consisting of x_2, x_3, x_5.
(c) Solve for a basis consisting of x_2, x_3, x_4.

Problem 2

$$2x_1 + x_2 \leq 9 \quad (1)$$
$$x_1 - 2x_2 \leq 2 \quad (2)$$
$$-3x_1 + 2x_2 \leq 3 \quad (3)$$
$$x_1 \geq 0, x_2 \geq 0$$

(a) Draw the feasible region on a graph.
(b) Indicate all the extreme points on the graph.
(c) Rewrite the constraint set with slack variables.
(d) (i) If Eqs (1) and (2) are active constraints, what is the extreme point on the graph? Call it $\boldsymbol{x}_B$.
(ii) Define the nonbasic and basic variable vectors $\boldsymbol{x}_B$ and $\boldsymbol{x}_N$.
(iii) Solve for the basic variable vector by matrix methods.
(iv) Compare (iii) with (i). The two solutions should be the same.

Problem 3

$$\text{Maximize } z = -2x_2 + x_1$$
$$x_2 \geq 1 + x_1$$
$$x_1 + x_2 \leq 7$$
$$x_1 \text{ unrestricted}, x_2 \geq 0$$

Since x_1 is unrestricted in sign, let $x_1 = x_1^+ - x_1^-$, where x_1^+ and x_1^- are *two new control variables* that replace x_1. They are nonnegative values satisfying the following constraints:

$$x_1^+ \geq 0 \qquad x_1^- \geq 0$$

Substitute $(x_1^+ - x_1^-)$ for x_1 in the given problem.
(a) Formulate the problem using x_1^+, x_1^-, and x_2 as control variables.
(b) Write in the standard form of linear programming.

Problem 4

An aggregate mixture contains a minimum of 30 percent sand and no more than 60 percent gravel nor 10 percent silt. Three sources of aggregate are available.

PIT	1	2	3
% Sand	5	30	100
% Gravel	60	70	—
% Silt	35	—	—
Cost/yd^3	\$2	\$10	\$8

Determine the mix of minimum unit cost, \$/yd^3

(a) Formulate a minimum cost model. Define all control variables with proper unit measures.
(b) Write in the standard form of linear programming. Define slack and surplus control variables and their units.

Problem 5

A sand and gravel mixture is to consist of exactly 50 percent sand and 50 percent gravel. Three pits are available. The percentage of sand and gravel and the cost of each mixture are:

PIT	1	2	3
% Sand	35	40	45
% Gravel	65	60	55
Unit cost/yd^3	\$8	\$9	\$11

Since it is impossible to obtain an exact 50–50 mixture of sand and gravel, gravel will have to be removed at a unit cost of \$4/yd^3, and sand will have to be added to the mixture at a unit cost of \$20/yd^3 of sand. A total mix of 10,000 yd^3 is required.

(a) Formulate a minimum cost model. Define the control variables and their units.
(b) Write in the standard form of linear programming. Define slack and surplus variables with proper unit measures.

Problem 6

Towns A and B produce 3 and 2 Mgal/day of wastewater, respectively. The BOD level of the wastewater is 200 mg/1. (See Figure 6.3.) Towns A and B belong to a regional water-treatment cooperative that operates water-treatment plants I and II. Plants I and II have a capacity of 3 and 4 Mgal/day and operate at a BOD removal efficiency of 90 and 75 percent, respectively.

The cooperative has an authority to assign the wastewater flow from towns A and B to either of plants I and II, as shown in the schematic diagram. Formulate a system-analysis model to minimize total annual treatment and pumping costs subject to the constraint that the total flow discharged into the river does not exceed 30 mg/l of BOD. The unit treatment and pumping costs

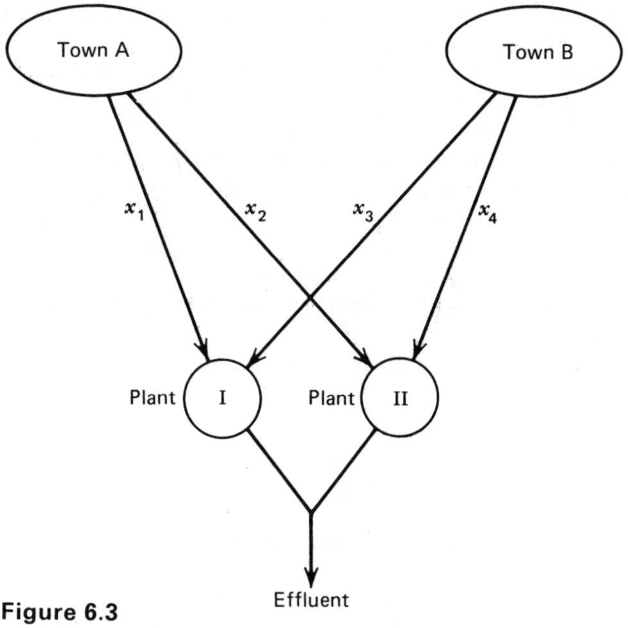

Figure 6.3

for each pipeline are given in the table. Let x_1, x_2, x_3, and x_4 be the wastewater flow in million gallons per day assigned to pipelines 1, 2, 3, and 4, respectively.

PIPELINE	UNIT ANNUAL TREATMENT AND PUMPING COSTS ($1000/Mgal/day—yr)
1	$46
2	$50
3	$55
4	$40

(a) Formulate a minimum cost model. Define the control variables with unit measures.
(b) Write in the standard form of linear programming. Define slack and surplus variables with proper unit measures.

Problem 7

A contractor will bid on a building construction project that consists of the following tasks:

TASK	PAYMENT SCHEDULE MONTH
1. Excavate site, remove 30,000 yd^3 of soil	3rd
2. Pour concrete foundation, 300 yd^3 of concrete	6th
3. Erect steel frame, 14,400 linear ft of steel	12th

A payment will be made after each task is completed. An estimate of the completion date of each task is indicated in the payment schedule.

The contractor has submitted a successful bid of \$400,000 on the entire project.

The contractor is a shrewd businessman, and he knows that it is better to charge larger sums of money for the earlier completed tasks and lesser amounts for later tasks. In other words, he would like to maximize the present worth of the money received. This type of bidding is called *unbalanced bidding*. The contractor will be able to accomplish his objective by adjusting the unit costs he charges for each task. His past experience shows he must charge between \$1.50/yd^3 and \$3.00 for soil excavation, at least \$100/yd^3 for cast-in-place concrete, and, but no more than \$100/ft for steel erection. Assume an interest rate of 12 percent per year.

(a) Formulate a mathematical model to meet the contractor's objective. Define control variables and their units.

(b) Write in the standard form of linear programming. Define slack and surplus variables with proper units.

6.2 THE SIMPLEX METHOD

The method will be used to determine the optimum solution of the following form:

$$\text{Minimize } z = \boldsymbol{c}'\boldsymbol{x}$$

$$\boldsymbol{A}\boldsymbol{x} = \boldsymbol{b}$$

$$\boldsymbol{x} \geq \boldsymbol{0}$$

where all elements of the vector $\boldsymbol{b}$ are positive, $\boldsymbol{b} \geq \boldsymbol{0}$.

Our development of the simplex method will be established to minimize the objective function, $z = \boldsymbol{c}'\boldsymbol{x}$. A maximization problem is solved by transforming the maximization problem to a minimization problem, and then, the minimization simplex algorithm is used to obtain the location of the optimum. Thus, we are able to solve both minimization and maximization problems with the same method. The simplex method is an iterative method to find the optimum solution of a linear mathematical model. It consists of the following steps:

1. Establish an initial candidate solution $\boldsymbol{x}^0$.
2. Test the initial candidate solution $\boldsymbol{x}^0$ to determine if it is the optimum solution $\boldsymbol{x}^*$.
3. If the test for the optimum solution fails, establish a new candidate solution $\boldsymbol{x}^1$, and repeat steps 2 and 3 until the optimality test is satisfied.

The superscripts 0 and 1 of $\boldsymbol{x}^0$ and $\boldsymbol{x}^1$ denote the initial estimate of $\boldsymbol{x}^*$ and the estimate of $\boldsymbol{x}^*$ for iteration 1. For an intermediate iteration k, the candidate solution of $\boldsymbol{x}^*$ is denoted as $\boldsymbol{x}^k$. If the optimality fails, $\boldsymbol{x}^k \neq \boldsymbol{x}^*$ (step 2), then a new candidate solution $\boldsymbol{x}^{k+1}$ is determined (step 3).

The simplex method involves a hill-climbing approach of moving from one extreme point to another one. New candidate solutions are established by comparing the relative values of the slopes of the objective function and selecting the point that not only minimizes the objective function the greatest amount, but also ensures that the new candidate is a feasible solution satisfying the constraint set.

Matrix algebra is used to explain the basic properties of the algorithm. Since the method is an iterative procedure, it is particularly suitable to computer solution.

Fundamental Relationships

If a candidate solution $\boldsymbol{x}$ satisfies the constraint set, $\boldsymbol{Ax} = \boldsymbol{b}$ and $\boldsymbol{x} \geq 0$, it is a *basic feasible solution* and a candidate for optimum solution of $z = \boldsymbol{c}'\boldsymbol{x}$. The superscript k will be dropped in this section because it does not add to the understanding of the procedure. A candidate solution will be designated as the partitioned vector, $\boldsymbol{x} = [\boldsymbol{x}_B\ \boldsymbol{x}_N]$. The basic candidate solution vector $\boldsymbol{x}_B$ may be written as a function of the nonbasic variables $\boldsymbol{x}_N$, or

$$\boldsymbol{x}_B = \boldsymbol{A}_B^{-1}[\boldsymbol{b} - \boldsymbol{A}_N\boldsymbol{x}_N]$$

Expanding this equation gives

$$\boldsymbol{x}_B = \boldsymbol{A}_B^{-1}\boldsymbol{b} - \boldsymbol{A}_B^{-1}\boldsymbol{A}_N\boldsymbol{x}_N$$

The value of z, written as a function of the nonbasic variables $\boldsymbol{x}_N$, is determined by substituting the candidate solution in the objective function:

$$z = \boldsymbol{c}_B'[(\boldsymbol{A}_B^{-1}\boldsymbol{b} - \boldsymbol{A}_B^{-1}\boldsymbol{A}_N\boldsymbol{x}_N)] + \boldsymbol{c}_N'\boldsymbol{x}_N$$

where $\boldsymbol{c}' = [\boldsymbol{c}_B'\ \boldsymbol{c}_N']$. Collecting the coefficients of $\boldsymbol{x}_N$, we may write z as

$$z = \boldsymbol{c}_B'\boldsymbol{A}_B^{-1}\boldsymbol{b} + [(\boldsymbol{c}_N' - \boldsymbol{c}_B'\boldsymbol{A}_B^{-1}\boldsymbol{A}_N)]\boldsymbol{x}_N$$

In a latter part of our discussion we refer to $[(\boldsymbol{c}_N' - \boldsymbol{c}_B'\boldsymbol{A}_B'\boldsymbol{A}_N)]$ as $\boldsymbol{c}_I'$, the *cost indicator*.

Since $\boldsymbol{x}_N$ are nonbasic variables and all elements of $\boldsymbol{x}_N$ are equal to zero, the values of the candidate solution $\boldsymbol{x}_B$ and candidate optimum solution z may be easily determined. Substitute $\boldsymbol{x}_N = \boldsymbol{0}$ and simply.

$$\boldsymbol{x}_B = \boldsymbol{A}_B^{-1}\boldsymbol{b}$$

$$z = \boldsymbol{c}_B'\boldsymbol{A}_B^{-1}\boldsymbol{b} = \boldsymbol{c}_B'\boldsymbol{x}_B$$

This analysis shows that a candidate solution $\boldsymbol{x}_B$ and the value of the objective function z are obtained by establishing a basic matrix $\boldsymbol{A}_B$.

In the simplex method, the choice of column vectors of $\boldsymbol{A}$, $\boldsymbol{a}_i$ that enter $\boldsymbol{A}_B$ will be selected to ensure that $\boldsymbol{x}_B$ is always a *basic feasible solution.* Furthermore, in each iteration the basic matrix is selected to ensure that new values of $\boldsymbol{x}_B$ will reduce the value of z or, at least, be equal to the value of z in the previous iteration. Thus, our discussion will begin by assuming that a basis feasible candidate solution $\boldsymbol{x}_B^k$ exists, where $k = 0, 1, 2, \ldots$. Recall that *all* basic feasible solutions are nonnegative vectors, $\boldsymbol{x}_B^0 \geq 0$, $\boldsymbol{x}_B^1 \geq 0$, $\boldsymbol{x}_B^2 \geq 0, \ldots$.

The Test for an Optimum Solution

A method for establishing if $z(\boldsymbol{x}^k) = z^k$ is the optimum minimum solution z^* must be developed. In other words, we must answer the question, is $\boldsymbol{x}^k = \boldsymbol{x}^*$ the location of the minimum z^*? We introduce the basic concepts by example.

Suppose the objective function is

$$z = 2x_1 + 3x_2 + 5x_3 - 2x_4$$

and the candidate solution $\boldsymbol{x}^k = [\boldsymbol{x}_B^k \; \boldsymbol{x}_N^K]$, is

$$\boldsymbol{x}_B^k = \begin{bmatrix} x_1^k \\ x_2^k \end{bmatrix} = \begin{bmatrix} 2 \\ 1 \end{bmatrix} \qquad \boldsymbol{x}_N^k = \begin{bmatrix} x_3^k \\ x_4^k \end{bmatrix} = \begin{bmatrix} 0 \\ 0 \end{bmatrix}$$

The value of the candidate solution is

$$z^k = 2 \cdot 2 + 3 \cdot 1 + 5 \cdot 0 - 2 \cdot 0 = 7$$

The candidate solution $\boldsymbol{x}^k$ may or may not be the location of the optimum minimum solution. We may determine if the current basic feasible solution $\boldsymbol{x}_B^k$ is the optimum solution by evaluating the elements of the cost indicators $\boldsymbol{c}_I^k$.

$$(z - z^k) = \boldsymbol{c}_I^{k\prime} \boldsymbol{x}_N$$

Substitute $z^k = 7$ and assume the objective function is a function of the nonbasic variables:

$$(z - 7) = 5x_3 - 2x_4$$

or

$$z = 7 + 5x_3 - 2x_4$$

Thus $\boldsymbol{c}_I' = [5 \; -2]$. If $\boldsymbol{x}_B^k$ is the location of the minimum solution, all other values of x in the feasible region will be greater than the candidate solution z^k or $z > z^k$. If $\boldsymbol{x}_B^k$ is not the location of the optimum solution, then a new basic vector $\boldsymbol{x}_B^{k+1}$ must be established. A nonbasic variable x_3 or x_4 must be entered into $\boldsymbol{x}_B^{k+1}$, and a basic variable x_1 or x_2 must be removed. In order for either x_3 or x_4 to enter $\boldsymbol{x}_B^{k+1}$, there must be positive values to satisfy the condition of a basic feasible solution $\boldsymbol{x}_B \geq \boldsymbol{0}$. This is an important restriction that will be included in our test for optimality.

First, assume x_3 is entered into the basis and x_4 remains a member of the nonbasis, $x_4 = 0$. Thus

$$z^{k+1} = 7 + 5x_3$$

Since x_3 is restricted to be a positive number, $x_3 \geq 0$, the value of $z^{k+1} = 7 + 5x_3$ will always be greater than $z^k = 7$. Thus, we can conclude that entering x_3 into $\boldsymbol{x}_B^{k+1}$ will not reduce z^k. The present basis $\boldsymbol{x}_B^{k\prime} = [x_1^k \; x_2^k]$, remains a candidate for the optimum minimum solution.

Next, x_4 is assumed to be entered into the basis $\boldsymbol{x}_B^{k+1}$ and x_3 remains in the nonbasis, $x_3 = 0$. The objective function is

$$z^{k+1} = 7 - 2x_4$$

Since x_4 must be a positive number, the value of z^{k+1} will be less than z^k for $x_4 > 0$. Thus, the candidate solution $z^k = 7$ may be reduced by entering x_4 into the basis. The current candidate solution $\boldsymbol{x}_B^{k\prime} = [x_1^k \; x_2^k]$ is not the minimum solution and a new candidate solution $\boldsymbol{x}^{k+1} = [\boldsymbol{x}_B^{k+1}\ \boldsymbol{x}_N^{k+1}]$ must be determined. We discuss the removal of x_1 or x_2 from $\boldsymbol{x}^{k+1}$ in the subsection entitled "Establishing a New Basis."

This example illustrates the essential steps utilized in testing a candidate solution as the optimum solution. We shall establish a more general test with the $\boldsymbol{c}_I^k$ vector. Since any one of the control variables of $\boldsymbol{x}_N^k$ may enter $\boldsymbol{x}^{k+1}$ and form a new basic feasible solution, the *sign* of coefficients of $\boldsymbol{c}_I^k$ will indicate if the solution of z^k will be increased or not. If *all* indicators $\boldsymbol{c}_I^k$ are positive, introducing *any* member of $\boldsymbol{x}_N$ into the basis will increase the value of z^k. If this condition occurs, then the location of the optimum point has been determined, $z^* = z^k$. We state this as a rule.

RULE 1 If the sign of all indicators of $\boldsymbol{c}_I^k$ are positive, then the current basic feasible solution $\boldsymbol{x}^k$ is the location of the optimal minimum solution, $\boldsymbol{x}^*$ and $z^k = z^*$.

Establishing a New Candidate Solution

If one or more of the indicators of $\boldsymbol{c}_I^k$ are negative, the candidate solution $\boldsymbol{x}_B^k$ is not an optimum, and the solution of z^k can be reduced. If there is only one value of $\boldsymbol{c}_I^k$ that is negative, then the corresponding nonbasic variable is entered into a new basis. Only one variable of $\boldsymbol{x}_N^k$ is entered into $\boldsymbol{x}_B^{k+1}$ during this step. If two or more of the indicators of $\boldsymbol{c}_I^k$ are negative, then the variable associated with the smallest negative value of $\boldsymbol{c}_I^k$ is entered into $\boldsymbol{x}_B^{k+1}$. For example, suppose that $z - z^k = \boldsymbol{c}^k\,\boldsymbol{x}_N^k$ is

$$(z - 8) = 3x_3 - 4x_4 - 8x_5$$

Since the coefficients of x_4 and x_5 are negative, entering either one of them into the basis will reduce z. Clearly, introducing x_5 into $\boldsymbol{x}_B^{k+1}$ will reduce z a larger amount; therefore, it is chosen as the variable to enter $\boldsymbol{x}_N^0$.

Entering the variable with smallest negative $\boldsymbol{c}_I^k$ coefficient does not ensure that the value of z^k is reduced the greatest amount, because the constraints may limit the range of the new basic variable. For example, consider the situation where x_5 can only increase to $x_5 = 1$ while x_4 may be increased to $x_4 = 3$; x_4 is thus the better choice. However, the gain is generally not considered with the extra computer programming effort, hence Rule 2.

RULE 2 If one or more of the indicators of $\boldsymbol{c}_I^k$ are negative, the current present solution can be reduced. The variable of $\boldsymbol{x}_N^k$ that enters the new basis, $\boldsymbol{x}_B^{k+1}$, denoted as $\boldsymbol{x}_p$, is the one associated with smallest negative value of $\boldsymbol{c}_I^k$.

We can interpret Rule 2 by tracing the possible paths from the current candidate solution at point $a = \boldsymbol{x}^0, (x_1^0, x_2^0)$, to the optimum solution at point $c = \boldsymbol{x}^*, (x_1^*, x_2^*)$ as shown in Figure 6.4. For the hypothetical situation, we assume that there are two negative indicators of $\boldsymbol{c}_I^k$. With two negative indicators, there are two possible paths to follow, from extreme point a to extreme points b or c, respectively. In following Rule 2, we find the smallest negative indicator and introduce the corresponding control variable into the basis. In Figure 6.4 this will be equivalent to moving from extreme point a to extreme point c. In this case, we find the minimum or optimum solution $\boldsymbol{x}^*$ in one iteration. If we choose the other path, extreme points b and d must be evaluated. These points are not the location of the optimum solution; therefore,

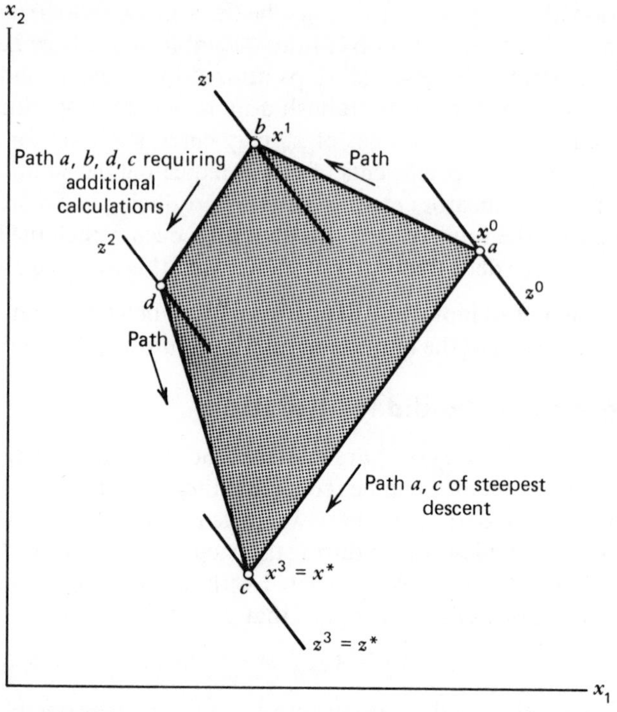

Figure 6.4 Hill-climbing approach and the search of extreme points.

two other iterations are required. Choosing the smallest negative value of c_I' may eliminate extra iterations.

By choosing the greatest negative value of c_I^k, we proceed to move down a path with the steepest descent. If we were maximizing, we would follow a path of steepest ascent; for this reason the simplex method is often called a hill-climbing procedure. By following Rule 2, we attempt to proceed to the minimum solution in the least number of iterations.

Establishing a New Basis

By Rule 2, we establish which control variable enters into a new candidate basis. We must determine which variable of the current basis x_B^k must be removed and placed in the nonbasis x_N^{k+1}. In essence, we are trading places. One control variable in the basis is being removed, and a control variable of the nonbasis is taking its place. The control variable that is removed from the basis is placed in the nonbasis. This approach satisfies the requirement that the number of control variables in the basis must be equal to the number of constraint equations. If there are m constraint equations, there must be m control variables in x_B^{k+1}.

For example, suppose we have a linear mathematical model consisting of three constraint equations $m = 3$, and five control variables $n = 5$. Furthermore, let us assume that the current basis contains

$$\boldsymbol{x}_B^k = \begin{bmatrix} x_1 \\ x_3 \\ x_4 \end{bmatrix} \qquad \boldsymbol{x}_N^k = \begin{bmatrix} x_2 \\ x_5 \end{bmatrix}$$

After completing the operations indicated by Rule 2, we find that x_5 is to be entered into the basis. If x_5 is entered into $\boldsymbol{x}_B^{k+1}$, one of the basic variables x_1, x_3, or x_4 must be removed and placed in $\boldsymbol{x}_N^{k+1}$. Suppose the constraint set is written as follows:

$$\begin{bmatrix} x_1 \\ x_3 \\ x_4 \end{bmatrix} = \begin{bmatrix} 1 \\ 5 \\ 3 \end{bmatrix} - \begin{bmatrix} -1 \\ 1 \\ 2 \end{bmatrix} \cdot x_5$$

or

$$x_1 = 1 + x_5$$

$$x_3 = 5 - x_5$$

$$x_4 = 3 - 2x_5$$

Since our goal is to remove one of the current control variables from $\boldsymbol{x}_B^k$, one of the control variables x_1, x_3, or x_4 must be entered in $\boldsymbol{x}_N^{k+1}$. This goal is accomplished by setting one of the variables of $\boldsymbol{x}_B^k$ equal to zero and determining if the remaining variables of $\boldsymbol{x}_B^k$ are basic feasible solutions.

Assume x_1 becomes a member of $\boldsymbol{x}_N^{k+1}$, or $x_1 = 0$. Solving for x_5, we find

$$x_1 = 0 = 1 + x_5$$

$$x_5 = -1$$

Since x_5 is negative, the basis $\boldsymbol{x}_B^{k+1'} = [x_5 \; x_3 \; x_4]$ cannot be a basic feasible solution.

Assume x_3 becomes a member of $\boldsymbol{x}_N^{k+1}$, or $x_3 = 0$.

$$x_3 = 0 = 5 - x_5$$

$$x_5 = 5$$

Since $x_5 > 0$, it is possible that x_3 can be a member of $\boldsymbol{x}_N^{k+1}$ if the remaining variables of $\boldsymbol{x}_B$, x_1, and x_4 are positive variables.

$$x_1 = 1 + x_5 = 1 + (5) = 6$$

$$x_4 = 3 - 2x_5 = 3 - 2(5) = -7$$

Since $x_4 < 0$, the basis $\boldsymbol{x}_B^{k+1'} = [x_1 \; x_5 \; x_4]$, is not a basic feasible solution.

Assume x_4 becomes a member of $\boldsymbol{x}_N$, $x_4 = 0$:

$$x_4 = 0 = 3 - 2x_5$$

$$x_5 = \tfrac{3}{2}$$

Determine x_1 and x_3:

$$x_1 = 1 + x_5 = 1 + (3) = 4$$

$$x_3 = 5 - x_5 = 5 - (3) = 2$$

Since all values x_1, x_3, and x_5 are positive, a new basic feasible solution has been found. The basic feasible and nonbasic vectors are

$$\boldsymbol{x}_B^{k+1} = \begin{bmatrix} x_1 \\ x_3 \\ x_5 \end{bmatrix} = \begin{bmatrix} 4 \\ 2 \\ 3 \end{bmatrix} \qquad \boldsymbol{x}_N^{k+1} = \begin{bmatrix} x_2 \\ x_4 \end{bmatrix} = \begin{bmatrix} 0 \\ 0 \end{bmatrix}$$

The same result may be obtained by observing that the variable associated with the minimum nonnegative ratio of $\boldsymbol{b}_i^k/\boldsymbol{a}_{i,p}^k$ is the variable that is removed from $\boldsymbol{x}_B^k$ and placed in $\boldsymbol{x}_N^{k+1}$. From the foregoing example $\boldsymbol{x}_B^{k+1} = \boldsymbol{b}^k - x_p^k \boldsymbol{a}_p^k$, where $x_p = x_5$ may be written as $\boldsymbol{x}_B^{k+1} + x_p^k \boldsymbol{a}_p^k = \boldsymbol{b}^k$; or

$$\begin{bmatrix} x_1 \\ x_3 \\ x_4 \end{bmatrix} + \begin{bmatrix} -1 \\ 1 \\ 2 \end{bmatrix} \cdot x_5 = \begin{bmatrix} 1 \\ 5 \\ 3 \end{bmatrix}$$

For simplicity in notation, the superscript k is removed. The ratios of $b_i/a_{i,p}$ may be written as

$$\frac{b_1}{a_{1,5}} = \frac{1}{-1} = -1$$

$$\frac{b_3}{a_{3,5}} = \frac{5}{1}$$

$$\frac{b_4}{a_{4,5}} = \frac{3}{2}$$

Since $b_1/a_{1,5} < 0$, x_1 cannot enter the new $\boldsymbol{x}_B$. Since $b_4/a_{3,5}$ is the *smallest positive ratio*, $x_q = x_4$ is removed from the basis.

This example gives the essential steps used in determining which control variable is removed from the current basis $\boldsymbol{x}_B^k$ and is placed in the new nonbasis $\boldsymbol{x}_N^{k+1}$. A general procedure for performing these steps may be establishing with the use of the equation:

$$\boldsymbol{x}_B^k + x_p^k \boldsymbol{a}_p^k = \boldsymbol{b}^k$$

The control variable x_p of $\boldsymbol{x}_B^k$ is defined to be the variable to be removed from $\boldsymbol{x}_B^k$ and placed in $\boldsymbol{x}_N^{k+1}$. The set of equations $\boldsymbol{x}_B^k + x_p^k \boldsymbol{a}^k = \boldsymbol{b}^k$ may be stated without the superscript k as

$$x_t + a_{k,p} x_p = b_t \qquad t = 1, 2, \ldots, m$$

The determination of x_q proceeds by selecting only those equations where the *sign of $a_{t,p}$ is positive, $a_{t,p} > 0$.* The control variable, $\boldsymbol{x}_p$, associated with the minimum value of $b_t/a_{t,p}$, will be the control variable that is removed from $\boldsymbol{x}_B$ and placed in $\boldsymbol{x}_N$. Since b_t *must be positive* by definition, $\boldsymbol{b} \geq 0$, all ratios must be positive. This procedure is stated as Rule 3 of the simplex method.

RULE 3 Calculate the ratio of $b_t/a_{t,p}$ for all $a_{t,p} > 0$ in the current basis,

$$\boldsymbol{x}_B + x_p \boldsymbol{a}_p = \boldsymbol{b}.$$

The control variable x_q that corresponds to the smallest positive value of $b_t/a_{t,p}$ is removed from the basis and placed in the nonbasis. The control variable x_p replaces x_q in the basis $k + 1$. The superscript k is not shown for simplicity in notation.

In case of a tie among minimum value ratios of $b_t/a_{t,p}$, any one of the control variables associated with the minimum ratio may enter the basis. If no variable satisfies Rule 3, the *domain is unbounded*, and no minimum value exists. When this result occurs, no practical real-world solution exists, because, most likely, the problem was incorrectly formulated. In this case, the mathematical model must be reformulated and solved again.

The Simplex Tableau

For hand calculations, the three rules of the simplex method and the method of successive elimination are generally used. The linear mathematical model may be written as

$$\text{Minimize } (z - z^k) = \boldsymbol{c}_I^{k\prime} \boldsymbol{x}_N^k + \boldsymbol{0}' \boldsymbol{x}_B^k$$

$$\boldsymbol{A}_N^k \boldsymbol{x}_N^k + \boldsymbol{I}\boldsymbol{x}_B^k = \boldsymbol{b}^k$$

$$\boldsymbol{x}_N^k = \boldsymbol{0}$$

$$\boldsymbol{x}_B^k \geq \boldsymbol{0}$$

where $\boldsymbol{x}_B^k$ and $\boldsymbol{x}_N^k$ represent a basic feasible candidate solution of the minimum of z. Special note should be made that the unit costs associated with $\boldsymbol{x}_B^k$ must be zero, $\boldsymbol{c}_B = \boldsymbol{0}$, and that all members of $\boldsymbol{b}^k$ must be positive. The candidate solution $\boldsymbol{x}_B^k$ may be written in an augmented matrix as

$$\begin{bmatrix} \boldsymbol{x}_B^k & \boldsymbol{A}_N^k & \boldsymbol{I} & \boldsymbol{b}^k & b/a \\ - & \boldsymbol{c}_I^{k\prime} & \boldsymbol{0}' & (z - z^k) & - \end{bmatrix}$$

The column vector $\boldsymbol{x}_B^k$ contains a list of the control variables that are current members of basis vector. It is understood that the variables that are not listed in $\boldsymbol{x}_B^k$ are members of the nonbasic vector $\boldsymbol{x}_N^k$ and, by definition, are equal to zero, $\boldsymbol{x}_N^k = \boldsymbol{0}$. The numerical values of the basic variables $\boldsymbol{x}_B^k$ and the objective function z^k are recorded in the tableau as $\boldsymbol{b}^k$ and $(z - z^k)$, respectively. The current values of the $\boldsymbol{A}_N^k$ matrix and the indicators of $\boldsymbol{c}_I^k$ are placed in the appropriate location of the augmented matrix. The column b/a is reserved for performing the calculation required by Rule 3.

In Section 6.3, the selection of the *initial basic feasible* solution $\boldsymbol{x}_B^0$, $k = 0$, will be discussed in greater detail. In this section we consider a special form of the linear model where the linear mathematical models consists of a set of constraint equations with less-than-or-equal-to constraints only.

$$\text{Minimize } z = \boldsymbol{c}'\boldsymbol{x}$$

$$\boldsymbol{a}_i'\boldsymbol{x} \leq b_i \qquad i = 1, 2, \ldots, m$$

$$\boldsymbol{x} \geq \boldsymbol{0}$$

Slack variables are introduced, and the model is written as

$$\text{Minimize } z = \boldsymbol{c}'\boldsymbol{x}$$

$$\boldsymbol{A}\boldsymbol{x} = \boldsymbol{b}$$

$$\boldsymbol{x} \geq \boldsymbol{0}$$

For this special case the easiest method to establish an initial basic feasible solution $\boldsymbol{x}_B^0$ is to assign all slack variables to $\boldsymbol{x}_B^0$. *No* calculations are needed for this initial guess.

Maximization Problems

The optimum solution of maximization problems of the form:

$$\text{Maximize } z = \boldsymbol{c}'\boldsymbol{x}$$

$$\boldsymbol{A}\boldsymbol{x} = \boldsymbol{b}$$

$$\boldsymbol{x} \geq \boldsymbol{0}$$

may be transformed to a minimization problem by multiplying the objective function by -1. Define the objective function as

$$a = (-1)z = -\boldsymbol{c}'\boldsymbol{x}$$

Finding the location of the minimum value of a is equivalent to determining the maximum value of z. This operation does not alter the shape of the objective function; it only changes the sign of the function. The transformed mathematical model is

$$\text{Minimize } a = -\boldsymbol{c}'\boldsymbol{x}$$

$$\boldsymbol{A}\boldsymbol{x} = \boldsymbol{b}$$

$$\boldsymbol{x} \geq \boldsymbol{0}$$

The minimum value of a may be determined with the simplex method. The optimum solution is designated at $\boldsymbol{x}^*$ with $\boldsymbol{a}^*$. The maximum value of z is located at $\boldsymbol{x}^*$ with maximum z equal to

$$z^* = -a^*$$

We illustrate the basic concept of solving a maximization problem with a one-dimensional model (see Figure 6.5.):

$$\text{Maximize } z = 2x_1$$

$$x_1 \leq 4$$

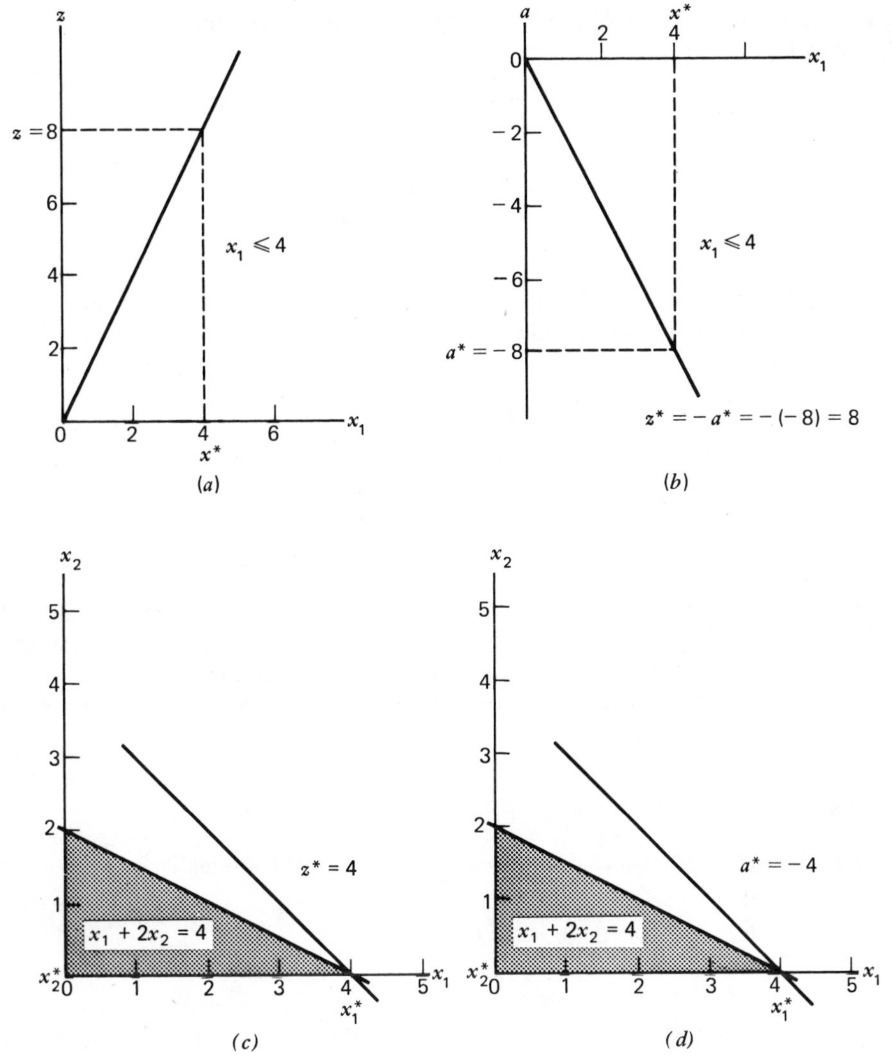

Figure 6.5 Solving a maximization problem as a minimization problem.

The optimum solution as shown in Figure 6.5*a* is $x_1^* = 4$ and $z^* = 8$. Solving this problem as a minimization problem with $a = -z$, the model is

$$\text{Minimize } a = -2x_1$$

$$x_1 \leq 4$$

The optimum solution as shown in Figure 6.5*b* is $x^* = 4$ and $a^* = -8$. Since $z = -a$, the maximum value of z is equal to $z^* = 8$. The location of the optimum solution x_1^* for the two problems is the same extreme point, $x_1^* = 4$.

The optimum solution of the following two-dimensional maximization model is shown in Figure 6.5*c*.

$$\text{Maximize } z = x_1 + x_2$$
$$x_1 + 2x_2 \le 4$$
$$x_1 \ge 0$$
$$x_2 \ge 0$$

The optimum solution is $\boldsymbol{x}^* = (4, 0)$ with $z^* = 4$. Transforming this model to a minimization model by multiplying z by -1 gives the same result as shown in Figure 6.5*d*:

$$\text{Minimize } a = -x_1 - x_2$$
$$x_1 + 2x_2 \le 4$$
$$x_1 \ge 0$$
$$x_2 \ge 0$$

The optimum solution $\boldsymbol{x}^*$ is (4, 0) with $a^* = -4$. Comparing the two solutions shows that the location of optimum solution is the same. Since $z = -a$, the optimum solution is $z^* = -a^* = 4$ and the same maximum value of z is obtained. The transformation $a = (-1)z$ will be used to solve multidimensional maximization problems with the simplex algorithm for minimization.

EXAMPLE 6.4 A Minimization Problem

(a) Use the simplex method to determine the solution of the following linear mathematical model:

$$\text{Minimize } z = -11x_1 - 4x_2$$
$$3x_1 + 5x_2 \le 15$$
$$5x_1 + 2x_2 \le 10$$
$$x_1 \ge 0,\ x_2 \ge 0$$

(b) Trace the path to the optimum solution on a graph.

Solution

(a) Slack variables x_3 and x_4 are added to the constraint equations to remove the inequality constraints:

$$\text{Minimize } z = -11x_1 - 4x_2$$
$$3x_1 + 5x_2 + x_3 \qquad = 15$$
$$5x_1 + 2x_2 \qquad + x_4 = 10$$
$$x_1 \ge 0,\ x_2 \ge 0,\ x_3 \ge 0,\ x_4 \ge 0$$

There are two constraint equations, $m = 2$; therefore, the initial basic feasible solution $\boldsymbol{x}_B^0$ must contain two variables. The easiest way to establish the initial basis is to assign all slack variables to $\boldsymbol{x}_N^0$:

$$\boldsymbol{x}_B^0 = \begin{bmatrix} x_3 \\ x_4 \end{bmatrix} = \begin{bmatrix} 15 \\ 10 \end{bmatrix}$$

Initial Solution

	$\boldsymbol{x}_B$	x_1	x_2	x_3	x_4	$\boldsymbol{b}$	b/a	
	x_3	3	5	1	0	15	$\frac{15}{3} = 5$	
	x_4	(5)	2	0	1	10	$\frac{10}{5} = 2$	←
Indicator row	c_I'	-11 ↑	-4	0	0	$(z - 0)$		

The simplex method may now proceed:

RULE 1 Since the c_I indicators of the variables of x_1 and x_2 are negative, the current solution is not a minimum.

RULE 2 The control variable that enters the basis is x_1 because $c_1 = -11$ is the smallest negative value of the indicator vector. An arrow is drawn under the x_1 column to indicate that it will enter the basis.

RULE 3 The control variable that leaves the basis is x_4 because $\frac{10}{5} = 2$ is the smallest positive b/a ratio. An arrow is drawn next to row 2 to indicate that x_4 is to be removed from the basis.

Iteration 1 Successive elimination is used to determine the next candidate solution; that is, $\boldsymbol{x}_{B'}' = [x_3\ x_1]$. A circle has been drawn around the pivot element of the $\boldsymbol{A}$ matrix. The pivot element will be reduced to 1 by dividing each element of the row by 5. The new row is written in a new simplex tableau. The remaining elements in column x_1, a_{11} and c_1, must be set equal to zero by making the appropriate row operations. In this case all elements of the x_1 row, row 2 in the newly established simplex tableau below, are multiplied by -3 and added to each element of the x_3 row, row 1 of the old tableau. In similar fashion, multiple each element of x_1, row 2 below, by 11 and add them to indicator row of the old tableau. The resulting simplex tableau is

First Iteration

$\boldsymbol{x}_b$	x_1	x_2	x_3	x_4	$\boldsymbol{b}$	b/a
x_3	0	$\frac{19}{5}$	1	$-\frac{3}{5}$	9	
x_1	1	$\frac{2}{5}$	0	$\frac{1}{5}$	2	
c_I'	0	$\frac{2}{5}$	0	$\frac{11}{5}$	$z + 22$	

The solution is an optimum solution because all indicator row values are positive:

$$x_1^* = 2, \quad x_3^* = 9 \quad \text{and} \quad z^* = -22$$

(b) The graphical solution is shown in Figure 6.6. The equation $5x_1 + 2x_2 = 10$ is an active constraint, and the equation $3x_1 + 5x_2 \leq 15$, is an inactive constraint as shown in the diagram.

The path of the simplex method shows the initial solution at point (0, 0). The candidate solution after one iteration and the optimum solution is (2, 0). The path of the solution is shown by the arrow.

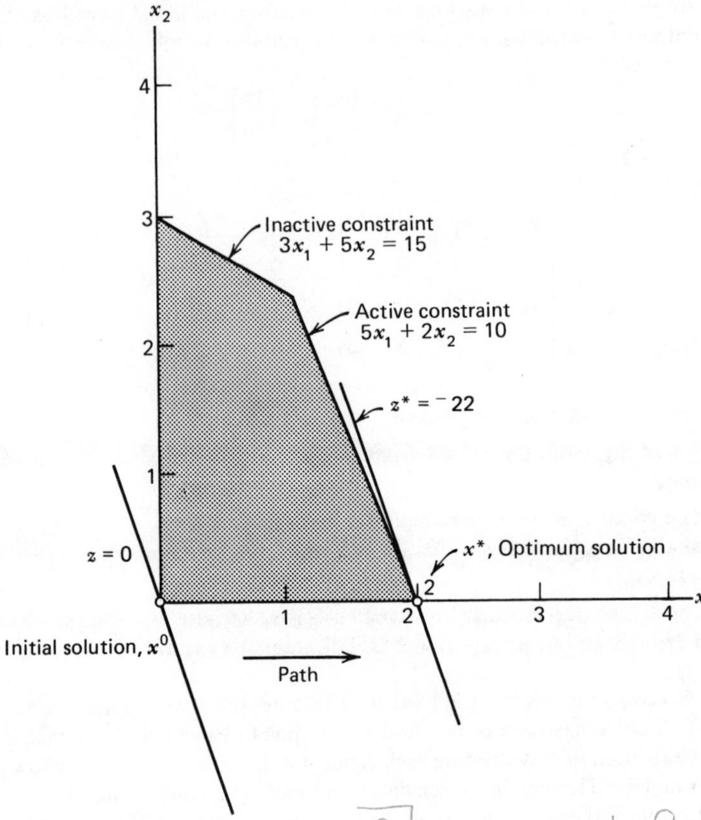

x_2
4
3
2
1
Inactive constraint
$3x_1 + 5x_2 = 15$
Active constraint
$5x_1 + 2x_2 = 10$
$z^* = -22$
$z = 0$
x^*, Optimum solution
2
1
2
3
4
x_1
Initial solution, x^0
Path

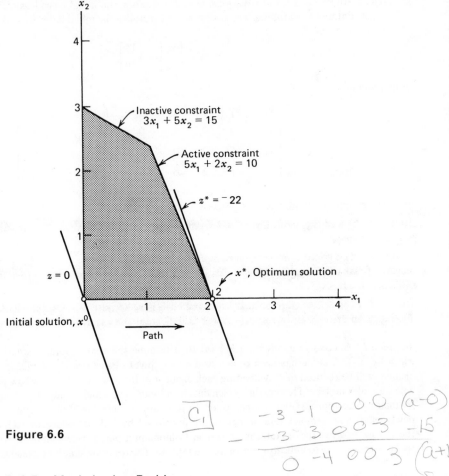

Figure 6.6

EXAMPLE 6.5 Maximization Problem

Determine the solution of the following linear mathematical model:

$$\text{Maximize } z = +3x_1 + x_2$$

$$x_1 - 2x_2 \leq 10$$

$$2x_1 + x_2 \leq 24$$

$$x_1 - x_2 \leq 5$$

$$x_1 \geq 0, x_2 \geq 0$$

(a) Transform the model from a maximization to a minimization problem.
(b) Solve part a by use of the simplex method.
(c) Trace the path to the optimum solution on a graph.
(d) On a graph show that the locations of optimum solutions x^* for the minimization and maximization problems are at the same extreme point.

Solution

(a) The simplex method utilized in this text requires that the objective function be a minimum. By multiplying the objective by -1, the objective function is inverted without changing the shape of the function. Define the transformed objective function as a.

$$a = (-1)z = -3x_1 - x_2$$

Thus, the objective function is:

C_1 → Minimize $a = -3x_1 - x_2$

(b) Slack variables $x_3, x_4,$ and x_5 are introduced to each constraint equation. The mathematical model is

x_1, x_2 ← non basic var's (given)

$$\text{Minimize } a = -3x_1 - x_2$$

x_3, x_4, x_5 ← basic vars

Eqn 1 $\quad x_1 - 2x_2 + x_3 \qquad\qquad = 10$

" 2 $\quad 2x_1 + x_2 + \quad + x_4 \quad = 24$

" 3 $\quad x_1 - x_2 + \qquad\quad + x_5 = 5$

$$x_1 \geq 0, x_2 \geq 0, x_3 \geq 0, x_4 \geq 0, x_5 \geq 0$$

The slack variables $x_3, x_4,$ and x_5 are assigned to the initial basis $x_B^0, x_3 = 10, x_4 = 24,$ and $x_5 = 5$. The simplex tableau is

Initial Solution "basic"

x_B	x_1	x_2	x_3	x_4	x_5	b	b/a
x_3	1	-2	1	0	0	10	$\frac{10}{1} = 10$
x_4	2	1	0	1	0	24	$\frac{24}{2} = 12$
x_5	①	-1	0	0	1	5	$\frac{5}{1} = 5$ ←
c_I	-3	-1	0	0	0	$(a - 0)$	

Eqn 1, 2, 3

The simplex method for minimization of a proceeds as follows.

RULE 1 Since the c_I indicators of x_1 and x_2 are negative, the initial solution is not a minimum.

RULE 2 The control variable x_1 enters the basis because $c_1 = -3$ is the smallest negative value of c_I. An arrow is drawn under the x_1 column to indicate that x_1 enters the basis.

RULE 3 The control variable x_5 leaves the basis because $b/a = 5$ is the smallest positive value of the b/a vector. An arrow indicates that x_5 leaves the basis. (non-neg.)

Iteration 1 Successive elimination is used to determine the next candidate solution, that is, $x_B' = [x_3 \ x_4 \ x_1]$. The result is

x_B	x_1	x_3	x_4	x_5	b	b/a
x_3	0	-1	1	0	-1	5	-5
x_4	0	③	0	1	-2	14	$\frac{14}{3}$ ←
x_1	1	-1	0	0	1	5	-5
c_I	0	-4	0	0	3	$(a + 15)$	

new basic var.

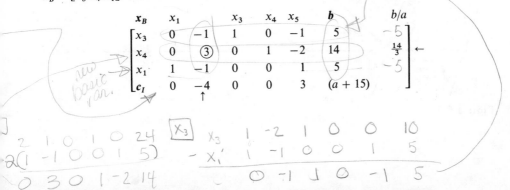

RULE 1 Since the c_I indicator of x_2 is negative, the candidate solution is not an optimum minimum solution.
RULE 2 The control variable x_2 enters the basis because it is the only negative indicator of c_I.
RULE 3 The control variable x_4 leaves the basis because $b/a = \frac{14}{13}$ is the only positive ratio of the b/a vector. Negative values of the element of $a_{i,p}$ are *not* considered in the b/a elevation.

A circle has been drawn around the pivot element of the A matrix. Successive elimination is performed. The new simplex tableau after the second iteration is

Second Iteration

x_B	x_1	x_2	x_3	x_4	x_5	b	b/a
x_3	0	0	1	$\frac{1}{3}$	$-\frac{5}{3}$	$\frac{29}{3}$	
x_2	0	1	0	$\frac{1}{3}$	$-\frac{2}{3}$	$\frac{14}{3}$	
x_1	1	0	0	$\frac{1}{3}$	$\frac{1}{3}$	$\frac{29}{3}$	
c_I	0	0	0	$\frac{4}{3}$	$\frac{1}{3}$	$\left(a + \frac{101}{3}\right)$	

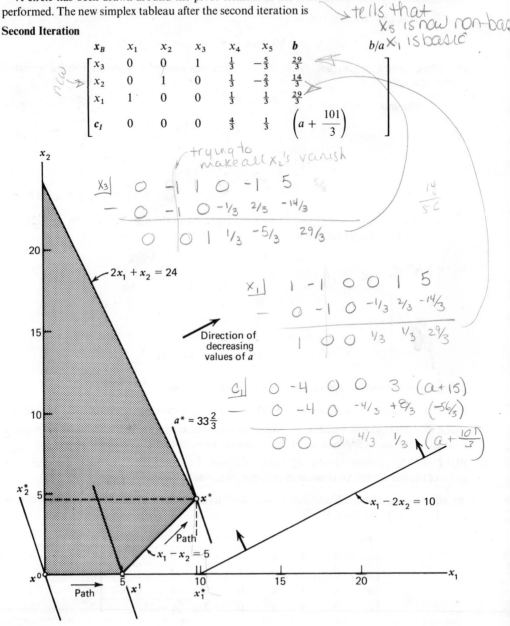

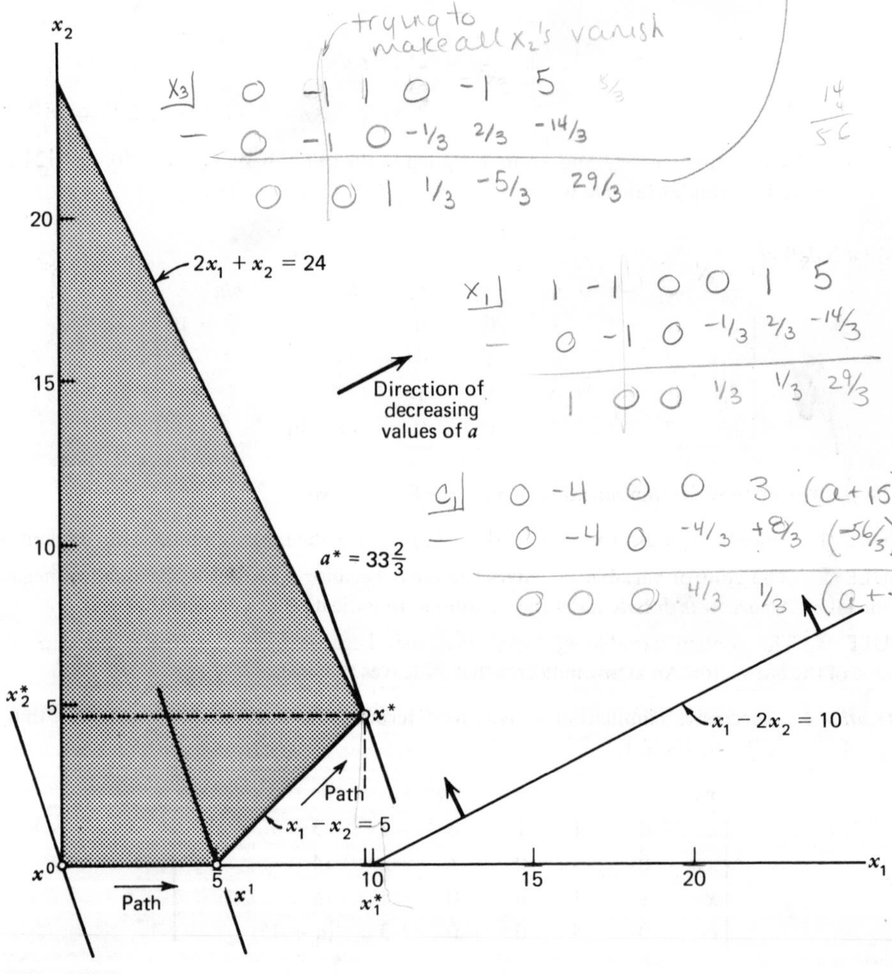

Figure 6.7

RULE 1 Since all indicators of c_I are positive, the candidate solution is the optimum minimum solution.

x_3, x_2, x_1 are basic variables
x_4, x_5 are non basic

The optimum location is

$$x_1^* = \tfrac{29}{3} = 9\tfrac{2}{3} \qquad x_2^* = \tfrac{14}{3} = 4\tfrac{2}{3} \qquad x_3^* = \tfrac{29}{3} = 9\tfrac{2}{3}$$

The optimum minimum of a is

$$a^* = \frac{-101}{3} = -33\tfrac{2}{3}$$

Since $z = -1a$, the maximum of z is

$$z^* = (-1)(-33\tfrac{2}{3}) = 33\tfrac{2}{3}$$

(c) The path to the minimum solution is shown in Figure 6.7. The path to the optimum solution initiates at (0, 0), proceeds to (5, 0), and then to optimum solution at ($9\frac{2}{3}$, $4\frac{2}{3}$).

(d) The feasibility region for the maximization of z or for the minimization of a is the same. The feasibility region is shown in Figure 6.8.

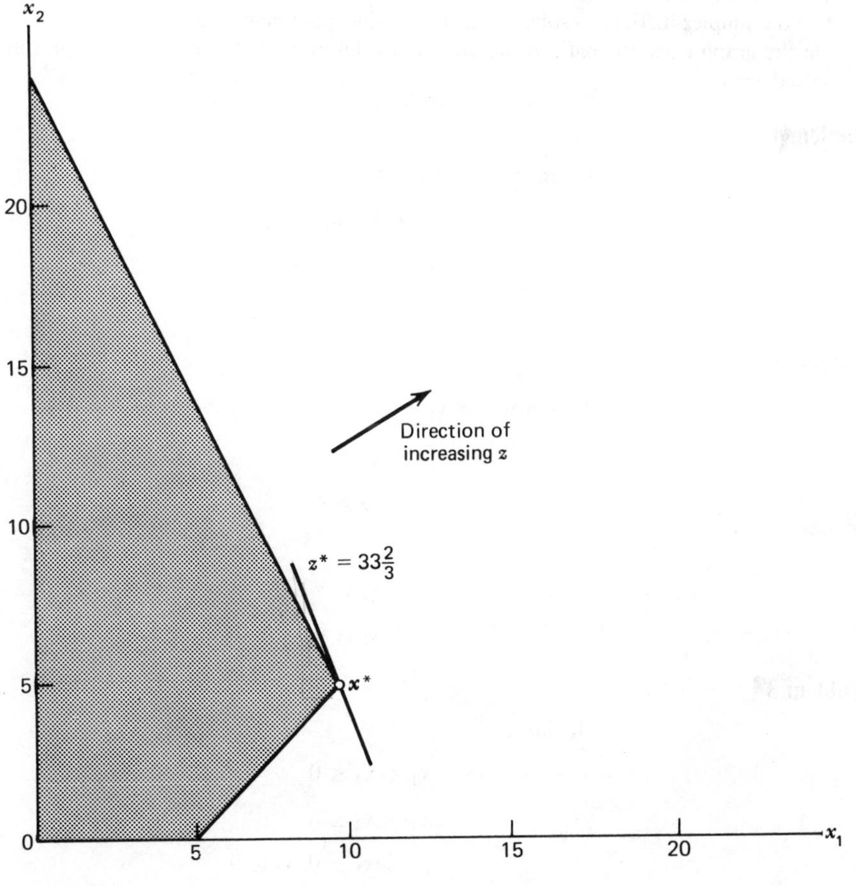

Figure 6.8

The maximum value of z is $33\frac{2}{3}$ as shown in the figure. As x_1 and x_2 are increased, the value of z increases because the unit costs c_1 and c_2 of the objective function are positive values; that is, $c_1 = 3$ and $c_2 = 1$. The minimum value a^* will occur at the same location as z^* because the value of a decreases with increasing values of x_1 and x_2. Since $a = -1z$, the unit costs of c_1 and x_2 of a are negative values; that is, $c_1 = -3$ and $c_2 = -1$. Multiplying the objective function of z by -1 does not change the shape of the linear function; therefore, the location of x^* will not change.

Remarks

The simplex method may be used for any multidimensional linear mathematical model. Since the original formulation of this problem contains only two control variables, we are able to point out a shortcoming of the mathematical approach as compared to the graphical approach. The constraint equation $x_1 - 2x_2 \leq 10$ will never be an active constraint, because it is not part of the feasible region. See Figure 6.7.

PROBLEMS

For each of the following problems:

(a) Solve by graphical means.

(b) Use the simplex method to solve the following linear mathematical models.

(c) On the graph trace the path to the optimum solution as followed by the simplex method solved above.

Problem 1

$$\text{Maximize } z = 11x_1 + 4x_2$$
$$3x_1 + 5x_2 \leq 15$$
$$5x_1 + 2x_2 \leq 10$$
$$x_1 \geq 0, x_2 \geq 0$$

Problem 2

$$\text{Minimize } z = x_1 - 2x_2$$
$$-x_1 + x_2 \leq 8$$
$$x_1 - x_2 \leq 8$$
$$x_1 + x_2 \leq 15$$
$$x_2 \leq 9$$
$$x_1 \geq 0, x_2 \geq 0$$

Problem 3

$$\text{Maximize } z = x_2$$
$$-x_1 + x_2 \leq 0$$
$$2x_1 + x_2 \leq 8$$
$$x_1 \geq 0, x_2 \geq 0$$

Problem 4

$$\text{Minimize } z = -5x_1 + 2x_2$$
$$2x_1 + x_2 \leq 9$$
$$x_1 - 2x_2 \leq 2$$
$$-3x_1 + 2x_2 \leq 3$$
$$x_1 \geq 0, x_2 \geq 0$$

Use the simplex method to solve the following problems:

Problem 5

$$\text{Maximize } z = 2x_1 + 2x_2 - x_3 - x_4$$
$$2x_1 + x_3 - 2x_4 \leq 6$$
$$-2x_1 + 2x_2 + x_3 - x_4 \geq -8$$
$$x_1 + x_2 + 2x_4 \leq 5$$
$$x_1 \geq 0, x_2 \geq 0, x_3 \geq 0, x_4 \geq 0$$

Problem 6

$$\text{Maximize } z = x_1 + x_2 + x_3 + 2x_4$$
$$x_1 + x_3 + x_4 \leq 4$$
$$x_1 + x_2 + 2x_4 \leq 8$$
$$x_1 + 2x_2 - x_3 - x_4 \leq 6$$
$$x_1 \geq 0, x_2 \geq 0, x_3 \geq 0, x_4 \geq 0$$

6.3 THE TWO-PHASE SIMPLEX METHOD

This discussion focuses upon the selection of the *initial basic feasible solution* $\mathbf{x}_B^0$ for linear mathematical models with $=$, $\geq$, and $\leq$ constraints. In the previous section, only $\leq$ constraints were considered. Slack variables were added to each equation, and the initial basic feasible solution could be easily determined by assigning all slack variables to $\mathbf{x}_B^0$. Unfortunately, if a constraint set of linear mathematical model contains an $=$ constraint or a $\geq$ constraint, the initial assignment of variables to $\mathbf{x}_B^0$ does give a basic feasible solution. For equations having greater-than constraints ($\geq$) surplus variables are added to the equation. This variable cannot be assigned to the set of basic variables $\mathbf{x}_B$, because it violates the condition that a basic variable must be a positive value, $\mathbf{x}_B \geq 0$. Equations with equality constraints offer a different problem in establishing an initial basic feasible solution.

The two-phase simplex method is straightforward because it avoids the need to make calculations in selecting the initial basis. The primary reason for utilizing this approach may appear to be solely for convenience. Not so, this approach is used because it eliminates the possibility of selecting an initial basis and, upon subsequent calculation, finding that it does not satisfy the requirement for a basic feasible solution.

Let us investigate the following linear programming problem and attempt to assign an initial basic feasible solution without performing any calculations:

$$\text{Minimize } z = 3x_1 + 4x_2 + 5x_3$$

$$\begin{aligned} x_1 + x_2 + x_3 &\le 5 \\ 2x_2 + x_3 &\ge 2 \\ x_1 + 3x_2 \quad &= 4 \\ x_1, x_2, x_3 &\ge 0 \end{aligned}$$

Add slack and surplus variables to the first and second constraint equations, respectively. The standard linear programming format is

$$\text{Minimize } z = 3x_1 + 4x_2 + 5x_3$$

$$\begin{aligned} x_1 + x_2 + x_3 + x_4 \qquad &= 5 \\ 2x_2 + x_3 \qquad - x_5 &= 2 \\ x_1 + 3x_2 \qquad\qquad &= 4 \\ x_1, x_2, x_3, x_4, x_5 &\ge 0 \end{aligned}$$

Since the goal is to assign an initial basis without performing any calculations, x_1, x_2, and x_3 will be assigned to $\boldsymbol{x}_N^0$. The control variables x_4 and x_5 are assigned to $\boldsymbol{x}_B^0$. Since there are three constraint equations, $m = 3$, the basis $\boldsymbol{x}_B^0$ must contain three variables in order to satisfy the requirement that

$$\boldsymbol{x}_B^0 = \boldsymbol{A}_B^{-1}[\boldsymbol{b} - \boldsymbol{A}_N \boldsymbol{x}_N^0]$$

Clearly, another variable x_1, x_2, or x_3 must be assigned to $\boldsymbol{x}_B^0$. Unfortunately, placing any one of these control variables in $\boldsymbol{x}_B$ will require calculations. In addition, it is possible that this assignment will not give a *feasible* solution, $\boldsymbol{x} \ge \boldsymbol{0}$. If this happens, another variable will have to be assigned to $\boldsymbol{x}_B^0$ and the calculations will have to be repeated. This method is unacceptable for models consisting of a large number of control variables. We shall investigate this example in greater detail and then show how artificial variables may be used to establish an *initial basic feasible solution.* The control variables x_1, x_2, and x_3 will be assigned to $\boldsymbol{x}_N$.

Each constraint equation will be investigated independently. Since x_1, x_2, and x_3 are members of $\boldsymbol{x}_N^0$ and equal to zero, the control variable x_4 from $x_1 + x_2 + x_3 + x_4 = 5$ is equal to

$$x_4 = 5$$

Since x_4 is positive and the associated unit cost of x_4 is equal to zero, x_4 is a viable choice for the initial basis. Thus, x_4 can be assigned to $\boldsymbol{x}_B^0$.

For $2x_2 + x_3 - x_5 = 2$, substituting zero for x_1, x_2, and x_3 results in x_5 being equal to

$$x_5 = -2$$

The surplus variable x_5 violates the requirement that the basic variable be positive. Therefore, x_5 cannot be assigned to a member of the initial basis $\boldsymbol{x}_B^0$.

Equation $x_1 + 3x_2 = 4$ presents a different type of problem. Since x_1 and x_2 are assumed to be members of $\boldsymbol{x}_N^0$ and equal to zero, the constraint equation cannot be satisfied.

$$(0) + 3(0) \neq 4$$

Obviously, both x_1 and x_2 cannot be members of the nonbasis $\boldsymbol{x}_B^0$.

This analysis shows that slack variables can be readily assigned to an $\boldsymbol{x}_B^0$. Equations with surplus variable or equality constraints present problems.

Artificial Variables

A new approach, called the *two-phase simplex method*, makes the *initial assignment* of control variables to $\boldsymbol{x}_B^0$ an easy task. The method makes use of a new control variable called the *artificial variable*. The use of artificial variables will be illustrated by introducing them to the constraint set where needed. Assign x_1, x_2, and x_3 to $\boldsymbol{x}_N^0$ and x_4; the slack variable from $x_1 + x_2 + x_3 + x_4 = 5$ is assigned to $\boldsymbol{x}_B^0$.

Add an artificial variable x_6 to $2x_3 + x_3 - x_5 = 2$.

$$2x_2 + x_3 - x_5 + x_6 = 2$$

The control variable x_5 is assigned to $\boldsymbol{x}_N^0$; thus, the value of the artificial variable is

$$x_6 = 2$$

It is a basic feasible solution that can be assigned to $\boldsymbol{x}_B$.

For $x_1 + 3x_2 = 4$ another artificial variable x_7 is introduced.

$$x_1 + 3x_2 + x_7 = 4$$

Since x_1 and x_2 are nonbasic variables, x_7 must be an artificial variable.

$$x_7 = 4$$

Since x_7 is positive, it may be assigned to $\boldsymbol{x}_B$.

As a result of adding artificial variables to the basis, the initial basic and nonbasic vectors are

$$\boldsymbol{x}_B^{0\prime} = [x_4 \quad x_6 \quad x_7] = [5 \quad 2 \quad 4]$$

$$\boldsymbol{x}_N^{0\prime} = [x_1 \quad x_2 \quad x_3 \quad x_5] = [0 \quad 0 \quad 0 \quad 0]$$

Since all members of $\boldsymbol{x}_B^0$ are positive values and all members of $\boldsymbol{x}_N^0$ are zero, we have established an initial basis and nonbasis that satisfies the requirement that $\boldsymbol{x}_B^0 \geq \boldsymbol{0}$ and $\boldsymbol{x}_N^0 = \boldsymbol{0}$. The constraint set becomes

$$\begin{aligned}
x_1 + x_2 + x_3 + x_4 \qquad\qquad\qquad &= 5\\
2x_2 + x_3 \qquad - x_5 + x_6 \qquad &= 2\\
x_1 + 3x_2 \qquad\qquad\qquad\qquad + x_7 &= 4
\end{aligned}$$

The Artificial Objective Function

The original model is still intact, and two extra variables, artificial variables, have been added to the constraint set. When x_6 and x_7 are members of the basis, the constraint set has *no* physical meaning. It is impossible to interpret these variables in terms of the original formulation. They offer only a mathematical convenience. When $x_6 = 0$ and $x_7 = 0$, the model reduces to its original form. Hence, our goal is to find some procedure to remove the artificial variables from $\mathbf{x}_B$ and place them in $\mathbf{x}_N$.

The two-phase simplex method systematically removes artificial variables from the basis and assigns them to the nonbasis and, concurrently, establishes a candidate solution comprised of variables from the original formulation. Finding an optimum minimum solution of the original mathematical model is achieved by introducing a second objective function to the model. The purpose of this second function is to remove the artificial variables from the model and, in the process, establish candidate solutions that satisfy the requirements for basic feasible solutions. This can be achieved by utilizing the rules of the simplex method discussed in Section 6.2.

The second objective function will be defined to be equal to the sum of the artificial variables:

$$w = \sum_i x_i$$

where the control variable x_i represents an artificial variable. From the foregoing example, the value of the artificial objective function w is expressed as

$$w = x_6 + x_7$$

The cost coefficients of x_6 and x_7 are both equal to one. The *cost coefficients associated with basic variables must be zero* for the condition that $(z - z^0) = \mathbf{c}_I' \mathbf{x}_N^0$ be satisfied. The second objective function will be rewritten in terms of the elements of $\mathbf{x}_N$. By utilizing the equations from the constraint set, x_6 and x_7 may be written as

$$x_6 = 2 - 2x_2 - x_3 + x_5$$

$$x_7 = 4 - x_1 - 3x_2$$

Substituting these equations into $w = x_6 + x_7$ gives

$$w = 6 - x_1 - 5x_2 - x_3 + x_5$$

or, equivalently,

$$(w - 6) = -x_1 - 5x_2 - x_3 + x_5$$

Since all artificial variables are initially members of $\mathbf{x}_B$ and, by definition, positive values, the value of w must always be positive. Our goal will be to minimize w until it is reduced to zero. When $w = 0$, all artificial variables will be equal to zero. When all artificial variables are members of $\mathbf{x}_N$, the model is a function of the originally formulated control variables. At this point, the mathematical model has been restored to its original physical meaning. The artificial objective function is introduced into the mathematical model as

$$\text{Minimize } (w - 6) = -x_1 - 5x_2 - x_3 + x_5$$
$$\text{Minimize } z = 3x_1 + 4x_2 + 5x_3$$
$$x_1 + x_2 + x_3 + x_4 = 5$$
$$2x_2 + x_3 - x_5 + x_6 = 2$$
$$x_1 + 3x_2 + x_7 = 4$$
$$x_1, x_2, \ldots \geq 0$$

The initial basis $\boldsymbol{x}_B^0$ for a two-phase linear mathematical model may be written in the following general form:

$$\text{Minimize } (w - w^0) = \boldsymbol{b}_I' \boldsymbol{x}_N^0$$
$$\text{Minimize } (z - 0) = \boldsymbol{c}_I' \boldsymbol{x}_N^0$$
$$\boldsymbol{A}_N^0 \boldsymbol{x}_N^0 + \boldsymbol{I}\boldsymbol{x}_B^0 = \boldsymbol{b}^0$$

where $\boldsymbol{d}_I = \boldsymbol{b}_I$ is defined as the unit costs of the artificial objective function. The simplex tableau for this model is

$$\begin{bmatrix} \boldsymbol{x}_B^0 & \boldsymbol{A}_N^0 & \boldsymbol{I} & \boldsymbol{b} & b/a \\ — & \boldsymbol{c}_I' & \boldsymbol{0} & (z - 0) & — \\ — & \boldsymbol{d}_I' & \boldsymbol{0} & (w - w^0) & — \end{bmatrix}$$

The Algorithm

The two-phase simplex method uses the same rules established in Section 6.2 for minimizing linear mathematical models. In the two-phase method, we must satisfy the requirements of two objective functions. During the first phase, the goal is to minimize the numerical value of w to zero. The cost coefficients of the w function $\boldsymbol{d}_I$ are used as the indicators during this stage. When $w = 0$, all artificial variables have been removed from the basis and the model is in terms of the original control variables only. At this point, the second phase of the method is begun. Since the goal is to minimize the z function, the cost coefficients of $\boldsymbol{c}_I$ are used as the indicators. During the second phase, the *artificial variables are never considered as possible entries to the basis*. Once they are removed, they are never reentered into the basis again.

EXAMPLE 6.6 Two-Phase Minimization Problem

(a) Use the two-phase simplex method to solve for the following linear mathematical model:

$$\text{Minimize } z = x_2$$
$$3x_1 + 4x_2 \geq 9$$
$$5x_1 + 2x_2 \leq 8$$
$$3x_1 - x_2 \leq 0$$
$$x_1 \geq 0, x_2 \geq 0$$

(b) Trace the path to the optimum solution of a graph.

Solution

(a) Slack and surplus variables are introduced into the model to remove the inequality constraints. The model becomes

$$\text{Minimize } z = x_2$$

$$3x_1 + 4x_2 - x_3 = 9 \quad (1)$$

$$5x_1 + 2x_2 + x_4 = 8 \quad (2)$$

$$3x_1 - x_2 + x_5 = 0 \quad (3)$$

We may establish the basic feasible solution by introducing an artificial variable to the first constraint equation:

$$3x_1 + 4x_2 - x_3 + x_6 = 9$$

The variable x_6 is the only variable necessary to establish an initial basic feasible solution. The artificial objective function is simply equal to

$$w = x_6$$

The value of w must be written in terms of $\boldsymbol{x}_N$ in order to qualify for the basis. From the constraint equation, we determine x_6 to be equal to

$$x_6 = 9 - 3x_1 - 4x_2 + x_3$$

The artificial objective function in terms of nonbasic variables $\boldsymbol{x}_N$ is

$$w = 9 - 3x_1 - 4x_2 + x_3$$

The mathematical model may be written as a two-phase linear mathematical model:

$$\text{Minimize } (w - 9) = -3x_1 - 4x_2 + x_3$$
$$\text{Minimize } (z - 0) = x_2$$
$$3x_1 + 4x_2 - x_3 + x_6 = 9$$
$$5x_1 + 2x_2 + x_4 = 8$$
$$3x_1 - x_2 + x_5 = 0$$

The initial basis and nonbasis vectors are

$$\boldsymbol{x}_B^{0\prime} = [x_4 \quad x_5 \quad x_6]$$
$$\boldsymbol{x}_N^{0\prime} = [x_1 \quad x_2 \quad x_3]$$

The simplex tableau for this model is

Initial Solution

$\boldsymbol{x}_B$	x_1	x_2	x_3	x_4	x_5	x_6	b	b/a	
x_6	3	(4)	−1	0	0	1	9	$\frac{9}{4} = 2.25$	←
x_4	5	2	0	1	0	0	8	$\frac{8}{2} = 4$	
x_5	3	−1	0	0	1	0	0		
$\boldsymbol{c}_I$	0	1	0	0	0	0	$(z - 0)$		
$\boldsymbol{d}_I$	−3	−4	1	0	0	0	$(w - 9)$		
		↑							

The rules of the simplex method are applied to the w indicator row until x_6, the artificial variable, is removed from the basis $\boldsymbol{x}_B$.

RULE 1 Since not all values of the indicator row $\boldsymbol{d}_I$ are positive, the initial solution is not a minimum.

RULE 2 The control variable x_2 enters the basis because $d_2 = -4$ is the smallest negative value of the $\boldsymbol{d}_I$ indicator vector. An arrow indicates that x_2 enters the basis.

RULE 3 The artificial variable x_6 leaves the basis because $b/a = 2.25$ is the smallest positive value of the b/a vector. An arrow indicates that b/a leaves the basis. Note that x_5 is not considered because the "a" coefficient is negative, $a = -1$.

Iteration 1 Successive elimination is used to determine the next candidate solution, $\boldsymbol{x}_B^{1\prime} = [x_2\ x_4\ x_5]$. A circle has been drawn around the pivot element of the $\boldsymbol{A}$ matrix. The results of successive elimination are as follows:

x_B	x_1	x_2	x_3	x_4	x_5	x_6	b	b/a	
x_2	0.75	1.00	−0.25	0.00	0.00	0.25	2.25	2.25/0.75 = 3	
x_4	3.50	0.00	0.50	1.00	0.00	−0.50	3.50	3.5/3.5 = 1	
x_5	(3.75)	0.00	−0.25	0.00	1.00	0.25	2.25	2.25/3.75 = 0.6	←
$\boldsymbol{c}_I$	−0.75	0.00	0.25	0.00	0.00	0.00	$(z - 2.25)$		
$\boldsymbol{d}_I$	0.00	0.00	0.00	0.00	0.00	1.0	$(w - 0.0)$		
	↑								

Since the artificial variable is no longer a member of the basis, the indicators of the z row are used as indicators.

RULE 1 Since not all values of $\boldsymbol{c}_I$ are positive, the candidate solution is not an optimum.

RULE 2 The control variable x_1 will be introduced to the basis because $c_1 = -0.75$ is the only negative indicator of $\boldsymbol{c}_I$.

RULE 3 The control variable x_5 is removed from the basis because $b/a = 0.6$ is the smallest positive value of the b/a vector.

Iteration 2 Since $w = 0$, the second indicator row of w and the column for x_6, the artificial variable is removed from further consideration. They are written in the following tableau.
Successive elimination is performed. The new candidate solution is

x_B	x_1	x_2	x_3	x_4	x_5	b
x_2	0.00	1.00	−0.20	0.00	−0.20	1.80
x_4	0.00	0.00	0.73	1.00	−0.93	1.40
x_1	1.00	0.00	−0.07	0.00	0.27	0.60
$\boldsymbol{c}_I$	0.00	0.00	0.20	0.00	0.20	$(z - 1.8)$

RULE 1 Since all $\boldsymbol{c}_I$ are positive, the optimum solution has been obtained.

The minimum is

$$x_1^* = 0.60 \qquad x_2^* = 1.80 \qquad x_4^* = 1.40$$

The minimum value of z is

$$z^* = 1.8$$

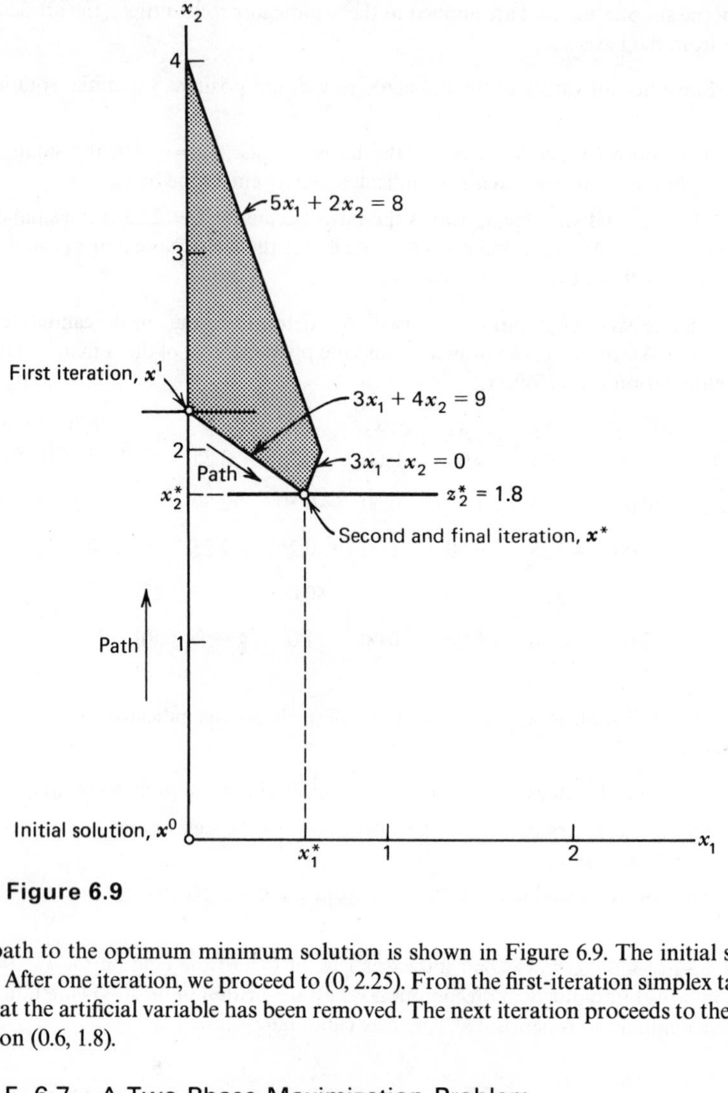

Figure 6.9

(b) The path to the optimum minimum solution is shown in Figure 6.9. The initial solution is (0, 0). After one iteration, we proceed to (0, 2.25). From the first-iteration simplex tableau, we see that the artificial variable has been removed. The next iteration proceeds to the optimum solution (0.6, 1.8).

EXAMPLE 6.7 A Two Phase Maximization Problem

(a) Use the two-phase simplex method to solve the following problem:

$$\text{Maximize } z = 2x_1 + x_2$$

$$2x_1 + 3x_2 \geq 6$$

$$2x_1 + x_2 \leq 6$$

$$x_2 \geq x_1$$

$$x_1 \geq 0, x_2 \geq 0$$

(b) Trace the path to the optimum solution on a graph.

Solution

(a) Add slack and surplus variables to the linear model. The negative transpose of the objective function, $a = -1z$, is required. The model becomes

$$\text{Minimize } a = -2x_1 - x_2$$

$$\begin{aligned} 2x_1 + 3x_2 - x_3 \quad &= 6 \\ 2x_1 + x_2 \quad + x_4 \quad &= 6 \\ -x_1 + x_2 \quad - x_5 &= 0 \end{aligned}$$

$$x_1 \geq 0, x_2 \geq 0, x_4 \geq 0, x_5 \geq 0$$

The basic feasible solution is established by introducing artificial variables x_6 and x_7 to the first and third constraint equations:

$$\begin{aligned} 2x_1 + 3x_2 - x_3 \quad + x_6 \quad &= 6 \\ -x_1 + x_2 \quad - x_5 \quad + x_7 &= 0 \end{aligned}$$

$x_6 = 6 - 2x_1 - 3x_2 + x_3$
$x_7 = x_1 + x_5 - x_2$

The artificial objective function is

$w = 6 - x_1 - 4x_2 + x_3 + x_5$

$$w = x_6 + x_7$$

or, in terms of $\boldsymbol{x}_N$ variables

$$w = 6 - x_1 - 4x_2 + x_3 + x_5$$

The two-phase linear mathematical model becomes

$$\begin{aligned} \text{Minimize } (w - 6) &= -x_1 - 4x_2 + x_3 + x_5 \\ \text{Minimize } (a - 0) &= -2x_1 - x_2 \end{aligned}$$

$$\begin{aligned} 2x_1 + 3x_2 - x_3 \quad + x_6 \quad &= 6 \\ 2x_1 + x_2 \quad + x_4 \quad &= 6 \\ -x_1 + x_2 \quad - x_5 \quad + x_7 &= 0 \end{aligned}$$

$$x_1, \ldots, x_7 \geq 0$$

The initial basis and nonbasis are

$$\boldsymbol{x}_B^{0\prime} = [x_6 \quad x_4 \quad x_7]$$

$$\boldsymbol{x}_N^{0\prime} = [x_1 \quad x_2 \quad x_3 \quad x_5]$$

Initial Solution

The simplex tableau is

$\boldsymbol{x}_B$	x_1	x_2	x_3	x_4	x_5	x_6	x_7	$\boldsymbol{b}$	b/a	
x_6	2	3	−1	0	0	1	0	6	$\frac{6}{3} = 2$	
x_4	2	1	0	1	0	0	0	6	$\frac{6}{1} = 6$	
x_7	−1	①	0	0	−1	0	1	0	$\frac{0}{1} = 0$	←
$\boldsymbol{c}_I$	−2	1	0	0	0	0	0	$(a - 0)$		
$\boldsymbol{d}_I$	−1	−4	1	0	1	0	0	$(w - 6)$		
		↑								

The simplex method proceeds as follows:

RULE 1 Since not all values of $\boldsymbol{d}_I$ are positive, the initial solution is not a minimum.

RULE 2 The control variable x_2 enters the basis because $d_2 = -4$ is the smallest negative value of $\boldsymbol{d}_I$.

RULE 3 The artificial variable x_7 is removed from the basis because $b/a = 0$ is the smallest value of b/a.

Iteration 1 The next candidate solution is

x_B	x_1	x_2	x_3	x_4	x_5	x_6	x_7	$\boldsymbol{b}$	b/a	
x_6	(5.00)	0.00	−1.00	0.00	3.00	1.00	−3.00	6.00	$\frac{6}{5} = 1.2$	←
x_4	3.00	0.00	0.00	1.00	1.00	0.00	−1.00	6.00	$\frac{6}{3} = 2$	
x_2	−1.00	1.00	0.00	0.00	−1.00	0.00	1.00	0.00		
$\boldsymbol{c}_I$	−1.00	0.00	0.00	0.00	1.00	0.00	1.00	$(a - 0.00)$		
$\boldsymbol{d}_I$	−5.00 ↑	0.00	1.00	0.00	−3.00	0.00	4.00	$(w - 6.00)$		

RULE 1 Since $\boldsymbol{d}_I$ contain negative values, the candidate solution is not an optimum.

RULE 2 The control variable x_1 is entered since $d_1 = -5$ is the smaller negative value of $\boldsymbol{d}_I$.

RULE 3 The control variable x_6 is removed because $b/a = 1.2$ is the smaller positive value of the b/a vector.

Iteration 2 The next candidate solution is

x_B	x_1	x_2	x_3	x_4	x_5	x_6	x_7	b	b/a	
x_1	1.00	0.00	−0.20	0.00	0.60	0.20	−0.60	1.20		
x_4	0.00	0.00	(0.60)	1.00	−0.80	−0.60	0.80	2.40	$2.4/0.60 = 4$	←
x_2	0.00	1.00	−0.20	0.00	−0.40	0.20	0.40	1.20		
$\boldsymbol{c}_I$	0.00	0.00	−0.60	0.00	0.80	0.60	−0.80	$(a + 3.60)$		
$\boldsymbol{d}_I$	0.00	0.00	0.00 ↑	0.00	0.00	1.00	1.00	$(w - 0)$		

RULE 1 Since all artificial variables have been removed from the basis, the $\boldsymbol{c}_I$ indicators are used to determine if this candidate solution is an optimum. Since all values of $\boldsymbol{c}_I$ are not positive, an optimum has not been found.

RULE 2 The control variable x_3 is entered in the basis because $c_3 = -0.60$ is the only negative value of $\boldsymbol{c}_N$. The cost coefficients c_6 and c_7 are not considered, because they are cost coefficients associated with the artificial variables x_6 and x_7.

RULE 3 The control variable x_4 is removed from the basis because $b/a = 2.4$ is the only positive value of the b/a vector.

Iteration 3 The next candidate solution is

$\boldsymbol{x}_B$	x_1	x_2	x_3	x_4	x_5	$\boldsymbol{b}$	b/a
x_1	1.00	0.00	0.00	0.33	0.33	2.00	
x_3	0.00	0.00	1.00	1.67	−1.33	4.00	
x_2	0.00	1.00	0.00	0.33	−0.67	2.00	
$\boldsymbol{c}_I$	0.00	0.00	0.00	0.00	1.00	$(a + 6)$	

RULE 1 Since all values of c_T are positive, an optimum solution is obtained.

Since $a^* = -6$, the maximum value of z is

$$z^* = -1(-6) = 6$$

The optimum solution occurs at

$$x_1^* = 2 \qquad x_2^* = 2 \qquad x_3^* = 4$$

(b) The graph of the feasible region and the optimum solution is shown in Figure 6.10.

Remarks

When a member of the basis is equal to zero, the solution is defined to be *degenerate*. If a degenerate solution occurs, it is possible that an optimum solution will not be obtained. In this unlikely situation, the solution of z and w will not be improved upon further iterations. When this occurs a control variable will be removed and reentered upon the subsequent iterations without reducing

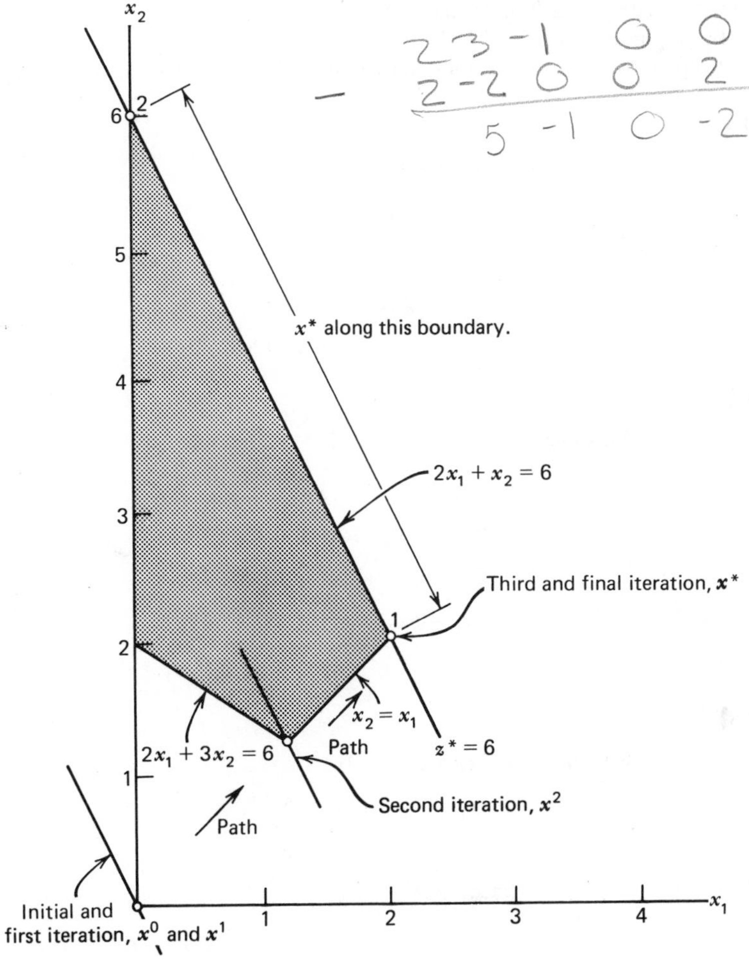

Figure 6.10

w or z. This situation is called *circling*. Since $x_7 = 0$ is in the initial basis and is being replaced by $x_6 = 0$, we have the possibility of circling, but it does not occur in this problem.

In the initial tableau, x_7 was removed from the basis $\boldsymbol{x}_B$ because $b/a = 0/1$ is the smallest value. Compare this tableau with tableau of iteration 1. In this case, x_6 was removed from $\boldsymbol{x}_B$ instead of x_2. The b/a for x_2 is equal to zero, $b/a = 0/-1 = 0$; however, the "a" coefficient is a negative value, thus violating rule 3. Consequently, x_2 is not considered for entry into $\boldsymbol{x}_B$.

The slope of objective function z is equal to the slope of the constraint equation, $2x_1 + x_2 = 6$. The optimum solution x^* consists of all points between points 1 and 2 as shown in the figure. Since the simplex method only considers extreme points as candidate solutions, points 1 and 2 are the only solutions that can be considered with the simplex algorithm. For the analysis conducted in part a, point (0, 6) was never considered as a candidate solution, because once an optimum solution is obtained, the procedure ends.

PROBLEMS

For Problem 1:

(a) Solve with the two-phase simplex method.
(b) Solve by graphical means.
(c) Trace the path to the optimum solution as followed by the simplex method of part a.

Problem 1

$$\text{Minimize } z = 4x_1 + 5x_2$$

$$x_1 + x_2 \geq 1$$

$$2x_1 + 4x_2 \geq 3$$

$$3x_1 + 7x_2 \leq 6$$

$$x_1 > 0, x_2 \geq 0$$

Problem 2

$$\text{Maximize } z = 2x_1 + 3x_2 - 2x_3$$

$$x_1 + 3x_2 + x_3 \geq 3$$

$$6x_1 - x_2 - x_3 \leq 6$$

$$2x_1 - 3x_2 + 2x_3 \leq 12$$

$$x_1 + 6x_2 - 4x_3 \leq 5$$

$$x_1 \geq 0, x_2 \geq 0, x_3 \geq 0$$

Problem 3

$$\text{Maximize } z = 3x_1 + 2x_2 + 3x_3$$

$$x_1 + x_2 + 2x_3 \geq 1$$

$$2x_1 \leq 4$$

$$x_3 \geq 2$$

$$x_1 \geq 0, x_2 \geq 0, x_3 \geq 0$$

Problem 4

$$\begin{aligned} \text{Maximize } z = \ & 2x_1 + 3x_2 \\ -& x_1 + x_2 \leq -1 \\ & 2x_1 \qquad \leq 1 \\ & x_1 > 0, x_2 \geq 0 \end{aligned}$$

There is no solution to this problem, the solution is infeasible.

(a) Show this result by graphical means.

(b) Use the two-phase simplex method. How can the unbounded solution be identified with the simplex method procedure?

(c) Will the observation made in part b be valid for the simplex method also?

Problem 5

Use the two-phase simplex method to solve Problem 4 of Section 6.1.

Problem 6

Use the two-phase simplex method to solve Problem 5 of Section 6.1.

Problem 7

Use the two-phase simplex method to solve Problem 6 of Section 6.1.

Problem 8

Use the two-phase simplex method to solve Problem 7 of Section 6.1.

6.4 NETWORK ANALYSIS

Many practical civil engineering problems have been formulated as linear resource allocation models. Example 6.1, A Construction Management Model, and Example 6.3, Minimum Cost Aggregate Mix Model, are two illustrations from this chapter. Network analysis, like resource allocation, has broad application to many disciplines of civil engineering including transportation and construction management. The critical path method, CPM, is probably the best known of the network analysis methods. In this section we show how CPM and other network analysis models are formulated as linear mathematical models.

Features of Network Analysis Model Formulation

There are several features common to network analysis models. One of the most useful ones is the ability to represent the models with simple network flow diagrams, as shown in Figure 6.11. These diagrams consist of directed arcs and nodes. Whether a network is comprised of single source and sink nodes or multiple source and sink nodes as shown in the figure, our goal is the same. It is to find the optimum assignment of arc flows x_i to satisfy a single minimization or maximization objective function and a set of constraint conditions. Depending upon the application, the control variables x_i are either continuous positive values, $x_i \geq 0$, or discrete integers, $x_i = 0, 1, 2, \ldots$.

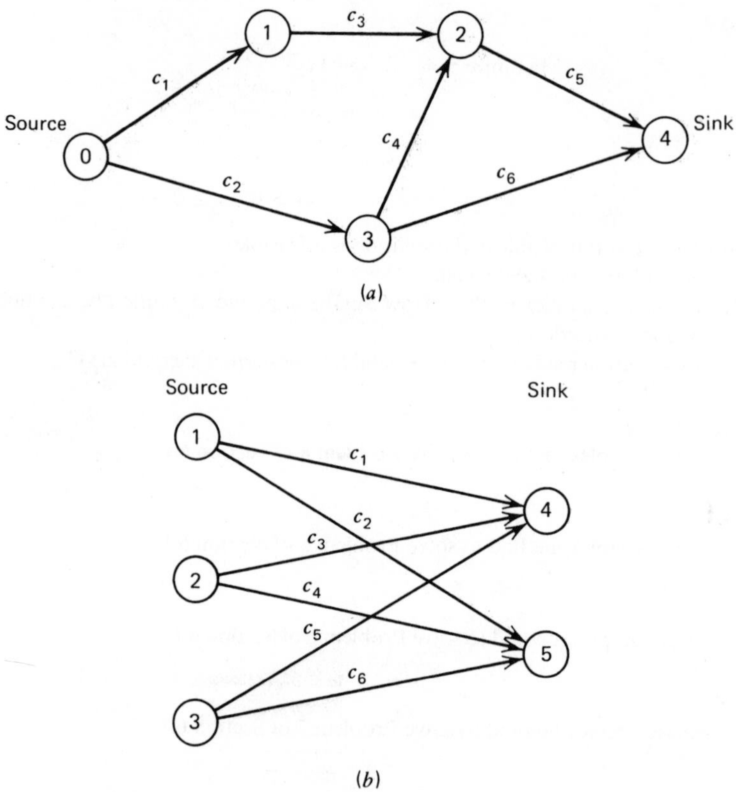

Figure 6.11 Examples of network flow diagrams. (*a*) Network with single source and sink nodes. (*b*) Network with multiple sources and sinks.

Unit costs c_i are assigned to each arc. The total cost for the entire network is equal to the sum of the individual cost assignments $c_i x_i$, or:

$$z = \sum_i c_i x_i = \mathbf{c}'\mathbf{x}$$

Another common feature is that the flow entering each node must be equal to the flow leaving it; that is:

$$\text{Flow in} = \text{Flow out}$$

If there are n nodes, there will be n constraint equations satisfying this conservation-of-flow condition. Other constraints may be imposed that will depend upon the application. For instance, in transportation networks a capacity constraint may be imposed on each arc. Thus, $x_i \leq c_i$, where c_i, the maximum flow in arc i, is introduced into the constraint set.

The formulation of network analysis problems gives a linear mathematical model that may be written in the standard form of linear programming.

$$z = \boldsymbol{c}'\boldsymbol{x}$$

$$\boldsymbol{A}\boldsymbol{x} = \boldsymbol{b}$$

$$\boldsymbol{x} \geq \boldsymbol{0}$$

where the $\boldsymbol{x}$ control vector can contain continuous and discrete variables.

Discrete linear programming models are given special names. When the $\boldsymbol{x}_i$ are integers, $x_i \geq 0, 1, 2, \ldots$, they are called *integer programming models*; when the x_i are restricted to zero or one, $x_i = 0, 1$, they are called *zero–one programming models*; and when the $\boldsymbol{x}$ vector consists of a set of discrete and continuous variables, they are called *mixed programming models.* In the following illustrative examples we see how these variables naturally arise in the formulation of models for different application.

Special Algorithms

The simplex method, as described in this textbook, may be used to solve any one of these network analysis linear programming model types. The reader should be aware that for problems, especially large network analysis problems consisting of many control variables, there are more computationally efficient algorithms than the simplex method. These algorithms take advantage of the unique mathematical properties of the model. The most notable of these algorithms, which are available in higher-level computer languages, such as FORTRAN, BASIC, and PASCAL (the latter two are popular programming languages for mini- and microcomputers) are called zero/one programming and integer programming. These programming algorithms are used to solve transportation, assignment, and CPM problems.

For some discrete linear programming problems, if a simple truncation procedure (rounding off to the nearest integer) is used, the simplex method may lead to a non-optimal solution or may violate one or more constraint equations. The *cutting plane method* is a popular integer programming method that addresses this problem. There are different versions of this method, but they all have the common characteristics of restructuring the constraint set to account for the restrictions placed upon it by the discrete variables and reformulating the model as a linear programming with continuous variables. With this procedure, the simplex method can be used to solve each reformulated linear programming model. Since many of these specialty algorithms use principles and concepts of the simplex method, they will not be described in this textbook. We limit our discussion to the formulation of different network analysis problems and assume that the simplex method will effectively lead to an optimum solution.

EXAMPLE 6.8 A Transportation Problem

A contractor must assign workers to four work sites each day. The travel time in minutes between each dispatch location and work site is shown on the directed areas of the network diagram in Figure 6.12. In order to maximize the number of productive work hours per day of each worker,

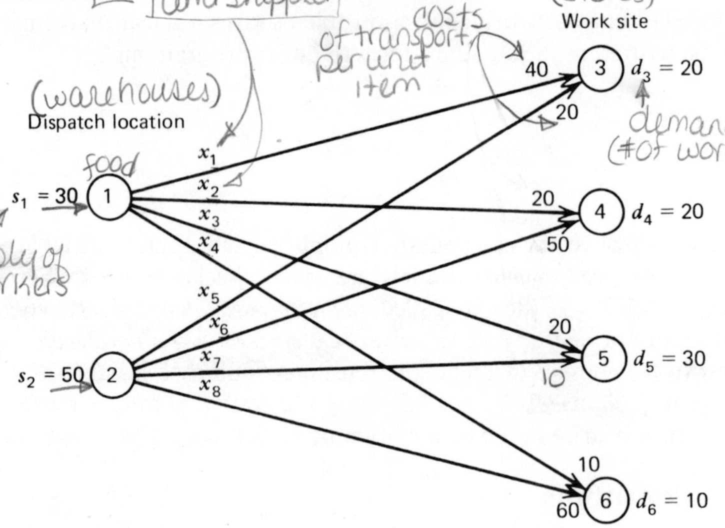

Figure 6.12

the contractor wishes to minimize the total worker travel time. Travel time is considered to be unproductive work time. The number of workers dispatched from locations 1 and 2 is 30 and 50 workers, $s_1 = 30$ and $s_2 = 50$, respectively. The numbers of workers required at each work site 3, 4, 5, and 6 are 20, 20, 30, and 10, respectively. They are shown on the network diagram as $d_3 = 20$, $d_4 = 20$, $d_5 = 30$, and $d_6 = 10$.

Formulate a mathematical model to minimize total travel time.

Solution

Control variables, $x_1, x_2, \ldots x_8$, are defined for each arc as shown in the figure. The total travel time T of all workers is the sum of the travel time for each arc. In terms of x, it is

$$T = 40x_1 + 20x_2 + 20x_3 + 50x_4 + 20x_5 + 50x_6 + 10x_7 + 60x_8$$

The conservation of flow condition must be observed at each node. At node 1, the supply of workers, $s_1 = 30$, is considered to be a flow into node 1 and work assignments are considered to be a flow out of the node 1. Thus:

$$\text{Flow in} = \text{Flow out}$$

$$30 = x_1 + x_2 + x_3 + x_4 \qquad \text{Node 1}$$

At node 2, a similar expression may be determined.

$$50 = x_5 + x_6 + x_7 + x_8 \qquad \text{Node 2}$$

At node 3, the number of workers assigned to work at node 3 must equal the demand for workers, $d_3 = 20$. Thus,

$$\text{Flow in} = \text{Flow out}$$

$$x_1 + x_5 = 20 \qquad \text{Node 3}$$

A similar approach may be used for nodes 4, 5, and 6. The resulting constraint equations are

$$x_2 + x_6 = 20 \qquad \text{Node 4}$$

$$x_3 + x_7 = 30 \qquad \text{Node 5}$$

$$x_4 + x_8 = 10 \qquad \text{Node 6}$$

The mathematical model is

$$\text{Minimize } T = 40x_1 + 20x_2 + 20x_3 + 50x_4 + 20x_5 + 50x_6 + 10x_7 + 60x_8$$

$$\begin{aligned}
x_1 + x_2 + x_3 + x_4 &= 30 \\
x_5 + x_6 + x_7 + x_8 &= 50 \\
x_1 + x_5 \qquad\qquad &= 20 \\
x_2 + x_6 \qquad\qquad &= 20 \\
x_3 + x_7 \qquad\qquad &= 30 \\
x_4 + x_8 \qquad\qquad &= 10 \\
\boldsymbol{x} &\geq 0
\end{aligned}$$

where each x_i is an integer for $i = 1, 2, \ldots, 8$

DEMAND

SUPPLY	3	4	5	6	TOTAL SUPPLY
1	40 X_1	20 X_2	20 X_3	10 X_4	30
2	20 X_5	50 X_6	10 X_7	60 X_8	50
TOTAL DEMAND	20	20	30	10	

IF TOT DEM ≠ TOT SUP, CREATE DUMMY SHIPMENT LOCATION
* see notes

Remarks

This in an integer programming problem. With the simplex method described in this textbook, artificial variables must be assigned to each constraint equation. Since there are six constraint equations, the total number of variables is 14. For larger networks, the number of control variables might make the simplex method too cumbersome. In Au and Stetson's book, they present a transportation algorithm that is a suitable hand calculation method for reasonably large problems.

In this problem the total supply equals the total demand.

$$s_1 + s_2 = d_1 + d_2 + d_3 + d_4 = 80$$

If the total supply and total demand are not equal, then the formulation will be different. Suppose the demand at work site 3 is reduced from 20 to 15 workers. The mathematical model becomes

$$\text{Minimize } T = 40x_1 + 20x_2 + 20x_3 + 50x_4 + 20x_5 + 50x_6 + 10x_7 + 60x_8$$

$$\begin{array}{llllllllr}
x_1 + x_2 + x_3 + x_4 & & & & & & & \leq 30 & \\
 & & & x_5 + x_6 + x_7 + x_8 & & & & \leq 50 & \text{Supply} \\
x_1 & & & x_5 & & & & = 15 & \\
 & x_2 & & & + x_6 & & & = 20 & \\
 & & x_3 & & & + x_7 & & = 30 & \text{Demand} \\
 & & & x_4 & & & + x_8 & = 10 & \\
 & & & & & & x & \geq 0 &
\end{array}$$

Since there is an oversupply of workers, the supply constraints at nodes 1 and 2 become an inequality constraint. Surplus variables x_9 and x_{10} may be introduced to the supply equations. Thus

$$x_1 + x_2 + x_3 + x_4 + x_9 = 30$$

$$x_5 + x_6 + x_7 + x_8 + x_{10} = 50$$

In this case, x_9 and x_{10} represent the number of workers who are not needed to complete the required work.

EXAMPLE 6.9 An Assignment Model

A contractor has three different types of excavation equipment that he wants to assign to three different excavation sites. He has rated the performance of each piece of equipment as to its ability to perform required tasks at the given sites. The ratings are on a scale of 0 to 10, where 0 represents the worst performance rating, and 10 is the best.

	JOB SITE		
MACHINE TYPE	1	2	3
1	8	3	7
2	0	10	3
3	6	5	4

The contractor desires to make the optimum assignment to maximize total performance.

(a) Draw a network diagram.

(b) Formulate the mathematical model optimally to assign the equipment to the excavation sites.

Solution

(a) The network source-sink diagram is shown in Figure 6.13. The supply of each machine is designated as $s_1 = 1$, $s_2 = 1$, and $s_3 = 1$. The demand for equipment at each site is designated as $d_1 = 1$, $d_2 = 1$, and $d_3 = 1$. The total supply equals the total demand.

$$s_1 + s_2 + s_3 = d_1 + d_2 + d_3 = 3$$

The costs in this case are designated on each arc in terms of the performance rating given in the above table. The control variables x_i, where $i = 1, 2, \ldots, 9$, are shown on each arc. The control variable is a zero–one integer variable, where

$$x_i = \begin{cases} 1, \text{ an assignment to arc } i \text{ is made.} \\ 0, \text{ no assignment to arc } i \text{ is made.} \end{cases}$$

(b) The overall performance rating z is

$$\text{Maximize } z = 8x_1 + 3x_2 + 7x_3 + 0x_4 + 10x_5 + 3x_6 + 6x_7 + 5x_8 + 4x_9$$

The conservation-of-flow condition must be satisfied at each node. For supply node 1:

$$\text{Flow in} = \text{Flow out}$$

$$1 = x_1 + x_2 + x_3 \qquad \text{Supply node 1}$$

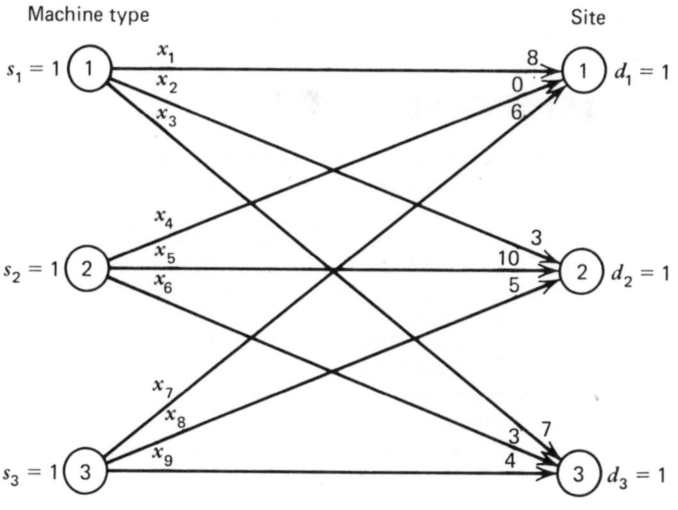

Figure 6.13

Similar expressions may be written for supply nodes 2 and 3 and demand nodes 1, 2, and 3.

Flow in = Flow out	
$1 = x_4 + x_5 + x_6$	Supply node 2
$1 = x_7 + x_8 + x_9$	Supply node 3
$x_1 + x_4 + x_7 = 1$	Demand node 1
$x_2 + x_5 + x_8 = 1$	Demand node 2
$x_3 + x_6 + x_9 = 1$	Demand node 3

The mathematical model is:

$$\text{Maximize } z = 8x_1 + 3x_2 + 7x_3 + 0x_4 + 10x_5 + 3x_6 + 6x_7 + 5x_8 + 4x_9$$

$$\begin{aligned}
x_1 + x_2 + x_3 \qquad\qquad\qquad\qquad\qquad\qquad &= 1 \\
x_4 + x_5 + x_6 \qquad\qquad\qquad &= 1 \\
x_7 + x_8 + x_9 &= 1 \\
x_1 \qquad + x_4 \qquad\qquad + x_7 \qquad &= 1 \\
x_2 \qquad + x_5 \qquad\qquad + x_8 \qquad &= 1 \\
x_3 \qquad + x_6 \qquad\qquad + x_9 &= 1
\end{aligned}$$

$$x_i = 0, 1 \qquad i = 1, 2, \ldots, 9$$

Remarks

Comparison between the transportation and assignment problems show conceptual differences, but the network diagram and mathematical model are identical in form. As a result, the same computer programs may be used to solve transportation and assignment problems. Since each supply and demand are limited to unity, the assignment model is a special case of the transportation model.

EXAMPLE 6.10 A Critical-Path Schedule

Constructing projects can be described as a set of work tasks or jobs. The estimated times to complete these tasks for the construction of a simple structure are

JOB	DESCRIPTION	TIME (WEEKS)
1	Excavation	1
2	Lay foundation	2
3	Erect walls	2
4	Install roof	1
5	Prefabricate walls	1
6	Prefabricate roof	1.5
7	Landscape grounds	1
8	Install heating system	2

A schedule of events is shown in Figure 6.14*a*. It shows that some tasks may proceed simultaneously while others must be completed in a specified sequence.

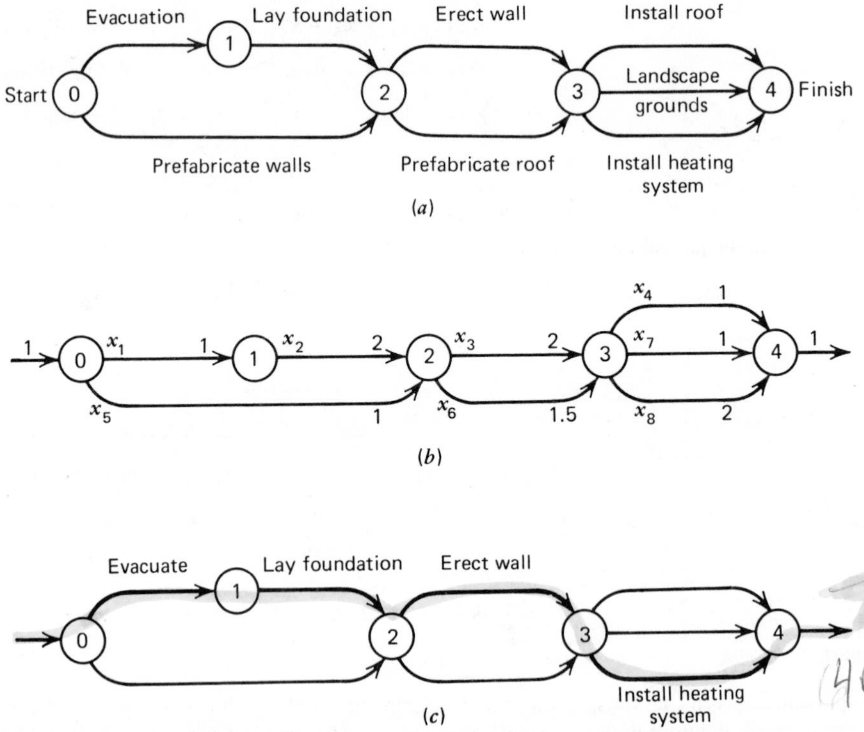

Figure 6.14 (*a*) Schedule diagram. (*b*) Network diagram. (*c*) Critical tasks.

Formulate a mathematical model to determine the critical path. Stated another way, formulate a model to determine the earliest time in which the project may be completed.

Solution

The critical path includes all jobs that must be completed without delay in order to have the project completed in the shortest possible construction time. The critical path is a contiguous linkage of arcs connecting the start and finish nodes. Jobs that do not fall on the critical path have *slack time.* A contractor has some flexibility in scheduling them. These tasks lie on a path that is considered to be parallel to the critical path. The total work time on a path of noncritical jobs will be less than the total time on a parallel critical path.

The schedule diagram has been redrawn as a network diagram in Figure 6.14*b* with the times to complete a task and the control variable x_i on the arcs. Unit flows are placed on the start and finish nodes, 0 and 4, respectively. The zero–one control x_i is used to indicate whether a job lies on the critical path or not. The control variables are

$$x_i = \begin{cases} 0 & \text{Job } i \text{ is on critical path} \\ 1 & \text{Job } i \text{ is not on critical path} \end{cases} \qquad i = 1, 2, \ldots, 8$$

Since the earliest time to complete the project is equal to the longest-time path connecting the start and finish nodes, our objective function is:

$$\text{Maximize } T = 1x_1 + 2x_2 + 2x_3 + 1x_4 + 1x_5 + 1.5x_6 + 1x_7 + 2x_8$$

Applying the conservation of flow condition at each node gives

$$\text{Flow in} = \text{Flow out}$$

$$1 = x_1 + x_5 \qquad \text{Node 0}$$

$$x_1 = x_2 \qquad \text{Node 1}$$

$$x_2 + x_5 = x_3 + x_6 \qquad \text{Node 2}$$

$$x_3 + x_6 = x_4 + x_7 + x_8 \qquad \text{Node 3}$$

$$x_4 + x_7 + x_8 = 1 \qquad \text{Node 4}$$

The model may be written in standard linear form.

$$\text{Maximize } T = 1x_1 + 2x_2 + 2x_3 + 1x_4 + 1x_5 + 1.5x_6 + 1x_7 + 2x_8$$

$$\begin{aligned} x_1 \qquad\qquad\qquad\qquad + x_5 \qquad\qquad\qquad\qquad &= 1 \\ x_1 - x_2 \qquad\qquad\qquad\qquad\qquad\qquad\qquad &= 0 \\ x_2 - x_3 \qquad + x_5 - x_6 \qquad\qquad &= 0 \\ x_3 - x_4 \qquad + x_6 - x_7 - x_8 &= 0 \\ x_4 \qquad\qquad + x_7 + x_8 &= 1 \end{aligned}$$

$$x_i = 0,1 \qquad i = 1, 2, \ldots, 8$$

Remarks

The solution of this problem may be determined by inspection. The critical path is shown in Figure 6.14*c*. The total time to complete the project is 7 weeks.

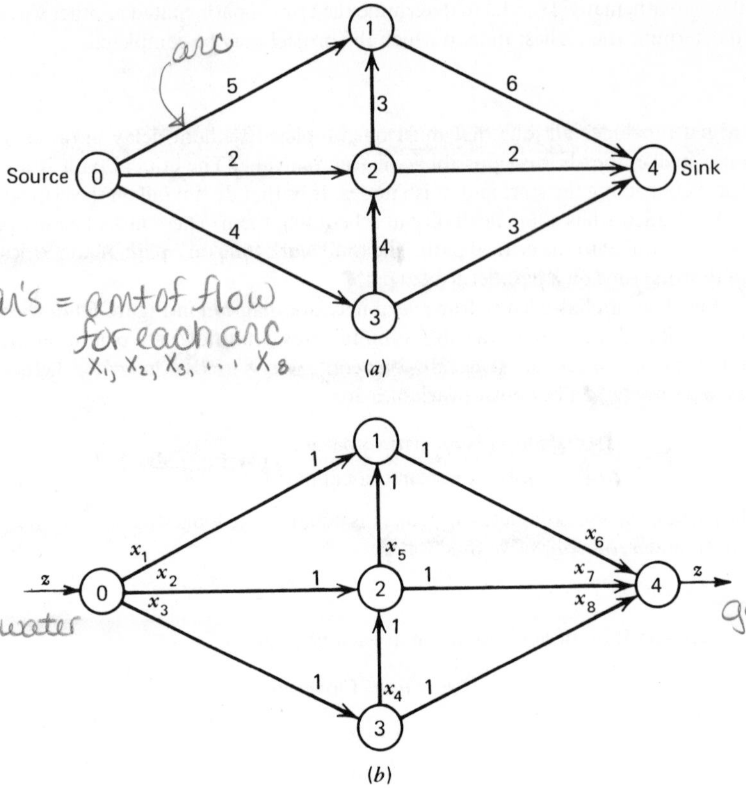

Figure 6.15

EXAMPLE 6.11 A Pipeline Capacity Model

Determine the capacity of the pipeline network system shown in Figure 6.15*a*. The flow capacity in million gallons per day is indicated on each directed arc. Formulate a mathematical model.

Solution

A network diagram is shown in Figure 6.15*b*. In this network, there are *unit costs* assigned to the directed arcs, $c_i = 1, i = 1, 2, \ldots, 8$. Furthermore, the capacity constraints will be included in the constraint set; they are not shown on the network diagram. The network diagram is only used to help formulate the problem with the conservation-of-flow condition.

The flow through each pipe segment of the network will be defined as x_i, where i is the pipe number. The maximum flow that may enter the source node and exit the sink node, nodes 0 and 4, is designated as the variable z. Thus, the objective function is

$$z = x_1 + x_2 + x_3 = x_6 + x_7 + x_8$$

The objective function may be written in terms of x_1, x_2, and x_3 or x_6, x_7, and x_8. Here $z = x_6 + x_7 + x_8$ is chosen.

The capacity constraint for each pipe segment is:

$$x_1 \leq 5 \quad x_2 \leq 2 \quad x_3 \leq 4 \quad x_4 \leq 4 \quad x_5 \leq 3 \quad x_6 \leq 6 \quad x_7 \leq 2 \quad x_8 \leq 3$$

The conservation of flow condition must be satisfied for nodes 1 through 3.

$$\text{Flow in} = \text{Flow out}$$

$$x_1 + x_5 = x_6 \qquad \text{Node 1}$$

$$x_2 + x_4 = x_5 + x_7 \qquad \text{Node 2}$$

$$x_3 = x_4 + x_8 \qquad \text{Node 3}$$

Nodes 0 and 4 are not included in the constraint set, because they were used to establish the objective function. Including them in the constraint set will introduce dependent equations.

The mathematical model is

$$\begin{aligned}
\text{Maximize } z = \quad & x_6 + x_7 + x_8 \\
& x_1 \le 5 \\
& x_2 \le 2 \\
& x_3 \le 4 \\
& x_4 \le 4 \\
& x_5 \le 3 \\
& x_6 \le 6 \\
& x_7 \le 2 \\
& x_8 \le 3 \\
& x_1 + x_5 - x_6 = 0 \\
& x_2 + x_4 - x_5 - x_7 = 0 \\
& x_3 - x_4 - x_8 = 0 \\
& \mathbf{x} \ge 0
\end{aligned}$$

PROBLEMS

Note. Many of these problems may be solved by inspection. Answer them assuming that they are more complex and that simplifying assumptions are not obvious.

Problem 1

Three employees were given three work tasks to complete. Their performance was measured in the time in hours it took to complete each task. The results of the time tests are as follows:

	TASK		
EMPLOYEE	1	2	3
1	3	2	5
2	4	1	6
3	3	4	7

(a) Draw a network flow diagram to obtain the optimum assignment to minimize total cost.
(b) Formulate a mathematical model, and define the control variables.
(c) Classify the model as a continuous, integer, zero–one, mixed programming model.

Problem 2

A contractor has four different construction jobs. He may assign 10 small or 5 large trucks to these sites. A performance rating that ranges from 0 to 10 is used to indicate the suitability of small and large trucks to each construction site.

	CONSTRUCTION SITES 1	2	3	4
Small truck	2	5	0	4
Large truck	8	5	10	6

The total demand for trucks exceeds the total supply. Twenty trucks are demanded, but only 15 are available. The demand for trucks by construction site is

SITE	DEMAND
1	6
2	5
3	5
4	4

The contractor wants to assign his trucks to maximize performance.
(a) Draw a network flow diagram.
(b) Formulate a mathematical model, and define all control variables.
(c) Classify the model as a continuous, integer, zero–one, mixed programming model.

Problem 3

The time in hours to travel between cities is shown in Figure 6.16. Supply depots exist in New York City and Portland, Maine. Ten trucks may be dispatched from New York City and 20 from Portland. The demand for supplies in terms of trucks or truck loads is

	TRUCK LOADS
Burlington, Vermont	6
Albany, New York	5
Springfield, Massachusetts	14
Boston, Massachusetts	5

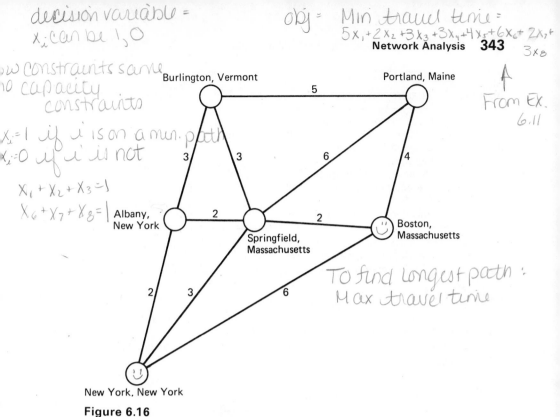

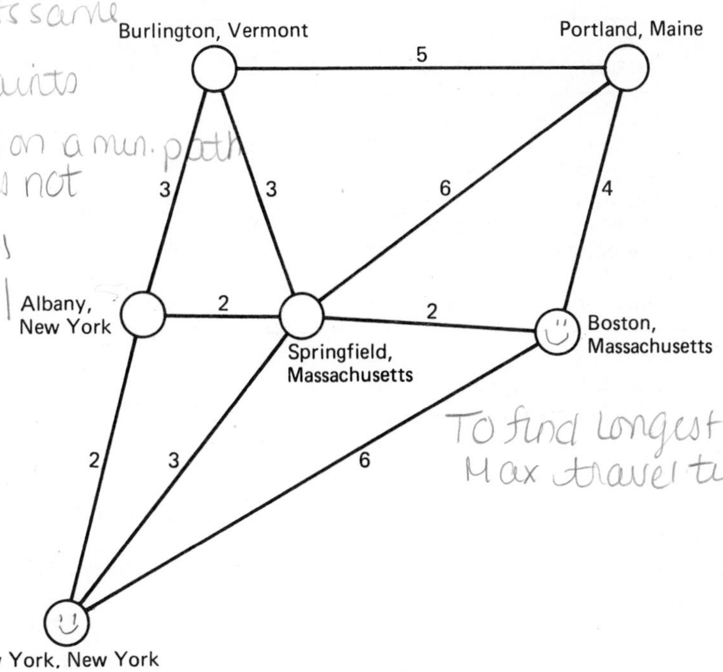

Figure 6.16

Treat this problem as a transportation problem to minimize total travel time.

(a) Draw a network flow diagram.

(b) Formulate as a linear programming problem. Define all control variables.

Problem 4

Repeat Problem 3 assuming that there are 30 trucks that can be dispatched from New York City and 30 trucks from Portland, Maine.

Problem 5

The time in hours to travel between cities is shown in Figure 6.16. A driver wants to determine the shortest route between New York City and Portland, Maine.

(a) Draw a network flow diagram using directed arcs.

(b) Formulate a mathematical model, and define all control variables.

(c) Classify the model as a continuous, integer, zero–one, mixed programming model.

Problem 6

Repeat Problem 5. In this problem, the driver will begin his trip in New York and end in Portland, but he must pass through each city named on the map, that is, Albany, Burlington, Springfield, and Boston. It is permissible for the driver to pass through the same city more than once.

Problem 7

A construction job consists of the following activities. A critical path is to be determined.

ACTIVITY	ACTIVITY DESCRIPTION	PREDECESSOR ACTIVITY	DURATION (DAYS)
1	Clear site of trees, etc.	—	3
2	Order and deliver material for foundation	—	14
3	Excavate foundation	1	5
4	Lay foundation	3	7
5	Order and deliver material for structure	—	14
6	Prefabricate structural members	5	5
7	Erect structure	4, 6	5
8	Install water and sewer lines	4	8

(a) Draw a network flow diagram.
(b) Formulate a mathematical model to determine the critical path.

Problem 8

Determine the maximum flow through the roadway network. The maximum flow on each roadway in vehicles per hour is labeled on the directed arcs in Figure 6.17.

Formulate a mathematical model to determine the roadway capacity of this system from points A to B.

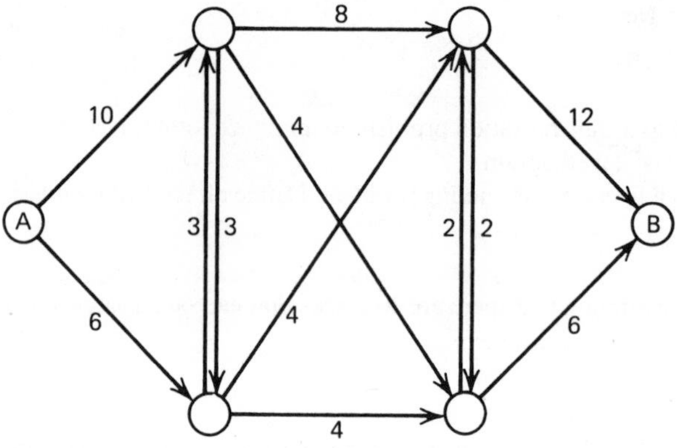

Figure 6.17

Summary

In this chapter the simplex algorithm was developed for minimization models of the standard form:

$$\text{Minimize } z = \boldsymbol{c}'\boldsymbol{x}$$

$$\boldsymbol{A}\boldsymbol{x} = \boldsymbol{b}$$

$$\boldsymbol{x} \geq \boldsymbol{0}$$

We have seen that strict adherence to this form is essential. Maximization problems and problems with various equality and inequality constraints can be written in this form. As a result, we can solve a broad range of linear optimization models with the same algorithm.

The simplex method is an iterative procedure consisting of three steps: (1) test for an optimum minimum solution; (2) if this test fails, determine a new control variable to enter the feasible solution; and (3) determine a control variable to be removed from the feasible solution. The procedure employs a hill-climbing approach that moves from one extreme point to another one. Artificial variables and the two-phase simplex method were used to solve problems where an initial basic feasible could not be easily identified. Resource allocation and network analysis problems were formulated and discussed.

Bibliography

Tung Au and Thomas Stelson. *Introduction to Systems Engineering.* Addison Wesley, Reading, Mass. 1969.

Richard de Neufville and Joseph Stafford. *Systems Analysis for Engineers and Managers.* McGraw-Hill, New York, 1971.

Donald P. Gaver and Gerald L. Thompson. *Programming and Probability Models in Operations Research.* Brooks/Cole, Monterey, Cal. 1973.

Gilbert Strang. *Linear Algebra and Its Applications.* Academic Press, New York. 1976.

Harvey Wagner. *Principles of Operations Research.* Prentice-Hall, Englewood Cliffs, N.J. 1969.

CHAPTER 7

Sensitivity Analysis and the Dual-Primal Relationships

After the completion of this chapter, the student should be able to

1. Evaluate the effect of change in unit cost and resource parameters on the optimum solution of a linear mathematical model by graphical and mathematical methods.
2. Utilize the primal–dual relationships for solving linear optimization problems.
3. Be familiar with economic terms of shadow price and opportunity cost for evaluating the performance of an optimum engineering design.

Thus far in our decision-making processes, we have assumed that resource costs; financial, physical, and institutional constraints; and the cost of money as well as the benefits are constants. If one or more of these factors is changed, the optimum allocation of resources and the selection of the best alternative may be affected. Investigating changes in any one of these parameters is called *sensitivity analysis*. In this chapter we limit our discussion of sensitivity analysis to linear mathematical models, or the so-called *primal models*, of the form:

$$\text{Minimize } z = \boldsymbol{c}'\boldsymbol{x}$$
$$\boldsymbol{A}\boldsymbol{x} = \boldsymbol{b}$$
$$\boldsymbol{x} \geq \boldsymbol{0}$$

Furthermore, our presentation will be limited to the study of changes in one or more of the parameters of the unit cost vector $\boldsymbol{c}$ and resource constraint vector $\boldsymbol{b}$. Since no mathematically efficient procedure exists for studying changes in $\boldsymbol{A}$, it will not be considered.

One possible method of performing a sensitivity analysis is to formulate a new mathematical model with the new cost and resource values. For large problems, this procedure may prove to be inefficient. For changes of $\boldsymbol{c}$ parameters, the simplex method may be employed. For changes in $\boldsymbol{b}$ parameters, a new linear mathematical model called the *dual* model will be formulated with the primal model and solved by the simplex method. The dual model utilizes a control vector set that is comprised of *opportunity costs* and new variables called *shadow prices*. We shall see that the solution of primal or dual problems will give the optimum solution z^*, the optimum resource allocation $\boldsymbol{x}^*$, plus opportunity-cost and shadow price information.

7.1 SENSITIVITY ANALYSIS

Our investigation of sensitivity analysis will begin by assuming the optimum solution to the following problem is known.

$$\text{Minimize } z = \boldsymbol{c}'\boldsymbol{x}$$
$$\boldsymbol{A}\boldsymbol{x} = \boldsymbol{b}$$
$$\boldsymbol{x} \geq 0$$

The optimum is designated as x^* and z^*. If the elements of either the unit cost vector $\boldsymbol{c}$ or the resource vector $\boldsymbol{b}$ are changed, the total cost z^* may be affected. The optimum basis $\boldsymbol{x}^*$ may change because the optimum solution may shift to a new extreme point.

Our discussion of sensitivity analysis will begin with the study of two-dimensional problems. At first we study the effects of the change in $\boldsymbol{c}$ and $\boldsymbol{b}$ by graphical means and learn that changes in these parameters *may* or *may not* affect the optimum solution $\boldsymbol{x}^*$ and z^*. After we obtain a visual interpretation of the problem, we utilize the simplex method to find the solutions of models where only changes in unit costs of $\boldsymbol{c}$ are made. In the following section changes in the resource constraint vector $\boldsymbol{b}$ are investigated with the use of dual formulation.

Changes in Unit Costs

Changes in unit costs $\boldsymbol{c}$ tend to shift and rotate the isocost line of the objective function. This may result in a change in total cost, a change in the location of the optimum solution, or both. Consider the graphical solution of the following two-dimensional minimum-cost model:

$$\text{Minimize } z = \$1.00x_1 + \$1.00x_2$$
$$2x_1 + x_2 \geq 6$$
$$x_1 + 2x_2 \geq 6$$
$$x_1 \geq 0, x_2 \geq 0$$

The optimum solution (2, 2) is shown in Figure 7.1*a*. The minimum cost solution is $z^* = \$4.00$.

Suppose that the unit cost is doubled and the constraint set is not altered. The optimization problem becomes

$$\text{Minimize } \hat{z} = \$2.00\hat{x}_1 + \$2.00\hat{x}_2$$
$$2\hat{x}_1 + 2\hat{x}_2 \geq 6$$
$$\hat{x}_1 + 2\hat{x}_2 \geq 6$$
$$\hat{x}_1 \geq 0, \hat{x}_2 \geq 0$$

The graphical solution to this problem is shown in Figure 7.1*b*. The solution is $\boldsymbol{x}^* = (2, 2)$ and $\hat{z}^* = \$8.00$.

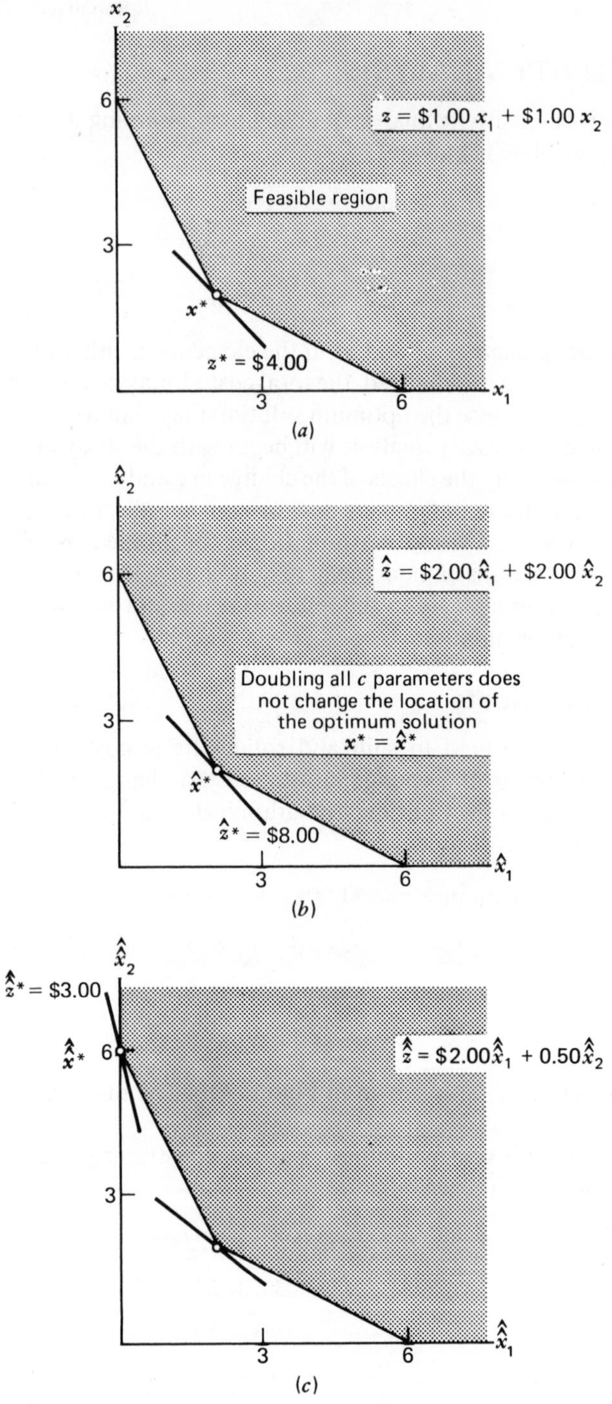

Figure 7.1 The effect of changing parameter in the cost vector ***C***.

Comparing the solution to the original and new problems, Figures 7.1*a* and 7.1*b* show the *location* of the optimum solution is the *same* for both problems. $\hat{\mathbf{x}}^* = \mathbf{x}^*$, but $\hat{z}^* > z^*$. The optimum solution of the new formulation is twice as large as the original problem. $\hat{z}^* = \$8.00$ and $z^* = \$4.00$. Uniform changes in both c_1 and c_2 will *translate* but not *rotate* the isocost line z. As a result, the location of the optimum solution will *not* change, only the magnitude of z^* will change. This conclusion is applicable to multidimensional problems too.

Consider the original formulation and doubling of the unit cost of c_1 from \$1.00 to \$2.00 and reducing the unit cost of c_2 from \$1.00 to \$0.50. The objective function of the new model becomes

$$\hat{\hat{z}} = \$2.00\hat{\hat{x}}_1 + 0.50\hat{\hat{x}}_2$$

Again, assume that the feasible region is unaltered. The graphical solution is shown in Figure 7.1*c*. The optimum solution is located at the extreme point $\hat{\hat{x}}^* = (0, 6)$ with $\hat{\hat{z}}^* = \$3.00$. Compare the graphical solutions of Figures 7.1*a* and 7.1*c*. The new isocost line has been translated and rotated in a clockwise direction, thus changing the location of x^*. For multidimensional problems, changes in the location of optimum solution $\mathbf{x}^*$ will depend upon the *relative changes* in all unit costs of $\mathbf{c}$. Remember that changes in $\mathbf{c}$ do not affect the feasible region.

In this section we used z and $\mathbf{x}$, $\hat{z}$ and $\hat{\mathbf{x}}$, and $\hat{\hat{z}}$ and $\hat{\hat{\mathbf{x}}}$ to distinguish among different mathematical models and their optimum solutions. In the remaining sections of the chapter we will distinguish between models as having differences in the unit cost parameters as $\mathbf{c}$ and $\hat{\mathbf{c}}$ and differences in the resource parameters as $\mathbf{b}$ and $\hat{\mathbf{b}}$. The control vector and objective function will be designated as $\mathbf{x}$ and $\mathbf{z}$. No distinguishing notation will be used for these variables.

Changes in Resource Constraints

Changes in parameters of the resource constraints $\mathbf{b}$ shift the isoquant lines and, in turn, change the feasible region. The optimum solution $\mathbf{x}^*$ and $\mathbf{z}^*$ may or may not be affected. Consider the minimum cost model:

$$z = \$2.00x_1 + 0.50x_2$$

$$2x_1 + x_2 \geq 6$$

$$x_1 + 2x_2 \geq 6$$

$$x_1 \geq 0, x_2 \geq 0$$

The optimum solution is $\mathbf{x}^* = (0, 6)$ with $z^* = \$3.00$ as shown in Figure 7.2*a*.

Any change in the active constraint $2x_1 + x_2 = 6$ will affect the location of the optimum solution (x_1^*, x_2^*) and the total cost z^*. Suppose the resource constraint $2x_1 + x_2 \geq 6$ is changed to $2x_1 + x_2 \geq 9$. The feasibility region is changed as shown in Figure 7.2*b*. The optimum solution is translated to (0, 9) with $z^* = \$4.50$.

Changes in the inactive constraint may or may not affect $\mathbf{x}^*$ and z^*. Suppose the inactive constraint $x_1 + 2x_2 \geq 6$ is changed to $x_1 + 2x_2 \geq 8$. The feasible set is

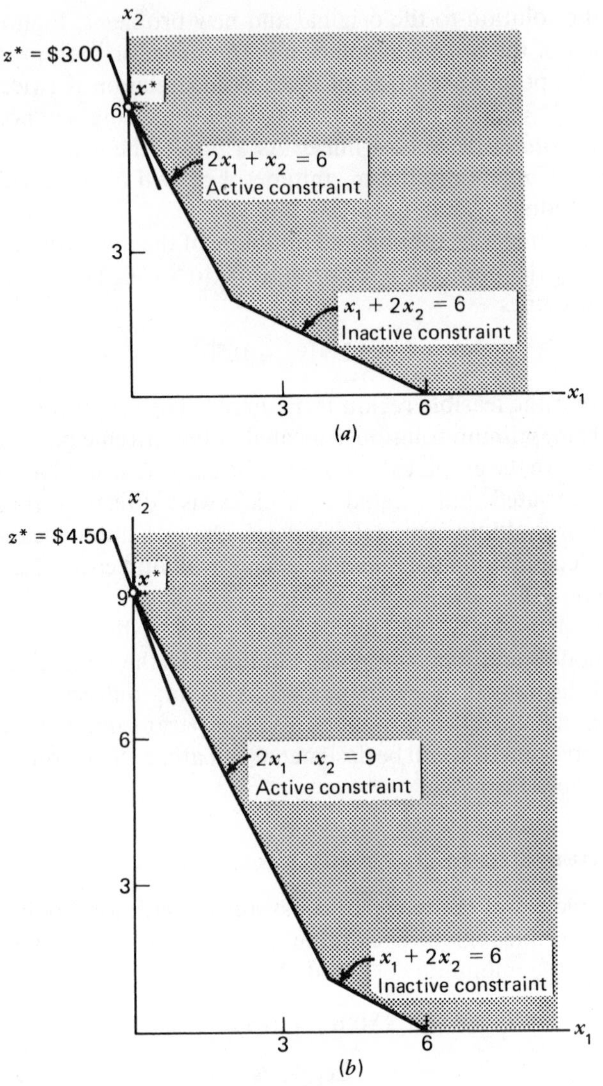

Figure 7.2 The effect of resource parameter changes of ***b***.

changed as shown in Figure 7.2*c*. The optimum solution, x^* and z^*, remains at (0, 6) with $z^* = \$3.00$.

Consider another change in an inactive constraint equation. Suppose the inactive constraint $x_1 + 2x_2 \geq 6$ is changed to $x_1 + 2x_2 \geq 15$. The feasible solution is shown in Figure 7.2*d*. The optimum solution $\boldsymbol{x}^*$ is (0, 7.5) with $z^* = \$3.50$. The optimum solution no longer lies on the isoquant $2x_1 + x_2 = 6$; it has become an inactive constraint. The active constraint becomes $x_1 + 2x_2 = 15$.

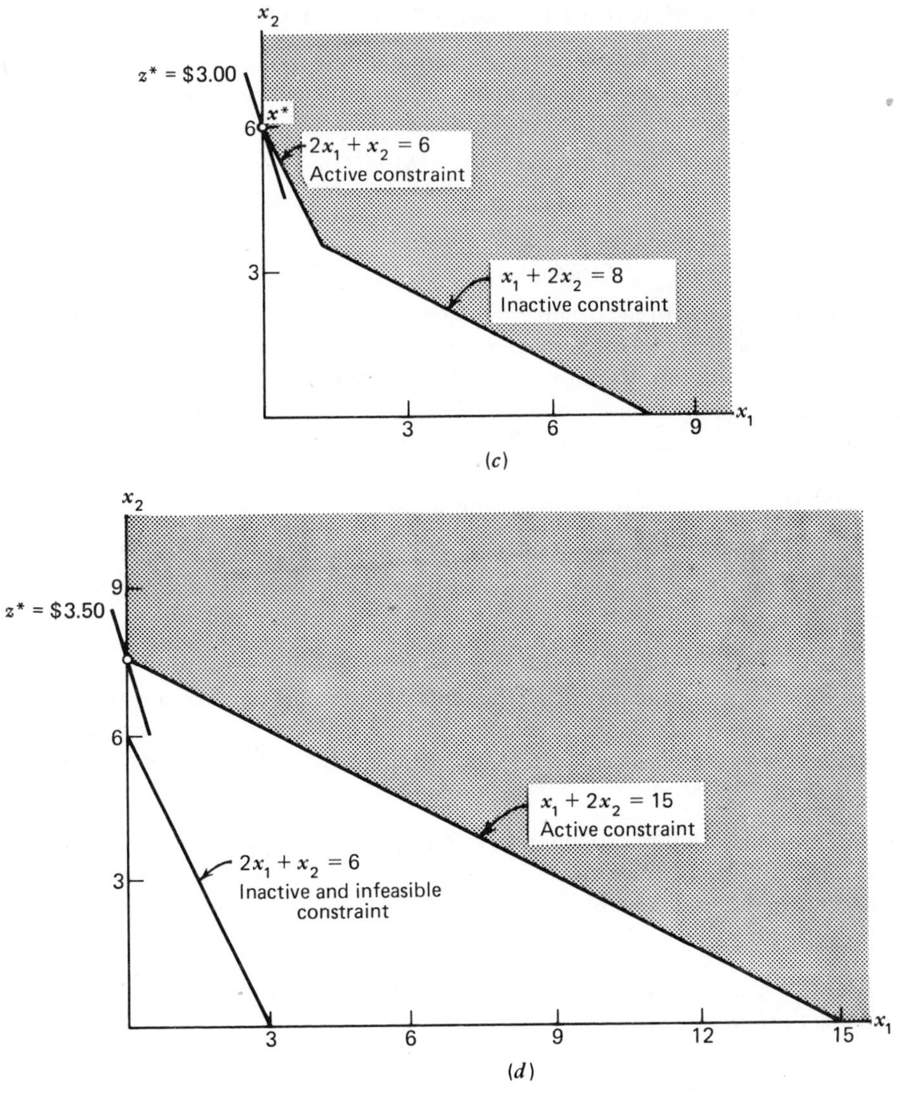

Figure 7.2 (Continued)

These examples shown in Figure 7.1 and 7.2 illustrate that the slope of the isocost and the boundary of the feasible region are extremely important in determining the location of the optimum solution $\boldsymbol{x}^*$. For two-dimensional problems, the graphical approach is most efficient in performing a sensivity analysis. For problems of higher order, the simplex method may be employed on the new formulation, or the final simplex tableau of the original mathematical model may be used. We take the second approach for changes in resource constraints.

The Simplex Method and Changes in Unit Costs

In order to distinguish between the original and new models and emphasize a particular difference between the models, the circumflex ^ will be used. For a change in $\boldsymbol{c}$, the original and new models in standard linear programming form are

Original	New
Minimize $z = \boldsymbol{c}'\boldsymbol{x}$	Minimize $z = \hat{\boldsymbol{c}}'\boldsymbol{x}$
$\boldsymbol{A}\boldsymbol{x} = \boldsymbol{b}$	$\boldsymbol{A}\boldsymbol{x} = \boldsymbol{b}$
$\boldsymbol{x} \geq \boldsymbol{0}$	$\boldsymbol{x} \geq \boldsymbol{0}$

The constraint sets of the original and new formulations are the same. The simplex method may be used to solve the new mathematical model. Instead of using this approach, the optimum solution $\boldsymbol{x}^*$ of the *final simplex iteration* from the original model will be used to establish a procedure for analyzing the new model. This approach should require less computational effort than solving the new model with the simplex method.

The approach consists of the following steps:

STEP 1 Assume the optimum solution $\boldsymbol{x}^*$ from the original mathematical model is the location for the new model, $\boldsymbol{x}^* = \boldsymbol{b}^*$. Calculate z^0, the initial estimate for the new model.

STEP 2 Establish a new indicator row $\hat{\boldsymbol{c}}_I$ for the new model.

STEP 3 Test for optimality. If all the unit cost elements of $\hat{\boldsymbol{c}}_I$ are positive, then the $\boldsymbol{x}^*$ is unchanged, (Rule 1 of the simplex method). If one or more elements of the indicator row $\hat{\boldsymbol{c}}_I$ are negative, find a new optimum by utilizing the simplex method Rules 1, 2, and 3.

Step 1 does not require explanation. For Step 2, $\hat{\boldsymbol{c}}_I$ will be obtained by utilizing the *final* simplex tableau of the original problem. It may be shown that the indicator row $\boldsymbol{c}_I$ in terms of the original cost vector $\boldsymbol{c}$ and final (optimum) simplex iteration is

$$\boldsymbol{c}_I' = \boldsymbol{c}_N' - \boldsymbol{c}_B'\boldsymbol{A}_N^*$$

It should be clear that the indicator row for the new model may be established by replacing the new unit costs of $\hat{\boldsymbol{c}}$ for $\boldsymbol{c}$:

$$\hat{\boldsymbol{c}}_I' = \hat{\boldsymbol{c}}_N' - \hat{\boldsymbol{c}}_B'\boldsymbol{A}_N^*$$

The value of z for the new model is simply

$$z^0 = \hat{\boldsymbol{c}}_B'\boldsymbol{b}^*$$

With this information we are ready to perform step 3.

The Simplex Tableau

The final iteration of the original simplex tableau, the optimum solution, may be written as the augmented matrix:

$$\begin{bmatrix} \hat{\boldsymbol{x}}_B^* & \boldsymbol{A}_N^* & \boldsymbol{I} & \boldsymbol{b}^* & b/a \\ - & \boldsymbol{c}_T^{*\prime} & \boldsymbol{0}' & (z - z^*) & - \end{bmatrix}$$

The simplex tableau for the new formulation will be established from the information contained in this matrix. Since $\boldsymbol{x}^*$ is considered the candidate solution for the new problem and the values of $\boldsymbol{x}_B^*$, $\boldsymbol{A}_N^*$, and $\boldsymbol{b}^*$ are unchanged, they will be introduced into a new simplex tableau without any calculations. A new indicator row is established with the use of equations $z^0 = \hat{\boldsymbol{c}}'\boldsymbol{b}^*$ and $\hat{\boldsymbol{c}}_I' = (\hat{\boldsymbol{c}}_N' - \hat{\boldsymbol{c}}_B'\boldsymbol{A}_N^*)$. Appropriate values of $\hat{\boldsymbol{c}}_I$ are placed in the $\hat{\boldsymbol{c}}_N$ and $\boldsymbol{0}$ submatrices of the simplex tableau.

$$\begin{bmatrix} \boldsymbol{x}_B^* & \boldsymbol{A}_N^* & \boldsymbol{I} & \boldsymbol{b}^* & b/a \\ - & \hat{\boldsymbol{c}}_N' & \boldsymbol{0}' & (z - \hat{z}^0) & - \end{bmatrix}$$

The simplex tableau for the new model is now complete, and it is ready for testing. The basis $\boldsymbol{x}_B^*$ is considered as a candidate solution, and the rules of the simplex method are used to test it for optimality. If the test fails, the simplex algorithm is applied until an optimum minimum solution is obtained.

EXAMPLE 7.1 A Change in Unit Cost Problem

Consider the two-dimensional linear mathematical model:

$$\text{Maximize } z = x_1 + 2x_2$$

$$x_1 + x_2 \le 2$$

$$x_1 \le 1$$

$$x_1 \ge 0, x_2 \ge 0$$

(a) Formulate and utilize the simplex method to determine the optimum solution $\boldsymbol{x}^*$.
(b) Determine if the optimum solution from part a will change when the unit cost c_2 is reduced to zero.
(c) Determine if the optimum solution from part a will change when the unit cost c_1 is increased to 3.
(d) On a graph show how the changes in $\boldsymbol{c}$ affect the magnitude of z^* and the locations of the optimum solution $\boldsymbol{x}_B^*$ found in parts a, b, and c.

Solution

(a) The problem will be solved as a minimization problem. Let $a = -1 \cdot z$. Introduce slack variables x_3 and x_4 to the constraint set. The model in standard form is

$$\text{Minimize } a = -x_1 - 2x_2$$

$$x_1 + x_2 + x_3 = 2$$

$$x_1 + x_4 = 1$$

$$x_1 \ge 0, x_2 \ge 0, x_3 \ge 0, x_4 \ge 0$$

The initial basis is $\boldsymbol{x}_B^{0\prime} = [x_3, x_4]$. The simplex tableau is

$\boldsymbol{x}_B$	x_1	x_2	x_3	x_4	$\boldsymbol{b}$	b/a	
x_3	1	(1)	1	0	2	$2/1 = 2$	←
x_4	1	0	0	1	1		
$\boldsymbol{c}_I$	−1	−2	0	0	$(a - 0)$		
		↑					

The rules of the simplex method are utilized. After one iteration the optimum solution is obtained:

$$\begin{array}{llllll} \boldsymbol{x}_B & x_1 & x_2 & x_3 & x_4 & \boldsymbol{b} \qquad\qquad b/a \end{array}$$
$$\begin{bmatrix} x_2 & 1 & 1 & 1 & 0 & 2 & \\ x_4 & 1 & 0 & 0 & 1 & 1 & \\ \boldsymbol{c}_I & 1 & 0 & 2 & 0 & (a+4) & \end{bmatrix}$$

The optimum solution is

$$x_1^* = 0 \qquad x_2^* = 2 \qquad x_3^* = 0 \qquad x_4^* = 1$$
$$a^* = -4 \qquad z^* = -1 \cdot -4 = 4$$

(b) For $c_2 = 0$, the new objective function is

$$\text{Maximize } z = x_1$$

The constraint set is the same as shown in part a. Let $a = -1 \cdot z$, thus the objective function in matrix notation becomes

$$\text{Minimize } a = \hat{\boldsymbol{c}}'\boldsymbol{x}$$

where $$[\hat{c}_1 \;\; \hat{c}_2 \;\; \hat{c}_3 \;\; \hat{c}_4] = [-1 \;\; 0 \;\; 0 \;\; 0]$$

From the final simplex tableau, we can identify

$$\boldsymbol{x}_B^* = \begin{bmatrix} x_2^* \\ x_4^* \end{bmatrix} \qquad \boldsymbol{x}_N^* = \begin{bmatrix} x_1^* \\ x_3^* \end{bmatrix} \qquad \boldsymbol{b}^* = \begin{bmatrix} 2 \\ 1 \end{bmatrix}$$

$$\boldsymbol{x}_N^{*\prime} = [x_1 \;\; x_3] = [0 \;\; 0]$$

and $$\boldsymbol{A}_N^* = \begin{bmatrix} a_{21} & a_{23} \\ a_{41} & a_{43} \end{bmatrix} = \begin{bmatrix} 1 & 1 \\ 1 & 0 \end{bmatrix}$$

Thus: $$\hat{\boldsymbol{c}}_B' = [\hat{c}_2 \;\; \hat{c}_4] = [0 \;\; 0]$$
$$\hat{\boldsymbol{c}}_N' = [\hat{c}_1 \;\; \hat{c}_3] = [-1 \;\; 0]$$

The value of $\hat{a}$ is

$$a^0 = \boldsymbol{c}_B'\boldsymbol{b}^* = [0 \;\; 0] \cdot \begin{bmatrix} 2 \\ 1 \end{bmatrix} = 0$$

The new indicator row is

$$\hat{\boldsymbol{c}}_I' = \hat{\boldsymbol{c}}_N' - \hat{\boldsymbol{c}}_B'\boldsymbol{A}_N^*$$

$$[\hat{c}_1 \;\; \hat{c}_3] = [-1 \;\; 0] - [0 \;\; 0] \cdot \begin{bmatrix} 1 & 1 \\ 1 & 0 \end{bmatrix}$$

$$\hat{\boldsymbol{c}}_I' = [\hat{c}_1 \;\; \hat{c}_3] = [-1 \;\; 0]$$

Since $\hat{c}_1$ is negative, the solution is not an optimum. The final simplex tableau from part a will be rewritten with $\hat{\boldsymbol{c}}_I$ and a. The simplex tableau becomes

$$\begin{array}{c|cccccc|c}
\boldsymbol{x}_B & x_1 & x_2 & x_3 & x_4 & \boldsymbol{b} & & b/a \\
x_2 & ① & 1 & 1 & 0 & 2 & & 2/1 = 2 \leftarrow \\
x_4 & 1 & 0 & 0 & 1 & 1 & & 1/1 = 1 \\
\hat{\boldsymbol{c}}_I & -1 & 0 & 0 & 0 & (a - 0) & & \\
 & \uparrow & & & & & &
\end{array}$$

Appropriate simplex method calculations are performed and new tableau is established

$$\begin{array}{c|cccccc|c}
\boldsymbol{x}_B & x_1 & x_2 & x_3 & x_4 & \boldsymbol{b} & & b/a \\
x_2 & 0 & 1 & 1 & -1 & 1 & & \\
x_1 & 1 & 0 & 0 & 1 & 1 & & \\
\hat{\boldsymbol{c}}_I & 0 & 0 & 0 & 1 & (a + 1) & &
\end{array}$$

Since Rule 1 of the simplex method is satisfied, the optimum solution is

$$x_1^* = 1 \qquad x_2^* = 1$$

$$a^* = -1$$

$$z^* = -1 \cdot a^* = -1(-1) = 1$$

(c) The approach is similar to the one used in part b. The new objective function is

$$\text{Maximize } z = 3x_1 + 2x_2$$

or the equivalent maximization objective function is

$$\text{Minimize } a = \hat{\boldsymbol{c}}\boldsymbol{x}$$

where
$$\hat{\boldsymbol{c}}' = [\hat{c}_1 \quad \hat{c}_2 \quad \hat{c}_3 \quad \hat{c}_4] = [-3 \quad -2 \quad 0 \quad 0]$$

Since x_1 is not of member of $\boldsymbol{x}_B^*$ from part a, changing c_1 will not affect the total cost, thus

$$a^0 = \hat{\boldsymbol{c}}'\boldsymbol{b}^* = -4$$

A new indicator $\hat{\boldsymbol{c}}_I$ is found with the relationship:

$$\hat{\boldsymbol{c}}_I = \hat{\boldsymbol{c}}_N' - \hat{\boldsymbol{c}}_B'\boldsymbol{A}_N$$

The vector matrix $\boldsymbol{A}_N^*$ was defined in part b. The vectors $\hat{\boldsymbol{c}}_N'$ and $\hat{\boldsymbol{c}}_B'$ are

$$\hat{\boldsymbol{c}}_N' = [\hat{c}_1 \quad \hat{c}_3] = [-3 \quad 0]$$

and
$$\hat{\boldsymbol{c}}_B' = [\hat{c}_2 \quad \hat{c}_4] = [-2 \quad 0]$$

Thus, the new indicator row is

$$[\hat{c}_1 \quad \hat{c}_3] = [-3 \quad 0] - [-2 \quad 0] \cdot \begin{bmatrix} 1 & 1 \\ 1 & 0 \end{bmatrix}$$

$$[\hat{c}_1 \quad \hat{c}_3] = [-1 \quad 2]$$

Since $\hat{c}_1$ is negative, the current basis $\boldsymbol{x}_B^* = [x_2 \quad x_4]$ is not optimum. The simplex tableau becomes

$\boldsymbol{x}_B$	x_1	x_2	x_3	x_4	$\boldsymbol{b}$	b/a	
x_2	1	1	1	0	2	$2/1 = 2$	
x_4	①	0	0	1	1	$1/1 = 1$	←
$\hat{c}_I$	-1	0	2	0	$(a + 4)$		
	↑						

Performing the necessary steps of the simplex method, the new tableau becomes

$\boldsymbol{x}_B$	x_1	x_2	x_3	x_4	$\boldsymbol{b}$	b/a
x_2	0	1	1	-1	1	
x_1	1	0	0	1	1	
$\hat{c}_I$	0	0	2	1	$(a + 5)$	

The optimum solution is

$$x_1^* = 1 \qquad x_2^* = 1 \qquad x_3^* = 0 \qquad x_4^* = 0$$

$$a^* = -5 \qquad z^* = -1 \cdot a^* = -1(-5) = 5$$

(d) The optimum solutions for parts a, b, and c are shown in Figure 7.3.

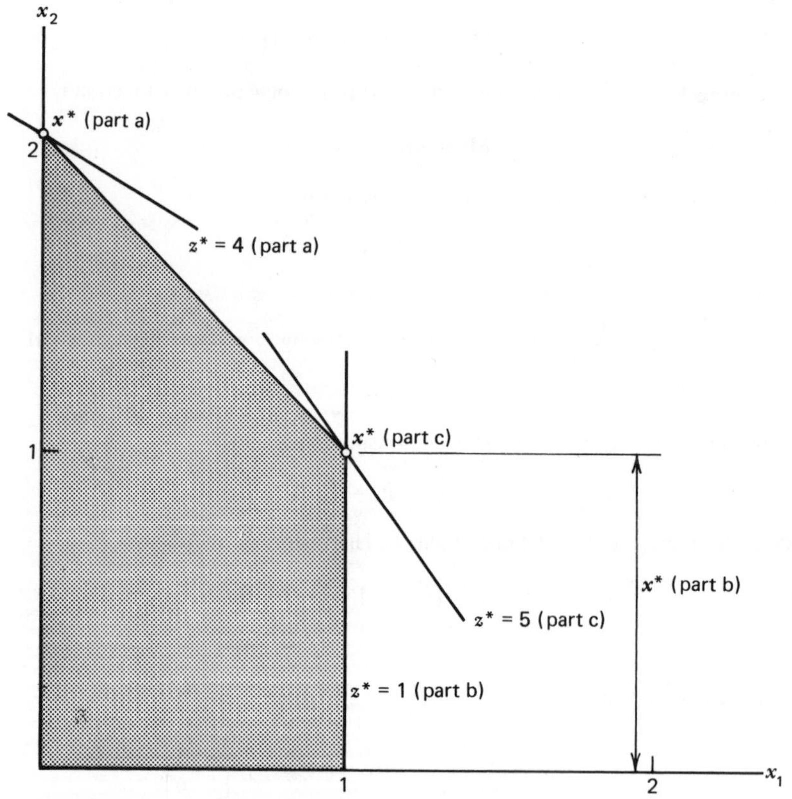

Figure 7.3

The graph shows that the change in unit costs of the objective function changes the slope of the isocost. If the change is large enough, the current optimum will change to a new extreme point. The original cost function is shown in the figure as $\hat{z}^* = 4$. This optimum solution is located at $\mathbf{x}^* = (0, 2)$. Reducing c_2 to zero causes the value of z to be reduced to $z^* = 1$ and the optimum solution to be shifted to (1, 1). This result is shown as $z^* = 1$ in the figure. The graphical solution shows optimum solution lies along the boundary of the feasible region $x_1 = 1$ from (1, 0) to (1, 1). This information was not gleaned from the solution by the simplex method. Likewise, increasing c_1 to 3 causes z to increase to 5 and changes the optimum solution to $x^* = (1, 1)$. This result is shown as $z^* = 5$ in the figure.

EXAMPLE 7.2 Roadway Aggregate Mix Problem: Changing Prices

The "most desirable" roadway aggregate has a grain size distribution as shown in Figure 7.4. The specification requires that the roadway aggregate meet the following limitations:

Roadway Aggregate Specification

	PERCENT VOLUME LOWER LIMIT	PERCENT VOLUME UPPER LIMIT
Gravel	5	20
Sand	65	85
Silt	—	15

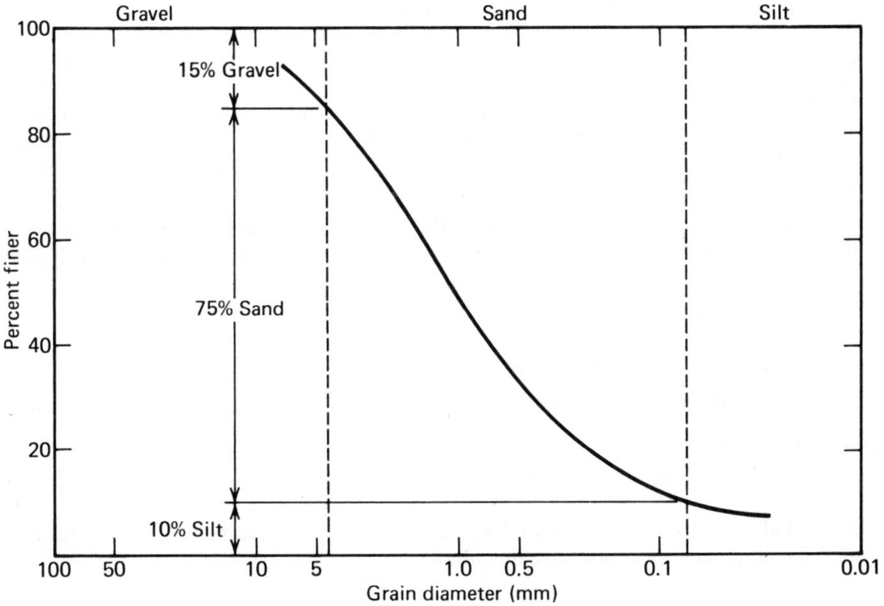

Figure 7.4

Three sources of material are in the vicinity of the construction site. The composition and the unit cost of the material delivered to the site are

SOURCES	% GRAVEL	% SAND	% SILT	$/YD³
River gravel	15	75	10	$6
Pea gravel	30	70	—	$8
Sand	—	100	—	$5

The roadway is to be built on glacial till. Sieve analysis shows that the soil contains 5 percent gravel, 30 percent sand, and 65 percent silt.

Supply 9000 yd^3 of aggregate meeting specification for a 3-mile length of road. Different sources of material may be mixed to form an aggregate mix. The on-site material may be used as the roadway aggregate if it meets specifications or mixed with other aggregates to form an acceptable mixture.

(a) Determine if any available source of aggregate meets specification without mixing.
(b) Formulate a minimum-cost aggregate mix model.
(c) Determine the minimum-cost solution that meets specification requirements.
(d) Determine if the solution will change if the unit cost of river gravel is reduced from $\$6.00/yd^3$ to $\$5.00/yd^3$.
(e) The solution from part d will show that river gravel will be substituted for pea gravel. The supplier of pea gravel wants to lower his selling price to become competitive again. Determine the maximum price that he may charge.
(f) In parts d and e, river gravel in a member of the optimum solution. Determine the highest price that the river gravel supplier may charge and still remain competitive.

Solution

(a) The river gravel is the only material that meets the aggregate specification. Furthermore, the river gravel satisfies the grain size distribution for the "most desirable" roadway aggregate as shown in the figure. In other words, we cannot find a better source of aggregate for the job. By mixing with other material, it may be possible to reduce the total cost of the aggregate.

Pea gravel and glacial till do not meet the specification requirements, because the pea gravel exceeds the acceptable limit for gravel and the glacial till contains too much silt. In order for these materials to be used they must be mixed with other materials.

(b) The control variables are defined as

x_1 = amount of river gravel (1000 yd^3).

x_2 = amount of pea gravel (1000 yd^3).

x_3 = amount of sand (1000 yd^3).

x_4 = amount of glacial till (1000 yd^3).

Let z equal the total cost of material delivered to the site. It is assumed the on-site glacial till material has no cost. The objective function is

$$\text{Minimize } z = \$6x_1 + \$8x_2 + \$5x_3$$

The total amount of aggregate required for the roadway is 9000 yd^3. The volume constraint is

$$x_1 + x_2 + x_3 + x_4 = 9$$

The acceptable ranges of gravel, sand, and silt are introduced into the model as a set of constraint equations. From the aggregate specification, the amount of gravel in the mix must be between 5 and 20 percent of the total mix. The upper- and lower-level gravel constraint equations are

$$0.15x_1 + 0.3x_3 + 0.05x_4 \leq 0.2(9) = 1.8$$

$$0.15x_1 + 0.3x_2 + 0.05x_4 \geq 0.05(9) = 0.45$$

The amount of sand in the mix must be between 65 and 85 percent of the entire mix. The upper- and lower-level sand constraint equations are

$$0.75x_1 + 0.7x_2 + 1x_3 + 0.3x_4 \leq 0.85(9) = 7.65$$

$$0.75x_1 + 0.7x_2 + 1x_3 + 0.3x_4 \geq 0.65(9) = 5.85$$

The amount of silt in the mix must not exceed 15 percent of the mix. The silt constraint equation is

$$0.1x_1 + 0.65x_4 \leq 0.15(9) = 1.35$$

The linear mathematical model is

$$\begin{array}{llr}
\text{Minimize } z = & 6x_1 + 8x_2 + 5x_3 & \\
& x_1 + x_2 + x_3 + x_4 = 9 & \text{Supply} \\
& 0.15x_1 + 0.3x_2 + 0.05x_4 \leq 1.8 & \text{Gravel} \\
& 0.15x_1 + 0.3x_2 + 0.05x_4 \geq 0.45 & \text{Gravel} \\
& 0.75x_1 + 0.7x_2 + x_3 + 0.3x_4 \leq 7.65 & \text{Sand} \\
& 0.75x_1 + 0.7x_2 + x_3 + 0.3x_4 \geq 5.85 & \text{Sand} \\
& 0.1x_1 + 0.65x_4 \leq 1.35 & \text{Silt} \\
& x_1, \ldots, x_4 \geq 0 &
\end{array}$$

(c) The optimum solution is calculated with the use of the simplex method. Slack variables x_5, x_6, and x_7 are introduced to constraint equations with less-than-or-equal-to constraints ($\leq$). Surplus variables x_8 and x_9 are introduced to constraint equations with greater-than-or-equal-to constraints ($\geq$). In order to establish an initial basic feasible solution, artificial variables x_{10}, x_{11}, and x_{12} are utilized for the equation with an equality constraint and, for the equations utilizing surplus variables, x_8 and x_9. The artificial objective function is

$$w = x_{10} + x_{11} + x_{12}$$

This equation is rewritten in terms of the nonbasic variables $\boldsymbol{x}_N$ and introduced into the model. The model is

$$\text{Minimize}\,(w - 15.3) = -1.90x_1 - 2x_2 - 2x_3 - 2x_4 + x_8 + x_9$$

$$\text{Minimize}\,(z - 0) = 6x_1 + 8x_2 + 5x_3$$

$$\begin{aligned}
x_1 + x_2 + x_3 + x_4 + x_{10} &= 9 \\
0.15x_1 + 0.3x_2 + 0.05x_4 + x_5 &= 1.8 \\
0.15x_1 + 0.3x_2 + 0.05x_4 - x_8 + x_{11} &= 0.45 \\
0.75x_1 + 0.7x_2 + x_3 \quad 0.3x_4 + x_6 &= 7.65 \\
0.75x_1 + 0.7x_2 + x_3 + 0.3x_4 - x_9 + x_{12} &= 5.85 \\
0.1\ x_1 + 0.65x_4 + x_7 &= 1.35
\end{aligned}$$

$$x_1, x_2, \ldots, x_{12} \geq 0$$

The simplex tableau for the initial solution is

$\boldsymbol{x}_B$	x_1	x_2	x_3	x_4	x_5	x_6	x_7	x_8	x_9	x_{10}	x_{11}	x_{12}	$\boldsymbol{b}$	
x_{10}	1.00	1.00	1.00	1.00	0.00	0.00	0.00	0.00	0.00	1.00	0.00	0.00	9.00	
x_5	0.15	0.30	0.00	0.05	1.00	0.00	0.00	0.00	0.00	0.00	0.00	0.00	1.80	
x_{11}	0.15	0.30	0.00	0.05	0.00	0.00	0.00	−1.00	0.00	0.00	1.00	0.00	0.45	
x_6	0.75	0.70	1.00	0.30	0.00	1.00	0.00	0.00	0.00	0.00	0.00	0.00	7.65	
x_{12}	0.75	0.70	(1.00)	0.30	0.00	0.00	0.00	0.00	−1.00	0.00	0.00	1.00	5.85	←
x_7	0.10	0.00	0.00	0.65	0.00	0.00	1.00	0.00	0.00	0.00	0.00	0.00	1.35	
$\boldsymbol{c}_I$	6.00	8.00	5.00	0.00	0.00	0.00	0.00	0.00	0.00	0.00	0.00	0.00	$(z - 0)$	
$\boldsymbol{d}_I$	−1.90	−2.00	−2.00	−1.40	0.00	0.00	0.00	1.00	1.00	0.00	0.00	0.00	$(w - 15.3)$	
			↑											

The initial basis is not a minimum solution. The arrows indicate that x_3 is entered and x_{12} is removed from the basis. Either x_2 or x_3 may be entered. The variable x_3 is chosen because it has the smallest unit cost, \$5.00, of these choices. It is possible that the optimum solution will be obtained in fewer iterations with this choice. After five iterations, the optimum solution is obtained. The optimum solution is shown in the final tableau. Since the artificial variables have no physical meaning and are members of $\boldsymbol{x}_N$, the columns x_{10}, x_{11}, and x_{12} are not shown in the final tableau.

Final Tableau (Optimum Solution)

$\boldsymbol{x}_B$	x_1	x_2	x_3	x_4	x_5	x_6	x_7	x_8	x_9	$\boldsymbol{b}$
x_9	0.00	0.00	0.00	0.00	0.00	0.00	−1.00	1.00	1.00	1.35
x_5	0.00	0.00	0.00	0.00	1.00	0.00	0.00	1.00	0.00	1.35
x_2	0.47	1.00	0.00	0.00	0.00	0.00	−0.26	−3.33	0.00	1.15
x_6	0.00	0.00	0.00	0.00	0.00	1.00	1.00	−1.00	0.00	0.45
x_3	0.37	0.00	1.00	0.00	0.00	0.00	−1.28	3.33	0.00	5.77
x_4	0.15	0.00	0.00	1.00	0.00	0.00	1.54	0.00	0.00	2.08
$\boldsymbol{c}_I$	0.38	0.00	0.00	0.00	0.00	0.00	8.46	10.00	0.00	$(z - 38.1)$
$\boldsymbol{d}_I$	0.00	0.00	0.00	0.00	0.00	0.00	0.00	0.00	0.00	$(w - 0.00)$

The optimum solution is to purchase the following amounts of material and mix them to meet specifications.

SOURCE	x_B^*	AMOUNT (yd³)	COST ($)
River gravel	—	0	0
Pea gravel	x_2	1,150	9,200
Sand	x_3	5,770	28,900
Glacial till	x_4	2,080	0

$z^* = \$38,100.$

It is interesting that river gravel, the "most desirable" roadway aggregate, is not purchased. Mixing pea gravel and sand with the glacial till that is on-site and is obtained at no cost produces the most cost-efficient mix to use. A savings of \$6900 is achieved by mixing the material, in lieu of using the river gravel exclusively. The unit cost of the glacial till, pea gravel, and sand mix is \$4.23/yd^3; river gravel is \$6/yd^3.

(d) Since the cost of river gravel c is reduced from \$6/yd^3 to \$5/yd^3, the cost function is changed from

$$z = 6x_1 + 8x_2 + 5x_3$$

to

$$z = 5x_1 + 8x_2 + 5x_3$$

The optimum solution from part c is

$$\boldsymbol{x}_B^{*\prime} = [x_9 \quad x_5 \quad x_2 \quad x_6 \quad x_3 \quad x_4]$$

$$\boldsymbol{x}_B^{*\prime} = [1.35 \quad 1.35 \quad 1.15 \quad 0.45 \quad 5.77 \quad 2.08]$$

The relationships $z^0 = \hat{\boldsymbol{c}}'_B\boldsymbol{b}^*$ and $\hat{\boldsymbol{c}}'_I = (\hat{\boldsymbol{c}}'_N - \hat{\boldsymbol{c}}'_B\boldsymbol{A}_N^*)$ are utilized to determine the new cost indicator row and test if the solution from part c remains the location of the optimum. The new cost $\hat{z}$ is determined with the expression:

$$z^0 = \hat{\boldsymbol{c}}'_B\boldsymbol{b}_B^* = \$38,100.00$$

The total cost is unchanged because the unit cost of the river gravel c_1 is not a member of the basis $\boldsymbol{x}_B^*$. The new indicator is established with

$$\hat{\boldsymbol{c}}'_I = \hat{\boldsymbol{c}}'_N - \hat{\boldsymbol{c}}'_B\boldsymbol{A}_N^*$$

The matrix $\boldsymbol{A}_N^*$ is obtained from the final tableau, where the nonbasic vector is

$$\boldsymbol{x}'_N = [x_1 \quad x_7 \quad x_8].$$

$$\boldsymbol{A}_N^* = \begin{bmatrix} a_{91}^* & a_{97}^* & a_{98}^* \\ a_{51}^* & a_{57}^* & a_{58}^* \\ a_{21}^* & a_{27}^* & a_{28}^* \\ a_{61}^* & a_{67}^* & a_{68}^* \\ a_{31}^* & a_{37}^* & a_{38}^* \\ a_{41}^* & a_{47}^* & a_{48}^* \end{bmatrix} = \begin{bmatrix} 0.00 & -1.00 & 1.00 \\ 0.00 & 0.00 & 1.00 \\ 0.47 & -0.26 & -3.33 \\ 0.00 & 1.00 & -1.00 \\ 0.37 & -1.28 & 3.33 \\ 0.15 & 1.54 & 0.00 \end{bmatrix}$$

Care must be used in establishing the cost vectors c'_B and c'_N. Particular attention is used to ensure that the unit costs are assigned to the proper location in the vector:

$$\hat{c}'_N = [\hat{c}_1 \quad \hat{c}_7 \quad \hat{c}_8] = [5 \quad 0 \quad 0]$$

$$\hat{c}'_B = [\hat{c}_9 \quad \hat{c}_5 \quad \hat{c}_2 \quad \hat{c}_6 \quad \hat{c}_3 \quad \hat{c}_4] = [0 \quad 0 \quad 8 \quad 0 \quad 5 \quad 0]$$

Perform the calculations as required:

$$\hat{c}'_N = [\hat{c}_1 \quad \hat{c}_7 \quad \hat{c}_8] = [-0.62 \quad 8.46 \quad 10.00]$$

Since not all members of $\hat{c}_I$ are positive, the current solution x_B^* is not an optimum solution.

By Rule 2, it is evident that control variable x_1 will be entered into the basis. It appears that the reduced price of river gravel is competitive with the other sources of material. The new optimum mix will be found with use of the simplex method. The indicator row of the final tableau in part c is modified as follows:

x_B	x_1	x_2	x_3	x_4	x_5	x_6	x_7	x_8	x_9	b	
x_9	0.00	0.00	0.00	0.00	0.00	0.00	−1.00	1.00	1.00	1.35	
x_5	0.00	0.00	0.00	0.00	1.00	0.00	0.00	1.00	0.00	1.35	
x_2	(0.47)	1.00	0.00	0.00	0.00	0.00	−0.26	−3.33	0.00	1.15	←
x_6	0.00	0.00	0.00	0.00	0.00	1.00	1.00	−1.00	0.00	0.45	
x_3	0.37	0.00	1.00	0.00	0.00	0.00	−1.28	3.33	0.00	5.77	
x_4	0.15	0.00	0.00	1.00	0.00	0.00	1.54	0.00	0.00	2.08	
$\hat{c}_I$	−0.62	0.00	0.00	0.00	0.00	0.00	8.46	10.00	0.00	$(z - 38.1)$	
	↑										

After one iteration, an optimum solution is obtained. The final tableau is

Final Tableau (Optimum Solution)

x_B	x_1	x_2	x_3	x_4	x_5	x_6	x_7	x_8	x_9	b
x_9	0.00	0.00	0.00	0.00	0.00	0.00	−1.00	1.00	1.00	1.35
x_5	0.00	0.00	0.00	0.00	1.00	0.00	0.00	1.00	0.00	1.75
x_1	1.00	2.11	0.00	0.00	0.00	0.00	−0.54	−7.03	0.00	2.43
x_6	0.00	0.00	0.00	0.00	0.00	1.00	1.00	−1.00	0.00	0.45
x_3	0.00	−0.78	1.00	0.00	0.00	0.00	−1.08	5.95	0.00	4.87
x_4	0.00	−0.32	0.00	1.00	0.00	0.00	1.62	1.08	0.00	1.70
$\hat{c}_I$	0.00	1.38	0.00	0.00	0.00	0.00	8.11	5.41	0.00	$(z - 36.5)$

The optimum solution is

SOURCE	x_B^*	AMOUNT (yd³)	COST ($)
River gravel	x_1^*	2,430	12,150
Pea gravel	—	0	0
Sand	x_3^*	4,870	24,350
Glacial till	x_4^*	1,700	0

$z^* = \$36,500.$

River gravel and sand are mixed with the glacial till to obtain the desired mix. Comparing the solution from part c with this one indicates that there is a savings of \$1600, less sand and glacial till required for this mix. This mix has a unit cost of \$4.06/yd^3.

(e) This approach is similar to the one used in part d except that we treat the unit cost of pea gravel $\hat{c}_2$ as a variable. The relationship $\hat{\boldsymbol{c}}_I' = \hat{\boldsymbol{c}}_N' - \hat{\boldsymbol{c}}_B' \boldsymbol{A}_N^*$ will be used to formulate a mathematical equation for price changes in pea gravel. The elements of $\boldsymbol{A}_N^*$, $\hat{\boldsymbol{c}}_N$, and $\hat{\boldsymbol{c}}_B$ are obtained from the final simplex tableau from part d.

$$\boldsymbol{A}_N^* = \begin{bmatrix} a_{92}^* & a_{97}^* & a_{98}^* \\ a_{52}^* & a_{58}^* & a_{58}^* \\ a_{12}^* & a_{17}^* & a_{18}^* \\ a_{62}^* & a_{67}^* & a_{68}^* \\ a_{32}^* & a_{37}^* & a_{38}^* \\ a_{42}^* & a_{47}^* & a_{48}^* \end{bmatrix} = \begin{bmatrix} 0.00 & -1.00 & 1.00 \\ 0.00 & 0.00 & 1.00 \\ 2.11 & -0.54 & -7.03 \\ 0.00 & 1.00 & -1.00 \\ -0.78 & -1.08 & 5.95 \\ -0.32 & 1.62 & 1.08 \end{bmatrix}$$

$$\hat{\boldsymbol{c}}_N' = [\hat{c}_2 \quad \hat{c}_7 \quad \hat{c}_8] = [c_2 \quad 0 \quad 0]$$

$$\hat{\boldsymbol{c}}_B' = [\hat{c}_9 \quad \hat{c}_5 \quad \hat{c}_1 \quad \hat{c}_6 \quad \hat{c}_3 \quad \hat{c}_4] = [0 \quad 0 \quad 5 \quad 0 \quad 5 \quad 0]$$

Utilize $\hat{\boldsymbol{c}}_I' = \hat{\boldsymbol{c}}_N' - \hat{\boldsymbol{c}}_B' \boldsymbol{A}_N^*$ to obtain

$$\hat{\boldsymbol{c}}_I' = [\hat{c}_2 \quad \hat{c}_7 \quad \hat{c}_8] = [(c_2 - 6.62) \quad 8.11 \quad 5.41]$$

Since $\hat{c}_2 = c_2 - 6.62$ is the only indicator that contains the variable c_2, its magnitude will determine if x_2 remains in the basis or not. The indicators $\hat{c}_7$ and $\hat{c}_8$ are independent of c_2; therefore, they are not affected by changes in the unit cost of pea gravel. The current cost of pea gravel is \$8.00/yd^3. Substituting this cost into $\hat{c}_2 = c_2 - 6.62$, gives $\hat{c}_2 = 1.38$. This is the same value of the indicator in the simplex tableau of part d. If the unit cost is reduced to \$6.62 or less, $\hat{c}_2 \leq 0$, x_2 or pea gravel will enter the basis and will become an ingredient of the minimum cost mix. The pea gravel supplier must lower his selling price to \$6.62 per cubic yard or less.

(f) The analysis of changes in unit cost of river gravel c_1 will be similar to part e. The unit cost parameter c_1 is treated as a variable. However, river gravel, in this case, is a member of the current basis; therefore, the total cost $\hat{z}$ is affected by unit cost changes in c_1. The analysis is more complicated than in part e is

$$z^0 = \hat{\boldsymbol{c}}_B' \boldsymbol{x}_B^*$$

where $$\hat{\boldsymbol{c}}_B' = [\hat{c}_9 \quad \hat{c}_5 \quad \hat{c}_1 \quad \hat{c}_6 \quad \hat{c}_3 \quad \hat{c}_4] = [0 \quad 0 \quad c_1 \quad 0 \quad 5 \quad 0]$$

and $$\boldsymbol{x}_B^{*\prime} = [x_9^* \quad x_5^* \quad x_1^* \quad x_6^* \quad x_3^* \quad x_4^*] = [1.35 \quad 1.35 \quad 2.43 \quad 0.45 \quad 4.87 \quad 1.70]$$

Substituting these vectors into the equation $z^0 = \hat{c}_B' x_B^*$ the relationship becomes

$$z^0 = 2.43c_1 + 24.35$$

This cost function is valid only if x_1 remains in the basis $\boldsymbol{x}_B^*$. The indicator vector is

$$\hat{\boldsymbol{c}}_I' = \hat{\boldsymbol{c}}_N' - \hat{\boldsymbol{c}}_B' \boldsymbol{A}_N^*$$

where $$\hat{\boldsymbol{c}}_N' = [\hat{c}_2 \quad \hat{c}_7 \quad \hat{c}_8] = [8 \quad 0 \quad 0]$$

and $$\hat{\boldsymbol{c}}_B' = [\hat{c}_9 \quad \hat{c}_5 \quad \hat{c}_1 \quad \hat{c}_6 \quad \hat{c}_3 \quad \hat{c}_4] = [0 \quad 0 \quad c_1 \quad 0 \quad 5 \quad 0]$$

The elements of $\hat{c}_I$ are

$$\hat{c}_2 = 11.9 - 2.11c_1$$

$$\hat{c}_7 = 5.40 + 0.54c_1$$

$$\hat{c}_8 = -29.8 + 7.03c_1$$

The optimum basis x_B^* will remain unchanged if all elements of the indicator row $\hat{c}_I$ remain nonnegative, $\hat{c}_2 \geq 0$, $\hat{c}_7 \geq 0$, and $\hat{c}_8 \geq 0$.

We may determine the range of unit costs of c_1 for which x_B^* remains unchanged. Let $\hat{c}'_I = 0$, or

$$\hat{c}_2 = 11.9 - 2.11c_1 = 0 \qquad \text{thus } c_1 = \$5.64/\text{yd}^3$$

$$\hat{c}_7 = 5.40 + 0.54c_1 = 0 \qquad \text{thus } c_1 < 0 \text{ is not feasible}$$

$$\hat{c}_8 = -29.8 + 7.03c_1 = 0 \qquad \text{thus } c_1 = \$4.24/\text{yd}^3$$

The selling price range is $\$4.24 \leq c_1 \leq \$5.64$. The total cost for changes in c_1, where x_B^* is unchanged, may be determined from $z = 2.43c_1 + 24.35$, or $\$34{,}700 \leq z \leq \$38{,}100$. When the price of river gravel is greater than \$5.64, pea gravel returns to the optimum solution. When the price of river gravel is reduced to a value below \$4.24, sand is removed from the optimum solution. Further analyses should show that a combination of glacial till, river gravel, and pea gravel will give the lowest cost.

Remarks

It is apparent from the solutions obtained in parts c and d that the glacial till is used in the mix to reduce cost. By adding glacial till, we shall never prepare a mix that is considered the "most desirable." We shall always utilize a mix that satisfies the specification limitations. This problem points out the importance of establishing engineering specifications. Since the optimum mixes meet the specification, the roadway should perform as well as the "most desirable" aggregate for the given life of the project. If it does not, the specification is not strict enough and should be changed.

PROBLEMS

Problem 1

There are two suppliers of pipe.

SOURCE	UNIT COST ($/linear ft)	SUPPLY (linear ft)
1	$100	100 maximum
2	$125	Unlimited

(a) Formulate a mathematical model to minimize the total cost of pipe. No less than 900 ft of pipe is required. Define control variables.
(b) By graphical means, find the optimum solution.
(c) Formulate a mathematical model where the price of source 2 material is \$90/linear ft.
(d) By graphical means, find the optimum solution to part c.

Problem 2

This problem is a continuation of Problem 1, part b.

(a) Formulate a mathematical model where the total supply of pipe from source 2 is limited to 800 ft.
(b) By graphical means, find the optimum solution to part a.
(c) Formulate a mathematical model where the total supply of pipe from source 2 is limited to 700 ft.
(d) By graphical means, find the optimum solution to part c if it exists.

Problem 3

A manufacturing company requires the daily consumption of 4 Mgal of water of a quality such that concentration of a certain pollutant must be kept below 100 mg/l. The water can be supplied from two sources: (1) purchase from a local water supply at the cost of \$100/Mgal, and (2) pump from a nearby stream at a cost of \$50/Mgal. The concentration of the pollutant from the first source is 50 mg/l, and that from the second source is 200 mg/l. The water from the two sources is completely mixed before it is used.

(a) Formulate the problem as a minimum-cost model.
(b) By graphical means, determine the amount of water to be purchased from each source to minimize total cost.
(c) By graphical means, determine the minimum-cost solution if the local water supply increases its unit cost by 50 percent.
(d) The company plans a new expansion that will at least double the daily consumption of water of the same quality. The local water company can supply up to 10 Mgal daily, but the quantity to be pumped from the nearby stream cannot exceed 2 Mgal daily. By graphical means, determine the quantities of water to be obtained from these two sources to meet the need of the new expansion such that cost will be minimized.

Problem 4

The original mathematical model in standard linear programming form is

$$\text{Minimize } z = -11x_1 - 4x_2$$

$$3x_1 + 5x_2 + x_3 \quad\quad = 15$$

$$5x_1 + 2x_2 \quad\quad + x_4 = 10$$

The final simplex tableau is

x_B	x_1	x_2	x_3	x_4	b
x_3	0	3.8	1	−0.6	9
x_1	1	0.4	0	0.2	2
	0	0.4	0	2.2	$(z + 22)$

A new mathematical model is formulated where the objective function is minimize $z = -11x_1 - 10x_2 + 3x_3 - 4x_4$. The constraint set is the same as shown above. Use the sensitivity analysis for changes in $\boldsymbol{c}$ to determine the optimum solution.

Problem 5

$$\text{Minimize } z = -2x_1 - x_2 - x_3$$

$$x_1 + 2x_2 + x_3 \leq 3$$

$$3x_1 + 5x_2 \leq 6$$

$$x_1, x_2, x_3, \geq 0$$

The final simplex tableau (optimal solution) is

x_B	x_1	x_2	x_3	x_4	x_5	b
x_3	0	0.33	1	1	−0.33	1
x_1	1	1.67	0	0	0.33	2
	0	2.67	0	1	0.33	$(z + 5)$

Use sensitivity analysis to determine that the location of x^* is the same as the one shown in the final simplex tableau for the following change in the objective function:

$$\hat{c}' = [-3 \quad -2 \quad 0]$$

Problem 6

$$\text{Maximize } z = x_1 + x_2$$

$$-x_1 + x_2 \leq 0$$

$$2x_1 + x_2 \leq 8$$

$$x_1 \geq 0, x_2 \geq 0$$

(a) Use the simplex method to determine the optimum solution of the problem.
(b) Solve part a by graphical means.
(c) Use sensitivity analysis and the final simplex tableau from part a to determine optimum solution for new unit costs of $c_1 = 2$ and $c_2 = 2$.
(d) Solve part c by graphical means.
(e) Use sensitivity analysis to determine the optimum solution for the new unit costs of $c_1 = 3$ and $c_2 = 0.5$.

Problem 7

In order to produce a certain steel alloy, it is required that the raw material must have no less than 1 lb of manganese, nor less than 3 lb of chromium nor more than 6 lb of copper. The raw material is drawn from two iron ores. The composite of the ores and its costs are

ORE	No. 1	No. 2
Manganese	1 lb/ton	1 lb/ton
Chromium	2 lb/ton	4 lb/ton
Copper	3 lb/ton	7 lb/ton
Cost	\$4/ton	\$5/ton

These ores may be mixed to meet specifications.

(a) Formulate a model to minimize total cost in dollars per ton.
(b) Solve by the simplex method.
(c) Excess chromium can be removed from the mix and sold for \$4.25/lb. Use the sensitivity analysis approach to determine if the optimum solution from part b is changed. If so, determine the new optimum solution and minimum cost.

7.2 DUAL–PRIMAL RELATIONSHIPS

Analysis of changes in the resource constraints requires the use of the dual–primal relationships for linear models. The dual formulation transposes the resource constraint vector $\boldsymbol{b}$ of the primal model into the objective function of the dual model. Thus, the dual model allows the analysis of changes in $\boldsymbol{b}$ to be performed in the same manner that was used in Section 7.1 for changes in $\boldsymbol{c}$. The dual–primal relationship utilizes new control variables called shadow prices and opportunity costs. The shadow price and opportunity cost give new economic insight and a different perspective of the meaning to resource allocation problems. The properties of the dual–primal relationships will be made by use of an example.

The Primal Model

Suppose a contractor requires at least 100 yd^3 of pure sand for a construction job. The contractor has two sources of soil from which she may extract the sand by screening. The soil properties and total refining costs are listed in the following table:

	SOURCE 1	SOURCE 2
Percent sand	60 percent	40 percent
Cost per cubic yard	\$4.80	\$4.00

This problem will be formulated as a resource allocation model with the control variables defined as

x_1 = amount of soil purchased and refined from source 1 in cubic yards.

x_2 = amount of soil purchased and refined from source 2 in cubic yards.

The total cost for purchasing and refining z is

$$z = \$4.80x_1 + \$4.00x_2$$

The amount of soil required from each source to obtain the minimum amount of pure sand is

$$0.6x_1 + 0.4x_2 \geq 100$$

where the amount of soil obtained from each source must be positive, $x_1 \geq 0$ and $x_2 \geq 0$. The mathematical or *primal model* is

$$\begin{aligned} \text{Minimize } z &= 4.80x_1 + 4.00x_2 \\ 0.6x_1 + 0.4x_2 &\geq 100 \qquad \text{Sand} \\ x_1 \geq 0,\ x_2 &\geq 0 \end{aligned}$$

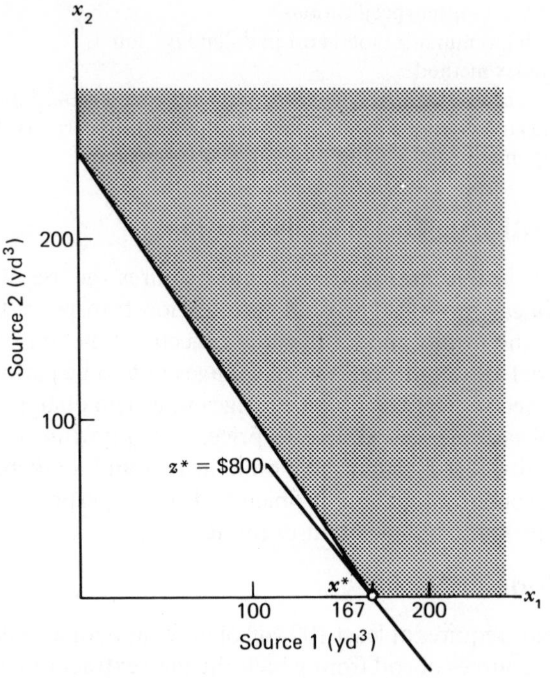

Figure 7.5 Solution to the buyer's problem.

The optimum solution of the primal or buyer's model is solved graphically as shown in Figure 7.5. The minimum cost to the buyer is \$800, that is, $z^* = \$800$. She will purchase 167 yd^2 from source 1, x_1^*, $= 167$ and $x_2^* = 0$.

The Dual Model

The *dual* is a transformation of the primal. Every linear mathematical model has a dual model. For analysis of changes in elements of the $\boldsymbol{b}$ vector, the dual is established by introducing a new set of control variables called *shadow prices*. We develop the concept of shadow price and establish a dual model by analyzing the sand-extraction example from a different point of view.

In the establishment of the primal model, we tacitly assumed that the contractor was a *buyer*. She will weigh the benefits and costs before deciding upon which source and amount of soil to purchase. The primal model is a summary of the decision-making strategy the contractor or buyer will utilize.

Let us introduce a mathematical model for the seller of pure sand. Assume that the conditions of perfect competition exist and that all sellers and buyers have the same information about the construction materials market. Therefore, the sand supplier knows that if he adjusts his price he can capture sales from the competition. Let the selling price of sand y_1 be a control variable, and establish a mathematical model to maximize the profit of the sand supplier. The total anticipated revenue from the sale of pure sand is

$$d = 100y_1$$

where y_1 equals the unit selling price of pure sand in units of dollars per cubic yard. Since the seller knows the unit selling prices of soil from sources 1 and 2, he realizes that if his selling price is too high, he will not sell his sand to the contractor. The seller can determine the unit cost of pure sand that can be extracted from the soil from sources 1 and 2 and, in turn, adjust his selling price y_1 such that it does not exceed the unit selling prices of his competition. The unit selling price of sand after it has been extracted from the soil is \$4.80/0.6 = \$8.00/yd³ for the source 1 and \$4.00/0.4 = \$10.00/yd³ for source 2. Thus, the sand supplier selling price will be restricted as follows:

$$y_1 \leq \$8.00 \qquad \text{Source 1}$$

$$y_2 \leq \$10.00 \qquad \text{Source 2}$$

The seller establishes the following maximum profit model for himself:

$$\begin{aligned} \text{Maximize } d &= 100y_1 \\ y_1 &\leq 8.00 \\ y_1 &\leq 10.00 \\ y_1 &\geq 0 \end{aligned}$$

Clearly, the supplier's optimum selling price will be $y_1^* = \$8.00/\text{yd}^3$, and his anticipated profit will be $d^* = \$800$.

In order to show a relationship between the maximum profit and minimum cost models for the seller and buyer, the maximum profit model will be written as

$$\begin{aligned} \text{Maximize } d &= 100y_1 \\ 0.6y_1 &\leq 4.80 \\ 0.4y_1 &\leq 4.00 \\ y_1 &\geq 0 \end{aligned}$$

This mathematical model, the seller's model, is defined to be the *dual.*

The optimum solutions to the buyer's and seller's models reveal interesting results. The maximum profit to the seller of pure sand equals the minimum cost to the buyer:

$$d^* = z^* = \$800.00$$

This result is a fundamental property of the dual–primal relationship and applies to all linear mathematical models. At the optimum, the *maximum solution of the dual equals the minimum solution of the primal,* or

$$d^* = z^*$$

If an optimum basic solution of the dual exists, then an optimum basic feasible solution of the primal must exist. If a solution of the dual does not exist, then no solution will exist for the primal.

Throughout this discussion, we have assumed that only two suppliers of sand exist. The sand supplier is not actually in the market! The mathematical model that is established for him is a planning tool that was used to estimate *anticipated revenues* if he enters the market. Since no sales transaction between the sand supplier and the contractor has taken place, the contractor will purchase 167 yd^3 of soil from source 1 and extract 100 yd^3 of pure sand from it. In essence, the seller of pure sand never existed in the mind of the buyer. The buyer, however, can use the dual solution to give her new economic insights. The shadow price y_1^* is one such entity.

Shadow Price

The *shadow price* is defined to be unit cost or the ratio of the change in total cost Δz to the change in a resource allocation:

$$\text{MC}_i = \frac{\Delta z}{\Delta b_i}$$

where i represents the ith resource constraint of $\boldsymbol{b}$. The shadow price of pure sand is \$8.00, or $y_1^* = \text{MC}_1 = \$8.00/\text{yd}^3$. If the buyer requires 10 additional cubic yards of pure sand, $\Delta b_1 = 10$, the additional cost is estimated determined to be

$$\Delta z = y_1^* \Delta b_1$$

$$\Delta z = \$8.00 \cdot 10 = \$80.00$$

The buyer's total purchase is estimated to be \$880.00, $z = z^* + \Delta z$. The shadow price is only used to estimate relatively small changes in the resource constraint. If relatively large changes in Δb_1 are made, it is possible that the estimate Δz may be wrong because the location of the optimum extreme point x^* may shift to a new location. Therefore, care must be taken in making this estimate. It might be helpful to the reader to review Figure 7.2 and the appropriate material in Section 7.1 on changes in $\boldsymbol{b}$.

We shall utilize the primal or buyer's model to calculate the change in the total cost z for an increase of 10 yd^3 extra of pure sand. Normally, this calculation is not performed if the shadow price is known. Here, we demonstrate that the buyer's total purchase as estimated with the shadow price is correct. We may determine the amount of soil she will purchase with the constraint resource equal to 110 instead of 100. Since she purchased all soil from source 1, her total purchase is

$$0.6x_1 + 0.4(0) = 110$$

$$x_1 = \$183.33$$

The total cost is

$$z = \$4.80(183.33) + \$4.00(0) = \$880.00$$

The buyer may use the shadow price to aid her in deciding where to purchase soil at minimum cost. Since the unit costs of sand from sources 1 and 2 are \$8.00 and 10.00 yd^3, respectively, she should purchase from source 1. The shadow price is equal to the minimum unit production cost of sand from source 1. Suppose that the supplier of sand decides to enter the market and sell pure sand at \$7.80/yd^3. The selling price of pure sand is less than the shadow price of \$8.00/yd^3; therefore, the contractor should purchase it from the supplier of pure sand. The total cost of this purchase is \$780.00 instead of \$800.00. Clearly, if the seller establishes his selling price of sand at \$8.00/yd^3 or more, there is no saving to the buyer, and she may continue to purchase from source 1 at a total cost of \$800.00.

The same interpretation of shadow price applies to the multidimensional linear mathematical model. When a primal model has m constraint equations, $\boldsymbol{Ax} = \boldsymbol{b}$, there will be m shadow prices. Each one of them can be associated with an element of the $\boldsymbol{b}$ vector and interpreted in terms of the context of the problem.

Dual–Primal Transformation

The primal and dual models in matrix form are

Primal

$$\text{Minimize } z = [4.80 \quad 4.00] \cdot \begin{bmatrix} x_1 \\ x_2 \end{bmatrix}$$

$$[0.6 \quad 0.4] \cdot \begin{bmatrix} x_1 \\ x_2 \end{bmatrix} \geq 100$$

$$\begin{bmatrix} x_1 \\ x_2 \end{bmatrix} \geq 0$$

and

Dual

$$\text{Maximize } d = [100] \cdot [y_1]$$

$$\begin{bmatrix} 0.6 \\ 0.4 \end{bmatrix} \cdot [y_1] \leq \begin{bmatrix} 4.80 \\ 4.00 \end{bmatrix}$$

$$[y_1] \geq 0$$

A pattern exists between the primal and dual formulations. The $\boldsymbol{c}$ and $\boldsymbol{b}$ vectors and $\boldsymbol{A}$ matrix of the models are the same for both models.

$$\boldsymbol{c}' = [4.80 \quad 4.00]$$

$$\boldsymbol{b}' = [100]$$

$$\boldsymbol{A}' = [0.6 \quad 0.4]$$

This pattern exists for all primal and dual models. This relationship is summarized by the following mathematical models:

Primal

$$\text{Minimize } z = \boldsymbol{c}'\boldsymbol{x}$$

$$\boldsymbol{Ax} \geq \boldsymbol{b}$$

$$\boldsymbol{x} \geq \boldsymbol{0}$$

Dual

$$\text{Maximize } d = \boldsymbol{b}'\boldsymbol{y}$$

$$\boldsymbol{A}'\boldsymbol{y} \leq \boldsymbol{c}$$

$$\boldsymbol{y} \geq \boldsymbol{0}$$

In order to establish a dual from a primal problem, it is necessary that the primal problem be a minimization problem with greater-than-or-equal-to constraints ($\geq$). The transpose of the primal must be a maximization problem with less-than-or-equal-to constraints ($\leq$) as required as shown. The optimum dual solution may be obtained by transforming it from a maximization problem to a minimization problem and by utilizing the simplex method.

Optimum Primal and Dual Solutions

Since the primal and dual models are linear transformations of one another, the optimum solution of both the primal $\boldsymbol{x}^*$ and dual $\boldsymbol{y}^*$ are obtained either from the final tableau of the primal or of the dual solutions. When the optimum solution of the dual $\boldsymbol{y}^*$ is obtained, the optimum solution of the primal $\boldsymbol{x}^*$ can be determined from the indicator row of the final simplex tableau of the dual model. See Table 7.1. Likewise, when the optimum solution of the primal $\boldsymbol{x}^*$ is determined, the optimum dual solution $\boldsymbol{y}^*$ is obtained from the indicator row of the final simplex tableau of the primal model. (See Table 7.2.)

The use of the seller's model introduces slack variables y_2 and y_3 into the dual

$$\text{Minimize } a = -100y_1$$

$$0.6y_1 + y_2 \qquad = \$4.80/\text{yd}^3$$

$$0.4y_1 \qquad + y_3 = \$4.00/\text{yd}^3$$

$$y_1 \geq 0 \quad \text{Shadow price}$$

$$y_2 \geq 0 \quad \text{Opportunity cost}$$

$$y_3 \geq 0 \quad \text{Opportunity cost}$$

Table 7.1 Optimum Solution of the Dual Model

BASIS	SHADOW PRICES	OPPORTUNITY COSTS	UNIT COSTS
	$y_1 \cdots y_m$	$y_{m+1} \cdots y_{m+n}$	
$\boldsymbol{y}_B^*$	$\boldsymbol{a}_1^* \cdots \boldsymbol{a}_m^*$	$\boldsymbol{a}_{m+1}^* \cdots \boldsymbol{a}_{m+n}^*$	$\boldsymbol{c}^*$
	$b_1^* \cdots b_m^*$	$b_{m+1}^* \cdots b_{m+n}^*$	$(d - d^*)$
	$x_{n+1}^* \cdots x_{n+m}^*$	$x_1^* \cdots x_n^*$	
	Slack and surplus variables	Primal variables	
	↓	↓	
Optimum Solution	$x_{n+1}^* = b_1^*$	$x_1^* = b_{m+1}^*$	
	$\vdots$	$\vdots$	
	$x_{n+m}^* = b_m^*$	$x_n^* = b_{m+n}^*$	

Table 7.2 Optimum Solution of the Primal Model

	PRIMAL VARIABLES			SLACK AND SURPLUS VARIABLES			RESOURCE CONSTRAINT
BASIS	x_1	$\cdots$	x_n	x_{n+1}	$\cdots$	x_{n+m}	
$\mathbf{x}_B^*$	$\mathbf{a}_1^*$	$\cdots$	$\mathbf{a}_n^*$	$\mathbf{a}_{n+1}^*$	$\cdots$	$\mathbf{a}_{n+m}^*$	$\mathbf{b}^*$
	c_1^*	$\cdots$	c_n^*	c_{n+1}^*	$\cdots$	c_{n+m}	$(z - z^*)$
	y_{m+1}^*	$\cdots$	y_{m+n}^*	y_1^*	$\cdots$	y_m^*	
	Opportunity costs			Shadow prices			
Optimum Solution	$y_{m+1}^* = c_1^*$			$y_1^* = c_{n+1}^*$			
	$\vdots$			$\vdots$			
	$y_{m+n}^* = c_n^*$			$y_m^* = c_{n+m}^*$			

The objective function of the dual has been rewritten as a minimization objective, $a = -100y_1$. Utilizing the augmented matrix and the rules of the simplex method we shall find the optimum dual solution. The initial augmented matrix is

$\mathbf{y}_B$	y_1	y_2	y_3	c	c/a
y_2	(0.60)	1.00	0.00	4.80	$4.80/0.6 = 8$ ←
y_3	0.40	0.00	1.00	4.00	$4.00/0.4 = 10$
	−100.0 ↑	0.00	0.00	$(a - 0)$	

After one iteration, the optimum dual solution is found.

$\mathbf{y}_B$	y_1 (Shadow price)	y_2 (Opportunity cost)	y_3 (Opportunity cost)	c	c/a
y_1	1.00	0.17	0.00	8.00	
y_3	0.00	−0.07	1.00	0.80	
	0.00	166.67	0.00	$(a + 800)$	
	↓ Surplus variable x_3	↓ Primal variables x_1	↓ Primal variables x_2		

The optimum dual solution is:

$$y_1^* = 8.00 \qquad y_2^* = 0.00 \qquad y_3^* = 0.80 \qquad a^* = -800$$

Since $d = (-1)a$, $d^* = \$800.00$.

The **x** variables are found in the indicator row. The primal variables x_1 and x_2 are located in indicator row under the slack variables columns of y_2 and y_3, thus $x_1^* = 166.67$ and $x_2^* = 0$. The surplus variable x_3 is located in the indicator row under the dual variable column y_1, thus, $x_3^* = 0$.

A check of this result will be determined by investigating the solution of the primal. A surplus variable x_3 is introduced:

$$\text{Minimize } z = 4.80x_1 + 4.00x_2$$
$$0.6x_1 + 0.4x_2 - x_3 = 100$$
$$x_1 \geq 0, x_2 \geq 0, x_3 \geq 0$$

An artificial variable x_4 is introduced, and an optimum solution is obtained by using the two-phase simplex method. The two-phase simplex model, where $w = x_4$, is

$$\text{Minimize } z = 4.80x_1 + 4.00x_2$$
$$\text{Minimize } (w - 100) = -0.6x_1 - 0.4x_2 + x_3$$
$$0.6x_1 + 0.4x_2 - x_3 + x_4 = 100$$
$$x_1 \geq 0, x_2 \geq 0, x_3 \geq 0, x_4 \geq 0$$

The simplex tableau is

$\mathbf{x}_B$	x_1	x_2	x_3	x_4	b	b/a	
x_3	(0.6)	0.4	−1	1	100	100/0.6 = 167	←
	4.8	4.0	0	0	$(z - 0)$		
	−0.6	−0.4	1	0	$(w + 100)$		
	↑						

The artificial variable x_4 is introduced to obtain a solution by the two-phase simplex method. Since it is an artificial variable, there is no corresponding variable in the primal model to which it can be related. After one iteration, the final simplex tableau is obtained:

$\mathbf{x}_B$	x_1	x_2	x_3	x_4	b	b/a
x_1	1	0.667	−1.67	1.67	167	
$\mathbf{c}_I$	0	0.800	8.00	−8.00	$(z - 800)$	
$\mathbf{d}_I$	0	0	0	1	$(w - 0)$	

This is the same result that was obtained from the dual solution. The optimum solution is

$$z^* = \$800.00$$
$$x_1^* = 167 \qquad x_2^* = 0, \quad x_3^* = 0$$

We ignore the artificial column of x_4 and the $\boldsymbol{d}_I$ row; they have no physical meaning

	Primal variables		Surplus variable	
	x_1	x_2	x_3	b
x_1	1	0.667	1.67	1.67
c_I	0	0.800	8.00	$(z - 800)$
Shadow price			↓ y_1^*	
Opportunity cost	↓ y_2^*	↓ y_3^*		

The shadow price is found in the indicator row under the surplus variable x_3, thus $y_1^* = 8.00$. The opportunity costs are located under the primal variable columns x_1 and x_2. Thus $y_2^* = 0$ and $y_3^* = 0.80$.

Opportunity Cost

In Chapter 3 the opportunity cost was described to be the inability for an investor who has limited capital to become involved in all investment opportunities. With the use of engineering economic methods, the so-called best alternative was selected, or, put another way, the opportunity cost was minimized. Only the best investments are considered viable alternatives. Here, the same principles are employed.

The *opportunity cost for resource allocation* problems is defined to be the unit cost or penalty for substituting a nonoptimum resource for an optimum one. It is defined as

$$\mathrm{MC}_j = \frac{\Delta z}{\Delta x_j}$$

where x_j is a representative resource variable from the nonbasic vector $\boldsymbol{x}_N$. The effects of utilizing nonoptimum resources will be demonstrated by use of the construction job example.

The optimum solution is $x_1^* = 167$, $x_2^* = 0$, and $x_3^* = 0$. If the minimum amount of sand is needed, the following equality constraint must be satisfied:

$$0.6x_1 + 0.4x_2 + x_3 = 100$$

Let us substitute 10 yd^3 of material from source 2 and determine the amount of soil from source 1.

$$0.6x_1 + 0.4(10) = 100$$

or

$$x_1 = 160 \text{ yd}^3$$

The total cost for this allocation is

$$z = \$4.80x_1 + \$4.00x_2$$

$$z = \$4.80(160) + 4.00(10) = \$808.00$$

Since the extra cost Δz is $\Delta z = z - z^* = \$808 - \$800 = \$8.00$. Thus, a penalty of \$8.00 is paid for using nonoptimum material.

By definition, opportunity cost is the unit cost

$$\frac{\Delta z}{\Delta x_2} = \frac{z - z^*}{\Delta x_2}$$

$$\frac{\Delta z}{\Delta x_2} = \frac{\$8.00}{10} = \$0.80/\text{yd}^3$$

The same result is obtained directly from the dual. We find that

$$y_2^* = \frac{\Delta z}{\Delta x_1} = \$0.00/\text{yd}^3$$

$$y_3^* = \frac{\Delta z}{\Delta x_2} = \$0.80/\text{yd}^3$$

Since $y_2^* = 0$, the first constraint equation is inactive, implying that x_1 is a member of basic solution. No penalty is paid to introduce more material from source 1 into the optimum. Since $y_3^* \neq 0$, the second constraint is inactive, implying that x_2 is not a member of the basis. For every cubic yard of material added from source 2, we pay a penalty of \$0.80/yd.^{3} As previously shown, to substitute 10 yd^3 of material 2 we pay a penalty of \$8.00. Using the definition of opportunity cost we obtain the same result.

$$\Delta z = y_3^* \Delta x_2 = (\$0.80/\text{yd}^3)(10) = \$8.00$$

Sensitivity Analysis and the Change in Resource Constraints

Since the transformation of the primal to the dual has placed the resource constraint parameters $\boldsymbol{b}$ in the objective function of the dual, we may employ the same approach for determining the effect of changing the resource constraints $\boldsymbol{b}$ in the dual formulation that we use for determining the effect of changing the unit costs $\boldsymbol{c}$ in the primal formulation of Section 7.1.

The procedure for determining the optimum solution and performing a sensitivity analysis with changes in $\boldsymbol{b}$ is as follows.

The original and new mathematical models are

Original	New
Minimize $z = \boldsymbol{c}'\boldsymbol{x}$	Minimize $z = \boldsymbol{c}'\boldsymbol{x}$
$\boldsymbol{A}\boldsymbol{x} \geq b$	$\boldsymbol{A}\boldsymbol{x} \geq \hat{\boldsymbol{b}}$
$\boldsymbol{x} \geq \boldsymbol{0}$	$\boldsymbol{x} \geq \boldsymbol{0}$

Since the constraint sets are not strict equalities, these models are not in standard linear programming form.

STEP 1 Establish a primal model with the following model form:

$$\text{Minimize } z = \boldsymbol{c}'\boldsymbol{x}$$
$$\boldsymbol{Ax} \geq \boldsymbol{b}$$
$$\boldsymbol{x} \geq \boldsymbol{0}$$

STEP 2 Establish a dual model utilizing the following model form:

$$\text{Maximize } d = \boldsymbol{b}'\boldsymbol{y}$$
$$\boldsymbol{A}'\boldsymbol{y} \leq \boldsymbol{c}$$
$$\boldsymbol{y} \geq \boldsymbol{0}$$

STEP 3 Formulate the dual model in the standard form of linear programming:

$$\text{Minimize } a = \boldsymbol{b}'\boldsymbol{y}$$
$$\boldsymbol{A}'\boldsymbol{y} = \boldsymbol{c}$$
$$\boldsymbol{y} \geq \boldsymbol{0}$$

The objective function has been transformed to a minimization objective, $a = -1d$. All elements of the cost constraint vector must be positive, $\boldsymbol{c} \geq 0$. Thus, all equations containing a negative c value must be multiplied by minus one and the inequality sign changed from $\leq$ to $\geq$. Slack and surplus variables are used to transform the constraint set from an inequality to equality constraint set.

$$\boldsymbol{A}'\boldsymbol{y} \{\leq, \geq\} \boldsymbol{c} \Rightarrow \boldsymbol{Ay} = \boldsymbol{c}$$

STEP 4 Solve the optimum solution of the dual by use of the simplex method. Use Table 7.1 to help identify $\boldsymbol{y}^*$ and $\boldsymbol{x}^*$.

STEP 5 Perform the sensitivity analysis for changes in $\boldsymbol{b}$ with the following relationships.

$$\boldsymbol{a}^0 = \boldsymbol{b}'_B \boldsymbol{c}^*_B$$

and

$$\hat{\boldsymbol{b}}'_I = [\hat{\boldsymbol{b}}'_N - \boldsymbol{b}'_B \boldsymbol{A}^*_N]$$

where $\hat{\boldsymbol{b}}$ are the resource constraints used in the new model.

For sensitivity analysis, these equations in step 5 are utilized in a manner similar to the approach used to analyze the change of $\boldsymbol{c}$ in Section 7.1. Once a^* is obtained, z^* may be calculated as $z^* = (-1)a^*$.

EXAMPLE 7.3 Changes in Resource Constraint

The following two-dimensional linear mathematical model is given:

$$\text{Minimize } z = -1x_1 - 2x_2$$

$x_1 + x_2 \leq 2$	Constraint 1
$x_1 \leq 2$	Constraint 2
$x_2 \leq 3$	Constraint 3

$$x_1 \geq 0, x_2 \geq 0$$

(a) Formulate the dual and determine the optimum solution, x^* and y^*, by the simplex method.
(b) With the use of the shadow prices and opportunity costs, estimate the change in z for a change in the resource of the first constraint equation from 2 to 2.5: $x_1 + x_2 \leq 2$ to $x_1 + x_2 \leq 2.5$. Estimate the penalty for introducing a unit of the nonoptimum variable into the optimum.
(c) Use the dual formulation to determine if the optimum solution x^* and y^* from part a will change when the resource constraint b_3 is reduced from 3 to $\frac{3}{4}$: $x_2 \leq 3$ to $x_2 \leq \frac{3}{4}$.
(d) On a graph determine if the change in b_2, part c, will affect the feasible region and the location of the optimum solution x^*. Compare this solution with the solutions from parts a and c.

Solution

(a) To formulate the dual properly, we must establish the primal model in the form:

$$\text{Minimize } z = \boldsymbol{c}'\boldsymbol{x}$$

$$\boldsymbol{Ax} \geq \boldsymbol{b}$$

$$\boldsymbol{x} \geq \boldsymbol{0}$$

Since all constraint equations contain less-than-or-equal-to constraints ($\leq$), they are transformed to greater-than-or-equal constraints ($\geq$) by multiplying each equation by (-1). The primal model written in matrix notation is

$$\text{Minimize } z = [-1 \quad -2] \cdot \begin{bmatrix} x_1 \\ x_2 \end{bmatrix}$$

$$\begin{bmatrix} -1 & -1 \\ -1 & 0 \\ 0 & -1 \end{bmatrix} \cdot \begin{bmatrix} x_1 \\ x_2 \end{bmatrix} \geq \begin{bmatrix} -2 \\ -1 \\ -3 \end{bmatrix}$$

$$\begin{bmatrix} x_1 \\ x_2 \end{bmatrix} \geq \begin{bmatrix} 0 \\ 0 \end{bmatrix}$$

The matrix $\boldsymbol{A}$ and vectors $\boldsymbol{b}$ and $\boldsymbol{c}$ are

$$\boldsymbol{A} = \begin{bmatrix} -1 & -1 \\ -1 & 0 \\ 0 & -1 \end{bmatrix} \qquad \boldsymbol{c}' = [-1 \quad -2] \qquad \boldsymbol{b}' = [-2 \quad -1 \quad -3]$$

The dual is

$$\text{Maximize } d = \boldsymbol{b}'\boldsymbol{y}$$

$$\boldsymbol{A}'\boldsymbol{y} \leq \boldsymbol{c}$$

$$\boldsymbol{y} \geq \boldsymbol{0}$$

or

$$\text{Maximize } d = [-2 \quad -1 \quad -3] \cdot \begin{bmatrix} y_1 \\ y_2 \\ y_3 \end{bmatrix}$$

$$\begin{bmatrix} -1 & -1 & 0 \\ -1 & 0 & -1 \end{bmatrix} \cdot \begin{bmatrix} y_1 \\ y_2 \\ y_3 \end{bmatrix} \leq \begin{bmatrix} -1 \\ -2 \end{bmatrix}$$

$$\begin{bmatrix} y_1 \\ y_2 \\ y_3 \end{bmatrix} \geq \begin{bmatrix} 0 \\ 0 \\ 0 \end{bmatrix}$$

To utilize the simplex method, the objective function must be transformed to a minimization objective function and all elements of the constraint vector $\boldsymbol{c}$ must be positive. Thus, the objective function is multiplied by (-1), or $a = -1 \cdot d$. Each constraint equation is multiplied by -1 to remove the negative values from the $\boldsymbol{c}$ vector. The dual model is

$$\text{Minimize } a = [2 \quad 1 \quad 3] \cdot \begin{bmatrix} y_1 \\ y_2 \\ y_3 \end{bmatrix}$$

$$\begin{bmatrix} 1 & 1 & 0 \\ 1 & 0 & 1 \end{bmatrix} \cdot \begin{bmatrix} y_1 \\ y_2 \\ y_3 \end{bmatrix} \geq \begin{bmatrix} 1 \\ 2 \end{bmatrix}$$

$$\begin{bmatrix} y_1 \\ y_2 \\ y_3 \end{bmatrix} \geq \begin{bmatrix} 0 \\ 0 \\ 0 \end{bmatrix}$$

Surplus variables y_4 and y_5 are introduced to each constraint equation. Artificial variables y_6 and y_7 must be used as the initial basic feasible solution. The artificial objective function will be

$$\text{Minimize } w = y_6 + y_7$$

The simplex tableau is

Initial Solution

$\boldsymbol{y}_B$	y_1	y_2	y_3	y_4	y_5	y_6	y_7	$\boldsymbol{c}$	b/a
y_6	①	1	0	-1	0	1	0	1	$1/1 = 1$ ←
y_7	1	0	1	0	-1	0	1	2	$2/1 = 2$
$\boldsymbol{b}_I$	2	1	3	0	0	0	0	$(a - 0)$	
$\boldsymbol{d}_I$	-2	-1	-1	1	1	0	0	$(w - 3)$	
	↑								

The rules of the two-phase simplex method are utilized. After two iterations the optimum solution is obtained.

Final Simplex Tableau

	Shadow prices			Opportunity costs		Artificial variables		
$\boldsymbol{y}_B$	y_1	y_2	y_3	y_4	y_5	y_6	y_7	$\boldsymbol{c}$
y_1	1	0	1	0	-1	0	1	2
y_4	0	-1	1	1	-1	-1	1	1
$\boldsymbol{b}_I$	0	1	1	0	2	0	-2	$(a - 4)$
$\boldsymbol{d}_I$	0	0	0	0	0	1	1	$(w - 0)$
Primal variables				x_1	x_2			
Slack variables	x_3	x_4	x_5					

Since the indicator $c_7 = -2$ is associated with an artificial variable and is not considered as an entry into the basis, the optimum solution is obtained.

The optimum solution is $z^* = -4$. The optimum $\boldsymbol{x}^*$ is obtained from the indicator of the dual solution:

$$\boldsymbol{x}^* = \begin{bmatrix} x_1^* \\ x_2^* \\ x_3^* \\ x_4^* \\ x_5^* \end{bmatrix} = \begin{bmatrix} 0 \\ 2 \\ 0 \\ 1 \\ 1 \end{bmatrix}$$

The shadow prices and opportunity costs are

$$\boldsymbol{y}^* = \begin{bmatrix} y_1^* \\ y_2^* \\ y_3^* \\ y_4^* \\ y_5^* \end{bmatrix} = \begin{bmatrix} 2 \\ 0 \\ 0 \\ 1 \\ 0 \end{bmatrix}$$

(b) Since $y_1^* = \Delta z/\Delta b_1 = 2$, the resource $x_1 + x_2 \leq 2$ is an active constraint. The change from the resource constraint of b_1 or $x_1 + x_2 \leq 2$ to $x_1 + x_2 \leq 2.5$ will affect z.

$$\Delta z = y_1^* \, \Delta b_1$$

Extreme care must be used by calculating Δz. The dual–primal relationship as shown in steps 1 and 2 of our procedure is strict. We must be particularly careful with inequality signs. Thus, the $x_1 + x_2 \leq 2$ and $x_1 + x_2 \leq 2.5$ will be rewritten as:

$$-x - x_2 \geq -2 \quad \text{and} \quad -x_1 - x_2 \geq -2.5$$

Thus
$$\Delta b = -2.5 - (-2.0) = -0.5$$

and
$$\Delta z = 2(-0.5) = -1$$

If this change is not made, the sign of Δz will be incorrect.

(c) The change of the $\boldsymbol{b}$ will be investigated with the following relationships.

$$a^0 = \hat{\boldsymbol{b}}'_B \boldsymbol{c}_B^*$$

$$\hat{\boldsymbol{b}}'_I = \hat{\boldsymbol{b}}'_N - \hat{\boldsymbol{b}}'_B \boldsymbol{A}'^*_N$$

and the minimization model of part a. The matrix $\boldsymbol{A}_N^*$ is equal to

$$\boldsymbol{A}_N^* = \begin{bmatrix} a_{12} & a_{13} & a_{15} \\ a_{42} & a_{43} & a_{45} \end{bmatrix} = \begin{bmatrix} 0 & 1 & -1 \\ -1 & 1 & -1 \end{bmatrix}$$

Since the artificial variables have no physical or economic meaning, the vectors $\boldsymbol{a}_6$ and $\boldsymbol{a}_7$ are not included in $\boldsymbol{A}'^*_N$. The values of the original $\boldsymbol{b}'_B$ and $\boldsymbol{b}'_N$ vectors are

$$\hat{\boldsymbol{b}}_N = [\hat{b}_2 \quad \hat{b}_3 \quad \hat{b}_5] = [1 \quad 3 \quad 0]$$

$$\hat{\boldsymbol{b}}_B = [\hat{b}_1 \quad \hat{b}_4] \quad = [2 \quad 0]$$

Since b_3 is changed from 3 to $\frac{3}{4}$ the $\boldsymbol{b}'_B$ vector becomes:

$$\hat{\boldsymbol{b}}'_I = [\hat{b}_2 \quad \hat{b}_3 \quad \hat{b}_5] = [1 \quad \tfrac{3}{4} \quad 0]$$

Substituting into $\hat{\boldsymbol{b}}'_I = \hat{\boldsymbol{b}}'_N - \hat{\boldsymbol{b}}'_B \boldsymbol{A}^*_N$ gives the new indicators:

$$\hat{\boldsymbol{b}}'_I = [\hat{b}_2 \quad \hat{b}_3 \quad \hat{b}_5] = [1 \quad -\tfrac{5}{4} \quad 2]$$

$$a^0 = [\hat{b}_1 \quad \hat{b}_4] \cdot \begin{bmatrix} c_1^* \\ c_4^* \end{bmatrix} = [2 \quad 0] \cdot \begin{bmatrix} 2 \\ 0 \end{bmatrix} = 4$$

The simplex tableau written in terms of dual variables with no artificial variables is

$$\begin{array}{c|cccccc|c} \boldsymbol{y}_B & y_1 & y_2 & y_3 & y_4 & y_5 & \boldsymbol{c} & c/a \\ \hline y_1 & 1 & 0 & 1 & 0 & -1 & 2 & 2/1 = 2 \\ y_4 & 0 & -1 & 1 & 1 & -1 & 1 & 1/1 = 1 \leftarrow \\ \boldsymbol{b}_I & 0 & 1 & -\tfrac{5}{4} & 0 & 2 & (a-4) & \\ & & & \uparrow & & & & \end{array}$$

The rules of the simplex method are utilized until an optimum solution is obtained. After two iterations, the optimum solution is obtained.

Final Dual Tableau

	Shadow prices			Opportunity costs		
$\boldsymbol{y}_B$	y_1	y_2	y_3	y_4	y_5	$\boldsymbol{c}$
y_2	1	1	0	-1	0	1
y_3	1	0	1	0	-1	2
$\boldsymbol{b}_I$	$\frac{1}{4}$	0	0	1	$\frac{3}{4}$	$(a - \frac{10}{4})$
Primal variables	↓	↓	↓	↓ x_1	↓ x_2	
Slack variables	x_3	x_4	x_5			

The optimum solution is $a^* = 10/4 = 2.5$, or $z^* = -a^* = 2.5$ where

$$\boldsymbol{x}^* = \begin{bmatrix} x_1^* \\ x_2^* \\ x_3^* \\ x_4^* \\ x_5^* \end{bmatrix} = \begin{bmatrix} 1 \\ \frac{3}{4} \\ \frac{1}{4} \\ 0 \\ 0 \end{bmatrix} \quad \text{and} \quad \boldsymbol{y}^* = \begin{bmatrix} y_1^* \\ y_2^* \\ y_3^* \\ y_4^* \\ y_5^* \end{bmatrix} = \begin{bmatrix} 0 \\ 1 \\ 2 \\ 0 \\ 0 \end{bmatrix}$$

(d) The optimum solution for part a is located at (0, 2) as shown in Figure 7.6 as $z^* = -4$. The constraint $x_2 \leq 3$ does not form the boundary of the feasible solution. Reducing the resource constraint from $x_1 \leq 3$ to $x_2 < \frac{3}{4}$ changes the shape of the feasible region. The optimum solution with new constraint is located at $(1, \frac{3}{4})$ and is shown as $z^* = -2.5$. These results agree with the solutions obtained in parts a and c.

Remarks

The shadow price may be used to estimate change in the objective function Δz for small changes in $\Delta \boldsymbol{b}$. From part a, the constraint $x_2 \leq 3$ is inactive, and $y_3^* = 0$. If we estimate Δz for part d, an incorrect estimate is obtained:

$$\Delta z = y_3^* \Delta b_3 = 0$$

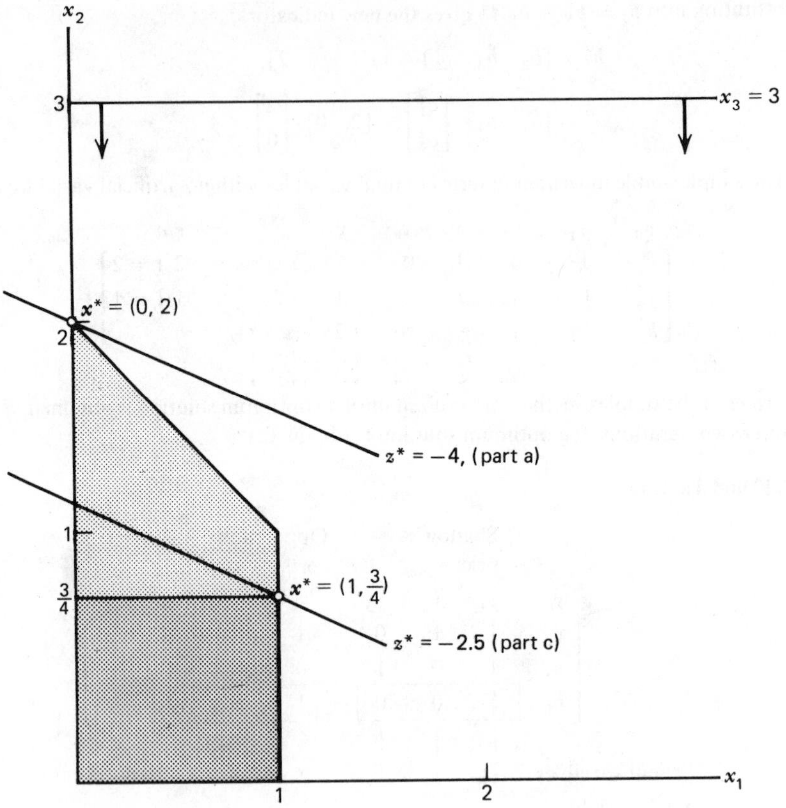

Figure 7.6

The actual change in z is $\Delta z = -2.5 - (-4.0) = 1.5$ as shown in the graph. In this case, even for small changes in $\Delta \boldsymbol{b}$, the estimate is incorrect. We can show that the basic feasible solution or extreme points have changed; thus the shadow price estimate will give an incorrect result.

When $x_2 \leq 3$, this equation lies outside the feasible region; therefore, points along this line will never become members of the basic feasible solution $\boldsymbol{x}_B$. For the optimum solution (0, 2), the active constraint is $x_1 + x_2 \leq 2$, and the inactive constraints are $x_1 \leq 1$ and $x_2 \leq 3$. When $x_2 \leq \frac{3}{4}$, the optimum solution is located at $(1, \frac{3}{4})$, the active constraints are $x_1 \leq 1$ and $x_2 \leq \frac{3}{4}$ and the inactive constraint $x_1 + x_2 \leq 2$. Comparing these two solutions shows that the optimum solution has shifted from one extreme point to another one.

EXAMPLE 7.4 Roadway Aggregate Mix Model: Limited Resources

The supply of sand for the roadway aggregate mix, Example 7.2, has been estimated to be as much as 6000 yd^3, or as little as 4000 yd^3. Determine the effect of limited sand supply upon the minimum cost solution. The aggregate must satisfy specification requirements as originally stated in Example 7.2. Utilize the original unit costs for material given in the statement of Example 7.2.

(a) Formulate the problem as a primal model with a supply limitation of 6000 yd^3 of sand. Formulate the dual and show that the dual optimum solution is the same as solution given in part a of Example 7.2.
(b) Determine the minimum cost solution for a supply limitation of 4000 yd^3 of sand.
(c) Determine the additional cost of substituting nonoptimum materials into the optimum solutions determined in parts a and b.
(d) Estimate the additional cost to supply 10 yd^3 of aggregate with 6000 yd^3 of sand and 4000 yd^3 of sand.

Solution

(a) The linear mathematical model formulated in part a of Example 7.2 is the primal model. The limited supply of sand is introduced to the model as the following constraint equation:

$$x_3 \le 6$$

The model becomes

$$\text{Minimize } z = 6x_1 + 8x_2 + 5x_3$$

$$\begin{array}{lr}
x_1 + x_2 + x_3 + x_4 = 9 & \text{Supply} \\
0.15x_1 + 0.3x_3 + 0.05x_4 \le 1.8 & \text{Gravel} \\
0.15x_1 + 0.3x_2 + 0.05x_4 \ge 0.45 & \text{Gravel} \\
0.75x_1 + 0.7x_2 + x_3 + 0.3x_4 \le 7.65 & \text{Sand} \\
0.75x_1 + 0.7x_2 + x_3 + 0.3x_4 \ge 5.85 & \text{Sand} \\
0.1x_1 + \quad + 0.65x_4 \le 1.35 & \text{Silt} \\
x_3 \le 6 & \text{Sand}
\end{array}$$

$$x_1, x_2, x_3, x_4 \ge 0$$

The primal requires that greater-than-or-equal-to constraints be utilized. The aggregate total-volume constraint equation, which is a strict equality constraint, will be written as two equations. One equation is written with a less-than-or-equal-to constraint; the other one is written with a greater-than-or-equal-to constraint. The equality constraint

$$x_1 + x_2 + x_3 + x_4 = 9$$

is replaced with the following two inequality constraints:

$$x_1 + x_2 + x_3 + x_4 \le 9$$

$$x_1 + x_2 + x_3 + x_4 \ge 9$$

All equations with less-than-or-equal-to constraints ($\le$) will be multiplied by (-1) to change the inequality to a greater-than-or-equal-to constraint ($\ge$). The primal model in matrix notation is

$$\text{Minimize } z = \boldsymbol{c}'\boldsymbol{x}$$

$$\boldsymbol{A}\boldsymbol{x} \ge \boldsymbol{b}$$

$$\boldsymbol{x} \ge \boldsymbol{0}$$

where: $\boldsymbol{c}' = [c_1 \;\; c_2 \;\; c_3 \;\; c_4] = [6.00 \;\; 8.00 \;\; 5.00 \;\; 0.00]$

$\boldsymbol{b}' = [b_1 \;\; b_2 \;\; b_3 \;\; b_4 \;\; b_5 \;\; b_6 \;\; b_7 \;\; b_8]$

$\boldsymbol{b}' = [-9.00 \;\; 9.00 \;\; -1.80 \;\; 0.45 \;\; -7.65 \;\; 5.85 \;\; -1.35 \;\; -6.00]$

and

$$\boldsymbol{A} = \begin{bmatrix} -1.00 & -1.00 & -1.00 & -1.00 \\ 1.00 & 1.00 & 1.00 & 1.00 \\ -0.15 & -0.30 & 0.00 & -0.05 \\ 0.15 & 0.30 & 0.00 & 0.05 \\ -0.75 & -0.70 & -1.00 & -0.30 \\ 0.75 & 0.70 & 1.00 & 0.30 \\ -0.10 & 0.00 & 0.00 & -0.65 \\ 0.00 & 0.00 & -1.00 & 0.00 \end{bmatrix}$$

The dual model is

$$\text{Maximize } d = \boldsymbol{b}'\boldsymbol{y}$$

$$\boldsymbol{A}'\boldsymbol{y} \le \boldsymbol{c}$$

$$\boldsymbol{y} \ge \boldsymbol{0}$$

or

$$\text{Maximize } d = [-9.00 \;\; 9.00 \;\; -1.80 \;\; 0.45 \;\; -7.65 \;\; 5.85 \;\; -1.35 \;\; -6.00] \cdot \boldsymbol{y}$$

$$\begin{bmatrix} -1.00 & 1.00 & -0.15 & 0.15 & -0.75 & 0.75 & -0.10 & 0.00 \\ -1.00 & 1.00 & -0.30 & 0.30 & -0.70 & 0.70 & 0.00 & 0.00 \\ -1.00 & 1.00 & 0.00 & 0.00 & -1.00 & 1.00 & 0.00 & -1.00 \\ -1.00 & 1.00 & -0.05 & 0.05 & -0.30 & 0.30 & -0.65 & 0.00 \end{bmatrix} \cdot \boldsymbol{y} \le \begin{bmatrix} 6.00 \\ 8.00 \\ 5.00 \\ 0.00 \end{bmatrix}$$

where

$$\boldsymbol{y}' = [y_1 \;\; y_2 \;\; y_3 \;\; y_4 \;\; y_5 \;\; y_6 \;\; y_7 \;\; y_8] \ge \boldsymbol{0}$$

Transform to a minimization problem.

$$\text{Minimize } a = -\boldsymbol{b}'\boldsymbol{y}$$

Since the $\boldsymbol{b}$ vector of the dual contains all positive values, the less-than-or-equal-to constraints require that slack variables be introduced to each constraint equation. They are defined to be y_9, y_{10}, y_{11}, and y_{12}.

The initial basis feasible solution will be $\boldsymbol{y}^{0'} = [y_9 \;\; y_{10} \;\; y_{11} \;\; y_{12}]$. The simplex tableau is

$\boldsymbol{y}_B$	y_1	y_2	y_3	y_4	y_5	y_6	y_7	y_8	y_9	y_{10}	y_{11}	y_{12}	$\boldsymbol{c}$
y_9	−1.00	1.00	−0.15	0.15	−0.75	0.75	−0.10	0.00	1.00	0.00	0.00	0.00	6.00
y_{10}	−1.00	1.00	−0.30	0.30	−0.70	0.70	0.00	0.00	0.00	1.00	0.00	0.00	8.00
y_{11}	−1.00	1.00	0.00	0.00	−1.00	1.00	1.00	−1.00	0.00	0.00	1.00	0.00	5.00
y_{12}	−1.00	1.00	−0.05	0.05	−0.30	0.30	−0.65	0.00	0.00	0.00	0.00	1.00	0.00
$\boldsymbol{b}_I$	9.00	−9.00	0.80	−0.45	7.65	−5.58	1.35	6.00	0.00	0.00	0.00	0.00	$(a - 0)$

After five iterations, the optimum solution is obtained. The artificial variables have been removed from the final tableau.

	Shadow prices								Opportunity costs				
y_B	y_1	y_2	y_3	y_4	y_5	y_6	y_7	y_8	y_9	y_{10}	y_{11}	y_{12}	c
y_9	0.00	0.00	0.00	0.00	0.00	0.00	0.00	0.37	1.00	−0.47	−0.37	−0.15	0.35
y_4	0.00	0.00	−1.00	1.00	1.00	−1.00	0.00	3.33	0.00	3.33	−3.33	0.00	10.00
y_7	0.00	0.00	0.00	0.00	−1.00	1.00	1.00	−1.28	0.00	0.26	1.28	−1.54	8.46
y_2	−1.00	1.00	0.00	0.00	−1.00	1.00	0.00	−1.00	0.00	0.00	1.00	0.00	5.00
b_I	0.00	0.00	1.35	0.00	0.45	1.35	0.00	0.23	0.00	1.15	5.77	2.08	$(a + 38.1)$
Primal variable									x_1	x_2	x_3	x_4	
Slack and surplus variables	x_5	x_6	x_7	x_8	x_9	x_{10}	x_{11}	x_{12}					

Since the objective function of the dual is a maximization objective function, it is transformed to a minimization function multiplying it by (−1). Thus, the objective function utilized in the simplex tableau is

$$a = -1 \cdot d$$

The optimum solution in terms of the primal control variables, x_1, x_2, x_3, and x_4 are located in the indicator row under the slack variable columns y_9, y_{10}, y_{11}, and y_{12}. The optimum solution is

$$x_1^* = 0.00 \qquad x_2^* = 1.15 \qquad x_3^* = 5.77 \quad x_4^* = 2.08 \qquad a^* = -38.1$$

Since $z = -1 \cdot a$, the optimum solution is

$$z^* = -1(-38.1) = 38.1$$

The optimum solution is summarized in the following table:

SOURCE	x_B^*	AMOUNT (yd³)	COST ($)
River gravel	—	0	0
Pea gravel	x_2	1,150	9,200
Sand	x_3	5,770	28,900
Glacial till	x_4	2,080	0

z^* = $38,100.

This solution is the same one obtained in Example 7.2.

(b) Since the optimum solution in part a requires that 5700 yd^3 of sand and only 4000 yd^3 of sand is available, the optimum solution from part a will be impacted. A new indicator will be established, and the optimum solution will be found with the final simplex tableau of part a. The new indicator row is calculated with use of

$$\hat{\boldsymbol{b}}_I' = \hat{\boldsymbol{b}}_N' - \hat{\boldsymbol{b}}_B' \boldsymbol{A}_N^*$$

The matrix A_N^* is obtained from the final tableau of part a:

$$A_N^* = \begin{bmatrix} a_{91}^* & a_{93}^* & a_{95}^* & a_{96}^* & a_{98}^* & a_{910}^* & a_{911}^* & a_{912}^* \\ a_{41}^* & a_{43}^* & a_{45}^* & a_{46}^* & a_{48}^* & a_{410}^* & a_{911}^* & a_{412}^* \\ a_{71}^* & a_{73}^* & a_{75}^* & a_{76}^* & a_{78}^* & a_{710}^* & a_{711}^* & a_{712}^* \\ a_{21}^* & a_{23}^* & a_{25}^* & a_{26}^* & a_{28}^* & a_{210}^* & a_{211}^* & a_{212}^* \end{bmatrix}$$

$$A_N^* = \begin{bmatrix} 0.00 & 0.00 & 0.00 & 0.00 & 0.37 & -0.47 & -0.37 & -0.15 \\ 0.00 & -1.00 & 1.00 & -1.00 & 3.33 & 3.33 & -3.33 & 0.00 \\ 0.00 & 0.00 & -1.00 & 1.00 & -1.28 & 0.26 & 1.28 & -1.54 \\ -1.00 & 0.00 & -1.00 & 1.00 & -1.00 & 0.00 & 1.00 & 0.00 \end{bmatrix}$$

The vectors $\hat{\boldsymbol{b}}_N$ and $\hat{\boldsymbol{b}}_B$ are

$$\hat{\boldsymbol{b}}_N' = [\hat{b}_1 \quad \hat{b}_3 \quad \hat{b}_5 \quad \hat{b}_6 \quad \hat{b}_8 \quad \hat{b}_{10} \quad \hat{b}_{11} \quad \hat{b}_{12}]$$
$$= [-9.00 \quad -1.80 \quad -7.65 \quad 5.85 \quad -4.00 \quad 0.00 \quad 0.00 \quad 0.00]$$

and $$\hat{\boldsymbol{b}}_B' = [\hat{b}_9 \quad \hat{b}_4 \quad \hat{b}_7 \quad \hat{b}_2] = [0.00 \quad 0.450 \quad -1.35 \quad 9.00]$$

Performing the matrix algebra as required gives a new indicator of

$$\hat{\boldsymbol{b}}_I' = [\hat{b}_1 \quad \hat{b}_3 \quad \hat{b}_5 \quad \hat{b}_8 \quad \hat{b}_{10} \quad \hat{b}_{11} \quad \hat{b}_{12}]$$
$$= [0.00 \quad 1.35 \quad 0.45 \quad 1.35 \quad -1.77 \quad 1.15 \quad 5.77 \quad 2.08]$$

The value of a^0 is

$$a^0 = \hat{\boldsymbol{b}}_B \boldsymbol{c}_B^*$$

The value of a^0 in this case is not changed, because the variable y_8 is not a member of $\boldsymbol{y}_B^*$ or $\boldsymbol{c}_B^*$. Thus, $\hat{a}$ is equal to -38.1 or $z^* = 38.1$. The simplex tableau is

$\boldsymbol{y}_B$	y_1	y_2	y_3	y_4	y_5	y_6	y_7	y_8	y_9	y_{10}	y_{11}	y_{12}	$\boldsymbol{c}$
y_9	0.00	0.00	0.00	0.00	0.00	0.00	0.00	0.37	1.00	−0.47	−0.37	0.15	0.35
y_4	0.00	0.00	−1.00	1.00	1.00	−1.00	0.00	3.33	0.00	3.33	−3.33	0.00	10.00
y_7	0.00	0.00	0.00	0.00	−1.00	1.00	1.00	−1.28	0.00	0.26	1.28	−1.54	8.46
y_2	−1.00	1.00	0.00	0.00	−1.00	1.00	0.00	−1.00	0.00	0.00	1.00	0.00	5.00
$\boldsymbol{b}_I$	0.00	0.00	1.35	0.00	0.45	1.35	0.00	−1.77	0.00	1.15	5.77	2.08	$(a + 38.1)$

After two iterations the optimum solution is obtained

$\boldsymbol{y}_B$	y_1	y_2	y_3	y_4	y_5	y_6	y_7	y_8	y_9	y_{10}	y_{11}	y_{12}	$\boldsymbol{c}$
y_8	0.00	0.00	−0.17	0.17	0.17	−0.17	0.00	1.00	1.18	0.00	−1.00	−0.18	2.09
y_{10}	0.00	0.00	−0.13	0.13	0.13	−0.13	0.00	0.00	−1.18	1.00	0.00	0.18	0.91
y_7	0.00	0.00	−0.18	0.18	−0.82	0.82	1.00	0.00	1.82	0.00	0.00	−1.82	10.91
y_2	−1.00	1.00	−0.17	0.17	−0.83	0.83	0.00	0.00	1.18	0.00	0.00	−0.18	7.09
$\boldsymbol{b}_I$	0.00	0.00	1.21	0.15	0.60	1.21	0.00	0.00	3.46	0.00	4.00	1.55	$(a + 40.7)$
Primal variables									x_1	x_2	x_3	x_4	
Slack and surplus variables	x_5	x_6	x_7	x_8	x_9	x_{10}	x_{11}	x_{12}					

The optimum solution is

$$x_1^* = 3.46 \quad x_2^* = 0.00 \quad x_3^* = 4.00 \quad x_4^* = 1.55 \quad a^* = -40.7$$

or

$$z^* = -1(-40.7) = \$40.70$$

The following purchase is recommended when 4000 yd^3 of sand is available:

SOURCE	x_B^*	AMOUNT (yd^3)	COST ($)
River gravel	x_1	3,460	20,700
Pea gravel	—	0	0
Sand	x_3	4,000	20,000
Glacial till	x_4	1,550	0

z^* = $40,700

Thus, if only 4000 yd^3 of sand is available, an additional $2600 is needed for purchasing aggregate material. See part (a), summary table. With this strategy, river gravel but no pea gravel is purchased; whereas when 5770 yd^3 of sand is available, pea gravel and no river gravel is purchased. The unit cost of this mix is $4.52/yd^3. The mix of part a has a unit cost of $4.23.

(c) The additional cost of substituting nonoptimum material for optimum material is the opportunity cost. The opportunity costs for the optimum solutions found in parts a and b can be determined from the final simplex tableaus. For part a the opportunity or penalty cost for utilizing river gravel in the mix is

$$y_9^* = MC_1 = \frac{\Delta z}{\Delta x_1} = \$0.35/\text{yd}^3 \text{ of river gravel}$$

Likewise, for part b, the opportunity cost for utilizing pea gravel in the mix is

$$y_{10}^* = MC_2 = \frac{\Delta z}{\Delta x_2} = \$0.91/\text{yd}^3 \text{ of pea gravel}$$

When there is a short supply of material, the penalty cost for using nonoptimum material is reflected in the opportunity costs. Here, $y_{10}^* > y_9^*$. There is a greater charge for substituting nonoptimal material when sand is limited to 4000 yd^3.

(d) The additional cost for changing the demand in roadway aggregate is estimated with the use of the shadow price. The material constraint is $x_1 + x_2 + x_3 + x_4 = 9$. Increasing the demand by 10 yd^3 of aggregate is equivalent to alternating the resource constraint. Since two constraint equations are used to represent this strict inequality, we use the shadow prices of $y_1 = \Delta z/\Delta b_1$ or $y_2 = \Delta z/\Delta b_2$.

With a supply of 6000 yd^3 of sand, we use the final simplex tableau for part a.

$$y_2^* = \frac{\Delta z}{\Delta b_2} = \$5.00/\text{yd}^3 \text{ of aggregate mix}$$

An additional 10 yd^3 is estimated to cost an additional $50.00.

With a supply of 4000 yd^3 of sand, we use the final simplex tableau of part b.

$$y_2^* = \frac{\Delta z}{\Delta b_2} = \$7.09/\text{yd}^3 \text{ of aggregate mix}$$

An additional 10 yd^3 of aggregate is estimated to cost an additional $70.90.

Remarks

This problem illustrates the financial penalties that must be paid when material is in short supply. When sand is in short supply, we have seen that the total cost, the opportunity cost, and shadow price are greater than when the sand supply is plentiful.

PROBLEMS

Problem 1

$$\text{Minimize } z = \boldsymbol{c}'\boldsymbol{x}$$

$$\boldsymbol{Ax} \geq \boldsymbol{b}$$

$$\boldsymbol{x} \geq \boldsymbol{0}$$

where

$$\boldsymbol{b}' = [2 \quad 3]$$

$$\boldsymbol{c}' = [8 \quad 9 \quad 10 \quad 2]$$

$$\boldsymbol{A} = \begin{bmatrix} 1 & -2 & 1 & -1 \\ 1 & -1 & 1 & -1 \end{bmatrix}$$

(a) Formulate the dual.
(b) Solve for $\boldsymbol{x}^*$ by utilizing the simplex method and the dual formulation of part a.
(c) Determine the shadow prices and opportunity costs. Use the marginal cost definitions.

Problem 2

$$\text{Minimize } z = 5x_1 - 10x_2 - 3x_3 + 2x_4$$

$$2x_1 + x_2 + x_3 + x_4 \leq 20$$

$$x_1 + 3x_2 + 2x_3 \geq 30$$

$$x_4 \geq 5$$

$$x_1 \geq 0, x_2 \geq 0, x_3 \geq 0, x_4 \geq 0$$

(a) Solve by the simplex method.
(b) Formulate the dual solution.
(c) Solve the dual by simplex method.
(d) Identify the shadow prices and opportunity costs.
(e) Is the solution of the primal equal to the solution of the dual?
(f) Identify the shadow prices and opportunity costs from the primal formulation and solution.

Problem 3

Consider the following primal model:

$$\text{Minimize } z = \boldsymbol{c}'\boldsymbol{x}$$

$$\boldsymbol{Ax} \leq \boldsymbol{b}$$

$$\boldsymbol{x} \geq \boldsymbol{0}$$

where

$$\boldsymbol{c}' = [-2 \quad 3], \boldsymbol{A} = \begin{bmatrix} -1 & 4 \\ 2 & 0 \end{bmatrix}, \boldsymbol{b}' = [3 \quad 1]$$

(a) Solve the primal by the simplex method.
(b) Use sensitivity analysis to determine if the optimum solution of part a is the same when c_2 is changed from 3 to -1.
(c) Formulate the dual of the primal model given above.
(d) Use the simplex to solve part c. Determine z^*, x^*, and y^*. Compare these answers with part a; they should agree.
(e) Determine shadow prices and opportunity costs from the final simplex tableau of the dual determined in part d.
(f) Use the shadow price definition $\Delta z/\Delta b_2$ to estimate the change in z due to the doubling of resource constraint b_2.
(g) Since the estimate of the change in z due to the change in b_2 (part d) may not be accurate, use sensitivity analysis to determine if your answer in part d is exact. If not exact, briefly state the reason for the inaccuracy.

Problem 4

$$\text{Minimize } z = -2x_1 + x_2 + x_3$$

$$5x_1 - 2x_2 \le 6$$

$$x_1 + x_3 \le 4$$

$$x_2 \le 5$$

$$x_1, x_2, x_3 \ge 0$$

The final simplex tableau of the dual (optimum) is [Note: maximum d = minimum $a = (-1)d$]:

y_B	y_1	y_2	y_3	y_4	y_5	y_6	b
y_1	1.0	0.2	0.0	−0.2	0.0	0.0	0.4
y_5	0.0	−0.4	−1.0	0.4	1.0	0.0	0.2
y_6	0.0	−1.0	0.0	0.0	0.0	1.0	1.0
b_i	0.0	2.8	5.0	1.2	0.0	0.0	$(a - 2.4)$

(a) By use of the definition of shadow price, estimate the change in z due to the changes in resource constraint from [6 4 5] to [7 6 10].
(b) Briefly describe how you ensure that the solution calculated in part a is an optimum solution by utilizing sensitivity analysis procedures.
(c) List all penalty charges for substituting nonoptimum activities of $\mathbf{x}$ into the optimum solution. Clearly define all of them and their magnitudes.

Problem 5

This problem was presented in Chapter 6 and is repeated here. See Section 6.1, Problem 6 and Section 6.3, Problem 7. Towns A and B produce 3 and 2 Mgal/day of wastewater, respectively. The BOD level of the wastewater is 200 mg/1. (See Figure 7.7.)

Towns A and B belong to a regional water-treatment cooperative that operates water-treatment plants I and II. Plants I and II have a capacity of 3 Mgal/day and 4 Mgal/day and operate at a BOD removal efficiency of 90 percent and 75 percent, respectively.

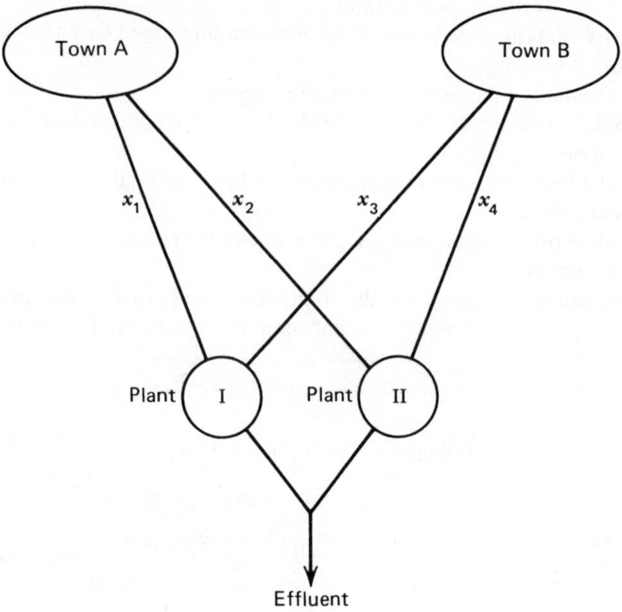

Figure 7.7

The cooperative has an authority to assign the wastewater flow from towns A and B to either plants I or II as shown in the schematic diagram. Formulate a system-analysis model to minimize total annual treatment and pumping costs subject to the constraint that the total flow discharged into the river does not exceed 30 mg/l of BOD. The unit treatment and pumping costs for each pipeline is given in the table below. Let x_1, x_2, x_3, and x_4 represent wastewater flow in million gallons per day assigned to pipelines 1, 2, 3, and 4, respectively.

PIPELINE NUMBER	UNIT ANNUAL TREATMENT AND PUMPING COSTS ($/Mgal/day–yr)
1	46
2	50
3	55
4	40

(a) Utilize the solution of the primal problem, Section 6.3, Problem 7, to determine z^*, $\mathbf{x}^*$, and identify all shadow prices and opportunity costs.
(b) Formulate the dual model.
(c) Solve for the optimum solution of the dual.
(d) Compare the results of parts a and c, and show they are the same.
(e) In 20 years the populations of the towns are expected to increase and are expected to produce 4 and 3 Mgal per day of wastewater, respectively. The total amount of waste equals the total capacity of the two treatment plants. Use the dual solution and sensitivity analysis to determine if an optimum solution exists at an effluent discharge equal to or less than 30 mg/l.

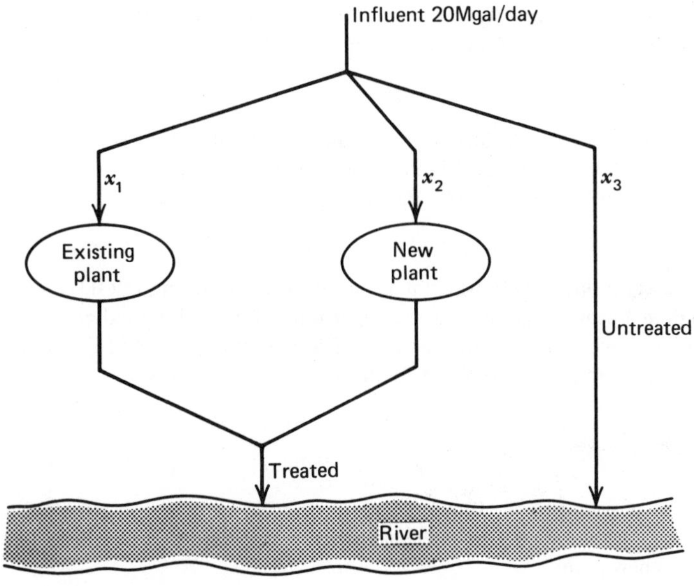

Figure 7.8

(f) If the effluent quantity constraint is changed to 50 mg/l at BOD, determine the optimum solution of part e.

Problem 6

A city is currently treating 8 Mgal/day of wastes in a treatment plant and dumping 12 Mgal/day of untreated waste into the river. The construction of a new plant has been proposed. A schematic diagram showing the overall treatment system is shown in Figure 7.8.

The control variables are

x_1 = wastewater treated in existing wastewater plant, million gallons per day.

x_2 = wastewater treated in proposed wastewater plant, million gallons per day.

x_3 = wastewater dumped into river, million gallons per day.

The existing plant has an 80 percent waste-removal efficiency rating. The proposed plant has a 90 percent efficiency rating. No more than 8 Mgal/day can be treated in the existing facility.

The Water Pollution Authority has recommended that a 15-Mgal plant be constructed. It is assumed that the plant has a 30-yr design life and a 15 percent annual construction loan can be secured. The Authority wants the water entering the river to have a concentration of 25 mg/l of BOD or less. The city has refused to build the new $43.2M facility because they claim the new plant is too expensive. Since the city has resisted, the Water Pollution Authority has imposed a $10,000 per day fine.

(a) Calculate the annual fine assuming the city pollutes the river 365 days per year.
(b) Calculate the annual construction cost of a new 15-Mgal facility.
(c) Compare the annual fine and construction costs. It is cheaper for the city to continue to pollute the river or to construct the plant?

(d) Determine the overall water quality of the 20 Mgal per day entering the river after it has been diluted with treated water from the existing plant. Assume $x_2 = 0$.

Problem 7

Use the information given in Problem 6. The Water Pollution Authority wants to determine the effect of the fine upon the town's decision to treat its waste.

(a) Show that the daily fine of \$10,000/day for untreated wastes is

$$c_3 = \$1.10 \times 10^{-6}/\text{mg BOD}$$

Assume 12 Mgal/day of wastewater at a concentration of 200 mg/l BOD enters the river.

(b) Formulate a linear programming model to minimize total annual construction and fine costs for the town. Consider the capacity of the existing plant and the effluent flow of 20 Mgal/day. The capital cost to construct the new plant is a function of x_2

$$c_2 = \$2.88 \times 10^6 x_2$$

$$(1 \text{ gal} = 3.785 \text{ liters})$$

(c) Solve for $\boldsymbol{x}^*$ of part d by the simplex method. Show that the fine is insufficient.

(d) Utilize the final solution of c to determine whether doubling the fine will persuade the city to construct a new plant.

(e) Why is not a water quality constraint introduced into the model of part b.

Problem 8

Use the information given in Problem 7, part b. The Water Pollution Authority wants to determine the effect of imposing different water quality restrictions on the construction cost of the new plant.

(a) Formulate a linear mathematical model to minimize annual construction costs only. Consider the capacity of the existing plant, the effluent flow of 20 Mgal/day, and the restriction that water entering the river must have a concentration of 25 mg/l or less. Do not consider this fine for nontreatment. In other words, let $c_3 = 0$.

(b) Formulate the dual for part a.

(c) Solve for the optimum solution $\boldsymbol{x}^*$ and the opportunity costs and shadow prices for part b.

(d) If the concentration of pollution is restricted to 20 mg/l or less, what are the size and cost of the new treatment facility?

Summary

Sensitivity analysis was used to determine the effect of changing the elements of the unit cost vector $\boldsymbol{c}$ and the resource constraint vector $\boldsymbol{b}$. Graphical and mathematical methods of analysis were used to solve these problems. The dual is not only used to perform sensitivity analysis of changing the $\boldsymbol{b}$ vector, it was used to give further insight into the decision-making process. With the dual and the use of the shadow price and opportunity cost, we can estimate the change in total cost z due to the change in unit costs $\boldsymbol{c}$ or resource constraint $\boldsymbol{b}$.

Bibliography

Refer to the Bibliography for Chapter 6.

CHAPTER 8

Classical Approach to Nonlinear Programming

After completion of this chapter, the student should be able to

1. Use the principles of calculus to locate the optimum solution of a mathematical model consisting of a nonlinear objective and a set of constraint functions.
2. Use the definition of Hessian matrix and other mathematical approaches to identify convex and concave functions and convex constraint sets.
3. Use the method of substitution and the Lagrange multiplier method to solve constrained optimization problems.
4. Identify the necessary and sufficient conditions for determining local and global optima.

In linear programming the search for an optimum solution is restricted to the evaluation of the extreme points of the feasible region. The special properties of linear equations are used to establish the simplex method for effectively locating the optimum solution. For nonlinear programming, the search for an optimum will not be as straightforward. The optimum solution in this case may occur at (1) an extreme point of the feasible region or constraint set, (2) along its boundary, or (3) at an interior point. If certain mathematical properties are present in the mathematical model, the search for the optimum solution may be narrowed to a workable and practical set of alternatives. Our goal then is to identify these special mathematical properties and utilize them for establishing a means for obtaining an optimum solution.

Unlike linear programming, an algorithm to solve these problems will not be established. The best we can achieve is to establish an *approach* for solving nonlinear problems. The actual method employed will depend upon its mathematical properties. These methods will be discussed here and in the next chapter. Generally, the approach consists of the following steps:

STEP 1 *Formulate* the model in the following form:

$$z = f(\mathbf{x})$$

$$\mathbf{g}(\mathbf{x})\{=, \leq, \geq\}\mathbf{b}$$

$$\mathbf{x} \geq \mathbf{0}$$

where

$$\mathbf{x}' = [x_1 \quad x_2 \quad \cdots \quad x_n]$$

STEP 2 *Locate* a candidate solution $\mathbf{x}^0$ for the optimum of z.

STEP 3 *Test* the candidate solution for the location of the global optimum point $\mathbf{x}^*$.

The function $f(\mathbf{x})$ and the functions of $\mathbf{g}(\mathbf{x})$ are assumed to be functions with continuous second derivatives.

The essential concepts for finding local and global optima of univariate and bivariate nonlinear functions are discussed in Chapter 2. These concepts will be extended and used for locating and testing candidate solutions for multivariate functions. The terminology, local and global optima, convex and concave functions, and necessary and sufficient conditions should be familiar. If not, the reader is encouraged to review the material before proceeding.

8.1 FUNDAMENTAL PRINCIPLES FOR FINDING GLOBAL OPTIMA

The type of the objective function $f(\mathbf{x})$ and the constraint set $\mathbf{g}(\mathbf{x})\{=, \leq, \geq\}\mathbf{b}$ plays an extremely important role in determining the approach to be used to locate the optimum solution. Consider the following two-dimensional optimization problem:

$$\text{Maximize } z = f(x_1, x_2)$$
$$\mathbf{g}(x_1, x_2) \leq \mathbf{b}$$
$$x_1 \geq 0, x_2 \geq 0$$

The problem is depicted in Figure 8.1. The global optimum or, equivalently, the global maximum may be determined by inspection. It is located at the highest point on the surface of z, $(\mathbf{x}^*, f(\mathbf{x}^*) = (x_1^*, x_2^*, f(x_1^*, x_2^*))$. The optimum point is an interior point of the constraint set $\mathbf{g}(\mathbf{x}) = \mathbf{g}(x_1, x_2) \leq \mathbf{b}$. For this illustration, the objective function is shown to be a *concave* function that possesses a *unique* maximum point at $\mathbf{x}^*$. For all other values of $\mathbf{x}$, $f(\mathbf{x}^*) > f(\mathbf{x})$. The geometric configuration of the z

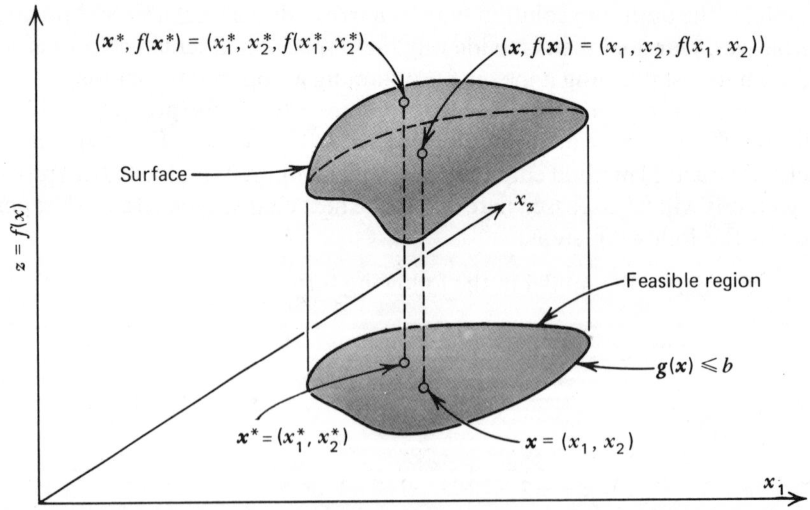

Figure 8.1 A global maximum, $f(\mathbf{x}^*) > f(\mathbf{x})$ for all $\mathbf{x}$ satisfying $\mathbf{g}(\mathbf{x}) \leq \mathbf{b}$.

surface is important in locating the optimum point. It is shown in later sections that the constraint set also affects determination of the optimum location. For now, we assume that the constraint set is well behaved and forms what is called a *convex set*.

If the objective function is *not* a concave function, the location of the global maximum may not be a unique point. If the surface z has a shape similar to a two-humped camel, two *local* maximum points exist. If both points are of equal height, the global maximum will exist at two different locations. If one point is higher than the other one, a global maximum exists at the higher one.

Obviously, the search for the maximum for the concave function, Figure 8.1, and the two-humped camel shape is relatively simple. Imagine that we are in a forest and cannot see the mountain top or tops. If we know that the mountain is a single-peaked mountain, a convex function, we can proceed up the incline of maximum ascent and be certain that the peak will be reached. If the mountain is camel backed or has two peaks, we are not sure we will reach the highest peak by climbing the steepest ascent. For problems of higher order when surfaces cannot be drawn, we shall use mathematics to help us determine if a function is concave or not. When searching for a minimum point, it is desirable to know if the objective function is a *convex function*. If this special property exists, the search for the minimum is easier to locate.

Now consider the search for a minimum of the concave function shown in Figure 8.1. Again, we take advantage of the fact that the objective function is a concave function. For this problem we have to investigate the boundary of the feasible region only. The global minimum cannot occur at an interior point if z is a concave function. Again, we are tacitly assuming the constraint set is a convex set. If this property is not satisfied, our search becomes more complicated. We begin our discussion by presenting mathematical theorems dealing with *convex* and *concave* functions.

The following mathematical theorems form the basis of locating the global optimum solution for multivariate mathematical models.

Global Minimum

THEOREM 1 If the objective function $f(\boldsymbol{x})$ is a *convex function* over a *convex constraint set* of $\boldsymbol{g}(\boldsymbol{x})\ \{=, \leq, \geq\}\ \boldsymbol{b}$, then any local minimum of $f(\boldsymbol{x})$ satisfying the constraint set will be a *global minimum.*

THEOREM 2 If the objective function $f(\boldsymbol{x})$ is a *concave function* over a *convex constraint set* of $\boldsymbol{g}(\boldsymbol{x})\ \{=, \leq, \geq\}\ \boldsymbol{b}$, then the *global minimum* will occur at one or more of the extreme points of the constraint set.

Global Maximum

THEOREM 3 If the objective function $f(\boldsymbol{x})$ is a *concave function* over a *convex constraint set* of $\boldsymbol{g}(\boldsymbol{x})\ \{=, \leq, \geq\}\ \boldsymbol{b}$, then any local maximum of $f(\boldsymbol{x})$ satisfying the constraint set will be a *global maximum.*

THEOREM 4 If the objective function $f(\boldsymbol{x})$ is a *convex function* over the *convex constraint set* of $\boldsymbol{g}(\boldsymbol{x})\ \{=, \leq, \geq\}\ \boldsymbol{b}$, then the *global maximum* will occur at one or more of the extreme points of the constraint set.

Clearly, methods for determining (1) whether the objective function $f(\boldsymbol{x})$ is a convex or concave or not and (2) whether a constraint set $\boldsymbol{g}(\boldsymbol{x})\ \{=, \geq, \leq\}\ \boldsymbol{b}$ is a convex set or not is essential to the determination of a global optimum.

Our discussion will deal with these determinations. These theorems give assurance that a global optimum will be obtained if the objective function is either a convex or concave function and the constraint set forms a convex set. If the objective function is neither convex nor concave or the constraint set is not a convex set, no assurances are made. Furthermore, if $\boldsymbol{g}(\boldsymbol{x})\ \{=, \geq, \leq\}\ \boldsymbol{b}$ is a convex set and $f(\boldsymbol{x})$ is a convex or concave function, the search for the optimum $\boldsymbol{x}^*$ will be narrowed. When these conditions are not satisfied, the search may be a tedious process. One of the aims of this chapter is to be able to identify these special types of objective functions and convex sets.

In this section and in Sections 8.2 and 8.3, the discussion focuses upon convex and concave functions and the determination of local optima of unconstrained mathematical models. When the determination of global optima of constrained mathematical models is sought, Theorems 1 through 4 will be employed. In these earlier sections, to simplify the discussion and understanding, the constraint set will be assumed to be a convex set. In Section 8.4 and in the remaining sections the effect of the constraint set upon the search process is investigated.

Local Minima and Maxima

A local minimum at the point $\boldsymbol{x}^\circ$, where $\boldsymbol{x}^{\circ\prime} = [x_1^\circ \quad x_2^\circ \quad x_3^\circ \quad \cdots \quad x_n^\circ]$, must satisfy the following inequality condition that $f(\boldsymbol{x}^\circ)$ must be less than equal to $f(\boldsymbol{x})$.

$$f(\boldsymbol{x}^\circ) \leq f(\boldsymbol{x})$$

The neighborhood of $\boldsymbol{x}^\circ$ is the set of points of the form to be $\boldsymbol{x} = \boldsymbol{x}^\circ + \Delta\boldsymbol{x}$ for small $\Delta\boldsymbol{x}$.

$$\begin{bmatrix} x_1 \\ x_2 \\ \vdots \\ x_n \end{bmatrix} = \begin{bmatrix} x_1^\circ \\ x_2^\circ \\ \vdots \\ x_n^\circ \end{bmatrix} + \begin{bmatrix} \Delta x_1 \\ \Delta x_2 \\ \vdots \\ \Delta x_n \end{bmatrix}$$

The Δx_i elements are positive or negative values.

A similar inequality exists for local maxima, $f(\boldsymbol{x}^\circ) \geq f(\boldsymbol{x})$. These inequalities will be used explicitly and implicitly to determine if a candidate for a local optimum satisfies the conditions for a local minimum, local maximum, or neither one.

Convex and Concave Functions

A function $f(\boldsymbol{x})$ is a convex function if it satisfies the following inequality.

$$f(\boldsymbol{x}) \leq \theta f(\boldsymbol{x}^1) + (1 - \theta) f(\boldsymbol{x}^2)$$

where $\boldsymbol{x} = \boldsymbol{x}(\theta)$ lies on the line segment between $\boldsymbol{x}^1$ and $\boldsymbol{x}^2$. That is, $\boldsymbol{x}(\theta)$ is of the form $\boldsymbol{x}(\theta) = \theta\boldsymbol{x}^1 + (1 - \theta)\boldsymbol{x}^2$ for some $\theta, 0 \leq \theta \leq 1$. Here, we choose two points in multidimensional space and convert this to a one-dimensional problem. In Chapter 2 a

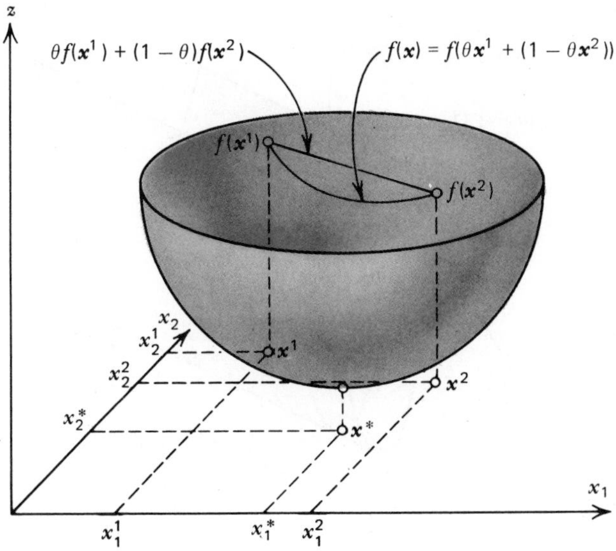

Figure 8.2 A unique minimum of a convex function.

similar definition was introduced for univariate functions. Figure 8.2 shows the essence of the test for a bivariate convex function. The minimum is located at x^*.

Choose any two points x^1 and x^2, and evaluate the values of $f(x^1)$ and $f(x^2)$. Draw a straight line between x^1 and x^2 and evaluate the nonlinear function $f(x)$ at a point $x(\theta)$ on line. The resulting inequality condition for a convex function is

$$f(\theta x^1 + (1 - \theta)x^2) \leq \theta f(x^1) + (1 - \theta) f(x^2)$$

As shown in Figure 8.2, this condition requires that the line segment between the points $(x^1, f(x^1))$ and $(x^2, f(x^2))$ in three dimensions lies above the three-dimensional surface. If for all values of θ between 0 and 1 this inequality condition is satisfied, the function is defined to be a convex function. All points along the straight line $\theta f(x^1) + (1 - \theta) f(x^2)$ must be greater than or equal to the corresponding points along the function $f(x) = f(\theta x^1 + (1 - \theta)x^2)$. A similar test may be established for a concave function. The definition of a concave function must satisfy the following inequality:

$$f(\theta x^1 + (1 - \theta)x^2) \geq \theta f(x^1) + (1 - \theta) f(x^2)$$

In Figures 8.2, 8.3, and 8.4 a strictly convex function, a convex function, and a function that is neither convex nor concave are shown. For Figure 8.2, two arbitrary points of x^1 and x^2 are chosen. Regardless of the values of x^1 and x^2, the following *strict* inequality constraint is satisfied.

$$f(\theta x^1 + (1 - \theta)x^2) < \theta f(x^1) + (1 - \theta) f(x^2)$$

The minimum point x^* for this function is a unique minimum.

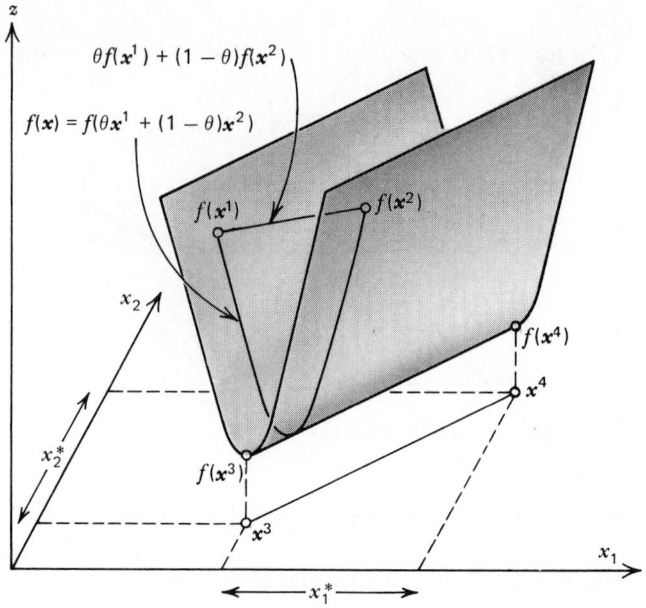

Figure 8.3 A nonunique minimum of a convex function.

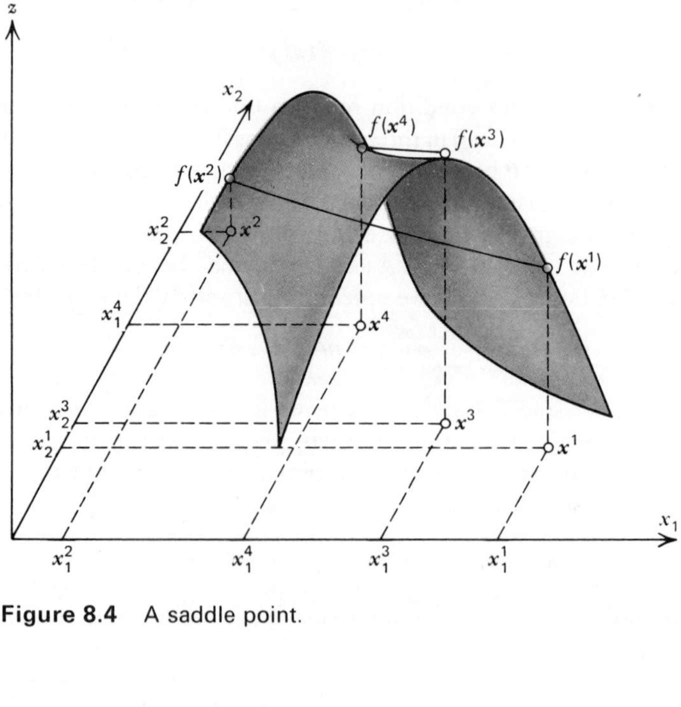

Figure 8.4 A saddle point.

For Figure 8.3 with arbitrary values of $\mathbf{x}^1$ and $\mathbf{x}^2$, the function $f(\mathbf{x})$ satisfies the following inequality constraint:

$$f(\theta\mathbf{x}^1 + (1 - \theta)\mathbf{x}^2) \leq \theta f(\mathbf{x}^1) + (1 - \theta)f(\mathbf{x}^2)$$

A relationship is not a strict inequality, because when $\mathbf{x}^1 = \mathbf{x}^3$ and $\mathbf{x}^2 = \mathbf{x}^4$, $f(\mathbf{x}) = \theta f(\mathbf{x}^1) + (1 - \theta)f(\mathbf{x}^2)$. The minimum of the function lies along a line between $\mathbf{x}^3$ and $\mathbf{x}^4$; therefore, it does not have a unique minimum point. The choice $\mathbf{x}^1$ and $\mathbf{x}^2$ plays a role in determining whether the function is a strictly convex function.

For Figure 8.4 the function does not satisfy the requirements of the definition of a convex function for all different combinations of two arbitrary points in the $x_1 - x_2$ plane; thus, it is not a convex function. It does not have a minimum, but a saddle point.

If the values of $\mathbf{x}$ are constrained to lie between certain points, the function shown in Figure 8.4 may satisfy the definition of a convex or a concave function. Consider the function $z = f(\mathbf{x})$ between $\mathbf{x}^1 \leq \mathbf{x} \leq \mathbf{x}^2$. For this range the function satisfies the condition for a strictly concave function:

$$f(\theta\mathbf{x}^1 + (1 - \theta)\mathbf{x}^2) > \theta f(\mathbf{x}^1) + (1 - \theta)f(\mathbf{x}^2)$$

But, for the range between $\mathbf{x}^3$ and $\mathbf{x}^4$,

$$f(\theta\mathbf{x}^3 + (1 - \theta)\mathbf{x}^4) < \theta f(\mathbf{x}^3) + (1 - \theta)f(\mathbf{x}^4)$$

the function $f(\mathbf{x})$ satisfies the condition for a strictly convex function. Thus, the function cannot be classified as either a convex or concave function for the region shown in the diagram.

EXAMPLE 8.1 A Cylindrical Tank

Consider the following mathematical model to maximize the volume of a cylindrical tank.

$$\text{Maximize } V = \pi r^2 l$$
$$l = 2r$$
$$0 \leq r \leq 10,$$
$$l \geq 0$$

where r is the tank radius and l is the tank length.

(a) Determine if function V is a convex function for the imposed constraint set.

(b) The constraint set forms a convex set. Determine the location of the global maximum.

Solution

(a) If $V = \pi r^2 l$ is a convex function, it must satisfy the following inequality equation:

$$f(\theta\mathbf{x}^1 + (1 - \theta)\mathbf{x}^2) < \theta f(\mathbf{x}^1) + (1 - \theta)f(\mathbf{x}^2) \quad (1)$$

where the $\mathbf{x}$ vector is a function of the variables r and l, the radius and length of the tank, respectively. Since $l = 2r$, and $\mathbf{x} = (x_1, x_2) = (r, l) = (r, 2r)$, the extreme points $\mathbf{x}^1$ and $\mathbf{x}^2$ are represented by the following vectors:

$$\mathbf{x}^1 = \begin{bmatrix} r^1 \\ l^1 \end{bmatrix} = \begin{bmatrix} 10 \\ 20 \end{bmatrix} \quad \text{and} \quad \mathbf{x}^2 = \begin{bmatrix} r^2 \\ l^2 \end{bmatrix} = \begin{bmatrix} 0 \\ 0 \end{bmatrix}$$

The term $\mathbf{x} = \theta\mathbf{x}^1 + (1-\theta)\mathbf{x}^2$ is

$$\mathbf{x} = \theta\begin{bmatrix} r^1 \\ l^1 \end{bmatrix} + (1-\theta)\begin{bmatrix} r^2 \\ l^2 \end{bmatrix} = \theta\begin{bmatrix} 10 \\ 20 \end{bmatrix}$$

Thus, $r = 10\theta$ and $l = 20\theta$. Substitute the variables r and l into Eq. (1); thus:

$$f(\theta\mathbf{x}^1 + (1-\theta)\mathbf{x}^2) = \pi r^2 l = \pi(10\theta)^2(20\theta) = 2000\pi\theta^3$$

and
$$\theta f(\mathbf{x}^1) + (1-\theta)f(\mathbf{x}^2) = \theta\pi(r^1)^2 l^1 + (1-\theta)\pi(r^2)^2 l^2$$
$$= \theta 2000\pi + 0$$

Equation (1) becomes
$$2000\pi\theta^3 < 2000\pi\theta$$

or
$$\theta^2 < 1$$

Since $0 \le \theta \le 1$, θ^2 is less than 1. The function $V = \pi r^2 l$ is a strict convex function over the specified constraint set.

(b) Since the objective function $f(\mathbf{x}) = V = 2\pi r^2 l$ is a convex function over the convex constraint set, by Theorem 3 the maximum volume will be located at an extreme point. The extreme points are

$$\mathbf{x}^1 = \begin{bmatrix} 10 \\ 20 \end{bmatrix} \quad \mathbf{x}^2 = \begin{bmatrix} 0 \\ 0 \end{bmatrix} \quad \mathbf{x}^3 = \begin{bmatrix} 0 \\ 20 \end{bmatrix} \quad \mathbf{x}^4 = \begin{bmatrix} 10 \\ 20 \end{bmatrix}$$

$V^1 = V^3$, $V^4 = \pi r^2 l = 0$, and $V^2 = \pi(10)^2(20) = 2000\pi$. Since $V^2 > V^1$, the global maximum occurs at $\mathbf{x}^2$; thus

$$V^* = 2000\pi$$

with
$$r^* = 10,\ l^* = 20$$

Remarks

Since the constraint $l = 2r$ is a strict equality constraint, we may simplify the solution of part a by substituting $l = 2r$ into the objective function.

$$V = 2\pi r^2(2r) = 2\pi r^3$$

Let $r = r'\theta + r^2(1-\theta)$, thus

$$r = 10\theta + 0(1-\theta) = 10\theta$$

and for $V = 2\pi r^3$ to be a convex function, the following inequality must hold

$$f(\theta r' + (1-\theta)r^2) < \theta f(r') + (1-\theta)f(r^2)$$

$$2\pi(10\theta^3) < \theta 2\pi(10)^3 + (1-\theta)2\pi(0)^3$$

which reduces
$$\theta^2 < 1$$

In this case, the method of substitution simplifies the analysis.

For part *b*, it should be clear that the maximum occurs at $r^2 = 10$. Theorem 3 is satisfied.

EXAMPLE 8.2 The Quadratic Function $f(\mathbf{x}) = x_1x_2$

The function $f(\mathbf{x}) = x_1x_2$ of two variables is a simple product that appears in many engineering problems. If x_1 and x_2 are defined as lengths of a rectangle, the product x_1x_2 is the area of the rectangle. The function cannot be classified as a convex or a concave function for all values of x_1

and x_2. The function may be either a convex or a concave function depending on the restrictions imposed upon it.

(a) Use the definitions of convex and concave functions and classify $f(\boldsymbol{x}) = x_1x_2$ as a convex, a concave, or neither a convex nor a concave function for the mathematical model:

$$\text{Maximize } z = f(x_1, x_2) = x_1x_2$$
$$0 \le x_1 \le 10$$
$$0 \le x_2 \le 20$$
$$x_1 \ge 0, x_2 \ge 0$$

Utilize lines drawn between the following extremes $\boldsymbol{x}^1$ and $\boldsymbol{x}^2$.

Line *A*. $\boldsymbol{x}^1$ and $\boldsymbol{x}^2$ between (1, 10) and (10, 1).

Line *B*. $\boldsymbol{x}^1$ and $\boldsymbol{x}^2$ between (1, 10) and (10, 5).

Line *C*. $\boldsymbol{x}^1$ and $\boldsymbol{x}^2$ between (1, 10) and (10, 10).

Line *D*. $\boldsymbol{x}^1$ and $\boldsymbol{x}^2$ between (1, 10) and (10, 20).

The constraint set forms a convex set.

(b) Use the definition of a convex function to show that when the function x_1x_2 is a convex function, it is an objective function in the mathematical model:

$$\text{Minimize } z = x_1x_2$$
$$x_2 = 2x_1$$
$$x_1 \ge 0, x_2 \ge 0$$

(c) Use the definition of a concave function to show that x_1x_2 is a concave function when the function is incorporated in a mathematical model as a constraint equation of the form $x_1x_2 \ge 10$.

Solution

(a) The definition of convex and concave functions are, respectively

$$f(\theta\boldsymbol{x}^1 + (1-\theta)\boldsymbol{x}^2) \le \theta f(\boldsymbol{x}^1) + (1-\theta)f(\boldsymbol{x}^2)$$

and

$$f(\theta\boldsymbol{x}^1 + (1-\theta)\boldsymbol{x}^2) \ge \theta f(\boldsymbol{x}^1) + (1-\theta)f(\boldsymbol{x}^2)$$

where $0 \le \theta \le 1$ and $\boldsymbol{x}^1$ and $\boldsymbol{x}^2$ are points in the x_1–x_2 plane.

$$\boldsymbol{x}^1 = \begin{bmatrix} x_1^1 \\ x_2^1 \end{bmatrix} \quad \text{and} \quad \boldsymbol{x}^2 = \begin{bmatrix} x_1^2 \\ x_2^2 \end{bmatrix}$$

The left-hand side of the inequalities $f(\boldsymbol{x})$ is the value of nonlinear function x_1x_2 evaluated at a point along the line $\theta\boldsymbol{x}^1 + (1-\theta)\boldsymbol{x}^2$ for a given value of θ. The right-hand side, $\theta f(\boldsymbol{x}^1) + (1-\theta)f(\boldsymbol{x}^2)$, is a linear equation for a given value of θ.

The value of $\boldsymbol{x}$ in vector notation as a function of θ is $\boldsymbol{x} = \theta\boldsymbol{x}^1 + (1-\theta)\boldsymbol{x}^2$. The values of the elements $\boldsymbol{x}_1$ and $\boldsymbol{x}_2$ of the vector $\boldsymbol{x}$ are:

$$\begin{bmatrix} x_1 \\ x_2 \end{bmatrix} = \theta\begin{bmatrix} x_1^1 \\ x_2^1 \end{bmatrix} + (1-\theta)\begin{bmatrix} x_1^2 \\ x_2^2 \end{bmatrix}$$

or, written in terms of two linear equations x_1 and x_2:

$$x_1 = \theta x_1^1 + (1 - \theta)x_1^2$$

$$x_2 = \theta x_2^1 + (1 - \theta)x_2^2$$

The right-hand side of the inequality is expressed as

$$\theta f(\mathbf{x}^1) + (1 - \theta)f(\mathbf{x}^2) = \theta f(x_1^1 x_2^1) + (1 - \theta)f(x_1^2 x_2^2)$$

The left-hand side of the inequality may be written in terms of θ by substituting linear equations of x_1 and x_2 into the $f(\mathbf{x}) = x_1 x_2$. The result is

$$f(\mathbf{x}) = x_1 x_2 = [\theta x_1^1 + (1 - \theta)x_1^2][\theta x_2^1 + (1 - \theta)x_2^2]$$

The various combination of lines A, B, C, and D are drawn between $\mathbf{x}^1$ and $\mathbf{x}^2$ as shown in the figure. The value of θ is varied from zero to one in steps of 0.1. The results are shown in Figure 8.5 and Table 8.1.

The evaluation of points along the lines A, B, C, and D shows that the function satisfies the definition of a convex, concave, or linear function, depending upon the value of the end points $\mathbf{x}_1$ and $\mathbf{x}_2$.

Line *A*. Concave function.

Line *B*. Concave function.

Line *C*. Linear function.

Line *D*. Concave function.

Thus, the function is neither a convex nor a concave function for the given feasible region.

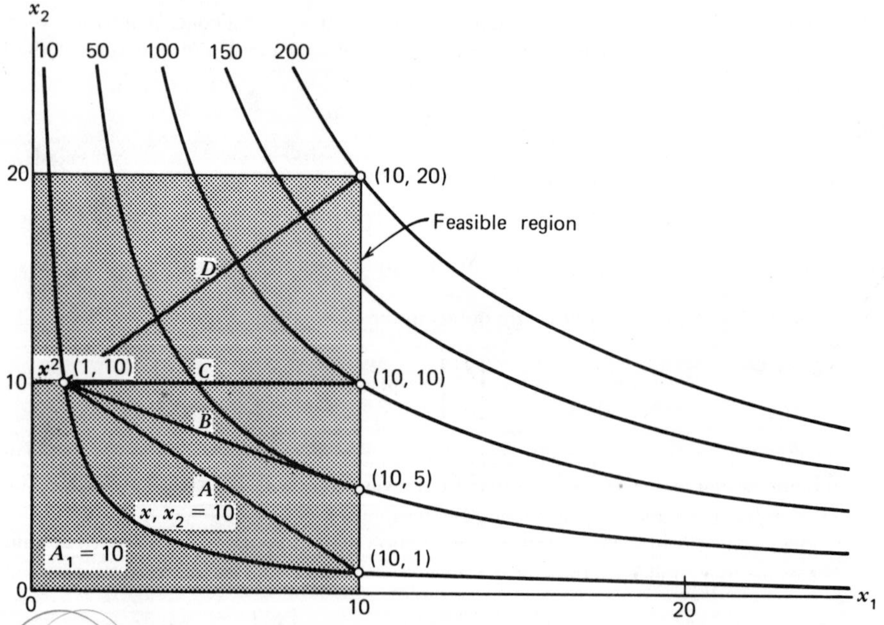

Figure 8.5

Table 8.1

	ALONG LINE *A*	ALONG LINE *B*	ALONG LINE *C*	ALONG LINE *D*
θ	$\mathbf{x}^1 = (1, 10)$ $\mathbf{x}^2 = (10, 1)$	$\mathbf{x}^1 = (1, 10)$ $\mathbf{x}^2 = (10, 5)$	$\mathbf{x}^1 = (1, 10)$ $\mathbf{x}^2 = (10, 10)$	$\mathbf{x}^1 = (1, 10)$ $\mathbf{x}^2 = (10, 20)$
	$f(\theta)$ R^*	$f(\theta)$ R	$f(\theta)$ R	$f(\theta)$ R
0	10 = 10	50 = 50	100 = 100	200 = 200
0.1	17.2 > 10	50.1 > 46	91 = 91	173 < 181
0.2	23 > 10	49.2 > 42	82 = 82	148 < 162
0.3	27.0 > 10	47.5 > 38	73 = 73	124 < 143
0.4	29.4 > 10	44.8 > 34	64 = 64	102 < 124
0.5	30.5 > 10	41.3 > 30	55 = 55	82.5 < 105
0.6	29.4 > 10	36.8 > 26	46 = 46	64.4 < 86
0.7	27.0 > 10	31.5 > 22	37 = 37	48.1 < 67
0.8	23.0 > 10	25.2 > 18	28 = 28	33.6 < 48
0.9	17.3 < 10	18.1 > 14	19 = 19	20.9 < 29
1.0	10 = 10	10 > 10	10 = 10	10 = 10
Classification	Concave function	Concave function	Linear function	Convex function

* $R = \theta f(\mathbf{x}^1) + (1 - \theta) f(\mathbf{x}^2)$.

(b) Select any two points along the line $x_2 = 2x_1$. For simplicity, we select $\mathbf{x} = (x_1, 2x_1)$ to have end points at $\mathbf{x}^1 = (10, 20)$ and at the origin $\mathbf{x}^2 = (0, 0)$ as shown in the figure. As a result, the range of $\mathbf{x}$ as a function of θ is:

$$\mathbf{x} = \theta \begin{bmatrix} x_1^1 \\ x_2^1 \end{bmatrix} + (1 - \theta) \begin{bmatrix} x_1^2 \\ x_2^2 \end{bmatrix} = \theta \begin{bmatrix} 10 \\ 20 \end{bmatrix} + (1 - \theta) \begin{bmatrix} 0 \\ 0 \end{bmatrix}$$

or

$$x_1 = 10\theta \quad \text{and} \quad x_2 = 20\theta$$

By substituting these equations into $x_1 x_2$, the function $f(\mathbf{x}) = x_1 x_2$ may be expressed as a function of θ, or $f(\mathbf{x}) = 200\theta^2$. Substitute $x_1 = 10\theta$, $x_2 = 20\theta$, and $f(\mathbf{x}) = 200\theta^2$ into the definition of convex function:

$$f(\mathbf{x}) < \theta f(\mathbf{x}^1) + (1 - \theta) f(\mathbf{x}^2)$$

$$200\theta^2 < (10)(20)\theta + (1 - \theta)(0)(0)$$

$$200\theta^2 < 200\theta$$

Simplifying, it becomes $\theta < 1$. Since θ lies between 0 and 1 and $\theta < 1$, we have shown that the function $f(\mathbf{x}) = x_1 x_2$ is a convex function along the line $x_2 = 2x_1$. The critical point $\mathbf{x}^\circ$ or minimum of $z = x_1 x_2$ must satisfy the relationship:

$$\frac{\partial z}{\partial x_1} = x_2 = 0$$

$$\frac{\partial z}{\partial x_2} = x_1 = 0$$

or $x_1^\circ = x_2^\circ = 0$. For all positive values of x_1 and x_2, $x_1 \geq 0$ and $x_2 \geq 0$; thus $\mathbf{x}^\circ$ is location of a local minimum. By Theorem 1, since $f(\mathbf{x})$ is a convex function along $x_2 = 2x_1$, $\mathbf{x}^\circ$ is the location of the global minimum $z^* = 0$, $x_1^* = 0$, $x_2^* = 0$.

(c) The results from part a will be used to show that $f(\mathbf{x}) = x_1 x_2$ is a concave function for the inequality $x_1 x_2 \geq 10$. The points $\mathbf{x}^1 = (1, 10)$ and $\mathbf{x}^2 = (10, 1)$ satisfy the equality $x_1 x_2 = 10$. A line drawn between them is shown as line A in Figure 8.5. The function $x_1 x_2$ is a concave function between (1, 10) and (10, 1) as shown in Table 8.1. The same conclusion is drawn for all other combinations of $\mathbf{x}^1$ and $\mathbf{x}^2$ that satisfy the condition $x_1 x_2 = 10$; thus, $x_1 x_2$ is a concave function for this restriction.

Remarks

The ability to classify functions is extremely important for utilizing Theorems 1 through 4. In part a the objective function cannot be classified as either a convex or a concave function; thus, Theorems 1 through 4 cannot be applied. For the simple model in part b, there is no problem of identifying $\mathbf{x} = (10, 20)$ as the location of the global maximum. Contrast the inspection method with the approach used in part b, where the local minimum is determined with calculus. A graphical solution is applicable also. For multidimensional problems, calculus will be used to find local optima whenever possible. In the following two sections, 8.2 and 8.3, our discussion investigates different methods of determining local optima.

PROBLEMS

Problem 1

(a) Plot the following functions:

$$z_1 = \ln x_1$$

$$z_2 = x_2^{-1}$$

(b) Use graphical methods and the definition of convex or concave function to prove that the functions are either convex or concave functions.

(c) Define the new function:

$$z = z_1 + z_2$$

Use the definition of a convex or concave function to prove z is neither a convex nor concave function.

Problem 2

By the use of the definition of a convex or concave function, prove that the linear function:

$$z = 3x_1 + 2x_2$$

satisfies the conditions for both a convex or concave function.

Problem 3

The number of highway miles M that can be paved is a function of labor and machine hours.

$$M = 0.5L^{0.2}K^{0.8}$$

where L is labor used in man hours and K is machine hours. Labor is paid \$10.00/hr and the machine operating cost is \$75.00/hr.

(a) Formulate a mathematical model to determine the maximum number of miles that can be paved given a budgetary constraint for total labor and machine operating costs of \$5000.
(b) Use a graphical means to determine the optimum solution.
(c) Assume that the constraint is active. Determine whether the objective function is a convex or concave function on the active constraint line.

Problem 4
Consider the following minimization problem:

$$\text{Minimize } z = (x_1 - 2)^2 + (x_2 - 3)^2$$
$$0 \leq x_1 \leq 5,$$
$$0 \leq x_2 \leq 6$$

The objective function is a convex function, and the constraint set forms a convex set.
(a) Draw the feasible region.
(b) Determine the location of critical solution $\boldsymbol{x}^\circ$ by graphical means.
(c) Prove that the solution is a global minimum, $\boldsymbol{x}^* = \boldsymbol{x}^\circ$.
(d) Consider the following objective function:

$$\text{Maximize } z = (x_1 - 2)^2 + (x_2 - 3)^2$$

Determine the critical solution $\boldsymbol{x}^\circ$ by evaluating $\partial z/\partial x_1 = 0$ and $\partial z/\partial x_2 = 0$.
(e) Prove that $\boldsymbol{x}^\circ$ is not the global maximum and that $\boldsymbol{x}^*$ is at extreme points.

8.2 QUADRATIC EQUATIONS AND LOCAL OPTIMA

In this section and the next one, methods for determining the location of critical points, points that are candidates for the location of the maximum or minimum of multivariate functions, are investigated. Calculus and the use of Taylor Series expansions are used.

To simplify our remarks, quadratic equations will be studied. The function $f(x_1x_2) = x_1^2 + 3x_2^2 + x_1x_2 - 6x_1 + 9$ is an example of a quadratic equation. It may be separated and written as a sum of quadratic, linear, and constant terms.

Quadratic equations have special mathematical properties that makes them particularly suitable to be written in matrix form, $f(\boldsymbol{x}) = a + b'\boldsymbol{x} + \boldsymbol{x}'\boldsymbol{C}\boldsymbol{x}$. The $\boldsymbol{x}$ vector consists of n control variables, where a is a constant, $\boldsymbol{b}$ is an n vector of constant coefficients, and $\boldsymbol{C}$ is an $n \times n$ symmetrical matrix of constant coefficients. The equation may be written in the following expanded form as

$$f(\boldsymbol{x}) = a + \boldsymbol{b}'\boldsymbol{x} + \boldsymbol{x}'\boldsymbol{C}\boldsymbol{x} = a + \sum_i b_i x_i + \sum_i \sum_j c_{ij} x_i x_j$$

The Quadratic Form

A multivariate function consisting exclusively of second-order terms $\boldsymbol{x}'\boldsymbol{C}\boldsymbol{x}$ is defined to be an expression in *quadratic form.* It can be written as follows:

$$\begin{aligned}\boldsymbol{x}'\boldsymbol{C}\boldsymbol{x} = \sum_i \sum_j c_{ij} x_i x_j &= (c_{11}x_1^2 + c_{12}x_1x_2 + \cdots + c_{1n}x_1x_n) \\ &+ (c_{21}x_2x_1 + c_{22}x_2^2 + \cdots + c_{2n}x_2x_n) \\ &+ \cdots + (c_{n1}x_nx_1 + c_{n2}x_nx_2 + \cdots + c_{nn}x_n^2)\end{aligned}$$

or, in matrix form, as

$$x'Cx = [x_1 x_2 \cdots x_n] \cdot \begin{bmatrix} c_{11} & c_{12} & \cdots & c_{1n} \\ c_{21} & c_{22} & \cdots & c_{2n} \\ \vdots & \vdots & & \vdots \\ c_{n1} & c_{n2} & \cdots & c_{nn} \end{bmatrix} \cdot \begin{bmatrix} x_1 \\ x_2 \\ \vdots \\ x_n \end{bmatrix}$$

The $\boldsymbol{C}$ matrix is a square symmetrical matrix, thus all off-diagonal terms must be equal, $c_{ij} = c_{ji}$.

Gradient Vectors

The first and second derivatives of $z = a + \boldsymbol{b}'\boldsymbol{x} + \boldsymbol{x}'\boldsymbol{C}\boldsymbol{x}$ will be required to determine the necessary and sufficient conditions for a local optimum. The gradient vector $\nabla(\boldsymbol{b}'\boldsymbol{x})$, the linear term of $\boldsymbol{b}'\boldsymbol{x}$ or $\boldsymbol{x}'\boldsymbol{b}$, is obtained by determining the partial derivatives with respect to each of the elements of $\boldsymbol{x}$

$$\nabla(\boldsymbol{b}'\boldsymbol{x}) = \frac{\partial}{\partial \boldsymbol{x}}(\boldsymbol{b}'\boldsymbol{x}) = \frac{\partial}{\partial \boldsymbol{x}}(\boldsymbol{x}'\boldsymbol{b}) = \frac{\partial}{\partial x_j}(b_1x_1 + b_2x_2 + \cdots + b_nx_n) \quad \text{for } j = 1, 2, \ldots, n$$

or

$$\nabla(\boldsymbol{b}'\boldsymbol{x}) = \begin{bmatrix} \frac{\partial}{\partial x_1}(b_1x_1 + b_2x_2 + \cdots + b_nx_n) \\ \frac{\partial}{\partial x_2}(b_1x_1 + b_2x_2 + \cdots + b_nx_n) \\ \vdots \\ \frac{\partial}{\partial x_n}(b_1x_1 + b_2x_2 + \cdots + b_nx_n) \end{bmatrix} = \begin{bmatrix} b_1 \\ b_2 \\ \vdots \\ b_n \end{bmatrix}$$

In matrix notation, the gradient of $\boldsymbol{b}'\boldsymbol{x}$ is simply equal to the column vector of $\boldsymbol{b}$

$$\nabla(\boldsymbol{b}'\boldsymbol{x}) = \boldsymbol{b}$$

The gradient of $\boldsymbol{x}'\boldsymbol{C}\boldsymbol{x}$ is:

$$\nabla(\boldsymbol{x}'\boldsymbol{C}\boldsymbol{x}) = \frac{\partial}{\partial \boldsymbol{x}}(\boldsymbol{x}'\boldsymbol{C}\boldsymbol{x}) = \frac{\partial}{\partial x_j}\left(\sum_i c_{ij}x_ix_j\right) \quad \text{for } j = 1, 2, \ldots, n$$

Performing the mathematical operations and simplifying reduces the expression to

$$\nabla(\boldsymbol{x}'\boldsymbol{C}\boldsymbol{x}) = \begin{bmatrix} 2(c_{11}x_1 + c_{12}x_2 + \cdots + c_{in}x_n) \\ 2(c_{21}x_1 + c_{22}x_2 + \cdots + c_{2n}x_n) \\ \vdots \\ 2(c_{n1}x_1 + c_{n2}x_2 + \cdots + c_{nn}x_n) \end{bmatrix}$$

In matrix form, the gradient of $\boldsymbol{x}'\boldsymbol{C}\boldsymbol{x}$ is simply equal to

$$\nabla(\boldsymbol{x}'\boldsymbol{C}\boldsymbol{x}) = 2\boldsymbol{C}\boldsymbol{x} = 2\boldsymbol{x}'\boldsymbol{C}$$

The second derivative of $x'Cx$ or $\partial^2(x'Cx)/\partial x_i \partial x_j$ for $i = 1, 2, \ldots, n$ and $j = 1, 2, \ldots, n$ is:

$$\nabla\nabla(x'Cx) = \nabla(2Cx) = \nabla(2x'C) = \frac{\partial}{\partial x_i}\left(\frac{\partial}{\partial x_j}\left(\sum_i \sum_j c_{ij} x_i x_j\right)\right)$$

This reduces to

$$\nabla\nabla(x'Cx) = 2 \cdot \begin{bmatrix} c_{11} & c_{12} & \cdots & c_{1n} \\ c_{21} & c_{22} & \cdots & c_{2n} \\ & \vdots & & \\ c_{n1} & c_{n2} & \cdots & c_{nn} \end{bmatrix} = 2C$$

The Necessary and Sufficient Conditions for Optima

The necessary and sufficient conditions for finding the location of a minimum or maximum will be determined with use of a Taylor series expansion for a multivariate function. This approach is an extension of the one used to find minima and maxima of univariate functions. The Taylor series expansion for a multivariate function about the point x° is

$$f(x) = f(x^\circ) + \nabla f(x^\circ)\,\Delta x + \tfrac{1}{2}\Delta x' \left[\frac{\partial^2 f(x^\circ)}{\partial x_i \partial x_j}\right] \Delta x + \cdots$$

where $x = x^\circ + \Delta x$ and Δx may be plus or minus values. For a quadratic function of n control variables, x, the higher-order terms are all zero. Suppose the point x° is a minimum point. The analysis for a maximum is similar to this one. The matrix of second partial derivatives $[\partial^2 f(x)/\partial x_i \partial x_j]$ is called the *Hessian matrix*. Evaluating each term of the Taylor expansion, we obtain

$$f(x^\circ) = a + b'x^\circ + x^{\circ\prime}Cx^\circ$$

$$\nabla f(x^\circ)\Delta x = (b' + 2Cx^\circ)\Delta x$$

$$\tfrac{1}{2}\Delta x \left[\frac{\partial^2 f(x^\circ)}{\partial x_i \partial x_j}\right] \Delta x = \Delta x' C \Delta x$$

Substituting these terms into $f(x)$ we obtain

$$f(x) = (a + b'x^\circ + x^\circ C x^\circ) + (b' + 2Cx^\circ)\,\Delta x + \Delta x' C \Delta x$$

If x° is the location of the minimum, then $f(x) > f(x^\circ)$ for all Δx. Since Δx may take on plus or minus values, the only possible way this requirement can be met is if $\nabla f(x^\circ) = 0$ and $\Delta x' C \Delta x$ is a positive quantity. In other words, the critical point must satisfy the following relationship.

$$b' + 2Cx^\circ = 0$$

or

$$x^\circ = -\tfrac{1}{2}C^{-1}b$$

if C^{-1} exists. This relationship will be used to determine the critical points whether the point is a minimum, a maximum, or a saddle point. This condition is called the *necessary condition for an optimum.* The sufficient condition for a minimum of the quadratic function is that the quadratic form $\Delta x' C \Delta x$ is always a positive value. If $\Delta x' C \Delta x$ is positive, then $f(x)$ is a convex function and a local minimum exists at x°. This condition will be investigated further after the following necessary conditions are summarized. The conditions for a local minimum are in the following theorems.

Local Minimum

THEOREM 5 A *sufficient condition* for $f(x)$ to have a local minimum at x°, where

$$\nabla f(x^\circ) = 0$$

is that $f(x)$ be a *convex* function.

THEOREM 6 A quadratic function, $f(x) = a + b'x + x'Cx$ is a convex function if C is *positive semidefinite* or *positive definite*. That is, for all nonzero x,

$$x'Cx \geq 0 \qquad \text{Positive semidefinite}$$

$$x'Cx > 0 \qquad \text{Positive definite}$$

Since $x = x^\circ + \Delta x$, if $x'Cx$ is positive definite or positive semidefinite, then $\Delta x' C \Delta x$ is positive definite or semidefinite also. Theorem 5 applies to quadratic functions and to other nonlinear functions as well. Theorems 5 and 6 used together form the sufficiency test for proving the critical point x° is the location of a minimum. Theorem 6 will be used to prove that $\Delta x' C \Delta x$ is a positive value.

Local Maximum

THEOREM 7 A *sufficient condition* for $f(x)$ to have a local maximum at x°, where

$$\nabla f(x^\circ) = 0$$

is that $f(x)$ be a concave function.

THEOREM 8 A quadratic function $f(x) = a + b'x + x'Cx$ is a concave function if C is a negative semidefinite or negative definite. That is, for all nonzero x:

$$x'Cx \leq 0 \qquad \text{Negative semidefinite}$$

$$x'Cx < 0 \qquad \text{Negative definite}$$

Positive and Negative Definite Matrices

The quadratic form $x'Cx$ is a scalar quantity that is either positive, negative, or zero for given values of an x vector. If $x'Cx$ is strictly positive or $x'Cx > 0$ for all values of x other than $x = 0$, the quadratic form is *positive definite*. If $x'Cx$ is strictly negative or $x'Cx < 0$ for all values of x other than $x = 0$, the quadratic form is *negative definite*. If $x'Cx \geq 0$ or $xCx \leq 0$, for all values of x other than $x = 0$, the quadratic form is *positive semidefinite* or *negative semidefinite*, respectively. A quadratic form that is positive for some values of x and negative for other values of x is called *indefinite*.

The following equations are examples of positive definite, positive semidefinite, and indefinite functions:

$$f(x) = x_1^2 + 2x_2^2 + 3x_3^2 \qquad \text{Positive definite}$$

$$f(x) = 2x_1^2 + 3(x_2 - x_3)^2 \qquad \text{Positive semidefinite}$$

$$f(x) = x_1^2 - 2x_2^2 + 3x_3^2 \qquad \text{Indefinite}$$

The function $f(x) = x_1^2 + 2x_2^2 + 3x_3^2$ is positive definite and a convex function because $f(x)$ is always positive for all positive and negative values of x except when $x = 0$. For a function $f(x) = 2x_1^2 + 3(x_2 - x_3)^2$, $f(x)$ is always positive for all positive and negative values of x except when $x_1 = 0$ and $x_2 = x_3$. When $x_2 = x_3$, $f(x)$ is equal to zero. This equation is positive semidefinite and a convex function. For the function $f(x) = x_1^2 - 2x_2^2 + 3x_3^2$, $f(x)$ will be either positive or negative depending on the values of the x vector. It is indefinite and neither a convex nor a concave function.

$$x_1^2 + 2x_2^2 + 3x_3^2 = [x_1 \quad x_2 \quad x_3] \cdot \begin{bmatrix} 1 & 0 & 0 \\ 0 & 2 & 0 \\ 0 & 0 & 3 \end{bmatrix} \cdot \begin{bmatrix} x_1 \\ x_2 \\ x_3 \end{bmatrix}$$

$$2x_1^2 + 3(x_2 - x_3)^2 = [x_1 \quad x_2 \quad x_3] \cdot \begin{bmatrix} 2 & 0 & 0 \\ 0 & 3 & -3 \\ 0 & -3 & 3 \end{bmatrix} \cdot \begin{bmatrix} x_1 \\ x_2 \\ x_3 \end{bmatrix}$$

$$2x_1^2 + 3x_2^2 - 3x_2x_3 + 3x_3^2$$

$$x_1^2 - 2x_2^2 + 3x_3^2 = [x_1 \quad x_2 \quad x_3] \cdot \begin{bmatrix} 1 & 0 & 0 \\ 0 & -2 & 0 \\ 0 & 0 & 3 \end{bmatrix} \cdot \begin{bmatrix} x_1 \\ x_2 \\ x_3 \end{bmatrix}$$

These equations may be written in quadratic form, $x'Cx$. A different method for classifying an equation as positive definite, positive semidefinite, or indefinite can be established by evaluating the C matrix only. A two-step evaluation is employed. First, all diagonals must be positive values. For a quadratic form of dimension n, all c_{ii} must be greater than zero, $c_{ii} \geq 0$ for $i = 1, 2, \ldots, n$. The main diagonal terms are equal to the second partial derivative of $f(x)$ with respect to x_i.

$$c_{ii} = \frac{\partial^2 f(x)}{\partial x_i^2} \geq 0$$

The first two equations satisfy this requirement. The third one does not. The second requirement is that the principal minors of the Hessian matrix, $[\partial^2 f(x)/\partial x_i \partial x_j] = C$, be all positive. The test for a positive definite matrix is

$$|c_{11}| > 0 \qquad \begin{vmatrix} c_{11} & c_{12} \\ c_{21} & c_{22} \end{vmatrix} > 0 \qquad \begin{vmatrix} c_{11} & c_{12} & c_{13} \\ c_{21} & c_{22} & c_{23} \\ c_{31} & c_{32} & c_{33} \end{vmatrix} > 0, \ldots$$

The test for a positive semidefinite matrix is

$$|c_{11}| \geq 0 \qquad \begin{vmatrix} c_{11} & c_{12} \\ c_{21} & c_{22} \end{vmatrix} \geq 0 \qquad \begin{vmatrix} c_{11} & c_{12} & c_{13} \\ c_{21} & c_{22} & c_{23} \\ c_{31} & c_{32} & c_{33} \end{vmatrix} \geq 0$$

For $f(\mathbf{x}) = x_1^2 + 2x_2^2 + 3x_3^2$, the principal minors are

$$|1| = 1 > 0 \qquad \begin{vmatrix} 1 & 0 \\ 0 & 2 \end{vmatrix} = 2 > 0 \qquad \begin{vmatrix} 1 & 0 & 0 \\ 0 & 2 & 0 \\ 0 & 0 & 3 \end{vmatrix} = 6 > 0$$

Since all principal minors are positive, the equation is a convex function. A unique minimum will exist that is analogous to the function illustrated in Figure 8.1. For $f(\mathbf{x}) = 2x_1^2 + 3(x_2 - x_3)^2$, the principal minors are

$$|2| = 2 > 0 \qquad \begin{vmatrix} 2 & 0 \\ 0 & 3 \end{vmatrix} = 6 > 0 \qquad \begin{vmatrix} 2 & 0 & 0 \\ 0 & 3 & -3 \\ 0 & -3 & 3 \end{vmatrix} = 0$$

Since the principal minors are either positive or equal to zero, the function is a convex function. A unique minimum will not exist for this case. This situation is analogous to one shown in Figure 8.2. For $f(\mathbf{x}) = x_1^2 - 2x_2^2 + 3x_3^2$, there is no need to determine the principal minors because the main diagonal terms of $\mathbf{C}$ have both positive and negative values; thus it cannot be either a positive or a negative definite matrix. It is indefinite.

For concave functions, $\mathbf{C}$ must be negative definite or negative semidefinite. The test for negative definite matrix is

$$|c_{11}| < 0 \qquad \begin{vmatrix} c_{11} & c_{12} \\ c_{21} & c_{22} \end{vmatrix} > 0 \qquad \begin{vmatrix} c_{11} & c_{12} & c_{13} \\ c_{21} & c_{22} & c_{23} \\ c_{31} & c_{32} & c_{33} \end{vmatrix} < 0, \ldots$$

The test for a negative semidefinite matrix is

$$|c_{11}| \leq 0 \qquad \begin{vmatrix} c_{11} & c_{12} \\ c_{21} & c_{22} \end{vmatrix} \geq 0 \qquad \begin{vmatrix} c_{11} & c_{12} & c_{13} \\ c_{21} & c_{22} & c_{23} \\ c_{31} & c_{32} & c_{33} \end{vmatrix} \leq 0, \ldots$$

For both negative definite and negative semidefinite matrices the signs will alternate starting with a negative sign.

EXAMPLE 8.3 Local Minimum of a Quadratic Equation

(a) Write the following expression in the matrix form: $z = c + \mathbf{b}'\mathbf{x} + \mathbf{x}'\mathbf{C}\mathbf{x}$.

$$z = x_1^2 + 2x_2^2 + 3x_3^2 - 2x_1x_2 - 2x_2x_3 - 9x_1 - 10x_2 - 12x_3 + 20$$

(b) Determine the critical point.
(c) Is the critical point a minimum?

Solution

(a) $z = [x_1 \quad x_2 \quad x_3] \cdot \begin{bmatrix} 1 & -1 & 0 \\ -1 & 2 & -1 \\ 0 & -1 & 3 \end{bmatrix} \cdot \begin{bmatrix} x_1 \\ x_2 \\ x_3 \end{bmatrix} + [-9 \quad -10 \quad -12] \begin{bmatrix} x_1 \\ x_2 \\ x_3 \end{bmatrix} + 20$

(b) The necessary condition for a critical point is

$$\nabla f(\boldsymbol{x}) = \boldsymbol{0} \quad \text{or} \quad \boldsymbol{x} = -\tfrac{1}{2}\boldsymbol{C}^{-1}\boldsymbol{b}$$

$$\boldsymbol{x} = \begin{bmatrix} x_1 \\ x_2 \\ x_3 \end{bmatrix} = -\tfrac{1}{2} \begin{bmatrix} 1 & -1 & 0 \\ -1 & 2 & -1 \\ 0 & -1 & 3 \end{bmatrix}^{-1} \cdot \begin{bmatrix} -9 \\ -10 \\ -12 \end{bmatrix} = -\tfrac{1}{2} \cdot \tfrac{1}{2} \begin{bmatrix} 5 & 3 & 1 \\ 3 & 3 & 1 \\ 1 & 1 & 1 \end{bmatrix} \cdot \begin{bmatrix} -9 \\ -10 \\ -12 \end{bmatrix}$$

$$\boldsymbol{x}^\circ = \begin{bmatrix} 87/4 \\ 69/4 \\ 31/4 \end{bmatrix}$$

(c) The critical point x° is a minimum if the Hessian matrix of $\boldsymbol{C}$ is positive definite. All main diagonal terms are positive, $c_{11} = 1, c_{22} = 2$, and $c_{33} = 3$. The principal minors are

$$\begin{vmatrix} 1 & -1 & 0 \\ -1 & 2 & -1 \\ 0 & -1 & 3 \end{vmatrix} = 2 > 0$$

$$\begin{vmatrix} 2 & -1 \\ -1 & 3 \end{vmatrix} = 5 > 0$$

$$|3| = 3 > 0$$

Therefore, $\boldsymbol{C}$ is positive definite and $\boldsymbol{x}^\circ$ is the location of the minimum.

EXAMPLE 8.4 Virtual Work

Use the principles of virtual work to determine the equilibrium position of the spring–mass system shown in Figure 8.6. Consider the dead weight of the blocks only.

(a) Use matrix algebra to solve for the equilibrium positions.

(b) Show the system is stable.

Solution

(a) The total potential energy V consists of the sum of potential energy of the system:

$$V = V_e + V_g$$

where V_e is the elastic energy of the springs, $\frac{1}{2}ke^2$; V_g is the gravitational energy of the weights, $-Wx$; e is the elongation of the spring; and x is the displacement from equilibrium position.

The total potential energy of the system is

$$V = \tfrac{1}{2}k_1x_1^2 + \tfrac{1}{2}k_2(x_2 - x_1)^2 - W_1x_1 - W_2x_2$$

$$V = \tfrac{1}{2}(2000)x_1^2 + \tfrac{1}{2}(500)(x_2^2 - 2x_1x_2 + x_1^2) - 2000x_1 - 4000x_2$$

In matrix notation

$$V = \tfrac{1}{2}[x_1 \quad x_2] \cdot \begin{bmatrix} 2000 & -500 \\ -500 & 500 \end{bmatrix} \cdot \begin{bmatrix} x_1 \\ x_2 \end{bmatrix} + [-2000 \quad -4000] \cdot \begin{bmatrix} x_1 \\ x_2 \end{bmatrix}$$

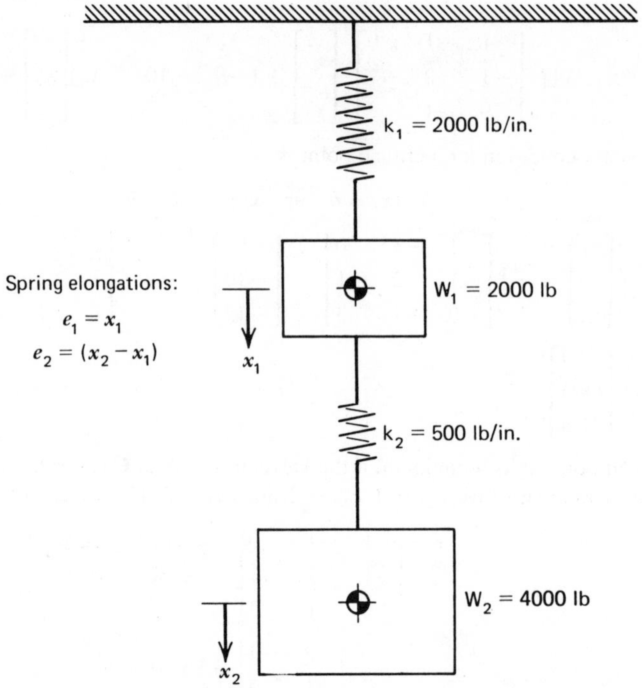

Figure 8.6

For equilibrium, the total energy is a minimum. Thus, the mathematical model is

$$\text{Minimize } V = V_e + V_g$$

The necessary condition is

$$\Delta_x V = 0$$

or

$$\tfrac{1}{2}(2) \cdot \begin{bmatrix} 2000 & -500 \\ -500 & 500 \end{bmatrix} \cdot \begin{bmatrix} x_1 \\ x_2 \end{bmatrix} + \begin{bmatrix} -2000 \\ -4000 \end{bmatrix} = \begin{bmatrix} 0 \\ 0 \end{bmatrix}$$

Solving for $\boldsymbol{x}$

$$\boldsymbol{x} = \begin{bmatrix} x_1 \\ x_2 \end{bmatrix} = \begin{bmatrix} 2000 & -500 \\ -500 & 500 \end{bmatrix}^{-1} \cdot \begin{bmatrix} 2000 \\ 4000 \end{bmatrix} = \begin{bmatrix} 4 \\ 12 \end{bmatrix}$$

The deflection of weight 1 is 4 in. and of weight 2 is 12 in.

(b) The system is stable if the total energy is a minimum. In other words, the sufficiency test for a minimum must be satisfied. It is sufficient to show that $\boldsymbol{C}$ is a positive semidefinite:

$$\boldsymbol{C} = \begin{bmatrix} 2000 & -500 \\ -500 & 500 \end{bmatrix}$$

All main diagonal terms are positive and each principal minor is positive.

$$2000 > 0 \qquad \begin{vmatrix} 2000 & -500 \\ -500 & 500 \end{vmatrix} = 750{,}000 > 0$$

We have proved that the system is stable.

PROBLEMS

Problems 1 through 3

For Problems 1 through 3, answer the following questions.

(a) Write the quadratic equations as a sum of a linear and quadratic forms, that is, $z = a + \boldsymbol{b}'\boldsymbol{x} + \boldsymbol{x}'\boldsymbol{C}\boldsymbol{x}$.

(b) Use the Hessian matrix to classify these functions as being positive definite, positive semidefinite, negative definite, negative semidefinite, or none of these.

(c) Classify each function as being a convex function, a concave function, or neither one.

Problem 1

$$z = 6 + 2x_1 - x_2 - x_1^2 - 2x_2^2 + x_1x_2 - x_3^2$$

Problem 2

$$z = 8 - 2x_1 - 3x_2 - x_1x_2 - x_2x_3 - x_1x_3$$

Problem 3

$$z = (x_1 - x_2)^2 + (x_2 - x_3)^2 + (x_1 - x_3)^2 + 4x_1^2 + 5x_2^2 + 6x_3^2$$

Problems 4 through 6

For Problems 4 through 6, answer the following questions.

(a) Write in quadratic form.

(b) Determine the critical point $\boldsymbol{x}^{\circ}$.

(c) Use Theorems 5 through 8 to determine if the critical point $\boldsymbol{x}^{\circ}$ is a local maximum or a local minimum.

Problem 4

$$f(\boldsymbol{x}) = x_1^2 + 3x_2^2$$

Problem 5

$$f(\boldsymbol{x}) = 2x_1^2 + x_2^2 - x_1x_2$$

Problem 6

$$f(\boldsymbol{x}) = x_1^2 - 2x_1x_2 + x_2^2$$

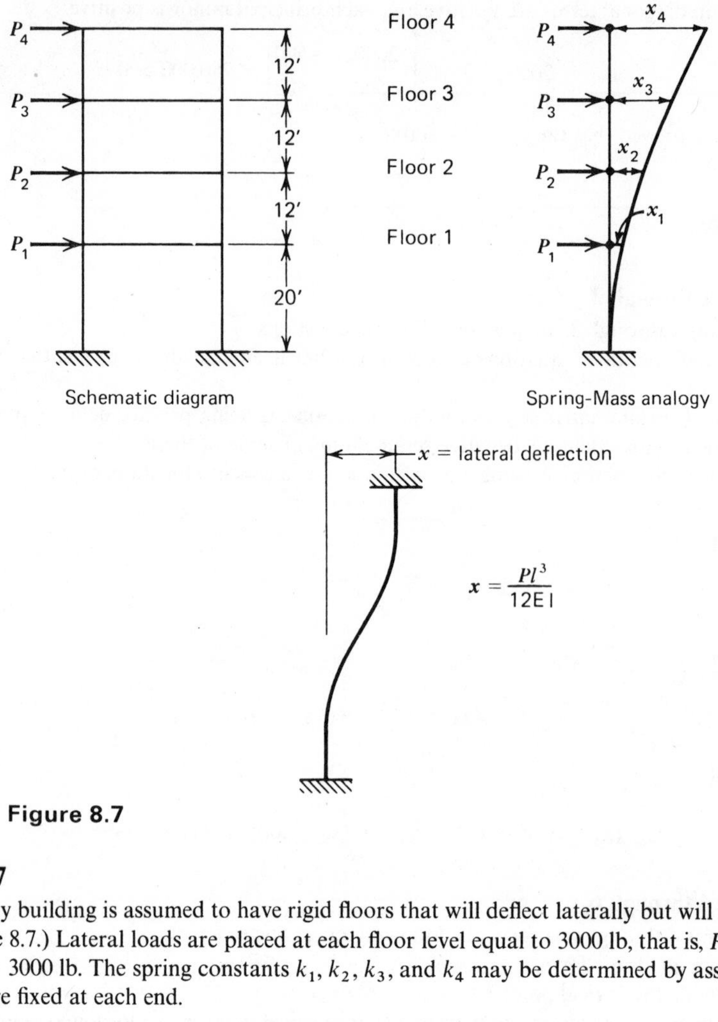

Figure 8.7

Problem 7

A four-story building is assumed to have rigid floors that will deflect laterally but will not rotate. (See Figure 8.7.) Lateral loads are placed at each floor level equal to 3000 lb, that is, $P_1 = P_2 = P_3 = P_4 = 3000$ lb. The spring constants k_1, k_2, k_3, and k_4 may be determined by assuming the columns are fixed at each end.

$$E = 30{,}000{,}000 \text{ psi} \qquad I = 500 \text{ in}^4$$

(a) Use the concepts of virtual work to formulate the problem to determine the lateral deflection of each floor.

(b) Determine the static deflections of each floor.

(c) Prove the structure is stable.

8.3 LOCAL OPTIMA OF NONLINEAR MULTIVARIATE FUNCTIONS

The concepts developed in Section 8.2 for quadratic equations are extended for a general nonlinear multivariate function. The necessary and sufficient conditions for finding a local minimum and a local maximum of a nonlinear function will be determined from calculus and the Taylor series expansion.

Local Minimum

The Taylor series expansion for the multivariate function $f(\boldsymbol{x})$ is

$$f(\boldsymbol{x}) = f(\boldsymbol{x}^\circ) + [\nabla f(\boldsymbol{x}^\circ)]'\Delta\boldsymbol{x} + \tfrac{1}{2}(\Delta\boldsymbol{x}')\boldsymbol{H}(\boldsymbol{x}^\circ)\Delta\boldsymbol{x} + \cdots$$

where $\nabla f(\boldsymbol{x}^\circ)$ is the gradient vector of $f(\boldsymbol{x})$ evaluated at $\boldsymbol{x} = \boldsymbol{x}^\circ$.

$$\nabla f(\boldsymbol{x}^\circ) = \begin{bmatrix} \dfrac{\partial f(\boldsymbol{x}^\circ)}{\partial x_1} \\ \dfrac{\partial f(\boldsymbol{x}^\circ)}{\partial x_2} \\ \vdots \\ \dfrac{\partial f(\boldsymbol{x}^\circ)}{\partial x_n} \end{bmatrix}$$

and $\boldsymbol{H}(\boldsymbol{x}^\circ)$ is the Hessian matrix evaluated at $\boldsymbol{x} = \boldsymbol{x}^\circ$.

$$\boldsymbol{H}(\boldsymbol{x}^\circ) = \begin{bmatrix} \dfrac{\partial^2 f(\boldsymbol{x}^\circ)}{\partial x_1^2} & \dfrac{\partial^2 f(\boldsymbol{x}^\circ)}{\partial x_1 \partial x_2} & \cdots & \dfrac{\partial^2 f(\boldsymbol{x}^\circ)}{\partial x_1 x_n} \\ \dfrac{\partial^2 f(\boldsymbol{x}^\circ)}{\partial x_2 \partial x_1} & \dfrac{\partial^2 f(\boldsymbol{x}^\circ)}{\partial x_2^2} & \cdots & \dfrac{\partial^2 f(\boldsymbol{x}^\circ)}{\partial x_2 \partial x_n} \\ \vdots & & & \vdots \\ \dfrac{\partial^2 f(\boldsymbol{x}^\circ)}{\partial x_n \partial x_1} & \dfrac{\partial^2 f(\boldsymbol{x}^\circ)}{\partial x_n \partial x_2} & \cdots & \dfrac{\partial^2 f(\boldsymbol{x}^\circ)}{\partial x_n^2} \end{bmatrix}$$

In order for $\boldsymbol{x}^\circ$ to be the location of the minimum, any point in the neighborhood of $\boldsymbol{x}^\circ$ must satisfy the inequality condition $f(\boldsymbol{x}^\circ) \le f(\boldsymbol{x})$. At a minimum, the gradient vector $f'(\boldsymbol{x}^\circ)$ must be equal to zero because each element of the $\Delta\boldsymbol{x}$ vector may consist of either plus or minus elements. The *necessary condition for a minimum* is

$$\boxed{\nabla f(\boldsymbol{x}^\circ) = \boldsymbol{0}}$$

The critical point $\boldsymbol{x}^\circ$, whether it be a location of a minimum, maximum, or point of inflection, must satisfy this equation. Each element of the gradient vector must be equal to zero:

$$\frac{\partial f(\boldsymbol{x}^\circ)}{\partial x_1} = 0$$

$$\frac{\partial f(\boldsymbol{x}^\circ)}{\partial x_2} = 0$$

$$\vdots$$

$$\frac{\partial f(\boldsymbol{x}^\circ)}{\partial x_n} = 0$$

If the third term of the Taylor expansion $\frac{1}{2}\Delta\boldsymbol{x}'[\boldsymbol{H}(\boldsymbol{x}^\circ)]\Delta\boldsymbol{x}$ is a positive value for all nonzero $\Delta\boldsymbol{x}$, then the condition $f(\boldsymbol{x}^\circ) \le f(\boldsymbol{x})$ is satisfied. In other words, if the following equality is satisfied

$$\Delta\boldsymbol{x}'\boldsymbol{H}(\boldsymbol{x}^\circ)\Delta\boldsymbol{x} > 0$$

then $\boldsymbol{x}^\circ$ is a local minimum of $f(\boldsymbol{x})$. Thus if the Hessian matrix $\boldsymbol{H}(\boldsymbol{x}^\circ)$ is positive definite, the sufficient condition for a minimum, $f(\boldsymbol{x}^\circ) < f(\boldsymbol{x})$, is satisfied. The Hessian matrix is positive definite matrix if (1) all main diagonal elements of $\boldsymbol{H}(\boldsymbol{x}^\circ)$ are greater than zero, and (2) the principle minors of $H(\boldsymbol{x}^\circ)$ are greater than zero. The first condition is satisfied if

$$\frac{\partial^2 f(\boldsymbol{x}^\circ)}{\partial x_1^2} > 0, \qquad \frac{\partial^2 f(\boldsymbol{x}^\circ)}{\partial x_2^2} > 0, \quad \ldots, \quad \frac{\partial^2 f(\boldsymbol{x}^\circ)}{\partial x_n^2} > 0$$

The second condition is satisfied if

$$\frac{\partial^2 f(\boldsymbol{x}^\circ)}{\partial x_1^2} > 0; \quad \begin{vmatrix} \dfrac{\partial^2 f(\boldsymbol{x}^\circ)^2}{\partial x_1^2} & \dfrac{\partial^2 f(\boldsymbol{x}^\circ)}{\partial x_1 \partial x_2} \\ \dfrac{\partial^2 f(\boldsymbol{x}^\circ)}{\partial x_2 \partial x_1} & \dfrac{\partial^2 f(\boldsymbol{x}^\circ)}{\partial x_2^2} \end{vmatrix} > 0; \; \ldots ; \; \begin{vmatrix} \dfrac{\partial f(\boldsymbol{x}^\circ)}{\partial x_1^2} & \dfrac{\partial^2 f(\boldsymbol{x}^\circ)}{\partial x_1 \partial x_2} & \cdots & \dfrac{\partial^2 f(\boldsymbol{x}^\circ)}{\partial x_1 \partial x_n} \\ \dfrac{\partial^2 f(\boldsymbol{x}^\circ)}{\partial x_2 \partial x_1} & \dfrac{\partial^2 f(\boldsymbol{x}^\circ)}{\partial x_2^2} & \cdots & \dfrac{\partial^2 f(\boldsymbol{x}^\circ)}{\partial x_2 \partial x_n} \\ \vdots & & & \\ \dfrac{\partial^2 f(\boldsymbol{x}^\circ)}{\partial x_n \partial x_1} & \dfrac{\partial^2 f(\boldsymbol{x}^\circ)}{\partial x_n \partial x_2} & \cdots & \dfrac{\partial^2 f(\boldsymbol{x}^\circ)}{\partial x_n^2} \end{vmatrix} > 0$$

The Hessian matrix is always a symmetrical matrix even for generalized nonlinear functions.

A nonlinear function that satisfies the condition of a positive definite Hessian matrix is a *convex function.*

THEOREM 9 A nonlinear function $f(\boldsymbol{x})$ is a convex function if the Hessian matrix $H(\boldsymbol{x}^\circ)$ evaluated at $\boldsymbol{x}^\circ$, where $\nabla f(\boldsymbol{x}^\circ) = \boldsymbol{0}$, is positive definite.

The case when the Hessian positive semidefinite is not considered.

Local Maximum

The necessary condition for a local *maximum* is the same as for a minimum:

$$\nabla f(\boldsymbol{x}^\circ) = \boldsymbol{0}$$

The critical point $\boldsymbol{x}^\circ$ satisfies the sufficient condition for a maximum when $\boldsymbol{H}(\boldsymbol{x}^\circ)$ is negative definite. (1) All main diagonal terms of the Hessian matrix must be negative values, and (2) the principal minors of the Hessian matrix must alternate in sign.

$$\frac{\partial^2 f(\boldsymbol{x}^\circ)}{\partial x_1^2} < 0; \quad \begin{vmatrix} \dfrac{\partial^2 f(\boldsymbol{x}^\circ)}{\partial x_1^2} & \dfrac{\partial^2 f(\boldsymbol{x}^\circ)}{\partial x_1 \partial x_2} \\ \dfrac{\partial^2 f(\boldsymbol{x}^\circ)}{\partial x_2 \partial x_1} & \dfrac{\partial^2 f(\boldsymbol{x}^\circ)}{\partial x_2^2} \end{vmatrix} > 0;$$

$$\begin{vmatrix} \dfrac{\partial^2 f(\boldsymbol{x}^\circ)}{\partial x_1^2} & \dfrac{\partial f^2(\boldsymbol{x}^\circ)}{\partial x_1 \partial x_2} & \cdots & \dfrac{\partial^2 f(\boldsymbol{x}^\circ)}{\partial x_1 \partial x_n} \\ \dfrac{\partial^2 f(\boldsymbol{x}^\circ)}{\partial x_2 \partial x_1} & \dfrac{\partial^2 f(\boldsymbol{x}^\circ)}{\partial x_2} & \cdots & \dfrac{\partial^2 f(\boldsymbol{x}^\circ)}{\partial x_2 \partial x_n} \\ \vdots & & & \vdots \\ \dfrac{\partial^2 f(\boldsymbol{x}^\circ)}{\partial x_n \partial x_1} & \dfrac{\partial^2 f(\boldsymbol{x}^\circ)}{\partial x_n \partial x_2} & \cdots & \dfrac{\partial f^2(\boldsymbol{x}^\circ)}{\partial x_n^2} \end{vmatrix} \begin{array}{l} > 0 \text{ if } n \text{ is even and} \\ < 0 \text{ if } n \text{ is odd} \end{array}$$

A nonlinear function that satisfies the condition of a negative definite Hessian matrix is a concave function.

THEOREM 10 A nonlinear function $f(\boldsymbol{x})$ is a concave function if the Hessian matrix evaluated at $\boldsymbol{x}^\circ$ where $\nabla f(\boldsymbol{x}^\circ) = \boldsymbol{0}$ is negative definite.

Separable Functions

When a function consists of a sum of separable functions, it is sometimes computationally easier to apply the following theorems for determining the convexity and concavity of the function.

THEOREM 11 If each $f(x_j)$ is a convex function over a convex set, then the sum of the functions, $f(\boldsymbol{x}) = \sum_j f(x_j)$ for $j = 1, 2, \ldots, n$ is a convex function.

THEOREM 12 If each $f(x_j)$ is a concave function over the convex set, then the sum of the functions, $f(\boldsymbol{x}) = \sum_j f(x_j)$, is a concave function.

The following are examples of separable and nonseparable equations:

$$f(\boldsymbol{x}) = x_1^2 + x_2^2 \qquad \text{Separable}$$

$$f(\boldsymbol{x}) = x_1^2 + x_2^2 - 4x_1x_2 \qquad \text{Nonseparable}$$

The first function may be separated into two parts with each part a function of x_1 and x_2, exclusively. Thus $f(\boldsymbol{x}) = f(x_1) + f(x_2)$, where $f(x_1) = x_1^2$ and $f(x_2) = x_2^2$. Since $f(x_1)$ and $f(x_2)$ are both convex functions, the sum of them is a convex function. It is not possible to separate the second function into two functions of x_1 and x_2 exclusively. Theorems 11 and 12 cannot be applied in this instance.

EXAMPLE 8.5 The Convexity of a Travel Demand Function

The total number of work trips by bus between business and residential zones Q is a function of the worker population in a business zone W, the worker population in a residential zone E, and the bus fare charged to travel the business and residential zones C. The total volume of work trips may be estimated with the function:

$$Q = 20W^{0.4}E^{0.3}C^{-0.2} \qquad \text{for } W \geq 0,\ E \geq 0, \text{ and } C \geq 0$$

(a) Utilize the Hessian test to determine if the function is a convex function, a concave function, or neither.

(b) Take the logarithm of the function and show that the function is a separable function. Confirm the result obtained in part a.

Solution

(a) The Hessian matrix with $x' = [W, E, C]$ is

$$H(x) = \begin{bmatrix} (-4.8W^{-1.6}E^{0.3}C^{-0.2}) & (2.4W^{-0.6}E^{-0.7}C^{-0.2}) & (-1.6W^{-0.6}E^{0.3}C^{-1.2}) \\ (2.4W^{-0.6}E^{-0.7}C^{-0.2}) & (-4.2W^{0.4}E^{-1.7}C^{-0.2}) & (-1.2W^{0.4}E^{-0.7}C^{-1.2}) \\ (-1.6W^{-1.4}E^{0.3}C^{-1.2}) & (-1.2W^{0.4}E^{-0.7}C^{-1.2}) & (4.8W^{0.4}E^{0.3}C^{-2.2}) \end{bmatrix}$$

For $W > 0$, $E > 0$, and $C > 0$, the main diagonal terms of the Hessian matrix are positive and negative values; therefore, the necessary condition for a convex or concave function is not satisfied. The function is neither a convex nor a concave function.

(b) The logarithmic transformation gives the separable function.

$$\ln Q = \ln 20 + 0.4 \ln W + 0.3 \ln E - 0.2 \ln C$$

The second derivatives of $\ln Q$ with respect to W, E, and C are

$$\frac{\partial(\ln Q)}{\partial W} = \frac{0.4}{W} = 0.4W^{-1} \qquad \frac{\partial^2(\ln Q)}{\partial W^2} = \frac{-0.4}{W^2} < 0 \qquad \text{for } W > 0$$

$$\frac{\partial(\ln Q)}{\partial E} = \frac{0.3}{E} = 0.3E^{-1} \qquad \frac{\partial^2(\ln Q)}{\partial E^2} = \frac{-0.3}{E^2} < 0 \qquad \text{for } E > 0$$

$$\frac{\partial(\ln Q)}{\partial T} = \frac{-0.2}{C} = -0.2C^{-1} \qquad \frac{\partial^2(\ln Q)}{\partial^2} = \frac{0.2}{C^2} > 0 \qquad \text{for } C > 0$$

The 0.4 ln W and 0.3 ln E terms are concave functions and the -0.2 ln C term is a convex function. The sum of convex and concave terms gives neither a convex nor a concave function because neither Theorem 11 nor 12 is satisfied.

EXAMPLE 8.6 Local Minimum of a Bivariate Model

Determine the location of the local minimum of

$$z = e^{x_1} - 5x_1 + (x_2 - 2)^2$$

Use the necessary and sufficient conditions for a local minimum as proof.

Solution

The necessary condition for an optimum solution is that all partial derivatives of $f(x)$ are equal to zero.

$$\nabla f(x) = 0$$

or

$$\frac{\partial f}{\partial x_1} = 0 \qquad \frac{\partial}{\partial x_1}(e^{x_1} - 5x_1 + (x_2 - 2)^2) = 0 \qquad e^{x_1} - 5 = 0$$

$$\frac{\partial f}{\partial x_2} = 0 \qquad \frac{\partial}{\partial x_2}(e^{x_1} - 5x_1 + (x_2 - 2)^2) = 0 \qquad 2(x_2 - 2) = 0 \tag{1}$$

Solving for x_1 and x_2,

$$x_1 = \ln 5 \qquad x_2 = 2$$

The critical point is $\boldsymbol{x}^\circ = (x_1^\circ, x_2^\circ) = (\ln (5), 2)$

The sufficient condition for a local minimum is that the Hessian matrix must be positive definite when evaluated at $\boldsymbol{x}^\circ$.

The Hessian matrix is

$$\boldsymbol{H}(\boldsymbol{x}) = \begin{bmatrix} \dfrac{\partial^2 f}{\partial x_1^2} & \dfrac{\partial^2 f}{\partial x_2^2} \\ \dfrac{\partial^2 f}{\partial x_1 \partial x_2} & \dfrac{\partial^2 f}{\partial x_2^2} \end{bmatrix} = \begin{bmatrix} e^{x_1} & 0 \\ 0 & 2 \end{bmatrix}$$

$\boldsymbol{H}(\boldsymbol{x})$ evaluated at $\boldsymbol{x}^\circ$ is

$$\boldsymbol{H}(\boldsymbol{x}^0) = \begin{bmatrix} e^{\ln 5} & 0 \\ 0 & 2 \end{bmatrix} = \begin{bmatrix} 5 & 0 \\ 0 & 2 \end{bmatrix}$$

The main diagonal elements of $\boldsymbol{H}(\boldsymbol{x}^\circ)$, 5 and 2, are greater than zero; therefore, the first part of the sufficiency test is satisfied. The determinant of the principal minors are positive.

$$|5| = 5 > 0 \qquad \begin{vmatrix} 5 & 0 \\ 0 & 2 \end{vmatrix} = 10 > 0$$

The objective function is a convex function, and the critical point is the location of a local minimum.

EXAMPLE 8.7 Determination of the Optimum Production Level

A company is considering manufacturing two types of products. The unit selling prices are \$100 and \$150 per unit for items 1 and 2, respectively. The costs to manufacture are $C_1 = 50q_1^{1.1}$ for item 1 and $C_2 = 110q_2^{1.05}$ for item 2. Assume that there is sufficient demand and that all goods produced will be sold.

(a) Determine the optimum production level to maximize total profit.
(b) Prove that the optimum solution is a local maximum.

Solution

(a) The mathematical model to maximize profit P is equal to revenue minus production cost, $P = R - C$. The total revenue is $R = 100q_1 + 150q_2$, and total cost is $C = 50q_1^{1.1} + 110q_2^{1.05}$.

$$P = \$100q_1 + 150q_2 - 50q_1^{1.1} - 110q_2^{1.05}$$

The necessary condition for a maximum is

$$\frac{\partial P}{\partial q_1} = 100 - 55q_1^{0.1} = 0$$

$$\frac{\partial P}{\partial q_2} = 150 - 115.5q_2^{0.05} = 0$$

Solving for q_1 and q_2 gives $q_1^\circ = (100/55)^{10} = 394$ and $q_2^\circ = (150/115.5)^{20} = 186$.

(b) If the Hessian matrix is negative definite, the profit function is a concave function and the critical point found in part a is a local maximum. The Hessian matrix evaluated at the critical point $q_1^\circ = 394$ and $q_2^\circ = 186$ is

$$\boldsymbol{H}(\boldsymbol{q}^\circ) = \begin{bmatrix} \dfrac{\partial^2 f}{\partial q_1^2} & \dfrac{\partial^2 f}{\partial q_1 \partial q_2} \\ \dfrac{\partial^2 f}{\partial q_2 \partial q_1} & \dfrac{\partial^2 f}{\partial q_2^2} \end{bmatrix} = \begin{bmatrix} -5.5(q_1^\circ)^{-0.9} & 0 \\ 0 & -5.78(q_2^\circ)^{-0.95} \end{bmatrix} = \begin{bmatrix} -0.03 & 0 \\ 0 & -0.04 \end{bmatrix}$$

The main diagonal elements of the Hessian matrix are negative, and the determinant principal minors alternate in sign.

$$|-0.03| = -0.03 > 0 \qquad \begin{vmatrix} -0.03 & 0 \\ 0 & -0.04 \end{vmatrix} = 0.0012 > 0$$

Thus the local maximum is located at $q_1^* = 394$ and $q_2^* = 186$. The maximum profit is $P^* = \$4920$.

PROBLEMS

Problem 1 through 3

For Problems 1 through 3 answer the following questions.

(a) Use the Hessian matrix to classify these functions as being positive definite, positive semidefinite, negative definite, negative semidefinite, or neither.

(b) Determine the critical point. What is the nature of the critical point?

Problem 1

$$z = \ln x_1 + x_2^2 \qquad \text{for } \boldsymbol{x} \geq 0$$

Problem 2

$$z = 6x_1^{0.2}x_2^{0.8} \qquad \text{for } \boldsymbol{x} \geq 0$$

Problem 3

$$z = e^{-x_1} + x_2^2 - \frac{x_1}{2} \qquad \text{for } \boldsymbol{x} \geq 0$$

Problems 4 and 5

For Problems 4 through 5, answer the following questions.

(a) Rewrite the function as a separable function.

(b) Use Theorem 11 or 12 to determine if the function is a convex function, a concave function, or neither one.

Problem 4

$$z = \ln x_1 - x_2^2 \qquad \text{for } \boldsymbol{x} \geq \boldsymbol{0}$$

Problem 5

$$z = e^{-x_1} + x_2^2 - \frac{x_1}{2} \qquad \text{for } \boldsymbol{x} \geq \boldsymbol{0}$$

Problem 6

The number of highway miles M that can be paved is a function of labor and machine hours.

$$M = 0.5L^{0.2}K^{0.8} \quad \text{for} \quad L > 0, K > 0.$$

where L is the labor used in man hours, and K is machine hours.

(a) Use Theorem 9 or 10 to determine if the function is a convex function, a concave function, or neither one.

(b) Take a logarithmic transfer of each side, and define a new separable function. Utilize Theorem 11 or 12 to confirm the results obtained in part a.

8.4 GLOBAL OPTIMA: LINEAR AND SIDE CONSTRAINTS

Thus far, our discussion has been directed primarily at finding local optima of mathematical models with no constraints. In this section, approaches to solving models with side and linear constraints are studied. The mathematical model has the following form:

$$z = f(\boldsymbol{x})$$

$$\boldsymbol{a}_i'\boldsymbol{x}\ \{=, \leq, \geq\}\ b_i \qquad i = 1, 2, \ldots, m \qquad \text{Linear constraints}$$

$$\boldsymbol{x} \leq \boldsymbol{b}, \boldsymbol{x} \geq \boldsymbol{0} \qquad \text{Side constraints}$$

This model consists of a set of linear constraint equations and side constraints. The number of side constraints may be equal to the number of control variables in the model.

Constraint sets of this type are convex constraint sets; therefore, one of the requirements for ensuring that a critical point is a global optimum is satisfied. See Theorems 1 through 4. In order to make this discussion complete, convex sets will be defined and constraint sets consisting of side and linear constraints will be shown to satisfy the requirements of these definitions.

Convex Sets

A constraint set is defined to be a convex set if a line joining any two points in the set falls within the boundary of the set. This is a general definition that is applicable to constraint sets consisting of side, linear, and nonlinear equations. The feasible

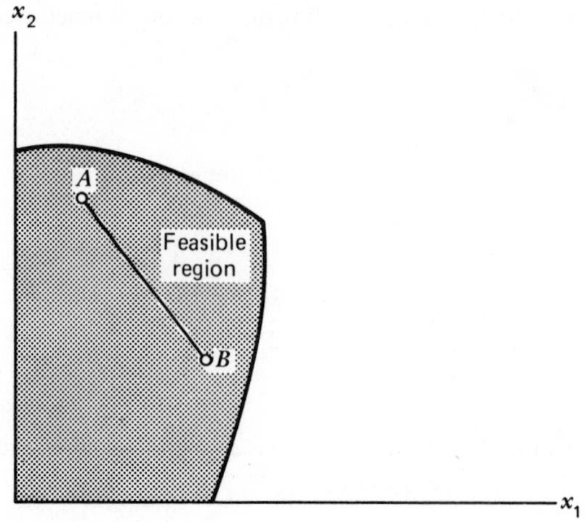

Figure 8.8 A convex set.

regions shown in Figures 8.8 and 8.9 are examples of convex and nonconvex sets. For problems of one and two dimensions, graphic methods of analyzing a constraint set are most useful. For models of higher order, mathematical theorems will be employed.

THEOREM 13 The intersection of convex sets is a convex set.

The half-plane $x_1 \geq 0$ is a convex set. The constraint set consisting of two half-planes, $x_1 \geq 0$ and $x_2 \geq 0$, is an example of an intersection of convex sets. By Theorem 13, the intersection of two convex sets, $x_1 \geq 0$ and $x_2 \geq 0$, forms a convex set. This

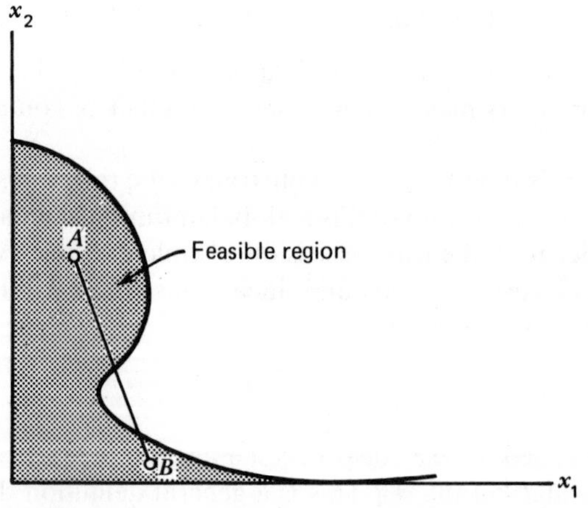

Figure 8.9 A nonconvex set.

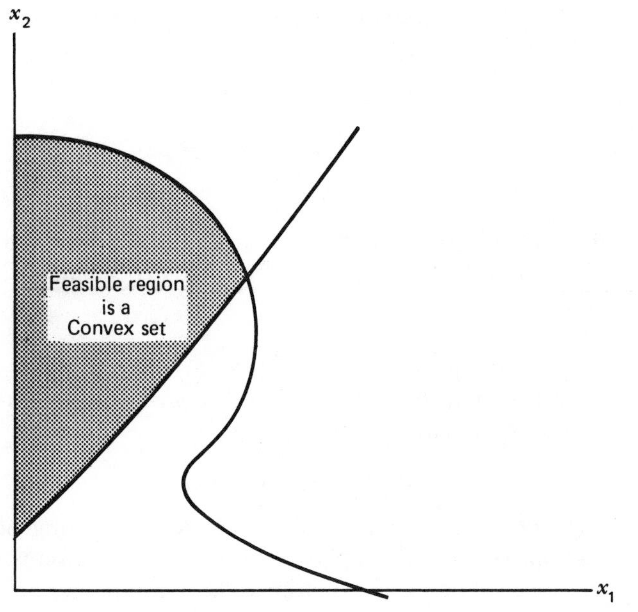

Figure 8.10 Intersection of nonconvex and convex sets.

result should be obvious. The power of this theorem lies in the fact that each equation can be analyzed separately. If each equation of the constraint set is found to be a convex, then the intersection of all of them will be a convex set.

It is not neccessarily true that *all* individual constraint equations be convex sets. The intersection of a nonconvex set with a convex set and nonconvex sets with another convex set may form a convex set. Figure 8.10 shows a convex formed by the intersection of a convex and a nonconvex set.

THEOREM 14 A hyperplane is a convex set. The linear equation defines a hyperplane $\boldsymbol{a}'\boldsymbol{x} = b$, and the linear inequalities $\boldsymbol{a}'\boldsymbol{x} \leq b$ and $\boldsymbol{a}'\boldsymbol{x} \geq b$ define half-spaces consisting of n control variables, $\boldsymbol{x}' = [x_1 x_2 \cdots x_n]$.

From Theorems 13 and 14, mathematical models restricted to side and linear constraints will be convex sets. As a result, for problems of this type global optima may be determined by evaluating the objective function $f(\boldsymbol{x})$ as a convex or a concave function.

The Method of Substitution

When side and linear constraints are introduced into the analysis as active constraints, the method of substitution is most applicable. If the original mathematical model of m control variables of $\boldsymbol{x}$, then the addition of each active constraint will reduce the order of the problem by 1. For example, consider the following optimization problem consisting of two control variables.

$$\text{Minimize } z = f(x_1, x_2)$$
$$g(x_1, x_2) = b$$
$$x_1 \geq 0, x_2 \geq 0$$

With the method of substitution, we select one variable in the constraint equation and let it be a dependent variable. Here we select x_2 as the dependent variable and solve for x_2 in terms of x_1.

$$x_2 = \phi(x_1)$$

Substitute $x_2 = \phi(x_1)$ into the objective function. The problem reduces to an objective function with a single unknown:

$$\text{Minimize } z = f(x_1, \phi(x_1))$$

This approach may be used for mathematical models consisting of equations of higher order. This approach is only applicable if the inverse relationship $x_2 = \phi(x_1)$ can be easily determined. For linear equations, the solution can be found without difficulty.

EXAMPLE 8.8 Effects of Plant Capacity on Production

In Example 8.7, "Determination of the Optimum Production Level," the production schedule that will result in maximum profit is to manufacture 394 and 186 units of items 1 and 2, respectively. Owing to limited floor space, the maximum number of units the company can produce at any one time is 500 units.

(a) Formulate a mathematical model to maximize profit.
(b) Determine the optimum production level, and show that it is a global maximum.

Solution

(a) The mathematical model from Example 8.7 may be used to establish a constrained optimization model.

$$\text{Maximize } P = 100q_1 + 150q_2 - 95q_1^{1.1} - 125q_2^{1.05}$$

The restriction that the production be limited because of the lack of floor space is

$$q_1 + q_2 \leq 500$$

The mathematical model is given by the following equation set:

$$\text{Maximize } P = 100q_1 + 150q_2 - 50q_1^{1.1} - 110q_2^{1.05}$$
$$q_1 + q_2 \leq 500$$
$$q_1 \geq 0, q_2 \geq 0$$

(b) From previous analysis of the unconstrained model, we know that the objective function is a concave function and the solution $q_1 = 394$ units and $q_2 = 186$ units violates the constraint $q_1 + q_2 \leq 500$.

Clearly, the constraint condition is active. By a method of substitution a local optimum may be found. If the constraint is active, then we may express the constraint condition as

$$q_1 + q_2 = 500$$

Let us solve for q_1 in terms of q_2 and substitute into the objective function.

$$q_1 = 500 - q_2$$

The objective function in terms of q_2 is

$$P = 100(500 - q_2) + 150q_2 - 50(500 - q_2)^{1.1} - 110q_2^{1.05}$$

The problem is reduced as a problem with a single variable. The necessary condition for a maximum is

$$\frac{dP}{dq_2} = -100 + 150 - 50(1.1)(500 - q_2)^{0.1}(-1) - (110)(1.05)q_2^{0.05} = 0$$

Simplifying

$$\frac{dP}{dq_2} = 50 + 55(500 - q_2)^{0.1} - 115.5q_2^{0.05} = 0$$

The solution is $q_2^\circ = 156$ and $q_1^\circ = 500 - q_2^\circ = 344$.

The second derivative of P with respect to q_2 is

(concave) by Hessian

$$\frac{\partial^2 P}{\partial q_2^2} = 55(0.1)(500 - q_2)^{-0.9}(-1) - 115.5(0.05)q_2^{-0.95}$$

This term is negative for all positive values of q_2; thus, the critical point where $q_2^0 = 156$ is a local maximum.

From Theorem 3, if the objective function is a concave function over a convex set, a local maximum point is a global maximum. We have shown that the critical point is a local maximum. All that remains is to show that the constraint set is a convex set. The constraint equation $q_1 + q_3 \leq 500$ is a linear function; therefore, it forms a convex set. The constraints $q_1 \geq 0$ and $q_2 \geq 0$ are both convex sets. Thus, the entire constraint set is an intersection of convex sets; therefore, it is a convex set also. Since all requirements of Theorem 3 are satisfied, the local maximum is a global maximum. No further analysis is necessary. The solution is

$$q_1^* = 344 \qquad q_2^* = 156 \qquad P^* = \$4870$$

EXAMPLE 8.9 Effects of Economy of Scale and Plant Capacity on Production

In Examples 8.7 and 8.8, "Determination of Optimum Production Level" and "Effect of Plant Capacity on Production," respectively, the company produced goods using a plant with no economy of scale and limited storage capacity. The company is contemplating replacing the old equipment and expanding the storage area to accommodate 800 units. The production cost functions for the equipment are $C_1 = 200q_1^{0.8}$ with a total production capacity of 500 units for item 1 and $C_2 = 160q_2^{0.09}$ with a production capacity of 600 units for item 2. Assume that the company will sell all items produced.

(a) Formulate a mathematical model to maximize profit.
(b) Determine the optimum production level and show that it is a global optimum.
(c) Compare the results obtained in part b with the optimum solutions of Example 8.7 and 8.8.

Solution

(a) As in previous examples and in this one, the control variables q_1 and q_2 will represent the production level for items 1 and 2, respectively. The net profit $P = R - C$ is equal to

$$P = 100q_1 + 150q_2 - 200q_1^{0.8} - 160q_2^{0.95}$$

The constraints on production are due to the storage and production capacities of the equipment. The storage capacity constraint is

$$q_1 + q_2 \leq 800$$

The production capacity constraints are

$$q_1 \leq 500 \qquad q_2 \leq 600$$

The mathematical model is

$$\text{Maximize } P = 100q_1 + 150q_2 - 200q_1^{0.8} - 160q_2^{0.95}$$

$$q_1 + q_2 \leq 800 \qquad \text{Storage}$$

$$q_1 \leq 500 \qquad \text{Production capacity}$$

$$q_2 \leq 600 \qquad \text{Production capacity}$$

$$q_1 \geq 0, q_2 \geq 0$$

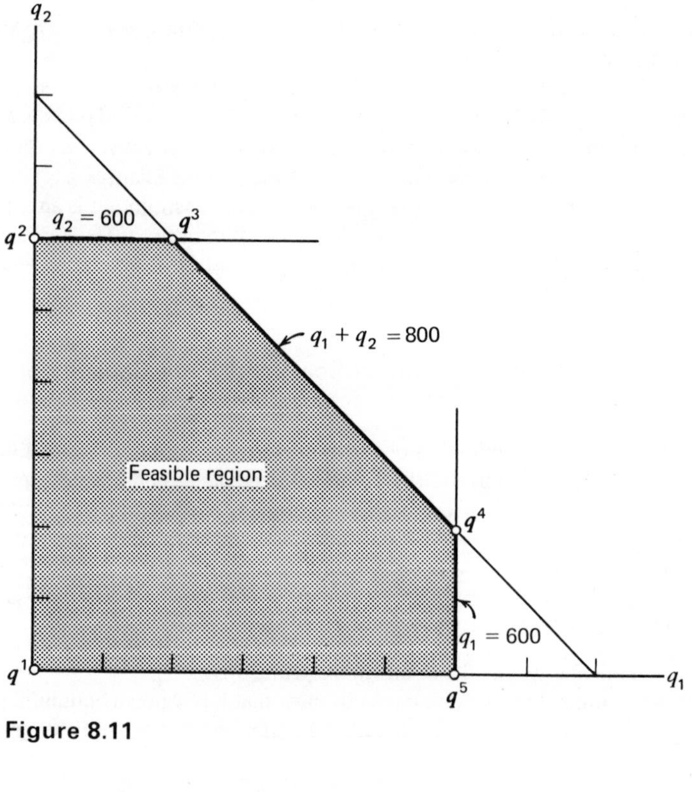

Figure 8.11

(b) As in other optimization examples, we investigate the nature of the objective function and the constraint set. Since the constraint set consists of a set of linear inequality constraints, it is a convex set by inspection. The Hessian test will be used to determine the nature of the objective function.

$$\mathbf{H}(\mathbf{q}) = \begin{bmatrix} \dfrac{\partial^2 P}{\partial q_1^2} & \dfrac{\partial^2 P}{\partial q_1 \partial q_2} \\ \dfrac{\partial^2 P}{\partial q_2 \partial q_1} & \dfrac{\partial^2 P}{\partial q_2^2} \end{bmatrix} = \begin{bmatrix} 32q_1^{-1.2} & 0 \\ 0 & 7.6q_2^{-1.05} \end{bmatrix}$$

The main diagonals of the Hessian matrix are positive for all positive values of q_1 and q_2. The determinants of the principal minors are

$$|32q_1^{-1.2}| = 32q_1^{-1.2} > 0 \qquad \text{for } q_1 \geq 0$$

$$\begin{vmatrix} 32q_1^{-1.2} & 0 \\ 0 & 7.6q_2^{-1.05} \end{vmatrix} = 243q_1^{-1.2}q_2^{-1.05} \geq 0 \qquad \text{for } q_1 \geq 0 \text{ and } q_2 \geq 0$$

Since the necessary and sufficient conditions for a positive definite matrix are satisfied, we conclude that the objective function is a convex function for value of q_1 and q_2 in the positive region.

Theorem 5 states that the global maximum will occur at an extreme point if the objective function is a convex function over a convex constraint set. Since these conditions are satisfied, the search for the optimum solution will require an evaluation of the extreme points only, They are shown in Figure 8.11. The following results are obtained:

	EXTREME POINTS		
POINT	q_1	q_2	PROFIT
1	0	0	\$0
2	0	600	\$20,800
3	200	600	\$26,400
4	500	300	\$30,000
5	500	0	\$21,100

The optimum solution is to

$$q_1^* = 500 \qquad q_2^* = 300 \qquad P = \$30{,}000$$

(c) The results from this example and the two previous two examples have been tabulated in the following table:

CASE	q_1^*	q_2^*	P^*	TOTAL OUTPUT $q_1^* + q_2^*$
1. Old equipment No storage restrictions	394	186	\$4,920	580
2. Old equipment Storage restrictions	344	156	\$4,870	500
3. New equipment Expanded plant	500	300	\$30,000	800

Quite clearly, for the data presented here, the best option for the company is to expand the plant with new equipment. If the company expands the plant to produce 580 units, the profit increase is marginal. The major impact upon profit is the introduction of new efficient equipment.

Remarks

For a mathematical point of view, we have examined three types of maximization problems here and in Examples 5.7 and 5.8, an unconstrained problem with a concave objective function, a constrained problem with a concave objective function, and a constrained problem with a convex objective function. Those problems illustrate most vividly the importance of the nature of the objective function and its effect upon the location of the optimum solution.

CASE	LOCATION OF OPTIMUM SOLUTION
1. Old equipment No storage equipment	Interior point where $q_1 > 0$ and $q_2 \geq 0$ are satisfied
2. Old equipment Storage restrictions	Boundary point where $q_1 + q_2 = 500$ is active
3. New equipment Expanded plant	Extreme point where $q_1 = 500$ and $q_1 + q_2 = 800$ are active

For case 1, we have called it an unconstrained problem. Strictly speaking, it is a constrained mathematical model because the constraints $q_1 \geq 0$ and $q_2 \geq 0$ must be satisfied.

Consider another case where the company decides to introduce only one piece of new equipment for producing item 1. The objective function is

$$P = \$100q_1 + 150q_2 - 200q_1^{0.8} - 110q_2^{1.05}$$

The Hessian matrix is

$$\boldsymbol{H}(\boldsymbol{q}) = \begin{bmatrix} \dfrac{\partial^2 P}{\partial q_1^2} & \dfrac{\partial^2 P}{\partial q_1 \partial q_2} \\[2ex] \dfrac{\partial^2 P}{\partial q_1 \partial q_2} & \dfrac{\partial^2 P}{\partial q_2^2} \end{bmatrix} = \begin{bmatrix} 32q_1^{-1.2} & 0 \\ 0 & -5.78q_2^{-0.95} \end{bmatrix}$$

The main elements of the Hessian matrix are neither all positive nor all negative values for $q_1 > 0$ and $q_2 > 0$. Thus, the objective function is neither a convex nor a concave function. Theorems 4 and 5 may not be utilized. The search becomes more complicated. The interior, boundary, and extreme points will have to be examined. There is the possibility of finding several local optima. These points will have to be carefully examined and compared before the global optimum can be identified. This procedure is more tedious than the ones used to find the global optima of well-behaved functions as illustrated in the examples. It will not be pursued here.

In part b, a graphical procedure was used to determine the locations of the extreme points. For a problem with only two control variables it is most appropriate. For higher-order problems, the extreme points may be determined with use of the equation

$$\boldsymbol{x}_B = \boldsymbol{A}_B^{-1}[\boldsymbol{b} - \boldsymbol{A}_N\boldsymbol{x}_N]$$

This equation was used in the determination of extreme points in solving linear programming problems. For part b, slack variables will be introduced and the constraint set will be

$$\begin{aligned} q_1 + q_2 + q_3 \qquad\qquad &= 800 \\ q_1 \qquad\quad + q_4 \qquad &= 500 \\ q_2 \qquad\quad + q_5 &= 600 \end{aligned}$$

The maximum number of possible combinations is:

$$\binom{n}{m} = \binom{5}{3} = \frac{5!}{3!2!} = 10$$

There are 10 possible combinations of the five control variables taken 3 at a time. For problems consisting of a large number of control variables, the combinations will become large.

PROBLEMS

Problems 1 through 4

For Problems 1 through 4 answer the following questions.

(a) Draw the feasible region.
(b) Does the feasible region form a convex set?
(c) Utilize Theorems 1 through 4 to determine if a global optimum may be found.
(d) Utilize the method of substitution to find the critical point $\boldsymbol{x}^\circ$ or global optimum $\boldsymbol{x}^*$.

Problem 1

$$\begin{aligned} \text{Maximize } z &= 3x_1 + 2x_2 \\ & x_1^2 + 4x_2^2 \leq 6 \\ & x_1 \geq 0,\ x_2 \geq 0 \end{aligned}$$

Problem 2

$$\begin{aligned} \text{Maximize } z &= x_1 + x_2 \\ & x_1^2 + (x_2 - 1)^2 \leq 1 \\ & x_1^2 + x_2^2 \leq 1 \\ & x_1 \geq 0,\ x_2 \geq 0 \end{aligned}$$

Problem 3

$$\begin{aligned} \text{Maximize } z &= x_1^{0.2} x_2^{0.8} \\ & x_1 + x_2 = 10 \\ & x_1 \geq 0,\ x_2 \geq 0 \end{aligned}$$

Problem 4

$$\begin{aligned} \text{Maximize } z &= 2x_1x_2 + 11x_2 - 3x_1^2 + 5x_2^2 \\ & x_1 + x_2 = 2 \\ & x_1 \geq 0,\ x_2 \geq 0 \end{aligned}$$

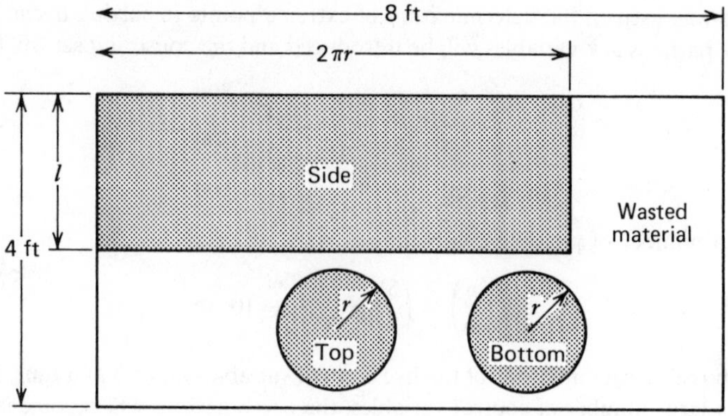

Figure 8.12

Problem 5

Design a cylinder of maximum volume. The cylinder will be constructed from a single 4 × 8 sheet of construction material. Define the control variables as

r = cylinder radius;
l = length of cylinder. (See Figure 8.12.)

(a) Formulate a mathematical model to meet the stated objective.
(b) Determine the optimum solution with the method of substitution, or any other search method where appropriate.
(c) If possible, prove your solution is a global optimum by utilizing one or more of the theorems presented in this chapter.

Problem 6

Repeat Problem 5 with the layout shown in Figure 8.13.

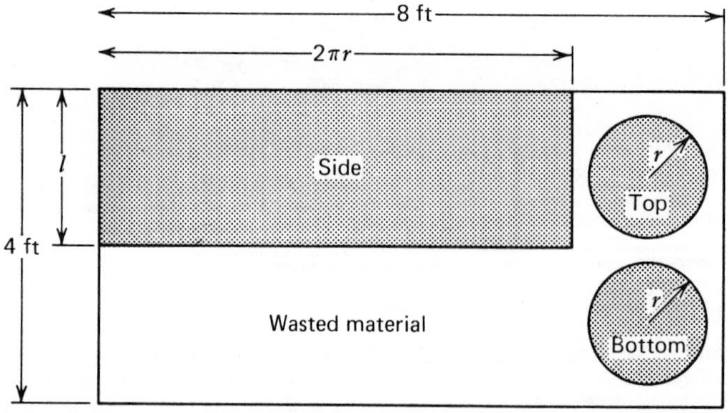

Figure 8.13

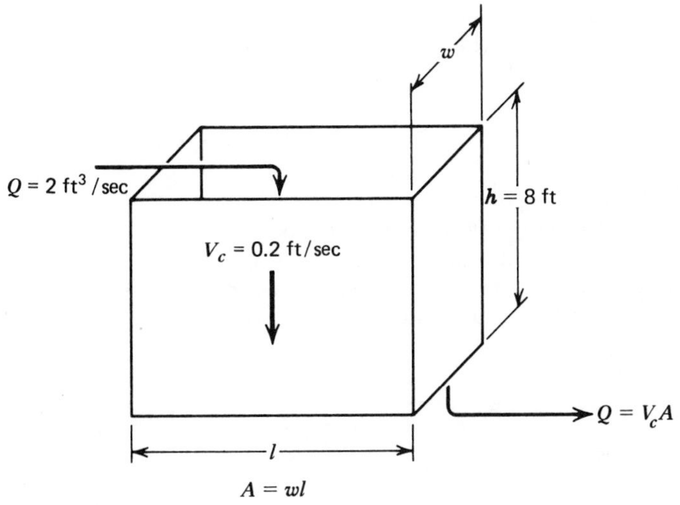

Figure 8.14

Problem 7

Consider the design of the grit chamber in Figure 8.14. The flow Q is a function of the overflow rate V_c and cross-sectional area of the tank A. Assume that the tank height is 8 ft, and the control variables are the tank length l and width w. Determine the minimum cost design. The unit construction cost is a function of the cross-sectional area of the walls, $c = \$2.50/\text{ft}^2$.

(a) Formulate a mathematical model to meet the stated objective. Clearly define control variables.
(b) Determine the optimum solution using the method of substitution.
(c) If possible, prove your solution is a global optimum by utilizing one or more of the theorems presented in this chapter.

Problems 8 and 9

For Problems 8 and 9 answer the following questions.

(a) Classify the objective function as a convex or a concave function. State the theorem utilized to prove your answer.
(b) Is the constraint set a convex set? Why?
(c) Find the global minimum. If appropriate, use the method of substitution to obtain your answer.
(d) Is your solution from part c, an extreme, boundary, or interior point?

Problem 8

$$\text{Minimize } C = 20x_1^{0.8} + 10x_2$$

$$x_1 + x_2 = 100$$

$$x_1 \geq 0, x_2 \geq 0$$

Problem 9

$$\text{Minimize } C = 20x_1^{1.2} + 10x_2$$
$$x_1 + x_2 = 100$$
$$x_1 \geq 0, x_2 \geq 0$$

Problems 10 and 11

For Problems 10 and 11 answer the following questions.

(a) Write the objective in matrix form: $z = a + \mathbf{bx} + \mathbf{x}'\mathbf{Cx}$.

(b) Determine whether the objective function is convex or concave. Why?

(c) Determine the global optimum. Utilize appropriate theorems to prove your answer.

Problem 10

$$\text{Minimize } z = x_1^2 + x_2^2 - 4x_1 - 2x_2 + 10$$
$$0 \leq x_1 \leq 3,$$
$$0 \leq x_2 \leq 3$$

Problem 11

$$\text{Maximize } z = x_1^2 + x_2^2 - 4x_1 - 2x_2 + 10$$
$$0 \leq x_1 \leq 3,$$
$$0 \leq x_2 \leq 3$$

8.5 THE LAGRANGE FUNCTION

The method of substitution is one method of solving optimization problems with active constraints. It is an efficient method; however, for some problems the approach is computationally impractical or impossible. The following constraint equation cannot be expressed as a single value of x_1 or x_2.

$$x_1 e^{x_1} + x_2 e^{x_2} = 6$$

The method of substitution cannot be used. The inverse relationships $x_1 = \phi(x_2)$ or $x_2 = \phi(x_1)$ cannot be found.

The Lagrange multiplier method is another method to solve optimization problems with active constraints. This method avoids the transformation and direct substitution into the objective function. Consider the following model with a single active constraint:

$$\text{Minimize } z = f(\mathbf{x})$$
$$g(\mathbf{x}) = b$$

The $\mathbf{x}$ vector is assumed to consist of n control variables. By introducing a new variable, the Lagrange multiplier λ, the constrained optimization model may be rewritten as an unconstrained optimization model. The revised model is called a Lagrange function, $L(\mathbf{x}, \lambda)$. The minimization problem with a single constraint will be written as

$$\text{Minimize } L(\mathbf{x}, \lambda) = f(\mathbf{x}) + \lambda(b - g(\mathbf{x}))$$

The methods for finding *local minima* may be used for the Lagrange function. The necessary conditions for a minimum or critical point are

$$\frac{\partial L}{\partial \boldsymbol{x}} = \boldsymbol{0}: \quad \frac{\partial L}{\partial x_1} = \frac{\partial f(\boldsymbol{x})}{\partial x_1} - \lambda \frac{\partial g(\boldsymbol{x})}{\partial x_1} = 0$$

$$\frac{\partial L}{\partial x_2} = \frac{\partial f(\boldsymbol{x})}{\partial x_2} - \lambda \frac{\partial g(\boldsymbol{x})}{\partial x_2} = 0$$

$$\vdots$$

$$\frac{\partial L}{\partial x_n} = \frac{\partial f(\boldsymbol{x})}{\partial x_n} - \lambda \frac{\partial g(\boldsymbol{x})}{\partial x_n} = 0$$

$$\frac{\partial L}{\partial \lambda} = 0: \quad b - g(\boldsymbol{x}) = 0 \quad \text{or} \quad g(\boldsymbol{x}) = b$$

If $\boldsymbol{x} \geq \boldsymbol{0}$ is introduced into the mathematical model, the Lagrange function may be written as $L(\boldsymbol{x}, \lambda)$ with $\boldsymbol{x} \geq \boldsymbol{0}$. The side constraints $\boldsymbol{x} \geq \boldsymbol{0}$ are not introduced into $L(\boldsymbol{x}, \lambda)$. There is no mathematical advantage to do so. This model is a constrained optimization model.

Maximization problems with a single constraint will have the same Lagrange function $L(\boldsymbol{x}, \lambda)$ and the same necessary conditions. Theorems 1 through 4 are the requirements for testing critical points for a global minimum and global maximum. The tests for determining the convexity or concavity of the objective and determining if the constraint set is a concave will have to be employed.

An Interpretation of the Lagrange Multiplier

Consider a mathematical model with two control variables x_1 and x_2 and with a single active constraint:

$$\text{Minimize } z = f(x_1, x_2)$$

$$g(x_1, x_2) = b$$

This model requires that the solution lie along the curve $g(x_1, x_2) = b$. The chain rule of differentiation will be used to develop the necessary conditions for a minimum. From this development the definition of the Lagrange multiplier will be determined.

Since $g(x_1, x_2) = b$ is an active constraint, the critical point must be a tangent point of $z = f(x_1, x_2)$ and $g(x_1, x_2) = b$ as shown in Figure 8.15. The functions $z = f(x_1, x_2)$ and $g(x_1, x_2) = b$ are tangent at the critical point (x_1°, x_2°). From the chain rule of differentiation, the objective equation is written as

$$\frac{dz}{dx_1} = \frac{\partial f(x_1, x_2)}{\partial x_1} + \frac{\partial f(x_1, x_2)}{\partial x_2}\left(\frac{dx_2}{dx_1}\right) = 0$$

or

$$\frac{dz}{dx_1} = \frac{\partial f}{\partial x_1} + \frac{\partial f}{\partial x_2}\left(\frac{dx_2}{dx_1}\right) = 0$$

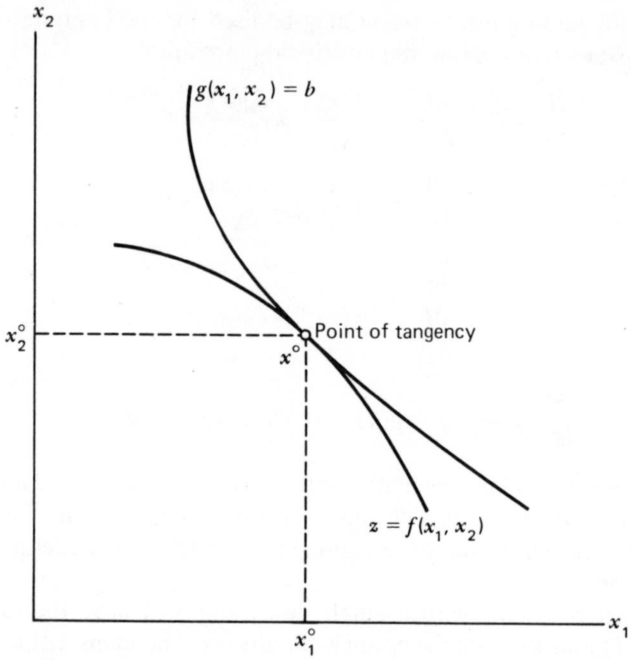

Figure 8.15 Optimality condition for a constrained mathematical model.

Utilizing the chain rule again, the constraint equation is written as

$$\frac{dg(x_1, x_2)}{dx_1} = \frac{\partial g(x_1, x_2)}{\partial x_1} + \frac{\partial g(x_1, x_2)}{\partial x_2}\left(\frac{dx_2}{dx_1}\right) = 0$$

or

$$\frac{dg(x_1, x_2)}{dx_1} = \frac{\partial g}{\partial x_1} + \frac{\partial g}{\partial x_2}\left(\frac{dx_2}{dx_1}\right) = 0$$

The total derivative dx_2/dx_1 appears in both equations. Rearranging the constraint equation, it becomes

$$\frac{dx_2}{dx_1} = \frac{-\partial g/x_1}{\partial g/\partial x_2}$$

Substitute it into the objective equation and reduce it to the following expression:

$$\frac{\partial f}{\partial x_1} - \frac{\partial f}{\partial x_2}\frac{\partial g/\partial x_1}{\partial g/\partial x_2} = 0$$

Rearrange the equation as

$$\frac{\partial f}{\partial x_1} - \frac{\partial f/\partial x_2}{\partial g/\partial x_2}\left(\frac{\partial g}{\partial x_1}\right) = 0$$

The ratio $\partial f/\partial x_2/\partial g/\partial x_2$, is called the Lagrange multiplier λ. Evaluated at the critical point (x_1°, x_2°) it is equal to

$$\lambda^\circ = \frac{\partial f(x_1^\circ, x_2^\circ)/\partial x_2}{\partial g(x_1^\circ, x_2^\circ)/\partial x_2}$$

Utilizing the expanded notation, the necessary condition is

$$\frac{\partial f(x_1^\circ, x_2^\circ)}{\partial x_1} - \lambda^\circ \frac{\partial g(x_1^\circ, x_2^\circ)}{\partial x_1} = 0$$

where $g(x_1^\circ, x_2^\circ) = b$, the active constraint, must be satisfied.

If the analysis is repeated, taking the total derivative with respect to x_2 instead of x_1, the result in expanded form is

$$\frac{\partial f(x_1^\circ, x_2^\circ)}{\partial x_2} - \mu^\circ \frac{\partial g(x_1^\circ, x_2^\circ)}{\partial x_2} = 0$$

where μ° is the Lagrange multiplier

$$\mu^\circ = \frac{\partial f(x_1^\circ, x_2^\circ)/\partial x_1}{\partial g(x_1^\circ, x_2^\circ)/\partial x_1}$$

The two Lagrange multipliers λ° and μ° are equal.

$$\lambda^\circ = \mu^\circ = \frac{\partial f(x_1^\circ, x_2^\circ)/\partial x_1}{\partial g(x_1^\circ, x_2^\circ)/\partial x_1} = \frac{\partial f(x_1^\circ, x_2^\circ)/\partial x_2}{\partial g(x_1^\circ, x_2^\circ)/\partial x_2}$$

Thus, the mathematical model

$$z = f(x_1, x_2)$$
$$g(x_1, x_2) = b$$

may be expressed as a Lagrange function, where x_1, x_2, and λ are variables.

In summary, the Lagrange function is

$$L(x_1, x_2, \lambda) = f(x_1, x_2) + \lambda(b - g(x_1, x_2))$$

The necessary condition for an optimum is

$$\frac{\partial L}{\partial \mathbf{x}} = \mathbf{0}: \quad \frac{\partial L}{\partial x_1} = \frac{\partial f(x_1^\circ, x_2^\circ)}{\partial x_1} - \lambda^\circ \frac{\partial g(x_1^\circ, x_2^\circ)}{\partial x_1} = 0$$

$$\frac{\partial L}{\partial x_2} = \frac{\partial f(x_1^\circ, x_2^\circ)}{\partial x_2} - \lambda^\circ \frac{\partial g(x_1^\circ, x_2^\circ)}{\partial x_2} = 0$$

$$\frac{\partial L}{\partial \lambda} = 0: \quad \frac{\partial L}{\partial x_2} = \frac{\partial f(x_1^\circ, x_2^\circ)}{\partial x_2} - \lambda^\circ \frac{\partial g(x_1^\circ, x_2^\circ)}{\partial x_2} = 0$$

$$g(x_1^\circ, x_2^\circ) = b$$

The Lagrange multiplier may be also interpreted as the change in z to the change in the resource constraint b, or

$$\lambda^\circ = \frac{dz^\circ}{db}$$

When $z = f(\boldsymbol{x})$ is assumed to be a cost function and $g(\boldsymbol{x}) = b$ is assumed to be a production constraint, the Lagrange multiplier is called a *shadow price*. The shadow price may be used to estimate the change in the production cost Δz due to a change in production level Δb, or $\Delta z = \lambda^\circ \Delta b$.

Problems with More Than One Active Constraint

These principles may be extended to multivariate models consisting of one or more active constraint equations:

$$z = f(\boldsymbol{x})$$

$$\boldsymbol{g}(\boldsymbol{x}) = \boldsymbol{b}$$

The vector $\boldsymbol{x}$ is assumed to consist of n control variables, and the constraint set consists of m constraint equations. The Lagrange function for this model will have the form:

$$L(\boldsymbol{x}, \boldsymbol{\lambda}) = f(\boldsymbol{x}) + \sum_i \lambda_i(b_i - g(\boldsymbol{x}))$$

or

$$L(\boldsymbol{x}, \boldsymbol{\lambda}) = f(\boldsymbol{x}) + \boldsymbol{\lambda}'[\boldsymbol{b} - \boldsymbol{g}(\boldsymbol{x})]$$

where

$$\boldsymbol{\lambda}' = [\lambda_1 \quad \lambda_2 \cdots \lambda_m]$$

There is a Lagrange multiplier associated with each constraint equation.

The necessary condition for a local optimum is

$$\nabla L_{\boldsymbol{x}} = \frac{\partial L}{\partial \boldsymbol{x}} = \boldsymbol{0}$$

$$\nabla L_{\boldsymbol{\lambda}} = \frac{\partial L}{\partial \boldsymbol{\lambda}} = \boldsymbol{0}$$

Since there are n control variables and m constraint equations, there are $n + m$ equations in the solution set.

$$\frac{\partial L}{\partial x_j} = \frac{\partial f(\boldsymbol{x})}{\partial x_j} - \sum_{i=1}^{m} \lambda_i \frac{\partial g_i(\boldsymbol{x})}{\partial x_j} = 0 \qquad j = 1, 2, \ldots, n$$

$$\frac{\partial L}{\partial \lambda_i} = b_i - g_i(\boldsymbol{x}) = 0 \qquad i = 1, 2, \ldots, m$$

The condition that $\partial L/\partial \boldsymbol{\lambda} = \boldsymbol{0}$ ensures that the set of all active constraint equations are satisfied.

A Search Method

From Theorems 1 through 4, the first step in determining the possible location of the optimum solution is to determine if the objective function is a convex function, a concave function, or neither. The theorems given in previous sections may be used to determine this. If the function is neither a convex nor a concave function, the optimum point or points may occur at the interior, boundary, or extreme point of the feasible region. As a result, a more intensive investigation of the candidate solutions will have to be made. On the other hand, if the objective function is a concave function and a minimum is sought, only the extreme points of the feasible region must be evaluated and compared. It is conceivable that a minimum point could occur at one or more extreme points. Theoretically, this is a straightforward procedure. Practically, for problems with great number of control variables and constraint equations, the search will be tedious.

For problems with an objective function that is a convex function, the use of necessary conditions for locating a local minimum is applicable. The optimum solution may occur at an interior point, a boundary point, or an extreme point. The following approach is recommended. There is no guarantee that the optimum solution will be found in the most expeditious manner. The steps are as follows:

STEP 1 Assume that all constraint equations are inactive constraints, and determine the local minimum. If the local minimum lies in the feasible region, a global optimum has been obtained because all conditions of Theorem 1 are satisfied.

STEP 2 If one or more constraints are violated, select a single-constraint equation and introduce the Lagrange multiplier method, or a method of substitution may be used. If the local minimum lies in the feasible region, it is a global optimum. If not, select another constraint equation, introduce it as an active constraint, and test it for optimality.

STEP 3 If the optimum is not found in Step 2, introduce two constraint equations as active constraints and test for optimality. Repeat if necessary by adding two or more constraints as active constraints. The Lagrange multiplier method will probably be easier to use than the method of substitution when several constraints are active.

One might ask, how does this approach differ from the problem with an objective function that is neither convex nor concave? The major difference for convex and concave objective functions is that once the local optimum is found, the global optimum is known by Theorems 1 and 3. For a nonconvex or a nonconcave objective function, the local optimum point may or may not be the global optimum. The only way to determine the global optimum is to compare all local optima and select the minimum point or points.

This procedure may be tedious for large problems with convex objective functions. Knowledge of the physical system may be utilized to eliminate analyses that will obviously violate the constraint condition.

EXAMPLE 8.10 A Fair Pricing Strategy

A transit company has determined the average cost to provide service on a two-bus route system is \$1.20 per customer. The company seeks to establish a fare schedule to account for the trip length of the rider and the ridership on each route. That is, patrons of the route with greater ridership may pay a fare that is less than the fare paid by patrons of the route with lower ridership. In a

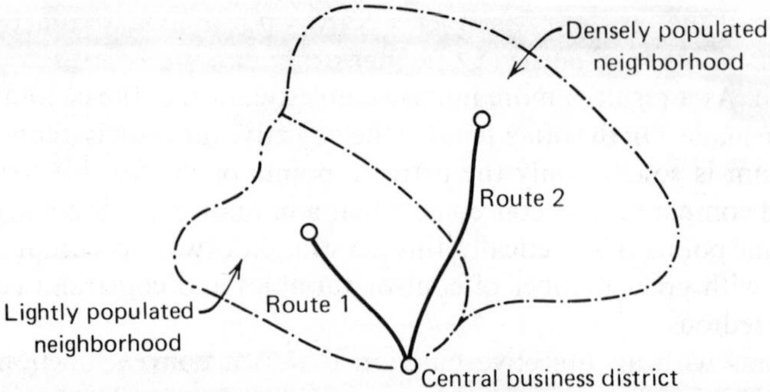

Figure 8.16

similar manner, patrons on the longer route may have to pay a greater fare than the patrons on the shorter route. (See Figure 8.16.)

ROUTE i	RIDERSHIP L_i, AVERAGE TRIP LENGTH	n_i, AVERAGE DAILY PATRONAGE
1	10 miles	600
$n = 2$	15 miles	1,400

A fair system is considered to be one that minimizes the variance of travel cost among its customers. The measure of effectiveness or weighted variance equation is

$$V = \frac{1}{N}\sum_{i=1}^{N} n_i(r_i - \bar{r})^2$$

where V is the variance of travel cost; r_i is the fare box revenue paid by each customer on route i; $\bar{r}$ is the \$1.20 average fare box revenue paid by each customer, or $\bar{r} = (1/N)\sum_i^N n_i r_i$; and N is the total number of customers.

(a) Formulate a mathematical model to determine on an equitable bus fare schedule that charges each customer on a cost per passenger-mile basis.

(b) Determine the global optimum by utilizing the Lagrange multiplier method.

Solution

(a) Define the price to be charged on each route as a control variable equal to the cost per passenger-mile on each route.

p_1 = unit fare as measured in cost per passenger-mile on route 1.
p_2 = unit fare as measured in cost per passenger-mile on route 2.

The measure of effectiveness and objective function of the mathematical model is equal to variance of travel cost among the customers:

$$V = \frac{n_1}{N}(r_1 - \bar{r})^2 + \frac{n_2}{N}(r_2 - \bar{r})^2$$

The relationship between fare box revenue and unit fare is

$$r_i = p_i L_i$$

or $r_1 = 10p_i$ and $r_2 = 15p_2$. Substituting this relationship and the ridership parameters from the above table gives a measure of effectiveness of

$$V = \frac{600}{2000}(10p_1 - 1.20)^2 + \frac{1400}{2000}(15p_1 - 1.20)^2$$

or

$$V = 0.3(10p_1 - 1.20)^2 + 0.7(15p_2 - 1.20)^2$$

Assume that the total fare box revenue R equals the total cost C to provide the service. The bus company is assumed to be a nonprofit operation:

$$\text{Fare box revenue} = \text{Total cost}$$

or

$$R = C$$

In terms of the control variables, the constraint equation is

$$p_1(n_1L_1) + p_2(n_2L_2) = \bar{r}N$$

or

$$p_1(600 \cdot 10) + p_2(1400 \cdot 15) = \$1.20 \cdot 2000$$

Simplifying,

$$6p_1 + 21p_2 = \$2.40$$

The mathematical model is

$$\text{Minimize } V = 0.3(10p_1 - 1.20)^2 + 0.7(15p_2 - 1.20)^2$$

$$6p_1 + 21p_2 = \$2.40$$

$$p_1 \geq 0,\ p_2 \geq 0$$

(b) The Lagrange function is

$$L = 0.3(10p_1 - 1.20)^2 + 0.7(15p_2 - 1.20)^2 + \lambda(2.40 - 6p_1 - 21p_2)$$

The necessary condition for a minimum is:

$$\frac{\partial L}{\partial p_1} = 0.6(10p_1 - 1.20) \cdot 10 - 6\lambda = 0 \quad (1)$$

$$\frac{\partial L}{\partial p_2} = 1.4(15p_2 - 1.20) \cdot 15 - 21\lambda = 0 \quad (2)$$

$$\frac{\partial L}{\partial \lambda} = 2.40 - 6p_1 - 21p_2 = 0 \quad (3)$$

Solving for p_1 and p_2 in terms of λ in Eqs. (1) and (2) gives:

$$p_1 = \frac{1.20 + \lambda}{10}$$

$$p_2 = \frac{1.20 + \lambda}{15}$$

Substituting p_1 and p_2 into Eq. (3) gives

$$6\left(\frac{1.20 + \lambda}{10}\right) + 21\left(\frac{1.20 + \lambda}{15}\right) = 2.40$$

Simplifying it gives $$(0.72 + 0.6\lambda) + (1.68 + 1.4\lambda) = 2.40$$

or $$\lambda^\circ = 0$$

Thus, substituting $\lambda^\circ = 0$ into Eqs. (1) and (2), the unit fares are

$$p_1^\circ = \$0.12 \text{ per passenger-mile}$$

$$p_2^\circ = \$0.08 \text{ per passenger-mile}$$

Theorem 1 will be used to prove the critical point as a location of a minimum. From Theorem 14, the linear constraint equation forms a convex set. The Hessian matrix of V is

$$H(\mathbf{p}^0) = \begin{bmatrix} \dfrac{\partial^2 V}{\partial p_1^2} & \dfrac{\partial^2 V}{\partial p_1 \partial p_2} \\ \dfrac{\partial^2 V}{\partial p_2 \partial p_1} & \dfrac{\partial^2 V}{\partial p_2^2} \end{bmatrix} = \begin{bmatrix} 60 & 0 \\ 0 & 315 \end{bmatrix}$$

The main diagonal elements are both positive as required for a convex function. The determinants of the principal minors are

$$|60| = 60 > 0 \qquad \begin{vmatrix} 60 & 0 \\ 0 & 315 \end{vmatrix} = 60 \cdot 315 > 0$$

The determinants of the principal minors are positive; therefore, the objective function is a convex function and the solution is a global minimum:

$$p_1^* = \$0.12 \text{ per passenger-mile}$$

$$p_2^* = \$0.08 \text{ per passenger-mile}$$

$$\lambda^* = 0$$

Remarks

The total average fare paid by the riders of each route are the same:

$$f_1 = p_1 L_1 = \$0.12 \times 10 = \$1.20 \text{ per trip}$$

$$f_2 = p_2 L_2 = \$0.08 \times 15 = \$1.20 \text{ per trip}$$

The fairness criterion places equal weight on each trip. Riders on routes 1 and 2 have different ridership patterns as measured by the average trip length and patronage figure. Nevertheless, the value of all these trips, no matter what route is selected, is treated equally.

These results may imply that all fares on routes 1 and 2 are fixed at \$1.20 per trip. This fare is the average price levied on patrons of routes 1 and 2 who travel an *average* distance of 10 miles on route 1 and 15 miles on route 2. The fare schedule may be established to charge the customers on a mileage basis, that is, \$0.12 and \$0.08 per passenger mile on routes 1 and 2, respectively. Thus, a patron of route 1 making a trip of 5 miles may be charged \$0.60. Over the long term, when all trip makers traveling various distances are considered, the expected revenue received will be equal \$1.20 per fare on route 1. The same reasoning applies on route 2.

Thus, we conclude that this pricing schedule tends to place equal value on all trips made. The total ridership and ridership distribution on the different routes do not alter the unit price charged on each individual route. Ridership will affect the average cost $\bar{c}$, which, in turn, will be reflected in the unit fares. Pricing schedules may be influenced by other factors including social, political, and other economic factors not considered here.

In the formulation of this problem the constraint condition that fare box revenue equals total cost is an active constraint. This example shows that the global optimum occurs along the constraint equation with a Lagrange multiplier equal to zero. It has been shown that the Lagrange multiplier is equal to the change in the value of the objective function divided by the change in the value of the constraint, or $\lambda = \partial z/\partial b = 0$.

For this problem, since $b = \bar{c}$, a change in the average service cost will not result in a change in the distribution of fares among the customers. If $\bar{c}$ is increased, then the extra cost or fare will be passed on to the riders as a proportional increase in fare to all users. The fare distribution will remain the same.

EXAMPLE 8.11 Maximum Volume of a Rectangular Box

The rectangular box in Figure 8.17 is supported along its edges by straight members. These members are cut from a single 12-ft rod. Determine the maximum volume of the box.
(a) Formulate the mathematical model to maximize volume.
(b) Use the Lagrange multiplier method to find the global optimum.

Solution

(a) The control variables are the length x_1, width x_2, and height x_3 of the box as shown in the figure. The volume of the box is $V = x_1 x_2 x_3$.

There are four length, four width, and four height members. The total length of these members is 12 ft:

$$4x_1 + 4x_2 + 4x_3 = 12$$

After simplification, the mathematical model becomes

$$\text{Maximize } V = x_1 x_2 x_3$$

$$x_1 + x_2 + x_3 = 3$$

$$x_1 \geq 0, x_2 \geq 0, x_3 \geq 0$$

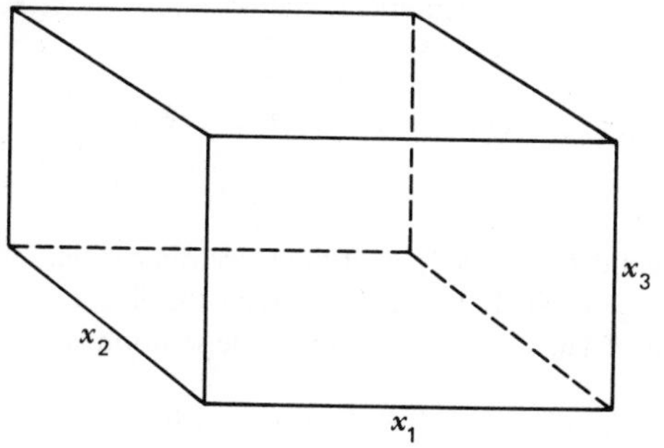

Figure 8.17 A box.

(b) The Lagrange function is

$$L(x_1, x_2, x_3, \lambda) = x_1x_2x_3 + \lambda(3 - x_1 - x_2 - x_3)$$

The necessary condition for a critical point is

$$\frac{\partial L}{\partial \boldsymbol{x}} = \boldsymbol{0}: \quad \frac{\partial L}{\partial x_1} = x_2x_3 - \lambda = 0 \tag{1}$$

$$\frac{\partial L}{\partial x_2} = x_1x_3 - \lambda = 0 \tag{2}$$

$$\frac{\partial L}{\partial x_3} = x_1x_2 - \lambda = 0 \tag{3}$$

$$\frac{\partial L}{\partial \lambda} = 0: \quad x_1 + x_2 + x_3 = 3 \tag{4}$$

Solving for x_2 and x_3 in Eqs. (1) and (2) gives

$$x_2 = \frac{\lambda}{x_3} \quad \text{and} \quad x_1 = \frac{\lambda}{x_3}$$

Thus

$$x_1 = x_2 = \frac{\lambda}{x_3} \tag{5}$$

Equation (3) may be rewritten as $x_1^2 - \lambda = 0$; thus

$$x_1 = x_2 = \lambda^{1/2}$$

From Eq. (5), $x_1 = \lambda^{1/2} = \lambda/x_3$; thus $x_3 = \lambda^{1/2}$. Likewise, $x_1 = x_2 = x_3 = \lambda^{1/2}$. Substitute into Eq. (4) and solve for λ.

$$x_1 + x_2 + x_3 = \lambda^{1/2} + \lambda^{1/2} + \lambda^{1/2} = 3 \qquad \text{or } \lambda^\circ = 1$$

The length of the sides is $x_1^\circ = x_2^\circ = x_3^\circ = 1$. The volume is constructed as a cube with equal sides of 1 ft.

The solution is a global maximum if $x_1x_2x_3$ is a concave function. This function is similar in form to the quadratic function, $f(\boldsymbol{x}) = x_1x_2$ stated in Example 8.2. The function x_1x_2 is concave along line A as shown in Table 8.1 and Figure 8.5 of Example 8.2. The line A is in the x_1–x_2 plane and has a mathematical relationship, $x_1 + x_2 = 10$. Note the similarity between the mathematical model of Example 8.2 and

$$f(\boldsymbol{x}) = x_1x_2 \qquad V = x_1x_2x_3$$

$$x_1 + x_2 = 10 \qquad x_1 + x_2 + x_3 = 3$$

In lieu of performing a detailed analysis of the three-dimensional model, we conclude that the function $V = x_1x_2x_3$ will be a concave along the line $x_1 + x_2 + x_3 = 3$. Since the sufficiency conditions of Theorem 3 are satisfied, the solution is a global optimum. Thus

$$x_1^* = x_2^* = x_3^* = 1 \text{ ft}$$

$$V^* = 1 \text{ ft}^3$$

EXAMPLE 8.12 A Minimum-Weight-Truss Problem

Determine the cross-sectional areas of the statically indeterminate truss members shown in Figure 8.18. Members in tension and compression are limited to a maximum stress of 20 ksi and 15 ksi, respectively. Members 1 and 3 are assumed to have the same cross-sectional areas, $A_1 = A_3$.

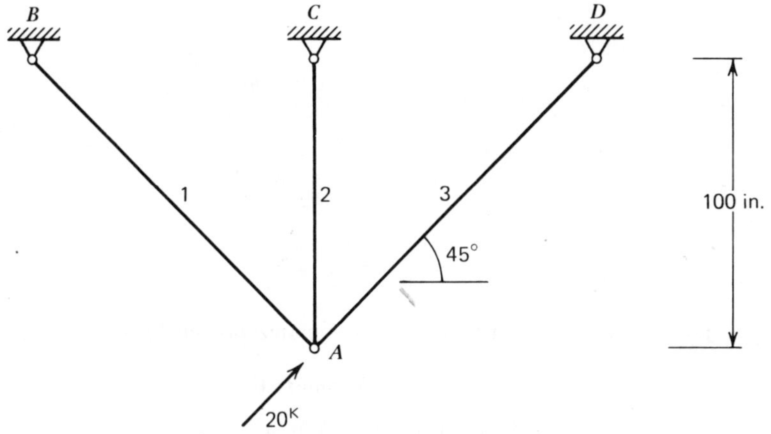

Figure 8.18 An indeterminate truss.

The member forces may be determined from two equations of equilibrium, $\sum F_x = 0$ and $\sum F_y = 0$ (the sum of the forces in the x and y directions must be equal to zero), and the condition of geometric capability. The axial stress is equal to tensile or compression axial forces divided by the cross-sectional area, $\sigma = F/A$. The member stresses have been determined as

$$\sigma_1 = \frac{10}{A_1} - \frac{10}{A_1 + \sqrt{2}A_2} \qquad \text{Member 1 in tension}$$

$$\sigma_2 = \frac{-20}{A_1 + \sqrt{2}A_2} \qquad \text{Member 2 in compression}$$

$$\sigma_3 = \frac{-10}{A_1} - \frac{10}{A_1 + \sqrt{2}A_2} \qquad \text{Member 3 in compression}$$

(a) Formulate a mathematical model to minimize the total weight or total volume of the structure.
(b) It will be found that the model found in part a is a constrained optimization model consisting of inequality constraints. Utilize the search method to determine the optimum solution.

Solution

(a) The objective function is

$$\text{Minimize } V = \frac{2\sqrt{2}}{2}\,100A_1 + 100A_2$$

or

$$\text{Minimize } V = 283A_1 + 100A_2 \tag{1}$$

where V equals the total volume of the truss members. The stress in the members must not exceed the allowable stress limits:

Tensile stress $\sigma_1 \leq 20$ ksi (2)

Compressive stress $\sigma_2 \leq -15$ ksi (3)

Compressive stress $\sigma_3 \leq -15$ ksi (4)

or

$$\frac{10}{A_1} - \frac{10}{A_1 + \sqrt{2}A_2} \leq 20 \tag{2'}$$

$$\frac{-20}{A_1 + \sqrt{2}A_2} \leq -15 \tag{3'}$$

$$\frac{-10}{A_1} - \frac{10}{A_1 + \sqrt{2}A_2} \leq -15 \tag{4'}$$

By appropriate mathematical manipulation, the optimization model may be written as

$$\text{Minimize } V = 283A_1 + 100A_2 \tag{1''}$$

$$A_1^2 + 1.414A_1A_2 - 0.707A_2 \geq 0 \tag{2''}$$

$$A_1 + 1.414A_2 \geq 1.333 \tag{3''}$$

$$-A_1^2 - 1.414A_1A_2 + 1.333A_1 + 0.943A_2 \geq 0 \tag{4''}$$

$$A_1 \geq 0, A_2 \geq 0$$

(b) The search method utilizes the approach of assuming that no constraints are active while solving the optimization problem, and then checking the constraint equations to determine if they are satisfied. If a constraint is violated, a single constraint is introduced, the constrained optimization problem is solved, and the constraints are checked again. This procedure is continued until a critical point is found that satisfies all constraint equation conditions.

The search proceeds as follows:

STEP 1 Assume no active constraints. The optimization model reduces to:

$$\text{Minimize } V = 283A_1 + 100A_2$$

The minimum solution of this model is

$$A_1^\circ = 0 \text{ in.}^2 \quad A_2^\circ = 0 \text{ in.}^2 \quad V^\circ = 0 \text{ in.}^3$$

This solution violates constraint (3″), thus it cannot be a global optimum.

STEP 2 Introduce constraint equation, Eq. (3″) into the formulation.

$$\text{Minimize } V = 283A_1 + 100A_2$$

$$A_1 + 1.414A_2 = 1.333$$

$$A_1 \geq 0, \quad A_2 \geq 0$$

Since the objective and constraint equations are linear equations, the optimum solution may be found by linear programming. Owing to the simplicity of this model, it should be evident that the solution will occur at the extreme point

$$A_1^\circ = 0 \text{ in.}^2 \quad A_2^\circ = 0.9427 \text{ in.}^2 \quad V = 94.27 \text{ in.}^3$$

Checking the constraint set above, we find constraints (2″) and (4″) are violated.

STEP 3 Introduce Eq. (2″). The mathematical model is

$$\text{Minimize } V = 283A_1 + 100A_2$$

$$A_1^2 + 1.414A_1A_2 - 0.707A_2 = 0$$

In this case, let us use the method of substitution. Rearranging the constraint equation we obtain

$$A_2 = \frac{A_1^2}{0.707 - 1.414A_1}$$

After substituting A_2 into the objective function and determining $dV/dA_1 = 0$, a quadratic equation is obtained. The solution of the quadratic equation is

$$A_1 = \frac{0.667 \pm \sqrt{0.445 - 2.668}}{2}$$

Only imaginary roots exist; thus no solution exists for this model.

STEP 4 Introduce Eq. (4″). The model is

$$\text{Minimize } V = 283A_1 + 100A_2$$

$$-A_1^2 - 1.414A_1A_2 + 1.333A_1 + 0.943A_2 = 0$$

The solution to this problem is

$$A_1^\circ = 1.052 \text{ in.}^2 \quad A_2^\circ = 0.544 \text{ in.}^2 \quad V^\circ = 354 \text{ in.}^3$$

A check of the constraints (2″) and (3″) show that they are satisfied; thus, this may be the location of the optimum solution. It can be shown that the constraint set is a nonconvex set. Thus, Theorems 1 and 2 are not applicable and the search continues.

STEP 5 Introduce Eqs. (2″) and (3″).

$$\text{Minimize } V = 283A_1 + 100A_2$$

$$A_1^2 + 1.414A_1A_2 - 0.707A_2 = 0$$

$$A_1 + 1.414A_2 = 1.333$$

Since the constraint set consists of two equations and two unknowns, they must be satisfied explicitly. We have no freedom to manipulate the control variables in this situation. Solving these equations gives

$$A_1^\circ = 0.364 \text{ in.}^2 \quad A_2^\circ = 0.686 \text{ in.}^2 \quad V^\circ = 171.6 \text{ in.}^3$$

The third constraint equation, Eq. (4″), is satisfied; thus, this solution may be the location of the minimum weight truss. Since the volume calculated in step 4 is greater than the one calculated here, 354 in.^{3} > 171.6 in.^{3}, it can be eliminated from further consideration.

STEP 6 Introduce Eqs. (3″) and (4″).

$$\text{Minimize } V = 283A_1 + 100A_2$$

$$A_1 + 1.414A_2 = 1.333$$

$$-A_1^2 - 1.414A_1A_2 + 1.333A_1 + 0.943A_2 = 0$$

The solution is

$$A_1^\circ = 1.333 \text{ in.}^2 \quad A_2^\circ = 0 \text{ in.}^2 \quad V^\circ = 377 \text{ in.}^3$$

This solution satisfies the inequality constraint of Eq. (2″); thus, it is the feasibility solution. However, since the volume is 377 in.^{3} and is greater than the volume obtained in step 5, this solution may be eliminated from further consideration.

No further analysis is needed. Introducing the three constraint equations in terms of two control variables A_1 and A_2 is not meaningful. Thus the optimum solution was obtained in step 5.

$$A_1^* = 0.364 \text{ in.}^2 \quad A_2^* = 0.686 \text{ in.}^2 \quad V^* = 171.6 \text{ in.}^3$$

A minimum-weight steel truss is 48.6 lb, where the density of steel is 489 lb/ft^3.

Remarks

The solution found in step 6, where $A_2 = 0$ in.^{2}, is the solution of a statically determinate truss. Clearly, a statically indeterminate truss found in step 5 is a superior design.

A graphical solution is shown in Figure 8.19, the location of the critical points found in steps 4, 5, and 6. We can see that each one satisfies the constraint set. The feasible region is not a convex set. Only by comparing the different solutions did we find the optimum location. Since there are only two control variables, it would have been much easier to utilize a graphical method of solution.

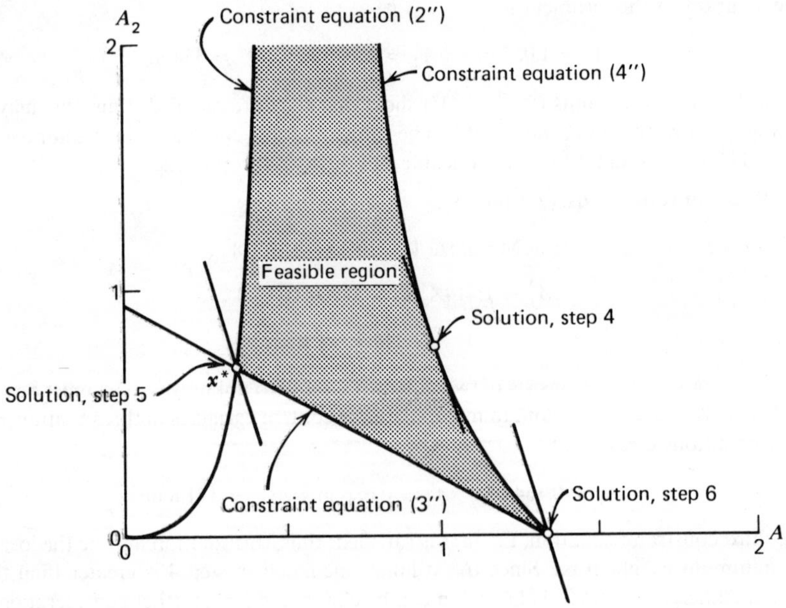

Figure 8.19

Even though this problem is simple, a considerable amount of computational effort is required by the search method. Moreover, if this problem consisted of a convex constraint set instead of a nonconvex constraint set, then the search would be ceased when one of the theorems for a minimum would be satisfied.

PROBLEMS

Problems 1 through 4

Problems 1 through 4 are the same ones given in Section 8.4. Questions labeled (a) through (c) in Section 8.4 will aid in the solution to the problems stated here.

Utilize the Lagrange multiplier method to find the critical point $\boldsymbol{x}^\circ$ or the global optimum $\boldsymbol{x}^*$.

Problem 1

$$\text{Maximize } z = 3x_1 + 2x_2$$
$$x_1^2 + 4x_2^2 \leq 6$$
$$x_1 \geq 0, x_2 \geq 0$$

Problem 2

$$\text{Maximize } z = x_1 + x_2$$
$$x_1^2 + (x_2 - 1)^2 \leq 1$$
$$x_1^2 + x_2^2 \leq 1$$
$$x_1 \geq 0, x_2 \geq 0$$

Problem 3

$$\text{Maximize } z = x_1^{0.2} x_2^{0.8}$$
$$x_1 + x_2 = 10$$
$$x_1 \geq 0, x_2 \geq 0$$

Problem 4

$$\text{Maximize } z = 2x_1x_2 + 11x_2 - 3x_1^2 + 5x_2^2$$
$$x_1 + x_2 = 2$$
$$x_1 \geq 0, x_2 \geq 0$$

Problems 5 and 6

For Problems 5 and 6 answer the following questions.

(a) Solve by the method of substitution.

(b) Solve by the Lagrange multiplier method.

Problem 5

$$\text{Minimize } z = x_1^2 + x_2^2 - 4x_1 - 2x_2 + 10$$
$$x_1 + 2x_2 = 3$$
$$x_1 \geq 0, x_2 \geq 0$$

Problem 6

$$\text{Minimize } z = e^{x_1} + \tfrac{1}{2}x_1x_2 + x_2^2$$

$$x_1 + x_2 = 3$$

$$x_1 \geq 0, x_2 \geq 0$$

Problem 7

Design a chemical reactor of minimum cost. The flow of liquid is steady at $Q = 2$ Mgal/day. The tank volume will be a function of flow rate Q and detention time t. The detention time t is the ratio of volume and flow.

$$t = \frac{V}{Q}$$

Assume the detention time is 90 min. The cost of construction is \$2.60/ft^2 of wall surface area and \$1.20/ft^2 of floor surface area. 1 gal $= 0.1337$ ft^3.

(a) Formulate a mathematical model to meet the stated objective. Clearly define the control variables.
(b) Determine the optimum solution, by utilizing the Lagrange multiplier method.
(c) If possible, prove your solution is an optimum by utilizing one or more of the theorems presented in this chapter.

Problem 8

The number of highway miles M that can be paved is a function of labor and machine hours.

$$M = 0.5L^{0.2}K^{0.8}$$

where L is the labor used in man hours and K is the machine hours. Determine the optimum combination of machine and labor to pave 16 miles of roadway. Labor is paid \$10.00/hr, and the machine operating cost is \$75/hr.

(a) Formulate a constrained mathematical model to minimize the paving cost subject to the paving requirement.
(b) Solve the problem by the use of the Lagrange multiplier method.
(c) Replace the production function $M = 0.5L^{0.2}K^{0.8}$ in the mathematical model of part a with its logarithmic transform, $\ln M = \ln 0.5 + 0.2 \ln L + 0.8 \ln K$. Use the Lagrange multiplier method and show the optimum solution is the same as part b.

Problem 9

Design a laminated wooden beam of minimum weight as shown in Figure 8.20. The bending stress $\sigma = Mc/I$ cannot exceed 1300 psi, the allowable bending stress for pine. The shear stress $\sigma = VQ/bI$ cannot exceed 120 psi for pine or 100 psi for the glue. The following notation is used:

M = bending moment.

V = shear force.

I = moment of inertia.

b = beam width.

$c = h/2 = \frac{1}{2}$ depth of the beam.

Q = first moment of the area.

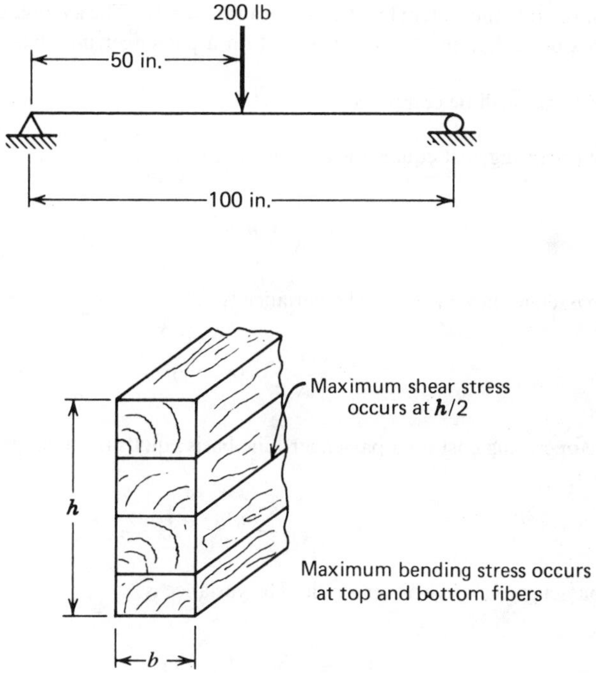

Figure 8.20

The beam weighs 24 lb/ft^3.

(a) Formulate a mathematical model to meet the stated objective. Clearly define the control variables.

(b) Determine the optimum solution by utilizing the search method.

(c) If possible, prove that your solution is an optimum by using one or more of the theorems presented in this chapter.

Problem 10

A nonprofit bus company desires to establish a fair pricing system for its customers. The ridership and trip length of each route of the system during the rush hour are

ROUTE	RIDERSHIP	AVERAGE TRIP LENGTH (Miles)
1	100	6
2	400	1.5
3	500	1.5
	$N = 1000$	$L = 9.0$

The total cost to operate the system for the rush hour is \$600.00. The average cost to operate the system $\bar{c}$ is 60 cents per rider, and the average cost on a passenger mile basis $\bar{p}$ is 31 cents per rider-mile.

Three pricing schemes will be compared.

(i) The average operating cost equals the average fare $\bar{r}$, where

$$\bar{c} = \frac{1}{N}\sum n_i r_i = \bar{r}$$

with r_i the passenger fare in cents. The variance is

$$V = \frac{1}{N}\sum n_i(r_i - \bar{r})^2$$

(ii) The average operating cost on a passenger-mile basis equals the average fare $\bar{r}$.

$$\bar{p} = \frac{1}{NL}\sum n_i l_i r_i = \bar{r}$$

with r_i the passenger fare in cents per mile. The variance is

$$V = \frac{1}{NL}\sum n_i l_i(r_i - \bar{r})^2$$

(iii) The average operating cost equals the average fare $\bar{r}$:

$$\bar{c} = \frac{1}{N}\sum n_i l_i r_i = \bar{r}$$

with r_i the passenger fare in cents per mile. The variance is

$$V = \frac{1}{N}\sum n_i(l_i r_i - \bar{r})^2$$

(a) Formulate a constrained optimization model to minimize the variance subject to the constraint that revenue equals costs for each pricing scheme.
(b) Utilize the Lagrange multiplier method to determine the optimum fare for each route.
(c) Complete the following table:

PRICING SCHEME	FOR ROUTE 1	FOR AVERAGE TRIP LENGTH 2	3
(i)			
(ii)			
(iii)			

(d) In your opinion, which pricing scheme is the best one?

8.6 THE KUHN–TUCKER CONDITIONS

Kuhn and Tucker have taken the concepts of the Lagrange multiplier for mathematical models with active constraints and extended them to mathematical models with active and inactive constraints. The mathematical model has the following form:

$$
\begin{aligned}
z &= f(\boldsymbol{x}) \\
g_i(\boldsymbol{x}) &\le b_i \qquad i = 1, 2, \ldots \\
g_k(\boldsymbol{x}) &\ge b_k \qquad k = 1, 2, \ldots \\
g_m(\boldsymbol{x}) &= b_m \qquad m = 1, 2, \ldots \\
\boldsymbol{x} &\ge 0
\end{aligned}
$$

where the $\boldsymbol{x}$ vector consists of n control variables, $\boldsymbol{x}' = [x_1 \quad x_2 \quad \cdots \quad x_n]$. Lagrange multipliers are introduced to transform the model into one with a single objective function, $L(\boldsymbol{x}, \boldsymbol{\lambda}, \boldsymbol{\mu}, \boldsymbol{v})$. The Lagrange multiplier is

$$L(\boldsymbol{x}, \boldsymbol{\lambda}, \boldsymbol{\mu}, \boldsymbol{v}) = f(\boldsymbol{x}) + \sum_i \lambda_i(b_i - g_i(\boldsymbol{x})) + \sum_k \mu_k(b_k - g_k(\boldsymbol{x})) + \sum_m v_m(b_m - g_m(\boldsymbol{x})),$$

where $\boldsymbol{\lambda}$, $\boldsymbol{\mu}$, and $\boldsymbol{v}$ are control variables associated with less-than-or-equal-to, greater-than-or-equal-to, and equality constraints, respectively. The necessary condition for a critical point is

$$\nabla_x L = \boldsymbol{0} \qquad \nabla_\lambda L = \boldsymbol{0} \qquad \nabla_\mu L = \boldsymbol{0} \qquad \nabla_v L = \boldsymbol{0}$$

The approach for establishing critical points is the same one utilized for the Lagrange multiplier method with active constraints only. However there are additional factors that must be considered.

The Kuhn–Tucker conditions for a global minimum and a global maximum are summarized in Tables 8.2 and 8.3. After a candidate solution or critical point has been obtained, the Kuhn–Tucker conditions may be used as a test to determine if it is a global optimum. The Kuhn–Tucker condition may be used as an alternate test for Theorems 1, 2, 3, and 4. Second, conditions 1 through 4 are the necessary conditions for a global optimum; thus one may use them to establish a critical point. If conditions 5 and 6 are satisfied, the critical point is a global optimum.

The Kuhn–Tucker conditions give a different insight to the nature of finding global optima. Consider a minimization problem of $z = f(\boldsymbol{x})$ subject to both active and inactive constraints. For a minimum solution, the Lagrange function $L(\boldsymbol{x}, \boldsymbol{\lambda}, \boldsymbol{\mu}, \boldsymbol{v})$ must be a minimum. Since it is a sum of terms, each term must be a minimum.

From Table 8.2 condition 2, the term $\lambda(b - g(\boldsymbol{x}))$ will be minimum when $\lambda = 0$ and $b - g(\boldsymbol{x}) \ge 0$, or when $\lambda \le 0$ and $b - g(\boldsymbol{x}) = 0$. When the constraint is inactive, the Lagrange multiplier will be equal to zero. If λ is equal to zero, the constraint equation will not influence the problem or its minimum value. When the constraint is active and the critical point is $\boldsymbol{x}^1$, the surface $z = f(\boldsymbol{x}) = f(\boldsymbol{x}^1)$ will be tangent to the surface of

Table 8.2 Kuhn–Tucker Conditions for a Global Minimum

Minimize $z = f(\mathbf{x})$

$$\begin{aligned} g_i(\mathbf{x}) &\le b_i & i &= 1, 2, \ldots \\ g_k(\mathbf{x}) &\ge b_k & k &= 1, 2, \ldots \\ g_m(\mathbf{x}) &= b_m & m &= 1, 2, \ldots \\ \mathbf{x} &\ge \mathbf{0} & \mathbf{x}'' &= [x_1 \quad x_2 \quad \cdots \quad x_n] \end{aligned}$$

Minimize $L(\mathbf{x}, \boldsymbol{\lambda}, \boldsymbol{\mu}, \mathbf{v}) = f(\mathbf{x}) + \sum_i \lambda_i(b_i - g_i(\mathbf{x})) + \sum_k \mu_k(b_k - g_k(\mathbf{x})) + \sum_m v_m(b_m - g_m(\mathbf{x}))$

The necessary and sufficient conditions for a global minimum are

1. $x_j \ge 0$ and $\partial L/\partial x_j \ge 0$.
2. (a) If $\lambda_i = 0$, then $b_i - g_i(\mathbf{x}) \ge 0$ (inactive constraint).
   (b) If $b_i - g_i(\mathbf{x}) = 0$, then $\lambda_i \le 0$ (active constraint).
3. (a) If $\mu_k = 0$, then $b_k - g_k(\mathbf{x}) \le 0$ (inactive constraint).
   (b) If $b_k - g_k(\mathbf{x}) = 0$, then $\mu_k \ge 0$ (active constraint).
4. v_m is unrestricted in sign, $-\infty \le v_m \le \infty$ and $b_m - g_m(\mathbf{x}) = 0$.
5. $f(\mathbf{x})$ is a convex function.
6. (a) $g_i(\mathbf{x})$ is a convex function
   (b) $g_k(\mathbf{x})$ is a concave function
   (c) $g_m(\mathbf{x})$ is a linear function.

Table 8.3 Kuhn–Tucker Conditions for a Global Maximum

Maximize $z = f(\mathbf{x})$

$$\begin{aligned} g_i(\mathbf{x}) &\le b_i & i &= 1, 2, \ldots \\ g_k(\mathbf{x}) &\ge b_k & k &= 1, 2, \ldots \\ g_m(\mathbf{x}) &= b_m & m &= 1, 2, \ldots \\ \mathbf{x} &\ge \mathbf{0} & \mathbf{x}' &= [x_1 \quad x_2 \quad \ldots \quad x_n] \end{aligned}$$

Maximize $L(\mathbf{x}, \boldsymbol{\lambda}, \boldsymbol{\mu}, \mathbf{v}) = f(\mathbf{x}) + \sum_i \lambda_i(b_i - g_i(\mathbf{x})) + \sum_k \mu_k(b_k - g_k(\mathbf{x})) + \sum_m v_m(b_m - g_m(\mathbf{x}))$

The necessary and sufficient conditions for a global maximum are

1. $x_j \ge 0$ and $\partial L/\partial x_j \le 0$.
2. (a) If $\lambda_i = 0$, then $b_i - g_i(\mathbf{x}) \ge 0$ (inactive constraint).
   (b) If $b_i - g_i(\mathbf{x}) = 0$, then $\lambda_i \ge 0$ (active constraint).
3. (a) If $\mu_k = 0$, then $b_k - g_k(\mathbf{x}) \le 0$ (inactive constraint).
   (b) If $b_k - g_k(\mathbf{x}) = 0$, then $\mu_k \le 0$ (active constraint).
4. v_m is unrestricted in sign, $-\infty \le v_m \le \infty$ and $b_m - g_m(\mathbf{x}) = 0$.
5. $f(\mathbf{x})$ is a concave function.
6. (a) $g_i(\mathbf{x})$ is a convex function
   (b) $g_k(\mathbf{x})$ is a concave function
   (c) $g_m(\mathbf{x})$ is a linear function.

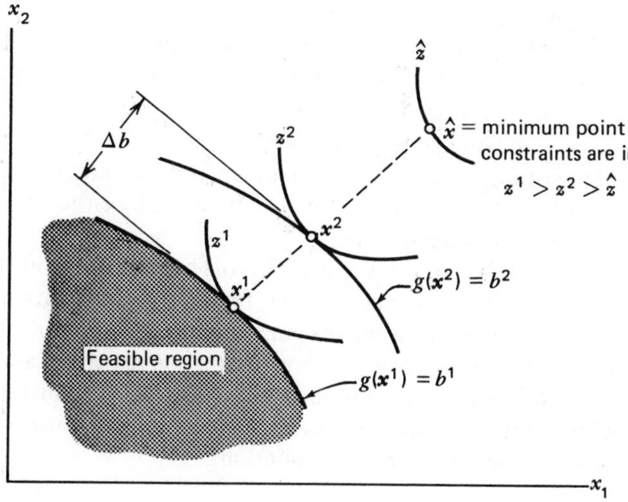

Figure 8.21 The $g(\mathbf{x}) \leq b$ constraint is active.

the constraint at $g(\mathbf{x}) = b$ at $\mathbf{x} = \mathbf{x}^1$ as shown in Figure 8.21. The minimum point with no constraints is located at $\hat{\mathbf{x}}$, it is outside the feasible region. If the constraint b is increased from b^1 to b^2, $\Delta b = b^2 - b^1 > 0$, and the value of z is to decrease from z^1 to z^2, $\Delta z = z^2 - z^1 < 0$, then $\lambda = \Delta z/\Delta b$ must be negative as stated in Table 8.2, condition 2(b).

A similar argument can be used to justify Table 8.2, condition 3. The Lagrange multiplier will be equal to zero, $\mu = 0$, if the constraint is inactive, $b - g(\mathbf{x}) \leq 0$, if the constraint is active, μ according to Kuhn Tucker conditions must be positive. Let the point $\hat{\mathbf{x}}$ be the minimum point if no constraint is imposed as shown in Figure 8.22.

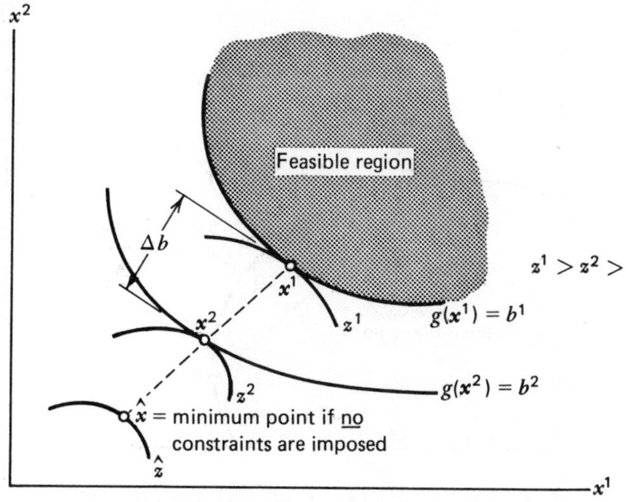

Figure 8.22 The $g(\mathbf{x}) \geq b$ constraint is active.

Consider the point x^1 where the $f(x)$ is tangent to $g(x^1) = b^1$. If the constraint is changed from b^1 to b^2 and z is to decrease, $\Delta z < 0$, the change in b must be negative, $\Delta b < 0$, as implied by the two-dimensional graph. A decrease in b from b^1 to b^2 or $\Delta b < 0$ will satisfy the condition $b^2 - g(x) \leq 0$, where $b^2 = b^1 + \Delta b$. Since $\mu = \Delta z/\Delta b$, the value of μ must be a positive value, $\mu > 0$.

For condition 4, Table 8.2 or 8.3, $g(x) = b$ is an active constraint and $f(x)$ must be tangent to it at all times. A change in b can be accompanied by either an increase or decrease in z. Thus, v is unrestricted in sign.

Conditions 5 and 6, Table 8.2 are directly related to Theorem 1. Condition 5 is the requirement that the objective function be a convex function. Condition 6 refers to the requirement that the constraint set be a convex constraint set. The conditions imposed by 6(a), 6(b) and 6(c) are alternative tests to show that a convex constraint set exists. They are particularly useful for a testing nonlinear function. The Hessian test for determining positive definite and negative definite matrices and, in turn, convex and concave functions may be used for nonlinear functions.

Convex Sets

Consider an equation of the inequality constraint $g(x) \leq b$. From condition 6(a), if $g(x)$ is a *convex function*, then $g(x) \leq b$ forms a convex set. The definition of convex function may be used to prove this statement. Consider two boundary points x^1 and x^2, as shown in Figure 8.23 as a two-dimensional model. A line joining x^1 and x^2 is expressed as

$$x = \theta x^1 + (1 - \theta)x^2$$

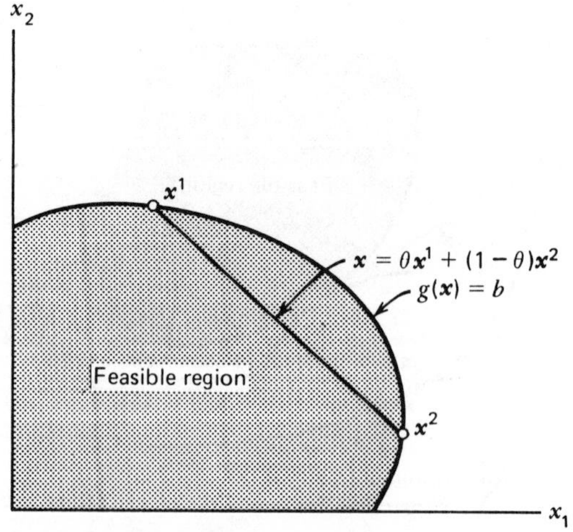

Figure 8.23 If $g(x)$ is a convex function, then $g(x) \leq b$ is a convex set.

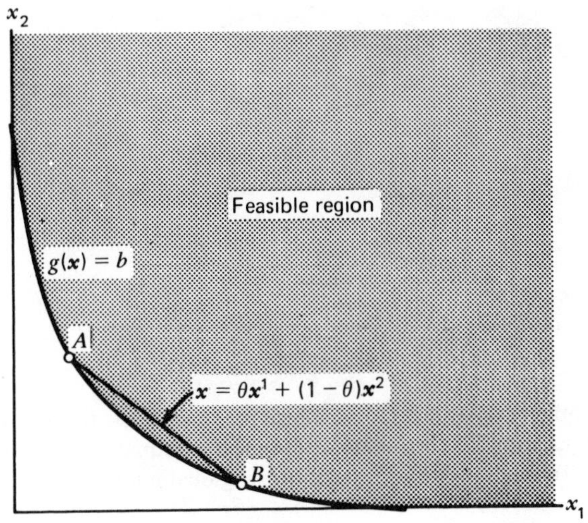

Figure 8.24 If $g(\mathbf{x})$ is a convex function, then $g(\mathbf{x}) \geq b$ is a convex set.

where $0 \leq \theta \leq 1$. All points of the line x lie inside the feasible region as required by the definition of a convex set. The definition of a convex function is

$$g(\mathbf{x}) = g(\lambda\mathbf{x}^1 + (1 - \lambda)\mathbf{x}^2) < \lambda g(\mathbf{x}^1) + (1 - \lambda)g(\mathbf{x}^2) \tag{1}$$

Since the points $\mathbf{x}^1$ and $\mathbf{x}^2$ are boundary points, the following conditions exist:

$$g(\mathbf{x}^1) = b \quad \text{and} \quad g(\mathbf{x}^2) = b$$

Substitute these relationships into the inequality equation and simplify:

$$g(\lambda\mathbf{x}^1 + (1 - \lambda)\mathbf{x}^2) \leq \lambda b + (1 - \lambda)b = b$$

or

$$g(\mathbf{x}) \leq b$$

This completes the proof. When $g(\mathbf{x})$ is a convex function, the inequality condition is satisfied and the constraint equation is a convex set.

A similar argument may be made to prove that $g(\mathbf{x}) \geq b$ is a convex set, condition 6(b), when $g(\mathbf{x})$ is a concave function. See Figure 8.24. Linear equations are special because they satisfy the conditions for a convex and concave function. As a result, for the equality constraint condition $g(\mathbf{x}) = b$, the only $g(\mathbf{x})$ function that satisfies will be a linear function.

EXAMPLE 8.13 Sizing of a Reactor Vessel

A chemical process for the treatment of industrial waste requires a detention time of at least 15 min. The reaction takes place in a continuous flow system having a flow rate of 2 ft^3/sec. Design a cylindrical tank of minimum cost. For simplicity, assume that the unit construction cost is a function of the tank surface area only. The tank will have a constant thickness.

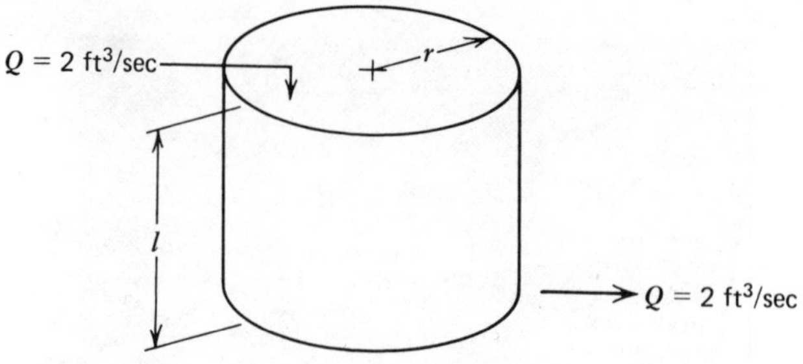

Figure 8.25

(a) Formulate a mathematical model to minimize the total construction cost.
(b) Use the Lagrange multiplier method to determine the optimum tank size.
(c) Use the Kuhn–Tucker conditions to test for the global minimum.

Solution

(a) Since construction cost is a function of tank surface area, minimizing the total surface area of the tank as a function of the tank radius and tank height will satisfy the objective of minimum cost. The control variables are r the tank radius in feet and l the tank height in feet. See Figure 8.25.

The total surface area A of the tank is equal to the surface areas of the base and wall.

$$A = 2\pi r l + \pi r^2$$

For steady-state flow, the rate of flow is equal to the tank volume V divided by the detention time T.

$$Q = \frac{V}{T}$$

The tank must be large enough to accommodate a reaction of at least 15 min. Thus, the constraint equation is of the form:

$$\frac{V}{Q} \geq T$$

where $Q = 2$ ft^3/sec, $T = 15$ min $= 900$ sec, and $V = \pi r^2 l$. Substituting and rearranging the expression gives

$$\pi r^2 l \geq 2 \cdot 900 = 1800$$

The mathematical model is

$$\text{Minimize } A = 2\pi r l + \pi r^2$$

$$\pi r^2 l \geq 1800$$

$$r \geq 0, l \geq 0$$

(b) The Lagrange function is

$$L = 2\pi r l + \pi r^2 + \mu(1800 - \pi r^2 l)$$

Assume the constraint is active. The necessary conditions for a minimum are

$$\frac{\partial L}{\partial r} = 2\pi l + 2\pi r - 2\pi r l \mu = 0 \tag{1}$$

$$\frac{\partial L}{\partial l} = 2\pi r - \pi r^2 \mu = 0 \tag{2}$$

$$\frac{\partial L}{\partial \mu} = 1800 - \pi r^2 l = 0 \tag{3}$$

From Eq. (2), we obtain $r = 2/\mu$. Substitute r into Eq. (1) and solve for l. The length is $l = 2/\mu$. The tank radius and height relationships are substituted into Eq. (3) and the result simplifies to

$$1800 - \pi\left(\frac{2}{\mu}\right)^2\left(\frac{2}{\mu}\right) = 0$$

$$\mu = \left(\frac{8\pi}{1800}\right)^{1/3}$$

The critical point is

$$r^\circ = l^\circ = \frac{2}{\mu} = \frac{2}{0.241} = 8.3 \text{ ft} \quad \text{with } \mu^\circ = 0.241$$

Now, assume that the solution may occur at an interior point and that $\pi r^2 l > 1800$ or inactive. Thus, $\mu = 0$ and the Lagrange function reduces to

$$L = 2\pi r l + \pi r^2$$

The necessary conditions for a minimum are

$$\frac{\partial L}{\partial r} = 2\pi l + 2\pi r = 0$$

$$\frac{\partial L}{\partial l} = 2\pi r = 0$$

The solution is $l = 0$ and $r = 0$. Clearly, this solution is an impractical one. The constraint $\pi r^2 l \geq 1800$ is violated also; therefore, this solution is infeasible. The critical point $r^\circ = l^\circ = 8.3$ ft satisfies the necessary condition for a minimum, and it will be tested as a global minimum.

(c) The necessary and sufficient conditions for a global minimum are obtained from Table 8.2. They are listed below. The numbers correspond to the condition numbers in Table 8.2. If a condition does not apply it is not listed.

1. If $r \geq 0$, then $\partial L/\partial r = 2\pi l + 2\pi r - 2\pi r l \mu \geq 0$.
If $l \geq 0$, then $\partial L/\partial l = 2\pi r - \pi r^2 \mu \geq 0$.

3(b). The constraint is assumed active; then $\mu \geq 0$ and $\partial L/\partial \mu = 1800 - \pi r^2 l = 0$.

5. The objective function $A = 2\pi r l + \pi r^2$ is a convex function.

6. $\pi r^2 l$ is a concave function.

Let us proceed to check the conditions, one at a time. Condition 1 is satisfied because in part b both r and l are positive values and $\partial L/\partial r$ and $\partial L/2l$ are equal to zero. Condition 3(b) is satisfied because the Lagrange multiplier is positive, $\mu^\circ = 0.241$, and the constraint is active, $\pi r^2 l = 1800$.

Condition 5 will be checked by utilizing Theorem 12. The sum of a set of convex functions is a convex function

$$A = A_1 + A_2$$

where

$$A_1 = 2\pi r l$$

$$A_2 = \pi r^2$$

Since $d^2 A_2/dr^2 = 2\pi > 0$, A_2 is positive definite for all values of r, thus A_2 is a convex function. Proving that $A_1 = 2\pi r l$ is a convex function is more difficult. The Hessian test gives

$$\boldsymbol{H}(\boldsymbol{x}) = \begin{vmatrix} 0 & 2\pi \\ 2\pi & 0 \end{vmatrix}$$

The main diagonals are equal to zero; therefore, the necessary condition for a convex function is not satisfied. Furthermore, $A_1 = 2\pi r l$ has the same form as the function $f(\boldsymbol{x}) = x_1 x_2$. In Example 8.2, we found that $x_1 x_2$ was not a convex function. However, if certain restrictions were placed upon $x_1 x_2$ we found it could be a convex function. Refer to column "along line D" of Table 8.3. In this problem the constraint $\pi r^2 l = 1800$ is active. Let us determine if A_1 is a convex when this constraint is imposed. Rewrite the constraint as $l = 1800/\pi r^2$. Substitute it into A_1 and simplify:

$$A_1 = \frac{3600}{r}$$

The second derivative is $d^2 A_1/dr^2 = 7200/r^2$. It is positive for all positive values of r, $d^2 A_1/dr^2 > 0$, thus A_1 is a convex function along the line $\pi r^2 l = 1800$ and $r \geq 0$. Since A_1 and A_2 are convex functions, the function $A = A_1 + A_2$ is convex. Condition 5 is satisfied.

Condition 6 is shown to be a convex set by utilizing a graphical procedure. All the points lying on the line connecting two arbitrary points on the boundary are inside the feasible region, thus the function $\pi r^2 l \geq 1800$ is a convex set. This condition will be satisfied only when $\pi r^2 l$ is a convex function. (See Figure 8.26.)

Since Kuhn–Tucker conditions are satisfied, the critical point $r^\circ = l^\circ = 8.3$ ft is the location of the global optimum:

$$r^* = l^* = 8.3\,\text{ft}$$

The tank volume and total surface area are

$$V^* = 1796\ \text{ft}^3$$

$$A^* = 2\pi r^* l^* + \pi r^{*2} = 649\ \text{ft}^2$$

Remarks

This problem, even for a reasonably straightforward problem, illustrates that it may be difficult to show that the Kuhn–Tucker necessary and sufficient conditions for a global optima are satisfied. Here the objective function was shown to be a convex function by assuming the constraint is active. Cooper and Steinberg state that the calculations become exceedingly complicated when the Hessian matrix is semidefinite. They also indicate that for problems in engineering, economics,

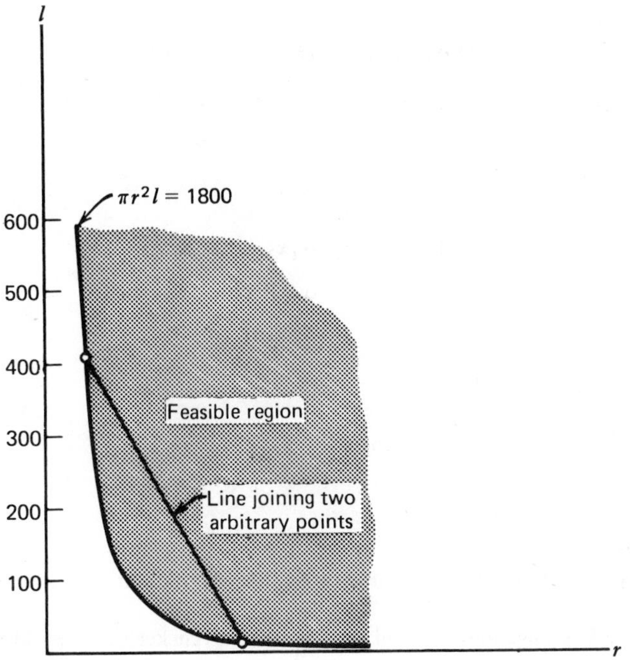

Figure 8.26 The line joining two arbitrary points on the boundary lies within the feasible region; therefore, $\pi r^2 l = 1800$ is a convex set.

and science the sufficiency test may not be worth performing because of complications in calculations. The nature of the critical point may be determined from physical conditions. These comments apply to the conditions for locating local maxima also.

PROBLEMS

Problems 1 through 3

For problems 1 through 3 answer the following questions.

(a) Is the objective function a convex function, a concave function, or neither one? Why?

(b) Does the constraint set form a convex set? Why?

(c) Determine the Lagrange function and the critical point $\mathbf{x}^\circ$.

(d) Are the Kuhn–Tucker necessary and sufficient conditions for a global maximum satisfied?

(e) Determine the location of the global maximum by graphical means.

Problem 1

$$\text{Maximize } z = (x_1 - 2)^2 + (x_2 - 1)^2 + 5$$

$$x_1^2 + x_2^2 = 1$$

$$x_1 \geq 0, x_2 \geq 0$$

Problem 2

$$\text{Maximize } z = 2x_1^2 + x_2^2 - 8x_1x_2$$
$$x_1 + x_2 \leq 1$$
$$x_1 \geq 0, x_2 \geq 0$$

Problem 3

$$\text{Maximize } z = x_1x_2$$
$$x_1^2 + x_2^2 = 12$$
$$x_1 \geq 0, x_2 \geq 0$$

Problem 4

$$\text{Maximize } z = x_1x_2$$
$$x_1 + x_2 = 1$$
$$x_1 \geq 0, x_2 \geq 0$$

(a) Solve by the method of substitution and prove that the solution is the location of the global optimum.
(b) Establish the Lagrange function and state the Kuhn–Tucker necessary and sufficient conditions for a global maximum. Determine the optimum solution and prove that it is the same as obtained in part a.
(c) Perform the Hessian test on the objective function and show that the Hessian matrix is not negative definite, and thus, that the objective function is not a concave function and the Kuhn–Tucker necessary and sufficient conditions for a global maximum are not satisfied.
(d) By taking the logarithm of the objective function the mathematical model becomes

$$\text{Maximize } A = \ln z = \ln x_1 + \ln x_2$$
$$x_1 + x_2 = 1$$
$$x_1 \geq 0, x_2 \geq 0$$

Solve the transformed mathematical model by the method of subsititution. Show that the optimum solution is the same as obtained in parts a and b.
(e) Establish the Lagrange function and state the Kuhn–Tucker conditions for a global maximum. Determine the optimum solution and show that it is the same as above.
(f) Show that the Kuhn–Tucker necessary and sufficient conditions for a global maximum are satisfied. (Comment: This problem has shown that the logarithmic transformation does not change the location of the optimum solution. Sometimes it is computationally easier to solve the separable function.)

Problem 5

Design an industrial storage building of maximum volume. The floor area of the building must be at least 900 ft^2. The maximum length of structural members will be limited to 200 ft. The rigid frame structure will consist of 4 columns and 4 beams. (See Figure 8.27.)
(a) Formulate a mathematical model for maximum volume.
(b) Use the Lagrange multiplier method to determine the critical point $\boldsymbol{x}^\circ$.

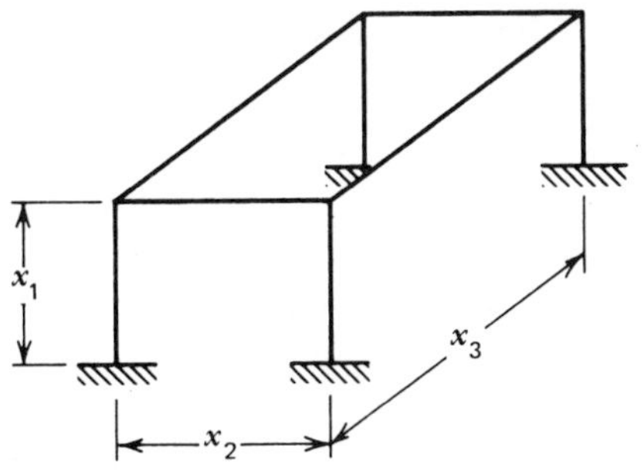

Figure 8.27

(c) Utilize the Kuhn–Tucker conditions and, if possible, show that critical point $\boldsymbol{x}^\circ$ is the global optimum solution $\boldsymbol{x}^*$.

Problem 6

The cost of building an unpaved highway will consist of a fixed cost and variable costs of paying labor L and renting machinery K. The weekly fixed cost is \$1000 (fixed costs consist of mortgage, utility bills, secretary salaries, etc.). Labor cost is \$400 per 1000 man hours. The weekly unit cost of renting machinery is a function of the number of machines rented. (See Figure 8.28.) The contractor wants to complete at least 3.6 miles of roadway per week. Production of roadway is represented by the function:

$$M = 1.8L^{0.2}K^{0.8}$$

where M is the number of miles paved per week and K is the number of machines used per week.

Because of a contractural agreement, the contractor must rent at least 4 machines per week. The supply of machines is limited to 20 per week.

(a) Formulate a mathematical model to meet the stated objective. Clearly define the control variables.

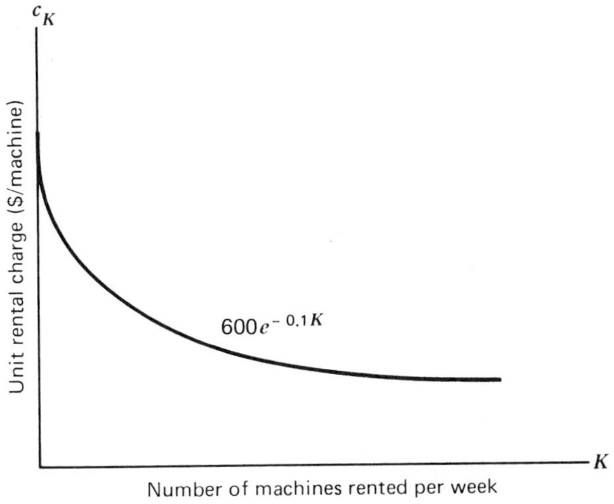

Figure 8.28

(b) Determine the critical solution $\boldsymbol{x}^\circ$ by utilizing the Lagrange method.
(c) Try to show that $\boldsymbol{x}^\circ$ is the global optimum point $\boldsymbol{x}^*$. Use the Kuhn–Tucker conditions.

Problem 7

Design a pressure vessel consisting of two spherical heads and a cylindrical section as shown in Figure 8.29. The volume of the tank must be large enough to hold 1000 gal or 1337 ft^3 of material at a pressure of 2000 psi. For a preliminary design, consider the hoop stresses in the walls of the cylinder and spherical head. The stress relationships are

$$\sigma = \frac{pr}{2t} \qquad \text{Hoop stress in cylinder}$$

$$\sigma = \frac{pr}{4t} \qquad \text{Hoop stress in sphere}$$

The allowable stress is 10,000 psi. Assume the tank wall thickness of spherical heads and cylinder wall are equal. The volume of a sphere is $V = \frac{4}{3}\pi r^3$. The surface area of a sphere is $A = 4\pi r^2$. Minimize the total volume of material needed for construction. The control variables are

r = Radius of the sphere and cylinder;

l = Length at the cylinder section;

t = Wall thickness of the sphere and cylinder.

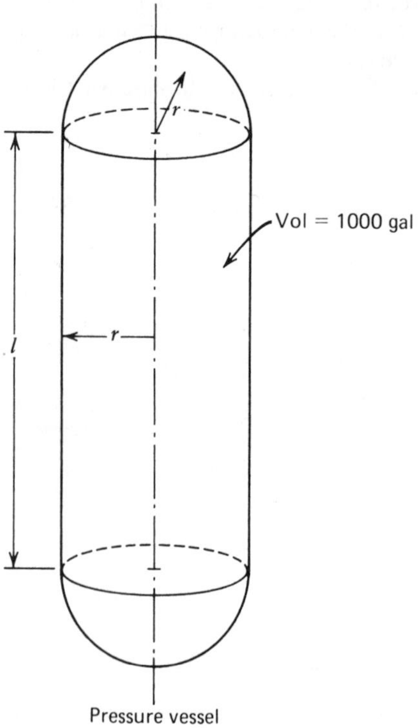

Pressure vessel

Figure 8.29

(a) Formulate a mathematical model.
(b) Determine the critical solution $\boldsymbol{x}^{\circ}$ by utilizing the Lagrange multiplier method.
(c) Try to show that $\boldsymbol{x}^{\circ}$ is the global optimum by using the Kuhn–Tucker conditions.

Summary

An optimization problem consisting of a convex or concave objective function and a convex constraint set is considered a well-behaved problem. For problems of this type, a global optimum can be found and is proved to exist. The amount of computational effort will depend upon the problem. The approach to finding candidates for the optimum solution will depend upon the type of the objective function. If a global maximum is sought and the objective function is a concave function, an evaluation of the extreme points is made. It is not necessary to search the interior or boundary of the feasible region. On the other hand, if the same conditions exist and a minimum is sought, a search of the interior, boundary and extreme points may have to be made. Although no mathematical algorithm exists, it is suggested that a search of the interior be made first, then the boundary, and the extreme points last. Fortunately, for well-behaved problems the search ceases when a local optimum is found. For problems that do not satisfy these special conditions, a more extensive search must be made.

The use of the method of substitution and the Lagrange function were to find critical points. Each method has certain theoretical and computational advantages. The Lagrange multiplier method is used to transform a mathematical model with strict equality constraints to one with a single objective function. The methods of calculus used for finding local optima of unconstrained problems may be used for Lagrange functions. Kuhn and Tucker have extended the Lagrange multiplier method to problems consisting of constraint equations with equality and inequality constraints.

Bibliography

Leon Cooper and David Steinberg. *Introduction to Methods of Optimization*. W. B. Saunders, Philadelphia, Penn. 1970.

R. H. Gallagher and O. C. Zienkiewicz. *Optimum Structural Design: Theory and Application*. John Wiley, New York. 1973.

Gilbert Strang. *Linear Algebra and Its Application*. Academic Press, New York. 1976.

G. Hadley. *Nonlinear and Dynamic Programming*. Addison-Wesley, Reading, Mass. 1964.

Ira Kirsch. *Optimum Structural Design*. McGraw-Hill, New York. 1980.

CHAPTER 9

Numerical Methods for Nonlinear Programming

After completion of this chapter, the student should be able to

1. Apply the basic principles of iterative search, the Newton and gradient projection methods, to solve nonlinear optimization problems and to obtain the roots of equations.
2. Appreciate the importance of graphical methods for estimating initial solutions of univariate functions, especially ones having multiple roots.
3. Write algorithms that solve for the roots of a given set of nonlinear equations by using the Newton method and the gradient projection method.

No numerical method is universally applicable to all kinds of nonlinear problems. A method of solution that proves highly efficient in solving one type of problem may fail or prove to be highly inefficient in solving another type. Almost every problem in engineering involving nonlinear equations requires special consideration in choosing or developing an algorithm to solve it. In linear programming, we found the simplex method to be a straightforward procedure for solving linear optimization problems because we knew the global optimum had to occur at an extreme point. In Chapter 8, we found that nonlinear problems are not as well behaved. Only unconstrained quadratic equations had features that made them straightforward to solve. Since engineering problems typically involve exponential, trigonometric, and polynomial functions, we present the basic principles of numerical analysis that are applicable to a broad range of functions. Since we do not take advantage of the special mathematical properties of any particular nonlinear function as we did with linear programming problems, these methods may not be computationally efficient for some problems. This is the price we must pay for generality. Our discussion focuses upon iterative methods of solution. These methods form the basis of many algorithms that are used to solve nonlinear problems.

In the preface of Forman Acton's textbook on numerical analysis, he states that "numerical equation solving is still largely an art," and he goes on to warn that "like most arts, it is learned with practice." Acton's words may appear to mean to the reader that numerical problem solving is hopelessly involved. It is our purpose to investigate iterative search methods as well as graphical procedures and demonstrate that even the most challenging problems may be solved.

The gradient search and gradient projection methods are used to solve unconstrained and constrained nonlinear optimization problems, respectively. The Newton method is another iterative search method that is used to solve for the *roots of equations*, not to solve optimization problems directly. Utilizing the necessary conditions for finding an optimum point, a set of n equations in n unknown is established. Since the Newton method is computationally efficient and easy to program on a computer, it has become a popular method. Sometimes the method will not converge or will converge at a point that is not of interest. We use graphical methods as well as the gradient search and gradient projection methods to overcome these difficulties.

Since the Newton method is applicable for solving a set of n equations with n variables, the Lagrange multiplier method is used to transform constrained optimization problems with active constraints into an unconstrained optimization problem. Thus, the following model

$$z = f(\boldsymbol{x})$$
$$\boldsymbol{g}(\boldsymbol{x}) = \boldsymbol{b}$$

will be transformed to

$$L(\boldsymbol{x}, \boldsymbol{\lambda}) = f(\boldsymbol{x}) + \boldsymbol{\lambda}'[\boldsymbol{b} - \boldsymbol{g}(\boldsymbol{x})]$$

where $\boldsymbol{x}$ and $\boldsymbol{\lambda}$ represent a set of n and m unknown variables x_i and λ_j for $i = 1, 2, \ldots, n$ and $j = 1, 2, \ldots, m$. The necessary condition for finding the critical point is

$$\nabla_{\boldsymbol{x}} L(\boldsymbol{x}, \boldsymbol{\lambda}) = \boldsymbol{0}$$
$$\nabla_{\boldsymbol{\lambda}} L(\boldsymbol{x}, \boldsymbol{\lambda}) = \boldsymbol{0}$$

This represents a set of $n + m$ nonlinear equations in $n + m$ variables. For an unconstrained problem the problem reduces to finding the critical values of the $\boldsymbol{x}$ vector of $\nabla_{\boldsymbol{x}} L(\boldsymbol{x}) = \nabla_{\boldsymbol{x}} f(\boldsymbol{x}) = \boldsymbol{0}$.

We cannot overemphasize the importance of the techniques and theorems of Chapter 8. The Newton and gradient projection methods will be used to *search for the critical points only*. They cannot distinguish between local optima, global optima, or saddle points. The theorems of Chapter 8 are employed for this purpose. Furthermore, before proceeding with the techniques described in this chapter, the objective and constraint set should be classified for its convexity properties. For instance, if Theorem 2 or 4 of Chapter 8 is applicable, there is no need to use the numerical methods of this chapter. Recall that the optimum solution will occur at one or more of the extreme points; thus, the search should be restricted only to investigating these points. The methods in this chapter are capable of finding the same result; however, they generally will require much more computational effort.

9.1 THE NEWTON METHOD

The Newton method utilizes an iterative approach to find the roots of a single univariate equation of the form:

$$h(x) = 0$$

and to find the roots of a set of n multivariate equations consisting of n variables of $\mathbf{x}$ of the form:

$$\boldsymbol{h}(\boldsymbol{x}) = \boldsymbol{0}$$

The functions $h(x)$ and $\boldsymbol{h}(\boldsymbol{x})$ represent a nonlinear function of x and $\mathbf{x}$, respectively. In this section our attention will focus upon the solution of univariate functions and finding the roots of nonlinear functions of the form $h(x) = 0$. In the next section, the roots of multivariate functions are discussed.

The Iterative Search Method

In general with any iterative search method, an initial guess of the solution is made. The superscript k will be used to designate the step, $k = 0, 1, 2, \ldots$. For the root of $h(x) = 0$, the initial guess will be represented as x^0, where $k = 0$.

If a value of x^k satisfies the expression $h(x) = 0$, it is a root of $h(x) = 0$, thus $h(x^k) = 0$. Generally speaking, x^k will not satisfy the strict equality condition, $h(x^k) = 0$. If x^k satisfies the inequality $|h(x^k)| \leq \varepsilon$, where ε represents a tolerance interval, then x^k is considered a root of $h(x)$.

If the initial guess x^0 does not satisfy the test for a root, the initial guess of x must be modified. The new estimate of x is x^1, or

$$x^1 = x^0 + \alpha(x^0)$$

where $\alpha(x^0)$ represents a correction factor. The new solution x^1 is tested. If $|h(x^1)| \leq \varepsilon$, then x^1 is a root. If the test fails, these steps are repeated. The $(k + 1)$th estimate of the root is represented in the iteration search equation as

$$x^{k+1} = x^k + \alpha(x^k)$$

This general procedure of testing and estimating is common to all iterative search methods. Newton used the Taylor series expansion to determine the correction factor $\alpha(x^k)$.

Roots of a Univariate Function

The essential steps of the Newton method are illustrated by the graph in Figure 9.1. The root is shown as x°. A candidate solution x^k will be considered a root in $h(x^k)$ lies within the given tolerance range between plus and minus ε as shown in the figure.

Assume that the initial guess x^0 is not a root of $h(x)$. That is, $|h(x^0)| > \varepsilon$ as shown in Figure 9.1. The new estimate x^1 will be determined by utilizing the Taylor series expansion of $h(x)$.

$$h(x^1) = h(x^0) + \frac{d}{dx} h(x^0)(x^1 - x^0) + \cdots$$

The new estimate of x^1 is assumed to be the root of the equation $h(x^1) = 0$. Thus, it can be found by setting the Taylor series equal to zero. If we ignore the higher-order terms we find

$$h(x^0) + \frac{d}{dx} h(x^0)(x^1 - x^0) = 0$$

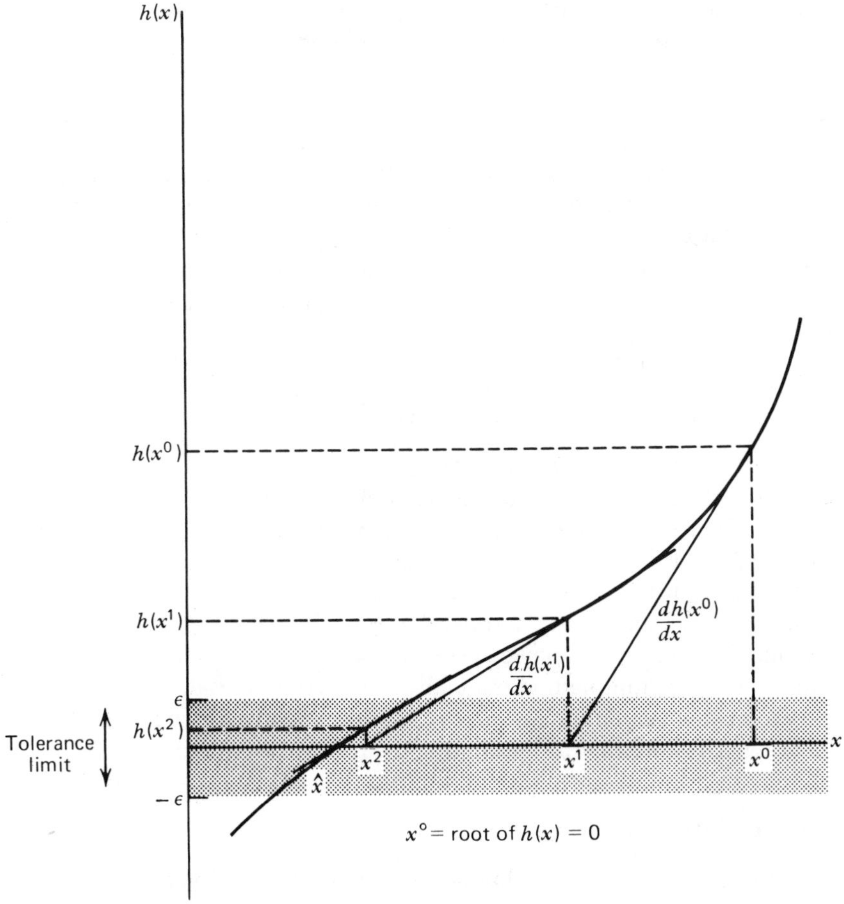

Figure 9.1 The Newton iterative approach to find the root of $h(x) = 0$.

Since x^0 and $dh(x^0)/dx$ exist, we obtain the next approximation to x^c of

$$x^1 = x^0 - \frac{h(x^0)}{dh(x^0)/dx}$$

or

$$\boxed{x^1 = x^0 - \frac{h(x^0)}{h'(x^0)}}$$

Here, as illustrated in Figure 9.1, the correction factor $\alpha(x^0)$ is the ratio $-h(x^0)/h'(x^0)$. The value of x^1 is the new estimate of the root of the equation. In Figure 9.1, the critical point is shown as $\hat{x}$. The candidate point x^1 is not the critical point, because $h(x^1)$ does not lie within the acceptance limit ε or $|h(x^1)| \leq \varepsilon$. The step is repeated with x^1 being the new approximation and x^2 the new estimate of the root of $h(x)$.

$$x^2 = x^1 - \frac{h(x^1)}{h'(x^1)}$$

In the example, the point x^2 is considered the root of $h(x)$ because $|h(x^2)| \leq \varepsilon$. The search is stopped. The root of x° is estimated to be x^2 or $x^\circ = x^2$. If ε were a smaller value, the search would continue.

The new method may be summarized by the following corrective equation and test. The $(k + 1)$th estimate of the root of $h(x)$ is estimated with the following iterative equation.

$$x^{k+1} = x^k - \frac{h(x^k)}{h'(x^k)} \qquad \text{for } k = 0, 1, 2, \ldots$$

The test for the root for x^{k+1} is $|h(x^{k+1})| \leq \varepsilon$.

This approach may be used to find roots of equations with multiple roots as illustrated in Figure 9.2. There are three roots, x^A, x^B, and x^C. The root that is found will depend upon the initial guess of x° as illustrated. In Figure 9.2*a*, the root x^A is estimated as x^2, and, in Figure 9.2*b*, x^C is estimated as x^2. Owing to the high degree of nonlinearity of this expression, the root x^B may be difficult to find. As a result, many different initial estimates of $\hat{x}$ may be necessary to find the root. If the proper estimate is not made, the Newton method may fail to converge to x^B. As a result, it is often necessary to make many initial estimates to ensure all roots are obtained. We recommend a graphical procedure for estimating the roots of a univariate function.

The Initial Estimate

A good initial estimate may be obtained through graphical methods. Suppose the roots of the following equation are sought.

$$h(x) = 2e^{-x/2} - \sin \pi x = 0$$

A root of $h(x)$ must satisfy the following equation.

$$h_1(x) = h_2(x)$$

where $h_1(x) = 2e^{-x/2}$ and $h_2(x) = \sin \pi x$. These two functions may be easily drawn or sketched. A sketch is usually sufficient to determine good initial guesses for the Newton method. Roots occur at each of the intersections shown in Figure 9.3 as points x^A, x^B, x^C, and x^D.

The Newton method is computationally efficient because it is quadratically convergent. If $\hat{x}$ is the root of $h(x) = 0$, then the error for δ^k, the estimate x^k, is the difference between the estimate and the root, $\delta^k = x^k - \hat{x}$. The error of the estimate of x^{k+1} is approximated as

$$\delta^{k+1} \sim (\delta^k)^2 \frac{h''(\hat{x})}{2h'(\hat{x})}$$

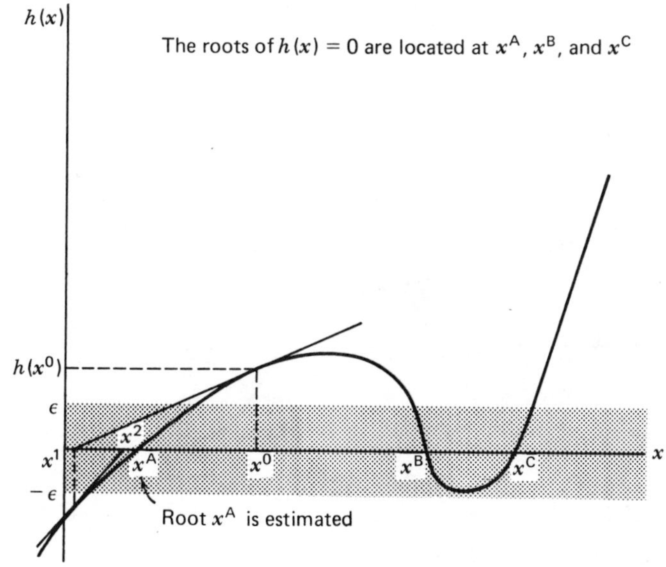

(a)

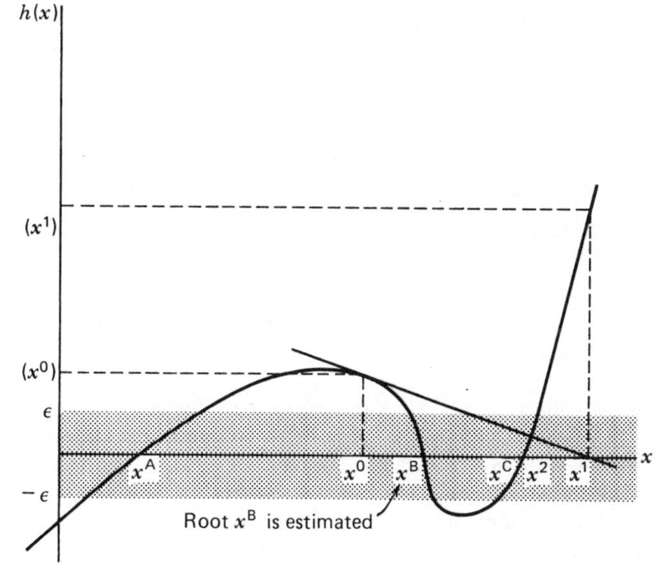

(b)

Figure 9.2 The importance of the mutual estimate of x^0.

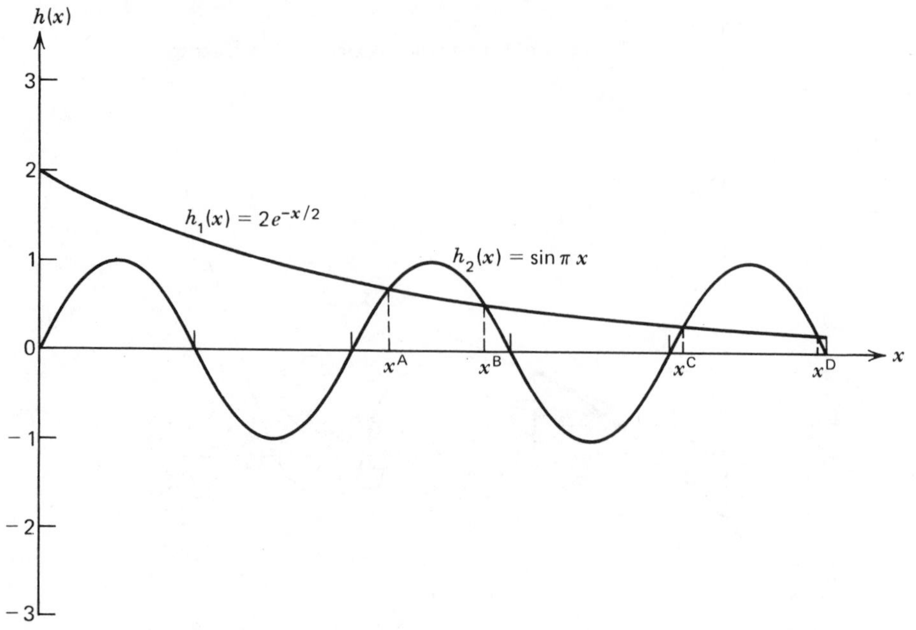

Figure 9.3 Graphical estimation of the roots of $h(x) = 2e^{-x/2} - 5 \sin \pi x = 0$.

where $\delta^{k+1} = x^k - x^\circ$ when x^k is sufficiently close to x°. Thus, squaring a small error at each iteration has the effect of doubling the number of significant figures. Even though the method is computationally efficient in converging to a root, the most important consideration is the initial estimate. If the initial estimate is outside a certain range of the root x°, the method will fail to find it.

EXAMPLE 9.1 A Minimum Weight Truss

Determine the angle α and the cross-sectional areas of the members 1, 2, and 3 of a minimum weight truss. (See Figure 9.4*a*.) Assume that the compression members, members 1 and 2, are of equal cross-sectional area and of equal length. The stress in all members must not exceed compressive or tensile limits of 20 kips/in^2.

(a) Formulate a mathematical model.
(b) Use graphical methods to estimate the optimum solution.
(c) Use the Newton method to search for the optimum solution.

Solution

(a) The principles of statics and strength of materials will be used to formulate the problem. During the course of the formulation observe that the geometric layout is the most important single factor in the design of this structure. The method of substitution will be used to reduce a multivariate model to one consisting of a single control variable α, as shown in the figure.

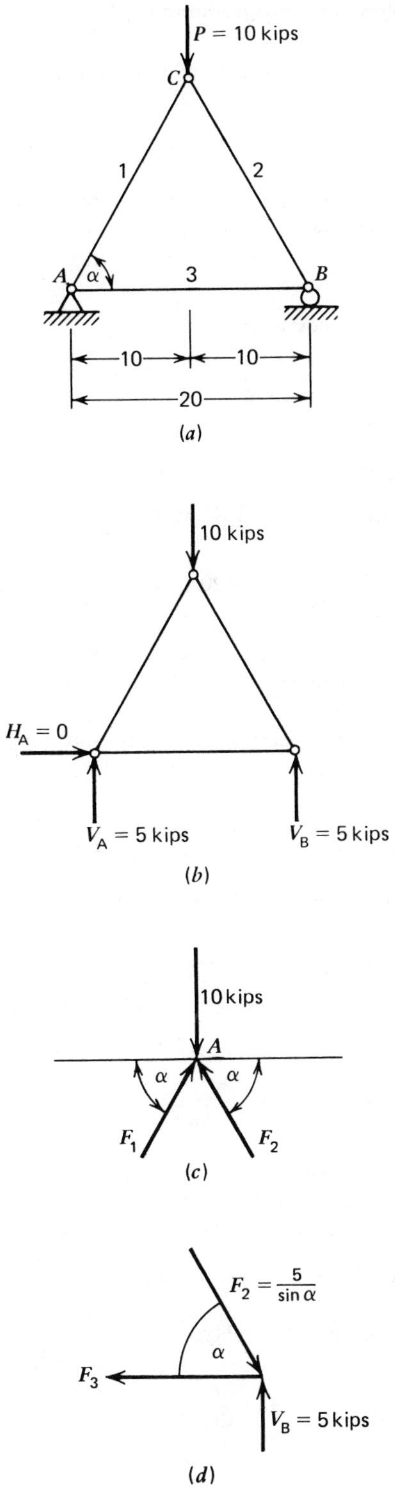

P = 10 kips
C
1
2
A
α
3
B
10
10
20
(a)
10 kips
H_A = 0
V_A = 5 kips
V_B = 5 kips
(b)
10 kips
A
α
α
F_1
F_2
(c)
F_2 = 5/sin α
α
F_3
V_B = 5 kips
(d)

Figure 9.4

The objective function to minimize the total weight, or equivalently, the total volume of the truss is

$$\text{Minimize } V = A_1 l_1 + A_2 l_2 + A_3 l_3$$

where A_1, A_2, and A_3 are cross-sectional areas of the members and l_1, l_2, and l_3 are the member lengths, respectively. The length of member 3 is known, $l_3 = 20$ ft. The lengths of members 1 and 2 are variables with $l_1 = l_2$. From trigonometry,

$$\cos\alpha = \frac{10}{l_1} = \frac{10}{l_2}$$

or

$$l_1 = l_2 = \frac{10}{\cos\alpha}$$

With this relationship and the assumption that $A_1 = A_2$, the objective function becomes

$$\text{Minimize } V = 2A_1\left(\frac{10}{\cos\alpha}\right) + 20A_3 = 20\left(\frac{A_1}{\cos\alpha} + A_3\right)$$

From statics, it may be shown that the vertical reactions at A and B are equal, $V_A = V_B = 5$ kips. The horizontal reaction of A is equal to zero. The free-body diagram for the truss is shown in Figure 9.4*b*.

The method of joints will be used to determine the member forces in the members. At joint C, the free-body diagram is as shown in Figure 9.4*c*. The sum of the vertical and horizontal forces is

$$\begin{aligned} \sum F_y = 0{:}\quad & 10 - F_1 \sin\alpha - F_2 \sin\alpha = 0 \\ \sum F_x = 0{:}\quad & F_1 \cos\alpha - F_2 \cos\alpha = 0 \end{aligned} \tag{1}$$

or

$$F_1 = F_2 \tag{2}$$

The member force is determined by substituting the results stated in Eq. (2) into Eq. (1). Simplifying, it becomes

$$F_1 = F_2 = \frac{5}{\sin\alpha} \tag{3}$$

Forces F_1 and F_2 are compressive forces.

The force in member 3 will be determined by analyzing a free-body diagram at point B. (See Figure 9.4*d*.) The sum of the horizontal foces is

$$\sum F_x = 0{:}\quad F_2 \cos\alpha - F_3 = 0$$

Substitute Eq. (2) into this expression and simplify:

$$F_3 = 5\,\frac{\cos\alpha}{\sin\alpha}$$

The force F_3 is a tensile force.

The stresses in members 1, 2, and 3 are limited to an allowable stress of 20 kips/in^2. Thus, the constraint set will consist of the following equations:

$$\frac{F_1}{A_1} = \frac{F_2}{A_2} \leq 20$$

$$\frac{F_3}{A_3} \leq 20$$

or

$$\frac{5}{A_1 \sin \alpha} \leq 20 \quad \text{or} \quad A_1 \geq \frac{1}{4 \sin \alpha}$$

$$\frac{5 \cos \alpha}{A_3 \sin \alpha} \leq 20 \quad \text{or} \quad A_3 \geq \frac{1}{4} \frac{\cos \alpha}{\sin \alpha}$$

The mathematical model becomes

$$\text{Minimize } V = 20\left(\frac{A_1}{\cos \alpha} + A_3\right)$$

$$A_1 \geq \frac{1}{4 \sin \alpha}$$

$$A_3 \geq \frac{1}{4} \frac{\cos \alpha}{\sin \alpha}$$

$$A_1 \geq 0, A_3 \geq 0$$

Both constraints must be active constraints; thus, A_1 and A_3 from the constraint set may be substituted into the objective function:

$$\text{Minimize } V = 20\left(\frac{1}{4 \cos \alpha \sin \alpha} + \frac{1}{4} \frac{\cos \alpha}{\sin \alpha}\right)$$

or

$$\text{Minimize } V = \frac{5}{\sin \alpha}\left[\frac{1}{\cos \alpha} + \cos \alpha\right]$$

(b) The critical point will occur at the point of zero slope, $dV/d\alpha = 0$, or

$$\frac{dV}{d\alpha} = \frac{-5 \cos \alpha}{\sin^2 \alpha}\left[\frac{1}{\cos \alpha} + \cos \alpha\right] + \frac{5}{\sin \alpha}\left[\frac{\sin \alpha}{\cos^2 \alpha} - \sin \alpha\right] = 0$$

Simplifying, we obtain the univariate equation:

$$h(\alpha) = \frac{dV}{d\alpha} = \left(\frac{-1}{\sin^2 \alpha} - \frac{\cos^2 \alpha}{\sin^2 \alpha}\right) + \left(\frac{1}{\cos^2 \alpha} - 1\right) = 0 \tag{4}$$

Multiply the equation by $\sin^2 \alpha$ and call each new term in the parentheses h_1 and h_2, respectively. Let $h_1(\alpha) = h_2(\alpha)$, thus, the optimum solution satisfies the following equation.

$$1 + \cos^2 \alpha = \tan^2 \alpha - \sin^2 \alpha \tag{5}$$

Graphing this equality as in Figure 9.5 shows that the approximate solution is $\alpha^\circ \simeq 55.^\circ$ and $V \simeq 14.1$ ft.-in.^{2} or $W^\circ = 48$ lb with the unit weight of steel equal to 490 lb/ft.^{3} of material. $A_1^\circ = A_2^\circ \simeq 0.31$ in.^{2} and $A_3^\circ = 0.35$ in^2.

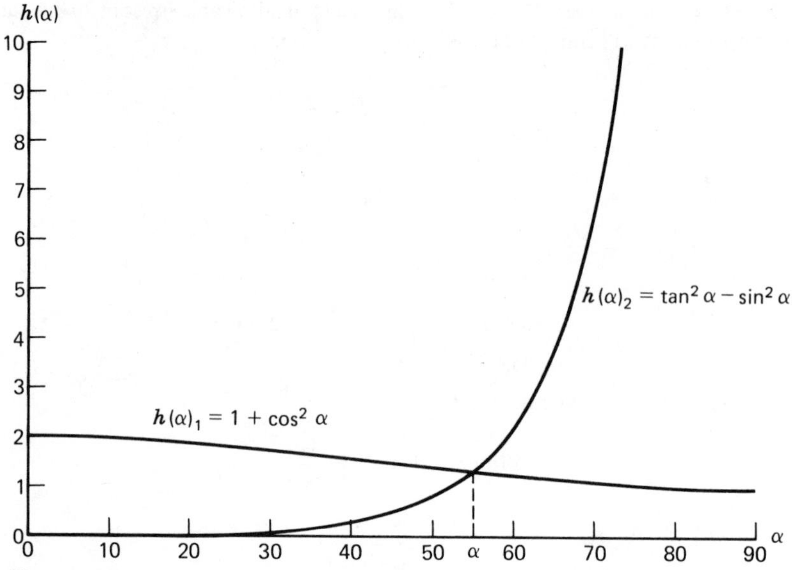

Figure 9.5

(c) The optimum solution will occur at the point where $h(\alpha) = 0$, see Eq. (4). The iteration equation of the Newton method is

$$\alpha^{k+1} = \alpha^k - \frac{h(\alpha^k)}{h'(\alpha^k)}$$

where α^{k+1} and α^k are equal to the new and old estimates of α, respectively, and where $h'(\alpha)$ is the first derivative of $h(\alpha)$. A computer program is utilized to solve for $h(\alpha) = 0$. The iteration is ceased when the following tolerance limit is satisfied:

$$|h(\alpha^{k+1})| \leq 0.01.$$

The iterations are as follows:

	α^k		
k	RADIANS	DEGREES	V^k
0	0.9599	55.000°	14.143 (Initial estimate)
1	0.9554	54.739°	14.142
2	0.9553	54.736°	14.142 (Tolerance limit is satisfied)

This table illustrates that the initial estimate is a good one, and the Newton method had little difficulty converging to a solution in a minimum number of iterations. The solution is a critical point.

$$\alpha^\circ = 54.736^\circ \qquad V^\circ = 14.418 \text{ ft.-in.}^2$$

Remarks

It may be shown that $h'(\alpha) = d^2V/d\alpha > 0$, and the critical point is the location of a minimum, thus

$$\alpha^* = \alpha^\circ = 54.736^\circ \qquad A_1^* = A_2^* = A_1^\circ = A_2^\circ = 0.31 \text{ in.}^2 \qquad A_3^* = A_3^\circ = 0.35 \text{ in.}^2$$

$$V^* = V^\circ = 14.418 \text{ ft.-in.}^2$$

In order to demonstrate the importance of the initial estimate, part c was solved again utilizing a different initial estimate. This time, suppose that the graphical solution of part a is not performed and an initial guess of $\alpha^0 = 15°$ is made. The output of the computer program is

k	α^k RADIANS	α^k DEGREES	V^k
0	0.2618	15.000°	38.660 (Initial estimate)
1	3.6189	207.345°	21.923
2	4.9406	283.074°	−23.853
3	5.0389	288.706°	−18.153
4	5.1553	295.413°	−15.275
5	5.2622	301.504°	−14.287
6	5.3176	304.678°	−14.145
7	5.3276	305.250°	−14.142
8	5.3279	305.264°	−14.142
9	5.3279	305.264°	−14.142 (Tolerance limit is satisfied)

The Newton method is able to converge to a root in a reasonable number of steps; the method is very efficient computationally. More important, the root here, α^9, $V^9 = 305.264°$ and $V^\circ = -14.1$ ft.-in^2, is not a solution. It does not agree with the one obtained in part c, α°, $V^\circ = 54.736°$ and $V^\circ = 14.4$ ft.-in.^{2}. From geometric considerations of the truss, the angle α must lie between 0 and 90 degrees, $0 \leq \alpha \leq 90°$. This root violates this condition. Furthermore, the negative volume does not make sense either. A plot of $h(\alpha)$ shows that at $\alpha^0 = 15°$ the slope $h'(\alpha^0)$ is very flat and close to zero. Since $h(\alpha^\circ)$ is a small numerical value, a drastic change in α takes place after the first iteration, $\alpha^1 = 207.345°$. Upon subsequent iterations, the procedure converges to $\alpha = 305.264°$, a root that is not of interest to us. (See Figure 9.6.)

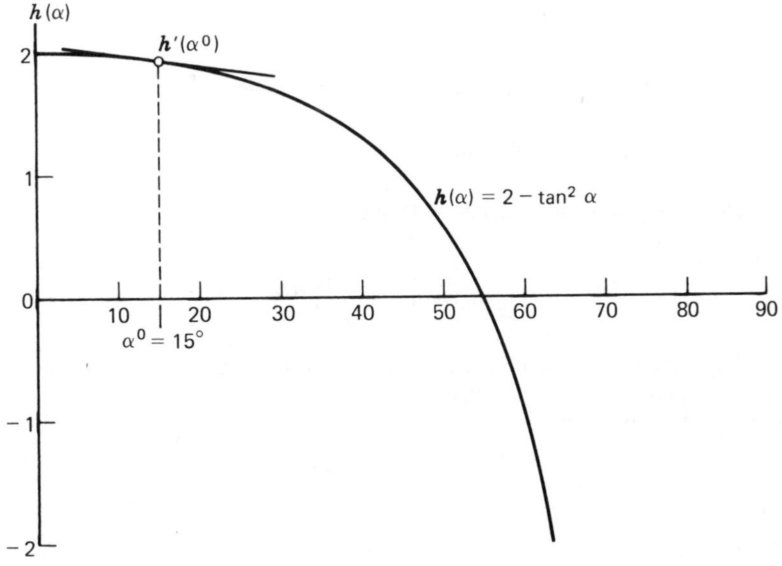

Figure 9.6

Sometimes it is possible to make a wild initial guess and find that the method will converge to a meaningful solution. This practice is not recommended. In this textbook, accurate drawings are presented in the Figures. Sketches may be drawn and crude estimates of the solution may be made. For this problem, an initial starting point of α between 45° and 65° will lead to the correct solution. See Figure Example 9.6. Even with the crudest of sketches, an estimate within this range should be achieved.

The observant reader will recognize in Eq. (5) the identity $\cos^2 \alpha + \sin^2 \alpha = 1$ and that it may be rewritten as

$$\tan^2 \alpha = 2$$

The optimum solution may be obtained without the graphical or Newton methods.

EXAMPLE 9.2 Maximum Fluid Flow Through a Pipe

The Manning equation is an empirically derived formula used to determine the flow in open channels, sewer lines, flumes, and so forth.

$$Q = \left(\frac{1.49}{n}\right) AR^{2/3} S_0^{1/2}$$

where Q is the flow in cubic feet per second; n is the channel roughness coefficient; A is the cross-sectional area of flow in square feet; P is the wetted perimeter in feet; R is the hydraulic radius equal to A/P, in feet; and S_0 is the slope of the channel. Consider a 3-ft-diameter concrete pipe ($n = 0.013$) on a slope of 0.0001.

(a) Formulate an optimization model to determine the maximum flow in the pipe.
(b) Use graphical methods to estimate the optimum solution.
(c) Use the estimate obtained in part b as an initial estimate and the Newton method to determine the optimum solution.

Solution

(a) (See Figure 9.7.) Our objective is to determine the maximum flow.

$$\text{Maximum } Q = \left(\frac{1.49}{n}\right) AR^{2/3} S_0^{1/2}$$

This equation may be rewritten in terms of the cross-sectional area A and wetted perimeter P. Thus

$$\text{Maximize } Q = \left(\frac{1.49}{n}\right) \frac{A^{5/3}}{P^{2/3}} S_0^{1/2}$$

where A and P will be written in terms of the control variable α, which is depicted in the diagram of the cross section of the pipe.

The wetted perimeter P is

$$P = \pi r + 2r\alpha = r(\pi + 2\alpha)$$

The area A is

$$A = \frac{\pi r^2}{2} + \int dA$$

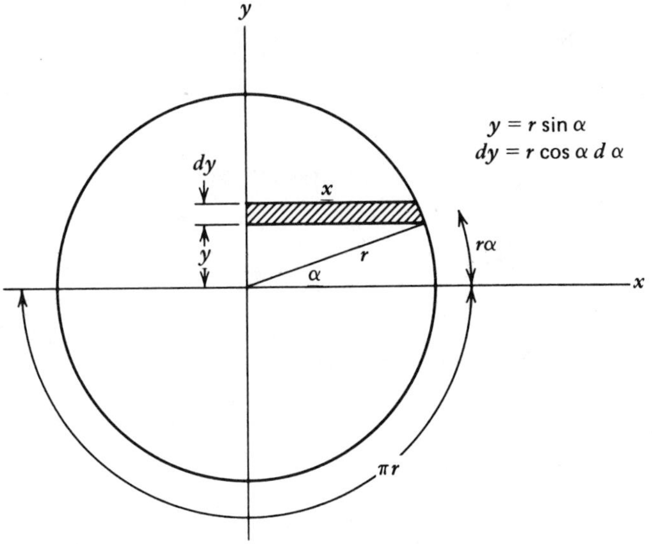

Figure 9.7

where dA is the incremental area shown in the figure, or

$$dA = x\,dy$$

but $$y = r \sin \alpha \text{ and } dy = r \cos \alpha \, d\alpha$$

and $$x^2 + y^2 = r^2 \quad \text{or} \quad x = (r^2 - y^2)^{1/2}$$

Substitute $y = r \sin \alpha$ for y; thus

$$x = (r^2 - r^2 \sin^2 \alpha)^{1/2}$$

$$A = \frac{\pi r^2}{2} + 2 \int_0^{\alpha} (r^2 - r^2 \sin^2 \alpha)^{1/2}\, r \cos \alpha \, d\alpha$$

Evaluating the integral gives

$$A = \frac{\pi r^2}{2} + r^2\left(\alpha + \frac{\sin 2\alpha}{2}\right)$$

The mathematical model in terms of the control variable α is

$$\text{Maximize } Q = \left(\frac{1.49}{n}\right) S_0^{1/2} \left(\frac{\pi r^2}{2} + r^2\left(\alpha + \frac{\sin 2\alpha}{2}\right)\right)^{5/3} (\pi r + 2r\alpha)^{-2/3}$$

$$0 \le \alpha \le \pi/2$$

(b) The optimum solution for maximum flow must satisfy the following condition:

$$\frac{dQ}{d\alpha} = 0$$

or

$$\frac{dQ}{d\alpha} = \left(\frac{1.49}{n}\right) S_0^{1/2} \left(\frac{5}{3}\right) \left(\frac{\pi r^2}{4} + r^2\left(\alpha + \frac{\sin 2\alpha}{2}\right)\right)^{2/3} (r^2(1 + \cos 2\alpha))(\pi r + 2r\alpha)^{-2/3}$$

$$+ \left(\frac{1.49}{n}\right) S_0^{1/2} \left(\frac{-2}{3}\right)(2r) \left(\frac{\pi r^2}{4} + r^2\left(\alpha + \frac{\sin 2\alpha}{2}\right)\right)^{5/3} (\pi r + 2r\alpha)^{-5/3} = 0$$

After substituting all known values of S_0, n, and r and simplifying, the expression reduces to

$$\frac{dQ}{d\alpha} = (\pi + 2\alpha)(3 + 5\cos 2\alpha) - 2\sin 2\alpha = 0$$

or

$$h_1(\alpha) = h_2(\alpha)$$

$$(\pi + 2\alpha)(3 + 5\cos 2\alpha) = 2\sin 2\alpha$$

A graph of $h_1(\alpha)$ and $h_2(\alpha)$ in Figure 9.8*a* shows that the intersection occurs at

$$\alpha = 1.1 \text{ rad}$$

(c) The iteration equation for the Newton method is

$$\boxed{\alpha^{k+1} = \alpha^k - \frac{h(\alpha^k)}{h'(\alpha^k)}}$$

where

$$h(\alpha) = \frac{dQ}{d\alpha} = (\pi + 2\alpha)(3 + 5\cos 2\alpha) - 2\sin\alpha$$

The Newton method with the initial estimate $\alpha^\circ = 1.1$ rad proceeds as follows:

		α^k	
k	Q^k (ft³/sec)	(radians)	(degrees)
0	7.189	1.100	63.025°
1	7.192	1.068	61.192°

The tolerance limit is assumed to be $|h(\alpha^{k+1}) \le 0.01$. The critical point is

$$\alpha^\circ = 61.192^\circ \quad \text{and} \quad Q^\circ = 7.192 \text{ ft}^3/\text{sec}$$

Remarks

It can be shown that $h'(\alpha) = d^2Q/d\alpha^2 < 0$ at $\alpha^\circ = 1.068$ and the location of the critical point is the location of maximum, thus $\alpha^\circ = \alpha^* = 1.068$.

To demonstrate the importance of the initial estimate, the following runs were made. The paths to convergence are:

	RUN 1	
k	Q^k (ft³/sec)	α^k (radians)
0	5.999	0.500
1	6.985	1.356
2	7.139	0.941
3	7.192	1.066
4	7.192	1.068 (Tolerance limit is satisfied)

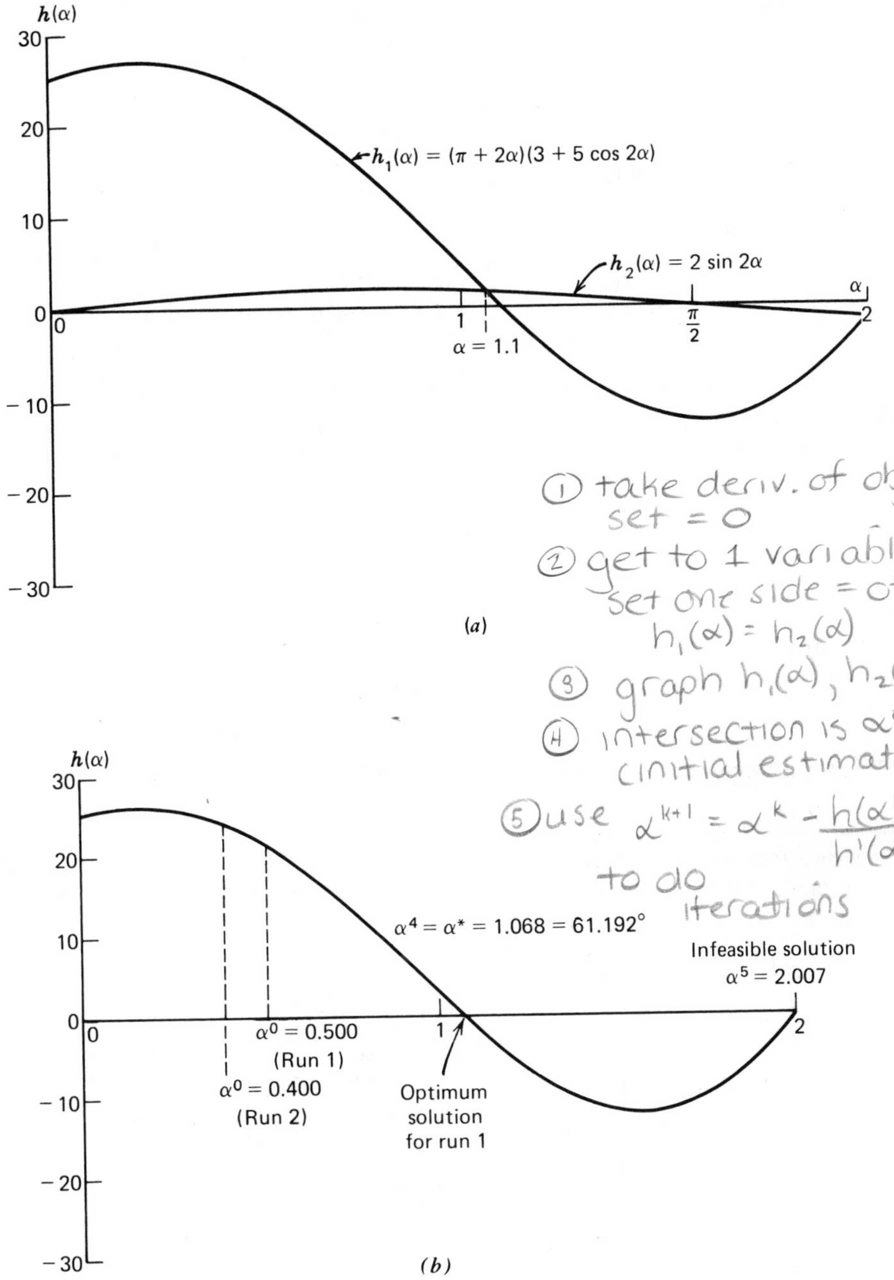

Figure 9.8

① take deriv. of obj. fcn
set = 0
② get to 1 variable
set one side = other side
$h_1(\alpha) = h_2(\alpha)$
③ graph $h_1(\alpha), h_2(\alpha)$
④ intersection is α^0
(initial estimate)
⑤ use $\alpha^{k+1} = \alpha^k - \dfrac{h(\alpha^k)}{h'(\alpha^k)}$
to do
iterations

This critical point is the optimum solution $\alpha^* = 1.068$ rad $= 61.192°$.

	RUN 2	
k	Q^k (ft^3/sec)	α^k (radians)
0	5.542	0.400
1	6.475	1.733
2	6.460	2.241
3	6.308	2.037
4	6.305	2.008
5	6.305	2.007 (Tolerance limit is satisfied)

This critical point is an infeasible solution, because $\alpha^5 = 2.007$ rads or 114.99° and α^* must be less than 90°.

These runs show that the Newton method converges rapidly. A graph of the function $h(\alpha)$ (Figure 9.8*b*) shows the initial points and points of convergence for runs 1 and 2. Note that $\alpha^5 = 114.99°$ is not a feasible solution.

PROBLEMS

Problem 1

The relationship between load P and deflection x is in Figure 9.9

$$P = kx$$

where k is the spring constant. Assume that the spring has nonlinear properties.

$$k = 2000e^{-x/20}$$

A load of 5000 lb is imposed. Determine the deflection.

(a) Establish a function of $h(x) = 0$ to solve for the deflection x. Utilize a graphical method to estimate x.

(b) Use the Newton method to determine x.

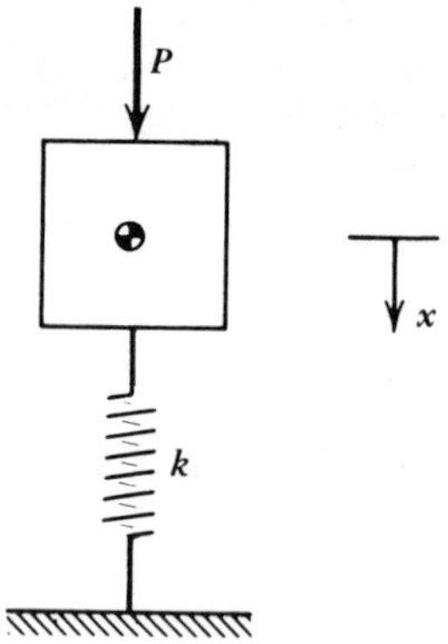

Figure 9.9

Problem 2

Consider a composite beam with the cross section shown in Figure 9.10a. The properties of the intervals are

	σ (psi)	ρ (lb/ft^3)	E(kips/in^2)
Steel	22,000	490	30,000
Plastic	1,200	40	2,000

where σ is the maximum allowable bending stress; ρ is the density; and E is the modulus of elasticity.

The beam will be subject to a maximum bending moment of 80,000 in.-lb. The beam has a 4 in. depth and 2-in. width. The objective is to determine a beam of minimum weight. The control variable is x, one-half depth of the plastic layer.

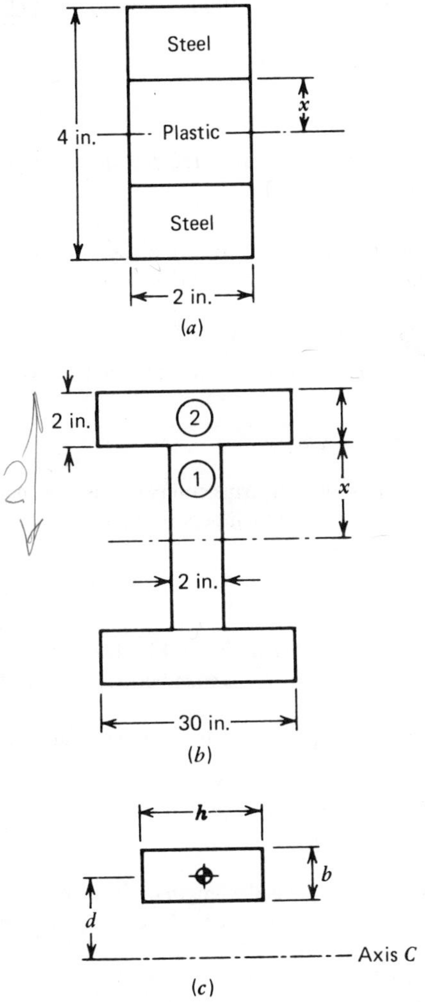

Figure 9.10

(a) Show that the moment of inertia of the composite member is

$$I = \frac{2}{12}(2x)^3 + 2\left(\frac{30}{12}(2 - x)^3 + 30(2 - x)\left(x + \frac{2 - x}{2}\right)^2\right)$$

where the composite member is treated as a member with the cross section shown in Figure 9.10*b*. The width of the steel portion is assumed to be equal to 30 ($2E_s/E_p = 30$, $E_s = 30{,}000$ kips/in.^{2} and $E_p = 2000$ kips/in.^{2}). The moment of inertia about the neutral axis is

$$I = \frac{bh^3}{12} + Ad^2$$

where b is the segment width; h is the segment depth; A is the segment area; and d is the distance to the segment centroid. (See Figure 9.10*c*.)

(b) Formulate a mathematical model to minimize the weight of the beam subject to the allowable bending stresses.

(c) Use the search method of Chapter 8 to solve the problem. Use the graphical method to estimate a solution.

(d) Use the Newton method to determine the solution.

Problem 3

The demand for prefabricated structural members is given by the demand function:

$$p = 10^{10} q^{-2}$$

where q is the total market demand and p is the unit selling price. One hundred firms are currently producing these members. The production function for each of these firms is

$$C_i = 400\, q_i^{1.2} \qquad i = 1, 2, \ldots, 100$$

A new firm enters the market with a new technology for producing prefabricated members. The cost function is

$$C_{101} = 450 q_{101}^{1.1}$$

(a) Determine the aggregate cost function for the 101 firms competing in this market.

(b) Determine the selling price p and the number of members sold at q at market equilibrium by use of graphical and the Newton methods.

Problem 4

Consider the design of a minimum weight truss as shown in Figure 9.11. The control variables are θ, as shown in the figure, and A, the cross-sectional area of the members. The members are circular members.

The allowable material stress is $\sigma_a = 40{,}000$ psi. The truss member must be designed for buckling. The critical buckling stress is

$$\sigma_{cr} = \frac{\pi^2 EI}{l^2 A}$$

where $E = 30{,}000{,}000$ psi, the modulus of elasticity; l is the length of the member; $A = \pi r^2$ is the cross-sectional area of a member; and $I = \pi r^4/4$ is the moment of inertia of a member.

(a) Formulate a mathematical model.

(b) Use the graphical and the Newton method to find the solution.

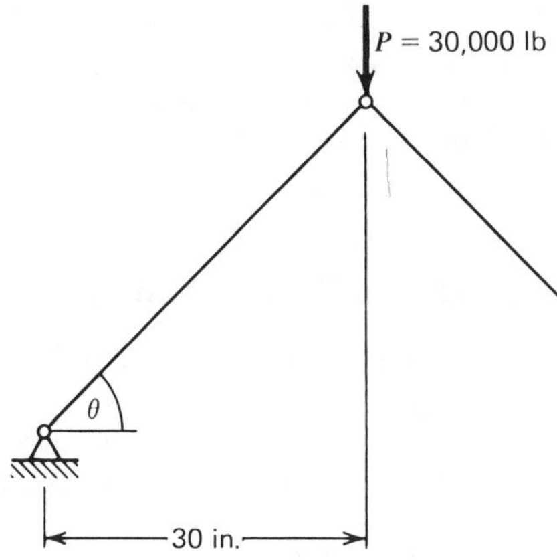

Figure 9.11

Problem 5

A manufacturer would like to construct a sphere that holds a gaseous substance at a pressure p of 10,000 psi. The larger the volume V, the more competitive his product will be. The volume of a sphere is $V = \frac{4}{3}\pi r^3$.

The total cost of sphere construction C is a function of the control variables t and the radius r of the sphere. The thicker the sphere wall, the more expensive it is to fabricate.

$$C = 2000\, t^{3/2} r$$

The selling price of the sphere must be equal to or less than \$10,000. Assume the selling price and cost C are equal.

In order for the sphere to be structurally sound, the tensile stress of the wall σ must not exceed the allowable stress of the wall material of 20,000 psi.

$$\sigma = \frac{pr}{2t}$$

The maximum wall thickness t is 1 in.

(a) Formulate a mathematical model.

(b) Use the search method of Chapter 8 to solve the problem. Use the graphical and the ~~Newton~~ methods to find the solution.

9.2 NEWTON AND GRADIENT PROJECTION METHODS

In the preceding discussion of the Newton method, we discussed the problem of establishing an initial estimate and advocated the use of a graphical method for this determination. Generally, for solving sets of nonlinear functions, the graphical method cannot be applied, and another approach must be employed. Making an educated guess is one possibility. We should not be surprised to find that this approach is somewhat haphazard and may prove to be unsuccessful. Gradient methods may be used in lieu of the Newton method. They do not converge to a solution as fast as the Newton method. As a result, we use the gradient projection method to estimate the location of the vicinity of the solution, and then the Newton method will be used to find the solution within a given tolerance interval. We expand the concept of the Newton method to finding the roots of nonlinear functions, introduce the gradient search and gradient projection methods, and finally combine these approaches into a single algorithm.

Newton Method for Finding the Root of an Equation Set

The Newton method may be used to find roots of a set of nonlinear equations:

$$\boldsymbol{h}(\boldsymbol{x}) = \boldsymbol{0}$$

where $\boldsymbol{x}' = [x_1 x_2 \cdots x_n]$ and $j = 1, 2, \ldots, n$. The set of n equations is

$$\begin{aligned} h_1(\boldsymbol{x}) &= 0 \\ h_2(\boldsymbol{x}) &= 0 \\ &\vdots \\ h_n(\boldsymbol{x}) &= 0 \end{aligned}$$

A Taylor series expansion may be written for each equation of the equation set. For equation i, the Taylor series expansion with an initial estimate $\boldsymbol{x}^0$ is

$$h_i(\boldsymbol{x}^1) = h_i(\boldsymbol{x}^0) + [\nabla_{\boldsymbol{x}} h_i(\boldsymbol{x}^0)]'(\boldsymbol{x}^1 - \boldsymbol{x}^0) + \cdots$$

where the vector:

$$[\nabla_{\boldsymbol{x}} h_i(\boldsymbol{x}^0)] = \begin{bmatrix} \dfrac{\partial h_i(\boldsymbol{x}^0)}{\partial x_1} \\ \dfrac{\partial h_i(\boldsymbol{x}^0)}{\partial x_2} \\ \vdots \\ \dfrac{\partial h_i(\boldsymbol{x}^0)}{\partial x_n} \end{bmatrix}$$

Neglecting the higher-order terms of the Taylor series expansion, the above equation set becomes

$$\begin{aligned} h_1(\boldsymbol{x}^1) &= h_1(\boldsymbol{x}^0) + [\nabla_{\boldsymbol{x}} h_1(\boldsymbol{x}^0)]'(\boldsymbol{x}^1 - \boldsymbol{x}^0) = 0 \\ h_2(\boldsymbol{x}^1) &= h_2(\boldsymbol{x}^0) + [\nabla_{\boldsymbol{x}} h_2(\boldsymbol{x}^0)]'(\boldsymbol{x}^1 - \boldsymbol{x}^0) = 0 \\ &\vdots \\ h_n(\boldsymbol{x}^1) &= h_n(\boldsymbol{x}^0) + [\nabla_{\boldsymbol{x}} h_n(\boldsymbol{x}^0)]'(\boldsymbol{x}^1 - \boldsymbol{x}^0) = 0 \end{aligned}$$

We rewrite this set of equations in a more compact form:

$$\boldsymbol{h}(\boldsymbol{x}^1) = \boldsymbol{h}(\boldsymbol{x}^0) + \boldsymbol{H}(\boldsymbol{x}^0)'(\boldsymbol{x}^1 - \boldsymbol{x}^0) = \boldsymbol{0}$$

where

$$h(x^0) = \begin{bmatrix} h_1(x^0) \\ h_2(x^0) \\ \vdots \\ h_n(x^0) \end{bmatrix}$$

and

$$H(x^0) = \begin{bmatrix} \frac{\partial h_1(x^0)}{\partial x_1} & \frac{\partial h_1(x^0)}{\partial x} & \cdots & \frac{\partial h_1(x^0)}{\partial x_n} \\ \frac{\partial h_2(x^0)}{\partial x} & \frac{\partial h_2(x^0)}{\partial x_2} & \cdots & \frac{\partial h_2(x^0)}{\partial x_n} \\ \vdots & \vdots & & \vdots \\ \frac{\partial h_n(x^0)}{\partial x_1} & \frac{\partial h_n(x^0)}{\partial x_2} & \cdots & \frac{\partial h_n(x^0)}{\partial x_n} \end{bmatrix}$$

Rearranging this equation we may solve for x^1:

$$x^1 = x^0 - [H(x^0)']^{-1} h(x^0)$$

If the root x^1 lies within the tolerance interval, then each equation of the set will satisfy the relationship $|h_i(x^1)| \leq \varepsilon$, for $i = 1, 2, \ldots, n$ and is considered a root of the equation set.

For the $(k + 1)$th iteration, the estimate of the root is

$$x^{k+1} = x^k - [H(x^k)']^{-1} h(x^k)$$

The test for the root is $|h_i(x^{k+1})| \leq \varepsilon$ for $i = 1, 2, \ldots, n$.

The Gradient Search Method for Unconstrained Models

The gradient search and gradient projection methods, like the Newton method, utilize an iterative search approach. These methods differ from the Newton method in the sense that they may be used to find the solution of optimization problems. An initial guess x^0 of the critical or global optimum is made. Since the global optimum implies that we satisfy certain conditions of optimality discussed in Chapter 8, we limit our discussion to finding the critical point x°. If x^0 satisfies a tolerance test, then the critical point is found. If not, a new estimate of x° is made with use of an iterative equation of the form:

$$x^1 = x^0 + msu^0$$

where u^0 is a direction vector and s is a scalar and m is a variable. The product ms represents the step size. The $k + 1$ estimate of x is

$$x^{k+1} = x^k + msu^k$$

In our discussion, the objective function for unconstrained and constrained optimization models will be limited to maximization problems:

$$\text{Maximize } z = f(x)$$

The tolerance test for locating the critical point of an optimization problem is different from the one used for roots of equations. The iteration procedure will cease when the relative difference between $f(\boldsymbol{x}^{k+1})$ and $f(\boldsymbol{x}^k)$ becomes null or meets the following criterion:

$$\left| \frac{f(\boldsymbol{x}^{k+1}) - f(\boldsymbol{x}^k)}{f(\boldsymbol{x}^k)} \right| \leq \varepsilon$$

If the step size is too large, the slope will become negative. In this case, the step size s is reduced and a new $\boldsymbol{x}^{k+1}$ is estimated. These estimations are repeated until the slope becomes small and meets the tolerance criterion.

The basic concept of the gradient search method is described most easily with the use of Figure 9.12. The local maximum of $z = f(\boldsymbol{x})$ is located at $\boldsymbol{x}^\circ = (x_1^\circ, x_2^\circ)$. Assume that $\boldsymbol{x}^\circ$ is an unknown. We use the gradient search method to demonstrate how $\boldsymbol{x}^\circ$ is estimated.

The overall approach is as follows. An initial guess of $\hat{\boldsymbol{x}}$ is made, call it $\boldsymbol{x}^0$ and then estimate $\boldsymbol{x}^1$ using a two-step procedure. First, the direction of steepest ascent $\boldsymbol{u}^0$ is calculated. Second, the maximum value of $f(\boldsymbol{x})$ in direction $\boldsymbol{u}^0$ from $\boldsymbol{x}^0$ is found by adjusting the step size ms. Next, a new direction $\boldsymbol{u}^1$ is calculated and a maximum of $f(\boldsymbol{x})$ in this direction is found. This approach is repeated until $\boldsymbol{x}^{k+1}$ is found to within a given limit of accuracy.

In Figure 9.12 $\boldsymbol{x}^\circ$ is found in three iterations, $\boldsymbol{x}^\circ = \boldsymbol{x}^3$. Points $\boldsymbol{x}^1$ and $\boldsymbol{x}^2$ are the points where $f(\boldsymbol{x})$ attains its maximum value on lines from $\boldsymbol{x}^0$ and $\boldsymbol{x}^1$ in directions $\boldsymbol{u}^0$ and $\boldsymbol{u}^1$, respectively. Since they do not satisfy the conditions for a maximum, they are intermediate points in the search.

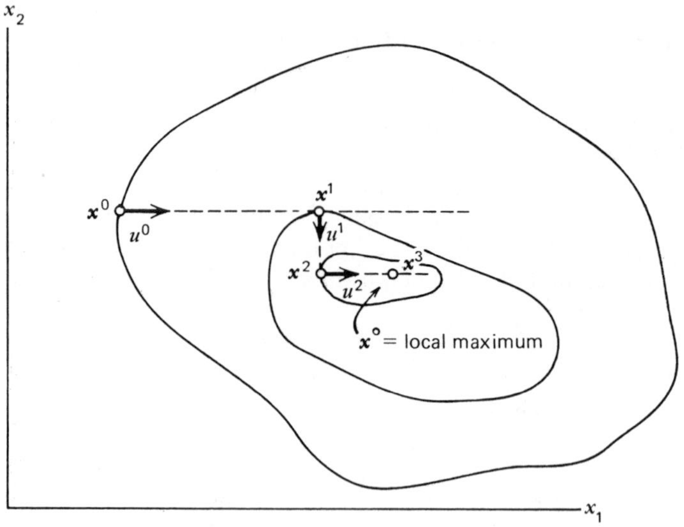

Figure 9.12 The iterative steps of the gradient search method.

The key to the gradient search method is the determination of the direction of steepest ascent. Before discussing this, the direction vector and direction cosines will be investigated.

Direction Cosines

In Figure 9.13 the direction of steepest ascent is shown as the vector $\boldsymbol{u}$ in the two-dimensional plane of x_1 and x_2. In order to avoid confusion, the superscript notation is dropped. Since $\boldsymbol{u}$ is a vector, we may express it as

$$\boldsymbol{u} = u_1\boldsymbol{i} + u_2\boldsymbol{j}$$

where u_1 and u_2 are direction cosines and $\boldsymbol{i}$ and $\boldsymbol{j}$ are unit vectors in the x_1 and x_2 directions. The direction cosines in terms of θ_1 and θ_2 are

$$u_1 = \cos\theta_1$$

$$u_2 = \cos\theta_2$$

The sum of the square of the direction cosines must be equal to one:

$$\cos^2\theta_1 + \cos^2\theta_2 = u_1^2 + u_2^2 = 1$$

In vector notation, this relationship is $\boldsymbol{u}'\boldsymbol{u} = |\boldsymbol{u}|^2 = 1$ and $|\boldsymbol{u}| = (\boldsymbol{u}'\boldsymbol{u})^{1/2}$. Similar relationships can be written for n dimensions where $\boldsymbol{u}' = [u_1 \quad u_2 \quad \cdots \quad u_n] = [\cos\theta_1 \quad \cos\theta_2 \quad \cdots \quad \cos\theta_n]$.

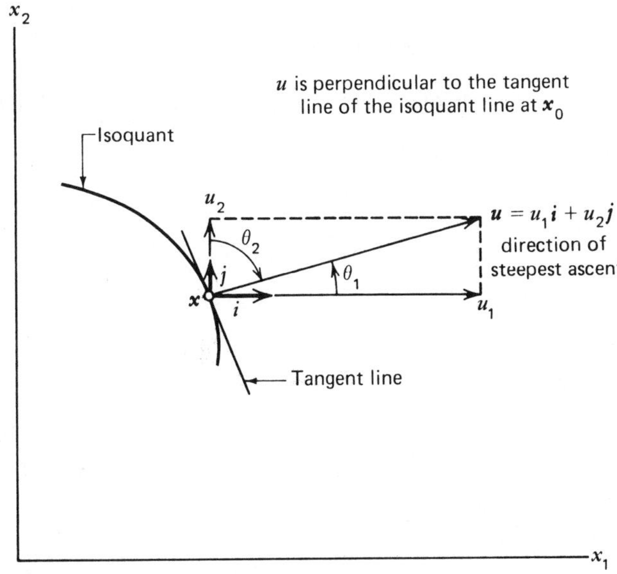

Figure 9.13 Direction of steepest ascent $\boldsymbol{u}^0$ at $\boldsymbol{x}^0$.

The Direction of Steepest Ascent

The direction of steepest ascent is calculated with the dot product at the point $\boldsymbol{x}$ as:

$$\nabla_{\boldsymbol{x}} f(\boldsymbol{x})\boldsymbol{u} = |\nabla_{\boldsymbol{x}} f(\boldsymbol{x})||\boldsymbol{u}|\cos\gamma$$

where angle γ is the angle between the gradient vector $\nabla_{\boldsymbol{x}} f(\boldsymbol{x})$ and the direction vector $\boldsymbol{u}$ as shown in Figure 9.14. Since $|\nabla_{\boldsymbol{x}} f(\boldsymbol{x})|$ is a constant at a given $\boldsymbol{x}$ and $|\boldsymbol{u}| = 1$, the control variable is assumed to be angle γ. The mathematical model is

$$\text{Maximize } z = |\nabla_{\boldsymbol{x}} f(\boldsymbol{x})||\boldsymbol{u}| \cos\gamma = |\nabla_{\boldsymbol{x}} f(\boldsymbol{x})| \cos\gamma$$

The maximum value of z occurs when $\cos\gamma = 1$ or $\gamma^* = 0°$.

Use the above equation to solve for $\boldsymbol{u}$. Let $\boldsymbol{g} = \nabla_{\boldsymbol{x}} f(\boldsymbol{x})$, $|\boldsymbol{g}| = |\nabla_{\boldsymbol{x}} f(\boldsymbol{x})|$ and $|u| = 1$. The direction of steepest ascent at $\boldsymbol{x}$ is

$$\boldsymbol{u} = \frac{\boldsymbol{g}}{|\boldsymbol{g}|} = \frac{\nabla_{\boldsymbol{x}} f(\boldsymbol{x})}{|\nabla_{\boldsymbol{x}} f(\boldsymbol{x})|}$$

The direction of steepest $\boldsymbol{u}$ is parallel to the gradient vector $\boldsymbol{g} = \nabla_{\boldsymbol{x}} f(\boldsymbol{x})$ because $\gamma^* = 0°$.

In superscript notation, the $k + 1$ estimate of the critical point $\boldsymbol{x}^\circ$ becomes

$$\boldsymbol{x}^{k+1} = \boldsymbol{x}^k + ms\boldsymbol{u}^k = \boldsymbol{x}^k + ms\frac{\boldsymbol{g}^k}{|\boldsymbol{g}^k|}$$

where $\boldsymbol{u}^k$ is evaluated at $\boldsymbol{x}^k$.

Now that the direction of steepest ascent $\boldsymbol{u}^k$ is known, the path to the maximum from $\boldsymbol{x}^k$ may be determined by adjusting the step size s,

$$\boldsymbol{x}(ms) = \boldsymbol{x}^k + ms\boldsymbol{u}^k$$

such that the maximum value of $f(\boldsymbol{x})$ at $\boldsymbol{x}(ms) = \boldsymbol{x}^{k+1}$ satisfies the tolerance test:

$$\left|\frac{f(\boldsymbol{x}^{k+1}) - f(\boldsymbol{x}^k)}{f(\boldsymbol{x}^k)}\right| < \varepsilon$$

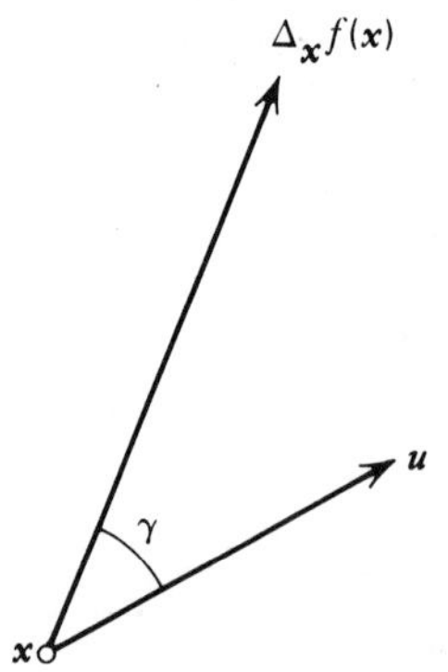

Figure 9.14 Dot product of two vectors.

There are many methods to determine the value of $\boldsymbol{x}^{k+1}$ in a given direction. The Newton method discussed in the previous section may be applied. Other methods, called direct search methods, do not require the evaluation of derivatives or gradients. They include the Fibonacci search method, the search method of Fletcher and Reeves, and many others. These methods are conceptually simple and easy to program on a computer. The method presented here uses an incremental search technique where a fixed step size s is assumed. The procedure continues to move toward $\boldsymbol{x}^{k+1}$ from $\boldsymbol{x}^k$ until the tolerance test is satisfied. If $\boldsymbol{x}^k$ is the starting point, the first estimate of $\boldsymbol{x}^{k+1}$ is $\boldsymbol{x}(s) = \boldsymbol{x}^k + s\boldsymbol{u}^k$. If $f(\boldsymbol{x}(s)) > f(\boldsymbol{x}^k)$ and the tolerance test fails the new estimate when a step size of $2s$ is used, $\boldsymbol{x}(2s) = \boldsymbol{x}^k + 2s\boldsymbol{u}^k$. If $f(\boldsymbol{x}(2s)) > f(\boldsymbol{x}^k)$ and the tolerance test fails, the process is continued. If we proceed past the maximum point $\boldsymbol{x}^{k+1}$, $f(\boldsymbol{x}(ms + s)) < f(\boldsymbol{x}(ms))$, where $(ms + s)$ is the last iteration and ms is the one preceding it. When this situation is encountered, a search between $\boldsymbol{x}(ms)$ and $\boldsymbol{x}(ms + s)$ is undertaken. Since the method does not utilize gradients or any property of the function being evaluated, it is not considered to be a computationally efficient method. However, it is an extremely cautious search procedure that should be successful in locating $\boldsymbol{x}^{k+1}$.

It can be shown that the directions of steepest ascent for sequential k and $k + 1$ steps are *orthogonal*. That is,

$$\boldsymbol{u}^k\boldsymbol{u}^{k+1} = 0$$

or, stated another way, the directions of steepest ascent for steps k and $k + 1$ are perpendicular to one another.

For ellipsoidal isoquants, the gradient search method will converge rapidly, $k = 2$, as shown in Figure 9.15*a*. Unfortunately, for some functions, once $\boldsymbol{x}^{k+1}$ is near $\boldsymbol{x}^\circ$ it tends to cycle about $\boldsymbol{x}^\circ$ and the convergence becomes very inefficient. For banana-shaped isoquants, the rate of convergence may be poor, as illustrated in Figure 9.15*b*. In these figures, the new directions are shown to be perpendicular to the last one.

The Gradient Projection Method for Constrained Models

The gradient projection method is a mathematical algorithm to solve constrained maximization problems of the form:

$$\text{Maximize } z = f(\boldsymbol{x})$$

$$\boldsymbol{g}(\boldsymbol{x}) = \boldsymbol{b}$$

$$\boldsymbol{x} \geq \boldsymbol{0}$$

The constraint set may consist of a set of linear and nonlinear equations. The reader should not confuse the gradient vector $\boldsymbol{g}$ with the function $\boldsymbol{g}(\boldsymbol{x})$. The algorithm utilizes the same iterative search technique employed in the gradient search method. Here, the direction of steepest ascent $\boldsymbol{u}$ must be adjusted in order to satisfy the constraint set. We develop the approach of calculating $\boldsymbol{u}$ for maximization problems with one linear, two linear, and nonlinear constraint sets, respectively.

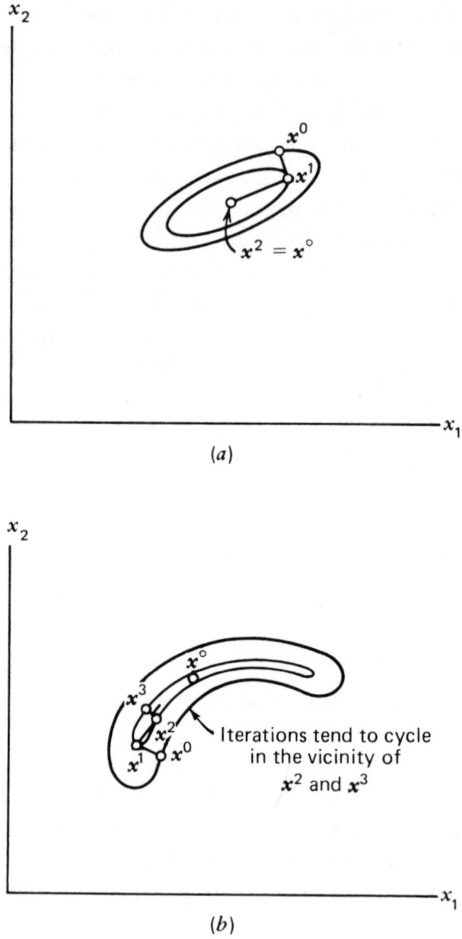

Figure 9.15 The efficiency of the gradient search method for ellipsoidal and banana-shaped isoquants. (*a*) Ellipsoidal isoquants. (*b*) Banana-shaped isoquants.

With a single linear active constraint, the equations involved in determining the maximum z at $\boldsymbol{x}^{k+1}$ from $\boldsymbol{x}^k$ are

$$\text{Maximize } z = f(\boldsymbol{x}^{k+1})$$
$$\boldsymbol{a}'\boldsymbol{x}^{k+1} = b$$
$$|\boldsymbol{u}^k| = 1$$

The model will be rewritten by introducing the Taylor series expansion for $f(\boldsymbol{x}^{k+1})$. After ignoring higher-order terms, the objective function becomes

$$f(\boldsymbol{x}^{k+1}) = f(\boldsymbol{x}^k) + ms[\nabla_{\boldsymbol{x}} f(\boldsymbol{x}^k)]'\boldsymbol{u}^k = f(\boldsymbol{x}^k) + ms\boldsymbol{g}^{k\prime}\boldsymbol{u}^k$$

Since $f(\boldsymbol{x}^k)$ is a constant, it will be dropped from further consideration. Since $\boldsymbol{a}'\boldsymbol{x}^{k+1} = b$, and $\boldsymbol{x}^{k+1} = \boldsymbol{x}^k + ms\boldsymbol{u}^k$ the constraint equation reduces to $\boldsymbol{a}'\boldsymbol{u}^k = 0$. The model becomes

$$\text{Maximize } z = \boldsymbol{g}^{k\prime}\boldsymbol{u}^k$$

$$\boldsymbol{a}'\boldsymbol{u}^k = 0$$

$$|\boldsymbol{u}^k| = 1$$

For simplicity, the superscripts will be dropped. The model becomes

$$\text{Maximize } z = \boldsymbol{g}'\boldsymbol{u}$$

$$\boldsymbol{a}'\boldsymbol{u} = 0$$

$$|\boldsymbol{u}| = 1$$

See Figure 9.16. The two-dimensional contour diagram of Figure 9.15 will be used to help explain the analysis. The gradient vector at point $\boldsymbol{x}$ is

$$\boldsymbol{g} = \nabla_{\boldsymbol{x}} f(\boldsymbol{x})$$

This vector is in the direction of maximum ascent if *no* constraints are imposed. It is perpendicular to isoquant surface $f(\boldsymbol{x})$ as shown. Clearly, making a step in direction

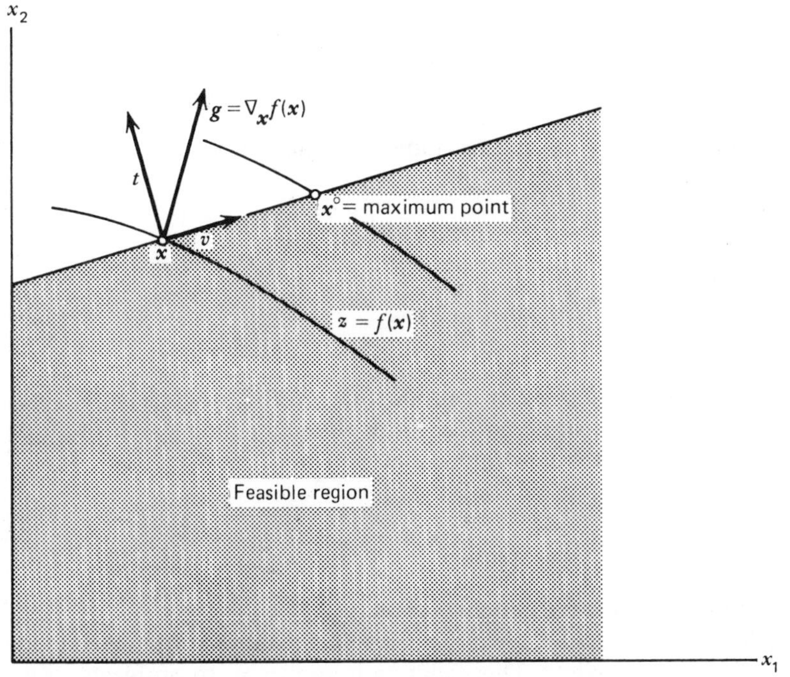

Figure 9.16 The direction of maximum feasible ascent.

$\boldsymbol{g}$ will violate the feasibility condition. The direction of maximum feasible ascent $\boldsymbol{v}$ is shown to be the vector tangent to the linear boundary line, $\boldsymbol{a}'\boldsymbol{x} = b$. The vector $\boldsymbol{t}$ is assumed to be perpendicular or orthogonal to $\boldsymbol{v}$, therefore parallel to $\boldsymbol{a}$. Utilizing vector addition, the direction vectors are related as follows:

$$\boldsymbol{g} = \boldsymbol{v} + \boldsymbol{t}$$

The direction of $\boldsymbol{t}$ is known, but its magnitude is not. However, since it is parallel to $\boldsymbol{a}$,

$$\boldsymbol{t} = \beta\boldsymbol{a}$$

Substitute the relationship for $\boldsymbol{t}$ into $\boldsymbol{g} = \boldsymbol{v} + \boldsymbol{t}$ and obtain

$$\boldsymbol{g} = \boldsymbol{v} + \beta\boldsymbol{a}$$

Multiply both sides of the equation by the transpose of $\boldsymbol{a}$.

$$\boldsymbol{a}'\boldsymbol{g} = \beta\boldsymbol{a}'\boldsymbol{v} + \beta\boldsymbol{a}'\boldsymbol{a}$$

In the optimization model, we noted that $\boldsymbol{a}'\boldsymbol{u} = 0$. Since $\boldsymbol{u}$ is the normalized direction vector of $\boldsymbol{v}$, or $\boldsymbol{u} = \boldsymbol{v}/|\boldsymbol{v}|$, it follows that $\boldsymbol{a}'\boldsymbol{v} = 0$. Thus, the expression simplifies to:

$$\boldsymbol{a}'\boldsymbol{g} = \beta\boldsymbol{a}'\boldsymbol{a} = \beta|\boldsymbol{a}|^2$$

The value of β is

$$\beta = \frac{\boldsymbol{a}'\boldsymbol{g}}{|\boldsymbol{a}|^2}$$

Substitute β into $\boldsymbol{g} = \boldsymbol{v} + \beta\boldsymbol{a}$ and rearrange the equation as

$$\boldsymbol{v} = \boldsymbol{g} - \boldsymbol{a}\left(\frac{\boldsymbol{a}'\boldsymbol{g}}{|\boldsymbol{a}|^2}\right)$$

$$\boldsymbol{v} = \left[\boldsymbol{I} - \frac{\boldsymbol{a}\boldsymbol{a}'}{|\boldsymbol{a}|^2}\right]\boldsymbol{g}$$

It should be noted that $\boldsymbol{a}'\boldsymbol{a} \neq \boldsymbol{a}\boldsymbol{a}'$. The product $\boldsymbol{a}'\boldsymbol{a}$ is scalar; whereas $\boldsymbol{a}\boldsymbol{a}'$ will be a square matrix. The direction vector $\boldsymbol{v}$ is parallel to the boundary line $\boldsymbol{a}'\boldsymbol{x} = b$. The normalized $\boldsymbol{u}$ vector, as imposed by the constraint equation $|\boldsymbol{u}| = 1$, is

$$\boldsymbol{u} = \frac{\boldsymbol{v}}{|\boldsymbol{v}|}$$

The elements of $\boldsymbol{u}$ are the directional cosines.

Throughout this discussion a two-dimensional diagram has been used to aid in the understanding of the method. This equation is applicable to problems with control variables of dimension n.

Consider the case with two active linear constraints. The same simplifications as shown for the single linear constraint is applicable here.

$$\text{Maximize } z = [\nabla f(\boldsymbol{x})]'\boldsymbol{u} = \boldsymbol{g}'\boldsymbol{u}$$

$$\boldsymbol{a}'_1\boldsymbol{u} = 0$$

$$\boldsymbol{a}'_2\boldsymbol{u} = 0$$

$$|\boldsymbol{u}| = 1$$

As in our previous development for one constraint, the direction vectors $\boldsymbol{g}$ and $\boldsymbol{t}$ are related as $\boldsymbol{g} = \boldsymbol{t} + \boldsymbol{v}$. In this case the direction vector $\boldsymbol{t}$ is assumed to be a linear combination of $\boldsymbol{a}_1$ and $\boldsymbol{a}_2$, the coefficients vector of the two linear constraint equations. Thus, the $\boldsymbol{t}$ vector is equal to

$$\boldsymbol{t} = \boldsymbol{a}_1\beta_1 + \boldsymbol{a}_2\beta_2 = \boldsymbol{M\beta}$$

where $\boldsymbol{M} = [\boldsymbol{a}_1\,\boldsymbol{a}_2]$ and $\boldsymbol{\beta}' = [\beta_1\,\beta_2]$. Substitute $\boldsymbol{t} = \boldsymbol{M\beta}$ into $\boldsymbol{g} = \boldsymbol{v} + \boldsymbol{t}$. It gives

$$\boldsymbol{g} = \boldsymbol{v} + \boldsymbol{M\beta}$$

Multiply both sides by $\boldsymbol{M}'$.

$$\boldsymbol{M}'\boldsymbol{g} = \boldsymbol{M}'\boldsymbol{v} + \boldsymbol{M}'\boldsymbol{M\beta}$$

Since
$$\boldsymbol{M}'\boldsymbol{v} = \begin{bmatrix} \boldsymbol{a}'_1\boldsymbol{v} \\ \boldsymbol{a}'_2\boldsymbol{v} \end{bmatrix}$$

$\boldsymbol{a}'_1\boldsymbol{u} = 0$ and $\boldsymbol{a}'_2\boldsymbol{u} = 0$ from the constraint set of the optimization model and $\boldsymbol{u} = \boldsymbol{v}/|\boldsymbol{v}|$, the equation reduces to

$$\boldsymbol{M}'\boldsymbol{g} = \boldsymbol{M}'\boldsymbol{M\beta}$$

Thus
$$\boldsymbol{\beta} = (\boldsymbol{M}'\boldsymbol{M})^{-1}\boldsymbol{M}'\boldsymbol{g}$$

Substitute into $\boldsymbol{g} = \boldsymbol{v} + \boldsymbol{M\beta}$

$$\boldsymbol{g} = \boldsymbol{v} + \boldsymbol{M}(\boldsymbol{M}'\boldsymbol{M})^{-1}\boldsymbol{M}'\boldsymbol{g}$$

or
$$\boldsymbol{v} = [\boldsymbol{I} - \boldsymbol{M}(\boldsymbol{M}'\boldsymbol{M})^{-1}\boldsymbol{M}']\boldsymbol{g}$$

This equation is a general expression that may be used for any number of active constraints. This expression will be used to find the direction vector for nonlinear active constraints. In normalized form, the direction cosines $\boldsymbol{u}$ are

$$\boldsymbol{u} = \frac{\boldsymbol{v}}{|\boldsymbol{v}|}$$

Consider the mathematical model consisting of a single nonlinear active constraint:

$$\text{Maximize } z = f(\boldsymbol{x}^{k+1}) = f(\boldsymbol{x}^k + s\boldsymbol{u}^k)$$

$$g(\boldsymbol{x}^{k+1}) = g(\boldsymbol{x}^k + s\boldsymbol{u}^k) = b$$

$$|\boldsymbol{u}^k| = 1$$

The Taylor series expansion of $f(\mathbf{x}^{k+1})$ and $g(\mathbf{x}^{k+1})$ will be used to transform the model. Ignoring the higher-order terms, the Taylor series expansions are subsituted into the mathematical model, giving

$$f(\mathbf{x}^{k+1}) = f(\mathbf{x}^k) + ms[\nabla_{\mathbf{x}} f(\mathbf{x}^k)]'\mathbf{u}^k$$

and

$$g(\mathbf{x}^{k+1}) = g(\mathbf{x}^k) + ms[\nabla_{\mathbf{x}} g(\mathbf{x}^k)]'\mathbf{u}^k = b$$

By dropping the superscript notation the mathematical model may be simplified as

$$\text{Maximize } z = \mathbf{g}'\mathbf{u}$$

$$\mathbf{m}'\mathbf{u} = 0$$

$$|\mathbf{u}| = 1$$

where $\mathbf{g} = \nabla_{\mathbf{x}} f(\mathbf{x})$ and $\mathbf{m} = \nabla_{\mathbf{x}} g(\mathbf{x})$. Do not confuse the vector $\mathbf{m}$ with the variable m.

The constraint equation $\mathbf{m}'\mathbf{u} = 0$ satisfies the orthogonality condition. The direction of maximum ascent is

$$\mathbf{v} = \left[\mathbf{I} - \frac{\mathbf{m}\mathbf{m}'}{|\mathbf{m}|^2}\right]\mathbf{g}$$

and

$$\mathbf{u} = \frac{\mathbf{v}}{|\mathbf{v}|}$$

For the two-dimensional case, $\mathbf{m}$ is shown in Figure 9.17 to be perpendicular to the direction of steepest ascent $\mathbf{v}$. A hyperplane orthogonal to $\mathbf{m}$ evaluated at $\mathbf{x}$ is called the *tangent hyperplane* at $\mathbf{x}$. In this example, the hyperplane is a tangent line.

Since $\mathbf{v}$ is a direction vector lying in the tangent hyperplane, it is possible that the new candidate point of $\mathbf{x}^\circ$ may not lie on the constraint surface. If $\mathbf{x}$ lies on the boundary as illustrated in Figure 9.17, a step in the direction of steepest ascent $\mathbf{u}$ will give an estimate of $\mathbf{x}^\circ$, which does not satisfy the surface constraint $g(\mathbf{x}) = b$. In other words, it lies in the infeasible region.

Since the estimate of $\mathbf{x}^\circ$ is in the vicinity of the constraint, the Newton method is an effective approach to bring it back to the constraint surface, $g(\mathbf{x}) = b$. The gradient projection and Newton methods may be combined into a single algorithm to find the roots of equations.

The direction of steepest ascent for m constraints, $i = 1, 2, \ldots, m$, is determined with the following expression.

$$\mathbf{v} = [\mathbf{I} - \mathbf{M}(\mathbf{M}'\mathbf{M})^{-1}\mathbf{M}']\mathbf{g}$$

where the matrix $\mathbf{M}$ consists of the gradient vectors of the active constraints $\nabla_{\mathbf{x}} h_i(\mathbf{x})$, or $\mathbf{M} = [\nabla_{\mathbf{x}} h_1(\mathbf{x}^k) \cdots \nabla_{\mathbf{x}} h_m(\mathbf{x}^k)]$. The direction cosines are the elements of the following vector.

$$\mathbf{u} = \frac{\mathbf{v}}{|\mathbf{v}|}$$

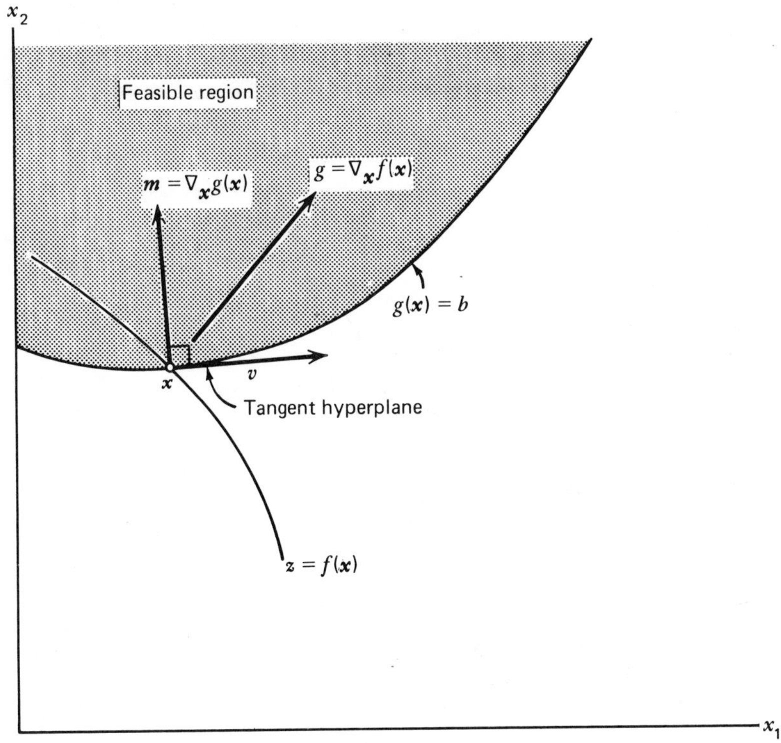

Figure 9.17 The direction of maximum feasible ascent does not lie on $g(\boldsymbol{x}) = b$.

The Gradient Projection–Newton Method

The Newton method may either lead to a root set of nonlinear equations with great efficiency or stray from the root with the same efficiency. If the initial solution $\boldsymbol{x}^\circ$ to the equation set is within a certain region of the solution $\boldsymbol{x}^\circ$, then the Newton method will give the desired result. The problem is to make a good initial guess. In many engineering problems, it is difficult to estimate $\boldsymbol{x}^\circ$. By alternately using the Newton and the gradient projection methods we should be able to find a solution. Since the method tends to be involved, it will be described with use of two-dimensional and three-dimensional nonlinear sets of equations.

The two-dimensional problem consists of two nonlinear equations in two unknowns

$$h_1(\boldsymbol{x}) = 0$$

$$h_2(\boldsymbol{x}) = 0$$

where $$\boldsymbol{x}' = [x_1 \; x_2]$$

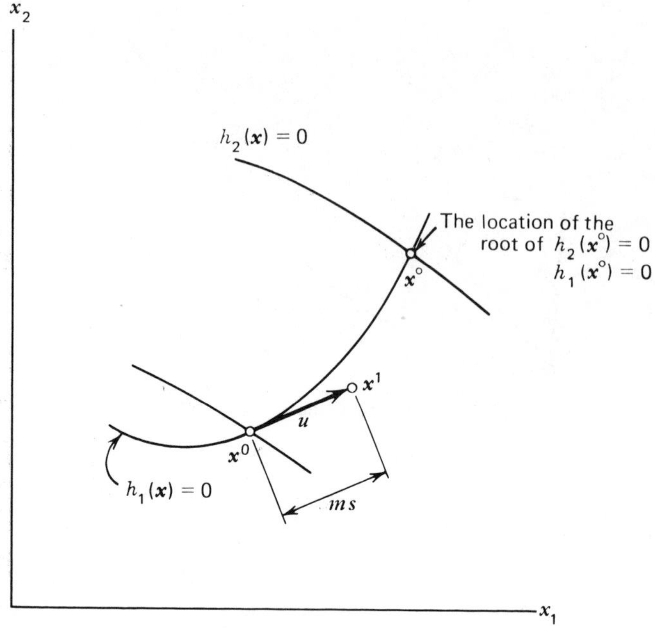

Figure 9.18 Finding the root of $h_2(\boldsymbol{x}) = 0$ given $\boldsymbol{x}^*$ satisfies $h_1(\boldsymbol{x}) = h_1(\boldsymbol{x}^k) = 0$.

An initial estimate, $\boldsymbol{x}^\circ = (x_1^0, x_2^0)$ of the first equation will be made.

$$h_1(\boldsymbol{x}^0) = 0$$

Since there are two variables in $\boldsymbol{x}$ and one equation, the solution $\boldsymbol{x}^0$ is considered a best guess solution. In this step, the relationship $h_2(\boldsymbol{x}) = 0$ is ignored. Unless we are very lucky, $\boldsymbol{x}^0$ will satisfy the equality condition of the second equation. In other words, $h_2(\boldsymbol{x}^0) \neq 0$. The situation is shown in Figure 9.18. The root of the equation set is shown as $\boldsymbol{x}^\circ$.

The gradient projection method will be used to make a cautious step toward the root $\boldsymbol{x}^\circ$. In this procedure a single step toward $\boldsymbol{x}^\circ$ is made without straying too far away from the boundary $h_1(\boldsymbol{x}) = 0$. The new estimate of the root $\boldsymbol{x}^1$ is

$$\boldsymbol{x}^1 = \boldsymbol{x}^0 + ms\boldsymbol{u}(\boldsymbol{x}^0)$$

where

$$\boldsymbol{u} = \boldsymbol{u}(x^0) = \frac{\boldsymbol{v}}{|\boldsymbol{v}|}$$

and

$$\boldsymbol{v} = \left[\boldsymbol{I} - \frac{\boldsymbol{m}\boldsymbol{m}'}{|\boldsymbol{m}|^2}\right]\boldsymbol{g}_2$$

$$\boldsymbol{g}_2 = \nabla_{\boldsymbol{x}} h_2(\boldsymbol{x}^0)$$

$$\boldsymbol{m} = \nabla_{\boldsymbol{x}} h_1(\boldsymbol{x}^0)$$

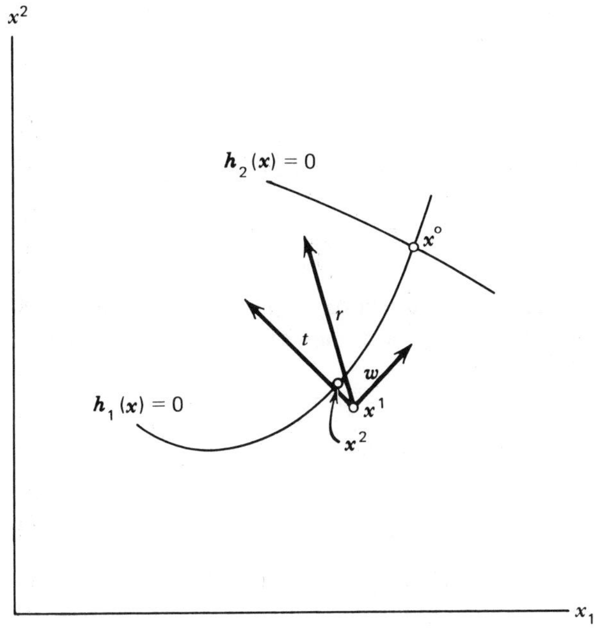

Figure 9.19 Constructing a vector $\boldsymbol{v}$ parallel to $h_1(\boldsymbol{x}) = 0$ at $\boldsymbol{x}^{k+1}$.

The new estimate $\boldsymbol{x}^1$ has strayed away from the nonlinear boundary $h_1(\boldsymbol{x}) = 0$ as shown in Figure 9.19. Newton's method will be used to correct this situation prior to taking another step toward $\boldsymbol{x}^\circ$. We seek to find a new location $\boldsymbol{x}^2$ such that

$$h_1(\boldsymbol{x}^2) = 0$$

We need two independent equations in two unknowns of $\boldsymbol{x}$ to solve this equation. From the Taylor series expansion and ignoring higher order terms, we obtain the first equation

$$h_1(\boldsymbol{x}^1) + \nabla_{\boldsymbol{x}} h_1(\boldsymbol{x}^1)(\boldsymbol{x}^2 - \boldsymbol{x}^1) = 0$$

The second equation will be determined by finding the vector $\boldsymbol{w}$ that is orthogonal to the vector $(\boldsymbol{x}^2 - \boldsymbol{x}^1)$.

$$\boldsymbol{w}'(\boldsymbol{x}^2 - \boldsymbol{x}^1) = 0$$

The vector $\boldsymbol{w}$ will be determined by introducing relationships among the vectors $\boldsymbol{w}$, $\boldsymbol{t}$, and $\boldsymbol{r}$ as shown in Figure 9.19. The vector $\boldsymbol{t}$ is assumed to be orthogonal to $\boldsymbol{w}$

$$\boldsymbol{t}'\boldsymbol{w} = 0$$

and the vector sum

$$\boldsymbol{w} + \boldsymbol{t} = \boldsymbol{r}$$

will be satisfied. The direction of $\boldsymbol{t}$ is known.

$$\boldsymbol{g}_1 = \nabla_{\boldsymbol{x}} h_1(\boldsymbol{x}^1)$$

The length of $\boldsymbol{t}$ is unknown, thus we assume

$$\boldsymbol{t} = \phi \boldsymbol{g}_1$$

where ϕ is an unknown constant. In order to find ϕ, we assume a vector $\boldsymbol{r}$ exists

$$\boldsymbol{r} = \begin{bmatrix} r_1 \\ r_2 \end{bmatrix}$$

Since the actual magnitude and direction of $\boldsymbol{r}$ are unimportant, r_1 and r_2 are randomly generated numbers. Since $\boldsymbol{t}$ and $\boldsymbol{w}$ are assumed to be orthogonal, it follows that $\boldsymbol{g}_1$ and $\boldsymbol{w}$ are orthogonal

$$\boldsymbol{g}_1' \boldsymbol{w} = 0$$

Since $\boldsymbol{t} = \phi \boldsymbol{g}_1$ and $\boldsymbol{r} = \boldsymbol{t} + \boldsymbol{w}$, we find

$$\boldsymbol{w} = \boldsymbol{r} - \phi \boldsymbol{g}_1$$

By substitution and simplification, we may find ϕ.

$\boldsymbol{g}_1' \boldsymbol{w} = 0$: $\quad \boldsymbol{g}_1'(\boldsymbol{r} - \phi \boldsymbol{g}_1) = \boldsymbol{g}_1' \boldsymbol{r} - \phi \boldsymbol{g}_1' \boldsymbol{g}_1 = \boldsymbol{g}_1' \boldsymbol{r} - |\boldsymbol{g}_1|^2 = 0$

or

$$\phi = \frac{\boldsymbol{g}_1' \boldsymbol{r}}{|\boldsymbol{g}_1|^2}$$

Now, the point $\boldsymbol{x}^2$ of the following equation set may be found with Newton's method.

$$h_1(\boldsymbol{x}^1) + \nabla \boldsymbol{h}_1(x^1)'(\boldsymbol{x}^2 - \boldsymbol{x}^1) = 0$$

$$\boldsymbol{w}'(\boldsymbol{x}^2 - \boldsymbol{x}^1) = 0$$

This set of equations may be written as $\boldsymbol{h}(\boldsymbol{x}^2) = \boldsymbol{h}(\boldsymbol{x}^1) + \boldsymbol{H}(\boldsymbol{x}^1)'(\boldsymbol{x}^2 - \boldsymbol{x}^1) = \boldsymbol{0}$ and solved as described in the section entitled "Newton Method for Finding the Root of an Equation Set." After the $\boldsymbol{x}^2$ is determined, we repeat the two step procedure by stepping toward the root by using gradient projection method and correcting the estimate by using Newton's method. This method is repeated until the solution $\boldsymbol{x}^k$ is considered sufficiently close to the root of $\boldsymbol{x}^\circ$. The Newton method with $\boldsymbol{x}^k$ as the initial estimate can now be used to solve

$$h_1(\boldsymbol{x}) = 0$$

$$h_2(\boldsymbol{x}) = 0$$

Consider another example with three equations and three unknown variables of $\boldsymbol{x}$:

$$h_1(\boldsymbol{x}) = 0$$

$$h_2(\boldsymbol{x}) = 0$$

$$h_3(\boldsymbol{x}) = 0$$

where $\quad \boldsymbol{x}' = [x_1 \quad x_2 \quad x_3]$

Here, we will not have the benefit of visualizing the approach with graphics as we did for the finding the roots of the two-dimensional problem. Our approach consists of making an initial estimate of $\boldsymbol{x}$; call it $\boldsymbol{x}^{0'} = [x_1^0 \quad x_2^0 \quad x_3^0]$, which satisfies the first equation. Thus

$$h_1(\boldsymbol{x}^0) = 0$$

For the case of three unknowns, make a guess of x_1^0 and x_2^0 and solve for x_3^0 such that $h_1(\boldsymbol{x}^0) = 0$.

Next, determine a new value of $\boldsymbol{x}$, $\boldsymbol{x}^1$, such that $h_2(\boldsymbol{x}^1) = 0$ without leaving the surface of the first equation $h_1(\boldsymbol{x}^1) = 0$. In other words,

Solve: $$h_2(\boldsymbol{x}) = 0$$

subject to the constraint: $$h_1(\boldsymbol{x}) = 0$$

with $\boldsymbol{x}^0$ as the starting point. The gradient search method will be utilized to find the location $\boldsymbol{x}^1$ that gives $h_2(\boldsymbol{x}^1) = 0$ such that it has not strayed too far away from the surface $h_1(\boldsymbol{x}) = 0$. If $h_2(\boldsymbol{x}^0) < 0$, the direction of steepest ascent is used; or

$$\boldsymbol{u} = \boldsymbol{u}(\boldsymbol{x}^0) = \frac{\boldsymbol{v}}{|\boldsymbol{v}|}$$

and $$\boldsymbol{v} = \left[\boldsymbol{I} - \frac{\boldsymbol{m}\boldsymbol{m}'}{|\boldsymbol{m}|^2}\right]\boldsymbol{g}_2$$

where $$\boldsymbol{g}_2 = \nabla_{\boldsymbol{x}} h_2(\boldsymbol{x}^0) \quad \text{and} \quad \boldsymbol{m} = [\nabla_{\boldsymbol{x}} h_1(\boldsymbol{x}^0)]$$

If $h_2(\boldsymbol{x}^0) > 0$ the direction of steepest descent $-\boldsymbol{u}(\boldsymbol{x}^0)$ is used. Only *one* step of size s is made along $\boldsymbol{u}(\boldsymbol{x}^0)$ or $-\boldsymbol{u}(\boldsymbol{x}^0)$. Call the new point $\boldsymbol{x}^1$, where $\boldsymbol{x}^1 = \boldsymbol{x}^0 + ms\boldsymbol{u}(\boldsymbol{x}^0)$. If $h_2(\boldsymbol{x}^1) = 0$ is not within a given tolerance limit, we may have stepped too far or we must continue to step in the direction toward $h_2(\boldsymbol{x}) = 0$. If $h_2(\boldsymbol{x}^1) > 0$ and $h_2(\boldsymbol{x}^0) < 0$, or if $h_2(\boldsymbol{x}^1) < 0$ and $h_2(\boldsymbol{x}^0) > 0$, the step sizes must be adjusted until $h_2(\boldsymbol{x}) = 0$ is satisfied. If $h_2(\boldsymbol{x}^1) > 0$ and $h_2(\boldsymbol{x}^0) > 0$ or $h_2(\boldsymbol{x}^1) < 0$ and $h_2(\boldsymbol{x}^0) < 0$, then we must continue the search for $h_2(\boldsymbol{x}) = 0$ by taking another step.

Before another step is taken we must ensure that we have not strayed too far from the surface $h_1(\boldsymbol{x}) = 0$. The Newton method with $\boldsymbol{x}^1$ as the initial point will be used to find a new location $\boldsymbol{x}^2$ that satisfies the following set:

$$\begin{aligned} h_1(\boldsymbol{x}^1) + \nabla h_1(\boldsymbol{x}^1)'(\boldsymbol{x}^2 - \boldsymbol{x}^1) &= 0 \\ \boldsymbol{w}_1'(\boldsymbol{x}^2 - \boldsymbol{x}^1) &= 0 \\ \boldsymbol{w}_2'(\boldsymbol{x}^2 - \boldsymbol{x}^1) &= 0 \end{aligned}$$

This equation set is a set of three independent equations in three unknowns, where

$$\begin{aligned} \boldsymbol{w}_1 &= \boldsymbol{r}_1 - \phi_1 \boldsymbol{g}_1 \\ \boldsymbol{w}_2 &= \boldsymbol{r}_2 - \phi_2 \boldsymbol{g}_1 \\ \boldsymbol{g}_1 &= \nabla_{\boldsymbol{x}} h_1(\boldsymbol{x}^1) \end{aligned}$$

with the $\boldsymbol{r}_1$ and $\boldsymbol{r}_2$ vectors consisting of random generated numbers

$$\boldsymbol{r}_1 = \begin{bmatrix} r_{11} \\ r_{12} \\ r_{13} \end{bmatrix} \qquad \boldsymbol{r}_2 = \begin{bmatrix} r_{21} \\ r_{22} \\ r_{23} \end{bmatrix}$$

Assume that $\boldsymbol{g}_1$ is orthogonal to $\boldsymbol{w}_1$ and $\boldsymbol{w}_2$, respectfully. Thus, the unknown constants ϕ_1 and ϕ_2 may be determined.

$$\boldsymbol{g}_1'\boldsymbol{w}_1 = 0: \qquad \boldsymbol{g}_1'\boldsymbol{r}_1 - \phi_1\boldsymbol{g}_1'\boldsymbol{g}_1 \quad \text{or} \quad \phi_1 = \frac{\boldsymbol{g}_1'\boldsymbol{r}_1}{|\boldsymbol{g}_1|^2}$$

$$\boldsymbol{g}_1'\boldsymbol{w}_2 = 0: \qquad \boldsymbol{g}_1'\boldsymbol{r}_2 - \phi_2\boldsymbol{g}_1'\boldsymbol{g}_1 \quad \text{or} \quad \phi_2 = \frac{\boldsymbol{g}_1'\boldsymbol{r}_2}{|\boldsymbol{g}_1|^2}$$

Now that all parameters in the equation set are known, the Newton method may be used to solve for $\boldsymbol{x}^2$. After $\boldsymbol{x}^2$ is determined, another step toward $h_2(\boldsymbol{x}) = 0$ is taken by utilizing the gradient projection and Newton methods. This procedure is continued until a value of $\boldsymbol{x}^k$ is found such that $h_1(\boldsymbol{x}^k) = 0$ and $h_2(\boldsymbol{x}^k) = 0$ within a given tolerance limit.

Now, the third equation is introduced.

Solve: $\qquad h_3(\boldsymbol{x}) = 0$

subject to the constraints: $\qquad h_1(\boldsymbol{x}) = 0$

$$h_2(\boldsymbol{x}) = 0$$

Basically, the procedure is repeated as previously described for two equations. Here, we highlight the important equations in gradient projection and Newton methods. Let us redefine $\boldsymbol{x}^k$ calculated above as $\boldsymbol{x}^0$, $\boldsymbol{x}^0 = \boldsymbol{x}^k$. For the gradient search, the direction of steepest ascent is

$$\boldsymbol{u} = \boldsymbol{u}(\boldsymbol{x}^0) = \frac{\boldsymbol{v}}{|\boldsymbol{v}|}$$

and

$$\boldsymbol{v} = [\boldsymbol{I} - \boldsymbol{M}(\boldsymbol{M}'\boldsymbol{M})^{-1}\boldsymbol{M}']\boldsymbol{g}_3$$

where $\qquad \boldsymbol{g}_3 = \nabla_{\boldsymbol{x}} h_3(\boldsymbol{x}^0)$

and $\qquad \boldsymbol{M} = [\nabla_{\boldsymbol{x}} h_1(\boldsymbol{x}^0) \;\; \nabla_{\boldsymbol{x}} h_2(\boldsymbol{x}^0)]$

For the Newton method, we require three independent equations in three unknowns. Assume an initial starting point $\boldsymbol{x}^1$; the location $\boldsymbol{x}^2$ is sought.

$$h_1(\boldsymbol{x}^1) + \nabla h_1(\boldsymbol{x}^1)'(\boldsymbol{x}^2 - \boldsymbol{x}^1) = 0$$

$$h_2(\boldsymbol{x}^1) + \nabla h_2(\boldsymbol{x}^1)'(\boldsymbol{x}^2 - \boldsymbol{x}^1) = 0$$

$$\boldsymbol{w}_1'(\boldsymbol{x}^2 - \boldsymbol{x}^1) = 0$$

where

$$w_1 = r - \phi_1 g_1 - \phi_2 g_2$$
$$g_1 = \nabla_x h_1(x^1)$$
$$g_2 = \nabla_x h_2(x^1)$$

with r, the random vector, is

$$r = \begin{bmatrix} r_1 \\ r_2 \\ r_3 \end{bmatrix}$$

Assume g_1 and g_2 to be orthogonal to w_1

$g'_1 w_1 = 0$: $\quad g'_1 r - \phi_1 g'_1 g_1 - \phi_2 g'_1 g_2 = 0$

$g'_2 w_1 = 0$: $\quad g'_2 r - \phi_1 g'_2 g_1 - \phi_2 g'_2 g_2 = 0$

Since we have two equations in two unknowns, we may solve for ϕ_1 and ϕ_2. Newton's method may now be used to find x^2. We continue this procedure until x^k is found to solve $h_3(x) = 0$ subject to $h_1(x) = 0$ and $h_2(x) = 0$.

After the location of x^k is found for $h_3(x) = 0$ subject to $h_1(x) = 0$ and $h_2(x) = 0$, the Newton method may be applied to the model:

$$h_1(x) = 0 \quad h_2(x) = 0 \quad h_3(x) = 0$$

where k is assumed to be initial estimate.

This procedure may be extended to finding the roots of n equations in n unknowns. It should converge for most engineering problems because we carefully proceed toward the root by never leaving the vicinity of any one surface of the equation set.

As previously stated, numerical analysis is largely an art; thus, there are other algorithms that may converge to a solution faster than this one. The reader should consult other textbook and computer program libraries for them.

EXAMPLE 9.3 A Biological Reactor System

Consider a biological reactor system consisting of reactor and clarifier tanks. A recycle line is used to keep a sufficient quantity of cells in the reactor. The cells consume the biological waste or food. A diffused air system is used to aerate the reactor vessel and remove the waste material. (See Figure 9.20.)

Design a minimum-construction-cost reactor system. Assume steady-state conditions; that is, the influent flow Q is constant, $Q = 1$ Mgal/day, and the biological waste of the influent is constant, $S_0 = 200$ mg/l. The construction cost of the reactor and clarifier tanks is an exponential cost function:

$$C = \$100{,}000 V^{0.6}$$

where V is the tank volume in million gallons. V_1 and V_2 will be used to designate the two tank volumes in million gallons. The cell concentration of the influent and sludge are, respectively, equal to $X_0 = 0$ mg/l and $X = 10{,}000$ mg/l. The rate of cell growth rate μ is 1.876/day, and the maximum yield coefficient Y is the ratio of the quantity of cells produced to the quantity of substrate utilized.

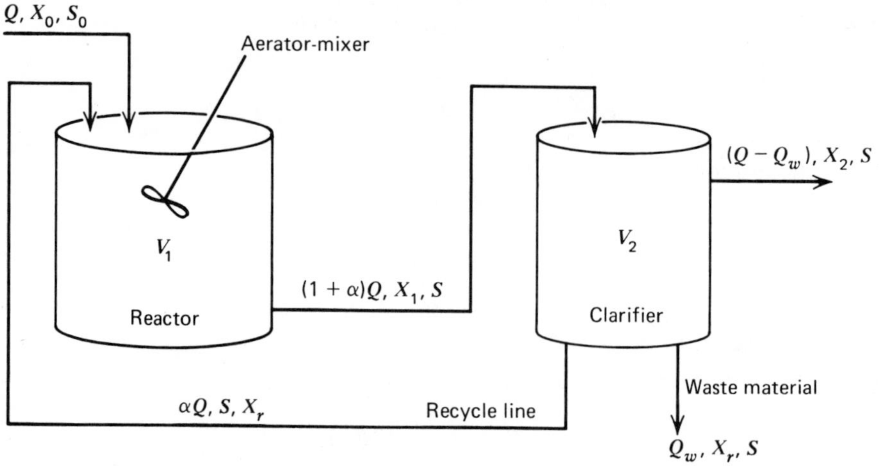

Figure 9.20 A schematic diagram of the system.

The clarifier volume V_2 is a function of the flow entering the tank, $(1 + \alpha)Q$, tank depth h and overflow rate Q_0:

$$V_2 = \frac{(1 + \alpha)Qh}{Q_0}$$

where α is the ratio of recycle flow Q_r to influent flow Q, or $\alpha = Q_r/Q$. The tank depth h is 8 ft, and the overflow rate Q_0 is 800 gal/day-ft^2.

(a) Formulate a mathematical model to minimize the cost of construction.

(b) Use the Lagrange multiplier method to formulate the constrained optimization model found in part a in terms of an unconstrained optimization model. State the necessary conditions for a critical point.

(c) Use the gradient projection—Newton method to establish a computer algorithm to solve the problem.

Solution

(a) The objective to minimize the construction cost C is

$$\text{Minimize } C = 100{,}000V_1^{0.6} + 100{,}000V_2^{0.6} \tag{1}$$

The constraint set will be determined by establishing mass balance equations for cells and biological waste in the reactor and a mass balance for biological waste in the clarifier. The mass balance on microorganisms, either cells or biological waste, is

$$\text{Accumulation} = \text{Inflow} - \text{Outflow} + \text{Net growth or decay}$$

The mass balance of cells in the reactor is

$$V_1 \frac{dX_1}{dt} = QX_0 + \alpha QX_r - (1 + \alpha)QX_1 + \mu X_1 V_1$$

or

$$\frac{dX_1}{dt} = \frac{X_0}{t_1} + \frac{\alpha X_r}{t_1} - \frac{(1 + \alpha)X_1}{t_1} + \mu X_1 \tag{2}$$

where $t_1 = V_1/Q$, the detention time for tank 1. The mass balance of biological waste in the reactor is

$$V_1 \frac{dS}{dt} = QS_0 + \alpha QS - (1 + \alpha)QS - \frac{\mu}{Y} X_1 V_1$$

Divide by V_1 and simplify with the substitution of t_1 where appropriate.

$$\frac{dS}{dt} = \frac{(S_0 - S)}{t_1} - \frac{\mu}{Y} X_1 \tag{3}$$

The mass balance of cells in the clarifier is

$$V_2 \frac{dX_2}{dt} = (1 + \alpha)QX_1 - (Q - Q_w)X_2 - (\alpha Q + Q_w)X_r \tag{4}$$

For steady-state conditions, the cell and biological waste concentration will not change over time. That is

$$\frac{dX_1}{dt} = 0 \qquad \frac{dS}{dt} = 0 \qquad V_2 \frac{dX_2}{dt} = 0$$

Thus, each constraint equation (2), (3), and (4) may be simplified to the following constraint set.

$$X_0 + \alpha X_r - (1 + \alpha)X_1 + \mu X_1 t_1 = 0 \tag{2'}$$

$$X_1 t_1 = \frac{Y}{\mu}(S_0 - S) \tag{3'}$$

$$\beta = [X_1(1 + \alpha) - \alpha X_r - X_2]/(X_r - X_2) \tag{4'}$$

where Eq. (4′) is obtained by setting $V_2(dX_2/dt) = 0$ and defining a new variable $\beta = Q_w/Q$. The unknown variables are α, t_1, X_1, and β.

Constraint (2′) may be simplified by substituting Eq. (3′) into Eqs. (2′) and is

$$[X_0 + \alpha X_r - Y(S_0 - S)]t_1 - (1 + \alpha)\frac{Y}{\mu}(S_0 - S) = 0 \tag{5}$$

Rewrite equation (2′) as

$$X_1(1 + \alpha) - \alpha Xr = X_0 - \mu X_1 t$$

and introduce it into equation (4′) and rewrite it as

$$\beta = \frac{X_0 - X_2 - \mu X_1 t}{X_r - X_2} \tag{2''}$$

substituting (3′) into (2″) gives

$$\beta = \frac{X_0 - X_2 + Y(S_0 - S)}{(X_r - X_2)} \tag{6}$$

The ratio β is independent of the control variables α, t_1, and X_1; therefore, Eq. (6) is removed from further consideration in the optimization process. Our model is reduced to a single constraint equation, Eq. (5), which is a function of t_1 and α only.

Thus far, we have considered the mass balance equations exclusively. The clarifier volume V_2 condition must be satisfied.

$$V_2 = \frac{(1+\alpha)Qh}{Q_0} \tag{7}$$

Equation (7) and $V_1 = Qt_1$ are substituted into Eq. (1), which reduces the objective to be a function of t_1 and α only.

$$\text{Minimize } C = 100{,}000\, Q^{0.6}t^{0.6} + 100{,}000\left(\frac{Qh}{Q_0}\right)^{0.6}(1+\alpha)^{0.6} \tag{8}$$

Substituting the values of Q, Q_0, S_0, S, h, X_r, and X_0 into Eqs. (5) and (8), the final model may be expressed as

$$\text{Minimize } C = 127.3t_1^{0.6} + 2110(1+\alpha)^{0.6}$$

$$(10{,}000\alpha_1 + 85)t_1 - 65{,}280\alpha = 65{,}280$$

$$\alpha \geq 0,\ t_1 \geq 0$$

Since the mass balance must be explicitly satisfied, the constraint is an active constraint.

(b) The Lagrange multiplier equation is

$$L(t_1, \alpha, \lambda) = 127.3t_1^{0.6} + 2110(1+\alpha)^{0.6} + \lambda[65{,}280 + 65{,}280\alpha - 10{,}000\alpha t_1 - 85t_1]$$

The necessary conditions for a critical point are

$$\frac{\partial L}{\partial \alpha} = 0: \qquad (2110)(0.6)(1+\alpha)^{-0.4} + \lambda(65{,}280 - 10{,}000t_1) = 0$$

$$\frac{\partial L}{\partial t_1} = 0: \qquad (127.3)(0.6)t_1^{-0.4} + \lambda(-10{,}000\alpha - 85) = 0$$

$$\frac{\partial L}{\partial \lambda} = 0: \qquad 65{,}280 + 65{,}280\alpha - 10{,}000\alpha t_1 - 85t_1 = 0$$

Our task is to determine the roots α, t_1, and λ of this set of equations.

(c) Since we utilize the Newton–gradient projection methods for solution, let us define

$$h_1(\mathbf{x}) = \frac{\partial L}{\partial \lambda} = 0$$

$$h_2(\mathbf{x}) = \frac{\partial L}{\partial t_1} = 0$$

$$h_3(\mathbf{x}) = \frac{\partial L}{\partial \alpha} = 0$$

with $\mathbf{x}' = [\alpha \quad t_1 \quad \lambda]$.

Since the gradient–Newton method is difficult to describe, only the numerical results are tabulated in Table 9.1.

Table 9.1

	t_1	α	λ	$h_1(\mathbf{x})$	$h_2(\mathbf{x})$	$h_3(\mathbf{x})$	COMMENT
Step 1	19.257	0.500	0.000	0.000	—	—	Initial estimates (1)
Step 2	19.257	0.500	0.000	0.000	1076.458	—	Initial estimate (2)
	19.257	0.500	−0.008	0.000	−0.002	—	Final estimate (2)
Step 3	19,257	0.500	−0.008	0.000	−0.002	−19.607	Initial estimate (3)
	25,048	0.339	−0.006	−316.469	4.863	0.000	Final estimate (4)
Step 4	25,048	0.339	−0.006	−316.469	4.863	0.000	Initial estimate (5)
	25,149	0.339	−0.006	0.000	0.000	0.000	Final solution

(1) Satisfy the relationship $h_1(\mathbf{x}) = 0$. Since $h_1(\mathbf{x})$ is a function of α and t_1 only, t_1 may be solved in terms of α.

$$t_1 = \frac{65{,}280(1 + \alpha)}{(10{,}000\alpha + 85)}$$

Assume an initial value of α equal to 0.5, $\alpha^0 = 0.5$, then $t_1^0 = 19.57$ min.
(2) Solve $h_2(\mathbf{x}) = 0$, subject to $h_1(\mathbf{x}) = 0$.
(3) Solve $h_3(\mathbf{x}) = 0$, subject to $h_1(\mathbf{x}) = 0$, $h_2(\mathbf{x}) = 0$.
(4) The constraint $h_3(\mathbf{x}) = 0$ is satisfied; however $h_1(\mathbf{x}) = 0$ and $h_2(\mathbf{x}) = 0$ are not satisfied. Since the Newton method converges rapidly in the vicinity of the root, this solution was tried and was found to be good initial estimate for step 4.
(5) Utilize the Newton method to determine the roots of $h_1(\mathbf{x}) = 0$, $h_2(\mathbf{x}) = 0$, $h_3(\mathbf{x}) = 0$.

The optimum solution is

$$C^* = \$3395$$

with $t_1^* = 25.15$ min, $\alpha^* = 0.339$, and $\lambda^* = -0.006$. The tank volumes are $V_1^* = 17{,}465$ gal and $V_2^* = 100{,}150$ gal.

Remarks

By the method of substitution we reduced the original formulation, which consisted of five control variables, α, V_1, V_2, Q_w, X_2 and four constraint equations, (2′), (3′), (4′) and (7), to a model consisting of two control variables α and t_1 and one constraint equation. It is important to simplify the problem as much as possible before utilizing numerical methods.

Since this optimization problem consists of two control variables and one constraint, graphical methods may be employed. The optimum solution and the solution path are shown in Figure 9.21.

PROBLEMS

Problem 1

$$z = x_1^2 - x_1x_2 + x_2^2 - 6x_1 + 9e^{x_3}$$

(a) Determine the direction of steepest ascent at the point:

$$\mathbf{x}^0 = [2 \quad 2 \quad 1 \quad 6]$$

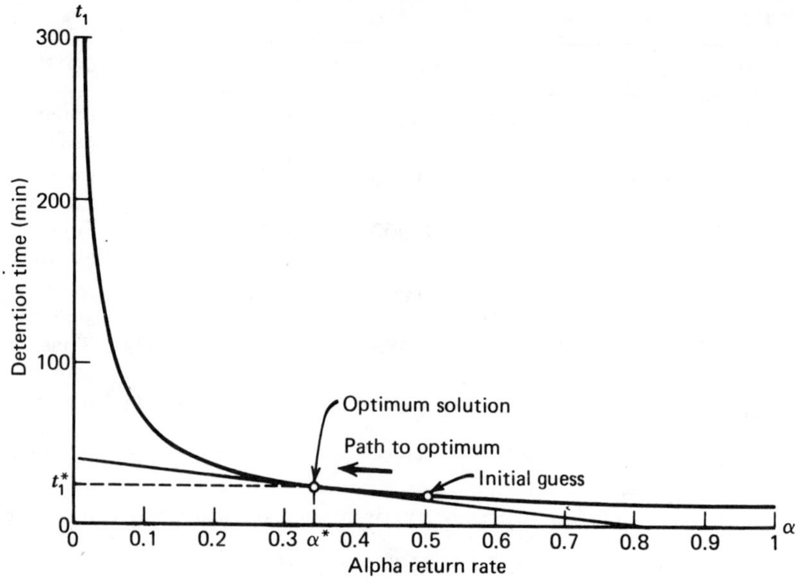

Figure 9.21

(b) Determine the direction cosines at

$$x^0 = [2 \quad 2 \quad 1 \quad 6]$$

on the boundary line:

$$x_1 + 2x_2 = 6$$

(c) Determine the direction cosines at

$$x^0 = [2 \quad 2 \quad 1 \quad 6]$$

on the boundary lines:

$$x_1 + 2x_2 = 6$$
$$x_2 + \quad x_4 = 8$$

Problem 2

$$\text{Minimize } z = (x_1 - 10)^2 + (x_2 - 5)^2$$
$$x_1^2 + x_2^2 \le 64$$
$$x_1 \ge 0, x_2 \ge 0$$

(a) By graphical means, find the optimum solution.
(b) Assume a starting point of $\boldsymbol{x}^0 = (8, 0)$. Determine the gradient projection vector $\boldsymbol{u}^0$ at that point.
(c) Let step size $s = 1$. Show that the step along the line in direction u^0 is in the infeasible region.
(d) Utilize the gradient projection—Newton method to solve for a feasible solution found in part c.

Problem 3

To determine the principal stresses σ_1 and σ_2 and the angle θ between the direction of σ_1 axis and the axis of stress meter 1 as shown in Figure 9.22 three stress meters are placed in the borehole at an angle 60° from one another. The following compressive stress readings were measured:

$$s_1 = 7722 \text{ psi}$$

$$s_2 = 6004 \text{ psi}$$

$$s_3 = 2775 \text{ psi}$$

The stress meters record only compressive and tensile stresses. Determine the principal stresses σ_1 and σ_2 and the angle θ with the use of the following set of empirical relationships:

$$-1.8[\sigma_1 \sin^2\theta + \sigma_2 \cos^2\theta] + 4[\sigma_1 \cos^2\theta + \sigma_2 \sin^2\theta] = s_1$$

$$-1.8[\sigma_1 \sin^2(\theta + 60°) + \sigma_2 \cos^2(\theta + 60°)] + 4[\sigma_1 \cos^2(\theta + 60°) + \sigma_3 \sin^2(\theta + 60)] = s_2$$

$$-1.8[\sigma_1 \sin^2(\theta + 120°) + \sigma_2 \cos^2(\theta + 120°)] + 4[\sigma_1 \cos^2(\theta + 120°) + \sigma_2 \sin^2(\theta + 120°)] = s_3$$

Utilize the Newton method to determine σ_1, σ_2, and θ.

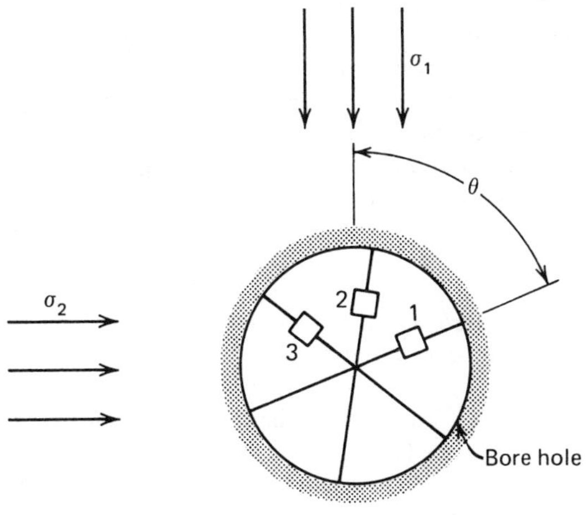

Figure 9.22 Stress meter marked as 1, 2, 3.

(a) Give the iterative equation.
(b) Assume an initial guess of

$$\sigma_1 = 1000 \text{ psi}$$
$$\sigma_2 = 500 \text{ psi}$$
$$\theta = 0°$$

and show the principal stresses and θ are

$$\sigma_1 = 3000 \text{ psi}$$
$$\sigma_2 = 2000 \text{ psi}$$
$$\theta = 20°$$

Assume ε, the tolerance level, is 0.01.

Summary

In this chapter we promoted the use of the Newton method because it is computationally efficient. If the initial guess is a poor one, it is possible that a solution may not be found. For univariate problems, a graphical method of estimation was presented. For multivariate problems, the gradient projection method was discussed and introduced into the Newton–gradient projection method.

The Newton–gradient projection method is a cautious procedure that should be successful in finding the solution to a variety of nonlinear equation sets. Other methods, that may take advantage of the mathematical properties of the nonlinear functions may prove to be more computationally efficient than the Newton–gradient projection method. For example, quadratic and geometric programming are notable examples of optimization methods used for quadratic and exponential functions, respectively. Separable programming utilizes the properties of separable functions to transform a nonlinear optimization problem into a linear programming problem, where the simplex method may be used to *estimate* the solution of the nonlinear problem. The cutting plane method is another method that transforms nonlinear problems to linear formulations. The methods described in this chapter require the calculation of derivatives. For some problems, this may be a difficult task to perform. Separable and dynamic programming are procedures that do not require the calculation of derivatives.

The methods discussed in this chapter contain fundamental concepts that are used in many of the methods available for solving nonlinear optimization problems and roots of equations. The reader is urged to read other books and technical journals and to refer to computer applications libraries to find the most appropriate method for his other particular problem.

Bibliography

Forman S. Acton. *Numerical Methods That Work*, Harper & Row, New York. 1970.

Leon Cooper and David Steinberg. *Introduction of Methods of Optimization*. W. B. Saunders, Philadelphia, Penn. 1970.

R. Fletcher and C. M. Reeves. "Function Minimization of Conjugate Gradients." *Computer Journal*, Volume 7, 1964, pp. 149–154.

G. Hadley. *Nonlinear and Dynamic Programming*. Addison-Wesley. Reading, Mass. 1964.

James L. Kuester and Joe H. Mize. *Optimization Techniques with FORTRAN*. McGraw-Hill, New York. 1973.

J. B. Rosen. "The Gradient Projection Method for Nonlinear Programming. Part I. Linear Constraints," *Journal of Industrial Applied Mathematics*, Volume 8, Number 1, March 1960, pp. 181–217.

J. B. Rosen. "The Gradient Projection Method for Nonlinear Programming. Part II. Nonlinear Constraints." *Journal of Industrial Applied Mathematics*, Volume 9, Number 4, December 1961, pp. 514–532.

Harvey M. Wagner. *Principles of Operations Research*. Prentice Hall, Englewood Cliffs, N.J. 1969.

CHAPTER 10

Curve Fitting and the Method of Least Squares

After completion of this chapter, the student should be able to:

1. Formulate empirically derived prediction equations using graphical and statistical methods.
2. Estimate unknown parameters of these prediction equations with the method of least squares.
3. Use these equations for forecasting and establishing confidence levels.

Throughout this book, empirical relationships have been used to give estimates of costs, production levels, and other technical factors. Here, special attention is given to the derivation of relationships from observed data. These relationships are called *regression equations*. In engineering and science, empirically derived relationships sometimes provide a satisfactory alternative to ones derived from physical laws. Regression analysis has been used to study complex problems where the technical process is not fully understood and where derivations from fundamental principles lead to complex models that are impractical to use. With linear regression analysis, dependency among variables may be identified and incorporated in a regression equation.

The following construction cost function for pumping stations is an example of an empirically derived cost function:

$$C = 33{,}200q^{1.26}$$

Given a value of flow q, the cost of construction may be estimated by $\hat{C} = 33200q^{1.26}$. The estimate of cost is denoted as $\hat{C}$ instead of C to emphasize that the cost is an *estimate* or *prediction*. It is also called a conditional estimate because the value of $\hat{C}$ is a function of a given value of q. This equation was derived from construction bid data for pumping stations of similar design and function. Only when the pumping station is built will the *true* construction cost be known. During the preconstruction phase of planning and design, the cost function can be used to give the "best" estimate of cost. Thus, particular attention must be given to gathering the proper data to derive this function. Equally important is the manner in which these data are used to formulate a meaningful prediction equation.

Regression equations take many forms, including the exponential relationship represented by the construction cost of pumping stations. One of the simplest and most important forms is the linear first-order, or simple linear regression equation

$$y = b_0 + b_1x$$

where b_0 and b_1 are the intercept and slope terms, respectively. In this model, the predicted value of y, $\hat{y}$, is a function of a single *explanatory* variable x. The exponential cost function $C = aq^b$ may be transformed to the simple linear regression form of $y = b_0 + b_1 x$ by taking logarithms of each side of the equation. The analytical details are discussed later in this chapter. The important point is that the same methods used to derive an exponential function may be used to derive a linear equation. These methods will be extended to models with more than one explanatory variable, such as the following bivariate linear and exponential equations:

$$\hat{y} = b_0 + b_1 x_1 + b_2 x_2$$

$$\hat{y} = b_0 x_1^{b_1} x_2^{b_2}$$

The derivation of a simple or multiple linear regression equation requires the following steps.

1. Gather data by sampling.
2. Formulate a mathematical model by examining the relationship between a dependent variable and the explanatory variables.
3. Estimate the model parameters by the method of least squares.

With the use of statistical methods, the empirically derived parameters of the equation may be validated. Statistical validation methods are beyond the scope of this textbook and will not be covered. Our approach is limited to, and relies upon, the use of graphical procedures and simple statistics to estimate the mathematical form of the regression equation and the method of least squares for determining the unknown model parameters.

One of the most important aspects of this approach is the data-gathering step. Data may be obtained by performing laboratory experiments, by performing field experiments, or by conducting surveys. The pumping-station construction-cost model was estimated with construction cost bid data gathered in a national survey. Data gathered by survey or by experiment may be expensive. As a consequence, the extent of the effort is generally restricted by cost limitations. In this chapter, it is assumed that the data are available.

10.1 MODEL FORMULATION

The simple linear regression equation $y = b_0 + b_1 x$ is an estimate of the *true* linear mathematical model:

$$Y = \beta_0 + \beta_1 x + \varepsilon$$

The true model is assumed to be an *exact* mathematical representation of a physical, economic, or sociological process. The linear process of the random variable Y is defined by an intercept β_0, slope β_1, and a random variable ε.

Rarely, in the real world, the true model is never known. The best we can achieve is to assume a true model form and determine the validity of the assumption by observation. Suppose the true model is $Y = \beta_0 + \beta_1 x + \varepsilon$. Our approach is to collect data of Y and x, estimate the parameters β_0 and β_1 with b_0 and b_1, and compare properties of the simple linear regression equation with the properties of the true model. If the two models exhibit the same properties within certain statistical limits, the empirically derived model will be assumed to be the proper choice. If not, it is possible that the dependency relationship between x and y is nonlinear, or that no linear or nonlinear relationship exists between x and y. All these possibilities are investigated in this chapter. Our discussion will begin by discussing random variables and the following models:

$$Y = \mu + \varepsilon$$

and

$$Y = \beta_0 + \beta_1 x + \varepsilon$$

It might be helpful for the reader to review the material of Section 4.1 on random variables and the normal probability distribution before proceeding in this chapter.

Random Variables and Statistical Estimates

Suppose concrete compression test cylinders are prepared utilizing the same ingredients and are mixed and cured under identical conditions. Although each cylinder is prepared with the same exacting care, imperfections will be present that will affect the ultimate strength of each cylinder. The outcome of a compression test will be represented by a set of ultimate strengths, $y_1, y_2, \ldots, y_n$. The average strength of the concrete samples or *sample mean* $\bar{y}$ is estimated with the $\bar{y}$ statistic.

$$\bar{y} = \frac{1}{n} \sum_i y$$

The *sample variance* is estimated with the s_y^2 statistic.

$$s_y^2 = \frac{1}{n-1} \sum_i (y_i - \bar{y})^2$$

The standard deviation is equal to the square root of the variance, $s_y = \sqrt{s_y^2}$. The sample mean $\bar{y}$ is a statistical estimate of the true ultimate strength of the concrete μ and the standard deviation s_y is an estimate of the true standard deviation $\sigma_y = \sigma$.

The true strength is represented by the model:

$$Y = \mu + \varepsilon$$

where Y is a random variable of ultimate compressive strength, μ is the true mean of ultimate compressive strength, and ε is the random variable explaining the variation of strength about μ. The variation of strength σ is represented by the bell-shaped curve, the normal probability density distribution shown in Figure 10.1. We shall assume that a histogram of the y data has been plotted and that it closely resembles the shapes of the normal probability distribution of Y. The model shows that the

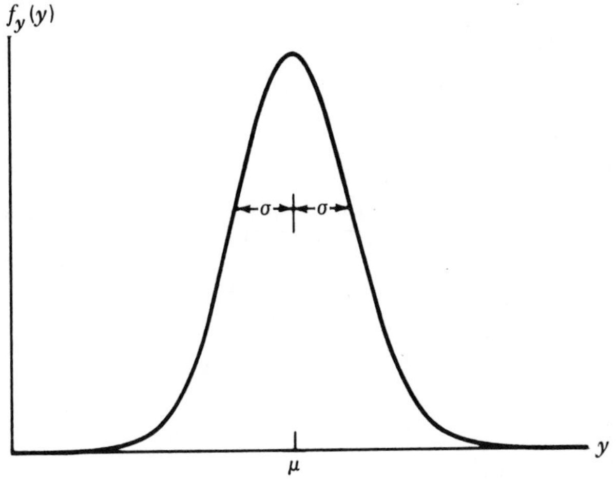

Figure 10.1 The normal probability density distribution of ultimate strength.

strength as represented by the random variable Y, from compression test to compression test.

The Normal Distribution

The properties of a random variable with a normal distribution $N(\mu, \sigma^2)$ are described by the mean μ, the standard deviation σ, and the following probability density function:

$$f_Y(y) = \frac{1}{\sqrt{2\pi\sigma}} \exp\left(\frac{-(y-\mu)^2}{2\sigma^2}\right)$$

The distribution of Y about the mean μ is shown in Figure 10.1.

The properties of the random variable Y as described by $N(\mu, \sigma_y^2)$ is identical to the true model, $Y = \mu + \varepsilon$. For the true model, the distribution of Y is assumed to be a sum of the constant μ and the random variation ε. The variation ε is assumed to be normally distributed with a mean of zero and variance σ_ε^2, $N(0, \sigma_\varepsilon^2)$. Thus, the random variable ε is described by the following probability density function:

$$f_\varepsilon(e) = \frac{1}{\sqrt{2\pi\sigma_\varepsilon}} \exp\left(\frac{-e^2}{2\sigma_\varepsilon^2}\right)$$

The normal probability $N(0, \sigma_\varepsilon^2)$ is shown in Figure 10.2. A plot of the bell-shaped curve in Figure 10.1, $Y = \mu + \varepsilon$, is identical to the one shown in Figure 10.2, except it is shifted to the right the distance μ. The variances σ^2 and σ_ε^2 are the same.

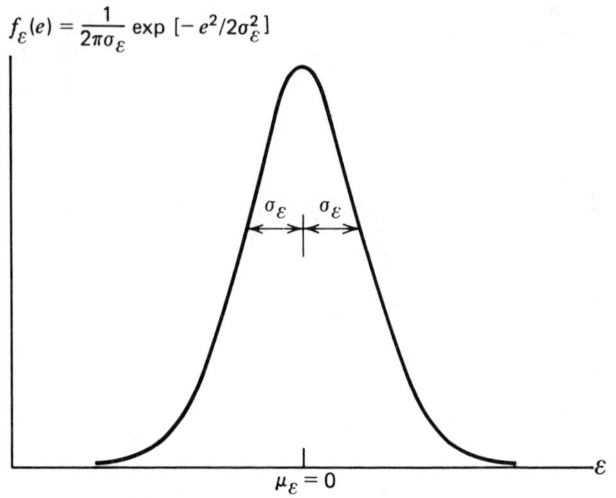

Figure 10.2 The PDF of the true model: $Y = \mu + \varepsilon$.

The Explanatory Variable

The true model $Y = \mu + \varepsilon$ illustrates the essential properties of a simple random process. Its properties are summarized as:

1. ε is normally distributed about the mean μ: $N(\mu, \sigma_\varepsilon^2)$.
2. The mean of ε is zero: $\mu_\varepsilon = 0$, thus $N(\mu, \sigma_\varepsilon^2) = N(0, \sigma_\varepsilon^2)$.
3. The variance of ε is constant: $V(\varepsilon) = \sigma_\varepsilon^2$.

The model $Y = \mu + \varepsilon$ is the sum of deterministic and random parts, μ and ε, respectively. The true model, $Y = \beta_0 + \beta_1 x + \varepsilon$, is similar in the sense that it is a sum of deterministic and random parts. The key difference is that its deterministic portion, $\beta_0 + \beta_1 x$, contains the *explanatory variable* x. Let us investigate the significance of the explanatory variable by example.

Suppose that another series ultimate strength tests are performed on concrete cylinders. Each cylinder is prepared with the same ingredients; however, for this series of experiments they are cured for different lengths of time and then tested for ultimate compressive strength. The ultimate strength is expected to increase with curing time; thus, the behavior will be represented by a true model

$$Y = \beta_0 + \beta_1 x + \varepsilon$$

where Y is a random variable of ultimate compressive strength, x is the curing time, β_0 and β_1 are constants, and ε is a normally distributed random variable about the line $\beta_0 + \beta_1 x$. The model is illustrated in Figure 10.3.

The properties of the true model, $Y = \beta_0 + \beta_1 x + \varepsilon$, are assumed to be as follows:

1. ε is normally distributed about the line, $Y = \beta_0 + \beta_1 x$: $N(0, \sigma_\varepsilon^2)$.
2. The mean of the random variable ε is zero: $\mu_\varepsilon = 0$.
3. The variance of ε is constant: $V(\varepsilon) = \sigma_\varepsilon^2$.

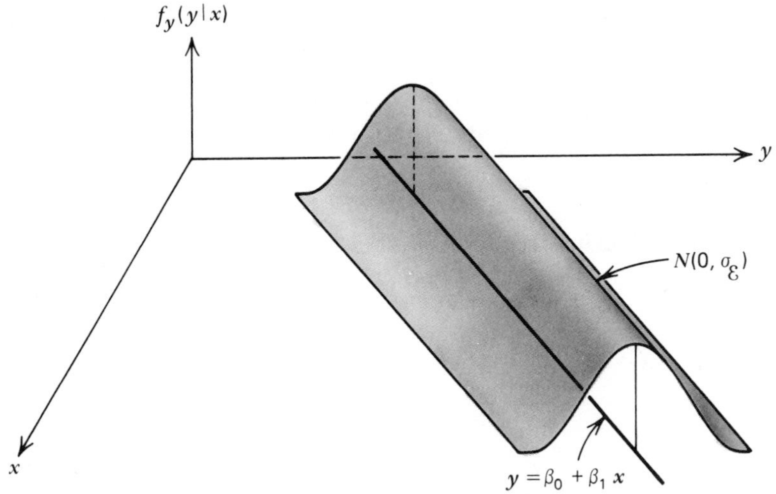

Figure 10.3 The PDF of the true model: $Y = \beta_0 + \beta_1 x_2 + \varepsilon$.

We focus attention to determining the strength of the linear relationship between observations of x and y and attempt to determine which true model, $Y = \mu + \varepsilon$, $Y = \beta_0 + \beta_1 x + \varepsilon$, or a nonlinear transform of $Y = \beta_0 + \beta_1 x + \varepsilon$, describes the actual process.

The Scattergram

One of the simplest and most powerful methods for determining the relationship between paired observations is the *scattergram*. A scattergram consists of plotting each pair of observations in the x–y coordinate. A relationship between the observations of x and y can be identified by inspection. In Figure 10.4 different linear relationships between x and y exist. Lines, so called *lines of best fit*, have been drawn through the data points to show the linear relationship more clearly. Although the lines do not fall through all the data points, they do summarize the pattern that exists between x and y.

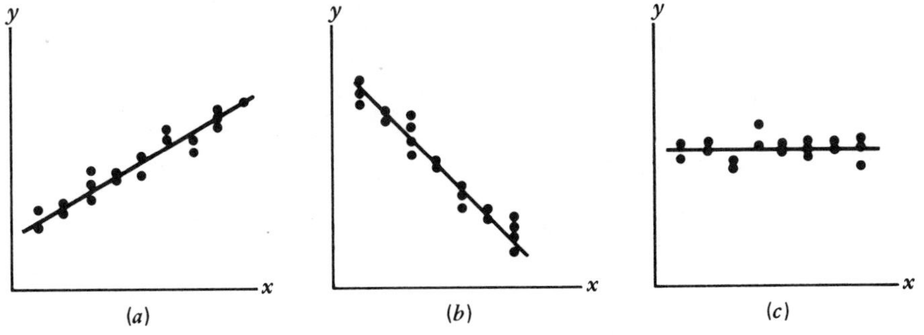

Figure 10.4 The scattergram for different samples of x and y.

If the scattergram has a pattern as shown by Figures 10.4*a* and 10.4*b*, it is reasonable to assume that the underlying relationship is represented by the true model:

$$Y = \beta_0 + \beta_1 x + \varepsilon$$

However, if a pattern as illustrated in Figure 10.4*c* exists, it might be better to assume a true model

$$Y = \beta_0 + \varepsilon$$

where β_0 is equal to the intercept of y, $\beta_0 = \mu$.

If no relationship exists, a random pattern as shown in Figure 10.5*a* may exist.

$$Y = \varepsilon$$

In Figure 10.5*b*, a nonlinear pattern exists that may be most appropriately represented as

$$Y = \beta_0 + \beta_1 x^2 + \varepsilon$$

Correlation Coefficient

The *correlation coefficient* is a statistical measure of the *linear* dependency between two variables. Data are sampled in pairs (x_i, y_i), where $i = 1, 2, \ldots, n$ and n equals the total number of observations. The correlation coefficient r_{xy} for variables x and y is estimated with the equation

$$r_{xy} = \frac{s_{xy}}{s_x s_y}$$

where s_{xy} is the estimated covariance of and x and y, $\sigma_{xy} = \mathrm{Cov}(x, y)$, s_x and s_y are the estimated standard deviations of x and y, σ_x and σ_y, respectively. The variance and covariance can be estimated with the following equations:

$$s_x^2 = \frac{1}{n-1} \sum_i (x_i - \bar{x})^2 = \frac{1}{n-1} \left(\sum_i x_i^2 - n\bar{x}^2 \right)$$

$$s_y^2 = \frac{1}{n-1} \sum_i (y_i - \bar{y})^2 = \frac{1}{n-1} \left(\sum_i y_i^2 - n\bar{y}^2 \right)$$

$$s_{xy} = \frac{1}{n-1} \sum_i (y_i - \bar{y})(x_i - \bar{x}) = \frac{1}{n-1} \left(\sum_i x_i y_i - n\bar{x}\bar{y} \right)$$

where $\bar{x}$ and $\bar{y}$ are the sample mean estimates of μ_x and μ_y, respectively. The standard deviations s_x and s_y are equal to the square root of the sample variances s_x^2 and s_y^2, respectively. Furthermore, observe that s_x^2 and s_y^2 are always positive; that is, $s_x^2 > 0$ and $s_y^2 > 0$, but s_{xy} may be positive or negative; that is, $-\infty \leq s_{xy} \leq \infty$. The mean estimates of x and y are determined by

$$\bar{x} = \frac{1}{n} \sum_i x_i$$

$$\bar{y} = \frac{1}{n} \sum_i y_i$$

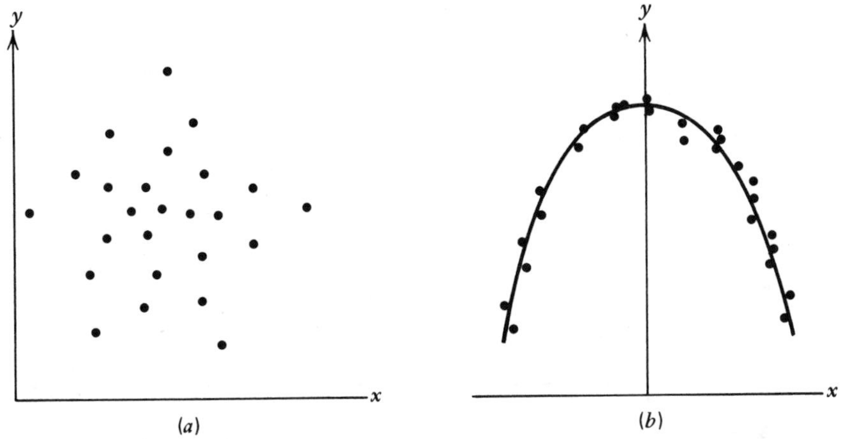

Figure 10.5 Random and nonlinear relationships.

In matrix notation, these equations are written as

$$s_x^2 = \frac{1}{n-1}(\mathbf{x}'\mathbf{x} - n\bar{x}^2)$$

$$s_y^2 = \frac{1}{n-1}(\mathbf{y}'\mathbf{y} - n\bar{y}^2)$$

$$s_{xy} = \frac{1}{n-1}(\mathbf{x}'\mathbf{y} - n\bar{x}\bar{y})$$

where $\mathbf{x}' = [x_1 \quad x_2 \quad \cdots \quad x_n]$ and $\mathbf{y}' = [y_1 \quad y_2 \quad \cdots \quad y_n]$ are vectors of the paired observations of x and y.

The correlation coefficient is limited to a range of values between plus and minus one.

$$-1 \leq r_{xy} \leq 1$$

If a perfect linear relationship exist, that is, all the points in the scattergram lie on a line, the value of r_{xy} will be equal to plus or minus one; that is, $r_{xy} = 1$ or -1. If $r_{xy} = 1$, the slope of the line is positive. If $r_{xy} = -1$, the slope of the line is negative. These concepts are shown in Figure 10.6. When a perfect linear relationship does not exist, the closer the value of r_{xy} approaches plus or minus one, $r_{xy} \sim 1$ or $r_{xy} \sim -1$, the stronger the linear dependency between the variables.

Since the correlation coefficient is a measure of the *linear* dependency between x and y only, when r_{xy} is equal to zero or approaches zero, $r_{xy} \sim 0$, we conclude that *no linear relationship exists* between x and y. This result does not exclude the possibility that a nonlinear relationship exists. The scattergrams shown in Figures 10.5*a* and 10.5*b* illustrates these ideas. The correlation coefficient for the random relationship shown in Figure 10.5*a* is approximately equal to zero; that is, $r_{xy} \sim 0$ and no functional relationship exists. On the other hand, the curve shown in Figure 10.5*b* is parabola. The correlation coefficient for this particular set of observed data pairs is equal to zero, $r_{xy} = 0$. Clearly, from the measure of correlation and the scattergram no linear

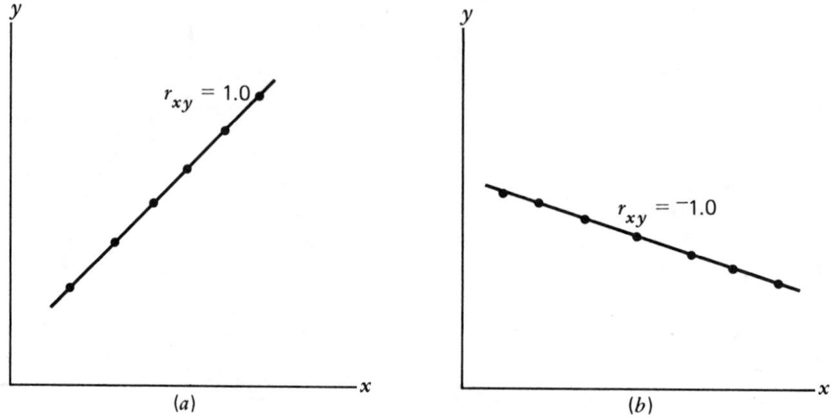

Figure 10.6 Perfect linear relationships.

relationship exists. Furthermore, it is apparent from the scattergram that a nonlinear relationship exists. The scattergram and correlation coefficient complement one another, and both will be used to establish the mathematical form of the predictive equation.

Transformation to Linear Form

Our method for establishing predictive equations *is restricted to linear equations* where $y = b_0 + b_1 x$ is an estimate of the true model, $Y = \beta_0 + \beta_1 x + \varepsilon$. Estimates b_0 and b_1 of the parameters of the intercept β_0 and the slope β_1 will be made by the method of least squares that is discussed in the following section. In order for this equation to have validity, there must be evidence that x and y are linearly dependent. The scattergram and the correlation coefficient are used in conjunction to determine the strength of this linear dependency.

Suppose an exponential relationship is assumed to be the equation of best fit:

$$y = b_0 x^{b_1}$$

By taking logarithms of each side we transform the nonlinear equation into a linear one:

$$\ln(y) = \ln(b_0) + b_1 \ln(x)$$

where

$$\mathbf{y} = \ln(y)$$

$$\mathbf{b}_0 = \ln(b_0)$$

$$b_1 = b_1$$

$$\mathbf{x} = \ln(x)$$

Thus, the linear relationship between y and x is represented as

$$\mathbf{y} = \mathbf{b}_0 + b_1 \mathbf{x}$$

Linear regression methods will be used to estimate $\mathbf{b}_0$ and b_1. This procedure, for instance, was used to estimate the cost function $\hat{c} = 33{,}200 q^{1.26}$.

Table 10.1 Nonlinear Model Transformations

NONLINEAR EQUATION	TRANSFORMATION	LINEAR EQUATION
1. $y = b_0 + b_1 x^2$	$\mathbf{x} = x^2$	$y = b_0 + b_1 \mathbf{x}$
2. $y = b_0 e^{b_1 x}$	$\mathbf{y} = \ln y$ $\mathbf{b}_0 = \ln b_0$	$\mathbf{y} = \mathbf{b}_0 + b_1 x$
3. $\frac{1}{y} = b_0 + b_1 \frac{1}{x}$	$\mathbf{y} = \frac{1}{y}$ $\mathbf{x} = \frac{1}{x}$	$\mathbf{y} = b_0 + b_1 \mathbf{x}$

Not all nonlinear equations may be transferred into a linear form.

Table 10.1 shows various transformations for models commonly encountered in engineering.

EXAMPLE 10.1 The Effect of the Random Variable ε

Consider the following four true models:

Model A. $Y = 2$ with no random variation, $\varepsilon = 0$.

Model B. $Y = 2 + \varepsilon$ with $\varepsilon = N(0, 1)$.

Model C. $Y = 1 + 2x$ with no random variation, $\varepsilon = 0$.

Model D. $Y = 1 + 2x + \varepsilon$ with $\varepsilon = N(0, 2)$.

For each model, a set of Y values have been randomly generated for each member of the **x** vector. The observations are as follows.

i	EXPLANATORY VARIABLE x	OBSERVATIONS OF y FOR A	B	C	D
1	1.0	2	3.4	3.0	2.8
2	2.2	2	1.1	5.4	2.9
3	2.4	2	2.4	5.8	5.9
4	4.2	2	0.3	9.4	10.3
5	3.8	2	1.7	8.6	9.0
6	7.8	2	1.5	16.6	16.0
7	10.0	2	1.3	21.0	16.2
8	9.6	2	0.7	20.2	19.4
9	9.0	2	0.0	19.0	21.5
10	8.5	2	3.4	18.0	16.8
11	2.9	2	1.4	6.8	8.9
12	6.4	2	1.5	13.8	11.0
13	0.9	2	2.2	2.8	3.9
14	1.0	2	1.6	3.0	3.5
15	0.7	2	2.2	2.4	2.8
16	5.1	2	3.0	11.2	12.9

Plot scattergrams of x and y, and calculate the mean and variance of x and y, the covariance of x and y, and the correlation coefficient for each model pair.

Solution

For each model the following statistics were determined.

Mean. $\bar{x} = \frac{1}{n}\sum_i x_i \qquad \bar{y} = \frac{1}{n}\sum_i y_i$

Variance. $s_x^2 = \frac{1}{n-1}(\mathbf{x}'\mathbf{x} - n\bar{x}^2) \qquad s_y^2 = \frac{1}{n-1}(\mathbf{y}'\mathbf{y} - n\bar{y}^2)$

Covariance. $s_{xy} = \frac{1}{n-1}(\mathbf{x}'\mathbf{y} - n\bar{x}\bar{y})$

Correlation Coefficient. $r_{xy} = \frac{s_{xy}}{s_x s_y}$

The estimates are

MODEL	$\bar{x}$	s_x^2	$\bar{y}$	s_y^2	s_{xy}	r_{xy}
A	4.72	11.42	2.00	0.00	0.00	0.00
B	4.72	11.42	1.73	1.00	−1.15	−0.34
C	4.72	11.42	1.04	45.7	22.85	1.00
D	4.72	11.42	10.2	40.2	20.6	0.96

The scattergrams for each of the models in Figure 10.7 show the linear dependency between x and y. For models A and B, the explanatory variable x is independent of y. For models C and D, the explanatory variable x is highly correlated with y.

Remarks

For practical engineering problems, a set of observations (x_i, y_i) will be gathered. The underlying true model will not be known. The purpose of this example is to illustrate the influence of independent variable x and the random variable ε upon the values of the observation random variables of Y.

EXAMPLE 10.2 Fitting a Nonlinear Function

The following paired observations have been made:

i	x	y
i	−3	18
2	−2	8
3	−1	2
4	0	0
5	1	2
6	2	8
$n = 7$	3	18

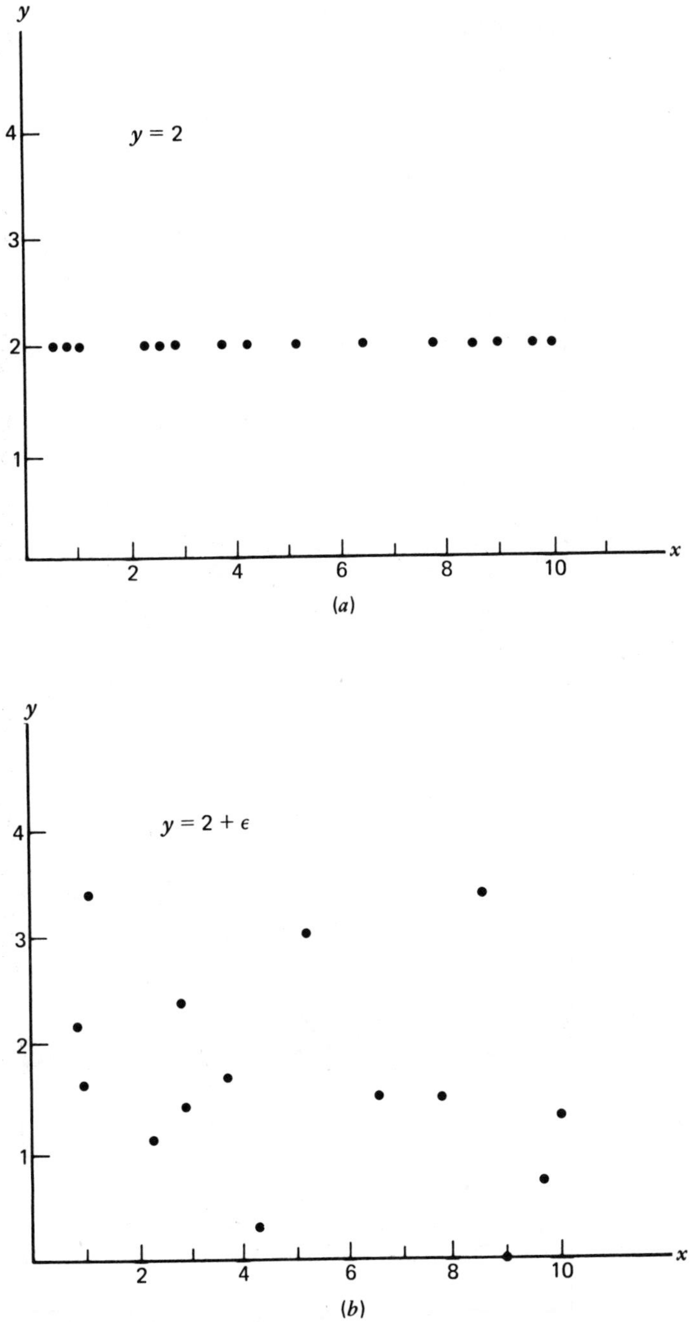

Figure 10.7 (*a*) Model A, $r_{xy} = 0$. The observations of x and y are linearly independent. (*b*) Model B, $r_{xy} = -0.34$. The observations of x and y appear to have a slight linear dependency.

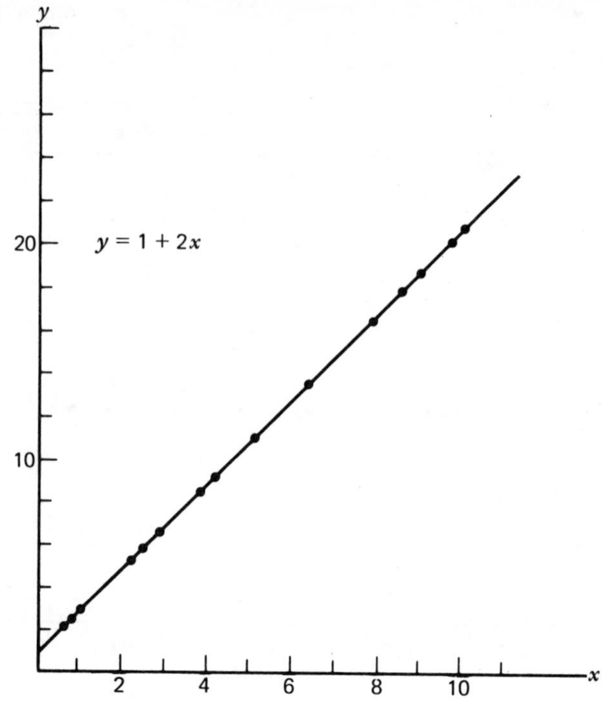

(c)

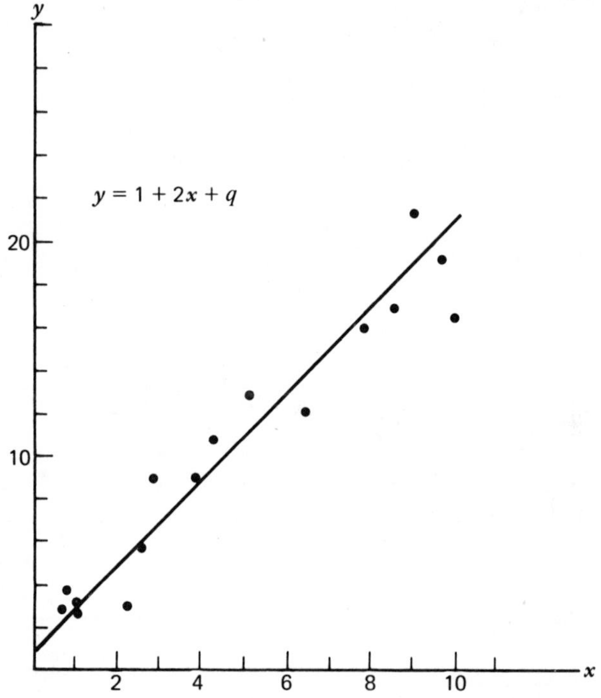

(d)

Figure 10.7 (*continued*) (*c*) Model C, $r_{xy} = 1$. Perfect linear dependency for the observations of x and y. (*d*) Model D, $r_{xy} = 0.96$. Strong linear dependency for the observations of x and y.

(a) Determine the correlation coefficient r_{xy}.
(b) Plot a scattergram.
(c) Utilize the scattergram to determine a transformation from the nonlinear to linear form.

Solution

(a) The correlation coefficient is defined to be $r_{xy} = s_{xy}/(s_x x_y)$. The following statistics have been obtained:

$$\bar{x} = 0.00 \qquad \bar{y} = 8.0$$
$$s_x = 2.16 \qquad s_y = 7.48$$
$$s_{xy} = 0.00$$

Thus

$$r_{xy} = \frac{s_{xy}}{s_x s_y} = \frac{0.00}{(2.16)(7.48)} = 0.00$$

There is no apparent linear dependency between the observations of x and y.

(b) The scattergram in Figure 10.8 shows that no linear relationship between x and y exists.

(c) Since the scattergram shows the relationship between y and x to be parabolic, assume the equation to have the form

$$y = b_0 x^2$$

where b_0 is an unknown constant. Define $\mathbf{x}$ as

$$\mathbf{x} = x^2$$

By utilizing the transformation $\mathbf{x} = x^2$ the data set for $\mathbf{x}$ becomes

i	$\mathbf{x}$	y
1	9	18
2	4	8
3	1	2
4	0	0
5	1	2
6	4	4
7	9	8

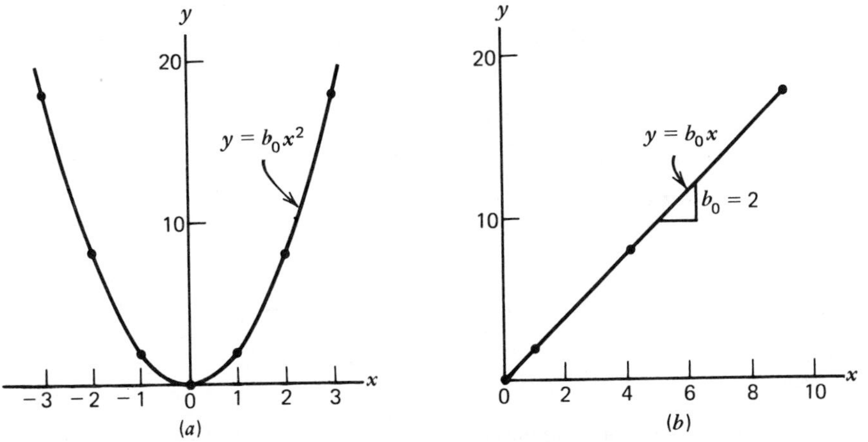

Figure 10.8 Each point is equivalent to two observations.

Plotting the corresponding pairs of observations of $\mathbf{x}$ and y, a linear relationship has been obtained. A perfect linear relationship between $\mathbf{x}$ and y exists. The correlation coefficient is equal to one, $r_{xy} = 1$. The slope of the line is easily determined to be equal to 2, $b_0 = 2$. The linear relationship becomes

$$y = 2\mathbf{x}$$

Since $\mathbf{x} = x^2$ the relationship becomes

$$y = 2x^2$$

Remarks

Clearly, for practical engineering problems we would not be so fortunate or so lucky as to transform a paired observation with zero correlation, $r_{xy} = 0$, to one with perfect correlation, $r_{xy} = 1$. This example, however, illustrates the essential steps taken in determining if the equation of best fit is a linear or nonlinear function. If a prediction equation is assumed to be nonlinear, an appropriate relationship must be identified to transfer the equation into a linear equation.

EXAMPLE 10.3 Chemical Reaction Rate

To design water and wastewater facilities, it is important to know how fast chemical reactions take place. Suppose a chemical reaction is

$$N + M \rightarrow Q$$

where N is the contaminant chemical, M is the reactant chemical, and Q is the product chemical. The rate of removal of the contaminant is defined as

$$k = \frac{-d[N]}{dt}$$

where $[N]$ is the concentration of N in milligrams per liter. For simplicity, the brackets will be dropped; thus $k = -dN/dt$. The following observations over time t, in minutes, were made.

t	N	
0	2.3	(Initial concentration, N_0)
200	2.1	
320	1.9	
520	1.7	
870	1.4	
1120	1.1	
1880	0.7	

(a) Solve the zero-, first-, and second-order reactions.

Zero order $$\frac{dN}{dt} = -k$$

First order $$\frac{dN}{dt} = -kN$$

Second order $$\frac{dN}{dt} = -kN^2$$

Transform the equations into linear form: $y = a + bx$.

(b) Utilize the scattergram and correlation coefficient to aid in the determination of the proper transformation.
(c) Estimate the reaction rate constant.

Solution

(a) The zero-order reaction is assumed to have the form:

$$\frac{dN}{dt} = -k$$

The solution to this differential equation gives $N = -kt + c$. When $t = 0$, the concentration is N_0; thus

$$N = N_0 - kt$$

The concentration N decreases at a constant rate k over time t.

The first- and second-order equations and their solutions are obtained in a similar manner.

$$\frac{dN}{dt} = -kN; \qquad N = N_0 e^{-kt} \quad \text{or} \quad \ln N = \ln N_0 - kt$$

$$\frac{dN}{dt} = -kN^2; \quad \frac{1}{N} = \frac{1}{N_0} + kt \quad \text{or} \quad N = \frac{N_0}{1 + N_0 kt}$$

A comparison of the three different equations is shown in Figure 10.9.

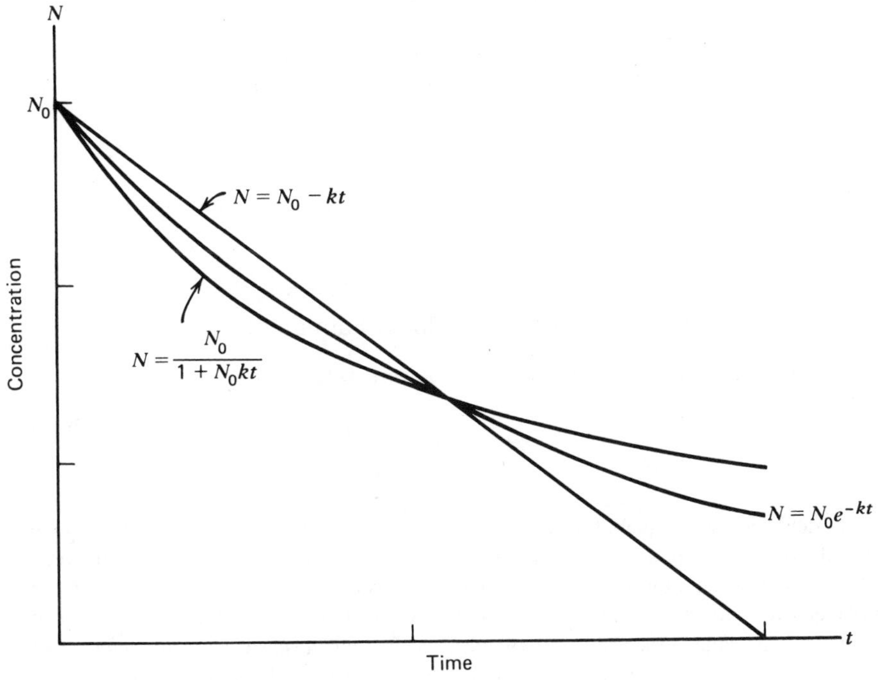

Figure 10.9

The transformations for each of the models are

ORDER	EQUATION	VARIABLE	CONSTANTS	LINEAR EQUATION
0	$N = N_0 - kt$	$y = N$ $x = t$	$a = N_0$ $b = -k$	$y = a + bx$
1	$\ln N = \ln N_0 - kt$	$y = \ln N$ $x = t$	$a = \ln N_0$ $b = -k$	$y = a + bx$
2	$\frac{1}{N} = \frac{1}{N_0} + kt$	$y = \frac{1}{N}$ $x = t$	$a = \frac{1}{N_0}$ $b = k$	$y = a + bx$

(b) Lines were drawn through the scattergram plots and the correlation coefficient was calculated for each case. See Figure 10.10. The scattergram and correlation coefficients give very little information, and no one model is a clear-cut choice. The principles of chemistry and other logical reasons, not just the scattergram and correlation coefficient, should be employed in choosing the best reaction equation form. In this example, the first-order reaction will be assumed.

(c) The intercept and slope of the line are estimated for $y = a + bt$ (see Figure 10.10a):

$$a = N_0 = 2.3 \text{ mg/l}$$

$$b = \frac{\ln(0.7) - \ln(2.3)}{1180 - 0} = \frac{-0.357 - 0.832}{1180} = -0.0010$$

$$k = -(-0.0010) = 0.0010$$

Thus, the first-order equation is $N = 2.3e^{-0.0010t}$ mg/l and the reactor rate is estimated to be $k = 0.0010$/min.

EXAMPLE 10.4 Traffic Flow Model: Model Formulation

The travel speed on a highway is assumed to vary linearly with traffic density

$$u = u_f\left[1 - \frac{k}{k_j}\right]$$

where the dependent and explanatory variables are: u is the travel speed in miles per hour (miles/hr) and k is the traffic density in vehicles per mile (vehicles/mile). The parameter u_f is the free mean speed or speed at which motorists will travel if the road is free of other travellers and independent of traffic density. This is $k = 0$. Geometric conditions, such as sight distance and no parking or parking along the edge of the roadway, will affect the free mean speed. The parameter k_j is called the jam density. At the jam density the traffic congestion is so severe that the traffic ceases to flow and the travel speed is equal to zero.

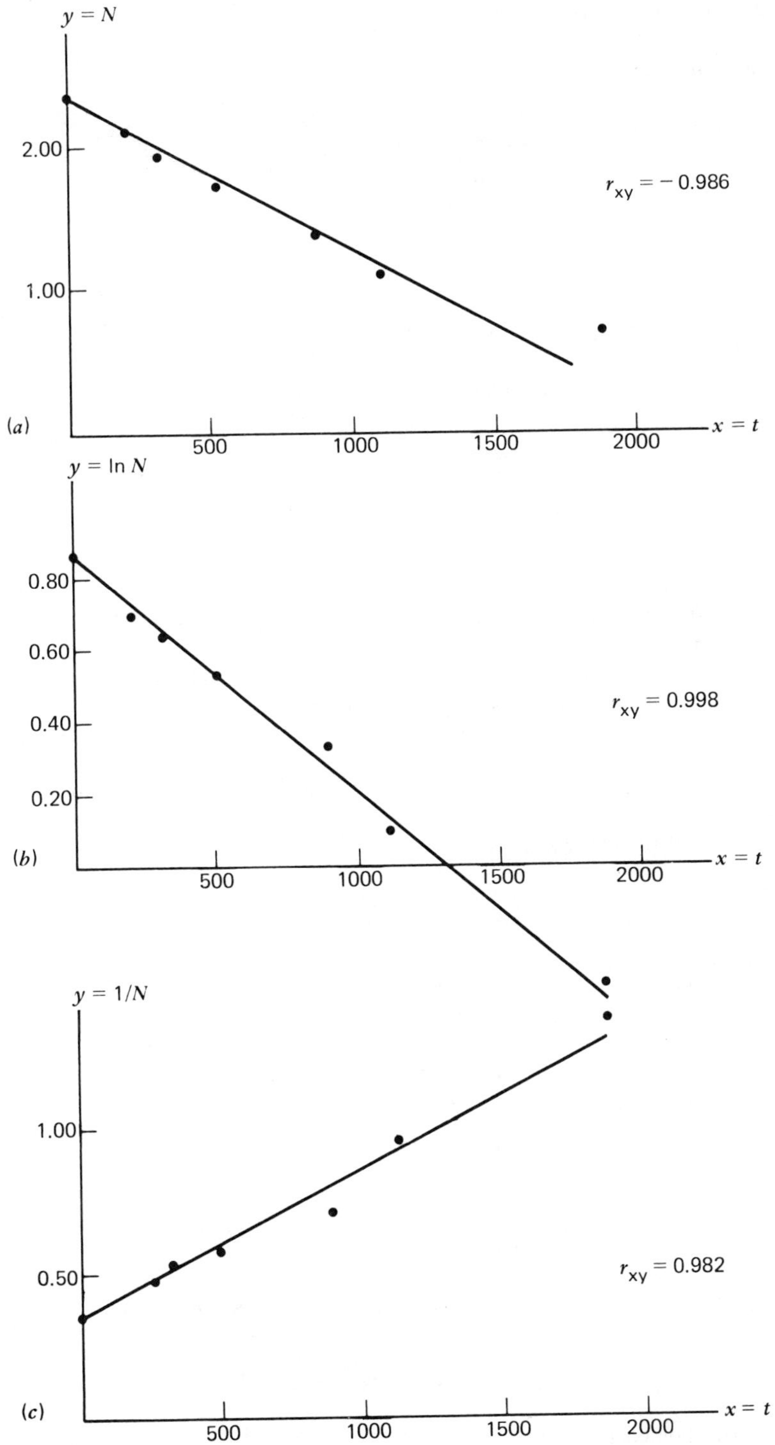

y = N
2.00
1.00
r_xy = −0.986
(a)
500
1000
1500
2000
x = t
y = ln N
0.80
0.60
0.40
0.20
r_xy = 0.998
(b)
500
1000
1500
2000
x = t
y = 1/N
1.00
0.50
r_xy = 0.982
(c)
500
1000
1500
2000
x = t

Figure 10.10

The following observations were made:

u = speed, miles/hr	k = traffic density, vehicles/mile
46.0	22.2
56.0	29.3
48.0	30.1
42.0	40.7
22.9	90.0
56.6	29.2
9.2	113.3
17.9	88.2
14.5	103.1
29.5	60.2
25.0	67.4
32.0	70.6
35.5	41.1
36.5	49.6

(a) Determine the transformations to estimate the model parameters of free mean speed and jam density.
(b) Determine the mean, variance and covariance of the speed and density.
(c) Plot the scattergram and determine the correlation coefficient.

Solution

(a) The free mean speed and jam density of $u = u_f[1 - k/k_j]$ will be estimated from the linear regression equation, $y = a + bx$. Rewrite the speed-density relationship and equate the parameters of the two equations:

$$u = u_f + \left(\frac{u_f}{k_j}\right)k$$

$$y = b_0 + b_1 x$$

Thus, $y = u$, $b_0 = u_f$, $b_1 = -u_f/k_j$, and $x = k$.

(b) Let $x = u$ and $y = k$; thus

$$\bar{u} = \bar{x} = \frac{1}{n}\sum_i x_i = 59.6 \text{ vehicles/mile}$$

$$\bar{k} = \bar{y} = \frac{1}{n}\sum_i y_i = 33.7 \text{ vehicles/mile}$$

$$s_u = s_x = \left[\frac{1}{n-1}(\mathbf{x'x} - n\bar{x}^2)\right]^{1/2} = 14.9 \text{ miles/hr}$$

$$s_k = s_y = \left[\frac{1}{n-1}(\mathbf{y'y} - n\bar{y}^2)\right]^{1/2} = 29.9 \text{ vehicles/mile}$$

$$s_{uk} = s_{xy} = \frac{1}{n-1}(\mathbf{x'y} - n\bar{x}\bar{y}) = -422.7$$

(c) The correlation coefficient is

$$r_{uk} = r_{xy} = \frac{s_{xy}}{s_x s_y} = \frac{-433.7}{(14.9)(29.9)} = -0.95$$

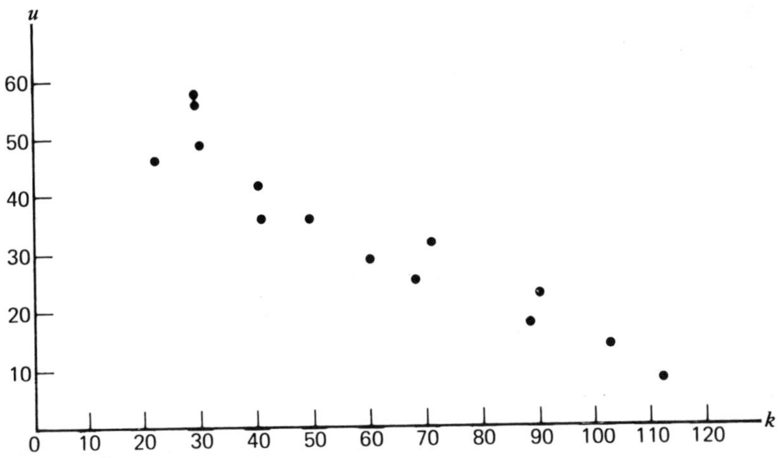

Figure 10.11

The scattergram and coefficient in Figure 10.11 indicate that a linear functional relationship between u and k is appropriate.

PROBLEMS

Problem 1

The effective stress at a given depth may be calculated by taking the difference in the overburden pressure and water pressure. The overburden pressure is equal to the sum of the products of soil

Table 10.2 London Clay Data

y SHEAR STRENGTH (kN/m²)	x EFFECTIVE STRESS (kN/m²)
7	10
14	32
18	47
35	88
37	92
41	65
42	107
51	108
41	122
46	121
55	140
55	153
63	182
71	188
75	182
80	215
80	243
92	261
98	260
101	270

density and soil strata thicknesses. This is a reasonably simple calculation to perform. Shear strength, on the other hand, must be obtained by field measurement. Our goal is to determine the form of the predictive equation to relate these variables. See Table 10.2.

(a) Draw a scattergram.
(b) Calculate the correlation coefficient.
(c) If a linear model $y = b_0 + b_1x$ is not appropriate, utilize a transformation and repeat parts a and b for the transformed data.

Problem 2

The amount of dissolved oxygen in a sample may be determined by the Winkler method. This chemical test method is reasonably time consuming and awkward to perform in the field. An electrical probe is a more efficient field instrument. A calibration curve or predictive equation is sought. See Table 10.3.

(a) Draw a scattergram.
(b) Calculate the correlation coefficient.
(c) If a linear model $y = b_0 + b_1x$ is not appropriate, utilize a transformation and repeat parts a and b for the transformed data.

Table 10.3 Calibration Data

y	x
DISSOLVED OXYGEN (mg/l)	ELECTRICAL PROBE (mV)
8.25	4.7
6.74	3.3
7.05	3.0
7.40	3.7
6.63	3.1
5.93	2.4
4.34	0.6
6.70	3.0
6.10	2.7
5.54	2.1
4.72	1.3
6.98	3.4
6.11	2.8
5.42	2.0
4.51	1.1
6.77	3.2
6.30	2.9
5.70	2.4
5.35	2.0
4.79	1.4

Problem 3

A national survey of the construction cost of sanitary sewers has been undertaken. A relationship between unit construction cost in dollars per linear foot and sewer size in discharge, million gallons per day, is sought. See Table 10.4.

Table 10.4 Sanitary Sewer Costs

y UNIT CONSTRUCTION COST ($/linear ft)	x DISCHARGE (Mgal/day)
50	0.23
52	0.30
70	1.10
96	1.20
80	0.70
150	4.0
188	5.0
260	10.0
300	15.0
315	15.0
370	20.0
340	25.0
315	22.0
325	24.0
325	20.0

(a) Draw a scattergram.
(b) Calculate the correlation coefficient.
(c) If a linear model $y = b_0 + b_1x$ is not appropriate, utilize a transformation and repeat parts a and b for the transformed data.

Problem 4

The flow–density q–k or y–x relationship is an important empirical relationship used by traffic engineers. Determine the relationship between these variables for the Holland Tunnel. See Table 10.5.

(a) Draw a scattergram.
(b) Calculate the correlation coefficient.
(c) If a linear model $y = b_0 + b_1x$ is not appropriate, utilize a transformation and repeat parts a and b for the transformed data.

Problem 5

(a) For the following predictive models, determine if the logarithmic or some other transformation to convert the nonlinear expression to a linear expression is possible. If no transformation exists, please indicate the reason.

(i) $y = b_0 2^{b_1 x}$

(ii) $y = \dfrac{b_0}{b_1 x^2}$

(iii) $y = \dfrac{1}{b_0 + b_1 x}$

(iv) $y = b_0 + b_1 2^{b_1 x}$

Table 10.5 Holland Tunnel Traffic Data

$q = y$ FLOW (vehicles/hr)	$k = x$ DENSITY (vehicles/mile)
7.5	19
7.8	18
8.1	20
8.3	21
8.5	20
9.5	26
9.7	30
7.8	28
10.8	34
10.8	37
11.3	37
12.0	42
12.3	50
12.5	46
12.6	52
12.8	57
12.7	60
12.3	61
12.5	69
12.0	71
12.1	78
11.5	80
11.0	85
10.2	90
9.5	93
9.2	105
8.1	110

(b) If a transformation exists, give the mathematical relationship for parameter and data conversion.

10.2 THE METHOD OF LEAST SQUARES

The line of best fit is the line that minimizes the sum of the square error, *SSE*, or the difference between the observed and predicted values of y:

$$e_i = y_i - \hat{y}_i$$

The parameters b_0 and b_1 of the simple regression equation $\hat{y} = b_0 + b_1 x$ will be estimated by establishing a mathematical model to minimize the sum of the square error, *SSE*:

$$\text{Minimize } z = SSE = \mathbf{e}'\mathbf{e} = e_1^2 + e_2^2 + \cdots + e_n^2$$

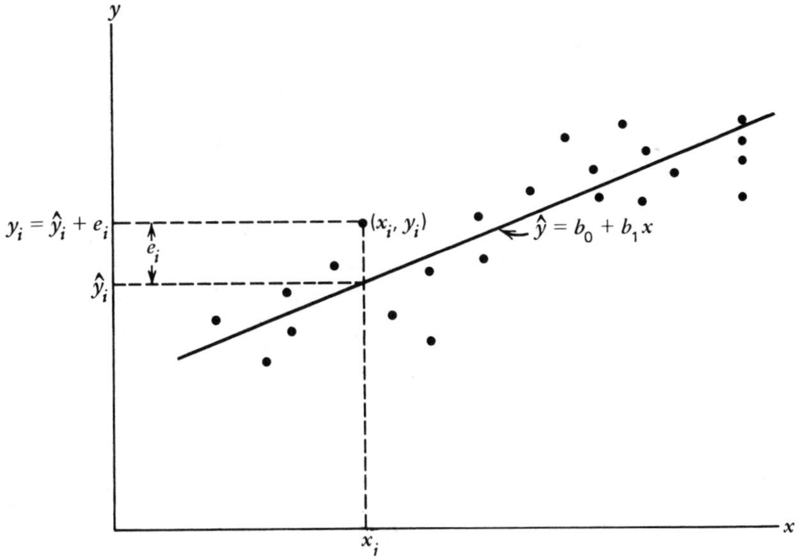

Figure 10.12 The line of best fit $\hat{y} = b_0 + b_1x$.

The value of e_i is depicted in Figure 10.12. At x_i, the predicted value of y is $\hat{y} = b_0 + b_1x_i$; thus

$$e_i = y_i - b_0 - b_1x_i$$

For n observations of x and y, the error vector $\mathbf{e}$ is

$$\mathbf{e} = \begin{bmatrix} e_1 \\ e_2 \\ \vdots \\ e_n \end{bmatrix} = \begin{bmatrix} y_1 - b_0 - b_1x_1 \\ y_2 - b_0 - b_1x_2 \\ \vdots \\ y_n - b_0 - b_1x_n \end{bmatrix} = \begin{bmatrix} y_1 \\ y_2 \\ \vdots \\ y_n \end{bmatrix} - \begin{bmatrix} 1 & x_1 \\ 1 & x_2 \\ \vdots & \vdots \\ 1 & x_n \end{bmatrix} \cdot \begin{bmatrix} b_0 \\ b_1 \end{bmatrix}$$

or $\mathbf{e} = \mathbf{y} - \mathbf{Xb}$. Thus, our model becomes

$$\text{Minimize } z = \mathbf{e}'\mathbf{e} = (\mathbf{y} - \mathbf{Xb})'(\mathbf{y} - \mathbf{Xb})$$

but

$$\text{Minimize } z = \mathbf{e}'\mathbf{e} = (\mathbf{y}' - \mathbf{b}'\mathbf{X}')(\mathbf{y} - \mathbf{Xb})$$

$$\text{Minimize } z = \mathbf{e}'\mathbf{e} = \mathbf{y}'\mathbf{y} - 2\mathbf{y}'\mathbf{Xb} + \mathbf{b}'\mathbf{X}'\mathbf{Xb}$$

This objective function is a quadratic equation in terms of the variable $\mathbf{b}$.

The condition for a minimum is $\nabla_b\, z = \mathbf{0}$.

$$\nabla_b\, z = -2\mathbf{X}'\mathbf{y} + 2\mathbf{X}'\mathbf{Xb} = \mathbf{0}$$

or the critical point is

$$\mathbf{b} = (\mathbf{X}'\mathbf{X})^{-1}\mathbf{Xy}$$

If the inverse $(\mathbf{X}'\mathbf{X})^{-1}$ exists, estimates of b_0 and b_1 can be found.

The critical point $\mathbf{b}$ is the location of the minimum if the Hessian matrix $\mathbf{X'X}$ is positive definite. Since $\mathbf{X} = [\mathbf{i}\ \mathbf{x}]$, where $\mathbf{i} = [1\ 1 \cdots 1]$ and $\mathbf{x}' = [x_1\ x_2 \cdots x_n]$, the product $\mathbf{X'X}$ is

$$\mathbf{X'X} = \begin{bmatrix} \mathbf{i} \\ \mathbf{x} \end{bmatrix} [\mathbf{i}\quad \mathbf{x}] = \begin{bmatrix} n & n\bar{x} \\ n\bar{x} & \mathbf{x'x} \end{bmatrix}$$

where $\mathbf{i'i} = n$ and $\mathbf{i'x} = \mathbf{x'i} = n\bar{x}$. Since the main diagonal terms are positive, $n > 0$ and $\mathbf{x'x}$, the necessary condition for a minimum is satisfied. The sufficient condition for a positive definite matrix is when the determinate of the principal minors of $\mathbf{X'X}$ are positive.

$$n > 0; \qquad \begin{vmatrix} n & n\bar{x} \\ n\bar{x} & \mathbf{x'x} \end{vmatrix} = n\mathbf{x'x} - n^2\bar{x}^2 \geq 0$$

However, the variance of x is defined as $s_x^2 = [1/(n-1)]\ (\mathbf{x'x} - n\bar{x}^2)$; thus, $n\mathbf{x'x} - n^2\bar{x}^2 = n(\mathbf{x'x} - n\bar{x}^2) = n(n-1)s_x^2$. Since s_x^2 is positive, the sufficient condition for a minimum is satisfied. The minimum is located at $\mathbf{b}$:

$$\mathbf{b} = (\mathbf{X'X})^{-1}\mathbf{Xy}$$

or

$$\begin{bmatrix} b_0 \\ b_1 \end{bmatrix} = \begin{bmatrix} n & n\bar{x} \\ n\bar{x} & \mathbf{x'x} \end{bmatrix}^{-1} \begin{bmatrix} \mathbf{i} \\ \mathbf{x}' \end{bmatrix} \cdot \mathbf{y}$$

$$\begin{bmatrix} b_0 \\ b_1 \end{bmatrix} = \frac{1}{n(\mathbf{x'x} - n\bar{x}^2)} \begin{bmatrix} \mathbf{x'x} & -n\bar{x} \\ -n\mathbf{x} & n \end{bmatrix} \cdot \begin{bmatrix} n\bar{y} \\ \mathbf{x'y} \end{bmatrix}$$

$$b_0 = \frac{\bar{y}\mathbf{x'x} - \bar{x}\mathbf{x'y}}{\mathbf{x'x} - n\bar{x}^2}$$

$$b_1 = \frac{\mathbf{x'y} - n\bar{x}\bar{y}}{\mathbf{x'x} - n\bar{x}^2}$$

Since $s_x^2 = [1/(n-1)](\mathbf{x'x} - n\bar{x}^2)$ and $s_{xy} = [1/(n-1)](\mathbf{x'y} - n\bar{x}\bar{y})$, the slope b_1 term may be written in terms of elementary statistics:

$$b_1 = \frac{s_{xy}}{s_x^2}$$

Furthermore, since $r_{xy} = s_{xy}/s_x s_y$, $b_1 = r_{xy} s_y / s_x$.

The sum of b_0 and $b_1 x$ is

$$b_0 + b_1\bar{x} = \frac{\bar{y}\mathbf{x'x} - n\bar{x}^2\bar{y}}{\mathbf{x'x} - n\bar{x}^2} + \frac{\bar{x}\mathbf{x'y} - n\bar{x}^2\bar{y}}{\mathbf{x'x} - n\bar{x}^2}$$

or

$$b_0 + b_1\bar{x} = \frac{\bar{y}(\mathbf{x'x} - n\bar{x}^2)}{(\mathbf{x'x} - n\bar{x}^2)} = \bar{y}$$

or

$$\bar{y} = b_0 + b_1\bar{x}$$

Thus, the estimates b_0 and b_1 are obtained from fundamental statistics and the relationships $\bar{y} = b_0 + b_1\bar{x}$ and $b_1 = s_{xy}/s_x^2$.

The *variance of error* about the regression line is defined as

$$s_e^2 = \frac{SSE}{(n-2)} = \frac{\mathbf{e}'\mathbf{e}}{(n-2)}$$

This is the estimate of the variance of the true model σ_ε^2, $Y = \beta_0 + \beta_1 x + \varepsilon$. Thus, the line of best fit $y = b_0 + b_1 x$ is the line that minimizes the variance of the residual error. The magnitude of s_e^2 has an important bearing upon the confidence that can be placed on a prediction, the smaller the magnitude of s_e^2, the greater the confidence in the prediction $\hat{y}$.

Predictions

The prediction of y for a given value of x is

$$\hat{y}_m = b_0 + b_1 x_m$$

where x_m is assumed to be a given value of x. Let us rewrite this expression in matrix form.

$$\hat{y}_m = [1 \quad x_m] \cdot \begin{bmatrix} b_0 \\ b_1 \end{bmatrix}$$

or

$$\hat{y}_m = \mathbf{x}'_m \mathbf{b}$$

where

$$\mathbf{x}'_m = [1 \quad x_m]$$

If ε is normally distributed, then the estimates b_0 and b_1 are normally distributed random variables. From estimation theory, it can be shown that a sum of a linear combination of normally and independently distributed random variables is a normal distribution. With this theorem, the variance of prediction $V(\hat{y}_m)$ may be estimated. It is

$$V(\hat{y}_m) = \mathbf{x}'_m(\mathbf{X}'\mathbf{X})^{-1}\mathbf{x}_m s_e^2$$

where s_e^2 is the residual square error. Expanding this term, the estimated variance $V(\hat{y}_m)$ is written as

$$V(\hat{y}_m) = s_e^2\left[\frac{1}{n^2} + \frac{(x_m - \bar{x})^2}{\sum_i (x_i - \bar{x})^2}\right]$$

From this expression, we see that the estimated variance can be less than the residual square error s_e^2. The confidence interval illustrated in Figure 10.13.

From this equation it is apparent that the data-gathering effort plays an important role in making reliable predictions. The size of the sample, that is, the number of observations n, and the range of x are important considerations. The estimated variance $V(\hat{y}_m)$ is reduced as the sample size n, and the range of observations of $\mathbf{x}$, is increased. In addition, the minimum variance for $\hat{y}$ occurs at the mean, that is, $x_m = \bar{x}$.

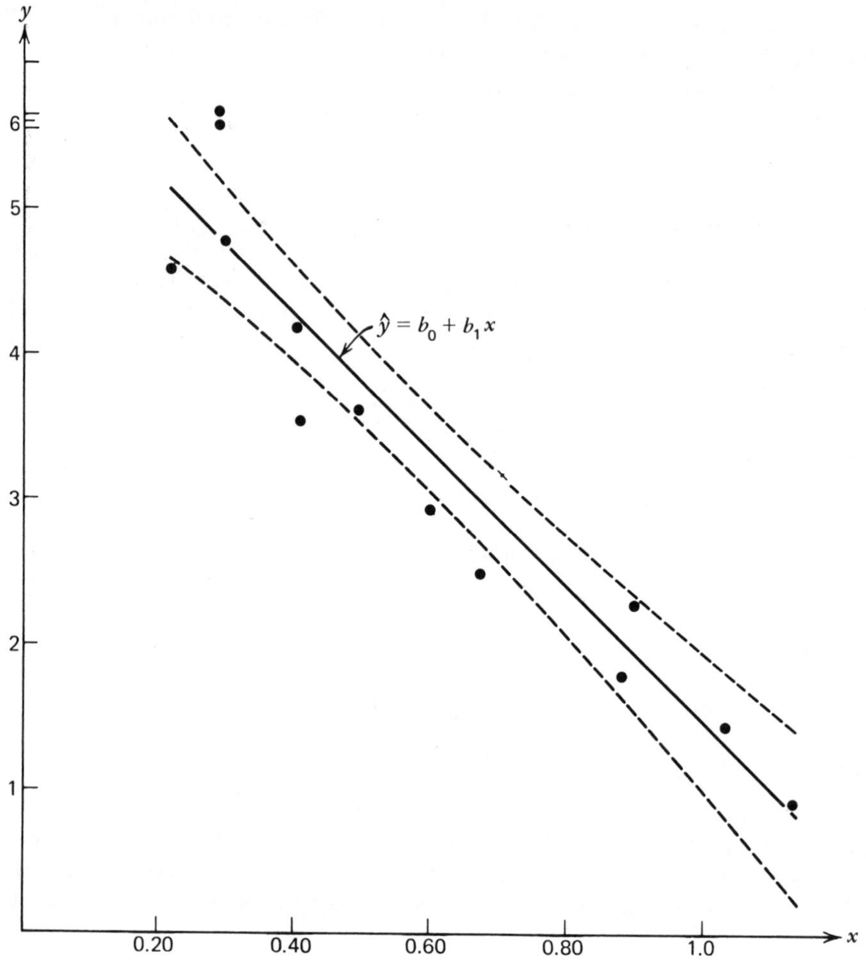

Figure 10.13 The confidence interval for $\hat{y} = b_0 + b_1 x$.

Multiple Regression

The parameters of a multiple linear regression $y = b_0 + b_1x_1 + b_2x_2 + \cdots + b_nx_n$ are also estimated by the method of least squares. The technique of the method of least squares is an extension of the method described for simple regression equations. The true model consisting of k linearly independent explanatory variables is

$$Y = \beta_0 + \beta_1x_1 + \beta_2x_2 + \cdots + \beta_kx_k + \varepsilon$$

All the properties of a simple linear regression are applicable to the multiple regression. Our discussion focuses on model formulation of the method of least squares and prediction.

All explanatory variables $x_1, x_2, \ldots, x_k$ that are expected to explain the outcome of Y are identified. This identification is not a trivial exercise of arbitrarily selecting variables that, by chance, may be correlated. This formulation requires a thorough understanding of how the process works. Each explanatory variable should be selected to reflect a principle of engineering, economics, or physical science. Once the explanatory variables have been selected and the data gathered, the strength of the linear relationship between y and each explanatory variable x_i may be determined by use of the correlation coefficient $r_{x_i y}$, $i = 1, 2, \ldots, k$, and the scattergrams between y and each x_i. These steps are identical to the one used for simple linear regression. If nonlinearities exist between any y and x_i pair, appropriate linear transformations may be applied.

In the model formulation stage, if any two explanatory variables x_i and x_j exhibit a strong linear relationship, *only one* of them is introduced into the prediction equation. If two explanatory are highly correlated, it is possible the estimated parameters of $b_0, b_1, \ldots, b_k$ may be distorted and may, in turn, lead to misleading predictions.

The method of least squares for multiple linear regression utilizes the same concept as utilized for simple linear regression. The objective will be to minimize the sum of the square error SSE.

$$\text{Minimize } z = SSE = \mathbf{e}'\mathbf{e} = e_1^2 + e_2^2 + \cdots + e_n^2$$

where $\mathbf{e} = \mathbf{y} - \hat{\mathbf{y}} = y - b_0 - b_1\mathbf{x}_1 - b_2\mathbf{x}_2 - \cdots - b_k\mathbf{x}_k$, or

$$\begin{bmatrix} e_1 \\ e_2 \\ \vdots \\ e_n \end{bmatrix} = \begin{bmatrix} y_1 - (b_0 + b_1x_{11} + b_2x_{12} + \cdots + b_kx_k) \\ y_2 - (b_0 + b_1x_{12} + b_2x_{22} + \cdots + b_kx_{k2}) \\ \vdots \\ y_n - (b_0 + b_1x_{1n} + b_2x_{2n} + \cdots + b_kx_{kn}) \end{bmatrix}$$

We may rewrite the expression in matrix form as

$$\mathbf{e} = \mathbf{y} - \mathbf{Xb}$$

where

$$\mathbf{y} = \begin{bmatrix} y_1 \\ y_2 \\ \vdots \\ y_n \end{bmatrix} \qquad \mathbf{X} = \begin{bmatrix} 1 & x_{11} & x_{21} & \cdots & x_{k1} \\ 1 & x_{12} & x_{22} & \cdots & x_{k2} \\ & & \vdots & & \\ 1 & x_{1n} & x_{2n} & \cdots & x_{kn} \end{bmatrix} \qquad \mathbf{b} = \begin{bmatrix} b_0 \\ b_1 \\ \vdots \\ b_k \end{bmatrix}$$

The $\mathbf{X}$ matrix may be represented as a set of column vectors;

$$\mathbf{X} = [\mathbf{i} \quad \mathbf{x}_1 \quad \mathbf{x}_2 \quad \cdots \quad \mathbf{x}_n].$$

The mathematical model becomes

$$\text{Minimize } z = SSE = \mathbf{y}'\mathbf{y} - 2\mathbf{y}'\mathbf{Xb} + \mathbf{b}'\mathbf{X}'\mathbf{Xb}$$

The condition for a minimum is satisfied when

$$\nabla_{\mathbf{b}} z = -2\mathbf{X}'\mathbf{y} + 2\mathbf{X}'\mathbf{Xb} = \mathbf{0}$$

or

$$\mathbf{b} = (\mathbf{X}'\mathbf{X})^{-1}\mathbf{Xy}$$

This is the same result obtained from simple linear regression. The major difference between simple and multiple linear regression is in the definition of the **X** matrix. As before, it can be shown that the Hessian matrix is positive definite and z or SSE is a minimum.

The prediction of y for given set of x values is

$$\hat{y}_m = b_0 + b_1 x_{1m} + b_2 x_{2m} + \cdots + b_k x_{km}$$

or, in matrix notation $\hat{y}_m = \mathbf{x}_m \mathbf{b}$, where $\mathbf{b}' = [b_0 \quad b_1 \quad b_2 \quad \cdots \quad b_m]$ and

$$\mathbf{x}_m = [1 \quad x_{1m} \quad x_{2m} \quad \cdots \quad x_{km}].$$

The variance of prediction $V(\hat{y}_m)$ is estimated with

$$V(\hat{y}_m) = \mathbf{x}_m'(\mathbf{X}'\mathbf{X})^{-1}\mathbf{x}_m s_e^2$$

where

$$s_e^2 = \frac{SSE}{n-k-1} = \frac{\mathbf{e}'\mathbf{e}}{n-k-1}$$

and e is the difference between the observed and predicted values of y, $e_i = y_i - \hat{y}_i$. Clearly, the variance of prediction is an extension of the simple regression case.

The *coefficient of determination* R^2 may be used to measure the significance of the regression equation in explaining the variance of y. R^2 is defined to be equal to the ratio of the variation of $\hat{y}$ (explained by the prediction equation) about the mean $\bar{y}$ to the variation of y (the unexplained or original observations of y) about the mean $\bar{y}$.

$$R^2 = \frac{SSR}{SSY}$$

where

$$SSR = \hat{\mathbf{y}}'\hat{\mathbf{y}} - n\bar{y}^2$$

$$SSY = \mathbf{y}'\mathbf{y} - n\bar{y}^2$$

For simple linear regression $R^2 = r_{xy}^2$, the coefficient of determination equals the correlation coefficient.

EXAMPLE 10.5 Traffic Flow Model: Method of Least Squares

(a) Utilize the method of least squares to estimate the parameters of free mean speed and jam density, or equivalently, the estimates of b_0 and b_1 of $y = b_0 + b_1 x$ as discussed in Example 10.4

(b) Utilize the following relationships:

$$\bar{y} = b_0 + b_1 \bar{x}$$

and

$$b_1 = \frac{s_{xy}}{s_x^2}$$

to estimate the parameters of b_0 and b_1.

(c) Determine the variance of prediction $V(\hat{u}_m)$ for $k_m = 50$ vehicles per mile.

Solution

(a) The parameters of the intercept b_0 and slope b_1 of $y = b_0 + b_1 x$ may be estimated by minimizing the sum of the square error, $e = y - \bar{y} = y - b_0 - b_1 x$, or

$$\text{Minimize } z = \sum_i e_i = \mathbf{e}'\mathbf{e}$$

where b_0 and b_1 are assumed to be unknowns. The critical point, as described in the text, is

$$\mathbf{b} = (\mathbf{X}'\mathbf{X})^{-1}\mathbf{X}\mathbf{y}$$

where $\mathbf{b}' = [b_0 \quad b_1]$ and the $\mathbf{b}$ and $\mathbf{X}$ matrices are

$$\mathbf{y} = \begin{bmatrix} 46.0 \\ 56.0 \\ 48.0 \\ 42.0 \\ 22.9 \\ 56.0 \\ 9.2 \\ 17.9 \\ 14.5 \\ 29.5 \\ 25.0 \\ 32.0 \\ 35.5 \\ 36.5 \end{bmatrix} \qquad \mathbf{X} = [\mathbf{i} \quad \mathbf{x}] = \begin{bmatrix} 1 & 22.2 \\ 1 & 29.3 \\ 1 & 30.1 \\ 1 & 40.7 \\ 1 & 90.0 \\ 1 & 29.2 \\ 1 & 113.3 \\ 1 & 88.2 \\ 1 & 103.1 \\ 1 & 60.2 \\ 1 & 67.4 \\ 1 & 70.6 \\ 1 & 41.1 \\ 1 & 49.6 \end{bmatrix}$$

Performing the necessary calculation, the products $\mathbf{X}'\mathbf{X}$ and $\mathbf{X}'\mathbf{y}$ are

$$\mathbf{X}'\mathbf{X} = \begin{bmatrix} n & n\bar{x} \\ n\bar{x} & \mathbf{x}\mathbf{x} \end{bmatrix} = \begin{bmatrix} 14 & 834.9 \\ 834.9 & 61390 \end{bmatrix}$$

$$\mathbf{X}'\mathbf{y} = \begin{bmatrix} \mathbf{i} \\ \mathbf{x} \end{bmatrix} \mathbf{y} = \begin{bmatrix} 471.6 \\ 22635 \end{bmatrix}$$

Thus

$$\mathbf{b} = (\mathbf{X}'\mathbf{X})^{-1}\mathbf{X}\mathbf{y}$$

$$\mathbf{b} = \begin{bmatrix} b_0 \\ b_1 \end{bmatrix} = \begin{bmatrix} 14 & 834.9 \\ 834.9 & 61390 \end{bmatrix}^{-1} \cdot \begin{bmatrix} 471.6 \\ 22635 \end{bmatrix}$$

$$= \begin{bmatrix} 0.378 & -0.0051 \\ -0.0051 & 8.62 \times 10^{-5} \end{bmatrix} \cdot \begin{bmatrix} 471.6 \\ 22634 \end{bmatrix} = \begin{bmatrix} 61.9 \\ -0.473 \end{bmatrix}$$

The estimates of free mean speed and jam density are

$$b_0 = u_f \quad \text{or} \quad u_f = 61.9 \text{ miles/hr}$$

$$b_1 = \frac{-u_f}{k_j} \quad \text{or} \quad k_j = \frac{-u_f}{b} = \frac{-61.9}{-0.473} = 131 \text{ vehicles/mile}$$

The linear regression model is $y = 61.9 - 0.473x$ or $u = 61.9 - 0.473k$. Equivalently, the model in terms of u and k is

$$u = 61.9\left[1 - \frac{k}{131}\right]$$

(b) For linear regression equations of the form $y = b_0 + b_1 x$, the parameters of b_0 and b_1 may be estimated with the following relationships:

$$b_1 = \frac{s_{xy}}{s_x^2} = \frac{-422.7}{(29.9)^2} = -0.473$$

As should be expected, $b_0 = \bar{y} - b_1\bar{x} = 33.7 - (-0.473)(59.6) = 61.9$. Thus,

$$y = 61.9 - 0.473x \quad \text{or} \quad u = 61.9 - 0.473k.$$

(c) The estimates of speed and residual error may be made with the following relationships:

$$\hat{\mathbf{u}} = \hat{\mathbf{y}} = 61.9 - 0.473k$$

$$\mathbf{e} = \mathbf{u} - \hat{\mathbf{u}} = \mathbf{y} - \hat{\mathbf{y}}$$

where

$$\mathbf{k} = \mathbf{x}$$

The speed estimates $\hat{\mathbf{u}}$ and $\mathbf{e}$ are

$$\hat{\mathbf{u}} = \begin{bmatrix} 51.4 \\ 48.0 \\ 47.7 \\ 42.6 \\ 19.3 \\ 48.1 \\ 8.34 \\ 20.2 \\ 13.1 \\ 33.4 \\ 30.0 \\ 28.5 \\ 42.5 \\ 38.4 \end{bmatrix} \qquad \mathbf{e} = \begin{bmatrix} 5.40 \\ -7.96 \\ -0.34 \\ 0.65 \\ -3.58 \\ -8.51 \\ -0.86 \\ 2.27 \\ -1.67 \\ 3.92 \\ 5.01 \\ -3.50 \\ 6.96 \\ 1.93 \end{bmatrix}$$

(d) The estimate of speed given, $k = 50$ vehicles per mile, is

$$\hat{u} = 61.9 - 0.473(50) = 38.3 \text{ miles/hr}$$

The estimate of variance is

$$V(\hat{y}) = V(\hat{u}) = \mathbf{x}_m'(\mathbf{X}'\mathbf{X})^{-1}\mathbf{x}_m\mathbf{s}_e^2$$

where $\mathbf{x}_m' = [1 \quad 50]$. Thus

$$V(\hat{u}) = [1 \quad 50] \cdot \begin{bmatrix} 0.378 & -8.0051 \\ -0.0051 & 8.62 \times 10^{-5} \end{bmatrix} \cdot \begin{bmatrix} 1 \\ 50 \end{bmatrix} (24.3) = 0.170$$

EXAMPLE 10.6 A Cost Model

A survey of the construction cost of wastewater clarifiers as a function of design flow has been tabulated. The data are

CONSTRUCTION COST, C (M$)	DESIGN FLOW, q (Mgal/day)
0.16	1.0
0.15	2.0
0.60	5.0
0.75	4.0
0.28	3.0
0.28	2.0
0.55	4.0
0.70	7.0

The scattergram in Figure 10.14 shows that a linear relationship, $C = cq$, or equivalently in terms of y and x, $y = b_1 x$, will be appropriate. The intercept is assumed equal to zero, thus $b_0 = 0$, in the model, $y = b_0 + b_1 x$.

Utilize the method of least squares to estimate the unit cost parameter c.

Solution

The sum of the square error SSE is

$$SSE = \mathbf{e}'\mathbf{e}$$

where $\mathbf{e} = \mathbf{y} - \hat{\mathbf{y}} = \mathbf{y} - \mathbf{b}\mathbf{x}$. For consistency notation, let $\boldsymbol{x} = \boldsymbol{x}$ even though $\boldsymbol{x}$ is a vector. The mathematical model to minimize SSE is

$$\text{Minimize } z = \mathbf{e}'\mathbf{e} = (\mathbf{y} - \mathbf{b}\boldsymbol{x})'(\mathbf{y} - \mathbf{b}\boldsymbol{x})$$

This model is identical to the one for the simple linear regression equation except $\mathbf{b} = [b_1]$ and $\mathbf{X} = \mathbf{x}$. The condition for a minimum z is

$$\nabla z = \mathbf{0}: \quad \mathbf{b} = (\mathbf{X}'\mathbf{X})^{-1}\boldsymbol{xy}$$

or simply,

$$\mathbf{b} = [b_1] = (\mathbf{x}'\mathbf{x})^{-1}\mathbf{xy} = \frac{\sum x_i y_i}{\sum x_i^2}$$

where

$$\mathbf{y} = \begin{bmatrix} 0.16 \\ 0.15 \\ 0.60 \\ 0.75 \\ 0.28 \\ 0.55 \\ 0.70 \end{bmatrix} \qquad \mathbf{x} = \begin{bmatrix} 1.0 \\ 2.0 \\ 5.0 \\ 4.0 \\ 3.0 \\ 2.0 \\ 4.0 \\ 7.0 \end{bmatrix}$$

The estimate of b_1 or the unit cost c is

$$b_1 = c = (124)^{-1} \cdot 12.3 = \frac{12.3}{124} = 0.10/\text{gal/day}$$

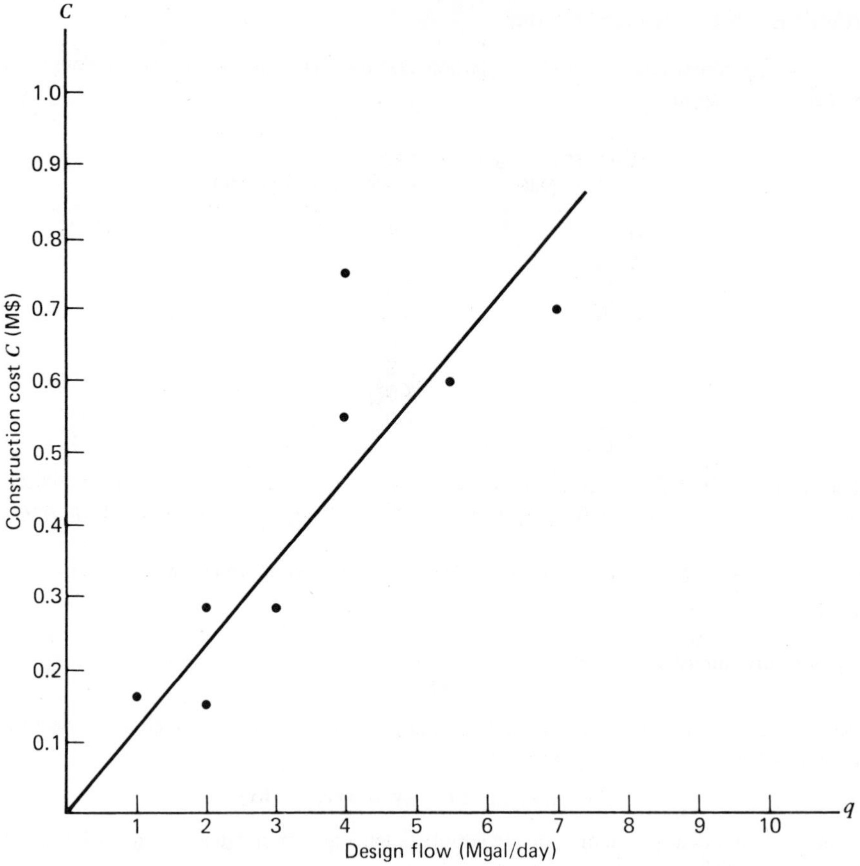

Figure 10.14

The cost model is

$$C = 0.10q$$

EXAMPLE 10.7 A Water Quality Model

One of the principal concerns of a water-treatment plant operator is the removal of color compounds from the water. The color is caused by a number of organic acid groups. Fulvic acid, a soluble material, turbidity, and insoluble materials are suspected to be the primary color-causing compounds. Three hundred and fifty-two samples were drawn from a river over a one-year period. A mathematical model to predict the overall water color quality is sought. The model variables are

y = Fulvic acid concentration in milligrams per liter.

x_1 = Absorbance, measured with a spectrophotometer at 350 nm.

x_2 = Turbidity concentration, FTU.

x_3 = Nessler tube color index, CU.

Absorbance and color measured with a Nessler tube are two different methods of measuring color. The spectrophotometer is considered to be a more precise measure because it uses a highly sensitive electrical device to measure the light passing through the sample. With a Nessler tube, a set of color standards is prepared and the water sample is matched to one of them. The following correlation coefficients were determined.

	y	x_1	x_2	x_3
y	1.00	0.93	0.44	0.90
x_1	—	1.00	0.091	0.96
x_2	—	—	1.00	0.081
x_3	—	—	—	1.00

Decide which explanatory variables should enter a multiple linear regression equation.

Solution

The correlation coefficient of x_1 and x_3 indicates that a strong linear relationship exists between them. A predictive equation relating x_1 and x_3 may be estimated. The models $x_3 = a_0 + a_1x_1$ or $x_1 = c_0 + c_1x_3$ may be determined by the method of least squares. Multiple regression equations may be distorted if they are both introduced. Since x_1 and x_3 are alternative methods of measuring the concentration of fulvic acid, only *one* of them is needed in the regression equation. Since the spectrophotometer is a more precise measuring device, the multiple regression model will be of the form:

$$y = b_0 + b_1x_1 + b_2x_2$$

Remarks

The correlation between fulvic acid and turbidity is $r_{yx} = 0.44$, showing a rather weak linear dependency between the two variables. The correlation between absorbance and fulvic acid, $r_{yx_1} = 0.93$, is much stronger. A simple linear regression model has been determined:

$$\hat{y} = 0.0280 + 0.0084x_1$$

Since $R^2 = r_{yx_1}^2$, or $R^2 = 0.86 = 86$ percent, it explains 86 percent of the unexplained variance of y. Although the linear relationship between absorbance and turbidity is weak, when it is introduced into the multiple linear regression equation,

$$\hat{y} = 0.0057 + 0.0081x_1 + 0.0075x_2$$

the combination of both x_1 and x_2 helps explain almost all the unexplained variance of absorbance, $R^2 = 98.7$ percent.

PROBLEMS

Problem 1

Utilize the method of least squares to establish mathematical relationships to estimate b_1 and b_2 of the multiple regression model.

$$y = b_1x + b_2x^2$$

Problem 2

The $\mathbf{X'X}$ matrix for the linear model, $\hat{y} = 14.0 - 0.08x$, is

$$\mathbf{X'X} = \begin{vmatrix} 25 & 1315 \\ 1315 & 76323.4 \end{vmatrix}$$

(a) Estimate the mean of x.
(b) Estimate the variance of x.
(c) Determine the prediction $\hat{y}_m$ and the variance of prediction for $x = 16$.

Problems 3 through 6

Refer to Problems 1 through 4 in Section 10.1 to answer Problems 3 through 6 below. Before a following problem can be attempted, the respective problem in Section 10.1 must be solved first.

Problem 3

(a) Use the method of least squares to determine the parameters of the model found in Problem 1, Section 10.1, part c.
(b) Predict the shear stress $\hat{y}_m$ and prediction variance for an effective stress of 100 kn/m^2.

Problem 4

(a) Use the method of least squares to determine the parameters of the model found in Problem 2, Section 10.1, part c.
(b) Compare the prediction variances for $x_m = 5$ and $x_m = 7$. Why is $V(\hat{y}_m)$ for $x = 7$ larger than $V(\hat{y}_m)$ for $x = 5$?

Problem 5

(a) Use the method of least squares to determine the parameters of the model found in Problem 3, Section 10.1, part c.
(b) Determine the prediction variance for a discharge of 10 Mgal/day.

Problem 6

(a) Use the method of least squares to determine the parameters of the model found in Problem 4, Section 10.1, part c.
(b) Predict the flow $\hat{y}_m$ and $V(\hat{y}_m)$ for a density of 55 vehicles/mile.

Problem 7

In Example 10.3, two points were used to estimate the slope N_0 and the reaction rate k of the model $N = N_0 e^{-kt}$. It was pointed out that the estimate of k and N_0 may be determined by the method of least squares; thus all observations are utilized. Here, since the initial concentration is known with confidence, assume that $N_0 = 2.3$ mg/l is known. Use the method of least squares to estimate the single parameter k of the model:

$$N = 2.3e^{-kt}$$

(a) Transform the model to a linear form and formulate a mathematical model to minimize the sum of the square error.
(b) Show that the estimate of $\mathbf{b}$ has the general form:

$$\mathbf{b} = (\mathbf{X}^1\mathbf{X})^{-1}\mathbf{X}\mathbf{y}$$

Appropriately define all vectors and matrices of this equation.

(c) Estimate the value of k.
(d) Predict the value of N for $t = 120$ sec. Determine the prediction variance at this point.

Problem 8

A transformation of

$$y = b_0 + b_1 e^{b_2 x}$$

to the simple regression form is not possible.

(a) Use the concepts of the method of least squares to establish an objective function to minimize the sum of square error SSE. Assume data gathering has been performed and an appropriate data set of y and x are available.
(b) What is the governing set of equations that must be solved to find b_0, b_1, and b_2.

Summary

Simple and multiple linear regression analysis, as described in this chapter, consists of three steps: data gathering, model formulation, and parameter estimation.

The scattergram and correlation coefficient were used to determine the linear dependency between dependent and explanatory variables and among explanatory variables. We sought to find strong correlations between the dependent and explanatory variables. For multiple regression, when a strong linear regression between a pair of explanatory variables was found, only one explanatory variable was entered into the model.

For model parameter estimation, the method of least squares was employed for both simple and multiple regression. The residuals or the difference between the observed and predicted values of the dependent variables were used to test the normal distribution assumptions of the prediction equation. Again, the same tests apply to both simple and multiple regression models.

By examining the residuals $\mathbf{e} = \mathbf{y} - \hat{\mathbf{y}}$ and performing hypothesis testing, the validity of a regression equation may be tested. These tests will give assurance that the regression equation and each estimated model parameter are statistically significant. This treatment is beyond the scope of our development and was not considered. For many engineering applications, the material presented here is sufficient.

Bibliography

Jack R. Benjamin and C. Allen Cornell. *Probability, Statistics and Decision for Civil Engineers.* McGraw-Hill, New York. 1970.

William H. Beyer, editor. *Handbook of Tables for Probability and Statistics.* The Chemical Rubber Co. 1966.

Richard de Neufville and Joseph Stafford. *Systems Analysis for Engineers and Managers.* McGraw-Hill, New York. 1971.

Norman Draper and Henry Smith. *Applied Regression Analysis.* John Wiley, New York. 1966.

Jan Kinenta. *Elements of Econometrics.* Macmillan, New York. 1971.

APPENDIX A

Matrix Methods

Matrix notation offers a convenient and compact way to represent large arrays of numbers. A matrix is always a rectangular array of numbers. A matrix will be designated with a bold italic upper case letter. The following rectangular array is simply designated as $\boldsymbol{A}$:

$$\boldsymbol{A} = \begin{bmatrix} a_{11} & a_{12} & \cdots & a_{1j} & \cdots & a_{1n} \\ a_{21} & a_{22} & \cdots & a_{2j} & \cdots & a_{2n} \\ \vdots & & & & & \\ a_{i1} & a_{i2} & \cdots & a_{ij} & \cdots & a_{in} \\ \vdots & & & & & \\ a_{m1} & a_{m2} & \cdots & a_{mj} & \cdots & a_{mn} \end{bmatrix}$$

The *order* or size of a matrix is designated by its number of rows m and its number of columns n. The order of matrix $\boldsymbol{A}$ is $m \times n$. A matrix with an equal number of rows and columns has order $n \times n$ and is called a *square* matrix. For example, the following matrix is a square 3×3 matrix.

$$\boldsymbol{A} = \begin{bmatrix} 1 & 2 & 3 \\ 4 & 5 & 6 \\ 7 & 8 & 9 \end{bmatrix}$$

The *main diagonal* of a square matrix is all the elements from the top left corner to the bottom right corner of the matrix. The main diagonal of $\boldsymbol{A}$ is

$$\boldsymbol{A} = \begin{bmatrix} \textcircled{1} & 2 & 3 \\ 4 & \textcircled{5} & 6 \\ 7 & 8 & \textcircled{9} \end{bmatrix}$$

All remaining elements are called the *off-diagonal* terms.

An *identity matrix* $\boldsymbol{I}$ is a special square matrix in which all main diagonal terms are equal to unity and all off-diagonal terms are equal to zero. A 4×4 identity matrix is

$$\boldsymbol{I} = \begin{bmatrix} 1 & 0 & 0 & 0 \\ 0 & 1 & 0 & 0 \\ 0 & 0 & 1 & 0 \\ 0 & 0 & 0 & 1 \end{bmatrix}$$

A *vector* is a special kind of matrix consisting of a single column of numbers. A vector is an $n \times 1$ matrix, more commonly called an n-vector. A vector will be designated with a lower case bold italic letter. Vector notation is used to designate a set of control variables, $x_1, x_2, \ldots, x_n$, as

$$\boldsymbol{x} = \begin{bmatrix} x_1 \\ x_2 \\ \vdots \\ x_n \end{bmatrix}$$

A set of unit costs, $c_1, c_{2_1}, \ldots, c_n$, may be represented with vector notation as $\boldsymbol{c}$, where

$$\boldsymbol{c} = \begin{bmatrix} c_1 \\ c_2 \\ \vdots \\ c_n \end{bmatrix}$$

ARITHMETIC OPERATIONS

To add or subtract matrices $\boldsymbol{A}$ and $\boldsymbol{B}$, they must be of the same order, $m \times n$. The addition of two matrices is the sum $\boldsymbol{C}$

$$\boldsymbol{C} = \boldsymbol{A} + \boldsymbol{B}$$

it is achieved by adding the corresponding matrix elements of $\boldsymbol{A}$ and $\boldsymbol{B}$.

$$c_{ij} = a_{ij} + b_{ij}$$

The order of addition does not affect the result, thus $\boldsymbol{A} + \boldsymbol{B} = \boldsymbol{B} + \boldsymbol{A}$. This is known as the commutative law of matrix addition. Subtraction of two matrices is similar to addition. It is done by subtracting corresponding elements:

$$\boldsymbol{C} = \boldsymbol{A} - \boldsymbol{B}$$

where

$$c_{ij} = a_{ij} - b_{ij}$$

In the *multiplication* of two matrices, $\boldsymbol{C} = \boldsymbol{AB}$, the matrices $\boldsymbol{A}$ and $\boldsymbol{B}$ must have common order. The number of columns of $\boldsymbol{A}$ must be equal to the number of rows of $\boldsymbol{B}$. If matrices $\boldsymbol{A}$ and $\boldsymbol{B}$ must have order $m \times n$ and $n \times p$, then the product matrix $\boldsymbol{C}$ has order $m \times p$:

$$\boldsymbol{C}_{(m \times p)} = \boldsymbol{A}_{(m \times n)} \boldsymbol{B}_{(n \times p)}$$

where the numbers in parentheses are the orders of the matrices. Generally, the product of two matrices is written as

$$\boldsymbol{C} = \boldsymbol{AB}$$

The elements of $\boldsymbol{C}$ are equal to

$$c_{ij} = a_{i1}b_{1j} + a_{i2}b_{2j} + \cdots + b_{in}b_{nj} = \sum_{k=1}^{n} a_{ik}b_{kj}$$

In other words, each element of $\boldsymbol{C}$, c_{ij}, is obtained by summing the product each row element $\boldsymbol{A}$ multiplied by each corresponding column element of $\boldsymbol{B}$. For example, the product of $\boldsymbol{A}$ and $\boldsymbol{B}$, $\boldsymbol{C} = \boldsymbol{AB}$, is

$$\boldsymbol{C} = \begin{bmatrix} 2 & 3 \\ 4 & 2 \\ -1 & 1 \end{bmatrix} \cdot \begin{bmatrix} 2 & -5 & 1 & 7 \\ 1 & 0 & 6 & -1 \end{bmatrix}$$

Since $\boldsymbol{A}$ is a 3×2 matrix and $\boldsymbol{B}$ is a 2×4 matrix, the product $\boldsymbol{C}$ must have order 3×4.

$$\boldsymbol{C} = \begin{bmatrix} (2 \cdot 2 + 3 \cdot 1) & (2 \cdot -5 + 3 \cdot 0) & (2 \cdot 1 + 3 \cdot 6) & (2 \cdot 7 + 3 \cdot -1) \\ (4 \cdot 2 + 2 \cdot 1) & (4 \cdot -5 + 2 \cdot 0) & (4 \cdot 1 + 2 \cdot 6) & (4 \cdot 7 + 2 \cdot -1) \\ (-1 \cdot 2 + 1 \cdot 1) & (-1 \cdot 5 + 1 \cdot 0) & (-1 \cdot 1 + 1 \cdot 6) & (-1 \cdot 7 + 1 \cdot -1) \end{bmatrix}$$

or

$$\boldsymbol{C} = \begin{bmatrix} 7 & -10 & 20 & 11 \\ 10 & -20 & 16 & 26 \\ -1 & -5 & 5 & -8 \end{bmatrix}$$

A SYMMETRICAL MATRIX

A *symmetrical matrix* is a square matrix in which all off-diagonal terms are equal to one another. That is

$$a_{ij} = a_{ji}$$

For example, the matrix $\boldsymbol{E}$ is a symmetrical matrix:

$$\boldsymbol{E} = \begin{bmatrix} 1 & 2 & 6 \\ 2 & 2 & 0 \\ 6 & 0 & 5 \end{bmatrix}$$

VECTOR AND MATRIX TRANSPOSE

A *transpose* of a matrix or vector is the interchange of columns with rows. The transpose of $\boldsymbol{A}$ is written as $\boldsymbol{A}'$. For example, the transpose of $\boldsymbol{A}$ is

$$\boldsymbol{A}' = \begin{bmatrix} 1 & 4 & 7 \\ 2 & 5 & 8 \\ 7 & 8 & 9 \end{bmatrix}' = \begin{bmatrix} 1 & 2 & 7 \\ 4 & 5 & 8 \\ 7 & 8 & 9 \end{bmatrix}$$

The transpose of the vectors $\boldsymbol{c}$ and $\boldsymbol{x}$ are written as

$$\boldsymbol{c}' = [c_1 \quad c_2 \quad \cdots \quad c_n]$$

and

$$\boldsymbol{x}' = [x_1 \quad x_2 \quad \cdots \quad x_n]$$

VECTOR PRODUCTS

The product of two vectors is a *scalar*. The vector may be written as

$$c = \boldsymbol{a}'\boldsymbol{b}$$

or, equivalently,

$$c = \boldsymbol{b}'\boldsymbol{a}$$

This result is a scalar. If $\boldsymbol{a}$ is an n-vector or $n \times 1$ matrix, then the transpose of $\boldsymbol{a}$ must be a $1 \times n$ row matrix. The product $c = \boldsymbol{a}'\boldsymbol{b}$ must be the product of a $1 \times n$ matrix by an $n \times 1$ matrix; therefore c contains only a single element. The scalar product is

$$c = \sum_{i=1}^{n} a_{1i}b_{i1} = a_{11}b_{11} + a_{12}b_{21} + \cdots + a_{1n}b_{n1}$$

For example, if $\boldsymbol{a}' = [3 \;\; 2 \;\; 1]$ and $\boldsymbol{b}' = [1 \;\; 0 \;\; 2]$, then

$$c = [3 \cdot 1 + 2 \cdot 0 + 1 \cdot 2] = 5$$

MATRIX PRODUCTS

In the foregoing, the products of two vectors $\boldsymbol{a}'\boldsymbol{b} = \boldsymbol{b}'\boldsymbol{a}$ are equal. The product of two square matrices is neither a scalar nor, generally, is it equal. That is,

$$\boldsymbol{AB} \neq \boldsymbol{BA}$$

This property can be illustrated by comparing the products of $\boldsymbol{AB}$ and $\boldsymbol{BA}$:

$$\boldsymbol{AB} = \begin{bmatrix} 2 & 0 \\ 0 & 3 \end{bmatrix} \cdot \begin{bmatrix} 1 & 2 \\ 3 & 4 \end{bmatrix} = \begin{bmatrix} 2 & 4 \\ 9 & 12 \end{bmatrix}$$

$$\boldsymbol{BA} = \begin{bmatrix} 1 & 2 \\ 3 & 4 \end{bmatrix} \cdot \begin{bmatrix} 2 & 0 \\ 0 & 3 \end{bmatrix} = \begin{bmatrix} 2 & 6 \\ 6 & 12 \end{bmatrix}$$

Obviously, the order of multiplication is very important in matrix algebra.

This example illustrates the effect of pre- and postmultiplying a diagonal matrix by a matrix also. Premultiplying $\boldsymbol{B}$ by the diagonal matrix $\boldsymbol{A}$ doubled the elements in the *first row* of $\boldsymbol{B}$ and tripled the elements in the *second row* of $\boldsymbol{B}$. Postmultiplying $\boldsymbol{B}$ by the diagonal matrix $\boldsymbol{A}$ doubled the element in the *first column* of $\boldsymbol{B}$ and tripled the element in the *second column* of $\boldsymbol{B}$.

QUADRATIC FORM

The pre- and postmultiplication of a matrix by vectors will result in a scalar product:

$$z = \boldsymbol{a}'\boldsymbol{C}\boldsymbol{b}$$

Using the rules of matrix multiplication, this equation may be written as

$$z \doteq \boldsymbol{a}'(\boldsymbol{C}\boldsymbol{b})$$

If matrix $\boldsymbol{C}$ is an $m \times n$ matrix, the vector $\boldsymbol{b}$ must be an $n \times 1$ matrix. The product $\boldsymbol{c} = (\boldsymbol{Cb})$ results in a vector or matrix of order $m \times 1$. Thus, the equation may be written as

$$z = \boldsymbol{a}'\boldsymbol{c}$$

The product of two vectors is a scalar.

When $\boldsymbol{C}$ is a square symmetrical $n \times n$ matrix and vectors $\boldsymbol{a}$ and $\boldsymbol{b}$ are equal to one another, $\boldsymbol{a} = \boldsymbol{b}$, a scalar product is formed:

$$z = \boldsymbol{a}'\boldsymbol{Ca}$$

If the equation is expanded, the product z may be written as

$$z = c_{11}a_1^2 + c_{22}a_2^2 + \cdots + c_{nn}a_n^2 + 2c_{12}a_1a_2 + 2c_{13}a_1a_3 + \cdots$$

The scalar product $z = \boldsymbol{a}'\boldsymbol{Ca}$ is called the *quadratic form.*

The quadratic form arises in many areas of mathematics and engineering. For instance, if a control vector $\boldsymbol{x}$ represents a set of measurement errors, $x_1, x_2, \ldots, x_n$, where the values of $\boldsymbol{x}$ can either be positive or negative, we may be interested in formulating a mathematical model to minimize the error. In order to weigh positive and negative errors on a equal basis, the objective will be to minimize the square errors. The mathematical model satisfying this objective is

$$\text{Minimize } z = \sum_i x_i^2 = x_1^2 + x_2^2 + \cdots + x_n^2$$

In matrix notation, the equation may be written as

$$\text{Minimize } z = \boldsymbol{x}'\boldsymbol{Ix} = \boldsymbol{x}'\boldsymbol{x}$$

where $\boldsymbol{I}$ is an $n \times n$ identity matrix.

COMMUTATIVE PROPERTIES

Contrast the important properties of matrix addition and multiplication. Matrix addition is commutative, and matrix multiplication is generally not commutative. That is,

$$\boldsymbol{A} + \boldsymbol{B} = \boldsymbol{B} + \boldsymbol{A}$$

$$\boldsymbol{AB} \neq \boldsymbol{BA}$$

The addition and multiplication of vectors is commutative in both cases:

$$\boldsymbol{a} + \boldsymbol{b} = \boldsymbol{b} + \boldsymbol{a}$$

$$\boldsymbol{a}'\boldsymbol{b} = \boldsymbol{b}'\boldsymbol{a}$$

Although matrices do not commute, vectors do; the rules of scalar algebra apply.

PARTITIONING

Matrix partitioning is a convenient way to analyze certain problems. Basically, matrix partitioning allows us to divide a vector or matrix into subvectors or submatrices, respectively. For example, the matrix $\boldsymbol{A}$ will be divided into four partition matrices, $\boldsymbol{C}$, $\boldsymbol{D}$, $\boldsymbol{E}$, and $\boldsymbol{F}$:

$$\boldsymbol{A} = \left[\begin{array}{ccc:c} a_{11} & a_{12} & a_{13} & a_{14} \\ a_{21} & a_{22} & a_{23} & a_{24} \\ \hdashline a_{31} & a_{32} & a_{33} & a_{34} \end{array}\right]$$

or

$$\boldsymbol{A} = \begin{bmatrix} \boldsymbol{C} & \boldsymbol{D} \\ \boldsymbol{E} & \boldsymbol{F} \end{bmatrix}$$

where

$$\boldsymbol{C} = \begin{bmatrix} a_{11} & a_{12} & a_{13} \\ a_{21} & a_{22} & a_{23} \end{bmatrix}$$

$$\boldsymbol{D} = \begin{bmatrix} a_{14} \\ a_{24} \end{bmatrix}$$

$$\boldsymbol{E} = [a_{31} \quad a_{32} \quad a_{33}]$$

$$\boldsymbol{F} = [a_{34}]$$

The rules of matrix algebra apply to partition matrices. In order to calculate the sum of $\boldsymbol{A}$ and $\boldsymbol{B}$, $\boldsymbol{A}$ and $\boldsymbol{B}$ must have the same order. Likewise, in the addition of $\boldsymbol{A}$ and $\boldsymbol{B}$ utilizing partition matrices, each corresponding partition matrix must have the same order. Thus, matrix $\boldsymbol{B}$ is partitioned in a manner that is similar to $\boldsymbol{A}$:

$$\boldsymbol{B} = \left[\begin{array}{ccc:c} b_{11} & b_{12} & b_{13} & b_{14} \\ b_{21} & b_{22} & b_{23} & b_{24} \\ \hdashline b_{31} & b_{32} & b_{33} & b_{34} \end{array}\right]$$

or

$$\boldsymbol{B} = \begin{bmatrix} \boldsymbol{G} & \boldsymbol{H} \\ \boldsymbol{J} & \boldsymbol{K} \end{bmatrix}$$

$$\boldsymbol{G} = \begin{bmatrix} b_{11} & b_{12} & b_{13} \\ b_{21} & b_{22} & b_{23} \end{bmatrix}$$

$$\boldsymbol{H} = \begin{bmatrix} b_{14} \\ b_{24} \end{bmatrix}$$

$$\boldsymbol{J} = [b_{31} \quad b_{32} \quad b_{33}]$$

$$\boldsymbol{K} = [b_{34}]$$

Since each matrix $\boldsymbol{A}$ and $\boldsymbol{B}$ and corresponding partition matrices have the proper order for addition we may calculate $\boldsymbol{A} + \boldsymbol{B}$ with the rules of matrix addition, or

$$\boldsymbol{A} + \boldsymbol{B} = \begin{bmatrix} (\boldsymbol{C} + \boldsymbol{G}) & (\boldsymbol{D} + \boldsymbol{H}) \\ (\boldsymbol{E} + \boldsymbol{J}) & (\boldsymbol{F} + \boldsymbol{K}) \end{bmatrix}$$

where

$$C + G = \begin{bmatrix} (a_{11} + b_{11}) & (a_{12} + b_{12}) & (a_{13} + b_{13}) \\ (a_{21} + b_{21}) & (a_{22} + b_{22}) & (a_{23} + b_{23}) \end{bmatrix}$$

$$D + H = \begin{bmatrix} (a_{14} + b_{14}) \\ (a_{24} + b_{24}) \end{bmatrix}$$

$$E + J = [(a_{31} + b_{31}) \quad (a_{32} + b_{32}) \quad (a_{33} + b_{33})]$$

$$F + K = [(a_{34} + b_{34})]$$

Multiplication in terms of partition matrices may be illustrated with the use of matrices $\boldsymbol{A}$ and $\boldsymbol{B}$ also.

$$\boldsymbol{AB'} = \begin{bmatrix} \boldsymbol{C} & \boldsymbol{D} \\ \boldsymbol{E} & \boldsymbol{F} \end{bmatrix} \cdot \begin{bmatrix} \boldsymbol{G} & \boldsymbol{H} \\ \boldsymbol{J} & \boldsymbol{K} \end{bmatrix}'$$

The transpose of $\boldsymbol{B}$ is

$$\boldsymbol{B'} = \begin{bmatrix} \boldsymbol{G'} & \boldsymbol{J'} \\ \boldsymbol{H'} & \boldsymbol{K'} \end{bmatrix}$$

Thus

$$\boldsymbol{AB'} = \begin{bmatrix} \boldsymbol{C} & \boldsymbol{D} \\ \boldsymbol{E} & \boldsymbol{F} \end{bmatrix} \cdot \begin{bmatrix} \boldsymbol{G'} & \boldsymbol{J'} \\ \boldsymbol{H'} & \boldsymbol{K'} \end{bmatrix}$$

and

$$\boldsymbol{AB'} = \begin{bmatrix} (\boldsymbol{CG'} + \boldsymbol{DH'}) & (\boldsymbol{CJ'} + \boldsymbol{DK'}) \\ (\boldsymbol{EG'} + \boldsymbol{FH'}) & (\boldsymbol{EJ'} + \boldsymbol{FK'}) \end{bmatrix}$$

The submatrices are

$(\boldsymbol{CG'} + \boldsymbol{DH'})$

$$= \begin{bmatrix} a_{11} & a_{12} & a_{13} \\ a_{21} & a_{22} & a_{23} \end{bmatrix} \cdot \begin{bmatrix} b_{11} & b_{21} \\ b_{12} & b_{22} \\ b_{13} & b_{23} \end{bmatrix} + \begin{bmatrix} a_{14} \\ a_{24} \end{bmatrix} \cdot [b_{14} \quad b_{24}]$$

$$= \begin{bmatrix} (a_{11}b_{11} + a_{12}b_{12} + a_{13}b_{13} + a_{14}b_{14}) & (a_{11}b_{21} + a_{12}b_{22} + a_{13}b_{23} + a_{13}b_{24}) \\ (a_{21}b_{11} + a_{22}b_{12} + a_{23}b_{13} + a_{24}b_{14}) & (a_{21}b_{21} + a_{22}b_{22} + a_{23}b_{23} + a_{24}b_{24}) \end{bmatrix}$$

$$(\boldsymbol{EG'} + \boldsymbol{FH'}) = [a_{31} \quad a_{32} \quad a_{33}] \cdot \begin{bmatrix} b_{11} & b_{21} \\ b_{12} & b_{22} \\ b_{13} & b_{23} \end{bmatrix} + [a_{34}] \cdot \begin{bmatrix} b_{14} \\ b_{24} \end{bmatrix}$$

$$= [(a_{31}b_{11} + a_{32}b_{12} + a_{33}b_{13} + a_{34}b_{14}) \quad (a_{31}b_{21} + a_{32}b_{22} + a_{33}b_{23} + a_{34}b_{24})]$$

$$(\boldsymbol{CJ'} + \boldsymbol{DK'}) = \begin{bmatrix} a_{11} & a_{12} & a_{13} \\ a_{21} & a_{22} & a_{23} \end{bmatrix} \cdot \begin{bmatrix} b_{31} \\ b_{32} \\ b_{33} \end{bmatrix} + \begin{bmatrix} a_{14} \\ a_{24} \end{bmatrix} \cdot [b_{34}]$$

$$= \begin{bmatrix} (a_{11}b_{31} + a_{12}b_{32} + a_{13}b_{33} + a_{14}b_{34}) \\ (a_{21}b_{31} + a_{22}b_{32} + a_{23}b_{33} + a_{24}b_{34}) \end{bmatrix}$$

and

$$(\boldsymbol{EJ'} + \boldsymbol{FK'}) = [a_{31} \quad a_{32} \quad a_{33}] \cdot \begin{bmatrix} b_{31} \\ b_{32} \\ b_{33} \end{bmatrix} + [a_{34}] \cdot [b_{34}]$$

$$= [a_{31}b_{31} + a_{32}b_{32} + a_{33}b_{33} + a_{34}b_{34}]$$

SIMULTANEOUS LINEAR EQUATIONS

Consider the set of linear equations:

$$\boldsymbol{Ax} = \boldsymbol{b}$$

The equation set consists of an $\boldsymbol{x}$ vector of order $n \times 1$ and an $\boldsymbol{A}$ matrix of order $m \times n$ where $n > m$. Let us subdivide the $\boldsymbol{x}$ vector into two partition vectors and call them $\boldsymbol{x}_B$ and $\boldsymbol{x}_N$, the basic and nonbasic vectors, respectively. Thus, $\boldsymbol{x}$ may be written as

$$\boldsymbol{x} = \begin{bmatrix} \boldsymbol{x}_B \\ \boldsymbol{x}_N \end{bmatrix}$$

where

$$\boldsymbol{x}'_B = [x_1 \quad x_2 \quad \cdots \quad x_m]$$

$$\boldsymbol{x}'_N = [x_{m+1} \quad \cdots \quad x_n]$$

Furthermore, to satisfy the multiplication requirement that $\boldsymbol{Ax}$, $\boldsymbol{A}$ must be subdivided into partition matrices $\boldsymbol{A}_B$ and $\boldsymbol{A}_N$, where $\boldsymbol{A}_B$ is a square $m \times m$ matrix and $\boldsymbol{A}_N$ is an $m \times (n - m)$ matrix. Thus, the equation set is

$$\boldsymbol{Ax} = [\boldsymbol{A}_B \ \boldsymbol{A}_N] \cdot \begin{bmatrix} \boldsymbol{x}_B \\ \boldsymbol{x}_N \end{bmatrix} = \boldsymbol{b}$$

Since the order of the submatrices $\boldsymbol{A}_B$ and $\boldsymbol{A}_N$ and partition vectors $\boldsymbol{x}_B$ and $\boldsymbol{x}_N$ were carefully chosen to obey the rules of matrix multiplication, it may be expressed as

$$\boldsymbol{A}_B\boldsymbol{x}_B + \boldsymbol{A}_N\boldsymbol{x}_N = \boldsymbol{b}$$

Suppose the set of equations consists of two constraint equations, $m = 2$, and four control variables, $n = 4$. Utilizing matrix partitioning, the constraint set may be written as

$$\begin{bmatrix} a_{11} & a_{12} \\ a_{21} & a_{22} \end{bmatrix} \cdot \begin{bmatrix} x_1 \\ x_2 \end{bmatrix} + \begin{bmatrix} a_{13} & a_{14} \\ a_{23} & a_{24} \end{bmatrix} \cdot \begin{bmatrix} x_3 \\ x_4 \end{bmatrix} = \begin{bmatrix} b_1 \\ b_2 \end{bmatrix}$$

where $\boldsymbol{x}'_B = [x_1 \quad x_2]$ and $\boldsymbol{x}'_N = [x_3 \quad x_4]$

DETERMINANTS

In the solution of $\boldsymbol{x}$ simultaneous equations, $\boldsymbol{Ax} = \boldsymbol{b}$, where $\boldsymbol{A}$ is a square matrix. The solution may be obtained by determining the inverse of $\boldsymbol{A}$ and premultiplying each side of the equation. The inverse of $\boldsymbol{A}$ is represented as $\boldsymbol{A}^{-1}$. To solve for $\boldsymbol{x}$ the following steps are performed. Premultiply both sides of the equation $\boldsymbol{Ax} = \boldsymbol{b}$ by the inverse of $\boldsymbol{A}$; thus

$$\boldsymbol{A}^{-1}\boldsymbol{Ax} = \boldsymbol{Ix} = \boldsymbol{x} = \boldsymbol{A}^{-1}\boldsymbol{b}$$

where the product $\boldsymbol{A}^{-1}\boldsymbol{A}$ is defined to be the identity matrix

$$\boldsymbol{A}^{-1}\boldsymbol{A} = \boldsymbol{I}$$

In order for this result to exist the inverse of $\boldsymbol{A}$ must exist, that is, $\boldsymbol{A}^{-1} \neq \boldsymbol{0}$. The solution of $\boldsymbol{Ax} = \boldsymbol{b}$ may be accomplished by several methods including matrix inversion and Cramer's rule. With these methods, the concept of a determinant is of prime importance.

The notation for a determinant of a matrix $\boldsymbol{A}$ is $|\boldsymbol{A}|$. The determinant of a 2×2 matrix is defined as

$$|\boldsymbol{A}| = \begin{vmatrix} a_{11} & a_{12} \\ a_{21} & a_{22} \end{vmatrix} = a_{11}a_{22} - a_{21}a_{21}$$

The determinant of $\boldsymbol{A}$ is equal to product of main diagonal terms a_{11} and a_{22} minus the product of the off-diagonal terms a_{21} and a_{22}, or

$$|\boldsymbol{A}| = \begin{vmatrix} a_{11} & a_{12} \\ a_{21} & a_{21} \end{vmatrix}$$

The determinate of a 3×3 matrix is

$$|\boldsymbol{A}| = \begin{vmatrix} a_{11} & a_{12} & a_{13} \\ a_{21} & a_{22} & a_{23} \\ a_{31} & a_{32} & a_{33} \end{vmatrix}$$

$$|\boldsymbol{A}| = a_{11}a_{22}a_{33} + a_{12}a_{23}a_{31} + a_{13}a_{21}a_{32} - a_{31}a_{22}a_{13} - a_{32}a_{23}a_{11} - a_{33}a_{21}a_{12}$$

The determinate of a 3×3 matrix may be obtained by taking the sum of the product of the main diagonal terms and the product of the terms parallel to it and subtracting the sum of the off-diagonal terms depicted as follows:

$$\begin{bmatrix} a_{11} & a_{12} & a_{13} \\ a_{21} & a_{22} & a_{23} \\ a_{31} & a_{32} & a_{33} \end{bmatrix} = a_{11}a_{22}a_{33} + a_{12}a_{23}a_{31} + a_{13}a_{21}a_{32}$$

minus

$$\begin{bmatrix} a_{11} & a_{12} & a_{13} \\ a_{21} & a_{22} & a_{23} \\ a_{31} & a_{32} & a_{33} \end{bmatrix} = a_{31}a_{22}a_{13} + a_{32}a_{23}a_{11} + a_{33}a_{21}a_{12}$$

or

$$|\boldsymbol{A}| = a_{11}a_{22}a_{33} + a_{12}a_{23}a_{31} + a_{13}a_{21}a_{32} - (a_{31}a_{22}a_{13} + a_{32}a_{23}a_{11} + a_{33}a_{21}a_{12})$$

A more general approach to solve for the determinant of $|\boldsymbol{A}|$ is to use the *method of cofactors*:

$$|\boldsymbol{A}| = \sum_{j=1}^{n} a_{ij}\boldsymbol{A}_{ij}$$

where $\boldsymbol{A}_{ij}$ is the cofactor of element a_{ij}. The cofactor is defined as

$$\boldsymbol{A}_{ij} = (-1)^{i+j}|\boldsymbol{M}_{ij}|$$

The term $|\boldsymbol{M}_{ij}|$ is the determinant of the minor of $\boldsymbol{A}_{ij}$. The minor is formed by eliminating the ith row and jth column of $\boldsymbol{A}$. Suppose we have a 3×3 matrix, the determinant of a minor $\boldsymbol{A}_{11}$ is

$$|\boldsymbol{M}_{11}| = \begin{vmatrix} a_{22} & a_{23} \\ a_{32} & a_{33} \end{vmatrix} = a_{22}a_{33} - a_{32}a_{23}$$

Likewise, the minor $\boldsymbol{M}_{12}$ is

$$\boldsymbol{M}_{12} = \begin{vmatrix} a_{21} & a_{23} \\ a_{31} & a_{33} \end{vmatrix} = a_{21}a_{33} - a_{31}a_{23}$$

Thus, the determinant of a 3×3 matrix is

$$|\boldsymbol{A}| = a_{11}(-1)^{1+1} \cdot \begin{vmatrix} a_{22} & a_{23} \\ a_{32} & a_{33} \end{vmatrix} + a_{12}(-1)^{1+2} \cdot \begin{vmatrix} a_{21} & a_{23} \\ a_{31} & a_{33} \end{vmatrix} + a_{13}(-1)^{1+3} \cdot \begin{vmatrix} a_{21} & a_{22} \\ a_{31} & a_{32} \end{vmatrix}$$

Simplifying, the determinant of $\boldsymbol{A}$ is

$$|\boldsymbol{A}| = a_{11}(a_{22}a_{33} - a_{32}a_{23}) - a_{12}(a_{21}a_{33} - a_{31}a_{23}) + a_{13}(a_{21}a_{32} - a_{31}a_{22})$$

or

$$|\boldsymbol{A}| = a_{11}a_{22}a_{33} + a_{12}a_{23}a_{31} + a_{13}a_{21}a_{32} - (a_{31}a_{22}a_{13} + a_{32}a_{23}a_{11} + a_{33}a_{21}a_{21})$$

CRAMER'S RULE

In lieu of presenting a rigorous development of Cramer's rule, a simple simultaneous equation consisting of two equations and two unknowns x_1 and x_2 will be solved. Algebraic row operations will be used to eliminate x_1 from the equation set. After obtaining the solution to x_2, it will be substituted back into the original equations and solved for x_1. The solution of x_1 and x_2 will be rearranged, and it will be seen that the solution is a ratio of determinants.

Let $\boldsymbol{Ax} = \boldsymbol{b}$ equal

$$\begin{bmatrix} a_{11} & a_{12} \\ a_{21} & a_{22} \end{bmatrix} \cdot \begin{bmatrix} x_1 \\ x_2 \end{bmatrix} = \begin{bmatrix} b_1 \\ b_2 \end{bmatrix}$$

Expand this to

$$a_{11}x_1 + a_{12}x_2 = b_1$$

$$a_{21}x_1 + a_{22}x_2 = b_2$$

Solve x_2 by multiplying all elements of the first equation by a_{21} and all elements of the second equation by $-a_{11}$.

$$(a_{21}a_{11})x_1 + (a_{21}a_{12})x_2 = a_{21}b_1$$

$$-(a_{11}a_{21})x_1 - (a_{11}a_{22})x_2 = -a_{11}b_2$$

Add the two equations. All x_1 terms will be eliminated and the equations will be a function of x_2 only.

$$(a_{21}a_{12} - a_{11}a_{22})x_2 = (a_{21}b_1 - a_{11}b_2)$$

Thus, the solution of x_2 is

$$x_2 = \frac{(a_{21}b_1 - a_{11}b_2)}{(a_{21}a_{12} - a_{11}a_{22})}$$

Substitute x_2 into the first equation and simplify to obtain a solution of x_1

$$a_{11}x_1 + a_{12}\frac{(a_{21}b_1 - a_{11}b_2)}{(a_{21}a_{12} - a_{11}a_{22})} = b$$

Solve for x_1:

$$x_1 = \frac{1}{a_{11}}\left[b_1 - \frac{a_{12}(a_{21}b_1 - a_{11}b_2)}{a_{21}a_{12} - a_{11}a_{22}}\right]$$

Combine the terms inside the brackets

$$x_1 = \frac{1}{a_{11}}\left[\frac{b_1a_{21}a_{12} - b_1a_{11}a_{22} - a_{12}a_{21}b_1 + a_{12}a_{11}b_2}{(a_{21}a_{12} - a_{11}a_{22})}\right]$$

and simplify to

$$x_1 = \frac{(a_{12}b_2 - a_{22}b_1)}{(a_{21}a_{12} - a_{11}a_{22})}$$

The numerator and denominator of x_1 and x_2 may be written as determinants:

$$x_1 = \frac{\begin{vmatrix} b_1 & a_{12} \\ b_2 & a_{22} \end{vmatrix}}{\begin{vmatrix} a_{11} & a_{12} \\ a_{21} & a_{22} \end{vmatrix}}$$

$$x_2 = \frac{\begin{vmatrix} a_{11} & b_1 \\ a_{21} & b_2 \end{vmatrix}}{\begin{vmatrix} a_{11} & a_{12} \\ a_{21} & a_{22} \end{vmatrix}}$$

These two equations show the essence of Cramer's method for a simple example. The following discussion utilizes vector notation to easily set up a problem for numerical calculation.

Represent the original problem $\boldsymbol{Ax} = \boldsymbol{b}$ as

$$[\boldsymbol{a}_1 \quad \boldsymbol{a}_2]\boldsymbol{x} = \boldsymbol{b}$$

where

$$\boldsymbol{a}_1 = \begin{bmatrix} a_{11} \\ a_{21} \end{bmatrix} \quad \boldsymbol{a}_2 = \begin{bmatrix} a_{12} \\ a_{22} \end{bmatrix} \quad \boldsymbol{x} = \begin{bmatrix} x_1 \\ x_2 \end{bmatrix} \quad \boldsymbol{b} = \begin{bmatrix} b_1 \\ b_2 \end{bmatrix}$$

The values of x_1 and x_2 are

$$x_1 = \frac{|\boldsymbol{b} \quad \boldsymbol{a}_2|}{|\boldsymbol{A}|}$$

$$x_2 = \frac{|\boldsymbol{a}_1 \quad \boldsymbol{b}|}{|\boldsymbol{A}|}$$

where $|\boldsymbol{A}|$, $|\boldsymbol{b} \quad \boldsymbol{a}_2|$, and $|\boldsymbol{a}_1 \quad \boldsymbol{b}|$ represent the determinants.

Cramer's rule can be extended to simultaneous equations of higher order. The overall procedure, which is the same one that is used for the two-dimensional case, is summarized by the following steps.

STEP 1 Calculate $|\boldsymbol{A}|$. If $|\boldsymbol{A}| = 0$, no solution exists.

STEP 2 Rewrite $\boldsymbol{A}$ as a matrix of column vectors:

$$\boldsymbol{A} = [\boldsymbol{a}_1 \quad \boldsymbol{a}_2 \quad \cdots \quad \boldsymbol{a}_n]$$

STEP 3 Calculate the solution vector $\boldsymbol{x}$, where each x_i is calculated as

$$x_1 = \frac{|\boldsymbol{b} \quad \boldsymbol{a}_2 \quad \cdots \quad \boldsymbol{a}_n|}{|\boldsymbol{A}|}$$

$$x_2 = \frac{|\boldsymbol{a}_1 \quad \boldsymbol{b} \quad \cdots \quad \boldsymbol{a}_n|}{|\boldsymbol{A}|}$$

$$\vdots$$

$$x_n = \frac{|\boldsymbol{a}_1 \quad \boldsymbol{a}_2 \quad \cdots \quad \boldsymbol{b}|}{|\boldsymbol{A}|}$$

MATRIX INVERSION

The solution to a simultaneous equation $\boldsymbol{Ax} = \boldsymbol{b}$ utilizes the inverse of $\boldsymbol{A}$ or $\boldsymbol{A}^{-1}$ directly. The inverse of $\boldsymbol{A}$ has the property that the product of $\boldsymbol{A}$ and $\boldsymbol{A}^{-1}$ is equal to the identity matrix, or

$$\boldsymbol{A}^{-1}\boldsymbol{A} = \boldsymbol{A}\boldsymbol{A}^{-1} = \boldsymbol{I}$$

Premultiply the left- and right-hand side of $\boldsymbol{Ax} = \boldsymbol{b}$ by $\boldsymbol{A}^{-1}$. Thus

$$\boldsymbol{A}^{-1}\boldsymbol{Ax} = \boldsymbol{Ix} = \boldsymbol{A}^{-1}\boldsymbol{b}$$

$$\boldsymbol{x} = \boldsymbol{A}^{-1}\boldsymbol{b}$$

The $\boldsymbol{A}^{-1}$ is equal to the transpose of the cofactor matrix divided by the determinant of $\boldsymbol{A}$.

$$\boldsymbol{A}^{-1} = \frac{[\boldsymbol{A}_{ij}]'}{|\boldsymbol{A}|}$$

METHOD OF SUCCESSIVE ELIMINATION

The solution of $\boldsymbol{Ax} = \boldsymbol{b}$ may be accomplished by another procedure utilizing the properties of an augmented matrix. The augmented matrix of the matrix equation $\boldsymbol{Ax} = \boldsymbol{b}$ is written as

$$[A \quad b]$$

The concept of successive elimination may be obtained by utilizing the properties of the inverse of $\boldsymbol{A}$. Premultiply the augmented matrix by $\boldsymbol{A}^{-1}$.

$$A^{-1} \cdot [A \quad b] = [(A^{-1}A)(A^{-1}b)]$$

Simplifying, the augmented matrix contains the solution of $\boldsymbol{x}$ to $\boldsymbol{Ax} = \boldsymbol{b}$:

$$[I \quad x]$$

where

$$x = A^{-1}b$$

With the method of successive elimination the inverse $\boldsymbol{A}^{-1}$ is not explicitly calculated as implied above. It involves the use of algebraic row operations. From algebra, the multiplication of all elements in the same row by a constant and addition of corresponding elements of any two rows do not change the mathematical meaning of the equations. Thus the objective of successive elimination is to reduce $\boldsymbol{A}$ of the augmented matrix $[A \quad b]$ to the identify matrix $\boldsymbol{I}$ by algebraic row operations.

NO SOLUTION

Thus far, it has been assumed that a unique solution to $\boldsymbol{Ax} = \boldsymbol{b}$ exists. If the inverse of $\boldsymbol{A}$ exists, then a unique solution exists:

$$x = A^{-1}b$$

If $\boldsymbol{A}^{-1}$ does not exist, no solution to $\boldsymbol{Ax} = \boldsymbol{b}$ can exist.

The determinant of the $\boldsymbol{A}$ matrix may be used as an indicator to determine if a unique solution exists. For methods of Cramer and matrix inversion, calculating $|A|$ first may save unnecessary calculation. For the method of successive elimination, if $\boldsymbol{A}^{-1}$ does not exist, $\boldsymbol{A}^{-1}\boldsymbol{A} = \boldsymbol{I}$ does not exist. In other words, since $\boldsymbol{A}^{-1}\boldsymbol{A} \neq \boldsymbol{I}$, the identity matrix $\boldsymbol{I}$ can not be found.

EXAMPLE A.1 Calculation of a 4 × 4 Determinant

Find the determinant of the matrix by use of the cofactor method.

$$|A| = \begin{vmatrix} 1 & 2 & 0 & 3 \\ 1 & 1 & 1 & 1 \\ 2 & 5 & 3 & 0 \\ 3 & 3 & 0 & 1 \end{vmatrix}$$

Solution

By the method of cofactors, the determinant of A is

$$|A| = \sum_{j=1}^{4} a_{1j}A_{1j}$$

$$|A| = (-1)^{1+1}\cdot\begin{vmatrix}1 & 1 & 1\\ 5 & 3 & 0\\ 3 & 0 & 1\end{vmatrix} + 2(-1)^{1+2}\cdot\begin{vmatrix}1 & 1 & 1\\ 2 & 3 & 0\\ 3 & 0 & 1\end{vmatrix} + 0(-1)^{1+3}\cdot\begin{vmatrix}1 & 1 & 1\\ 2 & 5 & 0\\ 3 & 3 & 1\end{vmatrix}$$

$$+ 3(-1)^{1+4}\cdot\begin{vmatrix}1 & 1 & 1\\ 2 & 5 & 3\\ 3 & 3 & 0\end{vmatrix}$$

$$|A| = \begin{vmatrix}1 & 1 & 1\\ 5 & 3 & 0\\ 3 & 0 & 1\end{vmatrix} - 2\cdot\begin{vmatrix}1 & 1 & 1\\ 2 & 3 & 0\\ 3 & 0 & 1\end{vmatrix} - 3\cdot\begin{vmatrix}1 & 1 & 1\\ 2 & 5 & 3\\ 3 & 3 & 0\end{vmatrix}$$

$$|A| = [(3 + 0 + 0) - (9 + 0 + 5)] - 2[(3 + 0 + 0) - (9 + 0 + 2)]$$
$$- 3[(0 + 9 + 6) - (15 + 9 + 0)]$$

$$|A| = -11 + 16 + 27 = 32$$

EXAMPLE A.2 Solution to Simultaneous Equations

Solve for x in the following problem by (a) Cramer's rule, (b) matrix inversion, and (c) successive elimination methods.

$$Ax = b$$

$$\begin{bmatrix}2 & 3 & 2\\ 0 & 3 & 2\\ 2 & 3 & 4\end{bmatrix}\cdot\begin{bmatrix}x_1\\ x_2\\ x_3\end{bmatrix} = \begin{bmatrix}14\\ 8\\ 16\end{bmatrix}$$

Solution

(a) The determinant of A is

$$|A| = \begin{vmatrix}2 & 3 & 2\\ 0 & 3 & 2\\ 2 & 3 & 4\end{vmatrix} = [(24 + 12 + 0) - (12 + 12 + 0)] = 12$$

Since $|A| \neq 0$, a solution exists. Rewrite A as a matrix of column vectors:

$$|A| = [a_1 \quad a_2 \quad a_3]$$

where

$$a_1 = \begin{bmatrix}2\\ 0\\ 2\end{bmatrix} \quad a_2 = \begin{bmatrix}3\\ 3\\ 3\end{bmatrix} \quad a_3 = \begin{bmatrix}2\\ 2\\ 4\end{bmatrix}$$

The solution of $\boldsymbol{x}$, $\boldsymbol{x}' = [x_1 \quad x_2 \quad x_3]$, is

$$x_1 = \frac{|\boldsymbol{b} \quad \boldsymbol{a}_2 \quad \boldsymbol{a}_3|}{|\boldsymbol{A}|}$$

$$x_1 = \frac{1}{|\boldsymbol{A}|} \cdot \begin{vmatrix} 14 & 3 & 2 \\ 8 & 3 & 2 \\ 16 & 3 & 4 \end{vmatrix} = \tfrac{36}{12} = 3$$

$$x_2 = \frac{|\boldsymbol{a}_1 \quad \boldsymbol{b} \quad \boldsymbol{a}_3|}{|\boldsymbol{A}|}$$

$$x_2 = \frac{1}{|\boldsymbol{A}|} \cdot \begin{vmatrix} 2 & 14 & 2 \\ 0 & 8 & 2 \\ 2 & 16 & 4 \end{vmatrix} = \tfrac{24}{12} = 2$$

$$x_3 = \frac{|\boldsymbol{a}_1 \quad \boldsymbol{a}_2 \quad \boldsymbol{b}|}{|\boldsymbol{A}|}$$

$$x_3 = \frac{1}{|\boldsymbol{A}|} \cdot \begin{vmatrix} 2 & 3 & 14 \\ 0 & 3 & 8 \\ 2 & 3 & 16 \end{vmatrix} = \tfrac{12}{12} = 1$$

The solution is $\boldsymbol{x}' = [3 \quad 2 \quad 1]$

(b) The solution by matrix inversion method, $\boldsymbol{x} = \boldsymbol{A}^{-1}\boldsymbol{b}$, is achieved by first calculating the inverse of $\boldsymbol{A}^{-1}$:

$$\boldsymbol{A}^{-1} = \frac{[A_{ij}]'}{|\boldsymbol{A}|}$$

Thus

$$\boldsymbol{A}^{-1} = \frac{1}{12} \cdot \begin{bmatrix} \begin{vmatrix} 3 & 2 \\ 3 & 4 \end{vmatrix} & (-1) \cdot \begin{vmatrix} 0 & 2 \\ 2 & 4 \end{vmatrix} & \begin{vmatrix} 0 & 3 \\ 2 & 3 \end{vmatrix} \\ (-1) \cdot \begin{vmatrix} 3 & 2 \\ 3 & 4 \end{vmatrix} & \begin{vmatrix} 2 & 2 \\ 2 & 4 \end{vmatrix} & (-1) \cdot \begin{vmatrix} 2 & 3 \\ 2 & 3 \end{vmatrix} \\ \begin{vmatrix} 3 & 2 \\ 3 & 2 \end{vmatrix} & (-1) \cdot \begin{vmatrix} 2 & 2 \\ 0 & 2 \end{vmatrix} & \begin{vmatrix} 2 & 3 \\ 0 & 3 \end{vmatrix} \end{bmatrix}'$$

$$\boldsymbol{A}^{-1} = \frac{1}{12} \cdot \begin{bmatrix} 6 & 4 & -6 \\ -6 & 4 & 0 \\ 0 & -4 & 6 \end{bmatrix}' = \frac{1}{12} \cdot \begin{bmatrix} 6 & -6 & 0 \\ 4 & 4 & -4 \\ -6 & 0 & 6 \end{bmatrix} = \begin{bmatrix} \frac{1}{2} & -\frac{1}{2} & 0 \\ \frac{1}{3} & \frac{1}{3} & -\frac{1}{3} \\ -\frac{1}{2} & 0 & \frac{1}{2} \end{bmatrix}$$

Thus

$$\boldsymbol{x} = \boldsymbol{A}^{-1}\boldsymbol{b}$$

$$\boldsymbol{x} = \begin{bmatrix} x_1 \\ x_2 \\ x_3 \end{bmatrix} = \begin{bmatrix} \frac{1}{2} & -\frac{1}{2} & 0 \\ \frac{1}{3} & \frac{1}{3} & -\frac{1}{3} \\ -\frac{1}{2} & 0 & \frac{1}{2} \end{bmatrix} \cdot \begin{bmatrix} 14 \\ 8 \\ 16 \end{bmatrix} = \begin{bmatrix} 3 \\ 2 \\ 1 \end{bmatrix}$$

The solution is

$$\boldsymbol{x}' = [3 \quad 2 \quad 1]$$

(c) The augmented matrix $[A \quad b]$ of $Ax = b$ is

$$\begin{bmatrix} 2 & 3 & 2 & 14 \\ 0 & 3 & 2 & 8 \\ 2 & 3 & 4 & 16 \end{bmatrix}$$

By row operations we will transform $[A \quad b]$ to $[I \quad x]$.

Iteration 1 Obtain a new matrix by

1. Divide row 1 by 2 and rewrite it.
2. Rewrite row 2 without any changes.
3. Obtain a new row 3 by multiplying row 1 of the new matrix by (-2) and adding it to row 3 of the above matrix.

The new matrix becomes

$$\begin{bmatrix} 1 & \frac{3}{2} & 1 & 7 \\ 0 & 3 & 2 & 8 \\ 0 & 0 & 2 & 2 \end{bmatrix}$$

Note all elements to the lower left of the main diagonal are zero.

Iteration 2

4. Obtain a new row 2 by dividing row 2 by (3).
5. Obtain a new row 1 by multiplying the new row 2 by $(-3/2)$ and adding it to row 1.
6. Since element $a_{32} = 0$, divide row 3 by (2) and rewrite it.

The new matrix is

$$\begin{bmatrix} 1 & 0 & 0 & 3 \\ 0 & 1 & \frac{2}{3} & \frac{8}{3} \\ 0 & 0 & 1 & 1 \end{bmatrix}$$

Iteration 3

7. Write rows 1 and 3 without change.
8. Multiply new row 3 by $(-\frac{2}{3})$ and add to row 2.

$$\begin{bmatrix} 1 & 0 & 0 & 3 \\ 0 & 1 & 0 & 2 \\ 0 & 0 & 1 & 1 \end{bmatrix}$$

An identity matrix has been obtained; therefore, a solution has been obtained. The solution is $x' = [3 \quad 2 \quad 1]$.

EXAMPLE A.3 Nonunique Solutions to a Set of Simultaneous Equations

Using the method of successive elimination, show that a unique solution does *not* exist for the following set of equations.

$$Ax = b$$

$$\begin{aligned} 2x_1 + 3x_2 + 2x_3 &= 14 \\ 3x_2 + 2x_3 &= 8 \\ 2x_1 \qquad\qquad &= 6 \end{aligned}$$

Solution

The augmented matrix $[A \quad b]$ is

$$\begin{bmatrix} 2 & 3 & 2 & 14 \\ 0 & 3 & 2 & 8 \\ 2 & 0 & 0 & 6 \end{bmatrix}$$

STEP 1 A new row 1 can be obtained by multiplying row 2 by (-1) and adding it to row 1. The new matrix is

$$\begin{bmatrix} 2 & 0 & 0 & 6 \\ 0 & 3 & 2 & 8 \\ 2 & 0 & 0 & 6 \end{bmatrix}$$

STEP 2 Since row 1 and 3 are identical we can eliminate row 3 by multiplying row 1 by (-1) and adding it to row 3.

$$\begin{bmatrix} 2 & 0 & 0 & 6 \\ 0 & 3 & 2 & 8 \\ 0 & 0 & 0 & 0 \end{bmatrix}$$

Since the third row of the augmented matrix is zero, a unique solution of $\boldsymbol{x}$ does not exist.

Remarks

The set of linear equations are not independent equations. Without using the method of successive elimination, we can see that equation (1) is equal to the sum of equations (2) and (3). Also notice that $|\boldsymbol{A}| = 0$; thus $\boldsymbol{A}^{-1}$ does not exist. Thus, we have several methods of identifying nonunique solution.

PROBLEMS

Problem 1

Consider the following matrices:

$$\boldsymbol{A} = \begin{bmatrix} 1 & 3 & -1 \\ 0 & 2 & 0 \\ 2 & 1 & 0 \\ 3 & 0 & 2 \end{bmatrix} \qquad \boldsymbol{B} = \begin{bmatrix} 1 & -1 & 2 & 4 \\ 2 & 0 & -1 & 0 \\ 0 & -1 & 3 & 0 \end{bmatrix}$$

$$\boldsymbol{C} = \begin{bmatrix} 0 & -1 & 0 & 2 \\ 2 & 0 & 2 & -1 \\ 1 & 0 & 3 & 1 \\ 0 & 1 & 4 & 0 \end{bmatrix} \qquad \boldsymbol{D} = \begin{bmatrix} 1 & 0 & 0 & 0 \\ 0 & 2 & 0 & 0 \\ 0 & 0 & -1 & 0 \\ 0 & 0 & 0 & 0 \end{bmatrix}$$

$$\boldsymbol{E} = \begin{bmatrix} 2 & 3 \\ 0 & 1 \\ 1 & 0 \\ 2 & -1 \end{bmatrix} \qquad \boldsymbol{x} = \begin{bmatrix} -1 \\ 2 \\ 0 \\ 1 \end{bmatrix} \qquad \boldsymbol{y} = \begin{bmatrix} 1 \\ -1 \\ 4 \\ 3 \end{bmatrix}$$

For each of the following operations state whether the operations are defined, and if so, perform the indicated operations.

(a) AB	(b) BA	(c) AC
(d) CA	(e) $A'C$	(f) $C'A$
(g) $(C'A)'$	(h) AD	(i) DA
(j) BE	(k) $E'B$	(l) Bx
(m) $x'y$	(n) xy	

Problem 2

Consider the following:

$$a' = [1 \quad 2 \quad 3] \qquad b' = [10 \quad 9 \quad 8]$$

$$A = \begin{bmatrix} 1 & 2 & -2 \\ 1 & 5 & 3 \\ -2 & 3 & 3 \end{bmatrix} \qquad B = \begin{bmatrix} 9 & 1 & 1 \\ 1 & 8 & 0 \\ 1 & 0 & 6 \end{bmatrix}$$

(a) What is the order of the vectors and matrices?
(b) Determine the vector products and show that $a'b = b'a$ are equal scalar quantities.
(c) Show the following products are scalars and that the following properties exist:

A) $a'Ab \neq b'Aa$
B) $a'Ab = b'A'a$
C) $a'Bb = a'B'b$

Problem 3

Calculate the determinate of A.

$$A = \begin{bmatrix} 2 & 0 & 3 & 4 \\ 0 & 1 & 0 & 0 \\ 2 & 1 & 5 & 0 \\ 0 & 0 & 0 & 3 \end{bmatrix}$$

Problem 4

Calculate the solution of x where $Ax = b$.

(i) $A = \begin{bmatrix} 1 & 1 & 1 \\ 0 & 2 & 3 \\ 3 & 1 & 1 \end{bmatrix} \qquad b = \begin{bmatrix} 3 \\ 4 \\ 3 \end{bmatrix}$

(ii) $A = \begin{bmatrix} 1 & 2 & 3 & 3 \\ 2 & 3 & 6 & 5 \\ 1 & 2 & 1 & 3 \\ 0 & 1 & 1 & 1 \end{bmatrix} \qquad b = \begin{bmatrix} 1 \\ 2 \\ 1 \\ 2 \end{bmatrix}$

(a) Cramer's rule.
(b) Matrix inversion.
(c) Method of successive elimination.

Problem 5

Solve for x_1, x_2, and x_3 in terms of x_4 for the following simultaneous equations by successive elimination:

$$\begin{aligned}
3x_1 + \quad\quad 3x_3 + 3x_3 &= 3 \\
2x_1 - x_2 + x_3 \quad\quad &= 3 \\
-x_1 - 2x_2 - 2x_3 - 5x_4 &= 1
\end{aligned}$$

Problem 6

Let $\boldsymbol{x}'_B = [x_1 \quad x_3]$ and $\boldsymbol{x}'_N = [x_4]$

$$\begin{aligned}
3x_1 + x_2 \quad\quad &= 6 \\
x_1 \quad\quad + x_3 &= 9
\end{aligned}$$

(a) Write the equation set in matrix form

$$\boldsymbol{A}_B \boldsymbol{x}_B + \boldsymbol{A}_N \boldsymbol{x}_N = \boldsymbol{b}$$

(b) Solve for $\boldsymbol{x}_B$ by use of matrix inversion.
(c) Solve for $\boldsymbol{x}_B$ by use of successive elimination.

Problem 7

Let $\boldsymbol{x}'_B = [x_1 \quad x_2 \quad x_3]$ and $\boldsymbol{x}'_N = [x_4 \quad x_5]$

$$\begin{aligned}
x_1 + x_2 - x_3 \quad\quad\quad\quad &= 1.5 \\
x_1 \quad\quad\quad + x_4 \quad\quad &= 4 \\
x_1 \quad\quad\quad\quad\quad - x_5 &= 1
\end{aligned}$$

(a) Write the equation set in matrix form, $\boldsymbol{A}_B \boldsymbol{x}_B + \boldsymbol{A}_N \boldsymbol{x}_N = \boldsymbol{b}$.
(b) Show that the solution for $\boldsymbol{x}_B$ is inconsistent. Use matrix inversion.
(c) Repeat part b by use of successive elimination.

Problem 8

Let $\boldsymbol{x}'_B = [x_1 \quad x_2 \quad x_3]$ and $\boldsymbol{x}'_N = [x_4]$.

$$\begin{aligned}
x_1 + x_2 + 2x_3 + 3x_4 &= 2 \\
-x_1 + 2x_2 - x_3 + x_4 &= 4 \\
-2x_1 + 4x_2 - 3x_3 - 2x_4 &= 2
\end{aligned}$$

(a) Write the equation set in matrix form: $\boldsymbol{A}_B \boldsymbol{x}_B + \boldsymbol{A}_N \boldsymbol{x}_N = \boldsymbol{b}$.
(b) Solve for $\boldsymbol{x}_B$ by use of matrix inversion.
(c) Solve for $\boldsymbol{x}_B$ by use of successive elimination.

APPENDIX B

Areas Under the Unit Normal Curve $N(0, 1)$[a]

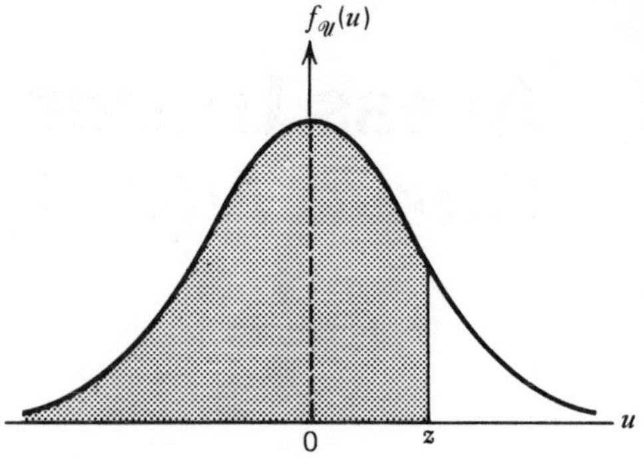

$$f_{\mathscr{U}}(u) = \frac{1}{\sqrt{2\pi}} e^{-1/2u^2}$$

$$F_{\mathscr{U}}(z) = \int_{-\infty}^{z} f_{\mathscr{U}}(u)\, du = \text{shaded area under the curve } f_{\mathscr{U}}(u)$$

u	0.00	0.01	0.02	0.03	0.04	0.05	0.06	0.07	0.08	0.09
−3.4	0.0003	0.0003	0.0003	0.0003	0.0003	0.0003	0.0003	0.0003	0.0003	0.0002
−3.3	0.0005	0.0005	0.0005	0.0004	0.0004	0.0004	0.0004	0.0004	0.0004	0.0003
−3.2	0.0007	0.0007	0.0006	0.0006	0.0006	0.0006	0.0006	0.0005	0.0005	0.0005
−3.1	0.0010	0.0009	0.0009	0.0009	0.0008	0.0008	0.0008	0.0008	0.0007	0.0007
−3.0	0.0013	0.0013	0.0013	0.0012	0.0012	0.0011	0.0011	0.0011	0.0010	0.0010
−2.9	0.0019	0.0018	0.0017	0.0017	0.0016	0.0016	0.0015	0.0015	0.0014	0.0014
−2.8	0.0026	0.0025	0.0024	0.0023	0.0023	0.0022	0.0021	0.0021	0.0020	0.0019
−2.7	0.0035	0.0034	0.0033	0.0032	0.0031	0.0030	0.0029	0.0028	0.0027	0.0026
−2.6	0.0047	0.0045	0.0044	0.0043	0.0041	0.0040	0.0039	0.0038	0.0037	0.0036
−2.5	0.0062	0.0060	0.0059	0.0057	0.0055	0.0054	0.0052	0.0051	0.0049	0.0048
−2.4	0.0082	0.0080	0.0078	0.0075	0.0073	0.0071	0.0069	0.0068	0.0066	0.0064
−2.3	0.0107	0.0104	0.0102	0.0099	0.0096	0.0094	0.0091	0.0089	0.0087	0.0084
−2.2	0.0139	0.0136	0.0132	0.0129	0.0125	0.0122	0.0119	0.0116	0.0113	0.0110
−2.1	0.0179	0.0174	0.0170	0.0166	0.0162	0.0158	0.0154	0.0150	0.0146	0.0143
−2.0	0.0228	0.0222	0.0217	0.0212	0.0207	0.0202	0.0197	0.0192	0.0188	0.0183
−1.9	0.0287	0.0281	0.0274	0.0268	0.0262	0.0256	0.0250	0.0244	0.0239	0.0233
−1.8	0.0359	0.0352	0.0344	0.0336	0.0329	0.0322	0.0314	0.0307	0.0301	0.0294
−1.7	0.0446	0.0436	0.0427	0.0418	0.0409	0.0401	0.0392	0.0384	0.0375	0.0367
−1.6	0.0548	0.0537	0.0526	0.0516	0.0505	0.0495	0.0485	0.0475	0.0465	0.0455
−1.5	0.0668	0.0655	0.0643	0.0630	0.0618	0.0606	0.0594	0.0582	0.0571	0.0559
−1.4	0.0808	0.0793	0.0778	0.0764	0.0749	0.0735	0.0722	0.0708	0.0694	0.0681
−1.3	0.0968	0.0951	0.0934	0.0918	0.0901	0.0885	0.0869	0.0853	0.0838	0.0823
−1.2	0.1151	0.1131	0.1112	0.1093	0.1075	0.1056	0.1038	0.1020	0.1003	0.0985
−1.1	0.1357	0.1335	0.1314	0.1292	0.1271	0.1251	0.1230	0.1210	0.1190	0.1170
−1.0	0.1587	0.1562	0.1539	0.1515	0.1492	0.1469	0.1446	0.1423	0.1401	0.1379
−0.9	0.1841	0.1814	0.1788	0.1762	0.1736	0.1711	0.1685	0.1660	0.1635	0.1611
−0.8	0.2119	0.2090	0.2061	0.2033	0.2005	0.1977	0.1949	0.1922	0.1894	0.1867
−0.7	0.2420	0.2389	0.2358	0.2327	0.2296	0.2266	0.2236	0.2206	0.2177	0.2148
−0.6	0.2743	0.2709	0.2676	0.2643	0.2611	0.2578	0.2546	0.2514	0.2483	0.2451
−0.5	0.3085	0.3050	0.3015	0.2981	0.2946	0.2912	0.2877	0.2843	0.2810	0.2776
−0.4	0.3446	0.3409	0.3372	0.3336	0.3300	0.3264	0.3228	0.3192	0.3156	0.3121
−0.3	0.3821	0.3783	0.3745	0.3707	0.3669	0.3632	0.3594	0.3557	0.3520	0.3483
−0.2	0.4207	0.4168	0.4129	0.4090	0.4052	0.4013	0.3974	0.3936	0.3897	0.3859
−0.1	0.4602	0.4562	0.4522	0.4483	0.4443	0.4404	0.4364	0.4325	0.4286	0.4247
−0.0	0.5000	0.4960	0.4920	0.4880	0.4840	0.4801	0.4761	0.4721	0.4681	0.4641

u	0.00	0.01	0.02	0.03	0.04	0.05	0.06	0.07	0.08	0.09
0.0	0.5000	0.5040	0.5080	0.5120	0.5160	0.5199	0.5239	0.5279	0.5319	0.5359
0.1	0.5398	0.5438	0.5478	0.5517	0.5557	0.5596	0.5636	0.5675	0.5714	0.5753
0.2	0.5793	0.5832	0.5871	0.5910	0.5948	0.5987	0.6026	0.6064	0.6103	0.6141
0.3	0.6179	0.6217	0.6255	0.6293	0.6331	0.6368	0.6406	0.6443	0.6480	0.6517
0.4	0.6554	0.6591	0.6628	0.6664	0.6700	0.6736	0.6772	0.6808	0.6844	0.6879
0.5	0.6915	0.6950	0.6985	0.7019	0.7054	0.7088	0.7123	0.7157	0.7190	0.7224
0.6	0.7257	0.7291	0.7324	0.7357	0.7389	0.7422	0.7454	0.7486	0.7517	0.7549
0.7	0.7580	0.7611	0.7642	0.7673	0.7704	0.7734	0.7764	0.7794	0.7823	0.7852
0.8	0.7881	0.7910	0.7939	0.7967	0.7995	0.8023	0.8051	0.8078	0.8106	0.8133
0.9	0.8159	0.8186	0.8212	0.8238	0.8264	0.8289	0.8315	0.8340	0.8365	0.8389
1.0	0.8413	0.8438	0.8461	0.8485	0.8508	0.8531	0.8554	0.8577	0.8599	0.8621
1.1	0.8643	0.8665	0.8686	0.8708	0.8729	0.8749	0.8770	0.8790	0.8810	0.8830
1.2	0.8849	0.8869	0.8888	0.8907	0.8925	0.8944	0.8962	0.8980	0.8997	0.9015
1.3	0.9032	0.9049	0.9066	0.9082	0.9099	0.9115	0.9131	0.9147	0.9162	0.9177
1.4	0.9192	0.9207	0.9222	0.9236	0.9251	0.9265	0.9278	0.9292	0.9306	0.9319
1.5	0.9332	0.9345	0.9357	0.9370	0.9382	0.9394	0.9406	0.9418	0.9429	0.9441
1.6	0.9452	0.9463	0.9474	0.9484	0.9495	0.9505	0.9515	0.9525	0.9535	0.9545
1.7	0.9554	0.9564	0.9573	0.9582	0.9591	0.9599	0.9608	0.9616	0.9625	0.9633
1.8	0.9641	0.9649	0.9656	0.9664	0.9671	0.9678	0.9686	0.9693	0.9699	0.9706
1.9	0.9713	0.9719	0.9726	0.9732	0.9738	0.9744	0.9750	0.9756	0.9761	0.9767
2.0	0.9772	0.9778	0.9783	0.9788	0.9793	0.9798	0.9803	0.9808	0.9812	0.9817
2.1	0.9821	0.9826	0.9830	0.9834	0.9838	0.9842	0.9846	0.9850	0.9854	0.9857
2.2	0.9861	0.9864	0.9868	0.9871	0.9875	0.9878	0.9881	0.9884	0.9887	0.9890
2.3	0.9893	0.9896	0.9898	0.9901	0.9904	0.9906	0.9909	0.9911	0.9913	0.9916
2.4	0.9918	0.9920	0.9922	0.9925	0.9927	0.9929	0.9931	0.9932	0.9934	0.9936
2.5	0.9938	0.9940	0.9941	0.9943	0.9945	0.9946	0.9948	0.9949	0.9951	0.9952
2.6	0.9953	0.9955	0.9956	0.9957	0.9959	0.9960	0.9961	0.9962	0.9963	0.9964
2.7	0.9965	0.9966	0.9967	0.9968	0.9969	0.9970	0.9971	0.9972	0.9973	0.9974
2.8	0.9974	0.9975	0.9976	0.9977	0.9977	0.9978	0.9979	0.9979	0.9980	0.9981
2.9	0.9981	0.9982	0.9982	0.9983	0.9984	0.9984	0.9985	0.9985	0.9986	0.9986
3.0	0.9987	0.9987	0.9987	0.9988	0.9988	0.9989	0.9989	0.9989	0.9990	0.9990
3.1	0.9990	0.9991	0.9991	0.9991	0.9992	0.9992	0.9992	0.9992	0.9993	0.9993
3.2	0.9993	0.9993	0.9994	0.9994	0.9994	0.9994	0.9994	0.9995	0.9995	0.9995
3.3	0.9995	0.9995	0.9995	0.9996	0.9996	0.9996	0.9996	0.9996	0.9996	0.9997
3.4	0.9997	0.9997	0.9997	0.9997	0.9997	0.9997	0.9997	0.9997	0.9997	0.9998

[a] Taken from Ronald E. Walpole and Raymond H. Myers, "Probability and Statistics for Engineers and Scientists," 2nd ed., Macmillan, 1978.

Index

ALTERNATIVE SELECTION MODELS

The net present worth NPW of a time stream of payments utilized the profit motive concept of maximizing $P = R - C$:

$$NPW = B_0 - A_0 - C_0$$

where B_0, A_0, and C_0 represent the present worth of future benefits B, operating and maintenance costs A, and capital investments C, respectively.

A cash-flow diagram for an uniform series of future benefits B, annual operating and maintenance costs A and an initial capital investment in year 0 C is

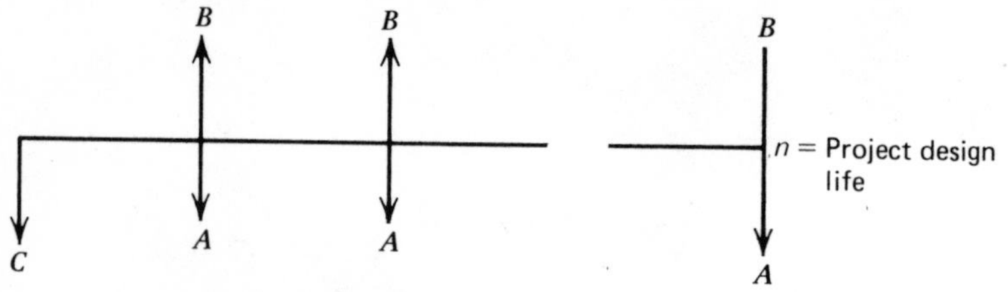

The decision-making process for selecting the best alternative among competing alternatives **A**, **B**, **C** is as follows:

1. Calculate $NPW = B_0 - A_0 - C_0$ for each alternative $= NPW^{\mathbf{A}}$, $NPW^{\mathbf{B}}$, $NPW^{\mathbf{C}}, \ldots$.
2. Consider only feasible alternatives satisfying the inequality condition, $NPW \geq 0$.
3. Select the feasible alternative with the maximum value of NPW.